Textbooks in Telecommunication Engineering

Era of the Internet and Network Science/Engineering

Series Editor

Tarek S. El-Bawab, Boston, MA, USA

Telecommunication and networks have evolved to embrace all aspects of our everyday life, including smart cities and infrastructures, healthcare, banking and businesses, manufacturing, space and aviation, meteorology and climate change, oceans and marine life, Internet of Things, defense, homeland security, education, research, social media, entertainment, and many others. Network applications and services continue to expand, virtually without limits. Therefore, specialized telecommunication and network engineering programs are recognized as a necessity to accelerate the pace of advancement in this field, and to prepare a new generation of engineers for imminent needs in our modern life. These programs need curricula, courses, labs, and textbooks of their own.

The IEEE Communications Society's Telecommunication Engineering Education (TEE) movement, led by Tarek S. El-Bawab- the editor of this Series, resulted in recognition of this field of engineering by the Accreditation Board for Engineering and Technology (ABET) on November 1, 2014. This Springer Series was launched to capitalizes on this milestone. The Series goal is to produce high-quality textbooks to fulfill the education needs of telecommunication and network engineering, and to support the development of specialized undergraduate and graduate curricula in this regard. The Series also supports research in this field and helps prepare its scholars for global challenges that lay ahead. The Series have published innovative textbooks in areas of network science and engineering where textbooks have been rare. It is producing high-quality volumes featuring innovative presentation media, interactive content, and online resources for students and professors.

Book proposals are solicited in all topics of telecommunication and network engineering including, but not limited to: network architecture and protocols; traffic engineering; network design, dimensioning, modeling, measurements, and analytics; network management and softwarization; cybersecurity; synchronization and control; applications of artificial intelligence in telecommunications and networks; applications of data science in telecommunications and networks; network availability, reliability, protection, recovery and restoration; wireless communication systems; cellular technologies and networks (through 5G, 6G, and beyond); satellite and space communications and networks; optical communications and networks; heterogeneous networks; broadband access and free-space optical communications; MSO/cable networks; storage networks; optical interconnects; and data centers; social networks; transmission media and systems; switching and routing (from legacy to today's paradigms); network applications and services; telecom economics and business; telecom regulation and policies; standards and standardization; and laboratories.

Proposals of interest shall be for textbooks that can be used to develop university courses, either in full or in part. They should include recent advances in the field while capturing whatever fundamentals that are necessary for students to understand the bases of the topic and appreciate its evolution trends. Books in this series will provide high-quality illustrations, examples, end-of-chapters' problems/exercises and case studies.

This series is indexed in Scopus. For further information and to submit proposals, please contact the Series Editor, Dr. Tarek S. El-Bawab, telbawab@ieee.org; or Mary James, Executive Editor at Springer, mary.james@springer.com.

An Liu · Ming-Min Zhao

Advanced Bayesian Signal Processing

From Foundations to Frontiers in Communication and Sensing

Springer

An Liu
Zhejiang University
Hangzhou, China

Ming-Min Zhao
Zhejiang University
Hangzhou, China

ISSN 2524-4345 ISSN 2524-4353 (electronic)
Textbooks in Telecommunication Engineering
ISBN 978-3-032-26655-2 ISBN 978-3-032-26656-9 (eBook)
https://doi.org/10.1007/978-3-032-26656-9

This Springer imprint is published by the registered company Springer Nature Switzerland AG
The registered company address is: Gewerbestrasse 11, 6330 Cham, Switzerland

Preface

Bayesian signal processing constitutes the theoretical foundation for many core components in modern telecommunications, including synchronization, channel estimation, signal detection, and channel decoding. Beyond these, next-generation standards such as 6G, Wi-Fi 7, and Wi-Fi 8 envision multi-functional wireless networks that integrate high-accuracy sensing capabilities, ranging from indoor positioning and activity recognition to radar sensing and Integrated Sensing and Communications (ISAC). These applications require accurate estimation of multiple target parameters such as distance, angle, and velocity in complex wireless environments, often leading to multi-dimensional non-convex estimation problems. Conventional methods like maximum likelihood (ML) and maximum-a-posteriori (MAP) estimation are prone to converge to poor local optima in such settings. In contrast, Bayesian signal processing excels in handling these challenges by incorporating prior information and modeling parameter uncertainty. Notably, compressive sensing, which is viewed as a class of Bayesian methods under sparse priors, has become a prominent research area, with extensive publications in major signal processing and wireless communications venues.

Against this backdrop, a textbook that systematically introduces Bayesian signal processing and its applications in wireless communications and sensing has become essential. While several books cover compressive sensing, few are suitable as textbooks for telecommunications students, and most lack comprehensive treatment of other Bayesian methodologies or in-depth wireless applications. This book fills that gap. It draws on the authors' graduate course "Compressive Sensing and Structured Sparse Recovery" at Zhejiang University, as well as their original research. It offers a systematic treatment of Bayesian signal processing from classical parameter estimation and sparse Bayesian inference to recent advanced methods, detailing their principles, design, performance, and trade-offs. Several key applications, including massive MIMO channel estimation, wireless localization and ISAC, illustrate how these techniques address real-world problems. Thus, the book serves both as an educational resource for students and a reference for researchers and practitioners. Some unique features that distinguish it from others on similar topics are list below:

1. A systematic introduction to Bayesian signal processing, covering parameter estimation, factor graphs, message passing, and sparse Bayesian inference.
2. Recent advanced methods such as Turbo sparse Bayesian inference with dynamic grid, subspace constrained variational Bayesian inference, and stochastic particle variational Bayesian inference.
3. A unified framework for comparing methods and algorithms, with a clear analysis of their strengths, weaknesses, and applicable scenarios.
4. Insights from the authors' own research on advanced Bayesian methods for emerging 6G technologies like massive MIMO and ISAC.
5. A companion MATLAB codebase implementing key algorithms discussed in the book.

This book is intended for senior undergraduates, graduate students, and industry researchers in telecommunications engineering, and can be used in courses on Bayesian signal processing, compressive sensing, and 6G wireless technologies. Readers are assumed to have a basic background in statistics, probability, linear algebra, signal detection, parameter estimation, and convex optimization. For those less familiar with the basics of parameter estimation, Part I provides necessary foundations, including expectation maximization and message passing. Those already proficient may start with Part II on sparse Bayesian inference, while readers interested in recent advances can proceed directly to Part III on advanced topics in future wireless networks. To aid lecturers in adopting this book, the accompanying materials of this book, including the solutions manual, lecture slides, and codebase, can be found on the website at https://github.com/biwcs-book.

Hangzhou, China
November 2025

An Liu
Ming-Min Zhao

Acknowledgements We wish to record our profoundest appreciation to Prof. Tarek El-Bawab. It was his initial invitation that set this book in motion, and his continued counsel and supportive actions have been a guiding light throughout this journey.

Sincere thanks are also due to our collaborators and friends, whose penetrating observations have left a distinct mark on this volume: Prof. Jisheng Dai, Prof. Xiaojun Yuan, Prof. Xiaoyan Kuai, Prof. Yinglei Teng, and Prof. Ye Xue. We would also like to extend our gratitude to the anonymous reviewers; their thoughtful and constructive comments helped us strengthen the work immeasurably.

We are especially indebted to our students, Yubo Wan, Zhixiang Hu, Wenkang Xu, Wei Xu, Yufan Zhou, Xiayu Zhu, Jingyi Li, Lebin Chen, Xiang Chen, Zhiyu Chen and Kangxin Hu. Their tireless efforts and close collaboration in compiling and shaping the material were the driving force that enabled us to meet our deadlines.

Finally, our hearts are filled with gratitude for our family members—our parents, wives, and children. Their love, patience, and steadfast support provided the bedrock upon which this endeavor was built, sustaining us through every challenge.

Competing Interests The authors have no competing interests to declare that are relevant to the content of this manuscript.

Note from the Series Editor

This Series has its roots in the Telecommunication Engineering Education (TEE) movement in the USA, which I had the honor of leading from 2008 to 2015. The movement resulted from the remarkable impacts telecom networks had on society and the many more impacts looming on the horizon at that time. Major changes were unfolding in the theoretical foundations of this field of engineering, its design strategies, and practical applications. The industries we knew as the Telecom, Cable, and Data-Communication industries through the twentieth century were undergoing innovative technical changes, repositioning themselves in the market, and converging; in preparation for the global economy of the twenty-first century.

The TEE movement was prompted by a personal experience I went through when moving from industry to academia. Telecom engineering, or network engineering—since the network has been central to the evolution telecom has undergone for three to four decades—became a distinct multidisciplinary field of study. This field blends ingredients from electrical engineering, computer engineering, computer science, and other disciplines, but it should not be taught merely as a branch of one of them. It became evident that society is better served by adding specialized undergraduate/graduate courses and programs in telecom/network engineering to our higher-education system, especially in the USA where this is uncommon.

The story of the TEE movement has been told in the literature (e.g., IEEE Communications Magazine, Vol. 53, No. 11, pp. 35–9, November 2015). The movement accomplished its main goal of adding telecom/network engineering to the ABET list of recognized engineering disciplines. ABET is the Accreditation Board of Engineering and Technology, Inc., which accredits engineering programs in the USA and more than 40 other countries. However, a lot of work had to be done to capitalize on that milestone. The literature of this discipline was rich in scholarly research papers, but not as rich in textbooks designed to support the development of new university courses and curricula. These textbooks are needed to facilitate rigorous academic training for a new generation of network engineers. The idea of this Series arose subsequently to help the telecom/networking community overcome this shortcoming. Since its inception, Springer Nature's TTE (Textbooks in Telecommunication Engineering) has been doing this job. The proximity of the acronyms herein—between TEE and TTE—is coincidental, but worth noting!

I am proud of the inroads TTE has made in publishing quality textbooks and in fulfilling the goals and objectives the Series has had since its launch. By mid-2026, we approach the milestone of having nearly 50 books, including 30 published and about 20 in the making (whether in the writing, peer-reviewing, or production phases). More than 65 authors have contributed to this Series so far and they come from nearly 25 different countries, cultures, and higher-education systems, scattered all over the world.

The topics covered by this Series so far span the entire landscape of telecom and network engineering, including wireless, optical, satellite; access, metro, and core; terrestrial, space, aerial, and under-water; devices, systems, and networks; signal processing; queuing theory, traffic engineering, and operations research; network management; Internet and Internet of Things (IoT); cybersecurity; applications of data science, artificial intelligence and machine learning in network engineering; applied electromagnetics and antennas; and economy/business aspects of telecom. The Series includes a communication-systems primer, a radio-receivers primer, and a cellular-generations primer, for undergraduate and graduate students. A synchronization primer, Wi-Fi primer, and a cybersecurity primer are close to finishing, and others are expected.

We have covered cutting-edge topics with modern textbooks designed to propose innovative academic courses and programs, bearing in mind sometimes that further breakthroughs are imminent in these topics. Many of our books are also ideal for short courses and training of practitioners and industry executives to catch up with the rapid progress in the field. These include books discussing the roles of edge and cloud computing; blockchains; and AI/ML applications in telecom and networks. Special topics such as railway telecom, navigation systems, semantic communications, low-altitude wireless and video analytics are covered. Other topics at the forefront of contemporary socioeconomic challenges, such as broadband access, the digital divide, industry 4.0, and applications of IoT in various aspects of modern life are addressed.

I have tried my best to make this Series a top global de-facto home for quality telecom/network engineering textbooks, and I hope that many in our community consider it as such. The fruits of the hard work put in by many people to achieve this goal are now ready for our community to reap.

I want to thank my wife Shahira and my daughter Nadine for their support of my endeavor with this Series over more than a decade so far. I have put a lot of time into this mission and they have understood how important this work is to me. I owe a great deal to all the authors of this Series, without whom we could not build up such a monumental body of knowledge. Words cannot describe my appreciation of their contributions and talent. I would like to also express my thanks to Springer Nature, and to Mary James in particular. Mary has been my main Springer partner throughout this journey, and I couldn't have thought of a better one. Finally, I thank all my colleagues in the field of telecom/network engineering and our readership for their continued support of this global project.

Tarek S. El-Bawab, Ph.D., FIEEE
Series Editor

Contents

Part I Bayesian Parameter Estimation

1 Bayesian Decision Theory (BDT): A Unified Framework for Parameter Estimation 3
1.1 Basic Concept of BDT 3
1.2 Tools for Solving Bayesian Decision Problem 5
1.2.1 Probability Representations 5
1.2.2 Properties of Probabilistic Inference 6
1.2.3 Bayesian Decision Rule 8
References 10

2 Point Estimators 11
2.1 Maximum A Posteriori (MAP) Estimation 11
2.2 Maximum Likelihood (ML) Estimation 12
2.2.1 Bias and Variance 14
2.2.2 Mean Squared Error 15
2.3 Least Squares Estimation 16
2.4 (Linear) Minimum Mean-Square Error Estimation 20
References 24

3 Bayesian Estimators 25
3.1 Basic Concepts 25
3.2 Bayesian Solution 26
3.3 Conjugate Priors 28
References 31

4 Expectation Maximization (EM) 33
4.1 Basic Concepts 33
4.2 EM Convergence 35
4.3 Geometric Interpretation 36
4.4 MAP Estimation via EM 38
References 40

5 Factor Graph and Message Passing 41
5.1 Motivation 41
5.2 Factor Graphs and the Sum-Product Algorithm 42
5.2.1 Computing a Single Marginal Function 43
5.2.2 Computing All Marginal Functions 44
5.2.3 The Max-Product Algorithm 46
5.2.4 Running Example for Repetition Code 46
5.3 Practicalities 47
5.3.1 Messages and Their Representations 47
5.3.2 The Sum-Product Algorithm for Factor Graphs with Cycles 50
5.4 Convergence Results for the Sum-Product Algorithm 51
References 54

6 **Subspace Methods** ... 57
6.1 MUSIC ... 57
6.1.1 Basic Principle ... 57
6.1.2 Root-MUSIC ... 59
6.1.3 2D MUSIC ... 60
6.2 ESPRIT ... 64
6.2.1 Basic Idea ... 66
6.2.2 Estimating the Subspace Rotation Operator ... 67
6.2.3 Dealing with ULAs ... 68
6.2.4 Multiple Invariance ESPRIT ... 70
References ... 79

Part II Sparse Bayesian Inference for Wireless Communications and Sensing

7 **Introduction to Compressive Sensing** ... 83
7.1 Introduction ... 83
7.1.1 Motivation ... 83
7.1.2 Core Idea ... 85
7.2 Theoretical Foundations ... 86
7.2.1 Sparsity ... 86
7.2.2 Incoherent Sampling ... 87
7.2.3 Undersampling ... 87
7.3 Robust Compressive Sampling ... 89
7.3.1 Restricted Isometry Property ... 89
7.3.2 Robust Signal Recovery from Noisy Data ... 91
References ... 99

8 **Optimization Based CS Recovery Algorithms** ... 101
8.1 Reweighted ℓ_1-Norm Minimization ... 101
8.1.1 Weighted ℓ_1-Norm Minimization ... 101
8.1.2 Reweighted ℓ_1-Norm Minimization ... 103
8.1.3 Connection to Majorization-Minimization ... 104
8.2 LASSO ... 106
8.2.1 LASSO Estimator ... 106
8.2.2 Solution to the LASSO Problem ... 107
8.2.3 Bayesian LASSO ... 110
8.2.4 Theoretical Insights ... 112
8.3 Generalized LASSO ... 112
8.3.1 Performance Analysis ... 113
8.3.2 Group LASSO ... 114
8.3.3 Burst LASSO ... 116
References ... 124

9 **Greedy CS Recovery Algorithms** ... 127
9.1 Orthogonal Matching Pursuit ... 127
9.1.1 Performance Analysis ... 128
9.1.2 Regularized OMP ... 130
9.1.3 Stagewise OMP ... 132
9.2 CoSaMP ... 134
References ... 146

10 Approximate Message Passing ... 149
10.1 Motivation for Approximate Message Passing ... 149
10.2 Derivation of AMP ... 149
10.2.1 Problem Setup ... 149
10.2.2 Factor Graph and Message Passing ... 150
10.2.3 From SPMP to AMP ... 151
10.3 Intuitive Interpretation of AMP ... 159
10.4 Denoising Function Design ... 160
10.4.1 Denoising Function for I.i.d. Bernoulli–Gaussian Distribution ... 160
10.4.2 Denoising Function for a General Independent Sparse Prior Distribution ... 160
10.4.3 Denoising Function for Unknown Independent Sparse Prior Distribution ... 161
10.5 State Evolution for AMP ... 162
10.5.1 A Concise Derivation of SE ... 163
10.5.2 Rigorous Results for SE ... 164
10.5.3 Connection Between the AMP Algorithm and State Evolution ... 165
10.6 Generalizations and Extensions of AMP ... 165
10.6.1 Complex-Valued AMP ... 166
10.6.2 Extensions to Different Sensing Matrices ... 166
10.6.3 Dependent Signal Structures and Non-separable Denoisers ... 166
10.6.4 Multiple Measurement Vector Problems ... 166
10.6.5 Bilinear Models ... 167
References ... 168

11 Sparse Bayesian Learning ... 171
11.1 Observation and Sparse Prior Models in SBL ... 171
11.2 Derivation of SBL Based on the EM Framework ... 173
11.2.1 Overview of the In-Exact Block MM Method ... 174
11.2.2 Detailed Implementations of the SBL-M Step ... 175
11.2.3 Convergence Analysis of SBL ... 177
11.3 Extension to Handle Group Sparsity ... 177
11.4 Application: FDD Massive MIMO Channel Estimation with Angular Reciprocity ... 179
11.4.1 Angular Reciprocity in FDD Systems ... 179
11.4.2 Joint Off-Grid Model for FDD Channel Estimation ... 180
11.4.3 Statistical Modeling with Angular Reciprocity for SBL Framework ... 182
11.4.4 SBL Algorithm for FDD Channel Estimation ... 182
11.4.5 Simulation Results and Performance Analysis ... 185
References ... 187

12 Variational Bayesian Inference ... 189
12.1 Motivation of VBI ... 189
12.2 Mathematical Foundations: The KLD and Mean-Field Approximation ... 190
12.2.1 The Optimization Goal: Minimizing the Reverse KLD ... 190
12.2.2 The Mean-Field Variational Family: The Factorization Assumption ... 191
12.2.3 Formalizing the Variational Optimization Problem ... 191
12.3 MFVBI: The General Framework ... 192
12.3.1 Derivation of the ELBO ... 192

12.3.2 The Closed-Form Solution for Optimal Variational Factors 192
12.3.3 The Coordinate Descent Algorithm 193
12.3.4 Properties, Convergence and Practical Challenges 193
12.4 Particle Variational Bayesian Inference 194
12.4.1 Motivation: Capturing Complex Posteriors with Particles 194
12.4.2 Problem Reformulation: Variational Optimization with Discrete Measures 194
12.4.3 Derivation of the Particle Weight Update Rule 195
12.4.4 Algorithm Summary, Complexity and Trade-Offs 195
12.5 Successive Linear Approximation VBI (SLAVBI) 196
12.5.1 Motivation: Enabling Closed-Form Updates in Nonlinear Models 196
12.5.2 Core Idea: Iterative Linearization and Gaussian Variational Posteriors 196
12.5.3 Derivation: Linearization and Variational Update 197
12.5.4 Algorithm Summary, Complexity and Connection to Extended Kalman Filtering 197
12.6 Sparse VBI for Compressive Sensing 198
12.6.1 Problem Setup: Linear Models with Uncertain Parameters 198
12.6.2 Revisiting the Hierarchical Sparse Prior 198
12.6.3 Derivation of Exact Variational Updates Under Mean-Field 199
12.6.4 Relationship to SBL 200
12.7 Application Examples: VBI in Wireless Localization 200
12.7.1 PVBI for Wireless Cooperative Localization 201
12.7.2 Sparse VBI for Massive MIMO Based Localization 203
References 209

Part III Advanced Topics in Future Wireless Networks

13 Structured Compressive Sensing with a Dynamic Grid 213
13.1 Motivation 213
13.2 Structured Sparsity Model and Structured CS Problem Formulation 214
13.2.1 Structured Sparsity Model 214
13.2.2 Structured CS Problem Formulation with a Dynamic Grid 218
13.3 Channel Estimation for Massive MIMO-OFDM Systems 219
13.3.1 Channel Model for Massive MIMO-OFDM Systems 220
13.3.2 Channel Estimation for FDD Systems 221
13.3.3 Channel Estimation for TDD-HBF Systems 225
13.4 Integrated Sensing and Communication 228
13.4.1 System Model 230
13.4.2 Formulation as a Structured CS Problem with a Dynamic Grid 233
References 239

14 Expectation Maximization-Based Turbo Sparse Bayesian Inference 241
14.1 Outline of EM-Turbo-SBI Framework 241
14.1.1 Problem Formulation 241
14.1.2 EM-Turbo-SBI Framework 242
14.1.3 M-Step: Inexact Block MM 243
14.1.4 EM-Based Surrogate Function Construction 244
14.1.5 Outline of the E-Step 245
14.2 Turbo-CS Algorithm 247
14.2.1 Outline of Turbo-CS Algorithm 247
14.2.2 LMMSE in Module A 249

14.2.3 Message Passing in Module B ... 250
14.2.4 Implementation Issues ... 253
14.3 Turbo-AMP Algorithm ... 255
14.3.1 Outline of Turbo-AMP Algorithm ... 256
14.3.2 AMP Algorithm for Independent BG Prior (Module A) ... 257
14.3.3 Implementation Issues ... 258
14.4 Turbo-VBI Algorithm ... 260
14.4.1 Outline of Turbo-VBI Algorithm ... 261
14.4.2 Sparse VBI Algorithm (Module A) ... 264
14.4.3 Implementation Considerations ... 267
14.5 Discussions and Comparative Analysis ... 269
14.5.1 Insight into Turbo-SBI Algorithms ... 269
14.5.2 Comparative Analysis of Turbo-SBI Algorithms ... 270
14.5.3 Guidance on Algorithm Selection ... 271
14.6 Applications ... 272
14.6.1 Turbo-CS for FDD Massive MIMO-OFDM Channel Estimation ... 272
14.6.2 Turbo-VBI for Joint Environment Sensing and Channel Estimation ... 275
References ... 279

15 Dynamic Sparse Bayesian Inference ... 281
15.1 Problem Formulation for Dynamic Sparse Bayesian Inference ... 281
15.1.1 System Model and Probabilistic Framework ... 281
15.1.2 Problem Formulation ... 283
15.2 Dynamic Turbo-CS Algorithm ... 284
15.2.1 Algorithm Overview ... 284
15.2.2 DTurbo-CS-E Step ... 285
15.2.3 Temporal Message Propagation ... 287
15.2.4 DTurbo-CS-M Step ... 289
15.3 Application Example: Channel Tracking in 5G NR Systems ... 290
15.3.1 Frame Structure and SRS Pattern ... 291
15.3.2 Channel Sounding Reference Signal in 5G NR Systems ... 292
15.3.3 Physical Channel Model ... 292
15.3.4 Sparse Channel Representation ... 293
15.3.5 Markov Channel Model for Dynamic Sparsity ... 294
15.3.6 Outline of the Channel Tracking Algorithm ... 296
15.3.7 Simulation Results ... 297
References ... 300

16 Advanced Variational Bayesian Inference Algorithms ... 303
16.1 The Computational Challenge in Structured Sparse Recovery ... 303
16.1.1 Revisiting the VBI Update: The Matrix Inversion Bottleneck ... 303
16.1.2 Two Philosophical Approaches to Complexity Reduction ... 304
16.2 Relaxation-Based Complexity Reduction: Inverse-Free VBI ... 305
16.2.1 Overview of the IF-VBI Approach ... 305
16.2.2 Compressive Observation and Sparse Prior Models ... 305
16.2.3 Variational Problem Formulation ... 306
16.2.4 Constructing a Relaxed Evidence Lower Bound ... 307
16.2.5 The IF-VBI Algorithm: An EM-Like Procedure ... 308
16.2.6 Convergence Behavior and Complexity Analysis ... 309
16.3 Constraint-Based Complexity Reduction: Subspace-Constrained VBI ... 309
16.3.1 Motivation: A Different Path to Efficiency ... 309
16.3.2 Problem Formulation with Constraints ... 310

16.3.3 Subspace-Constrained Mean Update ... 312
16.3.4 Updates for Remaining Variational Distributions ... 313
16.3.5 Convergence Analysis ... 315
16.4 Extending to Bilinear Models: Bilinear Subspace-Constrained VBI ... 315
16.4.1 Bilinear Sparse Bayesian Inference ... 315
16.4.2 BiSVBI Algorithm ... 317
16.5 Application I: Channel Estimation in Hybrid Beamforming Systems ... 320
16.5.1 Overview and Baseline Algorithms ... 320
16.5.2 Simulation Setup and Performance Metrics ... 321
16.5.3 Simulation Results ... 322
16.5.4 Influence of SNR ... 324
16.5.5 Influence of Number of Channel Paths ... 324
16.6 Application II: Joint Sensing and Data Recovery in ISAC Systems ... 325
16.6.1 Background: ISAC and the Bi-Static Sensing Challenge ... 325
16.6.2 System Model ... 326
16.6.3 Bilinear Sparse Bayesian Inference Formulation ... 327
16.6.4 The EM-BiSVBI Algorithm ... 329
16.6.5 Simulation Results ... 330
References ... 336

17 Bayesian Inference Based on Stochastic Optimization and Deep Learning ... 337
17.1 Review of Particle VBI for Non-Convex Parameter Estimation ... 337
17.1.1 Problem Formulation Based on VBI Framework ... 337
17.1.2 Particle-Based Variational Distributions ... 338
17.2 Stochastic Particle Variational Bayesian Inference ... 339
17.2.1 Core Ideas and Problem Formulation ... 339
17.2.2 SPVBI Algorithm Design Based on Parallel SSCA ... 340
17.2.3 Algorithm Summary ... 345
17.3 Deep-Unfolded SPVBI ... 345
17.3.1 Overview of the Deep Unfolding Technique ... 345
17.3.2 LSPVBI Network Architecture ... 346
17.3.3 Modular Architecture Design ... 347
17.3.4 Training Methodology ... 351
17.4 Application Example: Multiband WiFi Sensing ... 352
17.4.1 System Model and Problem Formulation ... 352
17.4.2 Performance Comparison ... 353
17.4.3 Computational Complexity Analysis ... 354
References ... 359

Index ... 361

Acronyms

ADC	Analog-to-Digital Converter
Adam	Adaptive Moment Estimation
AE-SC-VBI	Alternating Estimation-Based Subspace-Constrained Variational Bayesian Inference
AIC	Akaike Information Criterion
aNSE	Asymptotic Normalized Squared Error
AMP	Approximate Message Passing
AoA	Angle of Arrival
AoD	Angle of Departure
AO	Alternating Optimization
AWGN	Additive White Gaussian Noise
BC	Broadcast Channel
BDT	Bayesian Decision Theory
BDL	Bayesian Deep Learning
BER	Bit Error Rate
BG	Bernoulli-Gaussian
BGG	Bernoulli-Gamma-Gaussian
BiG-AMP	Bilinear Generalized Approximate Message Passing
BICM-ID	Bit-Interleaving Coded Modulation with Iterative Decoding
BiSVBI	Bilinear Subspace Variational Bayesian Inferenc
BNNs	Bayesian Neural Networks
BP	Belief Propagation/Back Propagation
BPSK	Binary Phase Shift Keying
BS	Base Station
BWP	Bandwidth Part
CDL	Clustered Delay Line
CFR	Channel Frequency Response
CGMT	Convex Gaussian Min-Max Theorem
CIR	Channel Impulse Response
CNN	Convolutional Neural Network
CoSaMP	Compressive Sampling Matching Pursuit
CP	Cyclic Prefix
CRB	Cramér-Rao Bound
CS	Compressive Sensing
CSI	Channel State Information
CSIT	Channel State Information at the Transmitter
DAD	Delay-Angular Domain
DiSouL	Direct Source Localization
DFT	Discrete Fourier Transform
DL	Deep Learning
DNN	Deep Neural Network
DoA	Direction-of-Arrival

DP	Downlink Pilot
DTurbo-CS	Dynamic Turbo CS
EKF	Extended Kalman Filtering
ELBO	Evidence Lower Bound
EM	Expectation-Maximization
EM-TDAMP	Expectation Maximization and Turbo Deep Approximate Message Passing
EM-Turbo-CS	Expectation-Maximization-Based Turbo Compressive Sensing
EM-Turbo-IF-VBI	Expectation-Maximization-Based Turbo Inverse-Free Variational Bayesian Inference
EM-Turbo-SBI	Expectation-Maximization-Based Turbo Sparse Bayesian Inference
EM-Turbo-VBI	Expectation-Maximization-Based Turbo Variational Bayesian Inference
ERP	Exact Reconstruction Principle
ESPRIT	Estimation of Signal Parameters via Rotational Invariance Techniques
FDD	Frequency Division Duplex
FER	Frame Error Rate
FIR	Finite Impulse Response
GG	Gamma-Gaussian
GNSS	Global Navigation Satellite Systems
GPS	Global Positioning System
HBF	Hybrid Beamforming
HMM	Hidden Markov Model
HyperNet	HyperNetwork
i.i.d.	Independent and Identically Distributed
IF-VBI	Inverse-Free VBI
IFSLA-VBI	Inverse-Free Successive Linear Approximation Variational Bayesian Inference
ISAC	Integrated Sensing and Communication
IS	Importance Sampling
IST	Iterative Soft Thresholding
KL	Kullback-Leibler
KLD	Kullback-Leibler Divergence
LASSO	Least Absolute Shrinkage and Selection Operator
LDPC	Low-Density Parity-Check
LLR	Log-Likelihood Ratio
LMMSE	Linear Minimum Mean Squared Error
LoS	Line of Sight
LP	Linear Program
LS	Least Squares
LSPVBI	Deep-Unfolded SPVBI
MAC	Multiple Access Channel
MAP	Maximum A Posteriori
MBAC	Memoryless Binary Asymmetric Channel
MCMC	Markov Chain Monte Carlo
MDL	Minimum Description Length
MFVBI	Mean-Field Variational Bayesian Inference
MGS	Modified Gram–Schmidt
MI	Multiple Invariance
MIMO	Multiple-Input Multiple-Output
MIMO-OFDM	Multiple-Input Multiple-Output Orthogonal Frequency-Division Multiplexing

MI-ESPRIT	Multiple Invariance Estimation of Signal Parameters via Rotational Invariance Techniques
MM	Majorization-Minimization
ML	Maximum Likelihood
MLP	Multilayer Perceptron
MMSE	Minimum Mean Squared Error
MRF	Markov Random Field
MRI	Magnetic Resonance Imaging
MS	Mobile Station
MSE	Mean Squared Error
MSMP	Min-Sum Message Passing
MU	Multi-User
MUSIC	MUltiple SIgnal Classification
MUs	Mobile Users
NLoS	Non-Line of Sight
NMSE	Normalized Mean Squared Error
NR	New Radio
NSE	Normalized Squared Error
OAMP	Orthogonal Approximate Message Passing
OFDM	Orthogonal Frequency-Division Multiplexing
OMP	Orthogonal Matching Pursuit
PAPR	Peak-to-Average Power Ratio
PO	Partially Orthogonal
PC-HBF	Partially-Connected Hybrid Beamforming
PDF	Probability Density Function
PL	Pseudo-Lipschitz
PSSCA	Parallel Stochastic Successive Convex Approximations
PVBI	Particle Variational Bayesian Inference
QP	Quadratic Program
QPSK	Quadrature Phase-Shift Keying
RBs	Resource Blocks
ReLU	Rectified Linear Units
RF	Radio Frequency
RIP	Restricted Isometry Property
RMSE	Root Mean Square Error
ROMP	Regularized Orthogonal Matching Pursuit
R-TST-MUSIC	Robust Time-Space-Time Multiple Signal Classification
SBI	Sparse Bayesian Inference
SBL	Sparse Bayesian Learning
SC-VBI	Subspace Constrained VBI
SE	State Evolution
SEMM	Single-E Multiple-M
SGD	Stochastic Gradient Descent
SISO	Single-Input Single-Output
SMC	Sequential Monte Carlo
SLAVBI	Successive Linear Approximation VBI
S-MUSIC	Spatial MUltiple SIgnal Classification
SNR	Signal-to-Noise Ratio
SPMP	Sum-Product Message Passing
SPVBI	Stochastic Particle Variational Bayesian Inference
SRSs	Sounding Reference Signals
SR-VBI	Super-Resolution Variational Bayesian Inference
SSCA	Stochastic Successive Convex Approximation

StOMP	Stagewise Orthogonal Matching Pursuit
STCS-DS	Structured Turbo Compressive Sensing with Delay Support
STCS-FS	Structured Turbo Compressive Sensing with Frequency Support
SVD	Singular Value Decomposition
TCS-IID	Turbo Compressive Sensing with i.i.d. Support
TDD	Time Division Duplex
TDD-HBF	Time Division Duplex Hybrid Beamforming
TDAMP	Turbo Deep Approximate Message Passing
TLS	Total Least-Squares
T-MUSIC	Temporal MUltiple SIgnal Classification
TST-MUSIC	Time-Space-Time MUltiple SIgnal Classification
Turbo-AMP	Turbo Approximate Message Passing
Turbo-CS	Turbo Compressive Sensing
Turbo-SBI	Turbo Sparse Bayesian Inference
Turbo-VBI	Turbo Variational Bayesian Inference
UAV	Unmanned Aerial Vehicle
ULA	Uniform Linear Array
UP	Uplink Pilot
UPA	Uniform Planar Array
USE	Uniform Spherical Ensemble
UT-AMP	Unitary Transformation Approximate Message Passing
UUP	Uniform Uncertainty Principle
VBI	Variational Bayesian Inference
VB-CESD	Variational Bayes Joint Channel Estimation and Symbol Decoding
WLS	Weighted Least-Squares
WR-MUSIC	Weighted Root-MUSIC
XR	Extended Reality
ZC	Zadoff-Chu

List of Figures

Fig. 1.1	Toy example of a binary classification problem, where the data points are generated using a Gaussian distribution	4
Fig. 2.1	Example to show the conjunction of the bias and variance of an estimator $\mathbb{E}\{\hat{\theta}\}$. **a** if the bias is large, decreasing the variance is useless, **b** if the variance is large, decreasing the bias is also useless	14
Fig. 3.1	A toy example of Bayesian estimation result, where the red curve indicates low uncertainty, while the magenta curve has higher uncertainty	26
Fig. 3.2	A toy example of different prior distributions	26
Fig. 4.1	Geometric interpretation of EM. Black curve: the log-likelihood of the observed data; red curve: the Q function; Blue dashed curve: the tangent to the observed log-likelihood at the point $P_X(D; \Psi^{(i)})$	37
Fig. 5.1	A factor graph for the product $f_A(x_1)f_B(x_1, x_2)f_C(x_1, x_3, x_4)$	43
Fig. 5.2	A rooted factor graph representation of the example given in Fig. 5.1 for computing a single marginal function $g_1(x_1)$	44
Fig. 5.3	Binary asymmetric channel model	46
Fig. 5.4	Factor graph for decoding a length 2 repetition over a MBAC	47
Fig. 5.5	General steps to solve an MAP estimation problem using the sum-product algorithm	48
Fig. 5.6	A numerical example for normalization and log-domain processing for decoding a length-2 repetition code in MBAC	49
Fig. 5.7	A schematic example of junction tree methods	50
Fig. 6.1	An example of the roots from half-wavelength spaced ULA measurements under different SNRs, with $M = 8$, $K = 3$, $T = 20$, $\theta_1 = -30°$, $\theta_2 = 20°$, and $\theta_3 = 60°$	60
Fig. 6.2	Evolution of the signal contents for the estimation of the parameters of channel path one	62
Fig. 6.3	Tree structure of the TST-MUSIC in solving the scenario shown in Fig. 6.2a	62
Fig. 6.4	Antenna array geometry for multiple source DoA estimation using ESPRIT (adapted from Fig. 3 in [20])	65
Fig. 6.5	Several ULA subarray choices for ESPRIT, **a** maximum overlap, **b** interleaved, **c** mixed (adapted from Fig. 1 in [23])	68
Fig. 6.6	Examples of arrays with multiple displacement invariances (adapted from Fig. 2 in [23])	71

Fig. 7.1 Illustration of Shannon's sampling theorem, **a** spectrum of a properly sampled bandlimited signal (blue) and the adjacent DFT images (red) that do not overlap; **b** a brick-wall low-pass filter, $H(f)$, removes the images, leaves the original spectrum, $X(f)$, and recovers the original signal from its samples; **c** the case that the Nyquist criterion is not satisfied, and adjacent copies overlap 84
Fig. 7.2 **a** Source 512×512 image with intensity values scaled to the unit interval [0, 1]. **b** Visual representation of 10^4 randomly selected wavelet coefficients. **c** Reconstruction using only the dominant 3×10^4 wavelet coefficients (remaining coefficients discarded), with intensity values constrained to the standard display range [0, 1]. As can be seen, natural images exhibit significant compressibility as their energy concentrates in limited coefficient subsets. Visual comparison reveals minimal perceptual degradation relative to the original 85
Fig. 7.3 **a** Original channel coefficients of a massive MIMO channel with 128 transmit antennas forming a ULA, and **b** DFT coefficients of the original channel. Relatively few DFT coefficients capture most of the channel energy; many such channels are highly compressible. **c** The reconstruction obtained by zeroing out all the coefficients in the DFT expansion but the 13 largest. The difference with the original channel coefficients is minor 86
Fig. 7.4 (Adapted from Fig. 3 in [2]) A signal x (horizontal axis) and its reconstruction $x^\star$ (vertical axis) obtained via solving problem (7.15). In this example, we set $N = 512$, $M = 256$, and $S = 64$. The sensing matrix has i.i.d. $\mathcal{N}(0, 1/M)$ entries and z is a Gaussian white noise vector adjusted so that $\|Ax\|_2/\|z\|_2 = 8$. Here, $\|x^\star - x\| \approx 1.95\epsilon$ 92
Fig. 8.1 (Adapted from Fig. 1 in [1]) An example of weighted ℓ_1-norm minimization for improving sparse signal recovery. (a) Sparse signal x_0, feasible set $y = A\tilde{x}$, and ℓ_1 ball of radius $\|x_0\|_1$. (b) There exists an $\tilde{x} = [1/3, 0, 1/3]^T \neq x_0$ for which $\|\tilde{x}\|_1 < \|x_0\|_1$. (c) Weighted ℓ_1 ball, and there exists no $\tilde{x} \neq x_0$ for which $\|W\tilde{x}\|_1 \leq \|Wx_0\|_1$ 103
Fig. 8.2 Near the origin, the log-sum penalty function $f_{\log,\epsilon}(t)$ approximates the canonical ℓ_0 sparsity measure $f_0(t)$ more closely than the convex ℓ_1 norm $f_1(t)$. As $\epsilon \to 0$, $f_{\log,\epsilon}(t)$ increasingly resembles $f_0(t)$, mimicking its steep slope at $t = 0$ to more aggressively penalize small non-zero values and enhance sparsity 106
Fig. 8.3 (Adapted from Fig. 2.4 in [6]) Soft-thresholding function $\mathcal{S}_\lambda(x) = \text{sign}(x)(|x| - \lambda)_+$ (blue dashed line). The black solid line represents the 45° reference line (identity function). The soft-thresholding operator shrinks values toward zero for $|x| \leq \lambda$ and shifts them linearly beyond this threshold 108
Fig. 8.4 (Adapted from Fig. 1 in [18]) Illustration of the system model and channel model for a massive MIMO system with the ULA of 128 antennas and half-wavelength inter-antenna distance at the base station (BS) 112

Fig. 8.5 (Adapted from Fig. 2 in [18]) Comparison of the block-sparse signal and burst-sparse signal. The example block-sparse signal has the dimension $N = 6$ and a single non-zero block of size $D = 3$. The example burst-sparse signal has the dimension $N = 6$ and a single non-zero burst of size $D = 3$ 117
Fig. 8.6 (Adapted from Fig. 3 in [18]) Illustration of the advantages of exploiting the burst-sparsity predicted by Corollary 8.1. Consider a massive MIMO system with the ULA of $N = 128$ antennas at the BS. The burst-sparse channel parameters are given in Fig. 8.4. The per-burst sparsity bound information at the user is $D = 8$ 119
Fig. 9.1 Schematic representation of the StOMP algorithm (Adapted from Fig. 1 in [11]) 132
Fig. 9.2 Simulation results of Example 9.1 (off-grid case) 139
Fig. 9.3 Simulation results of Example 9.1 (on-grid case) 139
Fig. 10.1 Factor graph corresponding to (10.3), where $N = 3, M = 2, p_n$ and f_m represent $p(x_n)$ and $\mathcal{N}(y_m; a_m^T x, \sigma_w^2)$, respectively 150
Fig. 10.2 Histogram of y_e^{10}. The red curve is the distribution of the effective AWGN observation $x + \tau_t Z$ 159
Fig. 10.3 Illustration of the soft-thresholding function $\eta^t(\cdot; \beta)$, where we set $\beta\tau_t^2 = 1$ 162
Fig. 11.1 An illustration of sparse prior and observation models in SBL 173
Fig. 11.2 An illustration of the massive MIMO system 179
Fig. 11.3 An illustration of the off-grid model 181
Fig. 11.4 (Adapted from Fig. 6 in [9]) NMSE of downlink channel estimate versus the number of training pilot symbols T. **a** SNR = 0dB; **b** SNR = 10dB 186
Fig. 12.1 Illustration of network node deployment in wireless cooperative localization 201
Fig. 12.2 Localization error over various prior location precisions 202
Fig. 12.3 A localization model in 5G massive MIMO system 203
Fig. 12.4 Average localization error versus the transmit power 207
Fig. 12.5 Probability of sub-meter precision versus the transmit power 207
Fig. 13.1 Illustration of the burst sparsity for a massive MIMO system with a ULA of 128 antennas and inter-antenna distance $\lambda/2$ at the BS, where λ is the wavelength of the carrier frequency, Ω_1 and Ω_2 are the index sets of the non-zero elements in the angular-domain channel (adapted from Fig. 1 in [1]) 214
Fig. 13.2 Illustration of the joint burst sparsity (adapted from Fig. 2 in [3]) 214
Fig. 13.3 Illustration of the common sparsity among the massive MIMO channels on different subcarriers 214
Fig. 13.4 Three-layer sparse prior model (adapted from Fig. 1 in [10]) 217
Fig. 13.5 Factor graph representation of the Markov chain $p(s)$ for modeling burst sparsity 218
Fig. 13.6 An illustration of massive MIMO-OFDM system 220
Fig. 13.7 Illustration of the BS's UPA steering vector 221
Fig. 13.8 An illustration of (**a**) the one-dimensional uniform grid for ULA when $Q = 32$ and (**b**) two-dimensional uniform grids for UPA when $N_1 = 8$ and $N_2 = 8$ 223

Fig. 13.9 Factor graph for the support-based two-layer sparse prior model of the DAD channel H_d. The factor nodes u_d^l and $w_{d,q}^l$ correspond to the probability distributions $p(t_d^l)$ and $p\left(s_{d,q}^l | s_{d,q-1}^l, t_d^l\right)$, respectively ... 225
Fig. 13.10 Architecture of the PC-HBF massive MIMO systems ... 226
Fig. 13.11 Illustration of the 2D burst sparsity for a massive MIMO system with the UPA ... 227
Fig. 13.12 Factor graph of the 2D Markov model ... 228
Fig. 13.13 Illustration for the applications of ISAC (adapted from Fig. 1 in [19]) ... 229
Fig. 13.14 An example of vehicle communication and sensing in ISAC scenarios. The BS sends downlink pilots to detect the target via the received echo signals, and the user sends uplink pilots for the BS to estimate the uplink communication channel ... 230
Fig. 13.15 Illustration of location domain radar and communication channels and their non-zero coefficients (Adapted from Fig. 1 in [29]) ... 230
Fig. 13.16 Frame structure of the ISAC system (adapted from Fig. 2 in [29]) ... 231
Fig. 13.17 Illustration of the three-layer sparse prior model for 2D joint burst sparsity (adapted from Fig. 3 in [29]) ... 234
Fig. 13.18 Factor graph of the 4-connected MRF model (adapted from Fig. 4 in [29]) ... 236
Fig. 14.1 Framework of the EM-Turbo-SBI algorithm ... 242
Fig. 14.2 Partitioning of the factor graph $\mathcal{G}$ in the Turbo-CS algorithm ... 248
Fig. 14.3 Partitioning of the factor graph $\mathcal{G}$ in the Turbo-AMP algorithm ... 256
Fig. 14.4 Partitioning of the factor graph $\mathcal{G}$ in the Turbo-VBI algorithm ... 262
Fig. 14.5 NMSE of channel estimation versus the number of iterations under SNR = 10dB ... 274
Fig. 14.6 NMSE of channel estimation versus SNR ... 274
Fig. 14.7 RMSE of target and scatterer localization versus SNR ... 276
Fig. 14.8 NMSE of radar and communication channel estimation versus SNR ... 277
Fig. 15.1 An illustration of factor graph $\mathcal{G}_B$ ... 286
Fig. 15.2 An illustration of the frame structure ... 291
Fig. 15.3 An illustration of the hopping SRS pattern (adapted from Fig. 1 in [18]) ... 291
Fig. 15.4 NMSE of the full-band channel versus time for $h_p = 4$ (adapted from Fig. 11 in [18]) ... 297
Fig. 15.5 NMSE of the first BWP channel versus time for $h_p = 4$ (adapted from Fig. 12 in [18]) ... 298
Fig. 16.1 NMSE of channel estimation versus the iteration number (adapted from Fig. 7 in [3]) ... 322
Fig. 16.2 NMSE performance of different channel estimation methods versus the compression ratio (Adapted from Fig. 8 in [3]) ... 323
Fig. 16.3 CPU time of different channel estimation methods versus the compression ratio (adapted from Fig. 9 in [3]) ... 323
Fig. 16.4 NMSE performance of different channel estimation methods versus the SNR (adapted from Fig. 10 in [3]) ... 324
Fig. 16.5 NMSE performance of different channel estimation methods versus the number of paths (adapted from Fig. 11 in [3]) ... 325
Fig. 16.6 Illustration of the system model, the location-domain channel, and its non-zero coefficients ... 326

Fig. 16.7 Convergence behavior of different algorithms (adapted from Fig. 2 in [20]) 331
Fig. 16.8 Channel estimation NMSE versus SNR (adapted from Fig. 3 in [20]) 331
Fig. 16.9 Data recovery NMSE versus SNR (adapted from Fig. 4 in [20]) 332
Fig. 16.10 Scatterer localization RMSE versus SNR (adapted from Fig. 5 in [20]) 332
Fig. 17.1 Illustration of particle approximation 339
Fig. 17.2 Positional update in a single iteration of the stochastic successive convex approximation (Adapted from Fig. 3 in [19]) 342
Fig. 17.3 Comparison of the parallel and non-parallel updating processes of SPVBI (Adapted from Fig. 4 in [19]) 344
Fig. 17.4 Deep-unfolding networks: combining the strengths of iterative algorithms and DNNs 346
Fig. 17.5 Network architecture of the deep-unfolding SPVBI algorithm (Adapted from Fig. 5 in [19]) 347
Fig. 17.6 An illustrative example of lightweight DNN approximation for discrete sampling (Adapted from Fig. 6 in [19]) 348
Fig. 17.7 An example demonstrating the sampling efficiency of different sampling methods (Adapted from Fig. 7 in [19]) 348
Fig. 17.8 Illustration of alternating projection 350
Fig. 17.9 Illustration of multiband WiFi sensing and spectrum distribution (Adapted from Fig. 2 in [19]) 352
Fig. 17.10 Iteration error curves with different particle numbers (Adapted from Fig. 12 in [19]) 354
Fig. 17.11 RMSE of line-of-sight (LoS) delay estimation with respect to the SNR (Adapted from Fig. 13 in [19]) 355

List of Tables

Table 3.1 Commonly used likelihood functions and the corresponding conjugate priors ... 29
Table 9.1 Comparison of optimization-based and greedy CS recovery algorithms ... 140
Table 9.2 Pros & Cons of optimization-based and greedy CS recovery algorithms ... 140
Table 10.1 AMP variable scalings in the large-system limit ... 154
Table 14.1 Factors, distributions, and functional forms in the factor graph $\mathcal{G}$ in Fig. 14.2a ... 248
Table 14.2 Factors, distributions, and functional forms in $\mathcal{G}_B$ of Fig. 14.2b ... 251
Table 14.3 Factors, distributions, and functional forms in the factor graph $\mathcal{G}$ in Fig. 14.4a ... 262
Table 14.4 Comparison of the Turbo-CS, Turbo-AMP, and Turbo-VBI, where PO stands for partially orthogonal ... 270
Table 14.5 System parameters for joint sensing and channel estimation Simulation ... 275
Table 15.1 Factors, distributions, and functional forms in Fig. 15.1 ... 286
Table 15.2 System parameters ... 297
Table 16.1 System configuration for channel estimation evaluation ... 321
Table 16.2 Simulation configuration parameters ... 330
Table 17.1 Comparison of the complexity order for different algorithms ... 355

Part I
Bayesian Parameter Estimation

Bayesian Decision Theory (BDT): A Unified Framework for Parameter Estimation

1

This chapter presents Bayesian decision theory (BDT) as a unified framework for optimal parameter estimation and decision-making under uncertainty. We introduce the core paradigm where an observer infers the state of the world (a random variable X) from correlated observations Y, leveraging their statistical relationship encoded in the joint distribution $P_{XY}(x, y)$. The BDT formalism hinges on defining a decision function $g(y)$ that maps observations y to state estimates $\hat{x}$, evaluated via a problem-specific loss function $L(\hat{x}, x)$ that quantifies decision costs. Optimality is achieved by minimizing the expected loss (risk), derived through probabilistic inference tools including likelihood models $P_{Y|X}(y|x)$, prior distributions $P(x)$, and Bayes' rule $P_{X|Y}(x|y) \propto P_{Y|X}(y|x)P(x)$. Key inference techniques, i.e., chain rule, marginalization, and independence assumptions, are detailed to compute posterior distributions and conditional risks. Practical applications span classification (e.g., 0–1 loss) and continuous estimation (e.g., quadratic loss), with examples illustrating medical diagnosis, signal processing, and communications. Besides, this chapter underscores BDT's strengths, including systematic integration of prior knowledge with observational data, causal interpretability, modularity via probability rules, and foundational optimality guarantees, and also acknowledges computational challenges in complex scenarios.

1.1 Basic Concept of BDT

Generally, the BDT is a very powerful tool to infer the state of a system from certain observations, or to compute the optimal decisions on a problem involving uncertainty [1, 2]. There are two basic concepts in BDT, i.e., world and observer. The world is also known as the environment, which contains many states or classes, drawn from a state or class random variable (or process) X, which can be a scalar, vector, or matrix. For example, if we are focusing on a fish classification problem that aims to categorize different types of fishes based on their characteristics, then X might be a bass or salmon, i.e., $X \in \{\text{bass, salmon}\}$. If a student grading problem is considered, i.e., we are evaluating and assigning a letter grade to each student's academic performance, then we have $X \in \{A, B, C, D, E\}$. Another example, if we are confronted with a medical diagnosis problem, then X might be expressed as $X \in \{\text{disease A, disease B}, \ldots\}$.

The observer usually refers to the entity or agent that makes observations or measurements, and it collects data through various means and uses this data to update their beliefs or estimates about an unknown quantity or parameter. The obtained observations or measurements are also drawn from a random variable (process), denoted by Y. Usually, the observation variables are statistically correlated with the state variables. This is intuitive since if these two processes are independent, then it would be impossible for us to obtain any information about the state from the observations. Because X and Y are correlated, i.e., there is a joint probability density function (PDF) of X and Y, denoted by $P_{XY}(x, y)$ (x and y denote some random realizations of X and Y, respectively), that cannot be simply decomposed as the product of two marginal PDFs $P_X(x)$ and $P_Y(y)$, we are able to draw some useful information about the system state X from the observations Y. In accordance with the abovementioned fish classification example, the observation random variable Y might be $Y \in \{\text{scale length, scale width}\} \in \mathbb{R}^2$ since the length and width information of a fish can be utilized as prior information for effective inference. For the student grading problem, the teacher may grade the students according to their scores obtained in each homework, thus a possible observation random variable might be $Y \in \{\text{HW}_1, \ldots, \text{HW}_N\} \in \mathbb{R}^N$. Similarly, for the medical diagnosis problem, we may have $Y \in \{\text{symptom}_1, \ldots, \text{symptom}_N\} \in \mathbb{R}^N$.

Based on the above discussions, we can see that two types of information are generally required, i.e., the prior information about the state and the observations. The BDT is a unified framework to effectively combine these two types of information, such that accurate inference/estimation can be achieved.

A. Liu and M.-M. Zhao, *Advanced Bayesian Signal Processing*, Textbooks in Telecommunication Engineering,
https://doi.org/10.1007/978-3-032-26656-9_1

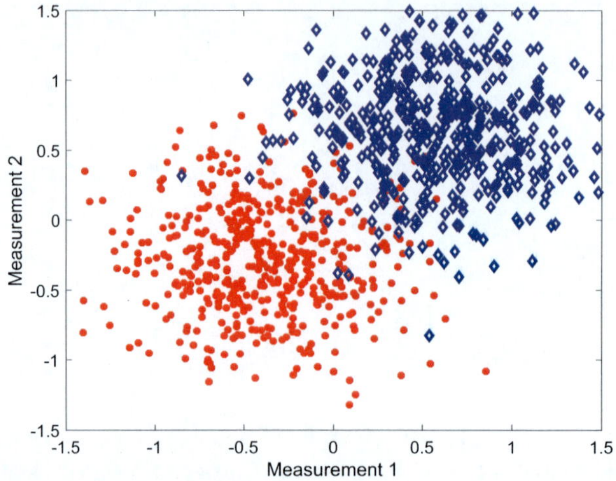

Fig. 1.1 Toy example of a binary classification problem, where the data points are generated using a Gaussian distribution

There is another important concept in BDT that must be carefully handled, which is known as the decision function. It can be viewed as a mapping that takes as input the observations and returns the state of the world x. Mathematically, the decision function, denoted by g, is defined as

$$g : \Omega \rightarrow \Psi, \tag{1.1}$$

where Ω denotes the observation space, i.e., $y \in \Omega$, and Ψ denotes the state space, i.e., $x \in \Psi$. The decision function g is supposed to provide an estimate of the state x based on the observation y, i.e.,

$$g(y) = \hat{x}, \tag{1.2}$$

where $\hat{x}$ is an estimate/prediction of x. To evaluate the performance of the decision function g, we resort to the loss function $L(\hat{x}, x)$, which is defined as the cost of deciding for $\hat{x}$ when the true state is x. If $x = \hat{x}$, then $L(\hat{x}, x)$ is usually set to 0, otherwise, $L(\hat{x}, x)$ should be a positive number. Therefore, the loss function quantifies the negative consequences of incorrect decisions, and it is used to evaluate the performance of different decision rules and to find the optimal decision strategy that minimizes the expected loss. The design of the loss function depends on the specific problem/application and the goals of the decision-making process.

To summarize, BDT provides a framework for making optimal decisions, i.e., seeking an optimal decision function, in the face of uncertainty, by minimizing a chosen loss function. In the following, we provide two toy examples to illustrate how BDT works for practical problems.

Example 1.1 (Classification Problem) The first example is a classification problem, where the observer's aim is to infer the state of the world from M possible states based on the received measurements, which can be expressed as

$$g(y) = \hat{x}, \quad \hat{x} \in \{1, 2, \ldots, M\}. \tag{1.3}$$

When $M = 2$, this problem reduces to the classical binary classification problem in the field of machine learning, where the goal is to classify data points into one of two possible classes, as shown in Fig. 1.1. In this case, we will mostly consider the following "0–1" loss function [3]:

$$L(g(y), x) = \begin{cases} 1, & g(y) \neq x, \\ 0, & g(y) = x, \end{cases} \tag{1.4}$$

which is very similar to the concept of bit error rate (BER) in communication systems.

Example 1.2 (Continuous Variable Estimation) The second example involves the estimation of a continuous variable, such as air temperature estimation and channel state estimation in wireless communications, where the observer tries to predict a continuous x according to

$$g(y) \in \mathbb{R}. \tag{1.5}$$

In this case, the squared error might be a suitable loss function, which can be expressed as

$$L(g(y), x) = \|g(y) - x\|^2. \tag{1.6}$$

The squared-error loss function (1.6) is a commonly used measure in wireless communications, sensing, machine learning, and statistics, e.g., the well-known minimum mean squared-error (MMSE) estimator in wireless communications is derived on the basis of this loss function [4].

1.2 Tools for Solving Bayesian Decision Problem

To solve a Bayesian decision problem, the most basic tool we use is the probability theory [5], where some probability representations (include joint distribution, likelihood distribution, prior probability, etc.) are used to represent the models we use in BDT, and some properties of probabilistic inference (such as chain rule of probability, marginalization, independence, and Bayes rule, etc.) are used to process these probabilities such that the posterior probability (or its approximations) of the state can be obtained eventually.

1.2.1 Probability Representations

In order to find the optimal decision function g, we need a probabilistic description of the problem. Given state x and observation y, we can statistically connect them by the likelihood function, which describes the probability of observing y given x, i.e., the conditional probability $P_{Y|X}(y|x)$. $P_{Y|X}(y|x)$ is also known as the observation model, as it is the model for the observations given the class or state of the world, i.e., it describes the relationship between the state x and observation y. In many applications, one of the most commonly seen observation model is the following linear observation model:

$$y = x + n, \tag{1.7}$$

where n is a Gaussian distributed noise with zero mean and variance σ^2, i.e., $n \in \mathcal{N}(0, \sigma^2)$. In this case, y is also a random variable that is the sum of the constant x and the Gaussian random variable n, thus the conditional probability $P_{Y|X}(y|x)$ can be expressed as

$$P_{Y|X}(y|x) = \frac{1}{\sqrt{2\pi\sigma^2}} e^{-\frac{(y-x)^2}{2\sigma^2}}. \tag{1.8}$$

Besides $P_{Y|X}(y|x)$, the probability distribution of the state x before observing any specific value of y, i.e., the prior probability $P(x)$, is also an important quantity that should be taken into consideration when solving a Bayesian decision problem as it can provide initial knowledge or assumptions about the possible values of the state x. The prior probability reflects a "prior belief" that, if all else is equal, the world will be in state x with a probability $P(x)$. In practice, x may follow a Gaussian distribution, or if we are focusing on a digital communication problem, then x can have two possible values, 0 and 1, usually with equal probabilities, i.e., $P(x = 0) = 0.5$ and $P(x = 1) = 0.5$. It is important to mention that in practice, it is possible that we may not know any prior information about x. In such cases, we usually make some assumptions or use non-informative priors, such as the uniform distribution [6].

Based on the likelihood function $P_{Y|X}(y|x)$ and prior information $P(x)$, we can now obtain the joint distribution $P_{X,Y}(x, y)$, which provides a comprehensive description of the Bayesian decision problem and also the relationship between x and y, i.e.:

$$P_{X,Y}(x, y) = P_{Y|X}(y|x)P(x). \tag{1.9}$$

Note that the concepts of likelihood, prior, and joint distributions are explored in great depth within the context of machine learning in [3].

1.2.2 Properties of Probabilistic Inference

The chain rule of probability, also known as the product rule of probability, is a fundamental concept in probability theory, and an important consequence of the definition of conditional probability. Recall (1.9), it tells us that the joint probability of x and y equals the product of the marginal probability $P(x)$ (also the prior probability if x has a specific physical meaning) and the conditional probability $P_{Y|X}(y|x)$. Thus, if we have a set of random variables $X_1, X_2, \ldots, X_N$, then by recursive application of (1.9) (or mathematical induction), the joint probability of these variables can be expressed as the product of conditional probabilities:

$$\begin{aligned} P_{X_1,X_2,\ldots,X_N}(x_1, x_2, \ldots, x_N) = & P_{X_1|X_2,\ldots,X_n}(x_1|x_2, \ldots, x_N) \\ & \times P_{X_2|X_3\ldots,X_N}(x_2|x_3, \ldots, x_N) \\ & \cdots \times P_{X_{N-1}|X_N}(x_{N-1}|x_N) P_{X_N}(x_N). \end{aligned} \tag{1.10}$$

Note that if we consider a Markov process and the future state (i.e., x_{i+1}) depends only on the current state (x_i) and not on the past states ($x_{i-1}, x_{i-2}, \ldots$), then the joint probability can be simplified as

$$\begin{aligned} P_{X_1,X_2,\ldots,X_N}(x_1, x_2, \ldots, x_N) = & P_{X_1}(x_1) \times P_{X_2|X_1}(x_2|x_1) \times P_{X_3|X_2}(x_3|x_2) \\ & \cdots \times P_{X_N|X_{N-1}}(x_N|x_{N-1}). \end{aligned} \tag{1.11}$$

The chain rule tells us that in the case when the conditional probabilities are more accessible or easier to estimate than the joint probability directly, we can determine the joint probability of all variables by decomposing it into a series of conditional probabilities. Thus, one of the main advantages of the chain rule is that it allows us to modularize interference problems.

Frequently, we have problems with multiple random variables, e.g., when visiting a doctor, a person can be characterized by a collection of random variables (x_1: temperature, x_2: blood pressure, x_3: weight, x_4: cough, etc.). We can summarize this as a vector $\boldsymbol{X}$ of N random variables, and let $P_{\boldsymbol{X}(x_1,\ldots,x_n)}$ denote the joint probability distribution. However, in many cases, we may only care about a subset of $\boldsymbol{X}$. For example, what if the doctor only wants to know if the patient has a cold or not? In this case, the probability of having a cold, denoted by $P(\text{cold})$, generally does not depend on blood pressure and weight, but on body temperature and cough, i.e., we need to know $P_{X_1,X_4}(x_1, x_4)$. Thus, given $P_{X_1,X_2,X_3,X_4}(x_1, x_2, x_3, x_4)$, we can obtain $P_{X_1,X_4}(x_1, x_4)$ by marginalizing it with respect to the selected subset of variables (i.e., X_1 and X_4 in the considered case). By summing (or integrating if continuous variables are considered) the other variables out, we have

$$P_{X_1,X_4}(x_1, x_4) = \sum_{x_2,x_3} P_{X_1,X_2,X_3,X_4}(x_1, x_2, x_3, x_4), \quad \text{(discrete case)} \tag{1.12}$$

$$P_{X_1,X_4}(x_1, x_4) = \int_{x_2}\int_{x_3} P_{X_1,X_2,X_3,X_4}(x_1, x_2, x_3, x_4)\mathrm{d}x_2\mathrm{d}x_3. \quad \text{(continuous case)} \tag{1.13}$$

It is worth mentioning that in some cases, there may exist both discrete and continuous variables, thus both summation and integral operations should be employed in the above marginalization process. However, only using integral operations is also feasible as we can express the PDF of the discrete variables as Dirac delta functions. Marginalization may seem trivial at first glance, but it is a major computational asset for probabilistic inference as there might be lots of variables which are irrelevant, especially in the case of large models. Besides, there are cases where direct evaluation of the joint probability may even be intractable, in which case employing marginalization plays an important role. Sometimes, we also combine it with the chain rule to explore independence relationships that will allow us to further reduce the complexity.

The chain rule and the marginalization rule are not merely algebraic identities; they are the fundamental engines of probabilistic inference. Together, they enable us to construct complex models from simpler components (via the chain rule) and to focus our inquiries on relevant variables by ignoring others (via marginalization). This modular approach is what makes probabilistic reasoning scalable to real-world problems. One may refer to [7] for further discussions about the profound importance of these rules in unifying information theory, inference, and machine learning.

Two random variables X and Y are independent if the probability distribution of one variable does not depend on the values of the other variable. This means that the joint probability distribution of X and Y can be expressed as the product of their marginal probability distributions, i.e.:

$$P_{X,Y}(x, y) = P_X(x)P_{Y|X}(y|x) = P_X(x)P_Y(y), \tag{1.14}$$

which also means that the distribution of Y is not affected by observing X. For example, let S and X denote two random variables that are discrete and continuous, respectively. Then, to compute $P_Y(y)$ based on the joint probability $P_{Y,X,S}(y, x, s)$, we can first employ the marginalization rule as follows:

$$P_Y(y) = \sum_s \int P_{Y,X,S}(y, x, s)\mathrm{d}x. \tag{1.15}$$

Then, according to the chain rule, we have

$$P_Y(y) = \sum_s \int P_{Y|X,S}(y|x, s)P_{S|X}(s|x)P_X(x)\mathrm{d}x. \tag{1.16}$$

By further applying the marginalization rule over S, it can be obtained that

$$\sum_s P_{Y|X,S}(y|x, s)P_{S|X}(s|x) = P_{Y|X}(y|x),$$

$$P_{Y(y)} = \int P_{Y|X}(y|x)P_{X(x)}\mathrm{d}x. \tag{1.17}$$

As can be seen, by dividing and grouping terms (also known as divide and conquer), the integral operation in (1.15) can be simplified.

Based on the abovementioned properties of probabilistic inference, we are ready to present the following Bayes rule, which is the central equation of Bayesian inference:

$$P_{X|Y}(x|y) = \frac{P_{X,Y}(x, y)}{P(y)} = \frac{P_{Y|X}(y|x)P(x)}{P(y)}, \tag{1.18}$$

which tells us how to infer the probability of the state x given the observation y. Note that the posterior probability $P_{X|Y}(x|y)$ has a strong physical meaning in practice, as the state x can be viewed as the inherent reason that cannot be directly observed, while the observation y is in general the consequence of this inherent reason, which is known. The Bayes rule given in (1.18) is extremely useful, because it allows us to "switch" the relation between the variables. Take the above medical diagnosis problem as the example, the doctor needs to know the disease that the patient has given his or her symptom, i.e., the following posterior probability is required to be calculated:

$$P_{X|Y}(\text{disease } x|\text{symptom } y). \tag{1.19}$$

This problem is very complicated because it is not causal, i.e., asking for the probability of cause given consequence is very difficult. However, by employing the Bayes rule, we can transform (1.19) into the probability of consequence given cause and some other "stuff," i.e.:

$$P_{X|Y}(\text{disease } x|\text{symptom } y) = \frac{P_{Y|X}(\text{symptom } y|\text{disease } x)P_X(\text{disease } x)}{P_Y(\text{symptom } y)}. \tag{1.20}$$

Note that obtaining the probability of having one particular symptom given the disease, i.e., $P_{Y|X}(\text{symptom } y|$ $\text{disease } x)$, is relatively easy, as we can get this information from many medical textbooks. Besides, the other "stuff" is also not difficult to obtain, because $P_X(\text{disease } x)$ does not depend on the patient and we can get it by collecting statistics over the entire population, and the marginal probability $P_Y(\text{symptom } y)$ can be obtained by resorting to the marginalization rule, i.e.:

$$P_Y(\text{symptom } y) = \sum_x P_{Y|X}(\text{symptom } y|\text{disease } x)P_X(\text{disease } x). \tag{1.21}$$

1.2.3 Bayesian Decision Rule

Recall that X, Y, $g(y)$, and $L(g(y), x)$ represent the state of the world, the observations, the decision function, and the loss of predicting x with $g(y)$. We are now ready to present the process of how to derive the optimal Bayesian decision rule. In this regard, we introduce the notion of risk, which is defined as the expected or average value of the loss, i.e.,

$$\text{Risk} = \mathbb{E}_{X,Y}\{L(g(Y), X)\} = \int\int P_{X,Y}(x, y)L(g(y), x)\mathrm{d}x\mathrm{d}y. \tag{1.22}$$

Because both X and Y are random variables, the expectation in (1.22) should be taken with respect to the joint distribution of X and Y. Thus, the optimal Bayesian decision rule is to minimize the above risk and then obtain the optimal decision function [8].

Specifically, based on the chain rule, (1.22) can be rewritten as

$$\begin{aligned}\text{Risk} &= \int P_Y(y)\int P_{X|Y}(x|y)L(g(y), x)\mathrm{d}x\mathrm{d}y \\ &= \int P_Y(y)R(y)\mathrm{d}y = \mathbb{E}_Y\{R(y)\},\end{aligned} \tag{1.23}$$

where $R(y) = \int P_{X|Y}(x|y)L(g(y), x)\mathrm{d}x$ is the conditional risk, given the observation y. Note that in (1.23), we did not decompose $P_{X,Y}(x, y)$ as $P_{X,Y}(x, y) = P_X(x)P_{Y|X}(y|x)$ as usual, but in the reverse order. As mentioned in Sect. 1.1, the loss function is by definition non-negative, i.e., $L(g(y), x) \geq 0, \forall x, y$, it follows that $R(y) \geq 0, \forall y$. Hence, Risk is minimum if we minimize $R(y)$ at all y, i.e., if we pick the decision function as follows:

$$g^*(y) = \arg\min_{g(y)} \int P_{X|Y}(x|y)L(g(y), x)\mathrm{d}x. \tag{1.24}$$

Therefore, if the posterior probability $P_{X|Y}(x|y)$ can be obtained by using the Bayes rule in (1.18), we are able to access the expression of the integral term in (1.24), according to which the optimal decision function $g^*(y)$ can be acquired. Once we have obtained $g^*(y)$, by plugging it into (1.22), the Bayes risk can be obtained as

$$R^* = \int\int P_{X,Y}(x, y)L(g^*(y), x)\mathrm{d}x\mathrm{d}y, \tag{1.25}$$

or

$$R^* = \int P_Y(y)\int P_{X|Y}(x|y)L(g^*(y), x)\mathrm{d}x\mathrm{d}y, \tag{1.26}$$

which is optimal (with respect to a particular loss function) and cannot be beaten.

Summary
To this end, it is important to mention that we only provide the abstract expressions of the optimal Bayesian decision rule and the Bayes risk. For some simple problems, we may be able to obtain their closed-form expressions, but in most cases, such closed-form expressions are not easily attainable, and some approximation skills and numerical methods should be employed to obtain a suboptimal solution. The complexity of Bayesian inference is mainly determined by the prior probability and the observation model, if they are relatively simple, then we may be able to obtain the closed-form

expression of the posterior probability or at least its numerical value; otherwise, the derivation would be very complex and difficult.

Generally, Bayesian decision theory or Bayesian methods have the following advantages:

1. Bayesian methods force us to be more objective and consider all available information. By incorporating prior knowledge and updating it with new data, we are encouraged to think critically about our results and not simply accept the first answer that comes along. In this way, Bayesian methods can help us make more informed decisions and avoid being misled by false assumptions or biases.
2. Bayesian models reflect the causal interpretation of the considered problem, and by adopting BDT, we can naturally decompose the posterior probability into "what we knew already" (prior) and "what data tell us" (likelihood). There is no need for heuristics to combine these two sources of information. This decomposition allows us to update our beliefs in a systematic and rational way based on new evidence.
3. Bayes rule, chain rule, and marginalization enable modularity and scalability to very complicated models and problems.

Homework Problems

1.1 (a) Define the following terms in the context of BDT: 1) state variable X; 2) observation variable Y; 3) decision function $g(y)$; 4) loss function $L(\hat{x}, x)$.

(b) Explain why statistical dependence between X and Y is essential for inference. Provide an example where independence would render inference impossible.

1.2 Consider a medical diagnosis system as follows:

- States: $X \in \{\text{healthy}, \text{diseased}\}$.
- Prior: $P(\text{diseased}) = 0.1$.
- Observation Y = blood test result (continuous).
- Likelihood:

$$P_{Y|X}(y|\text{healthy}) = \mathcal{N}(y; 100, 10), \quad P_{Y|X}(y|\text{diseased}) = \mathcal{N}(y; 150, 20). \tag{1.27}$$

(a) Calculate the joint distribution $P_{X,Y}(x, y)$ for x = diseased at $y = 140$.

(b) Compute the marginal distribution $P_Y(y)$ at $y = 140$.

1.3 A binary communication system transmits $X \in \{0, 1\}$ with prior $P(X = 0) = 0.6$ and observation model $Y = X + N$, where $N \sim \mathcal{N}(0, \sigma^2 = 1)$.

(a) Derive the posterior $P_{X|Y}(1|y)$ using Bayes' rule.

(b) Simplify the expression of $P_{X|Y}(1|y)$ for $y = 1.2$.

(c) Explain how causal reasoning (cause to effect) is reflected in your derivation.

1.4 For the system in Problem 1.3:

(a) Define the 0–1 loss function mathematically.

(b) Obtain the optimal decision rule $g^*(y)$ under 0–1 loss.

(c) Using (b), determine $g^*(y)$ at $y = 1.2$.

1.5 Consider a simple single-input–single-output (SISO) communication system with $H \sim \mathcal{CN}(\mu_h, \sigma_h^2)$ denoting the channel gain, and let $Y = HX + N$ denote the received signal with $N \sim \mathcal{CN}(0, \sigma_n^2)$ and $X = 1$.

(a) Given squared-error loss $L(\hat{h}, h) = |\hat{h} - h|^2$, show that the optimal estimator of H is $g^*(y) = \mathbb{E}\{H|Y = y\}$.

(b) Derive $\mathbb{E}\{H|Y = y\}$.
(c) Compute $g^*(y)$ for $\mu_h = 2, \sigma_h^2 = 4, \sigma_n^2 = 1, y = 3$.

1.6 A weather station estimates temperature X using sensor data Y, with prior $X \sim \mathcal{N}(25°\text{C}, 9)$ (i.e., $\mu_x = 25, \sigma_x^2 = 9$), likelihood $Y|X \sim \mathcal{N}(X, 4)$ (i.e., $\sigma_n^2 = 4$), and observation $y = 28°\text{C}$.

(a) Find $P_{X|Y}(x|y)$.
(b) Design the optimal estimator under squared-error loss.
(c) Calculate the Bayes risk R^* for this estimator.

1.7 (a) Explain why BDT requires computing $P_Y(y)$ via marginalization.
(b) Discuss one advantage of BDT in handling prior knowledge versus one computational challenge in high-dimensional spaces.
(c) Compare the 0–1 loss (classification) and squared-error loss (estimation) in terms of optimal decision behavior.

References

1. A. Wald, Statistical decision functions, in *Breakthroughs in Statistics: Foundations and Basic Theory* (Springer, New York, NY, 1950), pp. 342–357
2. J.O. Berger, *Statistical Decision Theory and Bayesian Analysis* (Springer Science & Business Media, 2013)
3. P.E. Hart, D.G. Stork, R.O. Duda, *Pattern Classification* (Wiley, Hoboken, 2001)
4. K.P. Murphy, *Machine Learning: A Probabilistic Perspective* (MIT Press, 2012),
5. E.T. Jaynes, *Probability Theory: The Logic of Science* (Cambridge University Press, 2003)
6. H. Raiffa, R. Schlaifer, *Applied Statistical Decision Theory* (Wiley, 2000)
7. D.J. MacKay, *Information Theory, Inference and Learning Algorithms* (Cambridge University Press, 2003)
8. C.P. Robert, *The Bayesian Choice: From Decision-theoretic Foundations to Computational Implementation* (Springer, New York, NY, 2007)

2 Point Estimators

Statistical inference can be broadly divided into two main categories, i.e., point estimation and Bayesian estimation. In point estimation, unknown parameters are treated as deterministic constants to be estimated from data, whereas Bayesian estimation considers them as random variables with prior distributions. This chapter comprehensively explores point estimation techniques for inferring unknown parameters from observed data, while the next chapter will delve into Bayesian estimation. Beginning with Bayesian approaches, we detail maximum a posteriori (MAP) estimation as the mode of the posterior distribution, highlighting its equivalence to the optimal decision rule under 0–1 loss. Maximum likelihood (ML) estimation is then presented as a frequentist alternative for learning distributions from data, we derive estimators through log-likelihood maximization and analyze their bias-variance tradeoffs via mean squared error (MSE). Least squares (LS) estimation is subsequently framed as a special case of ML under Gaussian noise, with closed-form solutions for linear models. Then, this chapter further delves into MMSE estimation, establishing the conditional mean $\mathbb{E}\{\boldsymbol{X}|\boldsymbol{Y}\}$ as the theoretically optimal estimator while acknowledging computational hurdles. To address practicality, linear MMSE (LMMSE) estimation is introduced, which leverages covariance matrices for tractable solutions, particularly effective under joint Gaussian assumptions. Moreover, connections between estimators (e.g., LS is equivalent to ML under Gaussianity, LMMSE as linear approximation of MMSE) are clarified, and the unifying role of the posterior distribution in Bayesian inference is emphasized, together with the inherent bias-variance-MSE tradeoffs governing estimator design.

2.1 Maximum A Posteriori (MAP) Estimation

In Bayesian statistics, an MAP estimate is an estimate of the unknown state, that equals the mode of the posterior distribution. The MAP can be used to obtain a point estimate of an unobserved state on the basis of observations, which can be expressed as

$$\begin{aligned} g_{\mathrm{MAP}}(y) &= \arg\max_x P_{X|Y}(x|y) \\ &= \arg\max_x \frac{P_{Y|X}(y|x)P_X(x)}{P_Y(y)} \\ &= \arg\max_x P_{Y|X}(y|x)P_X(x). \end{aligned} \tag{2.1}$$

Note that in the above derivation, the denominator of the posterior probability $P_Y(y)$ (i.e., the so-called marginal likelihood) is always positive and does not depend on X, and therefore plays no role in the optimization. MAP estimation is often used when the prior knowledge about the state/parameters is available and can be incorporated into the estimation process. It can also be used as a regularized estimation method, where the prior acts as a regularization term to prevent overfitting.

Questions

Consider again the binary classification problem in Chap. 1 and let us focus on the following "0–1" loss function:

$$L(g(y), x) = \begin{cases} 1, & g(y) \neq x, \\ 0, & g(y) = x. \end{cases} \tag{2.2}$$

A. Liu and M.-M. Zhao, *Advanced Bayesian Signal Processing*, Textbooks in Telecommunication Engineering,
https://doi.org/10.1007/978-3-032-26656-9_2

Prove that the optimal decision rule is the MAP rule, given by

$$g^*(y) = \arg\max_x P_{X|Y}(x|y), \tag{2.3}$$

and the associated risk is

$$R^* = \iint \mathbb{I}_{\{x \neq g^*(y)\}} p_{X,Y}(x, y)\, dx\, dy, \tag{2.4}$$

where the term $\mathbb{I}_{\{x \neq g^*(y)\}}$ denotes the indicator function, which is defined as follows:

$$\mathbb{I}_{\{x \neq g^*(y)\}} = \begin{cases} 1, & \text{if } x \neq g^*(y), \\ 0, & \text{if } x = g^*(y). \end{cases}$$

Based on this question, it would be interesting to mention that for the "0–1" loss function, the following two decision rules are both optimal and equivalent to the MAP rule given in (2.3):

$$g^*(y) = \arg\max_x P_{Y|X}(y|x) P_X(x), \tag{2.5}$$

$$g^*(y) = \arg\max_x \log P_{Y|X}(y|x) + \log P_X(x). \tag{2.6}$$

The derivation of (2.5) is straightforward according to the Bayes rule (1.17) because the marginal probability of y, i.e., $P_Y(y)$, is a fixed constant given the observation y. For this reason, (2.6) is also easy to obtain by taking the logarithm of the joint probability $P_{Y|X}(y|x)P_X(x)$. This step is very useful in practice, since we usually assume that the random variables therein obey exponential type distributions, such as Gaussian distribution, Laplace distribution, etc., the derivation would be much simplified in this case if we take the logarithm of $P_{Y|X}(y|x)P_X(x)$.

2.2 Maximum Likelihood (ML) Estimation

The Bayes risk is optimal only when the models are correct/accurate, i.e., it is guaranteed to achieve the minimum loss when we use the true probabilities. If the models are incorrect or do not accurately capture the true nature of the data or the decision problem, then the Bayes risk may not be optimal. In practice, it is almost impossible to obtain the true distribution of the state, in this case, we usually obtain its approximate prior distribution using data. In some cases, using this approximate prior information may even be worse than not using it at all. Therefore, there is a rather important issue here, which is how can we learn the distribution of a random variable from data? This is where the ML principle comes along.

Specifically, we first choose a parametric model for the distribution of a random variable X, i.e., we assume that the type of the distribution is known, but the parameters of this distribution, denoted by $\boldsymbol{\Theta}$, are unknown.[1] For example, if we assume that X follows Gaussian distribution, then there are two parameters in this distribution, which are its mean and variance, respectively. If X obeys Gamma distribution, then there are also two parameters, the parameter α is known as the shape parameter which determines the shape of the distribution curve, while the parameter β is called the inverse scale parameter, which mainly determines how steep the curve is. Conditioning on the known vector of parameters $\boldsymbol{\Theta}$, the likelihood distribution of X can be obtained by

$$P_X(x|\boldsymbol{\Theta}), \tag{2.7}$$

where $\boldsymbol{\Theta}$ is not a random variable (otherwise it would have to show up as a subscript). Note that in this context, we use the notations $P_X(D|\boldsymbol{\Theta})$ and $P_X(D; \boldsymbol{\Theta})$ interchangeably to represent the probability of data D given parameters $\boldsymbol{\Theta}$, with the semicolon emphasizing that $\boldsymbol{\Theta}$ are parameters rather than random variables. The second step of ML estimation is to assemble a collection of datasets $D = \{x_1, x_2, \cdots, x_N\}$, which is obtained by collecting N independent realizations of the random variable X. Then, the final step of ML estimation is to select the parameters to be the ones that maximize the probability of the data from that class, i.e.,

[1] Note that if we do not assume any distribution type for X, then it would become a variational Bayesian inference problem, which will be introduced in Chapter 12.

$$\begin{aligned}\mathbf{\Theta}^* &= \arg\max_{\mathbf{\Theta}} P_X(D;\mathbf{\Theta}) \\ &\overset{(a)}{=} \arg\max_{\mathbf{\Theta}} \prod_{i=1}^{N} P_X(x_i|\mathbf{\Theta}) \\ &\overset{(b)}{=} \arg\max_{\mathbf{\Theta}} \sum_{i=1}^{N} \log P_X(x_i|\mathbf{\Theta}),\end{aligned} \tag{2.8}$$

where in (a) we have used the fact that the points in D are generated independently, while (b) is because maximizing probabilities or their logarithms is equivalent, as in (2.6).

Example 2.1 (*ML Estimation in the Gaussian Case*). Let us consider the following Gaussian case:

$$f(X) = \frac{1}{\sqrt{2\pi\sigma^2}} e^{-\frac{(X-\bar{X})^2}{2\sigma^2}}, \tag{2.9}$$

where the parameters $\bar{X}$ and σ^2 denote the mean and variance, respectively. Given a collection of N independent samples $\{X_1, X_2, \cdots, X_N\}$, then the likelihood function can be written as

$$\begin{aligned}P_X(X_1, X_2, \cdots, X_N|\bar{X}, \sigma) &= \prod_{i=1}^{N} \left[\frac{1}{\sqrt{2\pi\sigma^2}} e^{-\frac{(X_i-\bar{X})^2}{2\sigma^2}}\right] \\ &= \frac{1}{(2\pi\sigma^2)^{N/2}} e^{-\frac{1}{2}\sum_{i=1}^{N}\left(\frac{X_i-\bar{X}}{\sigma}\right)^2}.\end{aligned} \tag{2.10}$$

Accordingly, the log-likelihood function can be expressed as

$$\Lambda = \ln P_X(X_1, X_2, \cdots, X_N|\bar{X}, \sigma) = -\frac{N}{2}\ln(2\pi) - N\ln\sigma - \frac{1}{2}\sum_{i=1}^{N}\left(\frac{X_i - \bar{X}}{\sigma}\right)^2. \tag{2.11}$$

To find the extreme points of Λ, we take its derivative with respect to $\bar{X}$ and set the corresponding derivative to zero, which yields

$$\frac{\partial\Lambda}{\partial\bar{X}} = \frac{1}{\sigma^2}\sum_{i=1}^{N}(X_i - \bar{X}) = 0, \tag{2.12}$$

or

$$\bar{X} = \frac{1}{N}\sum_{i=1}^{N} X_i, \tag{2.13}$$

and it is just the sample mean/average. Similarly, by setting $\frac{\partial\Lambda}{\partial\sigma} = 0$, we obtain

$$-\frac{N}{\sigma} + \frac{1}{\sigma^3}\sum_{i=1}^{N}(X_i - \bar{X})^2 = 0 \Rightarrow \hat{\sigma}^2 = \frac{1}{N}\sum_{i=1}^{N}(X_i - \bar{X})^2, \tag{2.14}$$

which is the sample variance.

When we talk about estimators, it is important to keep in mind that an estimate is a number, but an estimator is a random variable. An estimate is the value of the estimator for a given sample, and the estimator can be viewed as a mapping function from the given sample to the parameter $\mathbf{\Theta}$ that are required to be estimated, which can be generally expressed as

$$\hat{\mathbf{\Theta}} = f(X_1, X_2, \cdots, X_N). \tag{2.15}$$

This is intuitive since the data are random realizations of the random variable X and they are different in each experiment, as a result of these random realizations, the result of the estimator $\hat{\boldsymbol{\Theta}}$ should also be random. For example, if $D = \{x_1, x_2, \cdots, x_N\}$, when we say $\hat{\mu} = \frac{1}{N}\sum_{i=1}^{N} x_i$, what we mean is $\hat{\mu} = f(X_1, X_2, \cdots, X_N)|_{X_1=x_1,\cdots,X_N=x_N}$ with $f(X_1, X_2, \cdots, X_N) = \frac{1}{N}\sum_{i=1}^{N} X_i$, where each X_i is a random variable.

2.2.1 Bias and Variance

There are several ways to measure the quality of an estimator, among which the most commonly used metrics are bias and variance. Bias is the difference between the expected value of the estimator and the true value. Mathematically, letting $\hat{\theta} = f(X_1, \cdots, X_N)$ denote the estimator, then

$$Bias(\hat{\theta}) = \mathbb{E}_{X_1,\cdots,X_N}\{f(X_1, \cdots, X_N) - \theta\}. \tag{2.16}$$

Thus, if $Bias(\hat{\theta}) = 0$, the estimator is said to be unbiased, meaning its expectation equals the true value for any finite sample size. Moreover, it is often consistent, converging to the true value as the sample size grows large. Otherwise, we have a biased estimator, which systematically overestimates or underestimates the true value on average. In other words, an estimator that has bias will usually not converge to the perfect estimate θ, no matter how large the sample is, which means that it is not expressive enough to approximate the true value arbitrarily well. For example, if θ is negative and the estimator is $f(X_1, \cdots, X_N) = \frac{1}{N}\sum_{i=1}^{N} X_i^2$, then the bias is clearly non-zero. Therefore, the bias determines the asymptotic estimation performance when the sample size is asymptotically large. Next, let us consider this problem: assuming that the estimator converges to the true value, how many samples do we need? This question can be addressed by introducing the concept of variance, which is given by

$$Var(\hat{\theta}) = \mathbb{E}_{X_1,\cdots,X_N}\left\{\left(f(X_1, \cdots, X_N) - \mathbb{E}_{X_1,\cdots,X_N}\{f(X_1, \cdots, X_N)\}\right)^2\right\}. \tag{2.17}$$

The variance usually decreases as one collects more samples. It measures the spread or variability of the estimator's values around its expected value. In other words, it indicates how much the estimator's values can differ from one another when different samples are taken. A lower variance means that the estimator is more stable and reliable. Therefore, the performance of an estimator is generally ruled by the bias and variance, which are both important and usually a tradeoff exists between them, as shown in Fig. 2.1.

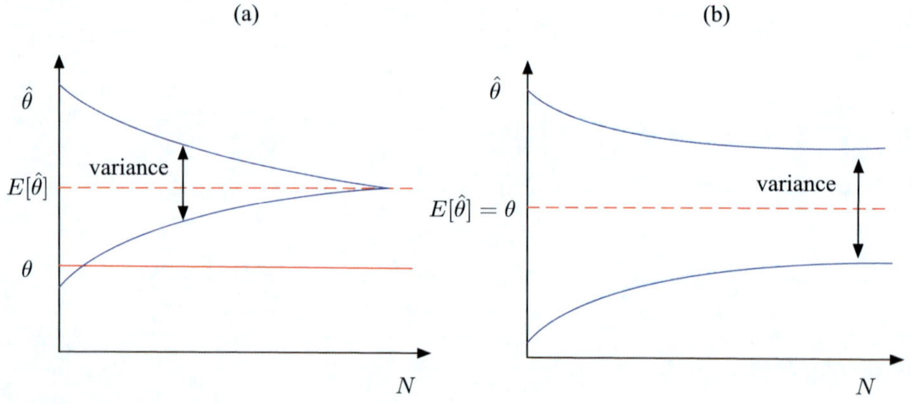

Fig. 2.1 Example to show the conjunction of the bias and variance of an estimator $\mathbb{E}\{\hat{\theta}\}$. **a** if the bias is large, decreasing the variance is useless, **b** if the variance is large, decreasing the bias is also useless

Example 2.2 (*Bias and Variance for ML Estimation in the Gaussian Case*). Consider a random variable X that follows Gaussian distribution $\mathcal{N}(\mu, \sigma^2)$, then the ML estimator for the mean of X can be expressed as $\hat{\mu} = \frac{1}{N}\sum_{i=1}^{N} X_i$ as discussed in Example 2.1. The bias of $\hat{\mu}$ is

$$\begin{aligned} Bias(\hat{\mu}) &= \mathbb{E}_{X_1,\cdots,X_N}\{\hat{\mu} - \mu\} \\ &= \frac{1}{N}\sum_{i=1}^{N} \mathbb{E}_{X_1,\cdots,X_N}\{X_i\} - \mu \\ &= \frac{1}{N}\sum_{i=1}^{N} \mathbb{E}_{X_i}\{X_i\} - \mu \\ &= \mu - \mu = 0, \end{aligned} \tag{2.18}$$

therefore the ML estimator is unbiased in this case. Further consider the variance of $\hat{\mu}$, we have

$$\begin{aligned} Var(\hat{\mu}) &= \mathbb{E}_{X_1,\cdots,X_N}\{(\hat{\mu} - \mathbb{E}_{X_1,\cdots,X_N}\{\hat{\mu}\})^2\} \\ &= \mathbb{E}_{X_1,\cdots,X_N}\{(\hat{\mu} - \mu)^2\} \\ &= \mathbb{E}_{X_1,\cdots,X_N}\left\{\left(\frac{1}{N}\sum_{i=1}^{N} X_i - \mu\right)^2\right\} \\ &= \frac{1}{N^2}\mathbb{E}_{X_1,\cdots,X_N}\left\{\sum_{i=1,j=1}^{N} (X_i - \mu)(X_j - \mu)\right\} = \frac{\sigma^2}{N}. \end{aligned} \tag{2.19}$$

Since σ^2 is finite, the variance of the estimator $\hat{\mu}$ will go to zero on the order of $1/N$ as $N \to \infty$.

2.2.2 Mean Squared Error

One possibility to account for both bias and variance is to minimize the MSE, which is defined as follows:

$$\mathrm{MSE}(\hat{\theta}) = \mathbb{E}_{X_1,\cdots,X_N}\left\{(f(X_1,\cdots,X_N) - \theta)^2\right\}, \tag{2.20}$$

where $\hat{\theta} = f(X_1, \cdots, X_N)$. By assuming that θ is an unknown but deterministic parameter, we have

$$\begin{aligned} \mathrm{MSE}(\hat{\theta}) &= \mathbb{E}\left\{(\hat{\theta} - \mathbb{E}\{\hat{\theta}\} + \mathbb{E}\{\hat{\theta}\} - \theta)^2\right\} \\ &= \mathbb{E}\left\{(\hat{\theta} - \mathbb{E}\{\hat{\theta}\})^2\right\} + 2\mathbb{E}\left\{(\hat{\theta} - \mathbb{E}\{\hat{\theta}\})(\mathbb{E}\{\hat{\theta}\} - \theta)\right\} + \mathbb{E}\left\{(\mathbb{E}\{\hat{\theta}\} - \theta)^2\right\} \\ &= \mathbb{E}\left\{(\hat{\theta} - \mathbb{E}\{\hat{\theta}\})^2\right\} + 2(\mathbb{E}\{\hat{\theta}\} - \mathbb{E}\{\hat{\theta}\})(\mathbb{E}\{\hat{\theta}\} - \theta) + (\mathbb{E}\{\hat{\theta}\} - \theta)^2 \\ &= Var(\hat{\theta}) + Bias^2(\hat{\theta}), \end{aligned} \tag{2.21}$$

where the subscript of the expectation operation, i.e., $X_1, \cdots, X_N$, is omitted here for brevity. Note that the above MSE is actually the conditional MSE, given that the true parameter is θ, and this concept is similar to the conditional risk introduced in Sect. 1.2.3. One can easily extend the MSE above to the case when θ is also a random variable. In this case, by further taking the expectation of $\mathrm{MSE}(\hat{\theta})$ with respect to θ, i.e., $\mathbb{E}_{\theta}\{\mathrm{MSE}(\hat{\theta})\}$, we can obtain the general average MSE.

To this end, we would like to mention that the above ML estimation is introduced from the perspective of learning the distribution of a random variable from data and a given parametric model. From the BDT perspective, i.e., to estimate the state of the world from given observations, the ML estimation can also be expressed as

$$\arg\max_{x} P_{Y|X}(y|x), \tag{2.22}$$

which can be viewed as a reduced form of the MAP rule under the condition that the prior probability $P_X(x)$ is unknown or X is simply assumed to obey uniform distribution.

2.3 Least Squares Estimation

The method of LS is a parameter estimation method in regression analysis to find the best-fitting line or curve to a set of data points by minimizing the sum of the squares of the offsets (i.e., "the residuals") of the points from the curve. LS problems mainly fall into two categories, i.e., linear LS and nonlinear LS, depending on whether or not the model functions are linear with respect to the unknown parameters. The linear LS problem occurs in statistical regression analysis, and it usually has a closed-form solution. The nonlinear problem is often solved by iterative refinement, where the system is approximated as a linear one at each iteration, and thus the core calculation is similar in both cases. There are interesting connections between ML estimation and LS methods. For example, in a general regression problem, we have two random variables X and Y, and a dataset of examples $D = \{(x_1, y_1), \cdots, (x_N, y_N)\}$. Note that in this problem, X and Y are both data, and X is not the state anymore. This is very common in practice as X and Y may denote the transmit and received pilot signals, respectively, which are both known in a typical channel estimation problem for wireless communication. Let us consider a parametric model of the form

$$y_n = f(x_n; \mathbf{\Theta}) + \epsilon_n, \ n = 1, \cdots, N, \tag{2.23}$$

where $\mathbf{\Theta}$ is a parameter vector, and ϵ_n's are independent and identically distributed (i.i.d.) random variable that accounts for noise, which are usually modeled as $\epsilon_n \sim \mathcal{N}(0, \sigma^2)$.

Under the i.i.d. Gaussian noise assumption, y_n is also Gaussian distributed with mean $f(x_n; \mathbf{\Theta})$, and the likelihood of the observations is proportional to

$$\exp\left(-\frac{1}{2\sigma^2}\sum_{n=1}^{N}(y_n - f(x_n; \Theta))^2\right). \tag{2.24}$$

Hence, the ML estimation of $\mathbf{\Theta}$ can be obtained by solving the following problem:

$$\arg\max_{\mathbf{\Theta}} e^{-\|\mathbf{y} - f(\mathbf{x}; \mathbf{\Theta})\|^2}, \tag{2.25}$$

where some constants are omitted here without loss of optimality. Problem (2.25) is further equivalent to the following distance-minimization problem:

$$\arg\min_{\mathbf{\Theta}} \|\mathbf{y} - f(\mathbf{x}; \mathbf{\Theta})\|^2, \tag{2.26}$$

which is known as LS. Therefore, LS estimation is really just the ML estimation under the assumption of i.i.d. Gaussian noises.

A particularly important and tractable case arises when the model is linear in the parameters, e.g., the following linear model:

$$f(x_n; \mathbf{\Theta}) = \sum_{i=0}^{K} \theta_i x_n^i, \ n = 1, \ldots, N.$$

Letting $\Pi_{ij} = x_i^{j-1}$ denote the element on the i-th row and $(j-1)$-th column of the matrix $\mathbf{\Pi} \in \mathbb{R}^{N\times(K+1)}$, i.e.,

$$\mathbf{\Pi} = \begin{bmatrix} 1 & x_1 & x_1^2 & \cdots & x_1^K \\ 1 & x_2 & x_2^2 & \cdots & x_2^K \\ \vdots & \vdots & & \ddots & \vdots \\ 1 & x_N & x_N^2 & \cdots & x_N^K \end{bmatrix}, \tag{2.27}$$

and putting the data $\{y_i\}_{i=1}^N$ in a vector $\mathbf{y}$, we can compute the LS as follows:

$$S(\boldsymbol{\Theta}) = \|\boldsymbol{y} - f(\boldsymbol{x}; \boldsymbol{\Theta})\|^2 = \|\boldsymbol{y} - \boldsymbol{\Pi}\boldsymbol{\Theta}\|^2$$
$$= \boldsymbol{y}^T\boldsymbol{y} - \boldsymbol{\Theta}^T\boldsymbol{\Pi}^T\boldsymbol{y} - \boldsymbol{y}^T\boldsymbol{\Pi}\boldsymbol{\Theta} + (\boldsymbol{\Theta}^T\boldsymbol{\Pi}^T\boldsymbol{\Pi}\boldsymbol{\Theta}). \tag{2.28}$$

The gradient of the LS is

$$\frac{\partial S(\boldsymbol{\Theta})}{\partial \boldsymbol{\Theta}} = -2\boldsymbol{\Pi}^T\boldsymbol{y} + 2\boldsymbol{\Pi}^T\boldsymbol{\Pi}\boldsymbol{\Theta}. \tag{2.29}$$

Setting this gradient to zero and solving for $\boldsymbol{\Theta}$, we can obtain

$$\boldsymbol{\Theta}^* = (\boldsymbol{\Pi}^T\boldsymbol{\Pi})^{-1}\boldsymbol{\Pi}^T\boldsymbol{y}, \tag{2.30}$$

which is the LS estimation of $\boldsymbol{\Theta}$.

In the general nonlinear case, we need to fit a set of N observed data points $\{(x_i, y_i)\}_{i=1}^N$ to a model that is nonlinear in K unknown parameters $\boldsymbol{\Theta} = [\theta_1, \theta_2, \ldots, \theta_K]^T$, where $N \geq K$. The objective is to find the parameter vector $\boldsymbol{\Theta}$ that minimizes the sum of squared residuals given by

$$S(\boldsymbol{\Theta}) = \sum_{i=1}^{N} r_i^2,$$

with the residuals defined as

$$r_i = y_i - f(x_i; \boldsymbol{\Theta}), \quad i = 1, \ldots, N.$$

Here, $f(\boldsymbol{x}; \boldsymbol{\Theta})$ is the nonlinear model function. Since the model is not linear in $\boldsymbol{\Theta}$, the function $S(\boldsymbol{\Theta})$ is generally non-quadratic and may possess multiple local minima. Thus, closed-form solutions are usually unavailable, and iterative numerical methods are employed. Typically, given an initial guess $\boldsymbol{\Theta}^{(k)}$ at iteration k, the model is approximated by a first-order Taylor expansion around $\boldsymbol{\Theta}^{(k)}$, given as follows:

$$f(x_i; \boldsymbol{\Theta}) \approx f(x_i; \boldsymbol{\Theta}^{(k)}) + \sum_{j=1}^{K} \frac{\partial f(x_i; \boldsymbol{\Theta}^{(k)})}{\partial \theta_j} \left(\theta_j - \theta_j^{(k)}\right). \tag{2.31}$$

Denoting the Jacobian matrix $\boldsymbol{J}^{(k)}$ with elements

$$J_{ij}^{(k)} = \frac{\partial f(x_i; \boldsymbol{\Theta}^{(k)})}{\partial \theta_j}, \tag{2.32}$$

and defining the shift vector $\Delta\boldsymbol{\Theta} = \boldsymbol{\Theta} - \boldsymbol{\Theta}^{(k)}$ and the residual vector $\Delta\boldsymbol{y}$ with components

$$\Delta y_i = y_i - f(x_i; \boldsymbol{\Theta}^{(k)}), \tag{2.33}$$

the linearized residuals become

$$r_i \approx \Delta y_i - \sum_{j=1}^{K} J_{ij}^{(k)} \Delta\theta_j. \tag{2.34}$$

In matrix form, we have

$$\boldsymbol{r} \approx \Delta\boldsymbol{y} - \boldsymbol{J}^{(k)}\Delta\boldsymbol{\Theta}, \tag{2.35}$$

where $\Delta\boldsymbol{y} = \boldsymbol{y} - f(\boldsymbol{x}; \boldsymbol{\Theta}^{(k)})$. Thus, the original nonlinear problem is replaced at each iteration by the following linear LS problem:

$$\arg\min_{\Delta\boldsymbol{\Theta}} \|\Delta\boldsymbol{y} - \boldsymbol{J}^{(k)}\Delta\boldsymbol{\Theta}\|^2. \tag{2.36}$$

To solve the linearized LS problem (2.36), the gradient of the linearized objective with respect to $\Delta\boldsymbol{\Theta}$ is first derived as

$$\frac{\partial}{\partial \Delta\boldsymbol{\Theta}} \|\Delta\boldsymbol{y} - \boldsymbol{J}^{(k)}\Delta\boldsymbol{\Theta}\|^2 = -2\boldsymbol{J}^{(k)T}\left(\Delta\boldsymbol{y} - \boldsymbol{J}^{(k)}\Delta\boldsymbol{\Theta}\right). \tag{2.37}$$

Setting this to zero yields the following normal equations:

$$J^{(k)T} J^{(k)} \Delta \Theta = J^{(k)T} \Delta y, \tag{2.38}$$

which are analogous to that in the linear LS case. Thus, the solution to problem (2.36) is

$$\Delta \Theta = \left(J^{(k)T} J^{(k)}\right)^{-1} J^{(k)T} \Delta y,$$

provided $J^{(k)T} J^{(k)}$ is invertible. These equations form the basis for the Gauss–Newton algorithm for a nonlinear LS problem, and the corresponding Gauss–Newton update step is

$$\Theta^{(k+1)} = \Theta^{(k)} + \Delta \Theta. \tag{2.39}$$

This process repeats until convergence, typically when $\|\Delta \Theta\|$ or the change in $S(\Theta)$ falls below a threshold. When measurements have different reliabilities, a weighted sum of squares is minimized, shown as follows:

$$S(\Theta) = \sum_{i=1}^{N} w_i r_i^2, \tag{2.40}$$

where w_i are positive weights, often chosen as the inverse of the measurement error variances. In matrix form, let $W = \text{diag}(w_1, \ldots, w_N)$. The weighted normal equations become

$$\left(J^{(k)T} W J^{(k)}\right) \Delta \Theta = J^{(k)T} W \Delta y. \tag{2.41}$$

All subsequent steps proceed analogously to the unweighted case. The weighted formulation is crucial in practice when observations have heterogeneous noise levels, ensuring that more reliable data points exert a stronger influence on the final estimate, thereby improving estimation accuracy.

Attention
In linear LS, the objective function $S(\Theta)$ is quadratic in Θ, and its contours with respect to any pair of parameters are concentric ellipses (provided the design matrix is full rank). The minimum lies at the center of these ellipses. For nonlinear LS, Θ is generally not quadratic. However, near a local minimum Θ^*, the second-order Taylor expansion yields

$$S(\Theta) \approx S(\Theta^*) + (\Theta - \Theta^*)^T (J^T J)(\Theta - \Theta^*), \tag{2.42}$$

so the contours are approximately elliptical in a small neighborhood of Θ^*. Far from the optimum, the contours deviate significantly from elliptical shapes. This explains why:

1. **Initialization Matters**: Starting values should be as close as possible to the true optimum to ensure convergence.
2. **Gauss–Newton May Diverge**: The algorithm relies on the local quadratic approximation; if the initial point is far from the minimum or the model is strongly nonlinear, the method may fail to converge.

Example 2.3 (*LS Channel Estimation*) Consider a SISO wireless communication system where a transmitter (e.g., base station (BS)) sends known pilot symbols to a receiver (e.g., mobile device) to estimate the channel impulse response (CIR). The system operates in a frequency-flat fading channel, where the channel can be modeled as a single complex coefficient h (due to narrowband assumption, $h = |h|e^{j\phi}$, where $|h|$ is the attenuation and ϕ is the phase shift). The received signal at time k is

$$y[k] = h \cdot s[k] + n[k], \quad k = 1, 2, \ldots, N$$

where $s[k]$ is the pilot symbol at time k (known at the receiver, e.g., for quadrature phase-shift keying (QPSK), $s[k] \in \{\pm 1 \pm j\}$), $n[k]$ is the additive white Gaussian noise (AWGN) with $n[k] \sim \mathcal{CN}(0, \sigma_n^2)$, $y[k]$ is the received signal observed at the receiver, and N denotes the number of pilot symbols. By stacking the N observations into a vector, we have

$$\underbrace{\begin{bmatrix} y[1] \\ y[2] \\ \vdots \\ y[N] \end{bmatrix}}_{\boldsymbol{y}} = \underbrace{\begin{bmatrix} s[1] \\ s[2] \\ \vdots \\ s[N] \end{bmatrix}}_{\boldsymbol{s}} \cdot h + \underbrace{\begin{bmatrix} n[1] \\ n[2] \\ \vdots \\ n[N] \end{bmatrix}}_{\boldsymbol{n}},$$

where $\boldsymbol{y} \in \mathbb{C}^{N\times 1}$, $\boldsymbol{s} \in \mathbb{C}^{N\times 1}$ and $h \in \mathbb{C}$.

The LS estimator minimizes the sum of squared residuals between received signals and hypothesized signals, i.e.,

$$\hat{h}_{\text{LS}} = \arg\min_{h} \|\boldsymbol{y} - \boldsymbol{s}h\|^2.$$

By expanding this cost function, we arrive at

$$J(h) = \|\boldsymbol{y} - \boldsymbol{s}h\|^2 = (\boldsymbol{y} - \boldsymbol{s}h)^H(\boldsymbol{y} - \boldsymbol{s}h),$$

and for complex variables, the Wirtinger derivative w.r.t. h^* (conjugate of h) is used, which gives

$$\frac{\partial J}{\partial h^*} = -\boldsymbol{s}^H(\boldsymbol{y} - \boldsymbol{s}h) = 0. \tag{2.43}$$

Solving (2.43) for h, we can obtain

$$\begin{aligned} \boldsymbol{s}^H(\boldsymbol{y} - \boldsymbol{s}\hat{h}_{\text{LS}}) =0 &\implies \boldsymbol{s}^H\boldsymbol{y} = \boldsymbol{s}^H\boldsymbol{s}\cdot\hat{h}_{\text{LS}} \\ &\implies \hat{h}_{\text{LS}} = (\boldsymbol{s}^H\boldsymbol{s})^{-1}\boldsymbol{s}^H\boldsymbol{y} \implies \hat{h}_{\text{LS}} = h + \underbrace{\frac{\boldsymbol{s}^H\boldsymbol{n}}{\|\boldsymbol{s}\|^2}}_{\text{error}}. \end{aligned} \tag{2.44}$$

Thus, the MSE of this LS channel estimator can be expressed as

$$\text{MSE} = \mathbb{E}\left\{\left|\hat{h}_{\text{LS}} - h\right|^2\right\} = \frac{\sigma_n^2}{\|\boldsymbol{s}\|^2},$$

from which we can observe that the MSE decreases with higher pilot energy ($\|\boldsymbol{s}\|^2$) or lower noise (σ_n^2).

Further consider a SISO system transmitting over a frequency-selective fading channel (e.g., multipath environment). The CIR has L taps, modeled as a finite impulse response (FIR) filter. The discrete-time received signal at sample index k is

$$y[k] = \sum_{l=0}^{L-1} h_l \cdot s[k-l] + n[k], \quad k = 1, 2, \ldots, N,$$

where h_l is the l-th tap of CIR (complex gain of path with delay l). Similarly, by stack N observations into a vector, we have

$$\underbrace{\begin{bmatrix} y[1] \\ y[2] \\ \vdots \\ y[N] \end{bmatrix}}_{\boldsymbol{y}} = \underbrace{\begin{bmatrix} s[1] & 0 & \cdots & 0 \\ s[2] & s[1] & \cdots & 0 \\ \vdots & \vdots & \ddots & \vdots \\ s[N] & s[N-1] & \cdots & s[N-L+1] \end{bmatrix}}_{\boldsymbol{S}} \underbrace{\begin{bmatrix} h_0 \\ h_1 \\ \vdots \\ h_{L-1} \end{bmatrix}}_{\boldsymbol{h}} + \underbrace{\begin{bmatrix} n[1] \\ n[2] \\ \vdots \\ n[N] \end{bmatrix}}_{\boldsymbol{n}},$$

where $\boldsymbol{S} \in \mathbb{C}^{N\times L}$ is the Toeplitz pilot matrix (structure from convolution), and the corresponding LS estimator and MSE are given by

$$\hat{\boldsymbol{h}}_{\text{LS}} = (\boldsymbol{S}^H\boldsymbol{S})^{-1}\boldsymbol{S}^H\boldsymbol{y},$$

and

$$\text{MSE} = \mathbb{E}\left\{\|\hat{\boldsymbol{h}}_{\text{LS}} - \boldsymbol{h}\|^2\right\} = \sigma_n^2 \cdot \text{tr}\left((\boldsymbol{S}^H\boldsymbol{S})^{-1}\right), \tag{2.45}$$

respectively. We can observe from (2.45) that the MSE fundamentally scales with the ratio of noise power to pilot energy, analogous to how estimation performance depends on the effective signal-to-noise ratio (SNR). In practical wireless system design, this underscores the importance of carefully balancing pilot resource allocation against expected noise conditions, i.e., more pilot symbols or higher transmission power may be needed in high-noise environments to maintain satisfactory channel estimation quality, while in cleaner channels, fewer resources might suffice. The optimization of this tradeoff directly impacts overall system efficiency and performance.

2.4 (Linear) Minimum Mean-Square Error Estimation

An MMSE estimator aims to find an estimate of an unknown parameter or signal that minimizes the MSE between the estimated value and the true value. Let $\boldsymbol{X}$ be an $N \times 1$ random vector (i.e., the state), and let $\boldsymbol{Y}$ be the $M \times 1$ measurement or observation vector. An estimator $\hat{\boldsymbol{X}}$ of $\boldsymbol{X}$ is any function of the measurement $\boldsymbol{Y}$, and the corresponding estimation error vector is given by $\hat{\boldsymbol{X}} - \boldsymbol{X}$ and its MSE is given by the trace of error covariance matrix as

$$\text{MSE} = \text{tr}\{\mathbb{E}\{(\hat{\boldsymbol{X}} - \boldsymbol{X})(\hat{\boldsymbol{X}} - \boldsymbol{X})^T\}\} = \mathbb{E}\{\|\hat{\boldsymbol{X}} - \boldsymbol{X}\|^2\}, \tag{2.46}$$

where the expectation is taken over the joint distribution of $\boldsymbol{X}$ and $\boldsymbol{Y}$. Then, the MMSE estimator is defined as the estimator that achieves the minimal MSE, i.e.,

$$\hat{\boldsymbol{X}}_{\text{MMSE}} = \arg\min_{\boldsymbol{X}} \text{MSE}. \tag{2.47}$$

In many cases, it is not possible to determine the analytical expression of the MMSE estimator, as it is very difficult to find the optimal decision function $\hat{\boldsymbol{X}}(\boldsymbol{Y})$. Therefore, in the following, we first focus on a relaxed form of the MMSE estimator by minimizing the MSE within the class of linear estimators.

Specifically, we postulate that the estimator is a simple linear function of $\boldsymbol{Y}$, i.e., $\hat{\boldsymbol{X}} = \boldsymbol{A}\boldsymbol{Y} + \boldsymbol{b}$, where $\boldsymbol{A}$ is a matrix and $\boldsymbol{b}$ is a vector. Therefore, essentially we are solving the following (convex) optimization problem:

$$\min_{\boldsymbol{A},\boldsymbol{b}} \mathbb{E}\{\|\boldsymbol{A}\boldsymbol{Y} + \boldsymbol{b} - \boldsymbol{X}\|^2\}. \tag{2.48}$$

First, notice that for any mean vectors $\bar{\boldsymbol{X}} = \mathbb{E}\{\boldsymbol{X}\}$ and $\bar{\boldsymbol{Y}} = \mathbb{E}\{\boldsymbol{Y}\}$ and any estimator $\hat{\boldsymbol{X}}$, we can always reduce problem (2.48) to a zero-mean case by considering $\boldsymbol{X}' = \boldsymbol{X} - \bar{\boldsymbol{X}}$ and $\boldsymbol{Y}' = \boldsymbol{Y} - \bar{\boldsymbol{Y}}$ as the new state and observation vectors, respectively. In other words, if $\hat{\boldsymbol{X}}'$ is the MMSE estimator for $\boldsymbol{X}'$ given $\boldsymbol{Y}'$, then $\hat{\boldsymbol{X}} = \hat{\boldsymbol{X}}' + \bar{\boldsymbol{X}}$ is the MMSE estimator for $\boldsymbol{X}$ given $\boldsymbol{Y}$. By restricting to the zero-mean case, i.e., assuming $\bar{\boldsymbol{X}} = \boldsymbol{0}$ and $\bar{\boldsymbol{Y}} = \boldsymbol{0}$, and then utilizing the orthogonality principle [1], we have

$$\mathbb{E}\{(\boldsymbol{X} - \hat{\boldsymbol{X}})^T\boldsymbol{Y}\} = tr\left(\mathbb{E}\{(\boldsymbol{X} - \hat{\boldsymbol{X}})\boldsymbol{Y}^T\}\right) = 0, \tag{2.49}$$

which is

$$tr\left(\mathbb{E}\{\boldsymbol{X}\boldsymbol{Y}^T\} - \boldsymbol{A}\mathbb{E}\{\boldsymbol{Y}\boldsymbol{Y}^T\} - \boldsymbol{b}\mathbb{E}\{\boldsymbol{Y}^T\}\right) = 0. \tag{2.50}$$

Therefore, the optimal $\boldsymbol{A}$ and $\boldsymbol{b}$ in the zero-mean case are

$$\boldsymbol{A} = \mathbb{E}\{\boldsymbol{X}\boldsymbol{Y}^T\}(\mathbb{E}\{\boldsymbol{Y}\boldsymbol{Y}^T\})^{-1}, \ \boldsymbol{b} = \mathbb{E}\{\boldsymbol{X}\} - \boldsymbol{A}\mathbb{E}\{\boldsymbol{Y}\} = \boldsymbol{0}. \tag{2.51}$$

Similarly, in the general case (non-zero mean), by letting $\boldsymbol{\Sigma}_{xy} = cov(\boldsymbol{X}, \boldsymbol{Y}) = \mathbb{E}\{(\boldsymbol{X} - \bar{\boldsymbol{X}})(\boldsymbol{Y} - \bar{\boldsymbol{Y}})^T\}$ and $\boldsymbol{\Sigma}_y = cov(\boldsymbol{Y}) = \mathbb{E}\{(\boldsymbol{Y} - \bar{\boldsymbol{Y}})(\boldsymbol{Y} - \bar{\boldsymbol{Y}})^T\}$ denote the cross-covariance matrix between $\boldsymbol{X}$ and $\boldsymbol{Y}$, and the auto-covariance matrix of $\boldsymbol{Y}$, respectively, we can obtain the LMMSE estimator (also known as the Wiener filter [2]) of $\boldsymbol{X}$ as

$$\hat{\boldsymbol{X}} = \boldsymbol{\Sigma}_{xy}\boldsymbol{\Sigma}_y^{-1}(\boldsymbol{Y} - \bar{\boldsymbol{Y}}) + \bar{\boldsymbol{X}}, \tag{2.52}$$

i.e., $\boldsymbol{A} = \boldsymbol{\Sigma}_{xy}\boldsymbol{\Sigma}_y^{-1}$ and $\boldsymbol{b} = \bar{\boldsymbol{X}} - \boldsymbol{A}\bar{\boldsymbol{Y}}$. Lastly, the error covariance matrix and minimum mean square error achievable by such an estimator are

$$cov(\boldsymbol{X} - \hat{\boldsymbol{X}}) = \boldsymbol{\Sigma}_x - \boldsymbol{\Sigma}_{xy}\boldsymbol{\Sigma}_y^{-1}\boldsymbol{\Sigma}_{xy}^T, \tag{2.53}$$

and

$$\mathbb{E}\{\|\boldsymbol{X} - \hat{\boldsymbol{X}}\|_2^2\} = tr(cov(\boldsymbol{X} - \hat{\boldsymbol{X}})), \tag{2.54}$$

respectively, where $\boldsymbol{\Sigma}_x = \mathbb{E}\{(\boldsymbol{X} - \bar{\boldsymbol{X}})(\boldsymbol{X} - \bar{\boldsymbol{X}})^T\}$.

It is important to point out that by presuming that the estimator is linear, we are able to simplify the overall estimation process. Instead of finding the optimal decision function $\hat{\boldsymbol{X}} = g(\boldsymbol{Y})$, only the parameters $\boldsymbol{A}$ and $\boldsymbol{b}$ are required to be optimized by adopting the linear estimator $\hat{\boldsymbol{X}} = \boldsymbol{A}\boldsymbol{Y} + \boldsymbol{b}$. In this way, a complex variational Bayesian inference problem can be simplified to a (convex) optimization problem. This idea is similar to the one we just introduced in Sect. 2.2, and is extremely useful in practice. For example, in modern machine learning or deep learning tasks, we usually use a deep neural network to represent the prior probability or likelihood function, and the network parameters can be optimized (through, e.g., backpropagation) to achieve good estimation performance.

With the same setting as above, we now seek an estimator $\hat{\boldsymbol{X}} = g(\boldsymbol{Y})$, in the space of all (measurable) functions of the observation $\boldsymbol{Y}$, and the result is given in the following theorem [1].

Theorem 2.1 *The MMSE estimator of $\boldsymbol{X}$ given $\boldsymbol{Y}$ is the following conditional mean:*

$$\hat{\boldsymbol{X}} = \mathbb{E}\{\boldsymbol{X}|\boldsymbol{Y}\}. \tag{2.55}$$

Proof Let $\boldsymbol{a}$ denote an estimate of $\boldsymbol{X}$. Then, if we have observed $\boldsymbol{Y} = \boldsymbol{y}$, the MSE is given by

$$h(\boldsymbol{a}) = \mathbb{E}\{\|\boldsymbol{X} - \boldsymbol{a}\|^2|\boldsymbol{Y} = \boldsymbol{y}\} = \mathbb{E}\{\|\boldsymbol{X}\|^2|\boldsymbol{Y} = \boldsymbol{y}\} - 2\boldsymbol{a}^T\mathbb{E}\{\boldsymbol{X}|\boldsymbol{Y} = \boldsymbol{y}\} + \|\boldsymbol{a}\|^2. \tag{2.56}$$

As can be seen, $h(\boldsymbol{a})$ is a quadratic function of $\boldsymbol{a}$, and we can find the minimizing value of $\boldsymbol{a}$ by differentiation:

$$\frac{\partial h(\boldsymbol{a})}{\partial \boldsymbol{a}} = -2\mathbb{E}\{\boldsymbol{X}|\boldsymbol{Y} = \boldsymbol{y}\} + 2\boldsymbol{a} = \boldsymbol{0}, \tag{2.57}$$

which leads to

$$\boldsymbol{a} = \mathbb{E}\{\boldsymbol{X}|\boldsymbol{Y} = \boldsymbol{y}\}. \tag{2.58}$$

Since $\boldsymbol{Y}$ is a random variable, the estimator $\hat{\boldsymbol{X}} = g(\boldsymbol{Y})$ is also a random variable. From the discussion above, we can conclude that the conditional expectation $\hat{\boldsymbol{X}} = \mathbb{E}\{\boldsymbol{X}|\boldsymbol{Y}\}$ has the lowest MSE among all other estimators $g(\boldsymbol{Y})$, which completes the proof.

Although the expression of the MMSE estimator, i.e., (2.55), appears simple at first glance, there are several reasons why we often make a linear assumption and employ the LMMSE estimator instead. First, in practical applications, the exact computation of the MMSE estimator can be extremely complex and computationally expensive. The MMSE estimator often requires the knowledge of the posterior probability $P_{\boldsymbol{X}|\boldsymbol{Y}}(\boldsymbol{x}|\boldsymbol{y})$, and then we need to perform an integral operation as $\int \boldsymbol{x} P_{\boldsymbol{X}|\boldsymbol{Y}}(\boldsymbol{x}|\boldsymbol{y})d\boldsymbol{x}$, which may be difficult, especially when $\boldsymbol{X}$ and $\boldsymbol{Y}$ are vector or matrix random variables. Second, the LMMSE estimator has some desirable properties that make it more suitable for many practical scenarios. For example, it is often easier to analyze and understand the behavior of a linear estimator. It also has certain robustness properties and can perform well even when the assumptions about the distributions are not perfectly met. Moreover, if $\boldsymbol{X}$ and $\boldsymbol{Y}$ are jointly Gaussian, then the LMMSE estimator and the optimal MMSE estimator coincide. This can be simply proved by resorting to the following theorem [3]:

Theorem 2.2 *Let $\boldsymbol{Z}$ follow a multivariate normal distribution, i.e., $\boldsymbol{Z} \sim \mathcal{N}(\boldsymbol{m}, \boldsymbol{\Sigma})$. Then, if $\boldsymbol{Z} = [\boldsymbol{X}^T, \boldsymbol{Y}^T]^T$, the conditional distribution of $\boldsymbol{X}$ given $\boldsymbol{Y}$, is also a multivariate normal distribution $\boldsymbol{X}|\boldsymbol{Y} \sim \mathcal{N}(\boldsymbol{m}_{x|y}, \boldsymbol{\Sigma}_{x|y})$, where the conditional mean (also the MMSE estimator $\hat{\boldsymbol{X}}$) and covariance are*

$$\begin{aligned} \boldsymbol{m}_{x|y} &= \bar{\boldsymbol{X}} + \boldsymbol{\Sigma}_{xy}\boldsymbol{\Sigma}_y^{-1}(\boldsymbol{Y} - \bar{\boldsymbol{Y}}), \\ \boldsymbol{\Sigma}_{x|y} &= \boldsymbol{\Sigma}_x - \boldsymbol{\Sigma}_{xy}\boldsymbol{\Sigma}_y^{-1}\boldsymbol{\Sigma}_{yx}, \end{aligned} \tag{2.59}$$

with block-wise mean and covariance defined as

$$\boldsymbol{m} = \begin{bmatrix} \bar{\boldsymbol{X}} \\ \bar{\boldsymbol{Y}} \end{bmatrix}, \quad \boldsymbol{\Sigma} = \begin{bmatrix} \boldsymbol{\Sigma}_x & \boldsymbol{\Sigma}_{xy} \\ \boldsymbol{\Sigma}_{yx} & \boldsymbol{\Sigma}_y \end{bmatrix}. \tag{2.60}$$

Besides, it is worth mentioning that the MMSE estimator $\hat{\boldsymbol{X}}$ and the estimation error vector $\boldsymbol{V} = \boldsymbol{X} - \hat{\boldsymbol{X}}$ are independent and satisfy

$$\hat{\boldsymbol{X}} \sim \mathcal{N}(\bar{\boldsymbol{X}}, \boldsymbol{\Sigma}_{xy}\boldsymbol{\Sigma}_y^{-1}\boldsymbol{\Sigma}_{yx}), \; \boldsymbol{V} \sim \mathcal{N}(\boldsymbol{0}, \boldsymbol{\Sigma}_{x|y}). \tag{2.61}$$

In the following chapters, we are essentially focusing on how to obtain (2.55) approximately, e.g., how to approximate the posterior probability $P_{\boldsymbol{X}|\boldsymbol{Y}}(\boldsymbol{x}|\boldsymbol{y})$ or the conditional/posterior mean $\mathbb{E}\{\boldsymbol{X}|\boldsymbol{Y}\}$. The main difference is that the observation and prior models are different, e.g., for compressive sensing problems, we usually assume sparse prior and linear observation model.

Summary

This chapter has established a comprehensive framework for point estimation techniques, demonstrating their theoretical foundations and practical implementations in uncertainty-driven decision problems. Beginning with Bayesian approaches, we derived the MAP estimator as the mode of the posterior distribution, and proved its equivalence to the optimal decision rule under 0–1 loss in classification tasks. The ML method was then introduced as a frequentist alternative for parameter learning, where estimators are obtained by maximizing the log-likelihood function. We analyzed key properties of ML estimators, including consistency, asymptotic normality, and the bias-variance trade-off, and highlighted their connection to MAP under uniform priors.

Subsequently, LS estimation was formulated as a special case of ML under i.i.d. Gaussian noise assumptions, yielding closed-form solutions for linear models. For Bayesian inference, the MMSE estimator was established as the conditional mean, which achieves theoretical optimality for quadratic loss but faces computational challenges in high dimensions due to posterior integration. To address this, LMMSE estimation was developed, which coincides exactly with MMSE under joint Gaussianity.

Key insights unifying these methods include

1. In BDT, the most important and useful element is the posterior probability (distribution), with which we are able to obtain all the estimators mentioned above. In other words, the posterior distribution actually contains all the information about the state that we can obtain from the received observations.
2. Equivalence exists among various point estimators, such as LS = ML under i.i.d. Gaussian noise, LMMSE = MMSE for jointly Gaussian states/observations, MAP = ML when priors are uniform, etc.
3. Computational tractability (LMMSE, LS) often trades against statistical optimality (MMSE, MAP), while bias reduction in estimators (e.g., ML's asymptotic unbiasedness) may increase variance in finite samples.

Homework Problems

2.1 Consider a medical diagnostic system for a disease with state $X \in \{0, 1\}$ (0: healthy, 1: diseased). The observation Y is a continuous biomarker level. Assume:

- Likelihoods: $Y|X = 0 \sim \mathcal{N}(\mu_0, \sigma^2)$, $Y|X = 1 \sim \mathcal{N}(\mu_1, \sigma^2)$ with $\mu_1 > \mu_0$.
- Prior: $P_X(1) = p$ (prevalence).

(a) Derive the MAP decision rule $g^*(y)$ in terms of μ_0, μ_1, σ, p. Explain that this MAP estimator minimizes the Bayesian risk for 0–1 loss. Under what condition does MAP become equivalent to ML? Specify the mathematical constraint on p.

(b) Instead of binary classification, consider estimating the probability of disease: $\theta = P(X = 1|Y)$. Show that $\theta = \frac{1}{1+e^{-(ay+b)}}$ for some a, b (express a, b in terms of μ_0, μ_1, σ, p). Besides, the true MMSE estimator for X is $\hat{X}_{\text{MMSE}} = \mathbb{E}\{X|Y\}$. Prove $\hat{X}_{MMSE} = \theta$ and compute $\text{MSE}(\hat{X}_{\text{MMSE}})$. Moreover, compare $\hat{X}_{\text{MMSE}}$ with the MAP decision: when would their outputs fundamentally differ? Which has higher MSE? Justify theoretically.

(c) Suppose the true model has heteroscedastic noise, i.e., $Y|X = 0 \sim \mathcal{N}(\mu_0, \sigma_0^2)$, $Y|X = 1 \sim \mathcal{N}(\mu_1, \sigma_1^2)$ but the diagnostician assumes homoscedastic noise (σ^2) as before. Derive the actual posterior $P_{X|Y}(x|y)$ under the true model. Analyze how this affects the MAP decision threshold, and the bias and variance of $\hat{\theta}$. Using $\mu_0 = 0, \mu_1 = 2, \sigma_0 = 1, \sigma_1 = 3, p = 0.2$, calculate decision thresholds for both models at $y = 1.5$ (for the homoscedastic noise, set $\sigma^2 = (\sigma_0^2 + \sigma_1^2)/2 = 5$).

(d) Define clinical utility as

$$\mathcal{U} = 100P(\text{TP}) - 50P(\text{FP}) - 10P(\text{FN}) \tag{2.62}$$

where TP=true positive, FP=false positive, FN=false negative. Express $\mathcal{U}$ in terms of a decision threshold T. Show that maximizing $\mathcal{U}$ is equivalent to a modified MAP rule with asymmetric loss. Prove the optimal threshold satisfies:

$$\frac{P_{Y|X}(T|1)}{P_{Y|X}(T|0)} = \frac{50}{110} \cdot \frac{1-p}{p}. \tag{2.63}$$

Interpret how prevalence (p) impacts this threshold vs. standard MAP.

2.2 Let $X_1, X_2, \ldots, X_N$ be i.i.d. samples from an exponential distribution with PDF:

$$f(x|\lambda) = \lambda e^{-\lambda x}, \quad x \geq 0, \lambda > 0.$$

(a) Derive the ML estimator $\hat{\lambda}_{\text{ML}}$ for λ.
(b) Compute the bias of $\hat{\lambda}_{\text{ML}}$. Is it asymptotically unbiased?
(c) Calculate the variance of $\hat{\lambda}_{\text{ML}}$.

2.3 Consider estimating the variance σ^2 of a Gaussian distribution $\mathcal{N}(\mu, \sigma^2)$ using N i.i.d. samples. Two estimators are proposed:

$$\hat{\sigma}^2_{\text{ML}} = \frac{1}{N}\sum_{i=1}^{N}(X_i - \bar{X})^2, \quad \hat{\sigma}^2_{\text{unbiased}} = \frac{1}{N-1}\sum_{i=1}^{N}(X_i - \bar{X})^2,$$

where $\bar{X} = \frac{1}{N}\sum_{i=1}^{N} X_i$.

(a) Prove that $\hat{\sigma}^2_{\text{ML}}$ is biased but asymptotically unbiased, while $\hat{\sigma}^2_{\text{unbiased}}$ is unbiased.
(b) Compute the MSE for both estimators. Which one has lower MSE when N is small?
(c) Explain the bias-variance tradeoff using these estimators.

2.4 Consider a dataset of N observations $\{(x_i, y_i)\}_{i=1}^{N}$ generated by the following linear model:

$$y_i = \theta_0 + \theta_1 x_i + \epsilon_i, \tag{2.64}$$

where the noise terms $\epsilon_i \overset{\text{i.i.d.}}{\sim} \mathcal{N}(0, \sigma^2)$ are Gaussian-distributed. Let $\boldsymbol{\Theta} = [\theta_0, \theta_1]^T$ denote the parameter vector.

(a) Show that the ML estimate of $\boldsymbol{\Theta} = [\theta_0, \theta_1]^T$ reduces to the following LS problem:

$$\min_{\theta_0,\theta_1} \sum_{i=1}^{N}(y_i - \theta_0 - \theta_1 x_i)^2.$$

(b) Derive the closed-form solution for $\hat{\theta}_0$ and $\hat{\theta}_1$ using matrix notation (as in Eq. (2.28)).

2.5 Let X and Y be jointly Gaussian random variables with known means μ_X, μ_Y, variances σ_X^2, σ_Y^2, and correlation coefficient ρ.

(a) Derive the LMMSE estimator $\hat{X}_{\text{LMMSE}}$ of X given Y.
(b) Compute the minimum MSE.
(c) Verify that in this case, $\hat{X}_{\text{LMMSE}} = \mathbb{E}\{X|Y\}$ (i.e., LMMSE = MMSE for jointly Gaussian variables).

2.6 An engineer uses the ML estimator for a parameter θ (assumed to be the mean, $\theta = \mu$) under a Gaussian model $N(\mu, \sigma^2)$, while the true data follows a Laplacian distribution $Lap(\theta, b)$ (with PDF $f(x; \theta, b) = \frac{1}{2b}\exp\left(-\frac{|x-\theta|}{b}\right)$, mean θ, and variance $2b^2$).

(a) Is the estimator still consistent (here, "consistent" means that as the sample size approaches infinity, the estimator converges in probability to the parameter being estimated)? Explain.
(b) Discuss potential impacts on bias, variance, and MSE compared to the case when the model is correct.

2.7 An engineer uses the ML estimator for the variance parameter σ^2 under a Gaussian model $N(\mu, \sigma^2)$, but the true data distribution is a uniform distribution $U(a, b)$ with variance $\frac{(b-a)^2}{12}$. Assume that the mean μ is known and matches the true mean of the uniform distribution, $\mu = \frac{a+b}{2}$.

(a) Is the ML estimator for σ^2 still consistent? Explain.
(b) Discuss the potential impacts on the bias, variance, and MSE of the estimator compared to when the model is correct.

2.8 Consider a linear observation model $\boldsymbol{y} = \boldsymbol{A}\boldsymbol{x} + \boldsymbol{n}$, where the observation $\boldsymbol{y}$ is an M by 1 vector, the parameter to be estimated $\boldsymbol{x}$ is an N by 1 vector, the known matrix $\boldsymbol{A}$ is M by N, and $\boldsymbol{n}$ is the noise vector. Assume $M \geq N$ and $\boldsymbol{A}^T\boldsymbol{A}$ is invertible.

(a) Suppose the elements of $\boldsymbol{x}$ and $\boldsymbol{n}$ have i.i.d. standard normal distributions. Derive the MAP, ML, LS, LMMSE, MMSE estimators of $\boldsymbol{x}$. What's the relationship between different estimators (e.g., which one is equivalent to which one? Which one has a better estimation performance in terms of the MSE?)?
(b) Suppose $\boldsymbol{x}$ still has i.i.d. standard normal distributed entries. $\boldsymbol{n}$ has i.i.d. entries but the PDF of each element of $\boldsymbol{n}$ is

$$p(n) = \begin{cases} Ce^{-\frac{n^2}{2}}, & \text{if } n \geq 0 \\ 0, & \text{otherwise} \end{cases},$$

where C is the normalization coefficient. Derive the MAP, ML, LS, LMMSE, MMSE estimators of $\boldsymbol{x}$.
(c) Suppose the elements of $\boldsymbol{x}$ and $\boldsymbol{n}$ have i.i.d. standard normal distributions. But the estimator has wrong information about the variance of $\boldsymbol{x}$, i.e., the variance σ_x^2 of each element of $\boldsymbol{x}$ input to the estimator is not equal to 1. Derive the MAP, ML, LS, LMMSE, and MMSE estimators of $\boldsymbol{x}$ under the wrong information σ_x^2. Discuss how the value of the input information σ_x^2 will affect the performance of different estimators. With the wrong information σ_x^2, is it possible that the MAP, MMSE, and LMMSE estimators are worse than the ML and LS estimators? If possible, when is this possible?

References

1. S.M. Kay, *Fundamentals of Statistical Signal Processing: Estimation Theory* (Prentice-Hall Inc., 1993)
2. N. Wiener, *Extrapolation, Interpolation, and Smoothing of Stationary Time Series* (The MIT Press, 1964)
3. C.R. Rao, C.R., Rao, M. Statistiker, M., C.R. Rao, C.R. Rao, *Linear Statistical Inference and Its Applications*, vol. 2 (Wiley, New York, 1973), pp. 263–270

3 Bayesian Estimators

This chapter introduces Bayesian estimation as a probabilistic framework where parameters are treated as random variables, in contrast with ML point estimation. It focuses on calculating the posterior distribution $P_{\boldsymbol{\Theta}|X}(\theta|D)$ to measure parameter uncertainty, achieved by combining the likelihood $P_{X|\boldsymbol{\Theta}}(x|\theta)$ and the prior $P(\theta)$ through Bayes' theorem. Different from ML estimation, Bayesian estimators provide full probability distributions rather than point estimates, enable natural uncertainty quantification (reducing with more data or lower noise), and retain all information from training data under correct models. Their limitations, such as sensitivity to inaccurate priors and computational complexity, are also discussed. A closed-form Gaussian solution is derived for mean estimation when the prior and likelihood are Gaussian, which demonstrates posterior updates via hyperparameters. The critical concept of conjugate priors (e.g., Beta prior for Bernoulli likelihood, Gaussian prior for Gaussian likelihood) is highlighted for analytical tractability, where posteriors retain the prior's parametric form, and thus sequential updates can be guaranteed. Examples illustrate conjugate pairs and practical trade-offs between prior informativeness and robustness.

3.1 Basic Concepts

Recall that in ML estimation, we usually choose a parametric model for probabilities, i.e., $P_X(x; \boldsymbol{\Theta})$, then we assemble a set of data samples, denoted by $D = \{x_1, \cdots, x_N\}$, which are drawn independently, and select the parameters that maximize the probability of the data, as given in (2.8). ML estimation does not explicitly consider prior knowledge and treats parameters, e.g., $\boldsymbol{\Theta}$, as deterministic but unknown values. However, in Bayesian estimation, $\boldsymbol{\Theta}$ is considered as a random variable. It updates the prior distribution of $\boldsymbol{\Theta}$ using the observed data to obtain a posterior distribution, which allows for uncertainty quantification and a more comprehensive understanding of the parameter space [1]. This paradigm shift from fixed parameters to random variables is the foundation of the Bayesian approach to statistical inference [2]. Specifically, the aim of Bayesian estimation is to compute the posterior distribution

$$P_{\boldsymbol{\Theta}|X}(\theta|D), \tag{3.1}$$

based on the probability density for observations given parameter, i.e., $P_{X|\boldsymbol{\Theta}}(x|\theta)$, and the prior distribution for parameter configurations that encodes the prior beliefs about them, i.e., $P_{\boldsymbol{\Theta}}(\theta)$.

There are a number of significant differences between Bayesian and ML estimates, which are given as follows:

1. ML estimator is a point estimator that produces a number, i.e., the "best" estimate, and to measure its quality, we need to measure its bias and variance [3]. This can also be done with repeated experiments. However, Bayes estimator produces a complete characterization of the parameter from the single dataset D.
2. Under Bayesian estimation, in addition to the most probable estimate, we can also obtain a characterization of the uncertainty by using the Bayes estimator, as illustrated in Fig. 3.1. It is noteworthy that the uncertainty obtained via Bayesian estimation is an objective quantity that tells us how good the estimator is, which is determined by the system and observation models. For example, if the number of observations is larger and the noises are smaller, then the uncertainty is generally lower.

A. Liu and M.-M. Zhao, *Advanced Bayesian Signal Processing*, Textbooks in Telecommunication Engineering,
https://doi.org/10.1007/978-3-032-26656-9_3

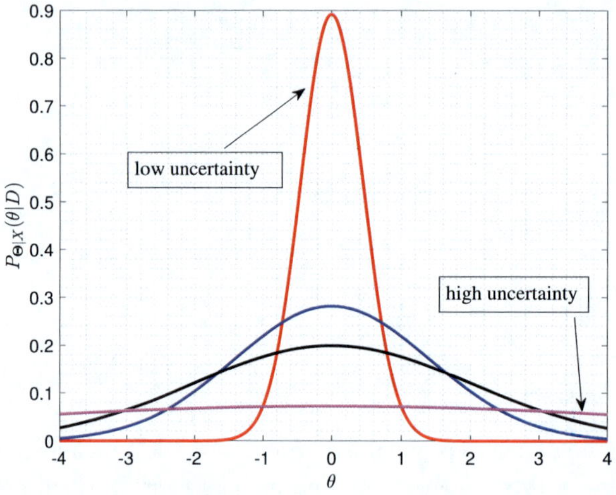

Fig. 3.1 A toy example of Bayesian estimation result, where the red curve indicates low uncertainty, while the magenta curve has higher uncertainty

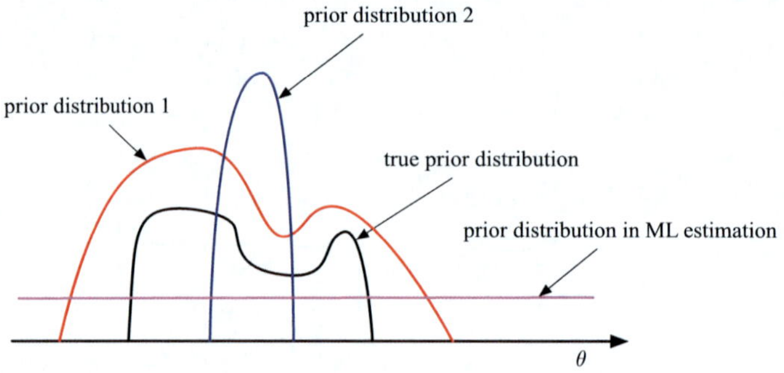

Fig. 3.2 A toy example of different prior distributions

3. In practice, we do not really care about the parameters themselves as they are required only in the sense that they allow us to build models that can be used to make predictions (e.g., the BDT). From this perspective, we can see that employing Bayesian estimation does not lose any information about the parameter that is contained in the training dataset, given the assumption that the models are correct.

Bayesian estimation also has its limitations. If the prior information in Bayesian estimation is inaccurate, then the posterior distribution, which is obtained by combining the prior distribution with the observed data, may not be a good representation of the true parameter values [4]. The estimates obtained from the posterior may be less reliable and could deviate significantly from the actual values. In extreme cases, a highly inaccurate prior might lead to overconfident or incorrect inferences. Additionally, the uncertainty quantification provided by Bayesian estimation might also be affected, potentially leading to incorrect estimates of the confidence intervals. As shown in Fig. 3.2, although prior distribution 1 (red) does not closely match the true prior distribution, the corresponding Bayes estimator can still achieve better performance than ML estimation generally as prior distribution 1 is more informative than the assumed uniform prior distribution (magenta) in ML. However, if the assumed prior distribution is prior distribution 2 (blue), then the result may be reversed, i.e., ML estimation might be better than Bayesian estimation in this case.

3.2 Bayesian Solution

In practice, Bayes estimator is relatively difficult to obtain as compared with the point estimators in Chap. 2, for the following reasons. First, it involves integrating over the parameter space, which can be a complex computational task. Second, determining the appropriate prior distribution and updating it with the observed data requires careful consideration and mathematical manipulation. Additionally, in many cases, the posterior distribution may not have a closed-form solution, making it even more challenging to obtain explicit expressions for the estimates and their properties [5].

In this section, we provide a detailed derivation of a Bayesian estimation solution [1]. We want to estimate the mean value of a Gaussian random variable X with unknown mean μ and known variance σ^2 from a set of N data samples $\{x_1, x_2, \cdots, x_N\}$. The prior distribution for the mean μ is also assumed to be Gaussian with known mean μ_0 and variance σ_0^2. In this case, the prior probability and likelihood function (observations) are both Gaussian, which are given by

$$P_\mu(\mu) = G(\mu, \mu_0, \sigma_0^2) \triangleq \frac{1}{\sqrt{2\pi\sigma_0^2}} e^{-\frac{(\mu-\mu_0)^2}{2\sigma_0^2}}, \tag{3.2}$$

and

$$P_{T|\mu}(D|\mu) = G(D, \mu, \sigma^2), \tag{3.3}$$

respectively, μ_0 and σ_0^2 are known hyper-parameters, and T is the random variable corresponding to the training dataset D. This means that both X and its mean μ are assumed to be Gaussian distributed. Then, we need to compute the posterior distribution for the mean μ, i.e.,

$$P_{\mu|T}(\mu|D) = \frac{P_{T|\mu}(D|\mu)P_\mu(\mu)}{P_T(D)}. \tag{3.4}$$

In (3.4), $P_{\mu|T}(\mu|D)$ is a probability density, thus we can ignore some constants, i.e., the terms that do not depend on μ, and normalize it when we are done. Accordingly, we only need to work with

$$P_{\mu|T}(\mu|D) \propto P_{T|\mu}(D|\mu)P_\mu(\mu) \propto \prod_i P_{X|\mu}(x_i|\mu)P_\mu(\mu), \tag{3.5}$$

where we have used the fact that the data samples $\{x_i\}$ are drawn independently. Then, by plugging the Gaussian prior and likelihood in (3.2) and (3.3) into (3.5), we obtain

$$\begin{aligned}
P_{\mu|T}(\mu|D) &\propto \prod_i G(x_i, \mu, \sigma^2)G(\mu, \mu_0, \sigma_0^2) \\
&\propto \exp\left\{-\sum_i \frac{(x_i-\mu)^2}{2\sigma^2} - \frac{(\mu-\mu_0)^2}{2\sigma_0^2}\right\} \\
&\propto \exp\left\{-\sum_i \frac{\mu^2 - 2x_i\mu + x_i^2}{2\sigma^2} - \frac{\mu^2 - 2\mu\mu_0 + \mu_0^2}{2\sigma_0^2}\right\} \\
&\propto \exp\left\{-\left(\frac{N}{2\sigma^2} + \frac{1}{2\sigma_0^2}\right)\mu^2 + 2\left(\frac{\sum_i x_i}{2\sigma^2} + \frac{\mu_0}{2\sigma_0^2}\right)\mu - \left(\frac{\sum_i x_i^2}{2\sigma^2} + \frac{\mu_0^2}{2\sigma_0^2}\right)\right\} \\
&\overset{(a)}{\propto} \exp\left\{-\left(\frac{N}{2\sigma^2} + \frac{1}{2\sigma_0^2}\right)\mu^2 + 2\left(\frac{\sum_i x_i}{2\sigma^2} + \frac{\mu_0}{2\sigma_0^2}\right)\mu\right\},
\end{aligned} \tag{3.6}$$

where (a) holds because ignoring the terms that do not depend on μ does not affect the probability density. From (3.6), we can see that $P_{\mu|T}(\mu|D)$ is a Gaussian distribution whose mean and variance can be obtained by putting it in the standard quadratic form. Using the "completing the squares" trick, i.e., $ax^2 + 2bx + c = a(x + b/a)^2 + c - b^2/a$, it follows that

$$P_{\mu|T}(\mu|D) \propto \exp\left\{-a(\mu - b/a)^2\right\} \tag{3.7}$$

where

$$a = \frac{N}{2\sigma^2} + \frac{1}{2\sigma_0^2}, \quad b = \frac{\sum_i x_i}{2\sigma^2} + \frac{\mu_0}{2\sigma_0^2}, \tag{3.8}$$

which further leads to

$$P_{\mu|T}(\mu|D) = G(\mu, \mu_N, \sigma_N^2), \tag{3.9}$$

where

$$\mu_N = \frac{\sigma_0^2 \sum_i x_i + \mu_0 \sigma^2}{\sigma^2 + N\sigma_0^2}, \quad \sigma_n^2 = \frac{\sigma^2 \sigma_0^2}{\sigma^2 + N\sigma_0^2}. \tag{3.10}$$

The Bayesian solution is a probability density, which is Gaussian in this particular case, and based on the above derivation, its mean is given by

$$\mu_N = \underbrace{\frac{N\sigma_0^2}{\sigma^2 + N\sigma_0^2}}_{\alpha_N} \mu_{ML} + \underbrace{\frac{\sigma^2}{\sigma^2 + N\sigma_0^2}}_{1-\alpha_N} \mu_0, \tag{3.11}$$

where μ_{ML} is the ML solution, i.e., the sample mean $\mu_{ML} = \frac{\sum_i x_i}{N}$ (as discussed in Example 2.1). As can be observed, μ_N is the weighted sum of μ_{ML} and μ_0, and α_N can be viewed as the weighting coefficient that is related to the number of data samples N, the observation noise power σ^2, and the prior variance of μ, i.e., σ_0^2. α_N satisfies $\alpha_N \in [0, 1]$, $\lim_{N\to\infty} \alpha_N = 1$ and $\lim_{N\to 0} \alpha_N = 0$, which is intuitive since as N increases, the information provided by the observations would be richer, and thus the proportion of ML estimation should be larger. Besides, α_N decreases as the observation noise power σ^2 increases, which means that we should lower the proportion of ML estimation if the information provided by the observations is less reliable. In addition, α_N increases with the increase of σ_0^2, i.e., if the prior information is not reliable, we should increase the proportion of ML estimation. This precision-weighted combination of prior and data estimates is a fundamental mechanism for sequential Bayesian learning [1, 5].

Furthermore, as $1/\sigma^2$ is usually called the precision and is denoted by τ, we can see that the posterior precision is, in this case, the sum of the prior precision and the data precision, i.e.,

$$\tau_{post} = \frac{\sigma^2 + N\sigma_0^2}{\sigma^2\sigma_0^2} = \underbrace{\frac{1}{\sigma_0^2}}_{\tau_{prior}} + \underbrace{\frac{N}{\sigma^2}}_{\tau_{data}}. \tag{3.12}$$

Combining (3.11) and (3.12), we can conclude that Bayesian estimation is able to combine the two sources of information which are from the observations and the prior, respectively.

3.3 Conjugate Priors

If there is no inherent reason to favor one prior probability distribution over another, a conjugate prior is sometimes selected for simplicity. A conjugate prior is defined as a prior distribution that belongs to a certain parametric family and for which the resulting posterior distribution also belongs to the same family [1], e.g., the aforementioned Gaussian likelihood has a Gaussian prior (conjugate prior). Whenever this is the case, we say that $P_\mu(\mu)$ is a conjugate prior for the likelihood $P_{X|\mu}(x|\mu)$ and the posterior $P_{\mu|T}(\mu|D)$ is the reproducing density. This is an extremely useful property because the Bayes estimator and its statistical properties (such as variance and confidence interval) can all be derived from the posterior distribution. Some commonly used likelihood functions and the corresponding conjugate priors are provided in Table 3.1.

Conjugate priors are particularly useful in sequential estimation. In this context, the posterior of the current observation is used as the prior in the next observation. In sequential estimation, if a conjugate prior is not used, the posterior distribution generally becomes more complex with each additional observation. In such cases, the Bayes estimator usually cannot be calculated without relying on numerical methods.

Example 3.1 (*Conjugate Prior (Bernoulli and Beta)*). Consider a Bernoulli random variable X, i.e., $X \sim Ber(\theta)$, which means that $P(X = 1) = \theta$ and $P(X = 0) = 1 - \theta$. Now, if we are given N realizations of this Bernoulli variable, i.e., $D = \{x_1, \cdots, x_N\}$, the likelihood function for θ is

$$P(D|\theta) = \prod_{i=1}^{N} P(X = x_i|\theta) = \prod_{i=1}^{N} \theta^{x_i}(1-\theta)^{1-x_i}. \tag{3.13}$$

Table 3.1 Commonly used likelihood functions and the corresponding conjugate priors

Likelihood	Conjugate prior
Bernoulli	Beta
Poisson	Gamma
Exponential	Gamma
Normal	Gamma
Normal	Normal

Let k be the number of x_i with value 1, i.e., $k = \sum_i x_i$, then the likelihood function $P(D|\theta)$ can be written in a more simple form as follows:

$$P(D|\theta) = \theta^k (1-\theta)^{N-k}. \tag{3.14}$$

Now, if we put a Beta distribution on the parameter θ of the Bernoulli likelihood, i.e., $\theta \sim Beta(\alpha, \beta)$, and its PDF is written as

$$P_\theta(\theta) = \frac{1}{B(\alpha, \beta)} \theta^{\alpha-1}(1-\theta)^{\beta-1}, \text{ for } \theta \in [0, 1], \tag{3.15}$$

where $B(\alpha, \beta)$ is the Beta function. Similar to (3.4), the posterior distribution for θ is given by

$$P_{\theta|T}(\theta|D) = \frac{P(D|\theta)P_\theta(\theta)}{P_T(D)} = \frac{1}{C}\theta^{\alpha+k-1}(1-\theta)^{\beta+N-k-1}, \tag{3.16}$$

where C is a normalizing constant. Notice that this posterior is also a Beta distribution with updated parameters, namely,

$$\theta|D \sim Beta(\alpha+k, \beta+N-k). \tag{3.17}$$

Therefore, the Bernoulli likelihood function has a conjugate prior that obeys the Beta distribution.

Summary

In summary, Bayesian estimation treats parameters as random variables, and computes the posterior distribution to integrate prior knowledge with observational data via Bayes' rule. It provides full probabilistic uncertainty quantification, which is different from point estimation (e.g., ML). We introduce two key tools for Bayesian estimation, i.e., (1) for Gaussian likelihoods with Gaussian priors, the posterior is Gaussian, with mean/variance analytically derivable; (2) conjugate priors (e.g., Beta for Bernoulli, Gamma for Poisson) yield posteriors in the same family, enabling tractable sequential updates.

The insights and implications of this chapter are listed as follows:

1. Bayesian estimation replaces point estimates with distributions, and it is able to quantify uncertainty intrinsically through posterior variance. In particular, posterior variance decreases with more data or lower noise, which enables us to measure estimate reliability objectively. During prior-data integration, posterior mean balances prior belief and data-driven ML estimate via precision weights.
2. Conjugate priors act as "plug-in" modules, enabling exact sequential inference by preserving distributional forms. They ensure tractability but are with limit flexibility, while non-conjugate cases require approximations.
3. For Bayesian estimation, performance hinges on prior accuracy, i.e., inaccurate priors skew posteriors, while weak priors revert to ML-like behavior.

Homework Problems

3.1 Suppose $X \sim \text{Poisson}(\lambda)$, and you observe a dataset $D = \{x_1, x_2, \ldots, x_N\}$ with $\sum_{i=1}^{N} x_i = S$. Use a Gamma prior $\lambda \sim \text{Gamma}(\alpha, \beta)$ (where β is the rate parameter).

(a) Write the likelihood $P(D|\lambda)$ and the prior $P(\lambda)$, and derive the posterior distribution $P(\lambda|D)$. Show that it is also a Gamma distribution, identifying the updated parameters.
(b) Discuss why the Gamma prior is conjugate for the Poisson likelihood, and explain how this property simplifies sequential updates if additional data is collected.

3.2 Consider a Poisson likelihood $X \sim \text{Poisson}(\lambda)$ with prior $\lambda \sim \text{Uniform}(0, 5)$. You observe data $D = \{2, 1, 3\}$.

(a) Write the unnormalized posterior $P(\lambda \mid D)$. Why is this prior non-conjugate?
(b) Compute the posterior mode (MAP estimate) numerically.
(c) Contrast this with a Gamma(2, 1) prior. Which prior gives higher posterior probability to $\lambda > 2$?

3.3 Conjugate priors are particularly useful in sequential estimation, as mentioned in Sect. 3.3. Suppose you are conducting an experiment where you observe a sequence of Bernoulli trials $X_i \sim \text{Ber}(\theta)$, and you use a Beta prior $\theta \sim \text{Beta}(\alpha, \beta)$. You collect data in two batches:

- Batch 1: $N_1 = 10$ trials with $k_1 = 4$ successes.
- Batch 2: $N_2 = 5$ trials with $k_2 = 3$ successes.

Start with a prior $\theta \sim \text{Beta}(2, 2)$.

(a) After observing Batch 1, compute the posterior distribution for θ. Use this posterior as the prior for Batch 2, and compute the final posterior after observing Batch 2.
(b) If you had observed all $N_1 + N_2 = 15$ trials at once with $k_1 + k_2 = 7$ successes, would the final posterior be the same? Explain why or why not, referencing the role of conjugate priors.

3.4 Suppose the lifetime X of a component follows an exponential distribution: $X \sim \text{Exp}(\lambda)$, where $P(X|\lambda) = \lambda e^{-\lambda x}$ for $x > 0$. You observe N independent lifetimes $D = \{x_1, \ldots, x_N\}$.

(a) Show that the Gamma distribution $\lambda \sim \text{Gamma}(\alpha, \beta)$ (with PDF $P(\lambda) \propto \lambda^{\alpha-1} e^{-\beta\lambda}$) is a conjugate prior for λ. Derive the posterior $P(\lambda|D)$.
(b) Given prior $\lambda \sim \text{Gamma}(3, 2)$ and data $D = \{1.2, 0.5, 2.1\}$, compute the posterior parameters and the posterior mean.
(c) Interpret the posterior mean as a weighted average of the prior mean and the ML estimator.

3.5 Suppose you are estimating the mean μ of a Gaussian likelihood $X \sim N(\mu, \sigma^2)$ with known variance $\sigma^2 = 1$, using a Gaussian prior $\mu \sim N(\mu_0, \sigma_0^2)$. You observe a dataset $D = \{x_1, x_2, \ldots, x_N\}$ with sample mean $\bar{x} = 5$, and $N = 10$.

(a) Consider two priors: (i) $\mu \sim N(0, 1)$, and (ii) $\mu \sim N(10, 1)$. Compute the posterior mean and variance for each prior using the formulas from Sect. 3.2.
(b) Discuss how the choice of prior affects the posterior mean in each case. If the true $\mu = 4$, which prior leads to a more accurate estimate, and why? Relate your answer to the concept of prior informativeness versus robustness.

References

1. C.M. Bishop, N.M. Nasrabadi, *Pattern Recognition and Machine Learning*, vol. 4, no. 4 (Springer, New York, 2006), p. 738
2. P.D. Hoff, *A First Course in Bayesian Statistical Methods*, vol. 580 (Springer, New York, 2009)
3. K.P. Murphy, *Machine Learning: a Probabilistic Perspective* (MIT Press, 2012)
4. A. Gelman, J.B. Carlin, H.S. Stern, D.B. Rubin, *Bayesian Data Analysis* (Chapman and Hall/CRC, 1995)
5. D. Barber, *Bayesian Reasoning and Machine Learning* (Cambridge University Press, Cambridge, 2012)

Expectation Maximization (EM) 4

In this chapter, we present the expectation-maximization (EM) algorithm as an iterative method for ML or MAP estimation in statistical models with unobserved latent variables. It addresses scenarios where direct likelihood optimization is intractable due to missing data or hidden structure. The EM framework alternates between: an E-step that computes the expected log-likelihood of complete data (the Q-function) using the posterior distribution of latent variables given current parameters; and an M-step that maximizes this expectation to update parameters. Theoretical analysis establishes EM's monotonic convergence via lower-bounding the observed-data log-likelihood (using Kullback-Leibler (KL) divergence) and proves convergence to stationary points under regularity conditions. Practical challenges, including computational complexity, non-convex optimization, and slow convergence, are discussed. Moreover, this chapter extends EM to MAP estimation by incorporating priors into the Q-function and introduces its relationship to the majorization-minimization (MM) algorithm, generalizing the E-step to surrogate function construction. A geometric interpretation illustrates EM as iterative bound optimization.

4.1 Basic Concepts

In statistics, the EM algorithm is an iterative method to find (local) ML or MAP estimates of parameters in statistical models, where the model depends on unobserved latent variables [7]. That is, either missing values exist among the data, or the model can be formulated more simply by assuming the existence of further unobserved data points. In this section, we consider ML estimation (the case with MAP estimation will be introduced in Sect. 4.4) and a problem with two types of random variables, i.e., the observed random variable X that we are interested in and the latent/hidden random variable Z that is not directly observed but is assumed to exist and influence the observed data. This means that we are not able to directly obtain the likelihood function $P_X(D; \Psi)$ (Ψ contains the unknown parameters and $D = \{x_1, \ldots, x_N\}$ is a given set of i.i.d. data samples), but only $P_Z(z; \Psi)$ and $P_{X|Z}(D|z; \Psi)$ are available. Note that the set D is called the incomplete data, while the set $D_c = \{(x_1, z_1), \ldots, (x_N, z_N)\}$ is called the complete data which cannot be seen, otherwise the problem would be a standard ML estimation problem, which has been studied before in Sect. 2.2. Usually, latent/hidden variables are introduced to capture the underlying structures or processes that are not directly measurable. They help to explain the relationships between the observed variables and provide a more comprehensive model of the data. Our goal is to find the unknown parameters Ψ^* that maximize the likelihood of D, i.e.,

$$\begin{aligned} \Psi^* &= \arg\max_{\Psi} P_X(D; \Psi) \\ &= \arg\max_{\Psi} \int P_{X|Z}(D|z; \Psi) P_Z(z; \Psi) dz \end{aligned} \tag{4.1}$$

However, calculating (4.1) is often intractable since Z is unobserved and the distribution of Z is unknown before attaining Ψ. The EM algorithm resolves this difficulty by alternating between an expectation (E) step and a maximization (M) step, which is shown as follows:

1. Start with an initial parameter estimate $\Psi^{(0)}$;
2. **E-step**: given the current parameters $\Psi^{(i)}$ (i denotes the iteration index) and observations in D, "guess" what the values of $\{z_i\}$ are, i.e., compute the conditional posterior probability $P(z|\Psi^{(i)}, D)$;

A. Liu and M.-M. Zhao, *Advanced Bayesian Signal Processing*, Textbooks in Telecommunication Engineering,
https://doi.org/10.1007/978-3-032-26656-9_4

3. **M-step**: with the new $\{z_i\}$, we have a complete data problem, solve this problem to obtain the updated parameters $\Psi^{(i+1)}$;
4. go to Step 2.

Another important point to note is that we generally do not know the complete data; thus, EM defines the Q-function as the expectation of the complete-data log-likelihood (taking the logarithm generally reduces the complexity) given observed data D evaluated using the current estimate for the parameters $\Psi^{(i)}$, which is known as the Q function [3], i.e.,

$$Q\left(\Psi;\, \Psi^{(i)}\right) = \mathbb{E}_{Z|X;\Psi^{(i)}}\left\{\log P_{X,Z}\left(D, \{z_1, \ldots, z_N\};\, \Psi\right)|D\right\}. \tag{4.2}$$

$Q\left(\Psi;\, \Psi^{(i)}\right)$ is the expected value of log-likelihood with respect to the complete data (joint X and Z), given that we observed the incomplete data X. Besides, it is important to mention that $Q\left(\Psi;\, \Psi^{(i)}\right)$ is a function of Ψ (i.e., the parameters that we want to determine), but to compute $Q\left(\Psi;\, \Psi^{(i)}\right)$, the parameter values from the previous iteration are required to be known, because we need a distribution for $Z|X$. Therefore, $Q\left(\Psi;\, \Psi^{(i)}\right)$ can be viewed as a local approximation of the original log-likelihood function $\log P_X(D;\, \Psi)$ at the current iterate $\Psi^{(i)}$. The above E-step and M-step can be more formally described as

1. **E-step**: given estimates $\Psi^{(i)}$, compute the expected log-likelihood of the complete data, i.e., $Q\left(\Psi;\, \Psi^{(i)}\right)$;
2. **M-step**: find the parameters that maximize $Q\left(\Psi;\, \Psi^{(i)}\right)$, i.e.,

$$\Psi^{(i+1)} = \arg\max_{\Psi} Q\left(\Psi;\, \Psi^{(i)}\right). \tag{4.3}$$

Finding an ML solution typically requires taking the derivatives of the likelihood function with respect to all the unknown values, the parameters and the latent/hidden variables, and simultaneously solving the resulting equations. In statistical models with latent variables, this is usually impossible as it results in interlocking equations. EM offers a way to solve this problem numerically, i.e., one can simply pick arbitrary values for one of the two sets of unknowns (parameters and latent/hidden variables), use them to estimate the other set, and keep alternating until the resulting values both converge to fixed points. After EM converges, we actually obtain two estimates, i.e., the (near)-ML estimate of Ψ and the Bayesian estimate of Z given the ML estimation Ψ. Although it is not obvious that this idea will work, it can be proven in this context as provided in the following section.

Attention

In practical applications, the EM algorithm may encounter the following difficulties:

1. In the E-step, we ought to compute the conditional posterior probability $P(z|\Psi^{(i)}, D)$, which may not be easy. Besides, the Q function as given in (4.2) is defined as the expected value of the log-likelihood function with respect to the conditional distribution of the latent variables given the observed data and the current estimate of the parameters, which might be difficult to calculate. Furthermore, even if we have obtained the Q function, it may not be a convex function which is relatively easy to handle and the resulting optimization problem in the M-step, i.e., (4.3), may not be a convex problem.[1]
2. In some cases, the convergence speed of the EM algorithm can be very slow, especially when dealing with large-scale data or complex models. This leads to long computation times and affects the practical efficiency of the algorithm.
3. For high-dimensional and/or large-scale data, the storage and computational complexity of the EM algorithm also become a problem. This is because in high-dimensional space, the number of parameters to be estimated increases greatly, and the calculation of the expectation and maximization steps also becomes more complex. While when the memory is limited, it may not be possible to store the entire dataset or perform large-scale matrix operations.

[1] In this case, we usually seek a stationary solution of problem (4.3), it can be proved that generally the resulting EM also converges a stationary solution of the original ML estimation problem [6].

Important

In (4.1), one can also ignore the integral operation and perform a joint ML estimation of z and Ψ, i.e., we can alternatively optimize z and Ψ, based on the joint likelihood function $P_{X,Z}(D, z; \Psi)$. Usually, such an approach can also converge to a stationary solution of the original ML estimation problem. Actually, the EM algorithm can also be viewed as an alternating optimization algorithm, where the E-step is to optimize z with given $\Psi^{(i)}$ such that a posterior estimation of z can be obtained (not a point estimation as in the abovementioned alternative ML estimation), while the M-step is to optimize Ψ with given z. If we make a decision of z in the E-step (i.e., obtaining its posterior mean) and calculate the Q function $Q\left(\Psi; \Psi^{(i)}\right)$ by assuming that the posterior distribution of z is a delta function, then the EM algorithm reduces to the well-known alternating optimization algorithm.

EM is preferred in some cases, such as when we want to obtain the posterior distribution of the hidden variables for some external reasons, and/or when this posterior distribution is relatively easy to calculate and by taking the estimation errors of the hidden variables into consideration, the estimation performance of Ψ can be improved. In practice, we usually divide the parameters into multiple groups, and flexibly determine the estimation methods (point estimation or Bayesian estimation) for each group of parameters according to the characteristics of the problem.

4.2 EM Convergence

We now establish the monotonicity of the observed-data log-likelihood under the EM updates. Recall that our goal is to maximize the log-likelihood function $\log P_X(D; \Psi)$. Using $P_{X,Z}(D, z; \Psi) = P_{Z|X}(z|D; \Psi)P_X(D; \Psi)$, we have

$$\log P_X(D; \Psi) = \log P_{X,Z}(D, z; \Psi) - \log P_{Z|X}(z|D; \Psi). \tag{4.4}$$

Then, by taking expectations on both sides of (4.4) over the conditional distribution $P_{Z|X}(\mathbf{Z}|X; \Psi^{(i)})$, and using the fact that the left-hand-side of (4.4) does not depend on Z, we obtain

$$\begin{aligned}\log P_X(D; \Psi) &= \mathbb{E}_{Z|X;\Psi^{(i)}}\left\{\log P_{X,Z}(D, z; \Psi)|D\right\}\\ &\quad - \mathbb{E}_{Z|X;\Psi^{(i)}}\left\{\log P_{Z|X}(z|D; \Psi)|D\right\}\\ &= Q\left(\Psi; \Psi^{(i)}\right) + H\left(\Psi|\Psi^{(i)}\right),\end{aligned} \tag{4.5}$$

where we have introduced

$$\begin{aligned}H\left(\Psi|\Psi^{(i)}\right) &= -\mathbb{E}_{Z|X;\Psi^{(i)}}\left\{\log P_{Z|X}(z|D; \Psi)|D\right\}\\ &= -\int P_{Z|X;\Psi^{(i)}}\left(z|D; \Psi^{(i)}\right)\log P_{Z|X}(z|D; \Psi)dz,\end{aligned} \tag{4.6}$$

which can be regarded as the difference between the log-likelihood function and the Q function.

To prove the convergence of EM, we turn to the following iterative improvement:

$$\begin{aligned}\log P_X\left(D; \Psi^{(i+1)}\right) - \log P_X\left(D; \Psi^{(i)}\right) &= Q\left(\Psi^{(i+1)}; \Psi^{(i)}\right) - Q\left(\Psi^{(i)}; \Psi^{(i)}\right)\\ &\quad + H\left(\Psi^{(i+1)}|\Psi^{(i)}\right) - H\left(\Psi^{(i)}|\Psi^{(i)}\right).\end{aligned} \tag{4.7}$$

By definition of the M-step in (4.3), it follows that

$$Q\left(\Psi^{(i+1)}; \Psi^{(i)}\right) \geq Q\left(\Psi^{(i)}; \Psi^{(i)}\right). \tag{4.8}$$

Besides, from the definition of $H\left(\Psi|\Psi^{(i)}\right)$ in (4.6), we have

$$\begin{aligned}&H\left(\Psi^{(i+1)}|\Psi^{(i)}\right) - H\left(\Psi^{(i)}|\Psi^{(i)}\right)\\ &= -\mathbb{E}_{Z|X;\Psi(i)}\left\{\log \frac{P_{Z|X}\left(z|D; \Psi^{(i+1)}\right)}{P_{Z|X}\left(z|D; \Psi^{(i)}\right)}\middle| D\right\},\end{aligned} \tag{4.9}$$

which further leads to

$$
\begin{aligned}
& H\left(\Psi^{(i+1)}|\Psi^{(i)}\right) - H\left(\Psi^{(i)}|\Psi^{(i)}\right) \\
& \overset{(a)}{\geq} -\log \mathbb{E}_{Z|X;\Psi(i)} \left\{ \left. \frac{P_{Z|X}\left(z|D;\Psi^{(i+1)}\right)}{P_{Z|X}\left(z|D;\Psi^{(i)}\right)} \right| D \right\} \\
& = -\log \int P_{Z|X;\Psi^{(i)}}\left(z|D;\Psi^{(i)}\right) \frac{P_{Z|X}\left(z|D;\Psi^{(i+1)}\right)}{P_{Z|X}\left(z|D;\Psi^{(i)}\right)} dz \\
& = -\log 1 = 0
\end{aligned}
\tag{4.10}
$$

where in (a), we have utilized the convexity of the negative log function and Jensen's inequality as given in the following [4].

Theorem 4.1 *Let X denote an integrable real-valued random variable and let ϕ denote a convex function, then we have*

$$
\phi(\mathbb{E}\{X\}) \leq \mathbb{E}\{\phi(X)\}. \tag{4.11}
$$

Based on (4.8), (4.10) and (4.7), it follows that

$$
\log P_X\left(D;\Psi^{(i+1)}\right) \geq \log P_X\left(D;\Psi^{(i)}\right), \forall i, \tag{4.12}
$$

i.e., the log-likelihood of the incomplete data is monotonically non-decreasing from iteration to iteration, hence EM converges. However, there is no general convergence theorem for the EM algorithm (with no bound on the convergence rate), because the convergence of the sequence $\{\Psi^{(i)}\}$ depends on the characteristics of $\log P_X(D;\Psi)$ and $H\left(\Psi|\Psi^{(i)}\right)$, and also the initial point $\Psi^{(0)}$. We can prove that $\{\Psi^{(i)}\}$ converges to a stationary point (not necessarily a local/global optimum, but usually a local optimum) of $\log P_X(D;\Psi)$, under certain regularity conditions. Please see [1, 3, 9, 11] for some detailed discussions.

4.3 Geometric Interpretation

One can also derive a geometric interpretation of EM from (4.5). According to the definition of $H\left(\Psi|\Psi^{(i)}\right)$ in (4.6), we can observe that $H\left(\Psi|\Psi^{(i)}\right)$ is of the form

$$
\begin{aligned}
H\left(\Psi|\Psi^{(i)}\right) &= -\int p_i(z)\log p(z)dz \\
&= \int p_i(z)\log\frac{p_i(z)}{p(z)}dz - \int p_i(z)\log p_i(z)dz \\
&= KL(p_i||p) + H(p_i)
\end{aligned}
\tag{4.13}
$$

where $p_i(z)$ and $p(z)$ represent $P_{Z|X;\Psi^{(i)}}\left(z|D;\Psi^{(i)}\right)$ and $P_{Z|X}(z|D;\Psi)$, respectively, $KL(p_i||p)$ the KL divergence [5] between p_i and p, and $H(p_i)$ is entropy of p_i. In essence, the KL divergence $KL(p_i||p)$ is a measure of how the reference probability distribution p_i is different from p, and it is always non-negative as a result known as Gibbs' inequality [2], with $KL(p_i||p) = 0$ if and only if the two distributions p_i and p are identical. Besides, the entropy $H(p_i)$ is also non-negative by definition. Therefore, the Q function $Q\left(\Psi;\Psi^{(i)}\right)$ satisfies

$$
Q\left(\Psi;\Psi^{(i)}\right) \leq \log P_X(D;\Psi), \tag{4.14}
$$

i.e., the Q function is a lower bound to the log-likelihood of the observed data. This allows us to interpret the EM steps geometrically, as shown in Fig. 4.1, i.e., the E-step is to lower-bound the observed log-likelihood, while the M-step is to maximize this lower bound. Next, consider the difference between the observed log-likelihood and the lower bound, i.e., $H\left(\Psi|\Psi^{(i)}\right)$, it can be easily obtained that[2]

[2] Actually, the difference between $\log P_X(D;\Psi)$ and $Q\left(\Psi;\Psi^{(i)}\right)$ is minimized at $\Psi^{(i)}$, since $H\left(\Psi^{(n)}|\Psi^{(i)}\right) - H\left(\Psi^{(i)}|\Psi^{(i)}\right) \geq 0, \forall n, i$ also holds, which implies that the gradients of $\log P_X(D;\Psi)$ and $Q\left(\Psi;\Psi^{(i)}\right)$ with respect to Ψ are identical at the point $\Psi^{(i)}$.

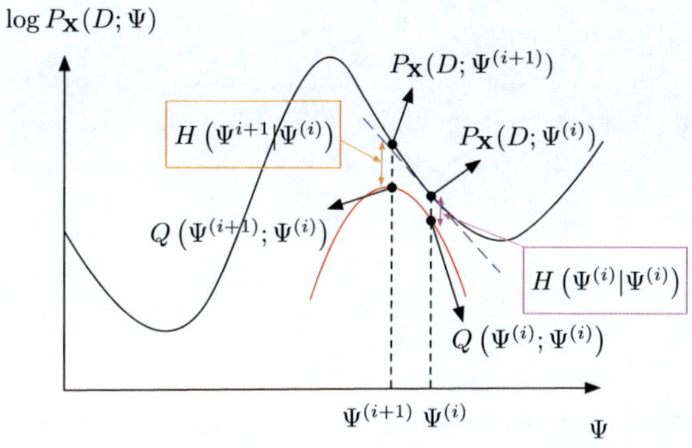

Fig. 4.1 Geometric interpretation of EM. Black curve: the log-likelihood of the observed data; red curve: the Q function; Blue dashed curve: the tangent to the observed log-likelihood at the point $P_X(D;\Psi^{(i)})$

$$\begin{aligned} H\left(\Psi^{(i+1)}|\Psi^{(i)}\right) - H\left(\Psi^{(i)}|\Psi^{(i)}\right) \\ &= KL(p_i||p_{i+1}) + H(p_i) - KL(p_i||p_i) - H(p_i) \\ &= KL(p_i||p_{i+1}) \geq 0,\ \forall i. \end{aligned} \tag{4.15}$$

Further based on the definition of the M-step, i.e., $Q\left(\Psi^{(i+1)};\Psi^{(i)}\right) \geq Q\left(\Psi^{(i)};\Psi^{(i)}\right)$, we can also have (4.12), which shows that EM converges iteratively without the need for step sizes.

Important

It is important to note that the famous EM algorithm is actually a special case of the MM algorithm [8], which has a long history that dates back to the 1970s. Specifically, MM generalizes EM by replacing the E-step, which calculates the conditional expectation of the log-likelihood of the complete data set, by a minorization step that finds a surrogate function, which keeps the key property of the E-step by being a lower bound of the objective function. As a consequence, MM shares most of the convergence results of EM. Compared to EM, which relies on a missing data interpretation of the problem, MM is easier to understand and has a wider scope of applications [10].

In general, the MM algorithm solves the following optimization problem:

$$\begin{aligned} &\min_{x} f(\boldsymbol{x}) \\ &\text{s.t. } x \in \mathcal{X}, \end{aligned} \tag{4.16}$$

where $\mathcal{X}$ is a nonempty closed set in $\mathbb{R}^n$ and $f:\mathcal{X}\to\mathbb{R}$ is a continuous function, which is assumed to go to infinity when $\boldsymbol{x}\in\mathcal{X}$ and $\boldsymbol{x}\to+\infty$. Given an initialized point $\boldsymbol{x}_0\in\mathcal{X}$, MM generates a sequence of feasible points $(\boldsymbol{x}_t)_{t\in\mathbb{N}}$ by the following induction. At point $\boldsymbol{x}_t$, a continuous surrogate function $g(\cdot|\boldsymbol{x}_t):\mathcal{X}\to\mathbb{R}$ satisfying the following upper bound property is constructed in the majorization step:

$$g(\boldsymbol{x}|\boldsymbol{x}_t) \geq f(\boldsymbol{x}) + c_t, \forall \boldsymbol{x}\in\mathcal{X}, \tag{4.17}$$

where $c_t = g(\boldsymbol{x}_t|\boldsymbol{x}_t) - f(\boldsymbol{x}_t)$, i.e., the difference of $g(\boldsymbol{x}|\boldsymbol{x}_t)$ and $f(\boldsymbol{x})$ is minimized at $\boldsymbol{x}_t$. Then, in the minimization step, $\boldsymbol{x}$ is updated as

$$\boldsymbol{x}_{t+1} \in \arg\min_{x\in\mathcal{X}} g(\boldsymbol{x}|\boldsymbol{x}_t). \tag{4.18}$$

The sequence $(f(\boldsymbol{x}_t))_{t\in\mathbb{N}}$ is non-increasing since

$$f(\boldsymbol{x}_{t+1}) \leq g(\boldsymbol{x}_{t+1}|\boldsymbol{x}_t) - c_t \leq g(\boldsymbol{x}_t|\boldsymbol{x}_t) - c_t = f(\boldsymbol{x}_t), \tag{4.19}$$

where the first inequality follows from (4.17), and the second inequality follows from (4.18). Thus, the objective value converges to a limit $f^\star$ by the assumption that f is bounded below. Moreover, $f^\star$ is guaranteed to be a stationary value and also the convergence of the sequence $(\boldsymbol{x}_t)_{t\in\mathbb{N}}$ can be ensured under some mild assumptions [10].

4.4 MAP Estimation via EM

EM can be equally applied to obtain MAP estimates with a straightforward extension. Recall that for MAP estimation, the goal is to maximize the posterior distribution, i.e.,

$$\begin{aligned}\Psi^* &= \arg\max_{\Psi} P_{\Psi|X}(\Psi|D) \\ &= \arg\max_{\Psi} P_X(D;\,\Psi)P_\Psi(\Psi),\end{aligned} \tag{4.20}$$

which is similar to ML estimation in (4.1) except in this case, the prior distribution of Ψ, i.e., $P_\Psi(\Psi)$, is taken into consideration. Problem (4.20) is still an estimation problem with incomplete data if the likelihood function $P_X(D;\,\Psi) = \int P_{X|Z}(D|z;\,\Psi)P_Z(z;\,\Psi)dz$ is not directly attainable. As we generally do not know the complete data posterior distribution $P_{\Psi|X,Z}(\Psi|D,z)$, EM creates a function for the expectation of the log-posterior given observed data D evaluated using the current estimate for the parameters $\Psi^{(i)}$, and the corresponding Q function can be derived as

$$\begin{aligned}&\mathbb{E}_{Z|X,\Psi}\left\{\log P_{\Psi|X,Z}(\Psi|D,z)|D,\,\Psi^{(i)}\right\} \\ &\quad= \underbrace{\mathbb{E}_{Z|X,\Psi}\left\{\log P_{X,Z|\Psi}(D,z|\Psi)|D,\,\Psi^{(i)}\right\}}_{\bar{Q}(\Psi|\Psi^{(i)})} + \underbrace{\mathbb{E}_{Z|X,\Psi}\left\{\log P_\Psi(\Psi)|D,\,\Psi^{(i)}\right\}}_{\log P_\Psi(\Psi)} \\ &\quad- \mathbb{E}_{Z|X,\Psi}\left\{\log P_{X,Z}(D,z)|D,\,\Psi^{(i)}\right\},\end{aligned} \tag{4.21}$$

where the equality can be easily obtained by resorting to the Bayes rule (1.18), and the last term can be dropped in the M-step as it does not depend on Ψ. Hence, the E-step and M-step for MAP estimation can be modified as follows:

1. **E-step**: given estimates $\Psi^{(i)}$, compute the expected log-posterior of the complete data (4.21), or equivalently, $\bar{Q}\left(\Psi|\Psi^{(i)}\right)$;
2. **M-step**: find the parameters that maximize (4.21), i.e.,

$$\Psi^{(i+1)} = \arg\max_{\Psi} \bar{Q}\left(\Psi|\Psi^{(i)}\right) + \log P_\Psi(\Psi). \tag{4.22}$$

Note that the M-step is typically a standard Bayesian estimation procedure.

Similar to Sect. 4.2, we can show the monotonicity of the modified EM algorithm for MAP estimation, and the details are omitted here for brevity.

Summary

This chapter introduced the EM algorithm as an iterative framework for ML or MAP estimation in statistical models with latent variables. EM addresses intractable likelihood optimization by alternating between: **E-step** that computes the expected complete-data log-likelihood (Q-function) using the posterior distribution of latent variables given current parameters, and **M-step** that updates parameters by maximizing the Q-function.

The core insights offered by this chapter can be summarized as follows.

1. EM monotonically increases the observed-data likelihood by constructing a tight lower bound (Q-function) via KL divergence properties. EM is a special case of the MM algorithm, where the E-step builds a surrogate function minorizing the objective.
2. EM converges to stationary points under regularity conditions, driven by non-negativity of KL divergence, and it can be extended to MAP estimation by incorporating priors into the Q-function.
3. There are also some disadvantages of EM, such as the E-steps in EM can be computationally intensive for complex posteriors, it is susceptible to local optima in non-convex problems, and it may exhibit slow convergence in high-dimensional spaces.

Homework Problems

4.1 Consider a dataset $D = \{x_1, x_2, \ldots, x_N\}$, where each x_i is generated from a mixture of two Bernoulli distributions. Specifically, for each x_i, there is a latent variable $z_i \in \{0, 1\}$ such that:

- If $z_i = 0$, then $x_i \sim \text{Ber}(\theta_0)$, i.e., $P(x_i \mid z_i = 0, \theta_0) = \theta_0^{x_i}(1-\theta_0)^{1-x_i}$,
- If $z_i = 1$, then $x_i \sim \text{Ber}(\theta_1)$, i.e., $P(x_i \mid z_i = 1, \theta_1) = \theta_1^{x_i}(1-\theta_1)^{1-x_i}$, and the prior probability of the latent variable is $P(z_i = 1) = \pi$, $P(z_i = 0) = 1 - \pi$. The parameters to estimate are $\Psi = (\theta_0, \theta_1, \pi)$.

(a) Write the complete-data log-likelihood $\log P_{X,Z}(D, z; \Psi)$.
(b) In the E-step, compute the posterior distribution $P(z_i \mid x_i, \Psi^{(i)})$ for each i.
(c) Derive the Q-function $Q(\Psi; \Psi^{(i)})$ for this model, showing all steps.

4.2 Consider a dataset $D = \{x_1, x_2, \ldots, x_N\}$, where each $x_i \sim N(\mu, \sigma^2)$, and σ^2 is known. The latent variable z_i indicates which of two means generated x_i: if $z_i = 0$, $x_i \sim N(\mu_0, \sigma^2)$; if $z_i = 1$, $x_i \sim N(\mu_1, \sigma^2)$. Assume a prior on the means $P(\mu_0) = N(0, \tau^2)$, $P(\mu_1) = N(0, \tau^2)$, and a prior on the mixing proportion $P(z_i = 1) = \pi$, with a Beta prior $\pi \sim \text{Beta}(\alpha, \beta)$.

(a) Write the modified Q-function for MAP estimation, incorporating the prior $P_\Psi(\Psi)$, where $\Psi = (\mu_0, \mu_1, \pi)$.
(b) Describe how the E-step and M-step differ from the standard EM algorithm for ML estimation. Derive the update for π in the M-step, considering the Beta prior.
(c) Discuss how the inclusion of priors might affect the convergence properties of the EM algorithm compared to the ML case.

4.3 The EM algorithm is a special case of the MM algorithm.

(a) Explain the key difference between the E-step in EM and the majorization step in MM. How does the MM framework generalize the EM algorithm?
(b) Consider a simple optimization problem: minimize $f(x) = x^2 + e^{-x}$. Construct a surrogate function $g(x|x_t)$ at the point $x_t = 0$ that satisfies the MM conditions (i.e., $g(x|x_t) \geq f(x)$ and equality holds at $x = x_t$). Hint: Use a first-order Taylor approximation for the non-quadratic term.

(c) Perform one iteration of the MM algorithm starting at $x_0 = 0$, and compute the next point x_1. Discuss how this process ensures a decrease in the objective function.

4.4 Consider a Gaussian mixture model with two components:

$$P(x_i|\Psi) = \pi_1 N(x_i|\mu_1, \Sigma_1) + \pi_2 N(x_i|\mu_2, \Sigma_2) \tag{4.23}$$

where $\Psi = \{\pi_k, \mu_k, \Sigma_k\}_{k=1,2}$, and latent variables $z_i \in \{1, 2\}$ indicate component membership.

(a) Derive the E-step: Compute $P(z_i = k|x_i, \Psi^{(t)})$ (responsibilities).
(b) Derive the M-step: Update $\pi_k^{(t+1)}, \mu_k^{(t+1)}, \Sigma_k^{(t+1)}$ by maximizing $Q(\Psi; \Psi^{(t)})$.
(c) Discuss how initialization affects convergence. Propose a robust initialization strategy.

4.5 For a hidden Markov model (HMM) with observed states $\{x_1, \ldots, x_T\}$ ($x_t \in \{1, \ldots, M\}$) and hidden states $\{z_1, \ldots, z_T\}$ ($z_t \in \{1, \ldots, N\}$), define A_{jk} as the transition probability $P(z_{t+1} = k \mid z_t = j)$, $B_j(x_t)$ as the emission probability $P(x_t|z_t = j)$, and π_j as the initial state probability $P(z_1 = j)$.

(a) Define the complete-data log-likelihood $\log P(X, Z|\Psi)$ with parameters $\Psi = \{A, B, \pi\}$.
(b) Derive the E-step (forward-backward algorithm) to compute $P(z_t|X, \Psi^{(t)})$ and $P(z_t, z_{t+1}|X, \Psi^{(t)})$.
(c) Derive the M-step updates for A, B, π.

References

1. R.A. Boyles, On the convergence of the EM algorithm. J. R. Stat. Soc. Ser. B Stat Methodol. **45**(1), 47–50 (1983)
2. P. Brémaud, *An Introduction to Probabilistic Modeling* (Springer Science & Business Media, 2012)
3. A.P. Dempster, N.M. Laird, D.B. Rubin, Maximum likelihood from incomplete data via the EM algorithm. J. Roy. Stat. Soc.: Ser. B (Methodol.) **39**(1), 1–22 (1977)
4. R. Durrett, Probability: Theory and Examples (Vol. 49). Cambridge University Press (2019)
5. S. Kullback, R.A. Leibler, On information and sufficiency. Ann. Math. Stat. **22**(1), 79–86 (1951)
6. G.J. McLachlan, T. Krishnan, *The EM Algorithm and Extensions* (Wiley, 2008)
7. X.L. Meng, D. Van Dyk, The EM algorithm—an old folk-song sung to a fast new tune. J. R. Stat. Soc. Ser. B Stat Methodol. **59**(3), 511–567 (1997)
8. J.M, Ortega W.C. Rheinboldt, Iterative Solution of Nonlinear Equations in Several Variables (Vol. 30). SIAM (1970)
9. R.A. Redner, H.F. Walker, Mixture densities, maximum likelihood and the EM algorithm. SIAM Rev. **26**(2), 195–239 (1984)
10. Y. Sun, P. Babu, D.P. Palomar, Majorization-minimization algorithms in signal processing, communications, and machine learning. IEEE Trans. Signal Process. **65**(3), 794–816 (2016)
11. C.J. Wu, On the Convergence Properties of the EM Algorithm. Ann. Stat., 95–103 (1983)

5 Factor Graph and Message Passing

This chapter introduces factor graphs as bipartite graphical representations of probabilistic model factorizations, which enables efficient computation of marginal posterior distributions via message-passing algorithms. Motivated by the computational intractability of direct MAP/MMSE estimation in high-dimensional problems (e.g., sequence detection in communications), factor graphs decompose complex joint distributions into local functions, exploiting conditional independencies. The core sum-product algorithm is detailed, i.e., it computes marginals through iterative message exchanges between variable and factor nodes, with rules governing message updates (products at variable nodes, summarization at factor nodes). Practical aspects include discrete/continuous message representations, addressing numerical stability via normalization, log-domain processing, and log-likelihood ratios (LLRs), and handling cyclic graphs via junction tree aggregation or loopy belief propagation, where convergence is analyzed for Gaussian/walk-summable models. The max-product variant for MAP estimation is derived, and a repetition code decoding example illustrates the framework. Key insights cover message-passing schedules (flooding vs. sequential), the role of LLRs in binary decision-making, and applications in error-correcting codes and multiple-input multiple-output (MIMO) detection. Theoretical guarantees for convergence and exact marginals in cycle-free graphs are contrasted with the empirical success of loopy propagation in cyclic structures approaching Shannon limits.

5.1 Motivation

Consider again a parameter estimation problem, and let $\boldsymbol{x} \in \mathcal{X}$ and $\boldsymbol{y} \in \mathcal{Y}$ denote the parameter to be estimated (with known prior $p_X(\boldsymbol{x})$) and the observation (with known likelihood function $p_{Y|X}(\boldsymbol{y}|\boldsymbol{x})$), respectively, then an estimator/detector can be expressed as

$$\boldsymbol{x}^*(\boldsymbol{y}) : \mathcal{Y} \rightarrow \mathbf{X}, \tag{5.1}$$

where for every possible $\boldsymbol{y}$, an estimate can be obtained, and $\boldsymbol{x}^*(\boldsymbol{y})$ is designed to minimize a certain loss function. Take the binary sequence estimation problem as an example (similar to the detection problem in wireless communications), i.e., the parameter to be estimated is a binary sequence $\boldsymbol{x} \in \{-1, +1\}^N$ (the transmitted symbol sequence with binary phase shift keying (BPSK) modulation), and the observation model can be expressed as

$$\mathbf{y} = \boldsymbol{x} + \boldsymbol{n}, \ \boldsymbol{n} \sim \mathcal{N}(\mathbf{0}, \sigma^2 \boldsymbol{I}), \tag{5.2}$$

where σ^2 denotes the noise variance. For this model, the MAP detector is optimal in the following sense: among all detectors, the MAP detector minimizes the probabilities of making an error. There are generally two variations of the MAP detector, the first one aims to minimize the error rate of the packet $\boldsymbol{x}$ (known as the frame error rate (FER)), i.e.,

$$\boldsymbol{x}^*(\boldsymbol{y}) = \arg\max_{\boldsymbol{x}} p_{X|Y}(\boldsymbol{x}|\boldsymbol{y}), \tag{5.3}$$

while the second one minimizes the error rate of the individual bits (known as the BER), i.e.,

$$x_k^*(\boldsymbol{y}) = \arg \max_{x_k \in \{-1,+1\}} p_{X_k|Y}(x_k|\boldsymbol{y}), \ \forall k. \tag{5.4}$$

A. Liu and M.-M. Zhao, *Advanced Bayesian Signal Processing*, Textbooks in Telecommunication Engineering,
https://doi.org/10.1007/978-3-032-26656-9_5

The above two MAP detectors are both useful (but different) since BER and FER are highly correlated performance metrics, usually a lower BER also leads to a lower FER. However, problem (5.3) maximizes the posterior distribution of $\boldsymbol{x}$ over all its possible combinations, which exhibits high computational complexity that is exponential in the length of $\boldsymbol{x}$. Employing problem (5.4) is able to alleviate this computational burden, but find the marginal posterior distribution $p_{X_k|Y}(x_k|\boldsymbol{y})$ is not easy.

For continuous variables, i.e., $\boldsymbol{x} \in \mathcal{R}^N$, we usually change the terminology from "detection" to "estimation", and different cost functions lead to different estimators, e.g., for the following "0-1" loss function:

$$c(\boldsymbol{x}, \boldsymbol{x}^*(\boldsymbol{y})) = \begin{cases} 0, & \|\boldsymbol{x} - \boldsymbol{x}^*(\boldsymbol{y})\| \leq \delta \\ 1, & \text{else} \end{cases}, \tag{5.5}$$

we use the MAP estimator (for $\delta \to 0$) $\boldsymbol{x}^*(\boldsymbol{y}) = \arg\max_{\boldsymbol{x}} p_{X|Y}(\boldsymbol{x}|\boldsymbol{y})$, while for the following MSE loss function:

$$c(\boldsymbol{x}, \boldsymbol{x}^*(\boldsymbol{y})) = \|\boldsymbol{x} - \boldsymbol{x}^*(\boldsymbol{y})\|^2, \tag{5.6}$$

the MMSE estimator $\boldsymbol{x}^*(\boldsymbol{y}) = \int x \, p_{X|Y}(x|y) \, dx$ is employed. In particular, the MMSE estimator is generally more difficult to obtain than the MAP estimator since a high-dimensional integral operation is required. If we can obtain the marginal posterior distribution of each parameter, i.e., $p_{X_k|Y}(x_k|\boldsymbol{y}), \forall k$, then the calculation of both the MAP and MMSE estimators can be simplified. However, similar to the discrete case, calculating $p_{X_k|Y}(x_k|\boldsymbol{y}), \forall k$ is very difficult because for each k, we should use an integration process to eliminate the parts that are associated with the other parameters.

This is the point where factor graph and message passing are brought into play. Generally, factor graphs are used to represent factorization of a probability distribution function, enabling efficient computations (in a distributed and parallel manner), such as the computation of marginal distributions through the message passing algorithm (the sum product algorithm is a popular kind) [1]. Using factor graphs reduces the complexity because instead of dealing with the entire high-dimensional joint posterior distribution directly, we can focus on the individual factors and their relationships. The structure of the factor graph reveals the conditional independencies among variables, and variables that are not directly connected through factors are conditionally independent given the other variables, which simplifies the problem of computing marginal posteriors. From another perspective, computing the marginal posterior distributions $p_{X_k|Y}(x_k|\boldsymbol{y}), \forall k$, separately involves many computational steps that are redundant, e.g., when computing $p_{X_1|Y}(x_1|\boldsymbol{y})$ and $p_{X_2|Y}(x_2|\boldsymbol{y})$, we need to perform integration over $\{x_3, x_4, \cdots\}$ in either case, and employing fact graph and message passing are able to effectively combine these redundant operations.

5.2 Factor Graphs and the Sum-Product Algorithm

Note that the posterior distribution $p_{X|Y}(\boldsymbol{x}|\boldsymbol{y})$ can be viewed as a certain joint distribution of $\{x_1, x_2, \ldots, x_N\}$, denoted by $f(x_1, x_2, \ldots, x_N)$, thus when discussing factor graph and message passing, we usually drop the conditioning on the observation $\boldsymbol{y}$. Throughout this chapter, we deal with $f(x_1, x_2, \ldots, x_N)$, and for this function, we are often interested in deriving its marginal functions, denoted $g_k(x_k)$. For each $a \in B_k$ (B_k denotes the domain of x_k), the value of $g_k(a)$ is obtained by summing the value of $f(x_1, \ldots, x_N)$ over all configurations of the variables that have $x_k = a$. This type of sum will be frequently used in this section, thus we define the following notations: if h is a function of three variables x_1, x_2 and x_3, then the "not-summary for x_2" is denoted by

$$\sum_{\sim\{x_2\}} h(x_1, x_2, x_3) \triangleq \sum_{x_1 \in B_1} \sum_{x_3 \in B_3} h(x_1, x_2, x_3). \tag{5.7}$$

In this notation, we have

$$g_k(x_k) \triangleq \sum_{\sim\{x_k\}} f(x_1, \ldots, x_N), \tag{5.8}$$

i.e., the kth marginal function associated with $f(x_1, \ldots, x_N)$ is the not-summary for x_k of f.

Suppose that $f(x_1, x_2, \ldots, x_N)$ factors into a product of several local functions, each having some subset of $\{x_1, x_2, \ldots, x_N\}$ as arguments, i.e., suppose that

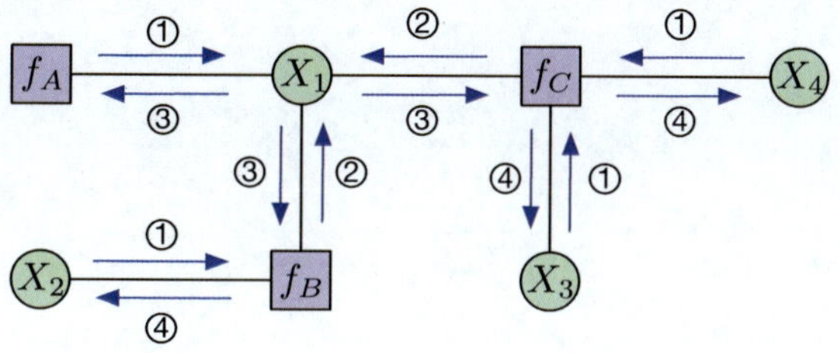

Fig. 5.1 A factor graph for the product $f_A(x_1)f_B(x_1, x_2)f_C(x_1, x_3, x_4)$

$$f(x_1, x_2, \ldots, x_N) = \prod_{k \in J} f_k(A_k), \tag{5.9}$$

where J is a discrete index set, A_k is a subset of $\{x_1, x_2, \ldots, x_N\}$, and $f_k(A_k)$ is a function having the elements of A_k as arguments. Then, a factor graph is formally defined as follows.

Definition: A factor graph is a bipartite graph that expresses the structure of the factorization (5.9). A factor graph has a variable node for each variable x_i, a factor node for each local function f_j, and an edge connecting variable node x_i to factor node f_j if and only if x_i is an argument of f_j.

Example 5.1 (*A Simple Factor Graph Example*). Let $f(x_1, x_2, x_3, x_4)$ be a function of four variables, and suppose that f can be expressed as a product

$$f(x_1, x_2, x_3, x_4) = f_A(x_1)f_B(x_1, x_2)f_C(x_1, x_3, x_4) \tag{5.10}$$

of three factors, so that $k \in J = \{A, B, C\}$, $A_A = \{x_1\}$, $A_B = \{x_1, x_2\}$ and $A_C = \{x_1, x_3, x_4\}$. The factor graph that corresponds to (5.10) is shown in Fig. 5.1, where squares denote factor nodes and circles represent variable nodes.

Note that $f(x_1, x_2, \ldots, x_N)$ is always decomposable, e.g., in the worst case, we can decompose it as

$$f(x_1, x_2, \ldots, x_N) = f_A(x_1)f_B(x_2|x_1)f_C(x_3|x_1, x_2)\ldots, \tag{5.11}$$

although no structural information is exploited here.

5.2.1 Computing a Single Marginal Function

For ease of understanding the sum-product algorithm,[1] we can imagine that there is a processor associated with each vertex of the factor graph, and the factor-graph edges represent channels by which these processors may communicate. In general, "messages" sent between processors are always simply some appropriate description of some marginal function [1].

We now describe a message-passing algorithm that will be temporarily called the "single-node sum-product algorithm", since it computes the marginal function $g_r(x_r)$ in a rooted cycle-free factor graph for a single value of k, with X_r taken as the root vertex. An example of a rooted cycle-free factor graph is provided in Fig. 5.2. The computation begins at the leaves of the factor graph, and mainly contains three phases, i.e., initialization, message computation and termination. In the **initialization phase**, each leaf variable node sends a trivial "identity function" message (i.e., the variables are uniformly distributed) to its parent, and each leaf factor node sends a description of f to its parent. For example, assuming that all the variables are defined over the set $\{0, 1\}$, then the functions can be represented as a vector of size 2, denoted by $[\mu(0), \mu(1)]$. In the **message computation phase**, each vertex waits for messages from all of its children before computing the message to be sent to its parent. A variable node simply sends the product of messages received from its children, while a factor node f_k with parent

[1] Since this algorithm operates by computing various sums and products, we refer to it as the sum-product algorithm.

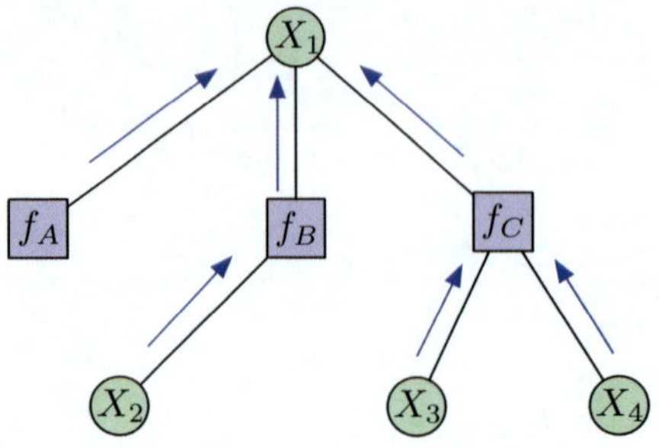

Fig. 5.2 A rooted factor graph representation of the example given in Fig. 5.1 for computing a single marginal function $g_1(x_1)$

X_i forms the product of f_k and the messages received from its children, and then operates on the result with a not-summary operator $\sum_{\sim\{x\}}$, which are given by

$$\mu_{X_i \to f_k}(x_i) = \prod_{h \in n(X_i) \setminus \{f_k\}} \mu_{h \to X_i}(x_i), \text{ (variable node to factor node)} \tag{5.12}$$

$$\mu_{f_k \to X_i}(x_i) = \sum_{\sim\{x_i\}} \left(f_k(A_k) \prod_{x_j \in A_k \setminus x_i} \mu_{X_j \to f_k}(x_j) \right), \text{ (factor node to variable node)} \tag{5.13}$$

respectively, where $h \in n(X_i)$ denote the set of neighbors of a given node X_i in a factor graph. By a "product of messages" we mean an appropriate description of the (point-wise) product of the corresponding functions. If the messages are parametrizations of the functions, then the resulting message is also the parametrization of the product function, not necessarily the numerical product of the messages. Similarly, the summary operator is applied to the functions, not necessarily literally to the messages themselves. Finally, in the **termination phase**, the marginal function $g_r(x_r)$ is obtained as the product of all messages received at X_r, which is

$$g_r(x_r) = \prod_{h \in n(X_r)} \mu_{h \to X_r}(x_r). \tag{5.14}$$

It is important to note that a message passed on the edge (X, f), either from variable X to factor f, or vice versa, is a single-argument function of x, the variable associated with the given edge. This follows since, at every factor node, not-summary operations are always performed for the variable associated with the edge on which the message is passed. Likewise, at a variable node, all messages are functions of that variable, and so is any product of these messages.

5.2.2 Computing All Marginal Functions

In practice, we may be interested in computing the marginal posterior distribution for more than one X. As mentioned before, message passing on factor graphs can be conducted in a distributed and parallel manner, i.e., computation of $g_k(x_k)$ for all k simultaneously can be efficiently accomplished by essentially "overlaying" on a single factor graph all possible instances of the single-node algorithm. In this case, no particular node is taken as a root node, and there is no fixed parent/child relationship among neighboring nodes. Instead, each neighbor w of any given node v is at some time regarded as a parent of v. The message passed from v to w is computed just as in the single-node algorithm,i.e., as if w were indeed the parent of v and all other neighbors of v were children.

Message passing is initiated at the leaves (those only connected to one neighbor). Each node v remains idle until messages have arrived on all but one of the edges incident on v. Just as in the single-node algorithm, once these messages have arrived, v is able to compute a message to be sent on the one remaining edge to its neighbor (temporarily regarded as the parent). Let us denote this temporary parent as node w. After sending a message to w, node v returns to the idle state, waiting for a "return message" to arrive from w. Once this message has arrived, the node is able to compute and send messages to each of its neighbors (other than w), each being regarded, in turn, as a parent.The algorithm terminates once two messages have been passed over every edge, one in each direction. At variable node x_k, the product of all incoming messages is the marginal function $g_k(x_k)$, just as in the single-node algorithm.

Note that the update rule at a variable node takes on the particularly simple form given by (5.12), because there is no local function to include. On the other hand, the update rule at a factor node given by (5.13) in general involves nontrivial function multiplications, followed by a not-summary operation. We also observe that variable nodes of degree two perform no computation: a message arriving on one (incoming) edge is simply transferred to the other (outgoing) edge.

Example 5.2 (*Sum-Product Algorithm Example*). Figure 5.1 shows the flow of messages that would be generated by the sum-product algorithm. The messages may be generated in four steps, as indicated with circles in Fig. 5.1. For simplicity, we assume that all the variables are defined over the set {0, 1}. Specifically, the messages are generated as follows.

Step 1:

$$\begin{aligned} \mu_{X_2\to f_B}(x_2) &= [1, 1], \\ \mu_{X_3\to f_C}(x_3) &= [1, 1], \\ \mu_{X_4\to f_C}(x_4) &= [1, 1], \\ \mu_{f_A\to X_1}(x_1) &= [f_A(0), f_A(1)]. \end{aligned} \tag{5.15}$$

Step 2:

$$\begin{aligned} \mu_{f_B\to X_1}(x_1) &= \sum_{x_2\in\{0,1\}} f_B(x_1, x_2)\mu_{X_2\to f_B}(x_2), \\ \mu_{f_C\to X_1}(x_1) &= \sum_{x_3,x_4\in\{0,1\}} f_C(x_1, x_3, x_4)\mu_{X_3\to f_C}(x_3)\mu_{X_4\to f_C}(x_4) \\ &= \sum_{x_3,x_4\in\{0,1\}} f_C(x_1, x_3, x_4). \end{aligned} \tag{5.16}$$

Step 3:

$$\begin{aligned} \mu_{X_1\to f_A}(x_1) &= \mu_{f_B\to X_1}(x_1)\mu_{f_C\to X_1}(x_1), \\ \mu_{X_1\to f_B}(x_1) &= \mu_{f_A\to X_1}(x_1)\mu_{f_C\to X_1}(x_1), \\ \mu_{X_1\to f_C}(x_1) &= \mu_{f_A\to X_1}(x_1)\mu_{f_B\to X_1}(x_1). \end{aligned} \tag{5.17}$$

Step 4:

$$\begin{aligned} \mu_{f_B\to X_2}(x_2) &= \sum_{x_1\in\{0,1\}} f_B(x_1, x_2)\mu_{X_1\to f_B}(x_1), \\ \mu_{f_C\to X_3}(x_3) &= \sum_{x_1,x_4\in\{0,1\}} f_C(x_1, x_3, x_4)\mu_{X_1\to f_C}(x_1)\mu_{X_4\to f_C}(x_4), \\ \mu_{f_C\to X_4}(x_4) &= \sum_{x_1,x_3\in\{0,1\}} f_C(x_1, x_3, x_4)\mu_{X_3\to f_C}(x_3)\mu_{X_1\to f_C}(x_1). \end{aligned} \tag{5.18}$$

Termination:

$$\begin{aligned} g_1(x_1) &= \mu_{f_A\to X_1}(x_1)\mu_{f_B\to X_1}(x_1)\mu_{f_C\to X_1}(x_1), \\ g_2(x_2) &= \mu_{f_B\to X_2}(x_2), \\ g_3(x_3) &= \mu_{f_C\to X_3}(x_3), \\ g_4(x_4) &= \mu_{f_C\to X_4}(x_4). \end{aligned} \tag{5.19}$$

Note that in the termination step, we compute $g_k(x_k)$ as the product of all messages directed toward X_k, and we ignore the messages with value [1, 1] for conciseness. Equivalently, since the message passed on any given edge is equal to the product of all but one of these messages, we may compute $g_k(x_k)$ as the product of the two messages that were passed (in opposite directions) over any single edge incident on X_k. Thus, for example, we may compute $g_1(x_1)$ in three different ways as given below:

$$\begin{aligned} g_1(x_1) &= \mu_{f_A \to X_1}(x_1)\mu_{X_1 \to f_A}(x_1) \\ &= \mu_{f_B \to X_1}(x_1)\mu_{X_1 \to f_B}(x_1) \\ &= \mu_{f_C \to X_1}(x_1)\mu_{X_1 \to f_C}(x_1). \end{aligned} \tag{5.20}$$

It is important to mention that in a cycle-free factor graph, accurate marginal posterior probabilities can be obtained through the sum-product algorithm [1]. This is because the cycle-free structure ensures that there will be no circular-dependency problems during the message-passing process. Messages can be updated in a certain order (for example, a two-way passing from leaf nodes to root nodes and then from root nodes to leaf nodes) and converge within a finite number of steps.

5.2.3 The Max-Product Algorithm

Instead of attempting to solve the marginal posterior distribution calculation problem, we may be interested in solving the related problem of maximization, and the goal here is to find the values of $\boldsymbol{x}$ that maximizes the global function $f(x_1, x_2, \cdots, x_N)$ (i.e., the most probable values in a probabilistic setting), and it can be defined as follows:

$$g_{X_k}^{\max}(x_k) = \max_{\sim\{x_k\}} f(x_1, x_2, \ldots, x_N). \tag{5.21}$$

An algorithm that solves this problem is nearly identical to the sum-product algorithm, with the sums replaced by maxima in the definitions.

5.2.4 Running Example for Repetition Code

Consider a channel coding problem in a memoryless binary asymmetric channel (MBAC) using a repetition code with code length 2, where the MBAC consists of binary input and output alphabets, i.e., $B, Y \in \{0, 1\}$, and two crossover probabilities $q_1 \triangleq P_{Y|B}(1|0)$ and $q_2 \triangleq P_{Y|B}(0|1)$, as shown in Fig. 5.3. The transmit codeword is given by $\boldsymbol{b} = [0, 0]^T$ or $\boldsymbol{b} = [1, 1]^T$. Assume that the transmitted message signal b is sent with equal probabilities for 0 and 1, and the received codeword is given by $\boldsymbol{y} = [0, 1]^T$, then the decoding problem is equivalent to find the MAP estimation of $\boldsymbol{b}$, i.e., we should find the value of b that maximizes $p_{B|\mathbf{Y}}(b|\boldsymbol{y})$.

Then, based on the MAP estimation in (2.1), we can focus on $p_{B,\mathbf{Y}}(b, \boldsymbol{y})$ and factorize it as

$$\begin{aligned} p_{B,\mathbf{Y}}(b, \boldsymbol{y}) &= p_{\mathbf{Y}|B}(\boldsymbol{y}|b)p_B(b) \\ &= p_{Y_1|B}(y_1|b)p_{Y_2|B}(y_2|b)p_B(b), \end{aligned} \tag{5.22}$$

where we have used the fact that Y_1 and Y_2 are independent for given b since the channel is memoryless, and the corresponding factor graph is shown in Fig. 5.4. According to the sum-product algorithm discussed above, the message-passing process over Fig. 5.4 mainly consists of two steps. In the first step, we calculate

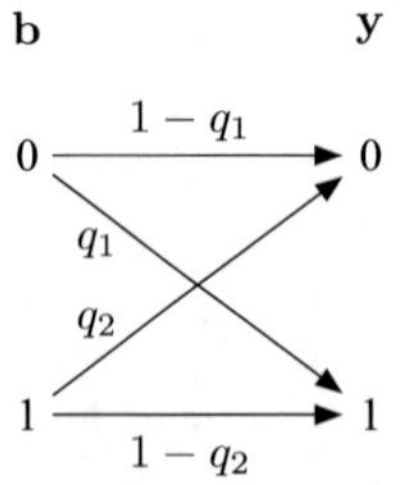

Fig. 5.3 Binary asymmetric channel model

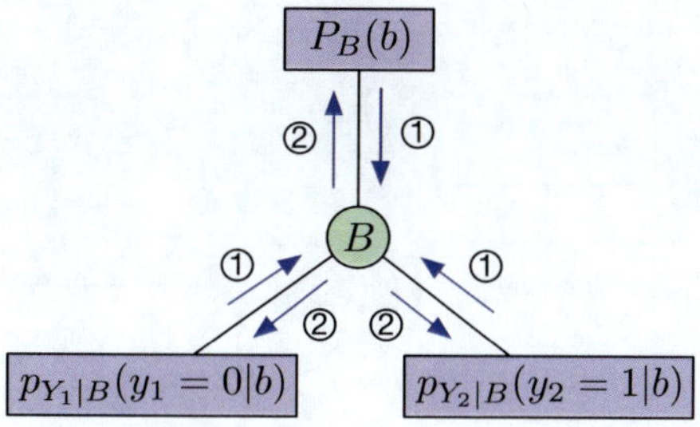

Fig. 5.4 Factor graph for decoding a length 2 repetition over a MBAC

$$
\begin{aligned}
\mu_{p_B \to B}(b) &= [1/2, 1/2]^T, \\
\mu_{p_{Y_1|B} \to B}(b) &= [1 - q_1, q_2]^T, \\
\mu_{p_{Y_2|B} \to B}(b) &= [q_1, 1 - q_2]^T,
\end{aligned}
\tag{5.23}
$$

while in th second step, we can obtain

$$
\begin{aligned}
\mu_{B \to p_B}(b) &= [q_1(1 - q_1), q_2(1 - q_2)]^T, \\
\mu_{B \to p_{Y_1|B}}(b) &= [\frac{1}{2}q_1, \frac{1}{2}(1 - q_2)]^T, \\
\mu_{B \to p_{Y_2|B}}(b) &= [\frac{1}{2}(1 - q_1), \frac{1}{2}q_2]^T.
\end{aligned}
\tag{5.24}
$$

Note that in the second step, calculating one message is enough for obtaining the marginal posterior distribution. Based on the above two steps, we are able to obtain

$$
\begin{aligned}
P_{B,\mathbf{Y}}(b = 0, \mathbf{y} = [0, 1]^T) &= \mu_{p_B \to B}(b = 0)\mu_{B \to p_B}(b = 0) = \frac{1}{2}q_1(1 - q_1), \\
P_{B,\mathbf{Y}}(b = 1, \mathbf{y} = [0, 1]^T) &= \mu_{p_B \to B}(b = 1)\mu_{B \to p_B}(b = 1) = \frac{1}{2}q_2(1 - q_2),
\end{aligned}
\tag{5.25}
$$

and then using the fact that $p(a|b) \propto p(a, b)$, it yields

$$
\begin{aligned}
P_{B|\mathbf{Y}}(b = 0|\mathbf{y} = [0, 1]^T) &= \frac{q_1(1 - q_1)}{q_1(1 - q_1) + q_2(1 - q_2)}, \\
P_{B|\mathbf{Y}}(b = 1|\mathbf{y} = [0, 1]^T) &= \frac{q_2(1 - q_2)}{q_1(1 - q_1) + q_2(1 - q_2)},
\end{aligned}
\tag{5.26}
$$

after normalization. It is worth mentioning that after obtaining the marginal posterior distribution $P_{B|\mathbf{Y}}$, we can also acquire the marginal distribution of Y, i.e., $p_{\mathbf{Y}}(\mathbf{y})$, as follows:

$$
p_{\mathbf{Y}}(\mathbf{y}) = \frac{p_B(b)p_{\mathbf{Y}|B}(\mathbf{y}|b)}{P_{B|\mathbf{Y}}(b|\mathbf{y})}.
\tag{5.27}
$$

To summarize, given an MAP estimation problem, we can follow the road map in Fig. 5.5.

5.3 Practicalities

5.3.1 Messages and Their Representations

As mentioned above, the messages that are passed along the edges are in essence functions, not some scalars or vectors in general. If the considered variables are discrete, then the messages can indeed be represented by scalars/vectors/matrices,

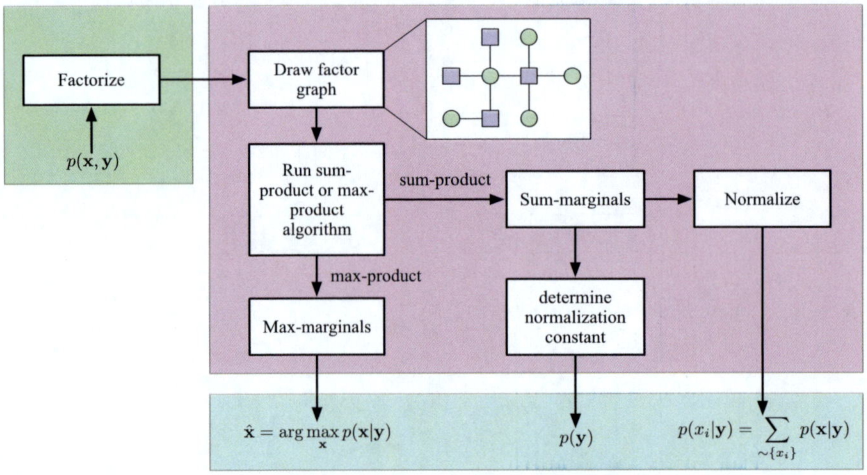

Fig. 5.5 General steps to solve an MAP estimation problem using the sum-product algorithm

which are relatively easy to handle. However, there may exist numerical stability issues during the message-passing process, especially when dealing with probabilities or other values that are multiplied or exponentiated. For example, when multiplying a large number of small probabilities (close to zero), the result may get smaller and smaller in magnitude and underflow to zero in a computer's floating-point representation. One effective approach is to use the normalization techniques, where messages can be scaled to a more manageable range. In each iteration of the message-passing process, the messages can be divided by a suitable scaling factor to prevent them from becoming too small. After the messages are passed and updated, they can be normalized to ensure that they still represent valid probability-like values. If the message is $\boldsymbol{m} = [a, b]^T$, then by interpreting $\boldsymbol{m}$ as a distribution, we can normalize it to $\boldsymbol{m} = [a/(a+b), b/(a+b)]^T$ with a normalization constant $a + b$, and the new message can be represented by $\boldsymbol{m}' = [a/(a+b), b/(a+b); a+b]^T$. Note that as long as we keep track of the normalization constants, the outcome of the sum-product algorithm will not be affected.

Another effective solution is to work in the log-domain. Instead of representing and manipulating the actual probabilities or messages, their logarithms are used. For example, when multiplying probabilities m_1 and m_2, instead of calculating $m_1 \times m_2$ (which may lead to underflow), we calculate $\log(m_1)+\log(m_2)$ in the log-domain. This addition operation is more numerically stable than multiplication. When the final result is needed, we can convert back from the log-domain using exponentiation. A numerical example for normalization and log-domain processing is shown in Fig. 5.6. Moreover, instead of go back and forth between the probability-domain and log-domain, one can utilize the Jacobian logarithm to simplify the calculation process, as given in the following theorem [2, 3]

Theorem 5.1 *For any given numbers a and b, the following equality holds:*

$$\mathbb{M}(a, b) \triangleq \ln(e^a + e^b) = \max(a, b) + \ln(1 + e^{-|b-a|}). \tag{5.28}$$

Proof Let us assume $a > b$, then we have

$$\begin{aligned}\ln(e^a + e^b) &= \ln(e^a(1 + e^b/e^a)\\ &= \ln(e^a(1 + e^{b-a}))\\ &= \ln(e^a(1 + e^{-(a-b)}) \quad \text{since } a > b\\ &= \ln(e^a) + \ln(1 + e^{-(a-b)})\\ &= a + \ln(1 + e^{-(a-b)}).\end{aligned} \tag{5.29}$$

Similarly, we can obtain $\ln(e^a + e^b) = b + \ln(1 + e^{-(b-a)})$ for the case when $b > a$. By combining the above results, it yields

$$\ln(e^a + e^b) = \max(a, b) + \ln(1 + e^{-|b-a|}), \tag{5.30}$$

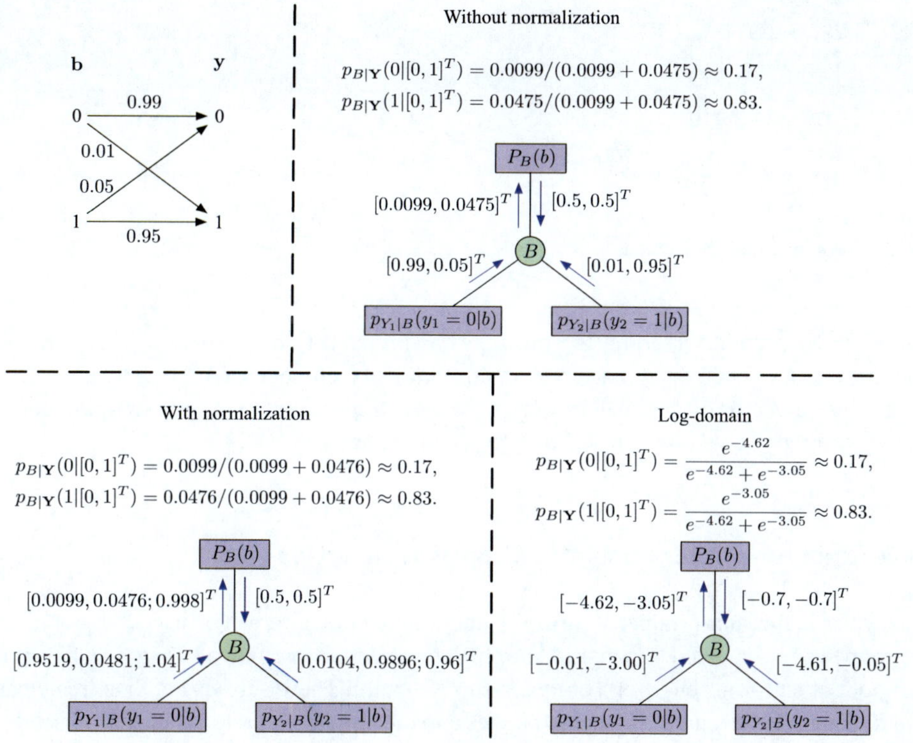

Fig. 5.6 A numerical example for normalization and log-domain processing for decoding a length-2 repetition code in MBAC

which completes the proof. □

Thus, instead of directly calculating the logarithm of the messages, e.g., the factor node to variable node message in (5.13), as in

$$L_{f_k \to X_i}(x_i) \triangleq \ln \mu_{f_k \to X_i}(x_i) = \ln \left\{ \sum_{\sim\{x_i\}} \left(f_k(A_k) \prod_{x_j \in A_k \backslash x_i} \mu_{X_j \to f_k}(x_j) \right) \right\}, \tag{5.31}$$

we can utilize the following update rule:

$$L_{f_k \to X_i}(x_i) = \mathbb{M}_{\sim\{x_i\}} \left(\ln f_k(A_k) + \sum_{x_j \in A_k \backslash x_i} L_{X_j \to f_k}(x_j) \right), \tag{5.32}$$

where we have used the recursive property of the operator $\mathbb{M}(\cdot, \cdot)$ (see Appendix A for the proof), i.e.,

$$\mathbb{M}(L_1, L_2, \ldots, L_L) = \mathbb{M}(L_1, \mathbb{M}(L_2, \ldots, L_L)). \tag{5.33}$$

Thus, the complex logarithm operation can be replaced by a max operation plus one table look-up, which is very efficient. Thus, the complex logarithm operation can be replaced by a max operation plus one table look-up, which is very efficient.

Furthermore, we can also combine the normalization and log-domain processing techniques, and use the LLR for message representation. This method is very suitable and popular for binary variables. For example, if the variable x is binary, i.e., $x \in \{0, 1\}$, then the LLR of x is defined as $\mathrm{LLR}(x) = \log P(x = 1)/P(x = 0)$. The LLR provides an intuitive interpretation of the evidence in favor of one state of a binary variable over the other. A positive LLR value indicates that the evidence leans more towards $x = 1$, while a negative LLR value indicates more evidence for $x = 0$. This interpretability helps in understanding the behavior of the message-passing algorithm and making decisions based on the passed messages. For example, in a decoding algorithm for error-correcting codes, the LLRs of the received bits can be used to make decisions about the most likely original codewords, and the magnitude and sign of the LLRs provide clear indications of the reliability of these decisions.

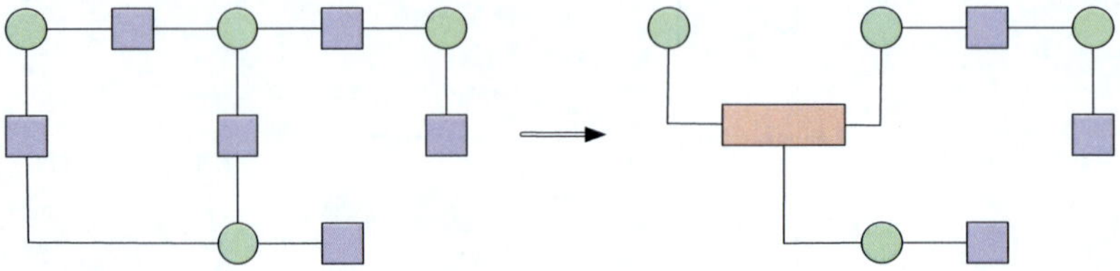

Fig. 5.7 A schematic example of junction tree methods

For continuous variables, we should interpret the messages as probability mass functions, and there are generally three ways to represent the messages: (1) make them discrete and use known techniques, (2) use a parametric representation (e.g., Gaussian mixtures), i.e., use a vector of parameters to represent each message, (3) use a non-parametric representation, in which case the messages are represented by a vector of (weighted) samples.

5.3.2 The Sum-Product Algorithm for Factor Graphs with Cycles

When there are loops/cycles in the sum-product algorithm, junction tree methods can be used [4]. In this method, the nodes in the factor graph are grouped or aggregated together to simplify the structure, as shown in Fig. 5.7. The process might involve identifying subsets of nodes that form a cycle or a complex sub-structure. These subsets are then treated as aggregated nodes. The relationships and messages between the aggregated nodes and the remaining nodes in the graph need to be re-defined. For example, the messages that were originally passed between individual nodes within the aggregated group might be combined or summarized into a single message representing the aggregated node. During the execution of the sum-product algorithm, the aggregated nodes are considered as a single entity for the purpose of message-passing and probability calculations. This helps in handling the cyclic dependencies more effectively, as it reduces the number of paths that messages can loop through and potentially allows for a more tractable approximation of the marginal probabilities or other inferences that the sum-product algorithm aims to compute. However, in some cases, the graph structure may be so complex or the loops may have such unique characteristics that aggregation does not simplify the problem enough or may even introduce additional complications, please see [1, 5] for more details.

A more modern approach is to ignore the cycles in the graph and simply apply the sum-product algorithm by following the same message propagation rules, since all updates are local. Due to the cycles in the factor graph, the message-passing process might be iterated multiple times. In some cases, the messages may converge to a fixed-point solution. Some of the most exciting applications of the sum-product algorithm, e.g., the decoding of turbo codes or low-density parity-check (LDPC) codes [1, 6], bit-interleaving coded modulation with iterative decoding (BICM-ID) [7], MIMO detection [8], multiuser detection [9], etc., occur precisely in circumstances where the underlying factor graph has cycles. Extensive simulation results (see, e.g., [10, 11]) indicate that when dealing with very long codes, the sum-product algorithm can attain remarkable performance very close to the Shannon limit on a Gaussian channel.

The sum-product algorithm is then sometimes called loopy belief propagation, because graphs typically contain cycles or loops, and the obtained results are not exactly the marginal posterior distribution, but some "beliefs" on this distribution. The initialization and scheduling of message updates must be adjusted slightly because the considered graphs might not contain any leaves. In this case, it helps to imagine that there is a global discrete-time clock and the messages are synchronized, with at most one message passed on any edge in any given direction at one time. A message sent from node v at time i will be a function of the local function at v (if any) and the (most recent) messages received at v prior to time i.

Different from cycle-free factor graphs, factor graph with cycles may have no nodes of degree one, thus the initialization phase should be modified accordingly. Usually, we initialize all variable messages to 1, thus every node is in a position to send a message at every time along every edge in the initialization phase. To this end, it is important to mention that the message-passing schedule in a factor graph, i.e., the specification of messages to be passed during each clock tick, will affect the final performance. A wide variety of message-passing schedules are possible, such as the flooding schedule, sequential schedule [12], etc. In the flooding schedule, messages are passed in a parallel fashion. At each iteration, all variable nodes send messages to all their adjacent factor nodes, and vice-versa. This means that the information spreads throughout the graph rapidly. The advantage of the flooding schedule is its simplicity and the fact that it can quickly disseminate information across the graph. However, it can be computationally expensive, especially for large graphs, as a large number of messages are updated in each iteration. In the sequential schedule, at most one message is passed anywhere in the graph at each clock tick. For instance, messages can be passed from one end of the graph to the other in a chain-like manner. The sequential schedule can be more efficient in terms of computation for certain graph structures, as it updates messages in a more controlled way. But it may take more iterations for the information to spread throughout the graph compared to the flooding schedule.

It would be useful to mention the notion "pending message" for factor graphs with cycles. Given an edge e, if a node v (connected to e) has received any messages on edges other than e after the transmission of the most previous message on v, we say a message is pending at v since the messages more recently received can affect the message to be sent on e. The receipt of a message at v from an edge e will create pending messages at all other edges connected to v. Only pending messages are required to be transmitted, since only pending messages can be different from the previous message sent on a given edge.

There is generally no universal guidelines for designing the best message-passing schedule, since it depends on many factors, such as the graphical structure, node property and some problem-specific requirements, etc. For example, in a simple chain-like graphical model, a sequential schedule from one end of the chain to the other and back might be intuitive and effective. However, in a more complex graph with loops and multiple branches, such a simple sequential schedule may not work well. Besides, different nodes in a factor graph may have different characteristics and functions. Some nodes may represent more stable or well-known variables, while others may be more uncertain or subject to frequent updates. The message schedule needs to take into account these differences. The nature of the problem being solved also dictates the message schedule. For instance, in a real-time decision-making problem, where quick responses are needed, a message schedule that prioritizes messages related to the most relevant variables for the decision at hand may be preferred. In contrast, in a problem focused on accurate posterior probability estimation with less time constraint, a more comprehensive and iterative message-passing schedule that explores all possible paths of information flow in the graph might be more suitable.

5.4 Convergence Results for the Sum-Product Algorithm

For general models, the convergence of the sum-product algorithm cannot be guaranteed, and even when converged, there is no guarantee on the quality of the converged marginals. However, for a Gaussian graphical model of arbitrary topology, i.e., when the joint probability can be written as

$$\begin{aligned} p(\boldsymbol{x}) &\propto \exp\left(-\boldsymbol{x}^T \boldsymbol{W} \boldsymbol{x} + \boldsymbol{b}^T \boldsymbol{x}\right) \\ &= \prod_{i,j} \exp\left(-w_{ij} x_i x_j\right) \times \prod_i \exp\left(b_i x_i\right), \end{aligned} \tag{5.34}$$

then the convergence of the sum-product algorithm can be guaranteed when $\boldsymbol{W}$ is diagonally dominant ($|w_{i,i}| > \sum_{j \neq i} |w_{i,j}|, \forall i$) [13]. Besides, if the sum-product algorithm converges, then the posterior marginal means calculated are exact, while the variances of the marginals are equal to the correct variances minus the summed conditional correlations between $\tilde{x}_1$ and all $\tilde{x}_j$ that are replicas of x_1, where $\tilde{x}_j$ denotes the unwrapped quantities in the unwrapped tree, and x_1 denotes the root node [13].[2] Moreover, the work [14] presented a walk-sum interpretation of inference in Gaussian graphical models, which holds for a wide class of models, called walk-summable. It was proved that in walk-summable models, the sum-product algorithm is guaranteed to converge, for both means and variances, and that upon convergence the means are correct, whereas the variances only capture walk-sums over back-tracking walks. For non-Gaussian models, only sufficient (not necessary) conditions for the convergence of the sum-product algorithm to a unique fixed point can be derived [15], and there is in general no guarantee on the quality of the converged solutions.

[2] More specifically, to analyze the behavior of the algorithm on a loopy graph, we imagine the process of a message being passed around a cycle. Each time a message completes a loop, it returns to a node with its own history "baked in", potentially leading to feedback or divergence. The "unwrapped tree" (also called the computation tree or universal cover) is a conceptual tool that captures this history. It is constructed by selecting a root node (i.e., x_1), and "unwrapping" the loopy graph into an infinite tree by replicating the nodes and their local connectivity. Each replica of a node in the original graph appears multiple times in the unwrapped tree, at different depths or in different branches. These replicas are denoted by $\tilde{x}_j$.

Summary
Factor graphs and message passing provide a powerful, unified framework for efficient probabilistic inference in high dimensions. They transform complex global calculations into distributed, local computations on a graph, enabling breakthroughs in applications like modern error-correcting codes (turbo, LDPC) and multi-antenna detection (MIMO) by effectively handling dependencies and leveraging problem structure, even in the presence of challenging cycles.

Below, we summarize the key insights and core concepts.

1. Factor graphs explicitly represent the factorization of a global function (e.g., joint probability) into local functions (factors), enabling a divide-and-conquer approach to inference. This modularity is crucial for managing complexity.
2. The sum-product algorithm computes marginals through iterative, local message exchanges between neighboring nodes: (1) for variable to factor node, product of incoming messages is performed, (2) for factor to variable node, sum (marginalization) over other arguments of the product of the local factor and incoming messages is conducted, and (3) after termination, marginal at each node is the product of all incoming messages.
3. While exact marginals are guaranteed only for cycle-free graphs, applying the same message-passing rules iteratively to graphs with cycles (loopy belief propagation) is highly effective in practice (e.g., turbo/LDPC decoding, MIMO detection), often approaching theoretical limits like Shannon capacity despite lacking universal convergence guarantees. Schedules (flooding, sequential) and concepts like "pending messages" manage updates.
4. There are some variants for the sum-product algorithm, e.g., the max-product algorithm that replaces sums with maximizations in the message rules for MAP estimation. For numerical issues in practice, we can address them via normalization, log-domain processing, and LLRs for binary variables, which enhance stability and interpretability.
5. By leveraging the graphical structure and conditional independencies, message passing avoids the exponential complexity of direct high-dimensional summation/integration or global maximization, making complex inference problems computationally feasible.
6. Convergence and correctness are provable for specific models like Gaussian graphical models under walk-summability or diagonal dominance. For general models, convergence is empirical but often yields excellent approximations, as demonstrated by groundbreaking applications in communications.

Appendix

Proof of the Recursive Property of the Jacobian Logarithm

Suppose that $\mathbb{M}(L_2, \cdots, L_L) = \ln(e^{L_2} + \cdots + e^{L_L})$ is known and define $\Delta \triangleq e^{L_2} + \cdots + e^{L_L}$, then we have

$$
\begin{aligned}
\mathbb{M}(L_1, \ldots, L_L) &= \ln(e^{L_1} + e^{L_2} + \cdots + e^{L_L}) \\
&= \ln(e^{L_1} + \Delta) \\
&= \max(L_1, \ln \Delta) + \ln(1 + e^{-|\ln \Delta - L_1|}) \\
&= \max(L_1, \mathbb{M}(L_2, \ldots, L_L)) + \ln(1 + e^{-|\mathbb{M}(L_2,\ldots,L_L) - L_1|}) \\
&= \mathbb{M}(L_1, \mathbb{M}(L_2, \ldots, L_L)).
\end{aligned}
\tag{5.35}
$$

By repeatedly substituting the above recursive formula into itself, the recursive property of the $\mathbb{M}(\cdot, \cdot)$ can be readily proved.

Homework Problems

5.1 (a) Define a factor graph formally. What do variable nodes, factor nodes, and edges represent?
(b) Consider the joint function

$$
f(x_1, x_2, x_3, x_4, x_5) = f_A(x_1, x_2) f_B(x_2, x_3) f_C(x_3, x_4) f_D(x_4, x_5) f_E(x_1, x_5).
$$

Draw the corresponding factor graph. Does this factor graph contain any cycles? Justify your answer.

5.2 Consider the factor graph shown in Fig. 5.1 representing $f(x_1, x_2, x_3, x_4) = f_A(x_1) f_B(x_1, x_2) f_C(x_1, x_3, x_4)$. Assume all variables are binary ($x_i \in \{0, 1\}$) and the local functions are defined as follows:

- $f_A(x_1)$: $[0.7, 0.3]^T$ (i.e., $f_A(0) = 0.7$, $f_A(1) = 0.3$)
- $f_B(x_1, x_2)$:

x_1	x_2	f_B
0	0	0.9
0	1	0.1
1	0	0.4
1	1	0.6

- $f_C(x_1, x_3, x_4)$:

x_1	x_3	x_4	f_C
0	0	0	0.8
0	0	1	0.2
0	1	0	0.5
0	1	1	0.5
1	0	0	0.1
1	0	1	0.9
1	1	0	0.3
1	1	1	0.7

(a) Initialization: Write down the messages sent from the leaf nodes (X_2, X_3, X_4, f_A) during Step 1 of the example in Fig. 5.1.
(b) Message to X_1: Calculate the messages $\mu_{f_B \to X_1}(x_1)$ and $\mu_{f_C \to X_1}(x_1)$ sent during Step 2 of the example in Sect. 5.2.
(c) Termination at X_1: Calculate the marginal function $g_1(x_1)$ at variable node X_1 using the termination rule (5.19). Normalize this result to represent the marginal posterior distribution $p_{X_1|Y}(x_1|\boldsymbol{y})$ (assuming $\boldsymbol{y}$ is conditioned upon).

5.3 Consider the repetition code ($N = 2$) decoding example over the MBAC channel described in Sect. 5.2.4. Assume the prior $p_B(b) = [0.5, 0.5]^\top$, crossover probabilities $q_1 = 0.2$, $q_2 = 0.3$, and the received vector $\boldsymbol{y} = [0, 1]^\top$.

(a) Write down the factorization of $p_{B,Y}(b, \boldsymbol{y})$ explicitly using the given q_1, q_2, and $\boldsymbol{y}$.
(b) Calculate the messages $\mu_{p_{Y_1|B} \to B}(b)$ and $\mu_{p_{Y_2|B} \to B}(b)$.
(c) Calculate the marginal posterior distribution $p_{B|Y}(b|\boldsymbol{y} = [0, 1]^\top)$. Based on this, what is the MAP estimate of the transmitted bit b?
(d) Compute the LLR for the bit b based on your result from (c). Interpret the sign and magnitude of the LLR.

5.4 Consider a communication system that uses a (6, 3) LDPC code to protect data transmitted over an AWGN channel. The parity-check matrix $\boldsymbol{H}$ for this code is given by

$$\boldsymbol{H} = \begin{bmatrix} 1\,0\,1\,0\,1\,0 \\ 1\,1\,0\,1\,0\,0 \\ 0\,0\,1\,1\,0\,1 \end{bmatrix},$$

which corresponds to the parity-check equations

$$\begin{cases} c_1 \oplus c_3 \oplus c_5 = 0, \\ c_1 \oplus c_2 \oplus c_4 = 0, \\ c_3 \oplus c_4 \oplus c_6 = 0. \end{cases}$$

The system employs BPSK modulation, where a codeword bit $c_i = 0$ is mapped to $+1$ and $c_i = 1$ is mapped to -1. Assume the a priori probability for each transmitted bit is equal, i.e., $p(c_i = 0) = p(c_i = 1) = 0.5$. The received vector after transmission over the AWGN channel with noise variance $\sigma^2 = 0.5$ is

$$\boldsymbol{y} = [0.9, -1.2, 0.4, 1.1, -0.8, -0.2].$$

(a) The goal of the decoder is to find the most probable codeword c given the received vector $\boldsymbol{y}$. Express the MAP decision rule for the entire codeword c. Using Bayes' theorem, show how this rule relates to the channel likelihood function $p(\boldsymbol{y}|\boldsymbol{c})$ and the prior probability $p(\boldsymbol{c})$. Explain why $p(\boldsymbol{c})$ is non-zero only for valid codewords that satisfy the parity-check equations.

(b) The first step in iterative decoding is to compute the initial LLR for each codeword bit based on the channel observation. Please derive the expression of the initial LLRs and calculate them for all six bits $c_1, \ldots, c_6$ using the given received vector $\boldsymbol{y}$ and noise variance σ^2.

(c) Draw the bipartite factor graph that represents the factorization of the joint probability distribution defined by this LDPC code. The graph should contain variable nodes (for codeword bits $c_1, \ldots, c_6$) and check nodes (for the parity-check equations defined by the rows of $\boldsymbol{H}$).

(d) The Sum-Product Algorithm (or Belief Propagation) is used to compute the a posteriori probabilities of the codeword bits. Describe the two main message-passing steps for one iteration of the algorithm in the LLR domain, i.e., Check Node to Variable Node (C2V) Update: Provide the update rule for the LLR message $L(f_i \rightarrow c_j)$ sent from a check node f_i to a variable node c_j. This rule is often expressed using the "box-plus" operator or the hyperbolic tangent function.
Variable Node to Check Node (V2C) Update: Provide the update rule for the LLR message $L(c_j ß f_i)$ sent from a variable node c_j to a check node f_i.

(e) The Min-Sum algorithm is a lower-complexity approximation of the Sum-Product algorithm. State the simplified update rule for the Check Node to Variable Node (C2V) message in the Min-Sum algorithm. Explain how this rule approximates the corresponding rule in the Sum-Product algorithm. (The V2C update rule remains the same).

(f) After several iterations of the Sum-Product or Min-Sum algorithm, how would you obtain the final a posteriori LLR for each bit and make a hard decision for the decoded codeword $\hat{\boldsymbol{c}}$? Describe a common stopping criterion for the iterative decoding process. If you were to plot the BER versus the signal-to-noise ratio (SNR) for this LDPC decoder, what is the characteristic shape of the performance curve you would expect to see, especially as the code length increases? Explain the concept of the "waterfall" region in this context.

5.5 Consider a factor graph with a cycle: $(x_1) - (f_A) - (x_2) - (f_B) - (x_3) - (f_C) - (x_1)$, where each (f_k) is a factor connecting two variables.

(a) Outline a possible message-passing schedule (e.g., flooding or sequential) and discuss how you would check for convergence.

(b) Explain the concept of "pending messages" and how they help in managing message updates in graphs with cycles.

References

1. F.R. Kschischang, B.J. Frey, H.A. Loeliger, Factor graphs and the sum-product algorithm. IEEE Trans. Inf. Theory **47**(2), 498–519 (2001)
2. P. Robertson, E. Villebrun, P. Hoeher, A comparison of optimal and sub-optimal MAP decoding algorithms operating in the log domain, in *Proceedings IEEE International Conference on Communications, ICC'95*, vol. 2 (IEEE, 1995), pp.1009–1013
3. W. Koch, A. Baier, Optimum and sub-optimum detection of coded data disturbed by time-varying intersymbol interference (applicable to digital mobile radio receivers), in *[Proceedings] GLOBECOM'90: IEEE Global Telecommunications Conference and Exhibition* (IEEE, 1990), pp. 1679–1684
4. F.V. Jensen, An introduction to Bayesian networks (1996)
5. S.M. Aji, R.J. McEliece, The generalized distributive law. IEEE Trans. Inform. Theory **46**(2), 325–343 (2000). https://doi.org/10.1109/18.825794
6. A.C. Reid, T.A. Gulliver, D.P. Taylor, Convergence and errors in turbo-decoding. IEEE Trans. Commun. **49**(12), 2045–2051 (2001)
7. X. Li, A. Chindapol, J.A. Ritcey, Bit-interleaved coded modulation with iterative decoding and 8 PSK signaling. IEEE Trans. Commun. **50**(8), 1250–1257 (2002)
8. C. Jeon, R. Ghods, A. Maleki, C. Studer, Optimality of large MIMO detection via approximate message passing, in *2015 IEEE International Symposium on Information Theory (ISIT)* (IEEE, 2015), pp. 1227–1231

9. S. Verdu, *Multiuser detection* (Cambridge University Press, 1998)
10. C. Berron, Near Shannon limit error-correcting coding and decoding: Turbo-codes, in *Proceedings of ICC'93* (1993)
11. D.J. MacKay, Good error-correcting codes based on very sparse matrices. IEEE Trans. Inf. Theory **45**(2), 399–431 (1999)
12. F.R. Kschischang, B.J. Frey, Iterative decoding of compound codes by probability propagation in graphical models. IEEE J. Sel. Areas Commun. **16**(2), 219–230 (1998)
13. Y. Weiss, W.T. Freeman, Correctness of belief propagation in Gaussian graphical models of arbitrary topology. Neural Comput. **13**(10), 2173–2200 (2001)
14. D.M. Malioutov, J.K. Johnson, A.S. Willsky, Walk-sums and belief propagation in Gaussian graphical models. J. Mach. Learn. Res. **7**, 2031–2064 (2006)
15. J.M. Mooij, H.J. Kappen, Sufficient conditions for convergence of the sum–product algorithm. IEEE Trans. Inf. Theory **53**(12), 4422–4437 (2007)

6 Subspace Methods

This chapter presents fundamental subspace-based algorithms for high-resolution parameter estimation in signal processing, focusing on MUSIC (MUltiple SIgnal Classification) and ESPRIT (Estimation of Signal Parameters via Rotational Invariance Techniques). MUSIC estimates parameters like direction-of-arrival (DoA) or frequency by exploiting the orthogonality between the signal subspace and the noise subspace derived from the eigendecomposition of the received signal's covariance matrix. It constructs a pseudospectrum whose peaks correspond to the estimated parameters. Variants like root-MUSIC (using polynomial rooting instead of spectral search) and 2D extensions are discussed, which address challenges like path clustering and computational efficiency. ESPRIT, in contrast, avoids spectral searches by leveraging a known translational invariance structure within the sensor array (e.g., identical subarrays displaced by a fixed vector). It directly estimates parameters by solving an (total) LS problem involving the signal subspaces of the subarrays. Extensions to uniform linear arrays (ULAs) using overlapping subarrays and the computationally more powerful multiple invariance (MI) ESPRIT for exploiting multiple displacement structures are detailed. Key aspects include signal number detection (e.g., using minimum description length (MDL)/Akaike information criterion (AIC)), computational complexity trade-offs, and practical considerations like noise robustness.

6.1 MUSIC

6.1.1 Basic Principle

The MUSIC algorithm is a powerful technique used mainly in the field of signal processing, and its objective is to estimate from measurements a set of constant parameters (such as the DoA) upon which the received signals depend. In many applications, such as radar, sonar, wireless communication, and microphone arrays, it is crucial to determine the direction from which the incoming signals are arriving. The MUSIC algorithm provides an efficient way to estimate these directions. For example, consider an array of M antennas receiving K narrowband signals. The received signal $\boldsymbol{y}(t) \in \mathbb{C}^{M\times 1}$ is given by

$$\boldsymbol{y}(t) = \boldsymbol{A}(\boldsymbol{\theta})\text{diag}(\boldsymbol{x}(t))\boldsymbol{s}(t) + \boldsymbol{n}(t), \tag{6.1}$$

where $\boldsymbol{s}(t) \in \mathbb{C}^{K\times 1}$ is the transmitted signal, $\boldsymbol{x}(t) = [x_1(t), \ldots, x_K(t)] \in \mathbb{C}^{K\times 1}$ is the complex gain, $\boldsymbol{A} = [\boldsymbol{a}(\theta_1), \ldots, \boldsymbol{a}(\theta_K)] \in \mathbb{C}^{M\times K}$ is the array response matrix with

$$\boldsymbol{a}(\theta) = \left[1, e^{-j2\pi\frac{d}{\lambda}\sin(\theta)}, \ldots, e^{-j2\pi\frac{(M-1)d}{\lambda}\sin(\theta)}\right]^T, \tag{6.2}$$

d denotes the antenna spacing, λ is the signal wavelength, and $\boldsymbol{n}(t) \in \mathbb{C}^{M\times 1}$ is the additive Gaussian noises.[1] It is important to mention that the number of source signals, K, should be less than the number of antennas, i.e., $K < M$.

The primary goal in this context is to estimate the DoAs of the K signals, as well as, in some cases, the signal waveform $\boldsymbol{s}(t)$ and the complex gain $\boldsymbol{x}(t)$. Note that from (6.1), it is easily seen that the estimation problem posed is scale-invariant

[1] For simplicity, we assume that the noises are spatially white in the following; however, this assumption is not mandatory; the extension to an arbitrary noise covariance $\sigma^2\boldsymbol{\Sigma}$ is not difficult, provided $\boldsymbol{\Sigma}$ is known.

A. Liu and M.-M. Zhao, *Advanced Bayesian Signal Processing*, Textbooks in Telecommunication Engineering,
https://doi.org/10.1007/978-3-032-26656-9_6

in the sense that absolute signal powers are not observable. For any nonsingular diagonal matrix, say $\boldsymbol{D}$, the data model is invariant with respect to the transformations $\text{diag}(\boldsymbol{x}(t))\boldsymbol{s}(t) \rightarrow \boldsymbol{D}^{-1}\text{diag}(\boldsymbol{x}(t))\boldsymbol{s}(t)$ and $\boldsymbol{A}(\boldsymbol{\theta}) \rightarrow \boldsymbol{A}(\boldsymbol{\theta})\boldsymbol{D}$. Thus, estimates of the signals and the associated array response vectors derived herein are to be interpreted modulo an arbitrary scale factor, unless knowledge of the gain pattern of one of the sensors is available.

The $M \times M$ autocorrelation matrix of $\boldsymbol{y}(t)$ is then given by

$$\boldsymbol{R}_y = \boldsymbol{A}(\boldsymbol{\theta})\boldsymbol{R}_x\boldsymbol{A}(\boldsymbol{\theta})^H + \sigma^2\boldsymbol{I}, \tag{6.3}$$

where σ^2 is the noise variance, $\boldsymbol{I}$ is an $M \times M$ identity matrix, and $\boldsymbol{R}_x$ is the $K \times K$ autocorrelation matrix of $\text{diag}(\boldsymbol{x}(t))\boldsymbol{s}(t)$. Since the exact $\boldsymbol{R}_y$ is in general difficult to obtain, it is traditionally estimated using the following sample correlation matrix:

$$\hat{\boldsymbol{R}}_y = \frac{1}{T}\sum_{t=1}^{T}\boldsymbol{y}(t)\boldsymbol{y}^H(t), \tag{6.4}$$

where $T > M$ is the number of vector observations (or known as snapshots). Given the estimate $\hat{\boldsymbol{R}}_y$, MUSIC estimates the DoA (or more generally frequency content) of the signal or autocorrelation matrix using an eigenspace method. Specifically, since $\hat{\boldsymbol{R}}_y$ is a Hermitian matrix, all of its M eigenvectors $\boldsymbol{U} \triangleq \{\boldsymbol{u}_1, \boldsymbol{u}_2, \cdots, \boldsymbol{u}_M\}$ are orthogonal to each other. Sort the eigenvalues of $\hat{\boldsymbol{R}}_y$ in decreasing order, then the eigenvectors $\{\boldsymbol{u}_1, \cdots, \boldsymbol{u}_K\}$ corresponding to the K largest eigenvalues (e.g., directions of largest variability) span the signal subspace $\mathcal{U}_S$. The remaining $M - K$ eigenvectors correspond to eigenvalue σ^2 and span the noise subspace $\mathcal{U}_N$, which is orthogonal to the signal subspace, i.e., $\mathcal{U}_S \perp \mathcal{U}_N$. Let $\boldsymbol{U}_S \triangleq [\boldsymbol{u}_1, \ldots, \boldsymbol{u}_K]$ and $\boldsymbol{U}_N \triangleq [\boldsymbol{u}_{K+1}, \ldots, \boldsymbol{u}_M]$, then we have

$$\hat{\boldsymbol{R}}_y = \boldsymbol{U}_S\boldsymbol{\Sigma}_S\boldsymbol{U}_S^H + \boldsymbol{U}_N\boldsymbol{\Sigma}_N\boldsymbol{U}_N^H, \tag{6.5}$$

where $\boldsymbol{\Sigma}_S$ and $\boldsymbol{\Sigma}_N$ are diagonal matrices that contain the largest K and remaining $M - K$ eigenvalues, respectively.

The general idea behind MUSIC is to use all the eigenvectors that span the noise subspace to improve the estimation performance. Since any signal vector $\boldsymbol{e}$ that resides in the signal subspace $\boldsymbol{e} \in \mathcal{U}_S$ must be orthogonal to the noise subspace, i.e., $\boldsymbol{e} \perp \mathcal{U}_N$, $\boldsymbol{e}$ must satisfy $\boldsymbol{e} \perp \boldsymbol{u}_i$ for all the eigenvectors $\{\boldsymbol{u}_i\}_{i=K+1}^{M}$ that spans the noise subspace $\mathcal{U}_N$. In order to measure the degree of orthogonality of an arbitrary vector $\bar{\boldsymbol{e}}$ with respect to all the $\boldsymbol{u}_i \in \mathcal{U}_N$, the MUSIC algorithm defines a squared norm:

$$d^2 = \|\boldsymbol{U}_N^H\bar{\boldsymbol{e}}\|^2 = \bar{\boldsymbol{e}}^H\boldsymbol{U}_N\boldsymbol{U}_N^H\bar{\boldsymbol{e}} = \sum_{i=K+1}^{M}|\bar{\boldsymbol{e}}^H\boldsymbol{u}_i|^2. \tag{6.6}$$

If $\bar{\boldsymbol{e}} \in \mathcal{U}_S$, then $d^2 = 0$ holds as implied by the orthogonality condition above. Taking a reciprocal of the squared norm expression creates sharp peaks at the signal DoAs (or frequencies, equivalently). The DoA estimation function for MUSIC (or the pseudospectrum) is

$$P_{\text{MUSIC}}(\boldsymbol{a}(\theta)) = \frac{1}{\boldsymbol{a}(\theta)^H\boldsymbol{U}_N\boldsymbol{U}_N^H\boldsymbol{a}(\theta)} = \frac{1}{\sum_{i=K+1}^{M}|\boldsymbol{a}(\theta)^H\boldsymbol{u}_i|^2}, \tag{6.7}$$

where $\boldsymbol{a}(\theta)$ is the candidate steering vector. The locations of the p largest peaks of the estimation function (obtained via spectral search) give the DoA estimates for the K signal components:

$$\hat{\theta} = \arg\max_{\theta}\ P_{\text{MUSIC}}(\boldsymbol{a}(\theta)). \tag{6.8}$$

In practice, however, a key issue in the above DoA estimation problem is the detection of the number of signals. One approach to this problem is based on the observation that the number of signals can be determined from the eigenvalues of the covariance matrix of the observation vector. In [4, 13], the authors developed a procedure, based on a nested sequence of hypothesis tests, to implement this approach. For each hypothesis, the likelihood ratio statistic is computed and compared to a threshold; the hypothesis accepted is the first one for which the threshold is crossed. The problem of this method is the subjective judgment required for deciding on the threshold levels. Later, in [28], a new approach was proposed to this problem that is based on the application of the information-theoretic criteria for model selection, i.e., AIC [1] and MDL [19, 28]. The advantage of this approach is that no subjective judgment is required in the decision process; the number of signals is determined as the value for which the AIC or the MDL criteria is minimized. In [29], an MDL principle based approach was proposed, which is applicable to any type of sources, including the case of sources which are fully correlated, referred to as the coherent signals case.

Attention

Based on the seminal paper by [28], the standard MDL criterion for estimating the number of signals K impinging on a sensor array is derived from information-theoretic principles and operates exclusively on the eigenvalues of the sample covariance matrix. This method is widely used for its consistency and ability to avoid subjective threshold settings. The MDL criterion is rooted in Rissanen's principle of MDL [19], which states that the best model is the one that minimizes the total code length required to describe the data. In the context of signal detection, the "model" is the number of signals K, while the "data" is the set of observations $\{\boldsymbol{y}(1), \ldots, \boldsymbol{y}(T)\}$.

The code length is composed of a term measuring how well the model fits the data (log-likelihood), and a penalty term for model complexity. For a candidate number of signals k, the MDL criterion is given by [28]

$$\mathrm{MDL}(k) = -\log \left(\frac{\prod_{i=k+1}^{M} \hat{\lambda}_i^{1/(M-k)}}{\frac{1}{M-k} \sum_{i=k+1}^{M} \hat{\lambda}_i} \right)^{(M-k)T} + \frac{1}{2} k(2M-k) \log T,$$

where $\hat{\lambda}_1 \geq \hat{\lambda}_2 \geq \cdots \geq \hat{\lambda}_M$ are the eigenvalues of the sample covariance matrix. The estimate of the number of signals is

$$\hat{K} = \arg \min_{k \in \{0,1,\ldots,M-1\}} \mathrm{MDL}(k).$$

6.1.2 Root-MUSIC

Compared with the conventional MUSIC algorithm, the root MUSIC algorithm [3] is to apply the polynomial root-finding method to replace the spectral search of zeros. Using the property that the steering vector of the signal is orthogonal to the noise, we wish to find the zeros of the following null spectrum:

$$f(z) = \boldsymbol{p}(z)^H \boldsymbol{U}_N \boldsymbol{U}_N^H \boldsymbol{p}(z), \tag{6.9}$$

where $\boldsymbol{p}(z) = [1, z^{-1}, \ldots, z^{-(M-1)}]^T$ and $z = e^{j2\pi \frac{d}{\lambda} \sin(\theta)}$. If we define the matrix $\boldsymbol{G} = \boldsymbol{U}_N \boldsymbol{U}_N^H \in \mathbb{C}^{M \times M}$, then $f(z)$ can be expanded into a polynomial in z, given by

$$f(z) = d_{-(M-1)} z^{-(M-1)} + \cdots + d_{M-1} z^{M-1}, \tag{6.10}$$

where $d_i = \sum_{m_1,m_2,(m_1-m_2)=i}^{M} \boldsymbol{G}_{m_1,m_2}$, $i \in [-(M-1), \ldots, M-1]$ are the sums along the diagonals of $\boldsymbol{G}$.

As can be seen, $f(z)$, $z \in \mathbb{C}$ is a polynomial of degree $2(M-1)$, and there are some useful properties of $f(z)$ which are given as follows: (1) $d_i = d_{-i}^*$ holds as $\boldsymbol{G}$ is Hermitian by construction; (2) as a result of $\mathcal{U}_S \perp \mathcal{U}_N$, K root pairs appear near the unit circle if channel noise exists, or on the unit circle if there is no noise [17]; (3) if one root exists at $\hat{z}$, another root will exist at $\frac{1}{z^*}$ because $f(z) = f^*(\frac{1}{z^*})$, i.e., the roots of $f(z)$ are mirror images of the unit circle. Therefore, by taking the phases of the K roots $\hat{z}_1, \hat{z}_2, \ldots, \hat{z}_K$ whose moduli are closest to 1 within the unit circle, we can have the DoA estimate as

$$\hat{\theta}_m = \arcsin\left(\frac{\lambda}{2\pi d} \arg\left(\hat{z}_m\right) \right), m = 1, \ldots, K. \tag{6.11}$$

Note that the roots can be found by treating d_i's as the polynomial coefficients representing the $(2M-2)$-degree characteristic polynomial of the $(2M-2) \times (2M-2)$ adjoint matrix $\boldsymbol{A}$ and then computing the eigenvalues of the adjoint matrix [16], where the matrix $\boldsymbol{A}$ is given by

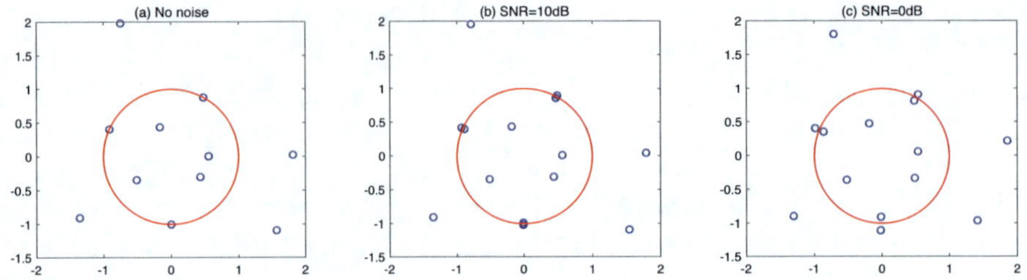

Fig. 6.1 An example of the roots from half-wavelength spaced ULA measurements under different SNRs, with $M = 8, K = 3, T = 20, \theta_1 = -30°$, $\theta_2 = 20°$, and $\theta_3 = 60°$

$$\boldsymbol{A} = \begin{bmatrix} -\frac{d_{M-2}}{d_{M-1}} & -\frac{d_{M-3}}{d_{M-1}} & \cdots & -\frac{d_{-(M-2)}}{d_{M-1}} & -\frac{d_{-(M-1)}}{d_{M-1}} \\ 1 & 0 & \cdots & 0 & 0 \\ 0 & 1 & \cdots & 0 & 0 \\ \vdots & & & & \vdots \\ 0 & 0 & \cdots & 1 & 0 \end{bmatrix}. \tag{6.12}$$

The computational complexity of this step is $O((2M-2)^3)$. The computational complexity of the spectral search in conventional MUSIC algorithm is $O(IM^2)$, where I is the number of searches. Therefore, if the search interval is approximately on the order of the reciprocal of the number of antennas, i.e., $1/M$, then the computational complexities of MUSIC and root-MUSIC are comparable. However, if we are aiming for super-resolution estimation, where the search interval is much less than $1/M$, then the computational complexity of MUSIC may become significantly higher than that of root-MUSIC.

Figure 6.1 depicts an example of the roots from half-wavelength spaced ULA measurements under different SNRs, with $M = 8$, $K = 3$, $T = 20$, $\theta_1 = -30°$, $\theta_2 = 20°$, and $\theta_3 = 60°$. As can be seen, for the no noise case, there are $2(M-1)$ solutions to the polynomial, of which K solutions (double roots) are on the unit circle, $(M-K-1)$ solutions are within the unit circle, and $(M-K-1)$ solutions are outside the unit circle [17]. Even when the measurements are severely corrupted by noise, the null spectrum can be used to obtain good estimates of the DoA locations.

6.1.3 2D MUSIC

The DoA-delay estimation is a classical problem encountered in radar, sonar, and geophysics. It also finds applications in source localization, accident reporting, cargo tracking, and intelligent transportation [18].

In a wireless communication system with multiple channel paths, one can obtain a better channel estimate by jointly exploring the path DoAs and the path propagation delays, thus significantly improving the system performance. Consider an MIMO orthogonal frequency division multiplexing (MIMO-OFDM) system, where the BS is equipped with M ULA antennas, and each single antenna user transmits uplink sounding reference signals (SRSs) to the BS for channel estimation and moves with low speed. The SRSs are usually obtained from sequences with good autocorrelation properties and low peak-to-average power ratio (PAPR), e.g., the SRSs in the 3GPP 5G NR Release 16 specification are obtained from Zadoff–Chu (ZC) sequence [24].

In a particular SRS symbol, the channel frequency response (CFR) at the nth $(0 \le n \le P-1)$ subcarrier is given by

$$\boldsymbol{h}[n] = \sum_{k=1}^{K} x_k e^{-j2\pi n \Delta_f \tau_k} \boldsymbol{a}(\theta_k), \tag{6.13}$$

where K is the number of propagation paths; Δ_f denotes the subcarrier spacing; x_k, θ_k, and τ_k denote the complex path gain, azimuth angle of arrival (AoA), and time delay of the kth path, respectively. Note that we have ignored various imperfection factors, such as Doppler phase rotation caused by user mobility, random phase noise, and the time offset, for simplicity. Accordingly, the received frequency-domain SRSs $\boldsymbol{Y} \in \mathbb{C}^{M \times P}$ in the considered SRS symbol is given by

$$\boldsymbol{Y} = \boldsymbol{H}\text{diag}(\boldsymbol{s}) + \boldsymbol{N}, \tag{6.14}$$

where $s \in \mathbb{C}^{P\times 1}$ is the frequency-domain SRSs transmitted from the user, $H \triangleq [h[0], h[1], \ldots, h[P-1]] \in \mathbb{C}^{M\times P}$ denotes the CFR matrix, and $N \in \mathbb{C}^{M\times P}$ is the AWGN with each element having zero mean and variance σ_n^2. Then, the received signal model (6.14) can be reformulated as

$$Y = A(\theta)\mathrm{diag}(x)B(\tau) + N, \tag{6.15}$$

where $B(\tau) = [b(\tau_1), b(\tau_2), \ldots, b(\tau_K)]^T \mathrm{diag}(s) \triangleq [\tilde{b}(\tau_1), \tilde{b}(\tau_2), \ldots, \tilde{b}(\tau_K)]^T \in \mathbb{C}^{K\times P}$ with $\tilde{b}(\tau_k) = \mathrm{diag}(s)b(\tau_k)$ and $b(\tau) = [1, e^{j2\pi\Delta_f\tau}, \ldots, e^{j2\pi(P-1)\Delta_f\tau}]^T$, and $x \triangleq [x_1, x_2, \ldots, x_K]^T$.

6.1.3.1 S-MUSIC and T-MUSIC

We first briefly review the spatial MUSIC (S-MUSIC) algorithm [21] and the temporal MUSIC (T-MUSIC) algorithm [12], respectively, which use the covariance matrices of the columns and the rows of Y, respectively, to estimate the DoAs and the delays. Note that the number of the channel paths K is assumed to be known a priori; otherwise, one may estimate K by thresholding the magnitude of the eigenvalues of the covariance matrices involved or by using the AIC and the MDL detection methods provided in [28].

We refer to the covariance matrix of the columns of Y as the spatial covariance matrix R^s, which is given by

$$R^s = \mathbb{E}\{YY^H\}. \tag{6.16}$$

Similarly, the rows of Y are the temporal sampling vectors of the received signals. We may express the temporal covariance matrix R^t as

$$R^t = \mathbb{E}\{Y^T Y^*\}. \tag{6.17}$$

Starting from the eigendecomposition of the covariance matrices, R^s and R^t can be, respectively, expressed as

$$\begin{aligned} R^s &= U_S^s \Sigma_S^s U_S^{sH} + U_N^s \Sigma_N^s U_N^{sH}, \\ R^t &= U_S^t \Sigma_S^t U_S^{tH} + U_N^t \Sigma_N^t U_N^{tH}. \end{aligned} \tag{6.18}$$

The column vectors of U_S^s and U_S^t corresponding to the K largest eigenvalues are the eigenvectors that span the signal subspace of R^s and R^t, respectively, while the column spaces spanned by the rest of the $M-K$ and $P-K$ eigenvectors of R^s and R^t are the orthogonal complement of the column spaces of U_S^s and U_S^t, respectively. Besides, both (Σ_S^s, Σ_N^s) and (Σ_S^t, Σ_N^t) are diagonal matrix pairs with the associated eigenvalues as their diagonal elements.

Then, the spectra of the S-MUSIC and T-MUSIC are, respectively, defined as

$$\begin{aligned} P_{\text{music}}^s(\theta) &= \frac{1}{a(\theta)^H \left(I - U_S^s U_S^{sH}\right) a(\theta)}, \\ P_{\text{music}}^t(\tau) &= \frac{1}{\tilde{b}(\tau)^H \left(I - U_S^t U_S^{tH}\right) \tilde{b}(\tau)}, \end{aligned} \tag{6.19}$$

and after searching θ and τ over the range of interest, the S-MUSIC and T-MUSIC estimate the K DoAs and path delays by

$$\begin{aligned} \hat{\theta} &= \arg\min_{\theta} a(\theta)^H \left(I - U_S^s U_S^{sH}\right) a(\theta), \\ \hat{\tau} &= \arg\min_{\tau} \tilde{b}(\tau)^H \left(I - U_S^t U_S^{tH}\right) \tilde{b}(\tau), \end{aligned} \tag{6.20}$$

here, the notations are slightly abused as we are actually searching for K peaks instead of a minimization. As mentioned in Sect. 6.1.1, since the covariance matrix of the received signal is usually unavailable, the implementation of the S-MUSIC and T-MUSIC thus employs the sample covariance matrices instead. For example, if T snapshots are observed, the sampled spatial covariance matrix $\hat{R}^s$ and the sampled temporal covariance matrix $\hat{R}^t$ are estimated, which can be expressed as

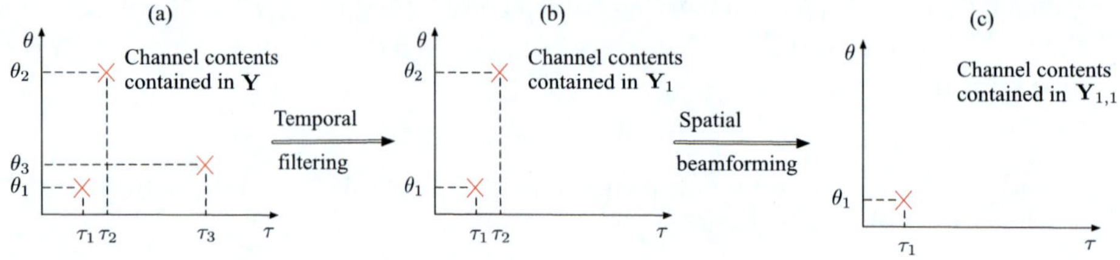

Fig. 6.2 Evolution of the signal contents for the estimation of the parameters of channel path one

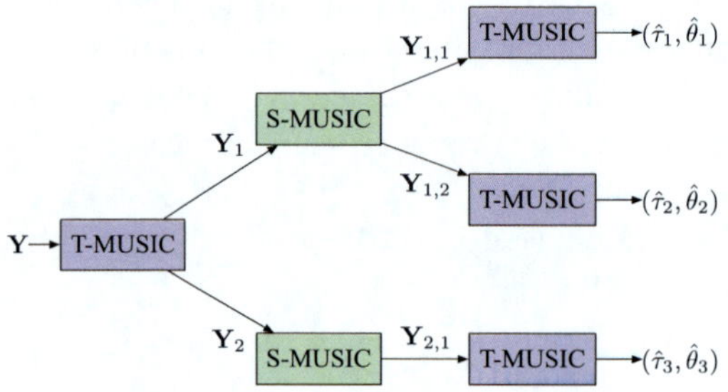

Fig. 6.3 Tree structure of the TST-MUSIC in solving the scenario shown in Fig. 6.2a

$$
\begin{aligned}
\hat{\boldsymbol{R}}^{s} &= \frac{1}{KT}\sum_{n=1}^{T} \boldsymbol{Y}^{(n)}\boldsymbol{Y}^{(n)H}, \\
\hat{\boldsymbol{R}}^{t} &= \frac{1}{MT}\sum_{n=1}^{T} \boldsymbol{Y}^{(n)T}\boldsymbol{Y}^{(n)*}.
\end{aligned} \tag{6.21}
$$

In scenarios where the incoming channel paths exhibit only mild contamination and are neither closely spaced in DoA nor significantly overlapped in delay, the S-MUSIC and T-MUSIC algorithms can, respectively, achieve reasonably accurate estimates of the path DoAs and delays. However, if path-clustering occurs either spatially or temporally, the associated signature matrix $\boldsymbol{A}(\boldsymbol{\theta})$ or $\boldsymbol{B}(\boldsymbol{\tau})$ may become ill-conditioned. In this case, the MUSIC spectrum may display a broadened "hump" rather than distinct peaks around each path-clustering region, thereby degrading the estimation resolution. In typical MIMO-OFDM wireless communication systems, such path-clustering challenges may be mitigated by leveraging the joint space-time characteristics of the propagation channel. In the following section, we elaborate on how the time-space-time MUSIC (TST-MUSIC) algorithm [27] refines these blurred estimates to resolve closely spaced channel paths through enhanced spatio-temporal processing.

6.1.3.2 TST-MUSIC

The rationale of the TST-MUSIC is to incorporate three 1D MUSIC (S-MUSIC and T-MUSIC) algorithms with beamforming techniques and the filtering techniques to group, isolate, and then estimate and pair the 2D parameters of a fading channel. To simplify the algorithm description, we first assume that there are only three rays present in the system. The general procedure of the TST-MUSIC is summarized at the end of this subsubsection.

As shown in Fig. 6.2, three channel paths are characterized by their temporal–spatial coordinates on the DoA-delay plane. Note that path 1 and path 2 possess close time delays ($\tau_1 \approx \tau_2$) but diverse DoAs ($\theta_1 < \theta_2$), whereas path 1 and path 3 are close in the DoAs ($\theta_1 \approx \theta_3$) but with far-apart delays ($\tau_1 < \tau_3$). The tree structure of the TST-MUSIC algorithm for this scenario is illustrated in Fig. 6.3. In addition, corresponding to Fig. 6.3, Fig. 6.2 shows the evolution of the data contents as the parameters of path 1 are estimated in the tree structure.

The TST-MUSIC treats those temporally close paths, which flatten the T-MUSIC spectrum, as a group. Therefore, path 1 and path 2 in Fig. 6.2a are regarded as one group, whereas path 3 is considered another group. By applying the T-MUSIC to the rows of $\boldsymbol{Y}$, the resulting group delays are estimated and denoted by $\hat{t}_1$ and $\hat{t}_2$, based on which we define the temporal filtering matrices $\boldsymbol{U}_i^t$ as

$$\begin{aligned} \boldsymbol{U}_1^t &= \boldsymbol{I} - \tilde{\boldsymbol{b}}(\hat{t}_1) \cdot \tilde{\boldsymbol{b}}(\hat{t}_1)^H, \\ \boldsymbol{U}_2^t &= \boldsymbol{I} - \tilde{\boldsymbol{b}}(\hat{t}_2) \cdot \tilde{\boldsymbol{b}}(\hat{t}_2)^H. \end{aligned} \tag{6.22}$$

Note that $\boldsymbol{U}_1^t$ (or $\boldsymbol{U}_2^t$) is also the complement projection matrix of $\tilde{\boldsymbol{b}}(\hat{t}_1)$ (or $\tilde{\boldsymbol{b}}(\hat{t}_2)$) with $\tilde{\boldsymbol{b}}(\hat{t}_1)^H \boldsymbol{U}_1^t = \boldsymbol{0}^T$ (or $\tilde{\boldsymbol{b}}(\hat{t}_2)^H \boldsymbol{U}_2^t = \boldsymbol{0}^T$). In the three-ray scenario shown in Fig. 6.2a, we have $\tau_1 \approx \tau_2 \approx \hat{t}_1 \leq \tau_3 \approx \hat{t}_2$, which implies that $\|\tilde{\boldsymbol{b}}(\tau_1)^H \boldsymbol{U}_1^t\| \approx \|\tilde{\boldsymbol{b}}(\tau_2)^H \boldsymbol{U}_1^t\| \approx 0 \ll \|\tilde{\boldsymbol{b}}(\tau_3)^H \boldsymbol{U}_1^t\|$ and $\|\tilde{\boldsymbol{b}}(\tau_1)^H \boldsymbol{U}_2^t\| \approx \|\tilde{\boldsymbol{b}}(\tau_2)^H \boldsymbol{U}_2^t\| \gg \|\tilde{\boldsymbol{b}}(\tau_3)^H \boldsymbol{U}_2^t\| \approx 0$, where the notation $\|\cdot\|$ denotes the 2-norm of a vector. With these facts, the TST-MUSIC postmultiplies $\boldsymbol{U}_i^t$ to $\boldsymbol{Y}$, which is referred to as the temporal filtering process to separate the paths with delays τ_1 and τ_2 from the path with delay τ_3. As a result, two group matrices, which are denoted as $\boldsymbol{Y}_1$ and $\boldsymbol{Y}_2$, are thus generated as

$$\boldsymbol{Y}_1 = \boldsymbol{Y}\boldsymbol{U}_2^t \approx [\boldsymbol{a}(\theta_1), \boldsymbol{a}(\theta_2)] \begin{bmatrix} \beta_1 & 0 \\ 0 & \beta_2 \end{bmatrix} \begin{bmatrix} \tilde{\boldsymbol{b}}(\tau_1)^T \boldsymbol{U}_2^t \\ \tilde{\boldsymbol{b}}(\tau_2)^T \boldsymbol{U}_2^t \end{bmatrix} + \boldsymbol{N}\boldsymbol{U}_2^t, \tag{6.23}$$

$$\boldsymbol{Y}_2 = \boldsymbol{Y}\boldsymbol{U}_1^t \approx \beta_3 \boldsymbol{a}(\theta_3)\, \tilde{\boldsymbol{b}}(\tau_3)^T \boldsymbol{U}_1^t + \boldsymbol{N}\boldsymbol{U}_1^t, \tag{6.24}$$

where "$\approx$" in (6.23) and (6.24) means that the residual signals from path 3 and from paths 1 and 2, respectively, are neglected. Discussions about the magnitude of the neglected residual signal will be given at the end of this section. Note that the transformed noise matrices $\boldsymbol{N}\boldsymbol{U}_2^t$ and $\boldsymbol{N}\boldsymbol{U}_1^t$ in (6.23) and (6.24) are still temporally and spatially white within the projected subspace, the detailed proof is given in Appendix 1. This property is essential for ensuring that the signal and noise subspaces can be efficiently separated.

Note that the two dominant rays contained in (6.23) have their DoAs $\theta_2 > \theta_1$. The DoA estimates $\hat{\theta}_1$ and $\hat{\theta}_2$ can thus be accurately obtained by applying the S-MUSIC to $\boldsymbol{Y}_1$. Similarly, $\hat{\theta}_3$ is obtained by applying the S-MUSIC to $\boldsymbol{Y}_2$. Note that $\hat{\theta}_1 \approx \hat{\theta}_3$, but the signals of path 1 and that of path 3 are separated into two different signal groups before the S-MUSIC is applied. Therefore, θ_1 and θ_3 can also be accurately estimated with the help of the temporal filtering following the first T-MUSIC. In addition, note that right after estimating $\{\theta_k\}$ in the S-MUSIC, the estimated array vectors $\{\boldsymbol{a}(\hat{\theta}_k)\}$ are determined and will be used in the spatial beamforming described below.

To further divide each group matrix into several single-path matrices, the spatial beamforming matrices $\boldsymbol{U}_i^s$ can be defined as

$$\boldsymbol{U}_1^s = \boldsymbol{I} - \boldsymbol{a}(\hat{\theta}_1)\boldsymbol{a}(\hat{\theta}_1)^H, \quad \boldsymbol{U}_2^s = \boldsymbol{I} - \boldsymbol{a}(\hat{\theta}_2)\boldsymbol{a}(\hat{\theta}_2)^H \tag{6.25}$$

for $\boldsymbol{Y}_1$, and

$$\boldsymbol{U}_3^s = \boldsymbol{I} - \boldsymbol{a}(\hat{\theta}_3)\boldsymbol{a}(\hat{\theta}_3)^H \tag{6.26}$$

for $\boldsymbol{Y}_2$, respectively. Note that $\boldsymbol{U}_i^s$ nulls the signal from path i as

$$\boldsymbol{U}_1^s \boldsymbol{a}(\hat{\theta}_1) = \boldsymbol{0}, \quad \boldsymbol{U}_2^s \boldsymbol{a}(\hat{\theta}_2) = \boldsymbol{0}, \quad \boldsymbol{U}_3^s \boldsymbol{a}(\hat{\theta}_3) = \boldsymbol{0}. \tag{6.27}$$

Similar to the temporal filtering process, the TST-MUSIC algorithm premultiplies $\boldsymbol{Y}_1$ by $\boldsymbol{U}_1^s$ and $\boldsymbol{U}_2^s$, which is referred to as the spatial beamforming process, to null the corresponding path. It follows that two single-path matrices $\boldsymbol{Y}_{1,1}$ and $\boldsymbol{Y}_{1,2}$ are formed, respectively, as

$$\boldsymbol{Y}_{1,1} = \boldsymbol{U}_2^s \boldsymbol{Y}_1 \approx \beta_1 \boldsymbol{U}_2^s \boldsymbol{a}(\theta_1)\, \tilde{\boldsymbol{b}}(\tau_1)^T \boldsymbol{U}_2^t + \boldsymbol{U}_2^s \boldsymbol{N}\boldsymbol{U}_2^t \tag{6.28}$$

$$\boldsymbol{Y}_{1,2} = \boldsymbol{U}_1^s \boldsymbol{Y}_1 \approx \beta_2 \boldsymbol{U}_1^s \boldsymbol{a}(\theta_2)\, \tilde{\boldsymbol{b}}(\tau_2)^T \boldsymbol{U}_2^t + \boldsymbol{U}_1^s \boldsymbol{N}\boldsymbol{U}_2^t. \tag{6.29}$$

Again, the residues of path 2 and path 1 are neglected in (6.28) and (6.29), respectively. The noise matrices in (6.28) and (6.29) are again still temporally and spatially white within the projected subspace. In (6.28) and (6.29), the single-path structure of $\boldsymbol{Y}_{1,1}$ and $\boldsymbol{Y}_{1,2}$ implies that the two paths with close path delays (path 1 and path 2) are separated into different subgroups by the spatial beamforming process. As a result, by applying the T-MUSIC algorithm to $\boldsymbol{Y}_{1,1}$ and $\boldsymbol{Y}_{1,2}$, path delay estimates $\hat{\tau}_1$ and $\hat{\tau}_2$ can then be accurately estimated, respectively. It also follows that the pairing of $(\hat{\tau}_1, \hat{\theta}_1)$ and $(\hat{\tau}_2, \hat{\theta}_2)$ is automatically achieved. Note that in the process of the second T-MUSIC algorithm, the temporal array vector should be $(\boldsymbol{U}_2^t)^T \tilde{\boldsymbol{b}}(\tau)$, as shown in (6.28) and (6.29), rather than $\tilde{\boldsymbol{b}}(\tau)$. Furthermore, since the group-delay information $\hat{t}_1$ is known, the searching region shrinks to the vicinity of $\hat{t}_1$ instead of the whole τ axis. For the other branch of the signal with respect to $\boldsymbol{Y}_2$, only the single path is found. Therefore, no spatial beamforming is needed, that is, $\boldsymbol{Y}_{2,1} = \boldsymbol{Y}_2$ and $\hat{t}_3 = \hat{t}_2$. Discussions about the magnitude of the neglected residual signal and why they can be ignored are provided in [27].

In addition, assume that the channel paths are not close in both of the DoAs and delays. The TST-MUSIC described above can be generalized as follows:

Step (1) **Grouping**: Applying the T-MUSIC algorithm to the received $\boldsymbol{Y}$, we obtain the group delays $\{\hat{t}_1, \ldots, \hat{t}_q\}$.

Step (2) **Temporal Filtering**: The temporal filtering matrices $\boldsymbol{U}^t_n, n = 1, \ldots, q$, are generated by

$$\boldsymbol{U}^t_n = \boldsymbol{I} - \tilde{\boldsymbol{b}}(\hat{t}_n)\tilde{\boldsymbol{b}}(\hat{t}_n)^H, \tag{6.30}$$

whereas the output of the nth temporal filter is given by

$$\boldsymbol{Y}_n = \boldsymbol{Y} \prod_{i=1, i\neq n}^{q} \boldsymbol{U}^t_i, \; n = 1, \ldots, q. \tag{6.31}$$

Step (3) **DoA Estimation**: Applying the S-MUSIC algorithm to each $\boldsymbol{Y}_n$, we estimate the DoAs of the paths in group n, as given by

$$\hat{\boldsymbol{\theta}}_n = [\hat{\theta}_{n,1}, \hat{\theta}_{n,2}, \ldots, \hat{\theta}_{n,r(n)}]^T, \tag{6.32}$$

for $n = 1, \ldots, q$, where $r(n)$ is the number of paths in the nth group, and $\sum_{n=1}^{q} r(n) = K$.

Step (4) **Spatial Beamforming**: The spatial beamforming matrix is obtained by

$$\boldsymbol{U}^s_{n,i} = \boldsymbol{I} - \boldsymbol{a}(\hat{\theta}_{n,i})\boldsymbol{a}(\hat{\theta}_{n,i})^H, \tag{6.33}$$

whereas the signal at the output of the mth spatial beamformer is given by

$$\boldsymbol{Y}_{n,m} = \prod_{i=1, i\neq m}^{r(n)} \boldsymbol{U}^s_{n,i} \boldsymbol{Y}_n, \tag{6.34}$$

for $n = 1, \ldots, q$ and $m = 1, \ldots, r(n)$.

Step (5) **Delay Estimation**: Apply the T-MUSIC algorithm again but with different temporal array manifolds $(\prod_{i=1, i\neq n}^{q} \boldsymbol{U}^t_i)^T \tilde{\boldsymbol{b}}(\tau)$ to each $\boldsymbol{Y}_{n,m}$. If single path is in the mth subgroup and $\hat{\tau}_{n,m}$ is obtained, pair it with $\hat{\theta}_{n,m}$; otherwise, pair $\hat{t}_n$ with $\hat{\theta}_{n,m}$.

6.2 ESPRIT

The ESPRIT algorithm is a high-resolution subspace-based method for estimating parameters such as DoAs, frequencies, or delays in signal processing [20]. Unlike MUSIC, which relies on spectral searches, ESPRIT exploits the rotational invariance inherent in the signal subspace structure to directly estimate parameters. This property allows ESPRIT to avoid computationally expensive grid searches, making it highly efficient for real-time applications. Note that one of the major breakthroughs afforded by the MUSIC algorithm as discussed in the previous section was the ability to handle arbitrary arrays of antennas (or sensors), but for ESPRIT, it retains most of the essential features of the arbitrary array of sensors, but achieves a significant reduction in computational complexity by imposing a constraint on the structure of the sensor array.

This constraint can be easily described by an example as follows. Consider a planar array of arbitrary geometry composed of $\frac{M}{2}$ sensor doublets as shown in Fig. 6.4. The elements in each doublet have identical sensitivity patterns and are translationally separated by a known constant displacement vector $\boldsymbol{\Delta}$. Other than the obvious requirement that each sensor has non-zero sensitivity in all directions of interest, the gain, phase, and polarization sensitivity of the elements in the doublet are arbitrary. Furthermore, there is no requirement that any of the doublets possess the same sensitivity patterns although, as discussed in [23], there are advantages to employing arrays with such characteristics. Besides, it is important to mention that by employing the doublet antenna structure, the ESPRIT algorithm can estimate the DoAs of $K \leq M/2$ narrowband sources. The MUSIC algorithm imposes the requirement $K < M$ and can therefore handle roughly twice as many sources as ESPRIT in general. For uniform linear arrays, however, ESPRIT can handle as many sources as MUSIC by employing overlapping subarrays, which will be discussed in detail later.

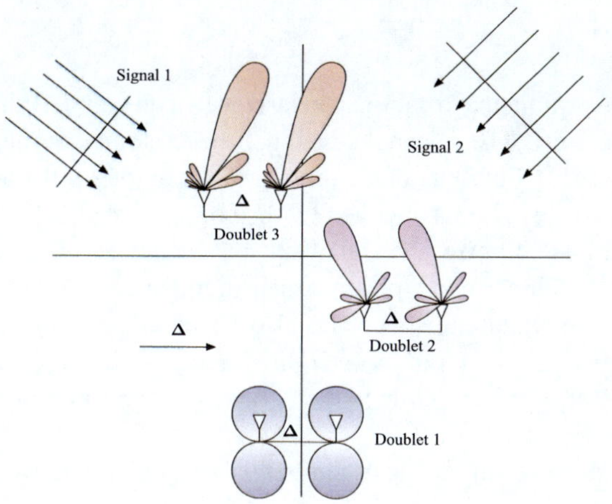

Fig. 6.4 Antenna array geometry for multiple source DoA estimation using ESPRIT (adapted from Fig. 3 in [20])

To describe mathematically the effect of the translational invariance of the sensor array, it is convenient to describe the array as being comprised of two subarrays, denoted by $\mathcal{Z}_x$ and $\mathcal{Z}_y$, which are identical in every respect although physically displaced (not rotated) from each other by a known displacement vector $\mathbf{\Delta}$ of magnitude Δ. The signals received at the ith doublet can then be expressed as

$$\begin{aligned} x_i(t) &= \sum_{k=1}^{K} s_k(t) a_i(\theta_k) + n_{x_i}(t) \\ y_i(t) &= \sum_{k=1}^{K} s_k(t) e^{j2\pi \frac{\Delta}{\lambda} \sin\theta_k} a_i(\theta_k) + n_{y_i}(t), \end{aligned} \tag{6.35}$$

where θ_k is now the direction-of-arrival of the kth source relative to the direction of the translational displacement vector $\mathbf{\Delta}$. Combining the outputs of each of the antennas in the two subarrays, the received data vectors can be written as follows:

$$\begin{aligned} \boldsymbol{x}(t) &= \boldsymbol{A}\boldsymbol{s}(t) + \boldsymbol{n}_x(t), \\ \boldsymbol{y}(t) &= \boldsymbol{A}\boldsymbol{\Phi}\boldsymbol{s}(t) + \boldsymbol{n}_y(t), \end{aligned} \tag{6.36}$$

where $\boldsymbol{A}$ is the array response matrix (or array manifold matrix), $\boldsymbol{s}(t)$ is the transmitted signal as defined in Sect. 6.1.1, the matrix $\boldsymbol{\Phi}$ is a diagonal $K \times K$ matrix of the phase delays between the doublet antennas for the K wavefronts, and is given by

$$\boldsymbol{\Phi} = \text{diag}\{e^{j\gamma_1}, \ldots, e^{j\gamma_K}\}, \tag{6.37}$$

with $\gamma_k \triangleq 2\pi \frac{\Delta}{\lambda} \sin\theta_k$. As can be seen, $\boldsymbol{\Phi}$ is a unitary matrix (operator) that relates the measurements from subarray $\mathcal{Z}_x$ to those from subarray $\mathcal{Z}_y$. $\boldsymbol{\Phi}$ is a simple scaling operator or rotation operator. The unitary nature of $\boldsymbol{\Phi}$ is a consequence of the narrowband plane wave assumption, an assumption that leads to unit-modulus sinusoidal signals in the spatial domain. In other applications, such as time series analysis, the diagonal elements of $\boldsymbol{\Phi}$ are potentially arbitrary complex numbers in which case $\boldsymbol{\Phi}$ could be an expansive or contractive operator.

Defining the total received signal as $z(t)$, the subarray outputs, $\boldsymbol{x}(t)$ and $\boldsymbol{y}(t)$, can be combined to yield

$$z(t) = \begin{bmatrix} \boldsymbol{x}(t) \\ \boldsymbol{y}(t) \end{bmatrix} = \bar{\boldsymbol{A}}\boldsymbol{s}(t) + \boldsymbol{n}_z(t),\ \bar{\boldsymbol{A}} = \begin{bmatrix} \boldsymbol{A} \\ \boldsymbol{A}\boldsymbol{\Phi} \end{bmatrix},\ \boldsymbol{n}_z(t) = \begin{bmatrix} \boldsymbol{n}_x(t) \\ \boldsymbol{n}_y(t) \end{bmatrix}. \tag{6.38}$$

Note that the DoA information is embedded in the diagonal elements of $\boldsymbol{\Phi}$, and it is the structure of $\bar{A}$ that is exploited to obtain estimates of the diagonal elements of $\boldsymbol{\Phi}$ without having to know $\boldsymbol{A}$.

6.2.1 Basic Idea

The basic idea behind ESPRIT is to exploit the rotational invariance of the underlying signal subspace induced by the translational invariance of the antenna array. The relevant signal subspace is the one that contains the outputs from the two subarrays described above, i.e., $\mathcal{Z}_x$ and $\mathcal{Z}_y$. Simultaneous sampling of the output of the arrays leads to two matrices, $\boldsymbol{U}_X$ and $\boldsymbol{U}_Y$, that span the same signal subspace (ideally, that spanned by the columns of $\boldsymbol{A}$).

As in the MUSIC algorithm, if there is no noise, the signal subspace can be estimated by collecting a sufficient number of measurements and finding any set of K linearly independent measurement vectors. These vectors span the K-dimensional subspace of $\mathbb{C}^{M/2\times 1}$ spanned by $\bar{\boldsymbol{A}}$. The signal subspace can also be obtained from knowledge of the covariance of the measurements, denoted by $\boldsymbol{R}_Z$. If $K \leq M/2$ (an assumption required in later developments), the $M-K$ smallest eigenvalues of $\boldsymbol{R}_Z$ are equal to σ^2. The K eigenvectors corresponding to the K largest eigenvalues are used to obtain $\boldsymbol{U}_S = [\boldsymbol{u}_1, \ldots, \boldsymbol{u}_K]$, where $span(\boldsymbol{U}_S) = span(\bar{\boldsymbol{A}})$.

Based on $span(\boldsymbol{U}_S) = span(\bar{\boldsymbol{A}})$, it can be inferred that there must exist a unique (recall $K \leq M/2$), nonsingular matrix $\boldsymbol{T}$ such that

$$\boldsymbol{U}_S = \bar{\boldsymbol{A}}\boldsymbol{T}. \tag{6.39}$$

Furthermore, the invariance structure of the array implies that $\boldsymbol{U}_S$ can be decomposed into $\boldsymbol{U}_X \in \mathbb{C}^{M/2\times K}$ and $\boldsymbol{U}_Y \in \mathbb{C}^{M/2\times K}$ (cf. $\mathcal{Z}_X$ and $\mathcal{Z}_Y$ subarrays) such that

$$\boldsymbol{U}_S = \begin{bmatrix} \boldsymbol{U}_X \\ \boldsymbol{U}_Y \end{bmatrix} = \begin{bmatrix} \boldsymbol{A}\boldsymbol{T} \\ \boldsymbol{A}\boldsymbol{\Phi}\boldsymbol{T} \end{bmatrix} \tag{6.40}$$

from which it is easily seen that

$$span(\boldsymbol{U}_X) = span(\boldsymbol{U}_Y) = span(\boldsymbol{A}). \tag{6.41}$$

Since $\boldsymbol{U}_X$ and $\boldsymbol{U}_Y$ share a common column space, the rank of

$$\boldsymbol{U}_{XY} \triangleq [\boldsymbol{U}_X, \boldsymbol{U}_Y] \tag{6.42}$$

is K, which implies that there exists a unique (recall $K \leq M/2$) rank K matrix $\boldsymbol{F} \triangleq [\boldsymbol{F}_X^T, \boldsymbol{F}_Y^T]^T \in \mathbb{C}^{2K\times K}$ with $\boldsymbol{F}_X, \boldsymbol{F}_Y \in \mathbb{C}^{K\times K}$ such that

$$\boldsymbol{0} = [\boldsymbol{U}_X, \boldsymbol{U}_Y]\boldsymbol{F} = \boldsymbol{U}_X\boldsymbol{F}_X + \boldsymbol{U}_Y\boldsymbol{F}_Y = \boldsymbol{A}\boldsymbol{T}\boldsymbol{F}_X + \boldsymbol{A}\boldsymbol{\Phi}\boldsymbol{T}\boldsymbol{F}_Y, \tag{6.43}$$

i.e., $\boldsymbol{F}$ spans the null space of $[\boldsymbol{U}_X, \boldsymbol{U}_Y]$. Defining[2]

$$\boldsymbol{\Psi} \triangleq -\boldsymbol{F}_X(\boldsymbol{F}_Y)^{-1}, \tag{6.44}$$

Equation (6.43) can be rearranged to yield

$$\begin{aligned} &\boldsymbol{A}\boldsymbol{T}\boldsymbol{F}_X\boldsymbol{F}_Y^{-1} + \boldsymbol{A}\boldsymbol{\Phi}\boldsymbol{T}\boldsymbol{F}_Y\boldsymbol{F}_Y^{-1} = -\boldsymbol{A}\boldsymbol{T}\boldsymbol{\Psi} + \boldsymbol{A}\boldsymbol{\Phi}\boldsymbol{T} = \boldsymbol{0} \\ &\Rightarrow \boldsymbol{A}\boldsymbol{T}\boldsymbol{\Psi} = \boldsymbol{A}\boldsymbol{\Phi}\boldsymbol{T} \\ &\Rightarrow \boldsymbol{A}\boldsymbol{T}\boldsymbol{\Psi}\boldsymbol{T}^{-1} = \boldsymbol{A}\boldsymbol{\Phi}. \end{aligned} \tag{6.45}$$

Assuming $\boldsymbol{A}$ to be full rank implies

$$\boldsymbol{T}\boldsymbol{\Psi}\boldsymbol{T}^{-1} = \boldsymbol{\Phi}. \tag{6.46}$$

Therefore, the eigenvalues of $\boldsymbol{\Psi}$ must be equal to the diagonal elements of $\boldsymbol{\Phi}$, and the columns of $\boldsymbol{T}$ are the eigenvectors of $\boldsymbol{\Psi}$. This is the key relationship in the development of ESPRIT and its properties. The signal parameters are obtained as nonlinear functions of the eigenvalues of the operator $\boldsymbol{\Psi}$ that maps (rotates) one set of vectors ($\boldsymbol{U}_X$) that span an $M/2$-dimensional signal subspace into another ($\boldsymbol{U}_Y$), as can be seen from (6.45).

[2] Here, we have used the fact that $\boldsymbol{F}_Y$ is invertible, and the detailed proof is provided in Appendix 2.

6.2.2 Estimating the Subspace Rotation Operator

In practical situations where only a finite number of noisy measurements are available, $\boldsymbol{U}_S$ is estimated from the covariance matrices of the measurements $\hat{\boldsymbol{R}}_Z$ or, equivalently, from the data matrix. The result is that $span(\boldsymbol{U}_S)$ is only an estimate of the signal subspace, and with probability one, $span(\boldsymbol{U}_S) \neq span(\bar{\boldsymbol{A}})$. Furthermore, $span(\boldsymbol{U}_X) \neq span(\boldsymbol{U}_Y)$. Thus, the objective of finding a $\boldsymbol{\Psi}$ such that $\boldsymbol{U}_X\boldsymbol{\Psi} = \boldsymbol{U}_Y$ is no longer achievable. A criterion for obtaining a suitable estimate must be formulated. The most commonly employed criterion for problems of this nature is the LS criterion, as discussed in Chap. 2.

The standard LS criterion applied to the model $\boldsymbol{AX} = \boldsymbol{B}$ to obtain an estimate of $\boldsymbol{X}$ assumes $\boldsymbol{A}$ is known and the error is to be attributed to $\boldsymbol{B}$. Assuming the set of equations is overdetermined, the columns of $\boldsymbol{A}$ are linearly independent, and the noise in the elements of $\boldsymbol{B}$ is zero mean and independent and identically distributed, the LS solution is

$$\hat{\boldsymbol{X}} = (\boldsymbol{A}^H\boldsymbol{A})^{-1}\boldsymbol{A}^H\boldsymbol{B}. \tag{6.47}$$

It is easily verified that the estimate is unbiased and minimum variance [20, 25]. The extension to arbitrary, but known covariance of the rows of $\boldsymbol{B}$ is straightforward and leads to the weighted least-squares (WLS) solution [22]. If both $\boldsymbol{A}$ and $\boldsymbol{B}$ are noisy, however, the LS solution is known to be biased.

Since it is not difficult to argue that the estimates $\boldsymbol{U}_X$ and $\boldsymbol{U}_Y$ are equally noisy, the LS criterion is clearly inappropriate. A criterion that takes into account noise on both $\boldsymbol{U}_X$ and $\boldsymbol{U}_Y$ is the total least-squares (TLS) criterion [11], which can be stated as finding residual matrices $\boldsymbol{R}_X$ and $\boldsymbol{R}_Y$ of minimum Frobenius norm, and $\hat{\boldsymbol{\Psi}}$ such that

$$(\boldsymbol{U}_X + \boldsymbol{R}_X)\hat{\boldsymbol{\Psi}} = \boldsymbol{U}_Y + \boldsymbol{R}_Y, \tag{6.48}$$

i.e.,

$$\begin{aligned} \min_{\boldsymbol{R}_X, \boldsymbol{R}_Y, \hat{\boldsymbol{\Psi}}} \quad & \|[\boldsymbol{R}_X, \boldsymbol{R}_Y]\|_F^2 \\ \text{s.t. } & (\boldsymbol{U}_X + \boldsymbol{R}_X)\hat{\boldsymbol{\Psi}} = \boldsymbol{U}_Y + \boldsymbol{R}_Y. \end{aligned} \tag{6.49}$$

By replacing $\hat{\boldsymbol{\Psi}}$ by $-\boldsymbol{F}_X(\boldsymbol{F}_Y)^{-1}$, regarding $\boldsymbol{F}$ as the optimization variable and appending a nontrivial constraint $\boldsymbol{F}^H\boldsymbol{F} = \boldsymbol{I}$ to eliminate the zero solution, problem (6.49) is shown to be equivalent to

$$\begin{aligned} \min_{\boldsymbol{R}_{XY}, \boldsymbol{F}} \quad & \|\boldsymbol{R}_{XY}\|_F^2 \\ \text{s.t. } & \boldsymbol{U}_{XY}\boldsymbol{F} = -\boldsymbol{R}_{XY}\boldsymbol{F}, \\ & \boldsymbol{F}^H\boldsymbol{F} = \boldsymbol{I}, \end{aligned} \tag{6.50}$$

where $\boldsymbol{R}_{XY} \triangleq [\boldsymbol{R}_X, \boldsymbol{R}_Y]$. Applying standard Lagrange techniques leads to a solution for $\boldsymbol{F}$ given by the eigenvectors corresponding to the K smallest eigenvalues of $\boldsymbol{U}_{XY}^H\boldsymbol{U}_{XY}$. The eigenvalues of $\hat{\boldsymbol{\Psi}}$ as defined above and calculated from the estimates $\boldsymbol{F}_X$ and $\boldsymbol{F}_Y$ are taken as estimates of the diagonal elements of $\boldsymbol{\Phi}$. The detailed derivation is provided in Appendix3.

To summarize, the TLS-ESPRIT algorithm based on a covariance formulation can be summarized as follows:

Step (1) Obtain an estimate of $\boldsymbol{R}_Z$, denoted by $\hat{\boldsymbol{R}}_Z$, from the measurements.

Step (2) Compute the eigendecomposition of $\hat{\boldsymbol{R}}_Z$

$$\hat{\boldsymbol{R}}_Z\bar{\boldsymbol{U}} = \bar{\boldsymbol{U}}\bar{\boldsymbol{\Lambda}}, \tag{6.51}$$

where $\bar{\boldsymbol{\Lambda}} = \text{diag}(\bar{\lambda}_1, \ldots, \bar{\lambda}_M)$ with $\bar{\lambda}_1 \geq \bar{\lambda}_2 \geq \cdots \geq \bar{\lambda}_M$, and $\bar{\boldsymbol{U}} = [\boldsymbol{u}_1, \ldots, \boldsymbol{u}_M]$.

Step (3) Estimate the number of sources $\hat{K}$.[3]

Step (4) Obtain the signal subspace estimate $\mathcal{U}_S = span(\boldsymbol{U}_S)$ and decompose it to obtain $\boldsymbol{U}_X$ and $\boldsymbol{U}_Y$, where

$$\boldsymbol{U}_S = [\boldsymbol{u}_1, \ldots, \boldsymbol{u}_{\hat{K}}] = \begin{bmatrix} \boldsymbol{U}_X \\ \boldsymbol{U}_Y \end{bmatrix}. \tag{6.52}$$

Step (5) Compute the eigendecomposition

[3] See [1, 19] for details on various techniques for estimating the number of sources.

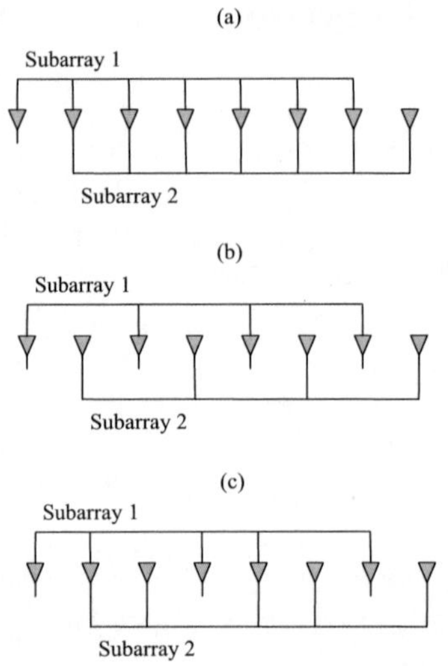

Fig. 6.5 Several ULA subarray choices for ESPRIT, **a** maximum overlap, **b** interleaved, **c** mixed (adapted from Fig. 1 in [23])

$$\boldsymbol{U}_{XY}^H \boldsymbol{U}_{XY} = \begin{bmatrix} \boldsymbol{U}_X^H \\ \boldsymbol{U}_Y^H \end{bmatrix} [\boldsymbol{U}_X, \boldsymbol{U}_Y] = \boldsymbol{U}\boldsymbol{\Lambda}\boldsymbol{U}^H, \tag{6.53}$$

where $\boldsymbol{\Lambda} = \text{diag}(\lambda_1, \ldots, \lambda_{2\hat{K}})$. Partition $\boldsymbol{U}$ into $\hat{K} \times \hat{K}$ submatrices,

$$\boldsymbol{U} = \begin{bmatrix} \boldsymbol{U}_{11} & \boldsymbol{U}_{12} \\ \boldsymbol{U}_{21} & \boldsymbol{U}_{22} \end{bmatrix}. \tag{6.54}$$

Step (6) Following the TLS solution where the matrix $\boldsymbol{F}$ (from the K smallest eigenvectors of $\boldsymbol{U}_{XY}^H \boldsymbol{U}_{XY}$) is partitioned as $\boldsymbol{F} = [\boldsymbol{F}_X^T, \boldsymbol{F}_Y^T]^T$, the operator $\boldsymbol{\Psi}$ satisfying $\boldsymbol{F}_Y = \boldsymbol{F}_X \boldsymbol{\Psi}$ is derived from the partition of $\boldsymbol{U}$. This yields the estimate $\hat{\boldsymbol{\Psi}} = -\boldsymbol{U}_{12}\boldsymbol{U}_{22}^{-1}$. Calculate its eigenvalues as

$$\hat{\phi}_k = \lambda_k(-\boldsymbol{U}_{12}\boldsymbol{U}_{22}^{-1}), \ \forall k = 1, \ldots, \hat{K}, \tag{6.55}$$

where $\lambda_k(\cdot)$ here denotes a function that extracts the kth largest eigenvalue of the argument.

Step (7) Estimate $\hat{\theta}_k = f^{-1}(\hat{\phi}_k)$, e.g., for DoA estimation as considered in this chapter, $\hat{\theta}_k = \sin^{-1}(\lambda \angle \hat{\phi}_k / (2\pi \Delta))$. Note that this expression assumes $\boldsymbol{\Delta}$ is collinear with the direction of wavefront progression. If $\boldsymbol{\Delta}$ has arbitrary orientation (e.g., planar arrays), the formula generalizes to

$$\hat{\theta}_k = \arcsin\left(\frac{\lambda \cdot \arg(\hat{\phi}_k)}{2\pi \|\boldsymbol{\Delta}\| \cos\alpha}\right), \tag{6.56}$$

where α is the angle between $\boldsymbol{\Delta}$ and the array aperture.

6.2.3 Dealing with ULAs

As mentioned above, an M-element antenna array is assumed to be composed of two identical $M/2$-element subarrays separated by a fixed displacement vector $\boldsymbol{\Delta}$, and the number of sources, K, is restricted to $K \leq M/2$. However, for ULAs, one can divide the array into two overlapping subarrays and thus increases the number of sources that can be handled. Some possible ULA subarray configurations are shown in Fig. 6.5. Note that the choice of subarray configuration also impacts the estimation accuracy and computational efficiency.

Let $\boldsymbol{J}_1$ and $\boldsymbol{J}_2$ be the $m \times M$ selection matrices that assign the elements of $z(t)$ to subarray 1 (the reference subarray) and subarray 2, respectively, where m denotes the number of elements in each subarray. For example, for the maximum overlapping subarray configuration, we have $m = M - 1$, $\boldsymbol{J}_1 = [\boldsymbol{I}_{M-1}, \boldsymbol{0}]$, and $\boldsymbol{J}_2 = [\boldsymbol{0}, \boldsymbol{I}_{M-1}]$, here, $\boldsymbol{I}_{M-1}$ is an identity matrix of size $M - 1$ and $\boldsymbol{0}$ is a vector of zeros. Define

$$\boldsymbol{J} \triangleq \begin{bmatrix} \boldsymbol{J}_1 \\ \boldsymbol{J}_2 \end{bmatrix}, \; \boldsymbol{G} = \boldsymbol{J}_1 \boldsymbol{A}, \tag{6.57}$$

then it can be easily obtained that $\boldsymbol{J}_2 \boldsymbol{A} = \boldsymbol{J}_1 \boldsymbol{A} \boldsymbol{\Phi}$, i.e.,

$$\boldsymbol{J}\boldsymbol{A} = \begin{bmatrix} \boldsymbol{G} \\ \boldsymbol{G}\boldsymbol{\Phi} \end{bmatrix}, \tag{6.58}$$

based on which we can model the received signal in a similar form as $z(t)$ in (6.38), and the ESPRIT directly applies. It is important to mention that for a ULA with M antennas, ESPRIT can resolve up to $M - 1$ sources by employing the maximum overlapping subarray configuration.

In many instances, it is preferable to avoid forming covariance matrices, and instead to operate directly on the measurement data. This approach leads to singular value decompositions (SVDs) of measurement data matrices, and in the following, an SVD variant of ESPRIT is discussed in detail. Specifically, let $\boldsymbol{Z}$ denote the measurement data matrix obtained by collecting a number (T) of measurement data vectors $z(t)$'s, i.e., $\boldsymbol{Z} = [z(1), \ldots, z(T)]$, then the SVD of $\boldsymbol{Z}$ is given as

$$\boldsymbol{Z} = \boldsymbol{U}\boldsymbol{\Sigma}\boldsymbol{V}^H, \tag{6.59}$$

where $\boldsymbol{U} \in \mathbb{C}^{M \times M}$ and $\boldsymbol{V} \in \mathbb{C}^{T \times T}$ are unitary matrices and $\boldsymbol{\Sigma}$ is a diagonal matrix of size $M \times T$ that holds the singular values from the largest (top left) in descending order. Let us assume that $T \geq M$. Notice that we have K sources or input signals. If there was no noise, there would only be K non-zero singular values. We assume that the K largest singular values stem from these input signals and other singular values are presumed to stem from noise. The matrices in the SVD of $\boldsymbol{Z}$ can be partitioned into submatrices, where some submatrices correspond to the signal subspace and some correspond to the noise subspace, i.e.,

$$\boldsymbol{U} = [\boldsymbol{U}_S, \boldsymbol{U}_N], \; \boldsymbol{\Sigma} = \begin{bmatrix} \boldsymbol{\Sigma}_S & \boldsymbol{0} & \boldsymbol{0} \\ \boldsymbol{0} & \boldsymbol{\Sigma}_N & \boldsymbol{0} \end{bmatrix}, \; \boldsymbol{V} = [\boldsymbol{V}_S, \boldsymbol{V}_N, \boldsymbol{V}_0], \tag{6.60}$$

where $\boldsymbol{U}_S \in \mathbb{C}^{M \times K}$ and $\boldsymbol{V}_S \in \mathbb{C}^{T \times K}$ contain the first K columns of $\boldsymbol{U}$ and $\boldsymbol{V}$, respectively, and $\boldsymbol{\Sigma}_S \in \mathbb{C}^{K \times K}$ is a diagonal matrix comprising the K largest singular values. Thus, the SVD of $\boldsymbol{Z}$ can be rewritten as

$$\boldsymbol{Z} = \boldsymbol{U}_S \boldsymbol{\Sigma}_S \boldsymbol{V}_S^H + \boldsymbol{U}_N \boldsymbol{\Sigma}_N \boldsymbol{V}_N^H, \tag{6.61}$$

where $(\boldsymbol{U}_S, \boldsymbol{\Sigma}_S, \boldsymbol{V}_S)$ and $(\boldsymbol{U}_N, \boldsymbol{\Sigma}_N, \boldsymbol{V}_N)$ represent the contributions of the input signal $z(t)$ and noise $\boldsymbol{n}_z(t)$ to $\boldsymbol{Z}$, respectively.

Attention

It is important to mention that when subarrays overlap in the ESPRIT context, the noise vector $\boldsymbol{n}_z(t)$ is not i.i.d. anymore. The presence of such non-i.i.d. noise in sensor arrays with overlapping subarrays is a direct consequence of the geometry itself. When subarrays share sensor elements, the noise measurements from those shared sensors are identical across the corresponding data vectors. This introduces perfect statistical correlation, violating the common assumption of spatially white noise (a covariance matrix of $\sigma^2 \boldsymbol{I}$) across the full array. This structured, non-i.i.d. noise covariance would be problematic for algorithms that assume simple, uncorrelated noise. However, ESPRIT elegantly circumvents this issue not by ignoring the correlation, but by embedding the array structure directly into its core mathematical formulation. This is achieved through the selection matrix $\boldsymbol{J}$, which is not an arbitrary construct but a precise mathematical representation of the physical array. It explicitly defines which sensors belong to which subarray, i.e., it inherently "knows" which sensors are shared and where the correlations occur. Instead of operating on the raw data, ESPRIT's derivation is based on the transformed signal subspace .

Furthermore, the performance of ESPRIT can be optimized through the use of a weighting matrix $\boldsymbol{W}$. While the foundational robustness to overlap is provided by the structure of $\boldsymbol{J}$, $\boldsymbol{W}$ refines the algorithm's asymptotic performance by whitening the errors in the estimated subspace vectors, giving less weight to more uncertain directions. One may refer to Sect. 6.2.4 for more details.

Hence, from the system model and based on the definitions $\boldsymbol{S} = [\boldsymbol{s}(1), \ldots, \boldsymbol{s}(T)]$ and $\boldsymbol{N} = [\boldsymbol{n}_z(1), \ldots, \boldsymbol{n}_z(T)]$, we can write $\boldsymbol{AS} = \boldsymbol{U}_S \boldsymbol{\Sigma}_S \boldsymbol{V}_S^H$ and $\boldsymbol{N} = \boldsymbol{U}_N \boldsymbol{\Sigma}_N \boldsymbol{V}_N^H$. Also, from the former, we can write

$$\boldsymbol{U}_S = \boldsymbol{AF}, \tag{6.62}$$

where $\boldsymbol{F} = \boldsymbol{S}\boldsymbol{V}_S \boldsymbol{\Sigma}_S^{-1}$. In the sequel, it is only important that there exists such an invertible matrix $\boldsymbol{F}$ and its actual content will not be important. Then, based on $\boldsymbol{J}_2\boldsymbol{A} = \boldsymbol{J}_1\boldsymbol{A}\boldsymbol{\Phi}$ and $\boldsymbol{U}_S = \boldsymbol{AF}$, we can further obtain

$$\begin{aligned} &\boldsymbol{J}_2\boldsymbol{A} = \boldsymbol{J}_1\boldsymbol{A}\boldsymbol{\Phi} \\ &\Rightarrow \boldsymbol{J}_2(\boldsymbol{U}_S\boldsymbol{F}^{-1}) = \boldsymbol{J}_1(\boldsymbol{U}_S\boldsymbol{F}^{-1})\boldsymbol{\Phi} \\ &\Rightarrow \boldsymbol{S}_2 = \boldsymbol{S}_1\boldsymbol{P}, \end{aligned} \tag{6.63}$$

where $\boldsymbol{S}_1 = \boldsymbol{J}_1\boldsymbol{U}_S$ and $\boldsymbol{S}_2 = \boldsymbol{J}_2\boldsymbol{U}_S$ denote the truncated signal subspaces, and

$$\boldsymbol{P} = \boldsymbol{F}^{-1}\boldsymbol{\Phi}\boldsymbol{F}. \tag{6.64}$$

The above equation has the form of an eigenvalue decomposition with $\boldsymbol{F}$ and $\boldsymbol{\Phi}$ containing the eigenvectors and eigenvalues, and the phases of the eigenvalues in the diagonal matrix $\boldsymbol{\Phi}$ are used to estimate the radial frequencies and the DoAs.

Thus, after solving for $\boldsymbol{P}$ in the relation $\boldsymbol{S}_2 = \boldsymbol{S}_1\boldsymbol{P}$, we would find the eigenvalues $\lambda_1, \lambda_2, \cdots, \lambda_K$ of $\boldsymbol{P}$, where $\lambda_k = \alpha_k e^{j\omega_k}$, and the radial frequencies $\omega_1, \omega_2, \ldots, \omega_K$ are estimated as the phases (argument) of the eigenvalues. In general, $\boldsymbol{S}_1$ is not invertible, thus in the step for solving $\boldsymbol{P}$, one can use the LS estimate $\boldsymbol{P} = (\boldsymbol{S}_1^H\boldsymbol{S}_1)^{-1}\boldsymbol{S}_1^H\boldsymbol{S}_2$, and another alternative would be the TLS estimate as discussed in Sect. 6.2.2.

Thus, the ESPRIT algorithm in this case is as follows:

Step (1) Compute the SVD of $\boldsymbol{Z} = \boldsymbol{U}\boldsymbol{\Sigma}\boldsymbol{V}^H$ and extract the signal subspace $\boldsymbol{U}_S \in \mathbb{C}^{M\times K}$ as the first K columns of $\boldsymbol{U}$.

Step (2) Compute $\boldsymbol{S}_1 = \boldsymbol{J}_1\boldsymbol{U}_S$ and $\boldsymbol{S}_2 = \boldsymbol{J}_2\boldsymbol{U}_S$.

Step (3) Solve for $\boldsymbol{P}$ in $\boldsymbol{S}_2 = \boldsymbol{S}_1\boldsymbol{P}$.

Step (4) Compute the eigenvalues $\lambda_1, \lambda_2, \cdots, \lambda_K$ of $\boldsymbol{P}$.

Step (5) The phases of the eigenvalues $\lambda_k = \alpha_k e^{j\omega_k}$ provide the radial frequencies ω_k and the DoAs.

It is noteworthy that the ESPRIT formulation presented here for ULAs represents the generic Standard ESPRIT framework. The solution method for the subspace rotation equation (e.g., $\boldsymbol{S}_2 = \boldsymbol{S}_1\boldsymbol{P}$) is not restricted to a specific type; either the LS or TLS criterion can be applied depending on the desired trade-off between computational simplicity and estimation accuracy.

6.2.4 Multiple Invariance ESPRIT

As discussed above, ESPRIT takes advantages of only a single displacement invariance in the antenna array. There are many situations, however, where the temporal or spatial samples possess several such invariances. Uniformly sampled time series and ULAs are prime examples. In such situations, the standard ESPRIT algorithm may be applied by first selecting any one of a number of pairs of subarrays that have the required displacement invariance, as shown in Fig. 6.5. The question naturally arises as to whether or not there is an optimal choice of subarrays, or, more fundamentally, whether or not there is a solution that exploits all the invariances simultaneously. There are the questions that the MI-ESPRIT addresses.

It should be noted that, though important in practice, ULAs are only a subset of the set of arrays possessing multiple invariances. A few examples of arrays with more general multiple displacement structure are shown in Fig. 6.6, where identical antennas are indicated by a common color. More general structures such as these would be useful, for example, in sonar applications where many identical sonobuoy subarrays are scattered over a large area, or in spaceborne applications where several identical antenna subarrays can be placed in orbit. Since the array manifold need not be known, these physical

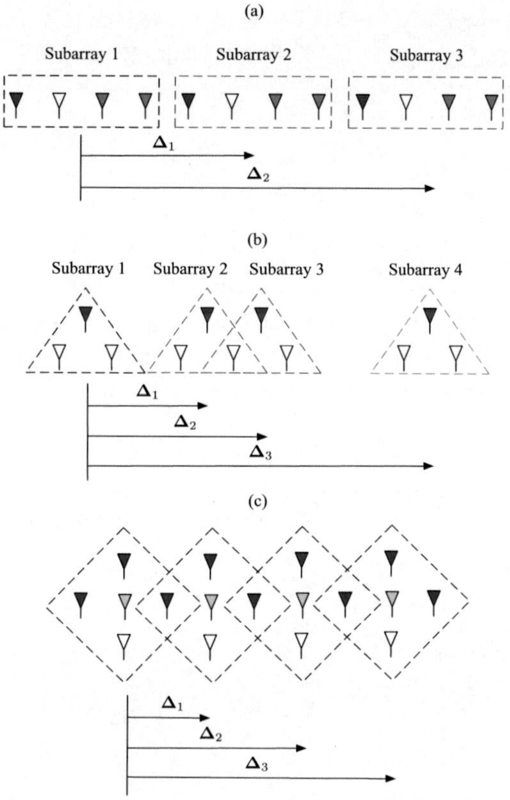

Fig. 6.6 Examples of arrays with multiple displacement invariances (adapted from Fig. 2 in [23])

subarrays can drift relative to one another (as long as their orientation remains the same) without significantly affecting the performance of the algorithm.

Assume the antenna array is composed of p identical subarrays of m antennas, each with its own displacement vector $\boldsymbol{\Delta}_i$ relative to the reference subarray. The displacement vectors are assumed to be collinear. As before, the total number of sensors M need not be equal to mp since overlapping subarrays are allowed. In such cases, we have $M < mp$. Let $\boldsymbol{J}_i$ be the $m \times M$ matrix that describes the ith subarray as in (6.57), and define

$$\boldsymbol{J} = \begin{bmatrix} \boldsymbol{J}_1 \\ \boldsymbol{J}_2 \\ \vdots \\ \boldsymbol{J}_p \end{bmatrix}. \tag{6.65}$$

Repost the received signal model as

$$\boldsymbol{z}(t) = \boldsymbol{A}\boldsymbol{s}(t) + \boldsymbol{n}(t), \tag{6.66}$$

then the eigendecomposition of the covariance matrix $\boldsymbol{R}_Z = \mathbb{E}\{\boldsymbol{z}(t)\boldsymbol{z}(t)^H\}$ has the form $\boldsymbol{R}_Z = \boldsymbol{U}_S\boldsymbol{\Sigma}_S\boldsymbol{U}_S^H + \sigma_n^2\boldsymbol{I}$. All subspace techniques are based on the observation that $span(\boldsymbol{U}_S) = span(\boldsymbol{A})$. This implies that there exists a full rank $K \times K$ matrix $\boldsymbol{T}$ satisfying $\boldsymbol{U}_S = \boldsymbol{A}\boldsymbol{T}$, which in turn implies that $\boldsymbol{J}\boldsymbol{U}_S = \boldsymbol{J}\boldsymbol{A}\boldsymbol{T}$. Consequently, analogous to (6.58), the structure inherent in $\boldsymbol{U}_S$ may be described by the following equation:

$$\boldsymbol{J}\boldsymbol{U}_S \triangleq \begin{bmatrix} \boldsymbol{U}_1 \\ \boldsymbol{U}_2 \\ \boldsymbol{U}_3 \\ \vdots \\ \boldsymbol{U}_p \end{bmatrix} = \begin{bmatrix} \boldsymbol{G} \\ \boldsymbol{G}\boldsymbol{\Phi} \\ \boldsymbol{G}\boldsymbol{\Phi}^{\delta_2} \\ \vdots \\ \boldsymbol{G}\boldsymbol{\Phi}^{\delta_{p-1}} \end{bmatrix} \boldsymbol{T}, \tag{6.67}$$

where $\delta_i \triangleq |\boldsymbol{\Delta}_i|/|\boldsymbol{\Delta}_1|$, and recall that $\boldsymbol{G} = \boldsymbol{J}_1\boldsymbol{A}$. The MI-ESPRIT problem can now be stated as follows.

Given p subspace estimates $\hat{\boldsymbol{U}}_1, \hat{\boldsymbol{U}}_2, \ldots, \hat{\boldsymbol{U}}_p \in \mathbb{C}^{m\times K}$, find a matrix $\boldsymbol{G} \in \mathbb{C}^{m\times K}$, a matrix $\boldsymbol{T} \in \mathbb{C}^{K\times K}$, and a diagonal matrix $\boldsymbol{\Phi} \in \mathbb{C}^{K\times K}$ to minimize

$$V = \left\| \begin{bmatrix} \hat{\boldsymbol{U}}_1 \\ \hat{\boldsymbol{U}}_2 \\ \hat{\boldsymbol{U}}_3 \\ \vdots \\ \hat{\boldsymbol{U}}_p \end{bmatrix} \boldsymbol{W}^{1/2} - \begin{bmatrix} \boldsymbol{G} \\ \boldsymbol{G\Phi} \\ \boldsymbol{G\Phi}^{\delta_2} \\ \vdots \\ \boldsymbol{G\Phi}^{\delta_{p-1}} \end{bmatrix} \boldsymbol{T} \right\|_F^2 . \tag{6.68}$$

Note that the displacements $\{\boldsymbol{\Delta}_i\}, i = 1, \ldots, p-1$ are not necessarily of equal magnitude. Also note that a weighting matrix $\boldsymbol{W} \in \mathbb{C}^{K\times K}$ not appearing before has been included in the cost function V for reasons given later. For the special case of uniformly spaced subarrays, we have $\boldsymbol{\Delta}_i = i\boldsymbol{\Delta}_1$, and consequently $\delta_i = i$. In the limiting case of a ULA, p is the number of antennas and $m = 1$.

To rewrite (6.68) in a more compact form, we rearrange the vertical stack of $m \times K$ matrices into a horizontal concatenation. Specifically, the first term in the Frobenius norm is a matrix of size $mp \times K$. By taking each of its p blocks $\hat{\boldsymbol{U}}_i\boldsymbol{W}^{1/2} \in \mathbb{C}^{m\times K}$ and placing them side by side, we obtain an $m \times pK$ matrix. Applying a similar rearrangement to the second term, we obtain

$$V = \|\boldsymbol{U}_W - \boldsymbol{B}\bar{\boldsymbol{\Phi}}\|_F^2, \tag{6.69}$$

where $\boldsymbol{U}_W$ and $\bar{\boldsymbol{\Phi}}$ are, respectively, defined as

$$\begin{aligned} \boldsymbol{U}_W &\triangleq [\hat{\boldsymbol{U}}_1\boldsymbol{W}^{1/2}, \hat{\boldsymbol{U}}_2\boldsymbol{W}^{1/2}, \ldots, \hat{\boldsymbol{U}}_p\boldsymbol{W}^{1/2}], \\ \bar{\boldsymbol{\Phi}} &\triangleq [\boldsymbol{T}, \boldsymbol{\Phi T}, \boldsymbol{\Phi}^{\delta_2}\boldsymbol{T}, \ldots, \boldsymbol{\Phi}^{\delta_{p-1}}\boldsymbol{T}], \\ \boldsymbol{B} &\triangleq \boldsymbol{G}, \end{aligned} \tag{6.70}$$

or

$$\begin{aligned} \boldsymbol{U}_W &\triangleq [(\boldsymbol{W}^{1/2})^T\hat{\boldsymbol{U}}_1^T, (\boldsymbol{W}^{1/2})^T\hat{\boldsymbol{U}}_2^T, \ldots, (\boldsymbol{W}^{1/2})^T\hat{\boldsymbol{U}}_p^T], \\ \bar{\boldsymbol{\Phi}} &\triangleq [\boldsymbol{G}^T, \boldsymbol{\Phi G}^T, \boldsymbol{\Phi}^{\delta_2}\boldsymbol{G}^T, \ldots, \boldsymbol{\Phi}^{\delta_{p-1}}\boldsymbol{G}^T], \\ \boldsymbol{B} &\triangleq \boldsymbol{T}^T. \end{aligned} \tag{6.71}$$

As explained shortly, the particular choice of definition is dictated by the desire for a minimal parameterization. Focusing on problem (6.69), it is noted that the objective function is separable in the parameter $\boldsymbol{B}$, and hence $\boldsymbol{B}$ can be directly obtained as a function of the remaining parameters, i.e., setting $\frac{\partial V}{\partial \boldsymbol{B}^*} = \boldsymbol{0}$ gives

$$\boldsymbol{B} = \boldsymbol{U}_W\bar{\boldsymbol{\Phi}}^H(\bar{\boldsymbol{\Phi}}\bar{\boldsymbol{\Phi}}^H)^{-1}. \tag{6.72}$$

By substituting the expression of $\boldsymbol{B}$ in (6.72) into (6.69), the MI-ESPRIT problem (6.68) may be reformulated as

$$\begin{aligned} \hat{\boldsymbol{\eta}} &= \arg\min_{\boldsymbol{\eta}} V \\ &= \arg\min_{\boldsymbol{\eta}} \left\| \boldsymbol{U}_W - \boldsymbol{U}_W\bar{\boldsymbol{\Phi}}^H\left(\bar{\boldsymbol{\Phi}}\bar{\boldsymbol{\Phi}}^H\right)^{-1}\bar{\boldsymbol{\Phi}} \right\|_F^2 \\ &= \arg\min_{\boldsymbol{\eta}} \|\boldsymbol{U}_W - \boldsymbol{U}_W\boldsymbol{P}_{\bar{\boldsymbol{\Phi}}}\|_F^2 \\ &= \arg\min_{\boldsymbol{\eta}} \|\boldsymbol{U}_W\boldsymbol{P}^{\perp}_{\bar{\boldsymbol{\Phi}}^H}\|_F^2 \\ &= \arg\min_{\boldsymbol{\eta}} \|\boldsymbol{r}(\boldsymbol{\eta})\|^2, \end{aligned} \tag{6.73}$$

where $\boldsymbol{\eta}$ is the parameter vector defined as[4]

[4] It is shown in [15] that this parameter vector $\boldsymbol{\eta}$ leads to a uniquely defined solution to (6.73).

$$\boldsymbol{\eta} = \left[\text{vec}\left(\tilde{\boldsymbol{I}}\Re\{\boldsymbol{G}\}\right)^T, \text{vec}\left(\tilde{\boldsymbol{I}}\Im\{\boldsymbol{G}\}\right)^T, \rho_1, \ldots, \rho_K, \theta_1, \ldots, \theta_K \right], \text{ if (6.70) is used,}$$
$$\boldsymbol{\eta} = \left[\text{vec}\left(\tilde{\boldsymbol{I}}\Re\{\boldsymbol{T}\}\right)^T, \text{vec}\left(\tilde{\boldsymbol{I}}\Im\{\boldsymbol{T}\}\right)^T, \rho_1, \ldots, \rho_K, \theta_1, \ldots, \theta_K \right], \text{ if (6.71) is used,} \tag{6.74}$$

where θ_i is the DoA of the ith path, ρ_i is the magnitude of the ith diagonal element of $\boldsymbol{\Phi}$[5], $\Re\{\boldsymbol{G}\}$ and $\Im\{\boldsymbol{G}\}$ are, respectively, the real and imaginary parts of $\boldsymbol{G}$, $\tilde{\boldsymbol{I}} = [\boldsymbol{0}_{(m-1)\times 1} \boldsymbol{I}]$ picks out all but the first row of the matrix to its right, and

$$\boldsymbol{r}(\boldsymbol{\eta}) \triangleq \text{vec}(\boldsymbol{U}_W \boldsymbol{P}^{\perp}_{\bar{\boldsymbol{\Phi}}^H}),$$
$$\boldsymbol{P}^{\perp}_{\bar{\boldsymbol{\Phi}}^H} \triangleq \boldsymbol{I} - \bar{\boldsymbol{\Phi}}^H (\bar{\boldsymbol{\Phi}}\bar{\boldsymbol{\Phi}}^H)^{-1} \bar{\boldsymbol{\Phi}}. \tag{6.75}$$

The choice of which set of definitions (i.e., (6.70) or (6.71)) to use is dictated by the dimension of $\boldsymbol{\eta}$. For the definitions in (6.70), $\boldsymbol{\eta} \in \mathbb{R}^{2mK}$, while for those in (6.71), $\boldsymbol{\eta} \in \mathbb{R}^{2K^2}$. Consequently, if $m > K$, the definitions in (6.71) are used, and vice versa. For notational convenience, define $m' = \min\{m, K\}$, so that in either case $\boldsymbol{\eta} \in \mathbb{R}^{2m'K}$.

It is clear from (6.37) that for DoA estimation problems, the diagonal elements of $\boldsymbol{\Phi}$ lie on the unit circle (i.e., $\rho_i = 1$). In such cases, the dimension of the parameter space may be further reduced by eliminating $\rho_i, i = 1, \cdots, K$, from the parameter vector $\boldsymbol{\eta}$. Doing so explicitly incorporates a unitary constraint on the diagonal matrix $\boldsymbol{\Phi}$ in the minimization of (6.73). However, for many time series applications (e.g., parameter estimation of damped exponentials, system identification, etc.) the unit-norm constraint does not apply, and the magnitude parameters must be retained in $\boldsymbol{\eta}$. To maintain generality in the discussion that follows, these terms will be included in $\boldsymbol{\eta}$.

The minimization of V over all $\boldsymbol{\eta} \in \mathbb{R}^{2m'K}$ is a nonlinear, computationally complex problem, and there exists no easily formulated solution as in the single invariance case. However, using a search technique to achieve $\partial V/\partial \boldsymbol{\eta} = 0$ is not a serious drawback since the single invariance version of ESPRIT can be applied to provide an excellent initial estimate of $\boldsymbol{\eta}$ with minimal computational cost. If the antenna array consists of equispaced subarrays, a maximum overlap configuration where $\bar{\boldsymbol{J}}_1 = [\boldsymbol{J}_1^T, \ldots, \boldsymbol{J}_{p-1}^T]^T$ and $\bar{\boldsymbol{J}}_2 = [\boldsymbol{J}_2^T, \ldots, \boldsymbol{J}_p^T]^T$ can be used. For the general case of unequal subarray separation, any suitable invariance can be employed by ESPRIT. One of the most popular and effective algorithms for solving nonlinear LS problems is Newton's method [5]. The method is based on a second-order expansion of the criterion function about the current parameter estimates, with the step in parameter space taken so as to achieve the local extremum of the approximating quadratic surface. The Newton iteration can be written as

$$\boldsymbol{\eta}(k+1) = \boldsymbol{\eta}(k) - \mu_k \boldsymbol{H}_k^{-1} \left. \frac{\partial V}{\partial \boldsymbol{\eta}} \right|_{\boldsymbol{\eta}=\boldsymbol{\eta}(k)}, \tag{6.76}$$

where μ_k is a step length (nominally 1) and $\boldsymbol{H}_k$ is the Hessian matrix at the kth iteration. The (i, j)th element of $\boldsymbol{H}_k$ is defined as

$$\{\boldsymbol{H}_k\}_{i,j} = \left. \frac{\partial^2 V}{\partial \eta_i \partial \eta_j} \right|_{\boldsymbol{\eta}=\boldsymbol{\eta}(k)}, \tag{6.77}$$

where η_i and η_j are, respectively, the ith and jth elements of $\boldsymbol{\eta}$. The derivatives $\partial V/\partial \eta_i$ are given by

$$\frac{\partial V}{\partial \eta_i} = 2\Re\{\boldsymbol{r}_i^H \boldsymbol{r}\}, \tag{6.78}$$

where $\boldsymbol{r}_i \triangleq \partial \boldsymbol{r}/\partial \eta_i$. Although the Newton algorithm achieves quadratic convergence when given a sufficiently accurate initial estimate, it requires computation of both first and second derivatives of the cost function. Since in the noise-free case the global minimum of V occurs when $\|\boldsymbol{r}\| = 0$ identically, it is expected that in the noisy case, $\|\boldsymbol{r}\|$ will be reasonably small in the vicinity of the solution. This fact, coupled with the generally high quality of the initial ESPRIT estimate, motivates the following approximation to the second-order derivative:

$$\{\boldsymbol{H}_k\}_{i,j} = 2\Re\{\boldsymbol{r}_i^H \boldsymbol{r}_j + \boldsymbol{r}_{ij}^H \boldsymbol{r}\} \approx 2\Re\{\boldsymbol{r}_i^H \boldsymbol{r}_j\}, \tag{6.79}$$

[5] The ESPRIT algorithm does not constrain the diagonal elements of $\boldsymbol{\Phi}$ to be on the unit circle, so $\rho_1, \ldots, \rho_K$ are implicitly parameters in the minimization.

where $\boldsymbol{r}_{ij} = \frac{\partial^2 \boldsymbol{r}}{\partial \eta_i \partial \eta_j}$.

The derivatives $\boldsymbol{r}_i$ may be computed by first noting that

$$\boldsymbol{r}_i = \text{vec}\left\{\boldsymbol{U}_W\left(\boldsymbol{P}^{\perp}_{\bar{\boldsymbol{\Phi}}^H}\bar{\boldsymbol{\Phi}}_i^H\bar{\boldsymbol{\Phi}}^{\dagger} + \bar{\boldsymbol{\Phi}}^{\dagger H}\bar{\boldsymbol{\Phi}}_i\boldsymbol{P}^{\perp}_{\bar{\boldsymbol{\Phi}}^H}\right)\right\}, \tag{6.80}$$

where $\bar{\boldsymbol{\Phi}}^{\dagger} \triangleq (\bar{\boldsymbol{\Phi}}\bar{\boldsymbol{\Phi}}^H)^{-1}\bar{\boldsymbol{\Phi}}$ and $\bar{\boldsymbol{\Phi}}_i \triangleq \partial\bar{\boldsymbol{\Phi}}/\partial\eta_i$. Expressions for the derivatives $\bar{\boldsymbol{\Phi}}_i$ using the definition of $\boldsymbol{\eta}$ in (6.74) are given below. In these expressions, $\boldsymbol{X} = \boldsymbol{T}^T$ if $m \geq K$, $\boldsymbol{X} = \boldsymbol{G}$ if $m < K$, $\boldsymbol{X}_{kl}$ is the (k, l)th element of $\boldsymbol{X}$, $\boldsymbol{I}_{kl}$ is a matrix with 1 in the (k, l)th position and zeros elsewhere, and $\Delta_1 \triangleq |\boldsymbol{\Delta}_1|$:

- $\eta_i = \Re\{\boldsymbol{X}_{kl}\}$: $\bar{\boldsymbol{\Phi}}_i = -[\boldsymbol{I}_{kl}, \boldsymbol{\Phi}\boldsymbol{I}_{kl}, \boldsymbol{\Phi}^{\delta_2}\boldsymbol{I}_{kl}, \cdots, \boldsymbol{\Phi}^{\delta_{p-1}}\boldsymbol{I}_{kl}]$;
- $\eta_i = \Im\{\boldsymbol{X}_{kl}\}$: $\bar{\boldsymbol{\Phi}}_i = -j[\boldsymbol{I}_{kl}, \boldsymbol{\Phi}\boldsymbol{I}_{kl}, \boldsymbol{\Phi}^{\delta_2}\boldsymbol{I}_{kl}, \cdots, \boldsymbol{\Phi}^{\delta_{p-1}}\boldsymbol{I}_{kl}]$;
- $\eta_i = \rho_k$: $\bar{\boldsymbol{\Phi}}_i = -\frac{1}{\rho_k}\left[\boldsymbol{0}, \boldsymbol{I}_{kk}\boldsymbol{\Phi}\boldsymbol{X}, \delta_2\boldsymbol{I}_{kk}\boldsymbol{\Phi}^{\delta_2}\boldsymbol{X}, \cdots, \delta_{p-1}\boldsymbol{I}_{kk}\boldsymbol{\Phi}^{\delta_{p-1}}\boldsymbol{X}\right]$;
- $\eta_i = \theta_k$: $\bar{\boldsymbol{\Phi}}_i = -(j2\pi\Delta_1\cos\theta_k/\lambda)\cdot\left[\boldsymbol{0}, \boldsymbol{I}_{kk}\boldsymbol{\Phi}\boldsymbol{X}, \delta_2\boldsymbol{I}_{kk}\boldsymbol{\Phi}^{\delta_2}\boldsymbol{X}, \cdots, \delta_{p-1}\boldsymbol{I}_{kk}\boldsymbol{\Phi}^{\delta_{p-1}}\boldsymbol{X}\right]$.

The derivatives of $\boldsymbol{r}$ are arranged so that

$$\left.\frac{\partial \boldsymbol{r}}{\partial \boldsymbol{\eta}}\right|_{\boldsymbol{\eta}=\boldsymbol{\eta}(k)} \triangleq \boldsymbol{r}_{\eta,k} \tag{6.81}$$

is an $mpK \times 2m'K$ matrix, and $\hat{\boldsymbol{H}}_k = \boldsymbol{r}_{\eta,k}^H\boldsymbol{r}_{\eta,k}$ is the expression for the approximate Hessian.

The Gauss–Newton implementation of the MI-ESPRIT algorithm then proceeds as follows:

Step (1) Using the ESPRIT algorithm to obtain initial estimates of $\theta_1, \ldots, \theta_K, \rho_1, \ldots, \rho_K$, $\boldsymbol{A}$ and $\boldsymbol{T}$.

Step (2) If $m \geq K$, set $\boldsymbol{X} = \boldsymbol{T}^T$; otherwise, set $\boldsymbol{X} = \boldsymbol{G}$. Form the initial parameter vector $\boldsymbol{\eta}(0)$ as $\boldsymbol{\eta}(0) = [\text{vec}(\tilde{\boldsymbol{I}}\boldsymbol{X})^T, \text{vec}(\tilde{\boldsymbol{I}}\tilde{\boldsymbol{X}})^T, \rho_1, \ldots, \rho_K, \theta_1, \ldots, \theta_K]$.

Step (3) Compute the quantities $\bar{\boldsymbol{\Phi}}$, $\boldsymbol{r}$, $\boldsymbol{r}_{\eta,k}$, as defined above.

Step (4) Update $\boldsymbol{\eta}(k+1) = \boldsymbol{\eta}(k) - \mu_k(\boldsymbol{r}_{\eta,k}^H\boldsymbol{r}_{\eta,k})^{-1}\boldsymbol{r}_{\eta,k}^H\boldsymbol{r}(\boldsymbol{\eta}(k))$.

Step (5) Evaluate $\|\partial V/\partial\boldsymbol{\eta}\|$, and if it is smaller than some prespecified tolerance, terminate the iterations; if not, repeat steps 3–5.

The bulk of the computation in the above algorithm occurs in forming the approximate Hessian and in taking its inverse at each update. For typical values of K, however, these computations are considerably less costly than the initial eigendecomposition needed to obtain $\hat{\boldsymbol{U}}_S$. In addition, simulation results indicate that only one or two iterations are required to achieve a solution sufficiently close to the global minimum.

In extremely difficult cases (when the initial estimate is severely degraded due to poor SNR or a small number of snapshots), it is often helpful to employ the Levenberg–Marquardt approach (see, for example, [10, 14]) and regularize the problem by adding a scaled identity matrix to the approximate Hessian. As the value of the criterion function decreases and the minimum is approached, the magnitude of the additive term is decreased to improve convergence speed. However, for most cases of interest, this type of regularization is not necessary and the above algorithm can be implemented as given.

Attention

The weighting factor $\boldsymbol{W}$ is introduced in the MI-ESPRIT cost function, i.e., Eq. (6.68), for the following key reasons:

1. **To achieve asymptotic efficiency (minimum variance estimates)**: The core theoretical reason is to minimize the asymptotic variance of the estimated parameters (e.g., DoAs). The analysis in [15, 23] shows that the parameter estimation error $\sqrt{T}(\hat{\boldsymbol{\eta}} - \boldsymbol{\eta}_0)$ (T denotes the number of snapshots used to compute the sample covariance matrix in practice) is asymptotically Gaussian distributed with zero mean and covariance matrix $\boldsymbol{C}(\boldsymbol{W})$ that depends on the weighting matrix $\boldsymbol{W}$. It is proven in [26] that a specific choice of $\boldsymbol{W}$, namely:

$$\boldsymbol{W}^{\star} = (\hat{\boldsymbol{\Sigma}}_S - \hat{\sigma}^2\boldsymbol{I})^2\hat{\boldsymbol{\Sigma}}_S^{-1}, \tag{6.82}$$

where $\hat{\sigma}^2$ is a consistent estimate of the noise variance and $\hat{\boldsymbol{\Sigma}}_S$ are the signal eigenvalues of the sample covariance, leads to parameter estimates that achieve the Cramér–Rao Bound (CRB) under certain assumptions (Gaussian signals, known array manifold parameterization). The CRB represents the theoretical lower limit on the variance of any unbiased estimator. In simpler terms, $\boldsymbol{W}^\star$ makes MI-ESPRIT the most statistically efficient (lowest possible variance) estimator possible asymptotically (as the number of snapshots T becomes large). Besides, $\boldsymbol{W}^\star$ weights the estimated signal subspace eigenvectors based on the quality of the estimated signal eigenvalues ($\hat{\boldsymbol{\Sigma}}_S$) relative to the noise floor ($\hat{\sigma}^2$). Eigenvectors corresponding to larger eigenvalues (stronger signal components or better estimates) are given more weight, while those corresponding to eigenvalues closer to the noise variance are given less weight. This optimally exploits the reliability information embedded in the eigenvalue decomposition of the sample covariance matrix.

2. **To improve performance in practical scenarios**: In practice, with finite data (T snapshots), the estimated signal subspace and consequently the subarray estimates are corrupted by noise. Treating all components of these estimates equally (i.e., using $\boldsymbol{W} = \boldsymbol{I}$, the identity matrix) is suboptimal. The optimal weighting $\boldsymbol{W}^\star$ accounts for the varying reliability of different components within the estimated signal subspace. Components associated with larger eigenvalues (higher SNR or more stable estimates) contribute more significantly to the fitting criterion, while components associated with eigenvalues near the noise level (more susceptible to noise) contribute less.
3. **To mitigate effects of overlapping subarrays**: The theoretical guarantee of asymptotic efficiency ($\boldsymbol{C}(\boldsymbol{W}^\star)$ achieving CRB) holds strictly under the condition $\boldsymbol{J}\boldsymbol{J}^T = \boldsymbol{I}$. This condition implies non-overlapping subarrays, meaning each sensor belongs to only one subarray. For overlapping subarrays (a common and often desirable configuration where sensors are shared between subarrays, e.g., Fig. 6.5a), the condition $\boldsymbol{J}\boldsymbol{J}^T = \boldsymbol{I}$ is violated. The analysis in the above-mentioned references does not strictly prove that $\boldsymbol{C}(\boldsymbol{W}^\star)$ is optimal in this overlapping case. However, numerical evaluations of the asymptotic covariance $\boldsymbol{C}(\boldsymbol{W})$ for overlapping scenarios consistently showed $\boldsymbol{C}(\boldsymbol{I}) > \boldsymbol{C}(\boldsymbol{W}^\star)$, meaning $\boldsymbol{W}^\star$ still provides lower variance than no weighting ($\boldsymbol{W} = \boldsymbol{I}$). Therefore, even when strict optimality is not proven (overlapping case), $\boldsymbol{W}^\star$ is strongly recommended and empirically superior. Besides, note that using overlapping subarrays exploits more of the array's inherent structure and is often necessary. $\boldsymbol{W}^\star$ helps maximize performance in these practical configurations.

While both MUSIC and ESPRIT are cornerstone subspace-based algorithms, their suitability depends on the application context. ESPRIT excels in structured environments with shift-invariant arrays and real-time requirements, whereas MUSIC offers greater flexibility for arbitrary geometries and scenarios where spectral visualization is beneficial. Understanding their trade-offs enables optimal algorithm selection in signal processing systems.

Below, we present a comparative analysis of the different ESPRIT algorithms, highlighting their key differences, advantages, and disadvantages.

1. Standard ESPRIT forms the foundational approach. Its primary advantage lies in computational efficiency achieved by exploiting a single translational invariance structure within the sensor array (e.g., doublets or overlapping subarrays in ULAs). This allows it to directly estimate parameters via eigenvalue decomposition of the subspace rotation operator, bypassing the computationally intensive spectral searches required by MUSIC. Consequently, it excels in real-time applications with structured arrays. However, standard ESPRIT exhibits significant limitations. Its performance is highly sensitive to subarray selection. While overlapping subarrays in ULAs enable handling up to $M-1$ sources (matching MUSIC's capacity), non-overlapping doublet configurations restrict it to $K \leq M/2$ sources. Furthermore, its reliance on LS estimation for solving the subspace relationship ($\boldsymbol{U}_X\boldsymbol{\Psi} = \boldsymbol{U}_Y$) introduces bias when noise corrupts both subarray measurements equally, leading to suboptimal accuracy, especially at lower SNRs. The basic formulation also lacks inherent robustness to coherent signals or challenging noise conditions compared to more advanced variants.
2. TLS-ESPRIT directly addresses a critical weakness of the standard LS approach. Recognizing that noise perturbs estimates of both subarray signal subspaces ($\boldsymbol{U}_X$ and $\boldsymbol{U}_Y$), TLS-ESPRIT employs the TLS criterion to minimize errors in both data matrices simultaneously. This fundamental shift provides superior estimation accuracy and noise robustness compared to LS-ESPRIT,[6] particularly under moderate-to-low-SNR conditions or with limited snapshots. It effectively mitigates the

[6] Here, LS-ESPRIT refers to the LS implementation of the ESPRIT algorithm.

bias inherent in the LS solution when noise is present on both sides of the equation. The trade-off is a moderate increase in computational complexity due to the requirement for an additional SVD or eigendecomposition of the augmented matrix $\boldsymbol{U}_{XY}^H \boldsymbol{U}_{XY}$ to solve the TLS problem. Despite this cost, TLS-ESPRIT is generally considered the practical variant of choice for most applications involving shift-invariant arrays due to its significantly improved performance over the basic LS version without an overwhelming computational burden.

3. MI-ESPRIT represents a significant advancement for scenarios where the array possesses multiple, known displacement invariances (e.g., ULAs, rectangular arrays, or distributed identical subarrays). Its core strength is its ability to exploit all available invariance structures simultaneously within a unified framework, rather than selecting just one pair of subarrays as in standard ESPRIT. This comprehensive utilization of array geometry information translates into theoretically optimal asymptotic estimation performance, as discussed above. However, these advantages come at a substantial computational cost. MI-ESPRIT requires solving a high-dimensional nonlinear optimization problem (minimizing the weighted fitting criterion over parameters related to $\boldsymbol{G}/\boldsymbol{T}$, magnitudes ρ_k, and angles θ_k). While efficient implementations using initial ESPRIT estimates and Gauss–Newton/Levenberg–Marquardt iterations exist, the computational load is significantly higher than standard or TLS-ESPRIT. Its implementation is also considerably more complex, demanding careful parameterization and iterative solvers. Therefore, MI-ESPRIT is primarily justified in applications demanding the highest possible estimation accuracy and where computational resources permit, or when dealing with complex array geometries exhibiting multiple invariances beyond simple ULAs.

While ESPRIT elegantly exploits a fixed translational invariance structure within the sensor array, recent advancements in hardware design have introduced the concept of movable antennas or fluid antennas, which can dynamically adjust their positions to create favorable array geometries [30, 31]. This paradigm shift, as explored in the context of wireless communication [7], sensing [8], or integrated sensing and communication (ISAC) [9], allows for the joint optimization of antenna positions and beamforming to improve the system performance, which demonstrates that the flexibility of movable antennas can significantly enlarge the performance region beyond what is achievable with conventional fixed-position arrays. Interested readers are encouraged to explore the growing body of literature on this topic for further advances in this direction.

Summary

This chapter explores subspace-based algorithms, specifically focusing on MUSIC and ESPRIT, which are critical for high-resolution parameter estimation in signal processing. In particular, MUSIC leverages the orthogonality between signal and noise subspaces, derived from the eigendecomposition of the signal covariance matrix, to construct a pseudospectrum where peaks indicate parameter estimates. Variants like root-MUSIC, which uses polynomial rooting, and TST-MUSIC, for joint DoA-delay estimation, enhance computational efficiency and address challenges like path clustering. ESPRIT, by contrast, exploits rotational invariance in structured arrays (e.g., uniform linear arrays) to directly estimate parameters without spectral searches, using techniques like TLS and MI-ESPRIT for improved performance.

Below is a summary of three important insights of subspace methods:

1. The fundamental principle of orthogonality between the signal and noise subspaces is extremely important, which lies at the core of subspace methods like MUSIC and ESPRIT. This property enables these algorithms to separate signal components from noise effectively, facilitating high-resolution parameter estimation, such as DoA or frequency, even in complex scenarios involving multiple signals or low SNRs. This capability makes subspace methods exceptionally powerful tools in applications like radar, sonar, and wireless communications.
2. Unlike MUSIC, which requires a spectral search to estimate parameters, ESPRIT leverages the rotational invariance inherent in structured arrays to bypass this step. This results in faster processing, making ESPRIT particularly well suited for real-time applications where rapid parameter estimation is critical.
3. There are several essential practical aspects of subspace methods. For example, accurately detecting the number of signals using techniques like AIC or MDL is very important. These methods ensure that the algorithms perform reliably in real-world conditions. Additionally, we explored variants like root-MUSIC, TST-MUSIC, TLS-ESPRIT, and MI-ESPRIT, which enhance the basic algorithms by addressing specific challenges, such as path clustering, computational complexity, and different invariance structures, demonstrating the adaptability and robustness of subspace techniques in diverse signal processing contexts.

Appendices

Appendix 1: Property of the Transformed Noise Matrices NU_2^t and NU_1^t

In this appendix, we prove that $\boldsymbol{NU}_2^t$ in (6.23) is spatially and temporally white with unchanged noise power within the row space of $\boldsymbol{U}_2^t$, i.e., within the projected subspace, where $\boldsymbol{U}_2^t$ is a projection matrix as defined in Sect. 6.1.3.2. The same argument also holds for $\boldsymbol{NU}_1^t$.

Let $\boldsymbol{N} = [\boldsymbol{n}_1, \ldots, \boldsymbol{n}_M]^T$, where $\boldsymbol{n}_k^T$ is a $1 \times P$ noise vector, and $\boldsymbol{U}_2^t = [\boldsymbol{u}_1, \ldots, \boldsymbol{u}_P]$, with n_{lm} and u_{lm} denoting the elements on the lth row and mth column of $\boldsymbol{N}$ and $\boldsymbol{U}_2^t$, respectively. Then, the element on the lth row and mth column of $\boldsymbol{NU}_2^t$ can be expressed as

$$\tilde{n}_{lm} = \boldsymbol{n}_l^T \boldsymbol{u}_m = \sum_{k=1}^{P} n_{lk} u_{km}, \tag{6.83}$$

and the correlation of $\tilde{n}_{lm}$ and $\tilde{n}_{pq}$ is given by

$$\begin{aligned}\mathbb{E}\left\{\tilde{n}_{lm}\tilde{n}_{pq}^H\right\} &= \mathbb{E}\left\{\left(\sum_{k=1}^{P} n_{lk}u_{km}\right)\left(\sum_{i=1}^{P} n_{pi}u_{iq}\right)^H\right\} \\ &= \sum_{k=1}^{P}\sum_{i=1}^{P} \sigma_n^2 \delta(l-p)\delta(k-i) u_{km} u_{iq}^H \\ &= \sigma_n^2 \delta(l-p) \boldsymbol{u}_q^H \boldsymbol{u}_m,\end{aligned} \tag{6.84}$$

where $\delta(\cdot)$ is the Kronecker delta function. With a slight abuse of notations, let $\tilde{b}_k$ denote the kth element of $\tilde{\boldsymbol{b}}(\hat{t}_2)$, then we have

$$\boldsymbol{U}_2^t = \begin{bmatrix} 0 & -\tilde{b}_1\tilde{b}_2^* & -\tilde{b}_1\tilde{b}_3^* & \cdots & -\tilde{b}_1\tilde{b}_P^* \\ -\tilde{b}_2\tilde{b}_1^* & 0 & -\tilde{b}_2\tilde{b}_3^* & \cdots & -\tilde{b}_2\tilde{b}_P^* \\ \vdots & \vdots & \vdots & \ddots & \vdots \\ -\tilde{b}_P\tilde{b}_1^* & -\tilde{b}_P\tilde{b}_2^* & \cdots & -\tilde{b}_P\tilde{b}_{P-1}^* & 0 \end{bmatrix}, \tag{6.85}$$

where the fact that the SRS signals are unit modulus has been used. Based on (6.85), we can further obtain

$$\boldsymbol{u}_q^H \boldsymbol{u}_m = (P-2)\tilde{b}_q\tilde{b}_m^*. \tag{6.86}$$

Thus, the correlation of $\tilde{n}_{lm}$ and $\tilde{n}_{pq}$ can be further simplified as

$$\mathbb{E}\left\{\tilde{n}_{lm}\tilde{n}_{pq}^H\right\} = -(P-2)\sigma_n^2 u_{qm}\delta(l-p). \tag{6.87}$$

Since u_{qm} is the (q, m)th element of the projection matrix $\boldsymbol{U}_2^t$ and satisfies $|u_{qm}| = 1$, it thus proves that the transformed noise matrix $\boldsymbol{NU}_2^t$ is spatially white. Furthermore, we can also see that the noise is temporally white with the same power within the projected subspace.

Appendix 2: Proof of the Invertibility of F_Y

By the rank-nullity theorem [2], the null space of $[\boldsymbol{U}_X, \boldsymbol{U}_Y]$ has dimension

$$\dim(\text{null}([\boldsymbol{U}_X, \boldsymbol{U}_Y])) = 2K - K = K.$$

Thus, the columns of $\boldsymbol{F}$ are linearly independent because they form a basis for the null space null$([\boldsymbol{U}_X, \boldsymbol{U}_Y])$. This implies that $\boldsymbol{F}$ is full column rank, i.e., rank$(\boldsymbol{F}) = K$. We now prove that $\boldsymbol{F}_Y$ is invertible by contradiction.

Assume that $\boldsymbol{F}_Y$ is not invertible, i.e., $\boldsymbol{F}_Y$ is singular. Then, there exists a non-zero vector $\boldsymbol{v} \in \mathbb{C}^{K\times 1}$ such that

$$\boldsymbol{F}_Y\boldsymbol{v} = \mathbf{0}.$$

Now, consider the product $\boldsymbol{F}\boldsymbol{v}$ given by

$$\boldsymbol{F}\boldsymbol{v} = \begin{bmatrix}\boldsymbol{F}_X\\ \boldsymbol{F}_Y\end{bmatrix}\boldsymbol{v} = \begin{bmatrix}\boldsymbol{F}_X\boldsymbol{v}\\ \boldsymbol{F}_Y\boldsymbol{v}\end{bmatrix} = \begin{bmatrix}\boldsymbol{F}_X\boldsymbol{v}\\ \mathbf{0}\end{bmatrix}.$$

Since $\boldsymbol{F}$ is a basis for the null space, we have

$$[\boldsymbol{U}_X, \boldsymbol{U}_Y]\boldsymbol{F}\boldsymbol{v} = \mathbf{0}. \tag{6.88}$$

Substituting $\boldsymbol{F} \triangleq \begin{bmatrix}\boldsymbol{F}_X\\ \boldsymbol{F}_Y\end{bmatrix}$ into (6.88) yields

$$[\boldsymbol{U}_X, \boldsymbol{U}_Y]\begin{bmatrix}\boldsymbol{F}_X\boldsymbol{v}\\ \mathbf{0}\end{bmatrix} = \boldsymbol{U}_X(\boldsymbol{F}_X\boldsymbol{v}) + \boldsymbol{U}_Y(\mathbf{0}) = \boldsymbol{U}_X(\boldsymbol{F}_X\boldsymbol{v}) = \mathbf{0}.$$

Since $\boldsymbol{U}_X$ has full column rank K (as $M/2 \geq K$ and it spans the signal subspace), its columns are linearly independent. Therefore, $\boldsymbol{U}_X(\boldsymbol{F}_X\boldsymbol{v}) = \mathbf{0}$ implies

$$\boldsymbol{F}_X\boldsymbol{v} = \mathbf{0},$$

which directly leads to

$$\boldsymbol{F}\boldsymbol{v} = \begin{bmatrix}\boldsymbol{F}_X\boldsymbol{v}\\ \boldsymbol{F}_Y\boldsymbol{v}\end{bmatrix} = \begin{bmatrix}\mathbf{0}\\ \mathbf{0}\end{bmatrix} = \mathbf{0}.$$

However, $\boldsymbol{v}$ is a non-zero vector, and $\boldsymbol{F}$ has full column rank K (because its columns form a basis for the null space). Therefore, $\boldsymbol{F}\boldsymbol{v} \neq \mathbf{0}$ holds for any non-zero $\boldsymbol{v}$. This is a contradiction. Hence, the initial assumption that $\boldsymbol{F}_Y$ is singular must be false. Therefore, $\boldsymbol{F}_Y$ is invertible.

Appendix 3: Lagrange Dual Method to Solve Problem (6.50)

To solve problem (6.50), first introduce Lagrange multiplier matrices $\boldsymbol{\Lambda} \in \mathbb{C}^{M/2\times K}$ (for $\boldsymbol{U}_{XY}\boldsymbol{F} = -\boldsymbol{R}_{XY}\boldsymbol{F}$) and $\boldsymbol{M} \in \mathbb{C}^{K\times K}$ (for $\boldsymbol{F}^H\boldsymbol{F} = \boldsymbol{I}$), then the Lagrangian can be expressed as

$$\mathcal{L} = tr(\boldsymbol{R}_{XY}\boldsymbol{R}_{XY}^H) + tr(\boldsymbol{\Lambda}^H(\boldsymbol{U}_{XY} + \boldsymbol{R}_{XY})\boldsymbol{F}) + tr(\boldsymbol{M}^H(\boldsymbol{F}^H\boldsymbol{F} - \boldsymbol{I})), \tag{6.89}$$

where $\boldsymbol{M}$ is a Hermitian matrix. Next, according to the first-order optimality condition [6], by taking derivatives of $\mathcal{L}$ with respect to $\boldsymbol{R}_{XY}$ and $\boldsymbol{F}$, and setting them to zeros, we have

$$\frac{\partial \mathcal{L}}{\partial \boldsymbol{R}_{XY}^*} = \boldsymbol{R}_{XY} + \boldsymbol{\Lambda}\boldsymbol{F}^H = \mathbf{0}, \tag{6.90}$$

$$\frac{\partial \mathcal{L}}{\partial \boldsymbol{F}^*} = (\boldsymbol{U}_{XY} + \boldsymbol{R}_{XY})^H\boldsymbol{\Lambda} + \boldsymbol{F}\boldsymbol{M}^H = \mathbf{0}. \tag{6.91}$$

By plugging $\boldsymbol{R}_{XY} = -\boldsymbol{\Lambda}\boldsymbol{F}^H$ into (6.91), we obtain

$$-\boldsymbol{F}\boldsymbol{\Lambda}^H\boldsymbol{\Lambda} + \boldsymbol{U}_{XY}^H\boldsymbol{\Lambda} + \boldsymbol{F}\boldsymbol{M}^H = \mathbf{0}. \tag{6.92}$$

Furthermore, plugging $\boldsymbol{R}_{XY} = -\boldsymbol{\Lambda}\boldsymbol{F}^H$ into the constraint $(\boldsymbol{U}_{XY} + \boldsymbol{R}_{XY})\boldsymbol{F} = \mathbf{0}$ yields

$$\boldsymbol{U}_{XY}\boldsymbol{F} - \boldsymbol{\Lambda}\boldsymbol{F}^H\boldsymbol{F} = \mathbf{0} \Rightarrow \boldsymbol{\Lambda} = \boldsymbol{U}_{XY}\boldsymbol{F}, \tag{6.93}$$

where the fact that $\boldsymbol{F}^H\boldsymbol{F} = \boldsymbol{I}$ is used. Thus, $\boldsymbol{R}_{XY}$ can be rewritten as $\boldsymbol{R}_{XY} = -\boldsymbol{U}_{XY}\boldsymbol{F}\boldsymbol{F}^H$. Next, substitute $\boldsymbol{R}_{XY}$ into the objective function of problem (6.50), we have

$$\|\boldsymbol{R}_{XY}\|_F^2 = \|\boldsymbol{U}_{XY}\boldsymbol{F}\boldsymbol{F}^H\|_F^2 = tr(\boldsymbol{F}\boldsymbol{F}^H\boldsymbol{U}_{XY}^H\boldsymbol{U}_{XY}\boldsymbol{F}\boldsymbol{F}^H) = tr(\boldsymbol{U}_{XY}^H\boldsymbol{U}_{XY}\boldsymbol{F}\boldsymbol{F}^H). \tag{6.94}$$

Therefore, to minimize $\|\boldsymbol{R}_{XY}\|_F^2$, $\boldsymbol{F}$ should be the eigenvectors corresponding to the K smallest eigenvalues of $\boldsymbol{U}_{XY}^H\boldsymbol{U}_{XY}$. Note that the dual matrix $\boldsymbol{M}$ can be obtained by plugging $\boldsymbol{\Lambda} = \boldsymbol{U}_{XY}\boldsymbol{F}$ into (6.91) and using the fact $\boldsymbol{F}^H\boldsymbol{F} = \boldsymbol{I}$.

Appendix 4: Derivation of (6.80)

Let $\boldsymbol{M} = \bar{\boldsymbol{\Phi}}^H\bar{\boldsymbol{\Phi}}^\dagger$, we have $\boldsymbol{P}^\perp_{\bar{\boldsymbol{\Phi}}^H} = \boldsymbol{I} - \boldsymbol{M}$. Then, by differentiating $\boldsymbol{M}$ with respect to η_i, it yields $\frac{\partial \boldsymbol{M}}{\partial \eta_i} = \frac{\partial \bar{\boldsymbol{\Phi}}^H}{\partial \eta_i}(\bar{\boldsymbol{\Phi}}\bar{\boldsymbol{\Phi}}^H)^{-1}\bar{\boldsymbol{\Phi}} + \bar{\boldsymbol{\Phi}}^H\frac{\partial(\bar{\boldsymbol{\Phi}}\bar{\boldsymbol{\Phi}}^H)^{-1}}{\partial \eta_i}\bar{\boldsymbol{\Phi}} + \bar{\boldsymbol{\Phi}}^H(\bar{\boldsymbol{\Phi}}\bar{\boldsymbol{\Phi}}^H)^{-1}\frac{\partial \bar{\boldsymbol{\Phi}}}{\partial \eta_i}$. Using the matrix-derivative formula $\frac{\partial(\bar{\boldsymbol{\Phi}}\bar{\boldsymbol{\Phi}}^H)^{-1}}{\partial \eta_i} = -(\bar{\boldsymbol{\Phi}}\bar{\boldsymbol{\Phi}}^H)^{-1}(\frac{\partial \bar{\boldsymbol{\Phi}}}{\partial \eta_i}\bar{\boldsymbol{\Phi}}^H + \bar{\boldsymbol{\Phi}}\frac{\partial \bar{\boldsymbol{\Phi}}^H}{\partial \eta_i})(\bar{\boldsymbol{\Phi}}\bar{\boldsymbol{\Phi}}^H)^{-1}$, substituting it into $\frac{\partial \boldsymbol{M}}{\partial \eta_i}$ and employing the definition of $\bar{\boldsymbol{\Phi}}^\dagger$ and the property $\boldsymbol{P}^\perp_{\bar{\boldsymbol{\Phi}}^H}\bar{\boldsymbol{\Phi}}^H = \boldsymbol{0}$ (from the definition of $\boldsymbol{P}^\perp_{\bar{\boldsymbol{\Phi}}^H}$), we obtain $\frac{\partial \boldsymbol{P}^\perp_{\bar{\boldsymbol{\Phi}}^H}}{\partial \eta_i} = \boldsymbol{P}^\perp_{\bar{\boldsymbol{\Phi}}^H}\bar{\boldsymbol{\Phi}}_i^H\bar{\boldsymbol{\Phi}}^\dagger + \bar{\boldsymbol{\Phi}}^{\dagger H}\bar{\boldsymbol{\Phi}}_i\boldsymbol{P}^\perp_{\bar{\boldsymbol{\Phi}}^H}$. Further using the property $\boldsymbol{r}_i = \frac{\partial \mathrm{vec}(\boldsymbol{U}_W\boldsymbol{P}^\perp_{\bar{\boldsymbol{\Phi}}^H})}{\partial \eta_i} = \mathrm{vec}\left(\boldsymbol{U}_W\frac{\partial \boldsymbol{P}^\perp_{\bar{\boldsymbol{\Phi}}^H}}{\partial \eta_i}\right)$, (6.80) can be readily derived.

Homework Problems

6.1 (a) Explain the principle behind the MUSIC algorithm for DoA estimation. How does it exploit the signal and noise subspaces?

(b) Consider a ULA with eight antennas spaced half a wavelength apart, receiving signals from two sources at DoAs of 30° and 60°. Assuming a large number of snapshots, sketch the expected MUSIC pseudospectrum. Label the axes and indicate where the peaks should appear.

6.2 Consider a ULA with eight antennas spaced $\lambda/2$ apart, where λ is the wavelength. Two signals arrive at DoAs of −20° and 20°.

(a) Write the steering vector $\boldsymbol{a}(\theta)$ for this array.

(b) Assuming the noise subspace is known, outline the steps to apply the root-MUSIC algorithm to estimate the DoAs. You do not need to compute numerical values, but describe the polynomial formed and how its roots are used to find the DoAs.

6.3 Consider a ULA with six antennas and maximum overlapping subarrays for ESPRIT.

(a) Define the selection matrices $\boldsymbol{J}_1$ and $\boldsymbol{J}_2$ for this configuration.

(b) Outline the steps to estimate the subspace rotation operator $\boldsymbol{\Psi}$ using the TLS criterion.

(c) If the estimated eigenvalues of $\boldsymbol{\Psi}$ are $\lambda_1 = e^{j\pi/3}$ and $\lambda_2 = e^{-j\pi/3}$, compute the corresponding DoAs assuming the antenna spacing is $\lambda/2$.

6.4 (a) For a channel with two paths clustered in delay ($\tau_1 \approx \tau_2$) but separated in DoA ($\theta_1 \neq \theta_2$), explain whether the first temporal filtering is necessary in the TST-MUSIC framework. How would you modify the TST-MUSIC procedure to estimate the parameters of these two paths?

(b) Compare S-MUSIC and T-MUSIC: under what channel conditions does TST-MUSIC outperform both?

References

1. H. Akaike, Information theory and an extension of the maximum likelihood principle, in *Selected Papers of hirotugu akaike* (Springer, New York, 1998), pp.199–213
2. S. Axler, *Linear Algebra Done Right* (Springer Nature, 2024), p. 390
3. A.J. Barabell, Improving the resolution performance of eigenstructure-based direction-finding algorithms, in *IEEE International Conference on Acoustics, Speech, and Signal Processing* (1983)
4. M.S. Bartlett, A note on the multiplying factors for various χ^2 approximations. J. Royal Stat. Soc. Ser. B (Methodol.) **16**(2), 296–298 (1954)
5. D.P. Bertsekas, Nonlinear programming. J. Oper. Res. Soc. **48**(3), 334–334 (1997)

6. S. Boyd, L. Vandenberghe, *Convex Optimization* (Cambridge University Press, UK, 2004)
7. L. Chen, M.M. Zhao, Q. Wu, M.J. Zhao, R. Zhang, A covariance-surrogate framework for movable-antenna enabled anti-jamming with unknown jammers (2025), arXiv preprint arXiv:2512.20380
8. X. Chen, M.-M. Zhao, M. Li, L. Li, M.-J. Zhao, J. Wang, Ambiguity function analysis and optimization of frequency-hopping MIMO radar with movable antennas. IEEE Internet Things J. **12**(12), 21836–21851 (2025)
9. L. Chen, M.-M. Zhao, M.-J. Zhao, R. Zhang, Antenna position and beamforming optimization for movable antenna enabled ISAC: optimal solutions and efficient algorithms. IEEE Trans. Signal Process. **73**, 3812–3828 (2025)
10. P.E. Gill, W. Murray, M.H. Wright, *Practical Optimization*. Society for Industrial and Applied Mathematics (2019)
11. G.H. Golub, C.F. Van Loan, *Matrix Computations* (Johns Hopkins University Press, 2013)
12. A. Jakobsson, A.L. Swindlehurst, P. Stoica, Subspace-based estimation of time delays and Doppler shifts. IEEE Trans. Signal Process. **46**(9), 2472–2483 (1998)
13. D.N. Lawley, Tests of significance for the latent roots of covariance and correlation matrices. Biometrika **43**(1/2), 128–136 (1956)
14. D.G. Luenberger, Y. Ye, *Linear and Nonlinear Programming*, vol. 2 (Addison-Wesley, Reading, MA, 1984)
15. B.E. Ottersten, M. Viberg, T. Kailath, Performance analysis of the total least squares ESPRIT algorithm. IEEE Trans. Signal Process. **39**, 1122–1135 (1991)
16. W.H. Press, *Numerical Recipes 3rd edition: The Art of Scientific Computing*, 3rd edn. (Cambridge University Press, 2007)
17. B.D. Rao, K.V.S. Hari, Performance analysis of Root-MUSIC. IEEE Trans. Acoust. Speech Signal Process. **37**(12), 1939–1949 (1989). https://doi.org/10.1109/29.45540
18. T.T. Rappaport, J.H. Reed, B.D. Woerner, Position location using wireless communications on highways of the future. IEEE Commun. Mag. **34**, 33–41 (1996)
19. J. Rissanen, A universal prior for integers and estimation by minimum description length. Ann. Stat. **11**(2), 416–431 (1983)
20. R. Roy, T. Kailath, ESPRIT-estimation of signal parameters via rotational invariance techniques. IEEE Trans. Acoust. Speech Signal Process. **37**(7), 984–995 (1989)
21. R. Schmidt, Multiple emitter location and signal parameter estimation. IEEE Trans. Antennas Propag. **34**(3), 276–280 (1986)
22. T. Strutz, *Data Fitting and Uncertainty: A Practical Introduction to Weighted Least Squares and Beyond* (Vieweg and Teubner, Wiesbaden, DEU, 2010)
23. A.L. Swindlehurst, B. Ottersten, R. Roy, T. Kailath, Multiple invariance ESPRIT. IEEE Trans. Signal Process. **40**(4), 867–881 (1992)
24. Technical Specification Group Radio Access Network; NR; Physical channels and modulation, document 11 V16.7.0, Release 16, 3GPP, September 2021
25. H. Theil, *Best Linear Unbiased Estimation and Prediction*. Principles of Econometrics (1971), pp.119–124
26. M. Viberg, B. Ottersten, Sensor array processing based on subspace fitting. IEEE Trans. Signal Process. **39**(5), 1110–1121 (1991)
27. Y.Y. Wang, J.T. Chen, W.H. Fang, TST-MUSIC for joint DOA-delay estimation. IEEE Trans. Signal Process. **49**(4), 721–729 (2001)
28. M. Wax, T. Kailath, Detection of signals by information theoretic criteria. IEEE Trans. Acoust. Speech Signal Process. **33**(2), 387–392 (1985)
29. M. Wax, I. Ziskind, Detection of the number of coherent signals by the MDL principle. IEEE Trans. Acoust. Speech Signal Process. **37**(8), 1190–1196 (1989)
30. K.K. Wong, A. Shojaeifard, K.F. Tong, Y. Zhang, Fluid antenna systems. IEEE Trans. Wirel. Commun. **20**(3), 1950–1962 (2020)
31. L. Zhu, W. Ma, R. Zhang, Movable antennas for wireless communication: opportunities and challenges. IEEE Commun. Mag. **62**(6), 114–120 (2023)

Part II

Sparse Bayesian Inference for Wireless Communications and Sensing

Introduction to Compressive Sensing

7

This chapter introduces compressive sensing (CS), a paradigm-shifting framework that enables efficient signal acquisition below the Nyquist rate by exploiting signal sparsity and incoherent measurements. Departing from Shannon's sampling theorem, CS addresses limitations in handling non-bandlimited signals, high sampling costs, and data redundancy. The core principles, i.e., sparsity (concise signal representations in bases like Fourier or wavelets) and incoherence (e.g., between spikes and sinusoids or noiselets and wavelets), ensure that sparse signals can be recovered from far than linear measurements than the signal dimension. Theoretical foundations establish that ℓ_1-minimization exactly recovers S-sparse signals when the sensing matrix satisfies incoherence conditions or the restricted isometry property (RIP). Robust extensions handle noisy data via least absolute shrinkage and selection operator (LASSO)-type optimization, with bounded error. CS is thus a universal, nonadaptive framework for sub-Nyquist sampling with applications in imaging, communications, and sensing.

7.1 Introduction

7.1.1 Motivation

Conventional approaches to sampling signals or images follow Shannon's celebrated theorem [1], i.e.,

Theorem 7.1 *If a function $x(t)$ contains no frequencies higher than B hertz, then it can be completely determined from its ordinates at a sequence of points spaced less than $1/(2B)$ seconds apart.*

A sufficient sample rate is therefore anything larger than $2B$ samples per second. Equivalently, for a given sample rate f_s, perfect reconstruction is guaranteed possible for a bandlimit $B < f_s/2$.

When the bandlimit is too high (or there is no bandlimit), the reconstruction exhibits imperfections known as aliasing. Modern statements of the theorem are sometimes careful to explicitly state that $x(t)$ must contain no sinusoidal component at exactly frequency B, or that B must be strictly less than one half the sample rate. The threshold $2B$ is called the Nyquist rate and is an attribute of the continuous-time input $x(t)$ to be sampled. The sample rate must exceed the Nyquist rate for the samples to suffice to represent $x(t)$. The threshold $f_s/2$ is called the Nyquist frequency and is an attribute of the sampling equipment. All meaningful frequency components of the properly sampled $x(t)$ exist below the Nyquist frequency. The condition described by these inequalities is called the Nyquist criterion, or sometimes the Raabe condition.

To illustrate this graphically, let us briefly recap the notations of Fourier transform and aliasing. When $x(t)$ is a function with a Fourier transform $X(f)$, i.e.,

$$X(f) \triangleq \int_{-\infty}^{\infty} x(t)\mathrm{e}^{-j2\pi ft}\mathrm{d}t, \tag{7.1}$$

then the samples $x[n]$ of $x(t)$ are sufficient to create a periodic summation of $X(f)$, denoted by

$$X_{1/T}(f) \triangleq \frac{1}{T}\sum_{k=-\infty}^{\infty} X(f - k/T) = \sum_{n=-\infty}^{\infty} x[n]\mathrm{e}^{-j2\pi fnT}, \tag{7.2}$$

A. Liu and M.-M. Zhao, *Advanced Bayesian Signal Processing*, Textbooks in Telecommunication Engineering,
https://doi.org/10.1007/978-3-032-26656-9_7

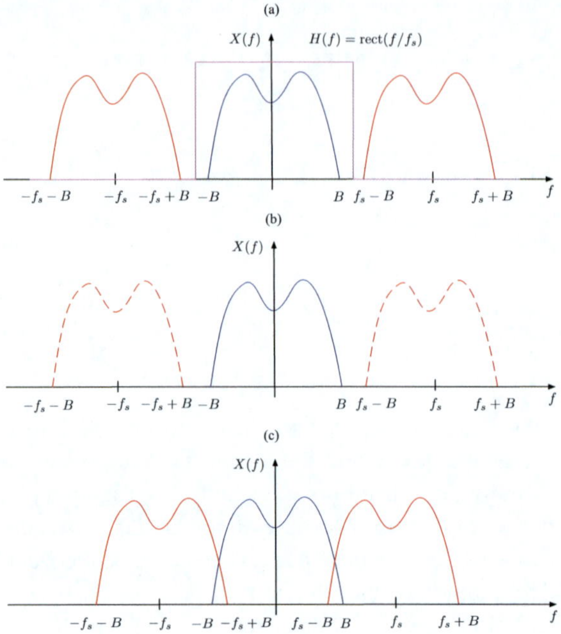

Fig. 7.1 Illustration of Shannon's sampling theorem, **a** spectrum of a properly sampled bandlimited signal (blue) and the adjacent DFT images (red) that do not overlap; **b** a brick-wall low-pass filter, $H(f)$, removes the images, leaves the original spectrum, $X(f)$, and recovers the original signal from its samples; **c** the case that the Nyquist criterion is not satisfied, and adjacent copies overlap

which is a periodic function that can be equivalently represented by a Fourier series, where the Fourier coefficients are denoted as $x[n]$. This function is also known as the discrete Fourier transform (DFT) of the sample sequence. As depicted in Fig. 7.1, copies of $X(f)$ are shifted by multiples of the sampling rate $f_s = 1/T$ and combined by addition. For a bandlimited function $X(f) = 0,$ for all $f \geq B$ and sufficiently large f_s, it is possible for the copies to remain distinct from each other. But if the Nyquist criterion is not satisfied, adjacent copies overlap, and it is not possible in general to discern an unambiguous $X(f)$. Any frequency component above $f_s/2$ is indistinguishable from a lower-frequency component, called an alias, associated with one of the copies. In such cases, the customary interpolation techniques produce the alias, rather than the original component. When the sample rate is predetermined by other considerations (such as an industry standard), $x(t)$ is usually filtered to reduce its high frequencies to acceptable levels before it is sampled. The type of filter required is a low-pass filter, and in this application it is called an anti-aliasing filter.

The Shannon sampling theorem, as articulated in Theorem 7.1, has governed signal acquisition for decades, dictating that a bandlimited signal must be sampled at least twice its highest frequency component to avoid aliasing. This principle has been foundational in designing systems for consumer audio and visual electronics, medical imaging devices, radio receivers, and so on, where analog-to-digital converter (ADC) uniformly samples signals at or above the Nyquist rate. However, this paradigm encounters critical limitations in scenarios where:

- **Signals are not naturally bandlimited**: For example, natural images, which exhibit sparsity in transform domains like wavelets (see Fig. 7.2), are not strictly bandlimited, requiring high-resolution sampling driven by spatial resolution rather than bandwidth.
- **High sampling rates are impractical**: In applications such as massive MIMO communication systems, ultra-wideband radar, magnetic resonance imaging (MRI), or sensor networks, hardware constraints (e.g., limited ADC speed, power, or sensor count) make Nyquist rate sampling infeasible or inefficient.
- **Redundant sampling is wasteful**: Traditional methods acquire far more data than necessary, as most real-world signals are compressible, i.e., their information can be represented by a few dominant coefficients in a suitable basis (e.g., transform domain wireless channel representation, wavelets for images and Fourier for sparse spectra, etc.).

Consider a signal $\boldsymbol{f} \in \mathbb{R}^N$ that is sparse or compressible: when expressed in an orthonormal basis $\boldsymbol{\Psi}$ (e.g., DFT, wavelets, curvelets), it admits a representation $\boldsymbol{f} = \boldsymbol{\Psi x}$ where most coefficients in $\boldsymbol{x}$ are zero or negligible. For such signals, the

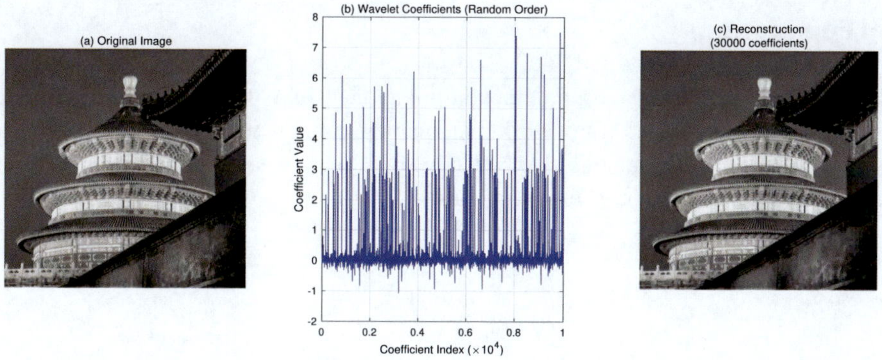

Fig. 7.2 **a** Source 512 × 512 image with intensity values scaled to the unit interval [0, 1]. **b** Visual representation of 10^4 randomly selected wavelet coefficients. **c** Reconstruction using only the dominant 3×10^4 wavelet coefficients (remaining coefficients discarded), with intensity values constrained to the standard display range [0, 1]. As can be seen, natural images exhibit significant compressibility as their energy concentrates in limited coefficient subsets. Visual comparison reveals minimal perceptual degradation relative to the original

Shannon–Nyquist framework is overly conservative: it samples uniformly across the entire domain, ignoring the fact that only a small subset of coefficients carries meaningful information. CS, a revolutionary paradigm emerging in the early 2000s, circumvents these limitations by leveraging two key principles: sparsity and incoherence. Unlike traditional sampling, which relies on bandlimiting, CS exploits the sparsity of signals in some transform domain and the incoherence between the sparsifying basis and a set of random sensing waveforms. This allows reconstructing the signal from $M \ll N$ linear measurements (here, M is the number of samples taken), directly capturing the signal's information content without redundant sampling.

7.1.2 Core Idea

The core idea of CS hinges on two intertwined principles: sparsity (signals have concise representations in some basis) and incoherence (sensing waveforms are dense in the sparsity domain), which together allow capturing the essential information of a signal with minimal measurements and reconstructing it via computational optimization [2].

- Sparsity expresses the idea that the "information rate" of a continuous time signal may be much smaller than suggested by its bandwidth, or that a discrete-time signal depends on a number of degrees of freedom which is comparably much smaller than its (finite) length. More precisely, CS exploits the fact that many natural signals are sparse or compressible in the sense that they have concise representations when expressed in the proper basis.
- Incoherence extends the duality between time and frequency and expresses the idea that objects having a sparse representation in $\boldsymbol{\Psi}$ must be spread out in the domain in which they are acquired, just as a Dirac or a spike in the time domain is spread out in the frequency domain. In other words, incoherence says that unlike the signal of interest, the sampling/sensing waveforms have an extremely dense representation in $\boldsymbol{\Psi}$.

The crucial observation is that one can design efficient sensing or sampling protocols that capture the useful information content embedded in a sparse signal and condense it into a small amount of data. These protocols are nonadaptive and simply require correlating the signal with a small number of fixed waveforms that are incoherent with the sparsifying basis. What is most remarkable about these sampling protocols is that they allow a sensor to very efficiently capture the information in a sparse signal without trying to comprehend that signal. Further, there is a way to use numerical optimization to reconstruct the full-length signal from the small amount of collected data. In other words, CS is a very simple and efficient signal acquisition protocol which samples in a signal independent fashion at a low rate and later uses computational power for reconstruction from what appears to be an incomplete set of measurements.

7.2 Theoretical Foundations

The theoretical underpinnings of CS rest on two fundamental pillars, i.e., sparsity and incoherence. This section formalizes these concepts, establishing the mathematical framework that enables signal recovery from sub-Nyquist measurements. We begin by defining sparsity in transform domains, then introduce incoherent sampling principles, and finally present the theoretical guarantees for exact recovery via convex optimization.

7.2.1 Sparsity

Many natural signals have concise representations when expressed in a convenient basis. Consider, for example, the coefficients of a three-path channel in Fig. 7.3a and its DFT transform in Fig. 7.3b. Although nearly all the channel coefficients have non-zero values, the DFT coefficients offer a concise summary: most coefficients are small, and the relatively few large coefficients capture most of the channel information.

Mathematically speaking, we have a vector $\boldsymbol{f} \in \mathbb{R}^N$ (such as the N channel coefficients in Fig. 7.3) which we expand in an orthonormal basis (such as the DFT basis when $\boldsymbol{f} \in \mathbb{C}^N$) $\boldsymbol{\Psi} = [\boldsymbol{\psi}_1, \boldsymbol{\psi}_2, \ldots, \boldsymbol{\psi}_N]$ as follows[1]:

$$\boldsymbol{f} = \sum_{i=1}^{N} x_i \boldsymbol{\psi}_i. \tag{7.3}$$

Generally, sparsity is a fundamental modeling tool which permits efficient signal processing, e.g., accurate statistical estimation and classification, efficient data compression, etc.

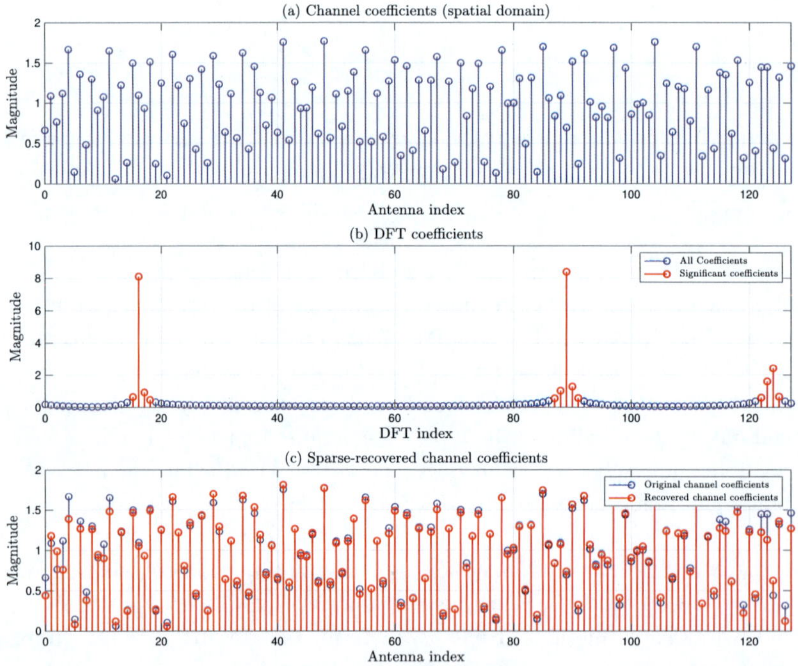

Fig. 7.3 **a** Original channel coefficients of a massive MIMO channel with 128 transmit antennas forming a ULA, and **b** DFT coefficients of the original channel. Relatively few DFT coefficients capture most of the channel energy; many such channels are highly compressible. **c** The reconstruction obtained by zeroing out all the coefficients in the DFT expansion but the 13 largest. The difference with the original channel coefficients is minor

[1] In general, since we need to process a continuous sequence of signals, the vectors $\boldsymbol{f}$ and $\boldsymbol{\psi}_i$ should properly be denoted as $\boldsymbol{f}(t)$ and $\boldsymbol{\psi}_i(t)$, respectively. However, for simplicity, we will consider only the sensing problem at a single time instant and therefore omit the time variable t throughout this chapter, unless otherwise specified.

7.2.2 Incoherent Sampling

Suppose the signal f is probed through a linear measurement system, where observable data are generated via inner products $\langle f, \boldsymbol{\phi}_k \rangle$ with predetermined sensing elements $\{\boldsymbol{\phi}_k\}$, i.e.,

$$y_k = \langle f, \boldsymbol{\phi}_k \rangle, \quad k = 1, \ldots, M, \tag{7.4}$$

where M denotes the number of measurements. That is, we simply correlate the object we wish to acquire with the waveforms $\boldsymbol{\phi}_k$. If the sensing waveforms are Dirac delta functions (spikes), for example, then $\boldsymbol{y} = [y_1, \ldots, y_M]$ is a vector of sampled values of f in the time or space domain. If the sensing waveforms are indicator functions of pixels, then $\boldsymbol{y}$ is the image data typically collected by sensors in a digital camera. If the sensing waveforms are sinusoids, then $\boldsymbol{y}$ is a vector of Fourier coefficients, and this is the sensing modality used in MRI. A plethora of analogous cases exists across diverse contexts.

Definition 7.1 The coherence between the sensing basis $\boldsymbol{\Phi}$ and the representation basis $\boldsymbol{\Psi}$ is

$$\mu(\boldsymbol{\Phi}, \boldsymbol{\Psi}) = \sqrt{N} \max_{1 \le k, i \le N} \left| \langle \boldsymbol{\phi}_k, \boldsymbol{\psi}_i \rangle \right|. \tag{7.5}$$

The coherence $\mu(\boldsymbol{\Phi}, \boldsymbol{\Psi}) \in [1, \sqrt{N}]$ measures the largest correlation between any two elements of $\boldsymbol{\Phi}$ and $\boldsymbol{\Psi}$. If they contain correlated elements, the coherence is large; otherwise, it is small.

Compressive sampling is mainly concerned with low-coherence pairs, and some examples of such pairs are given below:

- Choose $\boldsymbol{\Phi}$ to be the canonical or spike basis $\boldsymbol{\phi}_k(t) = \delta(t-k)$ and $\boldsymbol{\Psi}$ is the Fourier basis, i.e., $\boldsymbol{\psi}_i(t) = \frac{1}{\sqrt{N}} e^{j2\pi i t/N}$. Since $\boldsymbol{\Phi}$ is the sensing matrix, this corresponds to the classical sampling scheme in time or space. The time–frequency pair obeys $\mu(\boldsymbol{\Phi}, \boldsymbol{\Psi}) = 1$ and, therefore, we have maximal incoherence. Further, spikes and sinusoids are maximally incoherent not just in one dimension but in any dimension (in two dimensions, three dimensions, etc., see Appendix 1 for the details). This is quite intuitive since spikes/pulses are localized in the time domain while sinusoidal waves are localized in the frequency domain, their dot product (inner product) exhibits different behaviors at different positions and frequencies.
- The second example takes wavelet bases for $\boldsymbol{\Psi}$ and noiselets [3] for $\boldsymbol{\Phi}$. The coherence between noiselets and Haar wavelets is 1 and that between noiselets and Daubechies D4 and D8 wavelets is, respectively, about 2.2 and 2.9 across a wide range of sample sizes N. This extends to higher dimensions as well. The definitions of noiselets, Haar wavelet, and Daubechies D4 and D8 wavelets are provided in Appendix 2. Note that noiselets are also maximally incoherent with spikes and incoherent with the Fourier basis. The interest in noiselets comes from the facts that (1) they are incoherent with systems providing sparse representations of image data and other types of data and (2) they come with very fast algorithms; the noiselet transform runs in $O(NlogN)$ time, and just like the Fourier transform, the noiselet matrix does not need to be stored to be applied to a vector. This is of crucial practical importance for numerically efficient CS implementations.
- Random matrices typically exhibit low coherence with any fixed basis $\boldsymbol{\Psi}$, meaning they are largely incoherent. To form a random orthobasis $\boldsymbol{\Phi}$, one can independently sample N vectors uniformly from the unit sphere and then orthonormalize them, i.e., $\boldsymbol{\phi}_k \sim \mathcal{U}(\mathbb{S}), \mathbb{S} \triangleq \{\boldsymbol{\phi}_k \in \mathbb{R}^N : \|\boldsymbol{\phi}_k\|_2 = 1\}$; with high probability, the coherence between this random orthobasis and any fixed basis will be approximately $\sqrt{2 \log N}$. Extending this idea, random waveforms $\{\boldsymbol{\phi}_k(t)\}$ with independent identically distributed entries, such as Gaussian or ± 1 binary entries, also show very low coherence with any fixed representation $\boldsymbol{\Psi}$.

7.2.3 Undersampling

Ideally, one would like to measure all the N coefficients of f, but usually only a subset of f can be measured through the collected data $\{y_k\}$ as given in (7.4), i.e., $M < N$. With this information, the signal $\boldsymbol{x}$ might be recovered by ℓ_0 norm minimization, i.e., the reconstructed $f^\star$ can be expressed as $f^\star = \boldsymbol{\Psi}\boldsymbol{x}^\star$, where $\boldsymbol{x}^\star$ is the solution to the following problem:

$$\begin{aligned} &\min_{\tilde{\boldsymbol{x}} \in \mathbb{R}^N} \|\tilde{\boldsymbol{x}}\|_0 \\ &\text{s.t. } y_k = \langle \boldsymbol{\phi}_k, \boldsymbol{\Psi}\tilde{\boldsymbol{x}} \rangle, \quad k = 1, \ldots, M, \end{aligned} \tag{7.6}$$

where the operator $\|\cdot\|_0$ returns the number of non-zero entries of a given vector. As the ℓ_0 norm minimization problem in (7.6) is in general difficult to handle, we usually resort to the following ℓ_1 norm minimization problem:

$$\min_{\tilde{x}\in\mathbb{R}^N} \|\tilde{x}\|_1$$
$$\text{s.t. } y_k = \langle \phi_k, \Psi\tilde{x}\rangle, \quad k = 1, \ldots, M, \tag{7.7}$$

where $\|\tilde{x}\|_1 \triangleq \sum_i |x_i|$. Problem (7.7) implies that among all potential signals that satisfy $y_k = \langle \phi_k, \Psi\tilde{x}\rangle$, we pick that whose coefficient sequence has minimal ℓ_1 norm. Problem (7.7) is an ℓ_1-norm minimization problem subject to linear equality constraints, and it can easily be recast as a linear program (LP) (see Appendix 3) that can be efficiently solved by various algorithms, such as the simplex method and the interior-point method. Note that up to now we do not consider the effect of observation noises, the noisy case will be discussed later.

Given problem (7.7), a very important result asserts that when f is sufficiently sparse, the recovery via ℓ_1-minimization is provably exact (i.e., the solutions to (7.6) and (7.7) are equivalent), as provided in the following theorem.

Theorem 7.2 *([4]) Fix $f \in \mathbb{R}^N$ and suppose that the coefficient sequence x of f in the basis Ψ is S-sparse.*[2] *Select M measurements in the Φ domain uniformly at random. Then, if*

$$M \geq C\mu^2(\Phi, \Psi)S\log N \tag{7.8}$$

for some positive constant C, the solution to (7.7) is exact with overwhelming probability.[3]

Theorem 7.2 implies

- If there is no sparse constraint on x, the linear equation $y = \Phi\Psi x$ is underdetermined and has multiple solutions, which means that it is almost impossible to recover the signal x. However, if there is a sparse constraint on x, and the sparsity level S is sufficiently small, then the solution of this linear equation which also satisfies the sparse constraint is unique, and the signal can be perfectly recovered.
- The role of the coherence is completely transparent, i.e., the smaller the coherence, the fewer samples are needed. That's why we focused on low-coherence systems.
- There is no information loss by measuring just about any set of M coefficients, which might be far fewer than what the signal's size seems to require. If $\mu(\Phi, \Psi)$ is equal to or close to one, then approximately $S\log N$ samples are enough instead of N.
- We can exactly recover the signal f from our condensed dataset by minimizing a convex function. There is no requirement on any knowledge about how many non-zero coordinates x has, where these coordinates are, or what are their amplitudes, i.e., we assume all these are completely unknown in advance. We just run the algorithm, and if the signal is sparse enough, exact recovery happens.

Probability plays a critical role in ensuring effective signal recovery, and it is unrealistic to expect all measurement sets of size M to yield comparable outcomes. Actually, there exist sparse signals that nearly vanish in the Φ domain. For instance, one can find a sparse signal f and a very large subset (of size almost N, like $N - S$) where $y_k = \langle f, \phi_k\rangle = 0$ for all $k \in [1, M]$. An example is the Dirac comb as discussed in [5]. If such a subset is measured, the result is a stream of zeros, and no algorithm can reconstruct the signal. Despite this risk, Theorem 7.2 guarantees that the fraction of measurement sets leading to failed exact recovery is "negligible" (a very small probability, like a large negative power of N). Thus, the probability of failure is extremely low. For practical use, if the sampling size is large enough, this failure probability is treated as zero.

[2] Here, a vector $x \in \mathbb{R}^N$ is said to be S-sparse if it has at most S non-zero entries, i.e., $\|x\|_0 \leq S$.

[3] It is shown that the probability of success exceeds $1 - \delta$ if $M \geq C\mu^2(\Phi, \Psi)S\log(N/\delta)$. In addition, the result is only guaranteed for nearly all sign sequences x with a fixed support, see [4] for details.

7.3 Robust Compressive Sampling

Based on the above foundations, CS must also be equipped to manage two critical elements: nearly sparse signals and noise. First, the objects under consideration are typically not precisely sparse but rather approximately so. The core question here is whether accurate reconstructions of these objects can be derived from highly undersampled measurements. Second, in any real-world application, the measured data will always be corrupted by at least a minimal level of noise, as sensing devices do not possess infinite precision. Consequently, it is vital for CS to exhibit robustness against such nonidealities. At the very least, slight disturbances in the data should induce only minor disturbances in the reconstruction.

To proceed, we consider the following abstract problem of recovering a vector $\boldsymbol{x} \in \mathbb{R}^N$ from data:

$$\boldsymbol{y} = \boldsymbol{A}\boldsymbol{x} + \boldsymbol{z}, \tag{7.9}$$

where $\boldsymbol{A}$ is an $M \times N$ "sensing matrix" giving us information about $\boldsymbol{x}$, and $\boldsymbol{z}$ is a stochastic or deterministic unknown noise/error term. The setup of the previous section is of this form since with $\boldsymbol{f} = \boldsymbol{\Psi x}$ and $\boldsymbol{y} = \boldsymbol{R\Phi f}$ ($\boldsymbol{R}$ is the $M \times N$ matrix extracting the sampled coordinates in $[1, N]$), one can write $\boldsymbol{y} = \boldsymbol{A}\boldsymbol{x}$, where $\boldsymbol{A} = \boldsymbol{R\Phi\Psi}$. Hence, one can work with the abstract model (7.9) bearing in mind that $\boldsymbol{x}$ maybe the coefficient sequence of the object in a proper basis.

7.3.1 Restricted Isometry Property

We first introduce a key notion that has proved to be very useful to study the general robustness of CS, i.e., the so-called RIP [6].

Definition 7.2 For each integer $S = 1, 2, \ldots$, define the isometry constant δ_S of a matrix $\boldsymbol{A}$ as the smallest number such that

$$(1 - \delta_S)\|\boldsymbol{x}\|_2^2 \leq \|\boldsymbol{A}\boldsymbol{x}\|_2^2 \leq (1 + \delta_S)\|\boldsymbol{x}\|_2^2 \tag{7.10}$$

holds for all S-sparse vectors $\boldsymbol{x}$.

We can roughly state that a matrix $\boldsymbol{A}$ satisfies the RIP of order S when δ_S (the isometry constant for order S) is not very close to 1. When this property holds, $\boldsymbol{A}$ roughly maintains the Euclidean length (or ℓ_2-norm) of S-sparse signals. This, in turn, means that S-sparse vectors cannot lie in the null space of $\boldsymbol{A}$ (since if they did, their length after multiplying by $\boldsymbol{A}$ would be zero, contradicting the length preservation). An equivalent way to describe RIP is to say that every subset of S columns taken from $\boldsymbol{A}$ is nearly orthogonal.[4] To understand this equivalence, let us analyze both directions:

- **From near-orthogonality of S columns to norm preservation**: Suppose every subset of S columns of $\boldsymbol{A}$ is nearly orthogonal. For an S-sparse vector $\boldsymbol{x}$, $\boldsymbol{A}\boldsymbol{x} = \sum_{i \in S} x_i \boldsymbol{a}_i$, where S denotes the index set indicating the non-zero elements of $\boldsymbol{x}$ and $\{\boldsymbol{a}_i\}$ are the columns of $\boldsymbol{A}$. Then $\|\boldsymbol{A}\boldsymbol{x}\|_2^2 = \sum_{i \in S}\sum_{j \in S} x_i x_j \langle \boldsymbol{a}_i, \boldsymbol{a}_j \rangle$. Near-orthogonality implies $\langle \boldsymbol{a}_i, \boldsymbol{a}_j \rangle \approx 0$ for $i \neq j$, so $\|\boldsymbol{A}\boldsymbol{x}\|_2^2 \approx \sum_{i \in S} x_i^2 \langle \boldsymbol{a}_i, \boldsymbol{a}_i \rangle$. If $\langle \boldsymbol{a}_i, \boldsymbol{a}_i \rangle \approx 1$ (as $\boldsymbol{A}$ should preserve norms), then $\|\boldsymbol{A}\boldsymbol{x}\|_2^2 \approx \|\boldsymbol{x}\|_2^2$, and thus $(1 - \delta_S)\|\boldsymbol{x}\|_2^2 \leq \|\boldsymbol{A}\boldsymbol{x}\|_2^2 \leq (1 + \delta_S)\|\boldsymbol{x}\|_2^2$ holds with δ_S not close to 1.
- **From norm preservation to near-orthogonality of S columns**: Assume $\boldsymbol{A}$ roughly preserves the ℓ_2-norm of S-sparse signals. Then, we have $\|\boldsymbol{A}\boldsymbol{x}\|_2^2 = \sum_{i,j \in S} x_i x_j \langle \boldsymbol{a}_i, \boldsymbol{a}_j \rangle \approx \|\boldsymbol{x}\|_2^2 = \sum_{i \in S} x_i^2$. This implies $\sum_{i \neq j} x_i x_j \langle \boldsymbol{a}_i, \boldsymbol{a}_j \rangle$ is small. For this to hold for all $\boldsymbol{x}$, $\langle \boldsymbol{a}_i, \boldsymbol{a}_j \rangle \approx 0$ should hold for $i \neq j$, which implies that the S columns are nearly orthogonal.

To understand the connection between RIP and CS, let us consider a situation where we aim to obtain S-sparse signals using sensing matrix $\boldsymbol{A}$. First, focus on the case with $2S$-sparse signals. The RIP of order $2S$ ensures that for any $2S$-sparse vector $\boldsymbol{x}$, $(1-\delta_{2S})\|\boldsymbol{x}\|_2^2 \leq \|\boldsymbol{A}\boldsymbol{x}\|_2^2 \leq (1+\delta_{2S})\|\boldsymbol{x}\|_2^2$ holds according to (7.10). For two S-sparse vectors $\boldsymbol{x}_1, \boldsymbol{x}_2$, their difference $\boldsymbol{x}_1 - \boldsymbol{x}_2$ is at most $2S$-sparse. Applying RIP to $\boldsymbol{x}_1 - \boldsymbol{x}_2$, we directly obtain $(1 - \delta_{2S})\|\boldsymbol{x}_1 - \boldsymbol{x}_2\|_2^2 \leq \|\boldsymbol{A}\boldsymbol{x}_1 - \boldsymbol{A}\boldsymbol{x}_2\|_2^2 \leq (1+\delta_{2S})\|\boldsymbol{x}_1 - \boldsymbol{x}_2\|_2^2$. This preserves the pairwise distance between S-sparse signals in the measurement space $\boldsymbol{A}\boldsymbol{x}$. In CS, recovering S-sparse signals from $\boldsymbol{A}\boldsymbol{x} = \boldsymbol{y}$ requires distinguishing between different S-sparse signals. If $\delta_{2S} \ll 1$, the distance preservation ensures that distinct S-sparse signals $\boldsymbol{x}_1, \boldsymbol{x}_2$ map to distinct measurement vectors $\boldsymbol{A}\boldsymbol{x}_1, \boldsymbol{A}\boldsymbol{x}_2$. This prevents ambiguity in recovery, i.e., if

[4] They cannot be exactly orthogonal because there are more columns than rows in $\boldsymbol{A}$, making exact orthogonality impossible.

distances were not preserved, different signals might map to nearly identical $\boldsymbol{y}$, making reconstruction impossible. With RIP, algorithms (e.g., ℓ_1 minimization) can leverage these measurements to reliably discriminate and recover S-sparse signals, as the preserved distances provide a clear "separation" between signals in the measurement space. Thus, the condition $\delta_{2S} \ll 1$ guarantees the existence of efficient, robust algorithms for CS.

If the RIP holds, then the following linear program gives an accurate reconstruction:

$$\begin{aligned} &\min_{\tilde{\boldsymbol{x}} \in \mathbb{R}^N} \|\tilde{\boldsymbol{x}}\|_1 \\ &\text{s.t. } \boldsymbol{A}\tilde{\boldsymbol{x}} = \boldsymbol{y}. \end{aligned} \tag{7.11}$$

Theorem 7.3 *([7]) Assume that $\delta_{2S} < \sqrt{2} - 1$, then the solution $\boldsymbol{x}^\star$ to (7.11) (noiseless sparse recovery) obeys*

$$\begin{aligned} \|\boldsymbol{x}^\star - \boldsymbol{x}\|_2 &\le C_0 \frac{\|\boldsymbol{x} - \boldsymbol{x}_S\|_1}{\sqrt{S}}, \\ \|\boldsymbol{x}^\star - \boldsymbol{x}\|_1 &\le C_0 \|\boldsymbol{x} - \boldsymbol{x}_S\|_1, \end{aligned} \tag{7.12}$$

for some constant C_0, where $\boldsymbol{x}_S$ is the vector $\boldsymbol{x}$ with all but the largest S components set to 0.

The conclusions of Theorem 7.3 are stronger than those of Theorem 7.2. Specifically, if $\boldsymbol{x}$ is S-sparse, then $\boldsymbol{x} = \boldsymbol{x}_S$ holds (where $\boldsymbol{x}_S$ sets all but the largest S components of $\boldsymbol{x}$ to 0). In this case, both theorems yield exact recovery. However, if $\boldsymbol{x}$ is not S-sparse, Theorem 7.2 may not apply, but Theorem 7.3, through the inequalities in (7.12), guarantees that the recovered $\boldsymbol{x}^\star$ approximates $\boldsymbol{x}$ as well as if one knew the location of $\boldsymbol{x}$'s S largest components in advance (an "oracle" with full knowledge of $\boldsymbol{x}$). This means the reconstruction quality depends on how well $\boldsymbol{x}$ is approximated by $\boldsymbol{x}_S$, making Theorem 7.3 applicable to all signals, not just sparse ones. Besides, it is important to mention that Theorem 7.3 is deterministic, it depends solely on the sensing matrix $\boldsymbol{A}$ satisfying a deterministic hypothesis (e.g., a specific RIP condition $\delta_{2S} < \sqrt{2} - 1$). Once $\boldsymbol{A}$ meets this condition, the recovery properties are guaranteed without any probabilistic "chance" element, i.e., they will hold every time, unlike results where success is "with high probability."

In addition, it is important to distinguish between these two theoretical guarantees. Theorem 7.2 is based on the coherence $\mu(\boldsymbol{\Phi}, \boldsymbol{\Psi})$ between the sensing and sparsity bases (coherence is a simpler, pairwise property that is easy to compute). It offers a probabilistic recovery guarantee for sparse signals when measurements are drawn uniformly at random from an incoherent basis (e.g., random Fourier samples). Theorem 7.3, on the other hand, employs the RIP (verifying RIP for a given matrix is NP-hard in general), quantified by the constant δ_{2S}. RIP provides a deterministic and uniform guarantee, i.e., if $\delta_{2S} < \sqrt{2} - 1$, then ℓ_1-minimization recovers all S-sparse signals exactly, and it also handles approximately sparse signals and noise (see Theorem 7.4).

What is currently lacking is the relationship between S (the number of components that can be effectively recovered, which conforms to the hypothesis) and the number of measurements M (the number of rows of the sensing matrix $\boldsymbol{A}$). To derive more powerful results, we hope to find sensing matrices that satisfy the RIP and have S values close to M. Because if S is close to M, it means that in CS, the number of recoverable signal components is similar to the number of measurements, which can enhance the efficiency of signal recovery.

To construct a good sensing matrix, randomization is a powerful tool. Consider the following sensing matrices:

1. form $\boldsymbol{A}$ by sampling N column vectors uniformly at random on the unit sphere of $\mathbb{R}^M$;
2. form $\boldsymbol{A}$ by sampling i.i.d. entries from the normal distribution with mean 0 and variance $1/M$;
3. form $\boldsymbol{A}$ by sampling a random projection $\boldsymbol{P}$ and then normalize, i.e., $\boldsymbol{A} = \sqrt{N/M}\boldsymbol{P}$;
4. form $\boldsymbol{A}$ by sampling i.i.d. entries from a symmetric Bernoulli distribution ($\boldsymbol{P}(A_{i,j} = \pm 1/\sqrt{M}) = 1/2$) or other sub-Gaussian distributions.

With overwhelming probability, all these matrices obey the RIP (i.e., the condition in Theorem 7.3) provided that

$$M \ge C \cdot S \log(N/S), \tag{7.13}$$

where C is some constant depending on each instance. The claims for (1)–(3) use fairly standard results in probability theory (an example of the proof is provided in Appendix 4), arguments for (4) are more subtle, please see [8, 9]. In all these examples, when condition (7.13) holds, the probability of sampling a matrix that does not satisfy the RIP decreases exponentially with the increase of M. This indicates that under this condition, the possibility of randomly sampling a matrix that satisfies RIP is extremely high. Interestingly, there are no sensing matrices and no reconstruction algorithms that can draw the conclusions of Theorem 7.3 with substantially fewer samples than $C \cdot S \log(N/S)$ [10, 11]. This emphasizes that the number of samples required by Theorem 7.3 is a lower bound, and it is difficult for other methods to achieve the same effect with far fewer samples. In this sense, random matrices combined with ℓ_1 minimization is a near-optimal sensing strategy.

One can also establish the RIP for pairs of orthobasis as discussed in Sect. 7.2.2. With $\boldsymbol{A} = \boldsymbol{R\Phi\Psi}$, where $\boldsymbol{R}$ extracts M coordinates uniformly at random, it is sufficient to have [10, 12]

$$M \geq C \cdot S(\log N)^4 \tag{7.14}$$

for the property $\delta_{2S} < \sqrt{2} - 1$ to hold with large probability. If one wants the failure probability to be no more than $O(N^{-\beta})$ (where $\beta > 0$), then in Eq. (7.14), the best-known exponent is 5 rather than 4 (it is believed that Eq. (7.14) may actually hold with just $\log N$, but the best result now is that the exponent is 5). This proves that one can stably and accurately reconstruct nearly sparse signals from dramatically undersampled data in an incoherent domain.

Finally, the RIP can also hold for sensing matrices $\boldsymbol{A} = \boldsymbol{\Phi\Psi}$, where $\boldsymbol{\Psi}$ is an arbitrary orthobasis and $\boldsymbol{\Phi}$ is an $M \times N$ measurement matrix drawn randomly from a suitable distribution. If one fixes $\boldsymbol{\Psi}$ and populates $\boldsymbol{\Phi}$ as in (1)–(4), then with overwhelming probability, the matrix $\boldsymbol{A} = \boldsymbol{\Phi\Psi}$ obeys the RIP provided that (7.13) is satisfied, where again C is some constant depending on each instance. These random measurement matrices $\boldsymbol{\Phi}$ are in a sense universal [8], the sparsity basis need not even be known when designing the measurement system!

7.3.2 Robust Signal Recovery from Noisy Data

The previous subsection mainly addresses the first critical element, i.e., how to deal with nearly sparse signals. In practice, we are usually given noisy data as in (7.9). To address this issue, we usually use ℓ_1 minimization with relaxed constraints for reconstruction, which leads to the following optimization problem:

$$\begin{aligned} &\min_{\tilde{\boldsymbol{x}} \in \mathbb{R}^N} \|\tilde{\boldsymbol{x}}\|_1 \\ &\text{s.t. } \|\boldsymbol{A}\tilde{\boldsymbol{x}} - \boldsymbol{y}\|_2 \leq \epsilon, \end{aligned} \tag{7.15}$$

where ϵ bounds the amount of noise in the data. Problem (7.15) is commonly called basis pursuit denoising (BPDN), which is closely related to LASSO after [13, 14] (will be introduced in the next chapter), which was first proposed in [15]. This is again a convex problem (a second-order cone program) and can be solved efficiently.

For problem (7.15), we have the following theorem.

Theorem 7.4 *([7]) Assume that $\delta_{2S} < \sqrt{2} - 1$. Then the solution $\boldsymbol{x}^\star$ to (7.15) obeys*

$$\|\boldsymbol{x}^\star - \boldsymbol{x}\|_2 \leq C_0 \cdot \|\boldsymbol{x} - \boldsymbol{x}_S\|_1/\sqrt{S} + C_1 \cdot \epsilon \tag{7.16}$$

for some constants C_0 and C_1.

Theorem 7.4 shows that the reconstruction error is bounded by the sum of two terms. The first is the error which would occur if one had noiseless data. The second is just proportional to the noise level. Further, the constants C_0 and C_1 are typically small. With $\delta_{2S} = 1/4$ for example, $C_0 \leq 5.5$ and $C_1 \leq 6$. Figure 7.4 shows a reconstruction from noisy data. As can be observed, non-zero coefficients (points away from origin) cluster tightly around the ideal (red dashed) line, which implies that accurate recovery of significant components can be achieved. Besides, zero coefficients (points near origin) concentrate at (0, 0), thus effective noise suppression is possible. This result establishes CS as a practical and robust sensing mechanism. It works with all kinds of not necessarily sparse signals, and it handles noise gracefully.

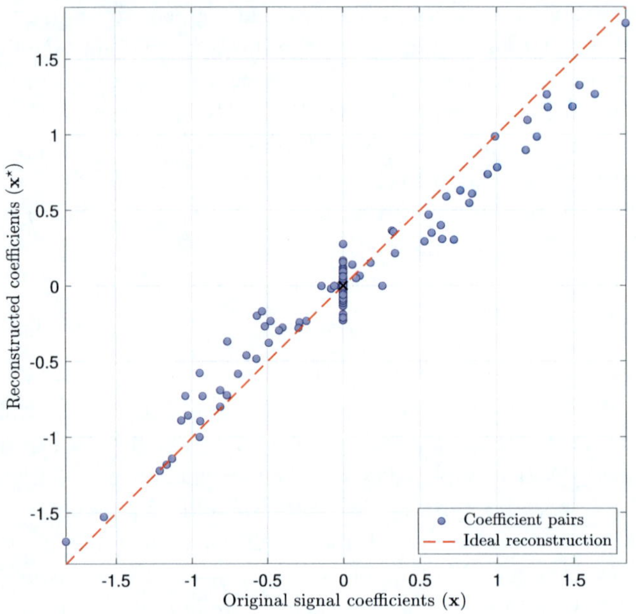

Fig. 7.4 (Adapted from Fig. 3 in [2]) A signal $\boldsymbol{x}$ (horizontal axis) and its reconstruction $\boldsymbol{x}^\star$ (vertical axis) obtained via solving problem (7.15). In this example, we set $N = 512$, $M = 256$, and $S = 64$. The sensing matrix has i.i.d. $\mathcal{N}(0, 1/M)$ entries and $\boldsymbol{z}$ is a Gaussian white noise vector adjusted so that $\|\boldsymbol{Ax}\|_2/\|\boldsymbol{z}\|_2 = 8$. Here, $\|\boldsymbol{x}^\star - \boldsymbol{x}\| \approx 1.95\epsilon$

Attention

In general, the coherence and RIP properties are different for sparse signal recovery, and their fundamental differences lies in their scope and strength, which are given as follows:

- Coherence is a local, pairwise property. It only looks at the correlation between two columns at a time, and it does not directly control how the matrix behaves on subsets of S columns. While RIP is a global property over all S-sparse subspaces. It constrains the behavior of $\boldsymbol{A}$ on every possible subset of S columns simultaneously. It ensures that every submatrix formed by S columns is well conditioned and near orthogonal.
- Coherence provides necessary conditions and non-uniform sufficient conditions. A low coherence is necessary for any recovery method to work. However, it does not guarantee that recovery will work equally well for all possible S-sparse signals under the same matrix $\boldsymbol{A}$ (non-uniform guarantee) [10]. RIP, however, provides strong uniform sufficient conditions. If $\boldsymbol{A}$ satisfies RIP with an appropriate δ_S, then recovery via convex optimization is guaranteed to work for all S-sparse signals $\boldsymbol{x}$ simultaneously (uniform guarantee), and it works robustly in the presence of noise and model mismatch.
- The coherence condition is in general easier to compute for a given matrix (only requires checking $N(N-1)/2$ inner products), and it is useful for designing matrices (e.g., ensuring low coherence) and providing quick checks or guarantees for simple recovery algorithms. In contrast, the RIP condition is very difficult to compute or verify for a given matrix (NP-hard in general [16]). It is primarily used as a powerful theoretical tool to prove that large classes of random matrices (e.g., Gaussian, Bernoulli, etc.) satisfy RIP with high probability when $M = O(S\log(N/S))$. This justifies the practical use of these matrices with convex optimization for robust recovery in noisy environments.

Summary

CS is a revolutionary sampling framework that fundamentally challenges the Nyquist–Shannon paradigm. It demonstrates that by leveraging inherent signal structure (sparsity) and designing measurements to be incoherent with this structure, the information bottleneck imposed by the Nyquist rate can be broken. Efficient computational recovery (e.g., ℓ_1-minimization) bridges the gap between incomplete measurements and accurate reconstruction, enabled by strong theoretical foundations (e.g., coherence, RIP). This paradigm shift enables efficient data acquisition where traditional sampling is impractical or wasteful.

The key innovations and findings are summarized as follows:

1. CS shifts the focus from signal bandwidth to signal sparsity. Real-world signals (communications channels, sensing targets, images, etc.) are often sparse or compressible when represented in a suitable domain (e.g., Fourier, Wavelets). This means their essential information can be captured by a small number of significant coefficients, discarding the rest with minimal loss. In other words, information content can be much lower than bandwidth suggests, and sampling should target this content, not the entire bandwidth.
2. CS relies on the principle of incoherence between the sensing basis ($\boldsymbol{\Phi}$, e.g., random projections, spikes) and the sparsity basis ($\boldsymbol{\Psi}$). High incoherence (low coherence μ) ensures that the sensing waveforms "spread out" the information of each sparse component across all measurements. Therefore, sensing in a domain maximally different from where the signal is sparse ensures each measurement captures a piece of global information, allowing dramatic undersampling $M \ll N$.
3. Recovering the sparse signal $\boldsymbol{x}$ from undersampled linear measurements $\boldsymbol{y} = \boldsymbol{\Phi}\boldsymbol{\Psi}\boldsymbol{x}$ is an ill-posed inverse problem. One important implication of CS is to solve this via ℓ_1-minimization, which promotes sparsity much more effectively than ℓ_2 (LS) and is computationally tractable (convertible to LP).
4. Performance of CS is rigorously guaranteed for sensing matrices that satisfy certain conditions. For the coherence condition, we can see from the above discussion that with high probability, exact recovery of S-sparse signals is possible from $M \geq C\mu^2(\boldsymbol{\Phi}, \boldsymbol{\Psi})S \log N$ random measurements in the noiseless case. While for the RIP condition, we know that matrices satisfying RIP (nearly preserving the norm of sparse vectors) ensure stable and robust recovery via ℓ_1-minimization, even for approximately sparse signals and in the presence of noise. Random sensing matrices (Gaussian, Bernoulli) satisfy RIP with $M \geq C \cdot S \log(N/S)$.
5. CS frameworks are able to handle real-world complexities. LASSO-type formulations (e.g., (7.15)) provide bounded error proportional to the noise level and the best S-term approximation error. Random sensing matrices ($\boldsymbol{\Phi}$) are often universal, meaning they work well with any fixed sparse basis ($\boldsymbol{\Psi}$), without requiring prior knowledge of the sparsity domain during acquisition.

Appendices

Appendix 1: Maximal Incoherence of Spikes and Sinusoids in Any Dimension

In the one-dimensional case, the inner product $\langle \boldsymbol{\phi}_k(t), \boldsymbol{\psi}_i(t) \rangle$ can be expressed as

$$\langle \boldsymbol{\phi}_k(t), \boldsymbol{\psi}_i(t) \rangle = \int \frac{1}{\sqrt{N}} \delta(t-k) \mathrm{e}^{j2\pi it/N} \mathrm{d}t = \frac{1}{\sqrt{N}} \mathrm{e}^{j2\pi ik/N}. \tag{7.17}$$

Thus, the magnitude is $|\langle \boldsymbol{\phi}_k(t), \boldsymbol{\psi}_i(t) \rangle| = \frac{1}{\sqrt{N}}$, which together with (7.5) leads to $\mu(\boldsymbol{\Phi}, \boldsymbol{\Psi}) = 1$.

In the two-dimensional space, the spike basis can be expressed as a two-dimensional Dirac delta function $\boldsymbol{\phi}_{k_x,k_y} = \delta(x-k_x, y-k_y)$, and the Fourier basis is $\boldsymbol{\psi}_{j_x,j_y} = \frac{1}{\sqrt{N}} \mathrm{e}^{j2\pi(j_x x + j_y y)/N}$. Their inner product is given by

$$\langle \boldsymbol{\phi}_{k_x,k_y}, \boldsymbol{\psi}_{j_x,j_y} \rangle = \iint \frac{1}{\sqrt{N}} \delta(x - k_x, y - k_y) \mathrm{e}^{j2\pi(j_x x + j_y y)/N} \mathrm{d}x\mathrm{d}y$$
$$= \frac{1}{\sqrt{N}} \mathrm{e}^{j2\pi(j_x k_x + j_y k_y)/N}, \tag{7.18}$$

and its absolute value is still $\frac{1}{\sqrt{N}}$. At this time, the coherence is also $\mu = \sqrt{N} \cdot \frac{1}{\sqrt{N}} = 1$. The same argument can be easily extended to higher dimensions. Therefore, the spike basis and Fourier basis are maximally incoherent in any dimension.

Appendix 2: Definitions of Haar, Daubechies D4/D8 Wavelets, and Noiselets

The Haar wavelet system is an orthonormal basis for $L^2(\mathbb{R})$ (the space of square-integrable functions) consisting of the scaling function $\phi(t)$ and the wavelet function $\psi(t)$, which are given by [17]

$$\phi(t) = \begin{cases} 1 & \text{if } 0 \le t < 1 \\ 0 & \text{otherwise,} \end{cases}$$

and

$$\psi(t) = \begin{cases} 1 & \text{if } 0 \le t < \frac{1}{2} \\ -1 & \text{if } \frac{1}{2} \le t < 1 \\ 0 & \text{otherwise,} \end{cases}$$

respectively. The full basis is generated by dyadic dilations and translations as follows:

$$\psi_{j,k}(t) = 2^{j/2}\psi(2^j t - k), \quad j, k \in \mathbb{Z},$$

where $j \in \mathbb{Z}$ is the scale parameter (dilates/compresses the wavelet) and $k \in \mathbb{Z}$ is the translation parameter (shifts the wavelet). This creates a set of basis functions that can analyze a signal at different resolutions (scales) and at different locations (translations).

The Haar transform is the process of projecting a function or a discrete signal onto the basis formed by the Haar wavelets and scaling functions. For a continuous function $f(t)$, the transform calculates the inner products (coefficients) of the function with the entire family of continuous wavelets $\psi_{j,k}(t)$ and scaling functions, which is more theoretical. For a discrete signal $\boldsymbol{x}$ of length N (e.g., a digital image or audio sample), the transform decomposes the signal into its wavelet coefficients, which is widely used in practical applications. The 1D discrete Haar transform can be computed by simply multiplying the signal vector by the Haar matrix, i.e.,

$$\boldsymbol{y} = \boldsymbol{H}\boldsymbol{x}, \tag{7.19}$$

where $\boldsymbol{H}$ is the $N \times N$ Haar matrix and $\boldsymbol{y}$ is the output vector of Haar coefficients. The Haar matrix $\boldsymbol{H}$ is a discrete, finite-sized matrix that implements the Haar Transform for signals of a specific length N, where N is a power of 2 (e.g., 2, 4, 8, 16, ...). It is an $N \times N$ orthogonal matrix, and its rows are formed by discretizing the scaled and translated Haar wavelet and scaling functions at N points. Specifically, its first row is always the scaling vector, which computes the global average of the entire signal (normalized). The subsequent rows are the wavelet vectors, which compute the differences between various segments of the signal at different scales. The Haar matrix can be structured hierarchically, from the coarsest scale (largest averages and differences) to the finest scale (smallest, most localized differences). In particular, the 2×2 Haar matrix (unnormalized $\boldsymbol{H}_2$) can be expressed as

$$\boldsymbol{H}_2 = \begin{bmatrix} 1 & 1 \\ 1 & -1 \end{bmatrix}, \tag{7.20}$$

and the general $2N \times 2N$ Haar matrix can be derived by the following equation:

$$\boldsymbol{H}_{2N} = \begin{bmatrix} \boldsymbol{H}_N \otimes [1, 1] \\ \boldsymbol{I}_N \otimes [1, -1] \end{bmatrix},$$

where $\boldsymbol{I}_N$ is the size-N identity matrix.

Daubechies wavelets are a family of orthogonal wavelets with compact support, defined by their filter coefficients [17]. For D4, the scaling function coefficients are expressible in exact form, involving square roots. These coefficients are

$$h_0 = \frac{1+\sqrt{3}}{4\sqrt{2}}, \quad h_1 = \frac{3+\sqrt{3}}{4\sqrt{2}},$$
$$h_2 = \frac{3-\sqrt{3}}{4\sqrt{2}}, \quad h_3 = \frac{1-\sqrt{3}}{4\sqrt{2}}.$$

The scaling function $\phi(t)$ satisfies the following dilation equation:

$$\phi(t) = \sqrt{2}\sum_{k=0}^{3} h_k \phi(2t-k),$$

and the wavelet function $\psi(t)$ is derived as

$$\psi(t) = \sqrt{2}\sum_{k=0}^{3} (-1)^k h_{3-k} \phi(2t-k).$$

For Daubechies D8, the scaling function has eight coefficients, which are typically provided numerically due to the complexity of their exact expressions. These coefficients are

$$h_0 \approx 0.23037781, \quad h_1 \approx 0.71484657,$$
$$h_2 \approx 0.63088076, \quad h_3 \approx -0.02798376,$$
$$h_4 \approx -0.18703481, \quad h_5 \approx 0.03084138,$$
$$h_6 \approx 0.03288301, \quad h_7 \approx -0.01059740.$$

The scaling function follows a similar dilation equation given by

$$\phi(t) = \sqrt{2}\sum_{k=0}^{7} h_k \phi(2t-k)$$

with the wavelet function defined analogously. The numerical values reflect the practical implementation in computational tools, where exact forms involving higher-order roots are less common and often computed via algorithms satisfying orthogonality and vanishing moment conditions.

Noiselets are functions designed for their incoherence with wavelet bases, particularly Haar wavelets, making them ideal for compressive sensing applications [3]. For noiselets, the mother bases function $\chi(x)$ is defined as

$$\chi(x) = \begin{cases} 1 & x \in [0, 1) \\ 0 & \text{otherwise,} \end{cases}$$

and the family of noiselets is constructed recursively as follows:

$$\begin{aligned} f_1(x) &= \chi(x) \\ f_{2n}(x) &= (1-i)f_n(2x) + (1+i)f_n(2x-1) \\ f_{2n+1}(x) &= (1+i)f_n(2x) + (1-i)f_n(2x-1). \end{aligned}$$

This recursive definition ensures unitarity and a fast transform with $O(n \log n)$ complexity, making noiselets useful as a sampling basis for signals that are sparse in the Haar domain.

Noiselet can be extended and discretized. The extended function $f_m(k, l)$ is defined as follows:

$$f_m(1,l) = \begin{cases} 1 & l = 0, \dots, 2^M - 1, \\ 0 & \text{otherwise}, \end{cases}$$

$$f_m(2k,l) = (1-i) f_m(k, 2l) + (1+i) f_m(k, 2l - 2^M),$$

$$f_m(2k+1,l) = (1+i) f_m(k, 2l) + (1-i) f_m(k, 2l - 2^M).$$

Using extended noiselet $f_m(k,l)$, we can generate the following $n \times n$ noiselet matrix N_n, where n is a power of two:

$$N_1 = [1],$$

$$N_n(k,:) = \frac{1}{2}[1-i, 1+i] \otimes N_{n/2}\left(\frac{k}{2}, :\right), \quad \text{when } k = 0, 2, 4, \dots, n-2,$$

$$N_n(k,:) = \frac{1}{2}[1+i, 1-i] \otimes N_{n/2}\left(\frac{k-1}{2}, :\right), \quad \text{when } k = 1, 3, \dots, n-1.$$

Suppose $2^m > n$, we can find that $N_n(k,l)$ is equal to $f_m\left(n+k, \frac{2^m}{n} l\right)$.

Appendix 3: Reformulation of ℓ_1-Norm Minimization with Equality Constraints as LP

This transformation is based on the linearity of the constraint and the decomposability of the ℓ_1 norm into non-negative components. Specifically, we start by decomposing each real-valued variable $\tilde{x}_i$ into the difference of two non-negative variables, i.e., $\tilde{x}_i = u_i - v_i$, where $u_i \geq 0$ and $v_i \geq 0$, and in the meantime let $|\tilde{x}_i| = u_i + v_i$. In this case, u_i and v_i cannot both be zeros, i.e., if $\tilde{x}_i \geq 0$, it must hold that $v_i = 0$ and $u_i = \tilde{x}_i$, while if $\tilde{x}_i < 0$, $u_i = 0$ and $v_i = -\tilde{x}_i$ should be satisfied.

Then, by substituting $\tilde{x}_i = u_i - v_i$ into the ℓ_1 norm, the objective function $\|\tilde{\boldsymbol{x}}\|_1$ becomes a linear function of u_i and v_i, i.e., $\|\tilde{\boldsymbol{x}}\|_1 = \sum_{i=1}^{N}(u_i + v_i) = \mathbf{1}^T\boldsymbol{u} + \mathbf{1}^T\boldsymbol{v}$, where $\boldsymbol{u} = [u_1, u_2, \dots, u_N]^T$, $\boldsymbol{v} = [v_1, v_2, \dots, v_N]^T$, and $\mathbf{1}$ is a vector of ones. Next, the linear equality constraint $y_k = \langle \boldsymbol{\phi}_k, \boldsymbol{\Psi}\tilde{\boldsymbol{x}} \rangle$ can be rewritten as $y_k = \langle \boldsymbol{\phi}_k, \boldsymbol{\Psi}(\boldsymbol{u} - \boldsymbol{v}) \rangle$. Therefore, problem (7.7) can be equivalently rewritten as the following LP problem:

$$\begin{aligned} \min_{\boldsymbol{u},\boldsymbol{v}\in\mathbb{R}^N} \ & \mathbf{1}^T\boldsymbol{u} + \mathbf{1}^T\boldsymbol{v} \\ \text{s.t. } & y_k = \langle \boldsymbol{\phi}_k, \boldsymbol{\Psi}(\boldsymbol{u} - \boldsymbol{v}) \rangle, \\ & \boldsymbol{u} \geq \mathbf{0}, \quad \boldsymbol{v} \geq \mathbf{0}. \end{aligned} \tag{7.21}$$

Appendix 4: Proof of RIP for i.i.d. Gaussian Sensing Matrices

To prove that $\delta_{2S} < \sqrt{2} - 1$ when $M \geq C \cdot S \log(N/S)$ for a sensing matrix $\boldsymbol{A}$ with i.i.d. $\mathcal{N}(0, 1/M)$ entries, follow the following three steps.

First, for any $2S$-sparse vector $\boldsymbol{x}$, the RIP states that $(1-\delta_{2S})\|\boldsymbol{x}\|_2^2 \leq \|\boldsymbol{A}\boldsymbol{x}\|_2^2 \leq (1+\delta_{2S})\|\boldsymbol{x}\|_2^2$. By normalizing $\|\boldsymbol{x}\|_2 = 1$, we ought to show $|\|\boldsymbol{A}\boldsymbol{x}\|_2^2 - 1| < \sqrt{2} - 1$, which then implies $\delta_{2S} < \sqrt{2} - 1$.

Second, let $Y = \|\boldsymbol{A}\boldsymbol{x}\|_2^2 = \sum_{i=1}^{M} y_i^2$, where $y_i = \sum_{j=1}^{N} A_{ij}x_j$. Since $A_{ij} \sim \mathcal{N}(0, 1/M)$, y_i follows $\mathcal{N}(0, \frac{\|\boldsymbol{x}\|_2^2}{M})$. Let $w_i = \frac{y_i}{\|\boldsymbol{x}\|_2/\sqrt{M}}$, we have $w_i \sim \mathcal{N}(0,1)$, and $Y = \frac{\|\boldsymbol{x}\|_2^2}{M}\sum_{i=1}^{M} w_i^2$. Using the Chernoff bound for Y, we have

$$\mathbb{P}(|Y-1| \geq \sqrt{2}-1) \leq 2\mathrm{e}^{-Mf(\sqrt{2}-1)}, \tag{7.22}$$

where $f(\delta) = \frac{1+\delta}{2}\log(1+\delta) - \frac{\delta}{2}$. For $\delta = \sqrt{2}-1$, $f(\sqrt{2}-1)$ is a positive constant.

Third, the number of $2S$-sparse vectors, denoted by $\binom{N}{2S}$, satisfies $\binom{N}{2S} \leq \left(\frac{eN}{2S}\right)^{2S}$. Applying the union bound, we can obtain

$$\mathbb{P}\left(\exists\, 2S\text{-sparse } \boldsymbol{x},\ |\|\boldsymbol{A}\boldsymbol{x}\|_2^2 - 1| \geq \sqrt{2}-1\right) \leq \left(\frac{eN}{2S}\right)^{2S} \cdot 2\mathrm{e}^{-Mf(\sqrt{2}-1)}. \tag{7.23}$$

Take the natural logarithm of $\left(\frac{eN}{2S}\right)^{2S} \cdot 2e^{-Mf(\sqrt{2}-1)} < 1$, it yields

$$2S \log\left(\frac{eN}{2S}\right) + \log 2 - Mf(\sqrt{2}-1) < 0. \tag{7.24}$$

By ignoring the constants e and 2, we can simplify $\log\left(\frac{eN}{2S}\right) \approx \log(N/S)$. Then, we have

$$2S \log(N/S) - Mf(\sqrt{2}-1) + \log 2 < 0. \tag{7.25}$$

Solving inequality(7.25) for M yields $M \geq \frac{2S\log(N/S)}{f(\sqrt{2}-1)}$. Let $C = \frac{2}{f(\sqrt{2}-1)}$, so $M \geq C \cdot S \log(N/S)$ ensures $\delta_{2S} < \sqrt{2}-1$ with high probability for all $2S$-sparse vectors. This completes the proof.

Homework Problems

7.1 Focus on the Shannon–Nyquist sampling theorem.

(a) Prove that the Fourier transform of an impulse train satisfies

$$\mathcal{F}\left\{\sum_{n=-\infty}^{\infty} \delta(t-nT_s)\right\} = \frac{1}{T_s}\sum_{k=-\infty}^{\infty} \delta\left(f - \frac{k}{T_s}\right),$$

where $\mathcal{F}\{x(t)\} = \int_{-\infty}^{\infty} x(t)e^{-j2\pi ft}dt$ and $\delta(\cdot)$ is the Dirac delta function. Use Fourier series properties in your proof.

(b) Consider a discrete impulse sequence

$$x[n] = \begin{cases} 1 & \text{for } n = 0, 4, 8, 12 \quad (0 \leq n \leq 15) \\ 0 & \text{otherwise,} \end{cases}$$

compute the 16-point DFT of $x[n]$ and show it equals

$$\frac{1}{4}[1, 0, 0, 0, 1, 0, 0, 0, 1, 0, 0, 0, 1, 0, 0, 0].$$

Explain how this verifies the discrete version of the result in (a). Compute the ℓ_∞ norm $\|X[k]\|_\infty$ of the DFT result.

(c) For the signal:

$$x(t) = \begin{cases} 1 & |t| \leq T_p \\ 0 & \text{otherwise,} \end{cases}$$

derive its Fourier transform $X(f)$. Prove energy conservation: $\int_{-\infty}^{\infty} |x(t)|^2 dt = \int_{-\infty}^{\infty} |X(f)|^2 df$. Sketch $|x(t)|$ and $|X(f)|$ for $T_p = 2$ ms. Calculate the ℓ_1, ℓ_2, and ℓ_∞ norms of $x(t)$.

(d) For the frequency response:

$$X(f) = \begin{cases} 1 & |f| \leq W \\ 0 & \text{otherwise,} \end{cases}$$

derive the inverse Fourier transform $x(t)$. Prove energy conservation. Sketch $|X(f)|$ and $|x(t)|$ for $W = 500$ Hz. Determine the time-domain ℓ_∞ norm $\|x(t)\|_\infty$.

(e) Using results from (a)–(d), derive the Nyquist–Shannon sampling theorem. Prove that perfect reconstruction requires $f_s > 2W$. For a bandlimited signal with $W = 100$ Hz, find the critical sampling rate. Calculate the ℓ_2 norm of the sinc reconstruction kernel and determine the ℓ_∞ norm of the sampling comb.

7.2 Consider the following underdetermined linear system:

$$y = [1\ 2]\begin{bmatrix} x_1 \\ x_2 \end{bmatrix} = 2. \tag{7.26}$$

Let $\boldsymbol{x} = [x_1, x_2]^T$ denote the solution vector.

(a) Define the ℓ_0, ℓ_1, ℓ_2, $\ell_{1/2}$, and ℓ_∞ norms mathematically.
(b) Sketch the boundaries of the ℓ_p balls for $p = \infty, 2, 1, 1/2$ in the 2D plane (label each clearly).
(c) Find all solutions for each minimization problem:

- $\min \|\boldsymbol{x}\|_0$ subject to $x_1 + 2x_2 = 2$,
- $\min \|\boldsymbol{x}\|_1$ subject to $x_1 + 2x_2 = 2$,
- $\min \|\boldsymbol{x}\|_{1/2}$ subject to $x_1 + 2x_2 = 2$,
- $\min \|\boldsymbol{x}\|_\infty$ subject to $x_1 + 2x_2 = 2$.

(d) For each solution in (c), plot it on the sketch from (b), explain derivation geometrically (intersection of constraint line $x_1 + 2x_2 = 2$ with ℓ_p unit ball scaled appropriately) and algebraically (using optimization techniques for each norm).
(e) Answer analytically. Is the ℓ_0 solution unique? Why? Is the ℓ_1 solution unique? Does it match ℓ_0? Is the ℓ_∞ solution unique? Under what conditions does it match ℓ_0?
(f) What geometric property makes ℓ_1 solutions coincide with ℓ_0 solutions for this system?

7.3 This problem focuses on sparsity and incoherence validation.

(a) Using MATLAB, generate a length $N = 256$ signal $\boldsymbol{x}$ with $S = 10$ non-zero entries (random positions, unit magnitudes).
(b) Compute its representation in: (i) Fourier basis $\boldsymbol{\Psi}_{\text{DFT}}$; (ii) Haar wavelet basis $\boldsymbol{\Psi}_{\text{Haar}}$.
(c) For each basis, plot the sorted magnitude of coefficients. Verify if $\boldsymbol{x}$ is sparse in these two bases.
(d) Compute coherence $\mu(\boldsymbol{\Phi}, \boldsymbol{\Psi})$ between (i) spike basis $\boldsymbol{\Phi}$ and $\boldsymbol{\Psi}_{\text{DFT}}$; (ii) noiselet basis $\boldsymbol{\Phi}_{\text{noiselet}}$ and $\boldsymbol{\Psi}_{\text{Haar}}$ using $\mu = \sqrt{N} \cdot \max_{i,j} |\langle \boldsymbol{\phi}_i, \boldsymbol{\psi}_j \rangle|$. Confirm $\mu \approx 1$ for spikes/Fourier and noiselet/Haar.

7.4 ℓ_1 versus ℓ_0 recovery (noiseless case)

(a) Design a Gaussian sensing matrix $\boldsymbol{\Phi} \in \mathbb{R}^{M \times N}$ with $M = 8$, $N = 32$, entries $\sim \mathcal{N}(0, 1/M)$.
(b) For $S = \{1, 3, 5, 7\}$, generate S-sparse $\boldsymbol{x}$ (random support, Gaussian amplitudes). Compute measurements $\boldsymbol{y} = \boldsymbol{\Phi x}$. Solve ℓ_0 minimization problem using exhaustive search over all S-sparse supports and ℓ_1 minimization problem using CVX [18], subject to $\boldsymbol{\Phi x} = \boldsymbol{y}$.
(c) Record success rate (exact recovery: $\|\boldsymbol{x} - \boldsymbol{x}^\star\|_2 < 10^{-6}$) over 100 trials. Plot S versus success rate for both methods.
(d) Verify Theorem 7.2: for $\mu = 1$, does $M \geq CS \log N$ hold empirically? Estimate C. Repeat this simulation for ℓ_1 minimization when i) $M = 32$, $N = 256$, and $S = \{1, 3, 5, 7\}$ and ii) $M = 128$, $N = 1024$, and $S = \{12, 16, 18, 20, 24, 32\}$.

7.5 Consider a frequency division duplex (FDD) massive MIMO system where the BS is equipped with $N_t = 128$ antennas (ULA with half-wavelength spacing), and the user equipment (UE) has a single antenna. For channel estimation, the BS transmits T pilot vectors $\{\boldsymbol{p}_1, \boldsymbol{p}_2, \ldots, \boldsymbol{p}_T\}$, where $\boldsymbol{p}_t \in \mathbb{C}^{N_t \times 1}$ and the UE receives

$$y_t = \boldsymbol{h}^H \boldsymbol{p}_t + n_t, \quad t = 1, 2, \ldots, T.$$

The following angle-domain sparse channel model is considered:

$$\boldsymbol{h} = \sum_{i=1}^{L_p} \alpha_i \boldsymbol{a}(\theta_i) \in \mathbb{C}^{N_t \times 1},$$

where $L_p = 6$ is the number of channel paths, $\alpha_i \sim \mathcal{CN}(0, 1)$ is the complex channel gain, $\theta_i \in [-60°, 60°]$ is the AoA, and $\boldsymbol{a}(\theta_i)$ is the ULA steering vector

$$[\boldsymbol{a}(\theta_i)]_m = \mathrm{e}^{-j2\pi m \frac{d}{\lambda} \sin\theta_i}, \quad d = \lambda/2, \quad m = 0, 1, \ldots, N_t - 1.$$

Consider the compressive regime, i.e., $T \ll N_t$ (e.g., $T = 10 < 128$), the corresponding measurement vector at the user is given by

$$\boldsymbol{y} = \begin{bmatrix} y_1 \\ \vdots \\ y_T \end{bmatrix} = \underbrace{\begin{bmatrix} \boldsymbol{p}_1^H \\ \vdots \\ \boldsymbol{p}_T^H \end{bmatrix}}_{\boldsymbol{\Phi} \in \mathbb{C}^{T \times N_t}} \boldsymbol{h} + \boldsymbol{n}.$$

(a) When is $\boldsymbol{h}$ sparse in the DFT basis $\boldsymbol{F}$, i.e.,

$$\boldsymbol{h} = \boldsymbol{F}\boldsymbol{x}, \quad \boldsymbol{x} \in \mathbb{C}^{N_t \times 1} \quad (L_p\text{-sparse}).$$

Rewrite measurements as

$$\boldsymbol{y} = \boldsymbol{\Phi}\boldsymbol{F}\boldsymbol{x} + \boldsymbol{n} = \boldsymbol{A}\boldsymbol{x} + \boldsymbol{n},$$

and answer what property should $\boldsymbol{\Phi}$ satisfy for CS recovery?

(b) Derive the LS solution for $\boldsymbol{h}$. Meanwhile, by assuming known channel covariance matrix $\boldsymbol{R}_{\boldsymbol{hh}} = \mathbb{E}\{\boldsymbol{h}\boldsymbol{h}^H\}$, derive the LMMSE solution for $\boldsymbol{h}$. Why is $\boldsymbol{R}_{\boldsymbol{hh}}$ difficult to estimate in practice?

(c) Use CS-based estimation method to solve

$$\min_{\boldsymbol{x}} \frac{1}{2}\|\boldsymbol{y} - \boldsymbol{A}\boldsymbol{x}\|_2^2 + \lambda\|\boldsymbol{x}\|_1$$

with $\lambda = \sigma_n\sqrt{2\log N_t}$.

(d) Simulations and analysis: given parameters $N_t = 128$, $L_p = 6$, $T = 10$, and $\mathrm{SNR} = 10$ dB, using Gaussian ($[\boldsymbol{p}_t]_i \sim \mathcal{CN}(0, 1/N_t)$) and DFT-based pilot design (random DFT columns), do the following experiments:

- vary $T \in \{6, 8, 10, 12, 14\}$: Plot NMSE for CS-based method, LS, LMMSE;
- vary SNR $\in \{0, 10, 20, 30\}$ dB: Compare all methods;
- beamspace visualization: plot $|\boldsymbol{x}|$ (true versus estimated)

utilizing the NMSE $\left(\mathbb{E}\left\{\|\boldsymbol{h} - \hat{\boldsymbol{h}}\|_2^2/\|\boldsymbol{h}\|_2^2\right\}\right)$ metrics.

References

1. C.E. Shannon, Communication in the presence of noise. Proc. IRE **37**(1), 10–21 (1949)
2. E.J. Candès, M.B. Wakin, An introduction to compressive sampling. IEEE Signal Process. Mag. **25**(2), 21–30 (2008)
3. R. Coifman, F. Geshwind, Y. Meyer, Noiselets. Appl. Comput. Harmon. Anal. **10**(1), 27–44 (2001)
4. E. Candès, J. Romberg, Sparsity and incoherence in compressive sampling. Inverse Prob. **23**(3), 969 (2007)
5. E.J. Candès, J. Romberg, T. Tao, Robust uncertainty principles: exact signal reconstruction from highly incomplete frequency information. IEEE Trans. Inf. Theory **52**(2), 489–509 (2006)
6. E.J. Candès, T. Tao, Decoding by linear programming. IEEE Trans. Inf. Theory **51**(12), 4203–4215 (2005)
7. E.J. Candès, J.K. Romberg, T. Tao, Stable signal recovery from incomplete and inaccurate measurements. Comm. Pure Appl. Math. **59**, 1207–1223 (2006)
8. R. Baraniuk, M. Davenport, R. DeVore, M. Wakin, A simple proof of the restricted isometry property for random matrices. Constr. Approx. **28**, 253–263 (2008)
9. S. Mendelson, A. Pajor, N. Tomczak-Jaegermann, Uniform uncertainty principle for Bernoulli and subgaussian ensembles. Constr. Approx. **28**, 277–289 (2008)
10. E.J. Candès, T. Tao, Near-optimal signal recovery from random projections: universal encoding strategies? IEEE Trans. Inf. Theory **52**(12), 5406–5425 (2006)
11. D.L. Donoho, Compressed sensing. IEEE Trans. Inf. Theory **52**(4), 1289–1306 (2006)
12. M. Rudelson, R. Vershynin, On sparse reconstruction from Fourier and Gaussian measurements. Comm. Pure Appl. Math. J. Issu. Courant Inst. Math. Sci. **61**(8), 1025–1045 (2008)
13. R. Tibshirani, Regression shrinkage and selection via the LASSO. J. R. Stat. Soc. Ser. B Stat Methodol. **58**(1), 267–288 (1996)
14. S.S. Chen, D.L. Donoho, M.A. Saunders, Atomic decomposition by basis pursuit. SIAM Rev. **43**(1), 129–159 (2001)
15. F. Santosa, W.W. Symes, Linear inversion of band-limited reflection seismograms. SIAM J. Sci. Stat. Comput. **7**(4), 1307–1330 (1986)

16. A.M. Tillmann, M.E. Pfetsch, The computational complexity of the restricted isometry property, the nullspace property, and related concepts in compressed sensing. IEEE Trans. Inf. Theory **60**(2), 1248–1259 (2013)
17. S. Mallat, *A Wavelet Tour of Signal Processing: The Sparse Way*, 3rd edn. (Academic Press, 2009)
18. CVX Research, Inc., CVX: Matlab software for disciplined convex programming, version 2.0. https://cvxr.com/cvx, April 2011

8 Optimization Based CS Recovery Algorithms

This chapter presents optimization-based algorithms for CS signal recovery. We first introduce reweighted ℓ_1-minimization, an iterative technique that enhances sparsity recovery by adaptively adjusting weights based on intermediate solutions, which significantly outperforms standard ℓ_1-minimization with fewer measurements. Theoretical connections to MM algorithms and the log-sum penalty function are established, and its improved performance is explained. Then, this chapter extensively covers the LASSO method, its formulation, Bayesian interpretation, coordinate descent solution, soft-thresholding operator, and theoretical guarantees (prediction consistency, sparsistency) are detailed. Further, we explore structured sparsity extensions, including group LASSO for group-wise variable selection and Burst LASSO tailored for burst-sparse signals encountered in applications like massive MIMO channel estimation. The burst LASSO employs a block-sparse lifting transformation and is analyzed using the convex Gaussian min-max theorem (CGMT), demonstrating substantial gains in pilot efficiency and phase transition behavior over standard methods. Performance metrics like asymptotic normalized squared error (aNSE) are analyzed for generalized formulations.

8.1 Reweighted ℓ_1-Norm Minimization

From the previous chapter, it is well understood that it is possible to reconstruct sparse signals exactly from what appear to be highly incomplete sets of linear measurements and that this can be done by constrained ℓ_1 minimization, as given in (7.11). In this section, a novel method is introduced, known as the reweighted ℓ_1-norm minimization, for sparse signal recovery that in many situations outperforms ℓ_1 minimization in the sense that substantially fewer measurements are needed for exact recovery.

8.1.1 Weighted ℓ_1-Norm Minimization

First consider the following "weighted" ℓ_1 minimization problem:

$$\begin{aligned} &\min_{\tilde{\boldsymbol{x}}\in\mathbb{R}^N} \sum_{i=1}^{N} w_i|\tilde{x}_i| \\ &\text{s.t. } \boldsymbol{y} = \boldsymbol{A}\tilde{\boldsymbol{x}}, \end{aligned} \tag{8.1}$$

where $w_1, w_2, \ldots, w_N$ are positive weights. Just like its "unweighted" counterpart in (7.11), this convex problem can be recast as a linear program as discussed in the appendix of Chap. 7. In the sequel, we denote the objective function $\sum_{i=1}^{N} w_i|\tilde{x}_i|$ by $\|\boldsymbol{W}\tilde{\boldsymbol{x}}\|_1$ where $\boldsymbol{W}$ is the diagonal matrix with $w_1, w_2, \ldots, w_N$ on the diagonal and zeros elsewhere.

The weighted ℓ_1 minimization problem (8.1) can be viewed as a relaxation of a weighted ℓ_0 minimization problem as given below

$$\begin{aligned} &\min_{\tilde{\boldsymbol{x}}\in\mathbb{R}^N} \|\boldsymbol{W}\tilde{\boldsymbol{x}}\|_0 \\ &\text{s.t. } \boldsymbol{y} = \boldsymbol{A}\tilde{\boldsymbol{x}}. \end{aligned} \tag{8.2}$$

Whenever the solution to

A. Liu and M.-M. Zhao, *Advanced Bayesian Signal Processing*, Textbooks in Telecommunication Engineering,
https://doi.org/10.1007/978-3-032-26656-9_8

$$\begin{aligned} \min_{\tilde{x} \in \mathbb{R}^N} \; & \|\tilde{x}\|_0 \\ \text{s.t. } & y = A\tilde{x} \end{aligned} \tag{8.3}$$

is unique, it is also the unique solution to (8.2) provided that the weights do not equal to zero. This is intuitive because the ℓ_0 norm counts the number of non-zero elements in a vector, and for the weighted ℓ_0 norm $\|W\tilde{x}\|_0$, where $w_i > 0, \forall i$, the weighting operation does not alter the sparsity pattern of $\tilde{x}$, i.e., weighting scales the magnitudes of non-zero entries but preserves their existence. However, the corresponding ℓ_1 relaxations (7.11) and (8.1) will have different solutions in general, because ℓ_1-norm penalizes all non-zero entries uniformly, while the weights $w_i > 0$ assign non-uniform penalties to different coordinates, and larger weights amplify the cost of non-zero entries, incentivizing sparser solutions at those positions. In other words, the weighting fundamentally alters the optimization landscape, leading to different minimizers. Hence, one may think of the weights $\{w_i\}$ as free parameters in the convex relaxation, whose values, if set wisely, could improve the signal reconstruction.

One possible use for the weights could be to counteract the influence of the signal magnitude on the ℓ_1 penalty function. Suppose, for example, that the weights were inversely proportional to the true signal magnitude, i.e.,

$$w_i = \begin{cases} \frac{1}{|x_{0,i}|}, & x_{0,i} \neq 0, \\ \infty, & x_{0,i} = 0, \end{cases} \tag{8.4}$$

where $x_0 = [x_{0,1}, x_{0,2}, \ldots, x_{0,N}]^T$ is the true signal. If x_0 is S-sparse, i.e., obeys $\|x_0\|_0 \leq S$, then (8.1) is guaranteed to find the correct solution with this choice of weights, assuming only that $M \geq S$ and no M-dimensional subspace defined by A contains "degenerate" geometric configurations, i.e., any subset of S columns of A is linearly independent (since $M \geq S$). Specifically, for indices i where $x_{0,i} = 0$, the weight $w_i = \infty$. This imposes an infinite penalty on any non-zero value at these positions in the minimization of

$$\sum_{i=1}^{N} w_i|\tilde{x}_i| = \sum_{x_{0,i}\neq 0} \frac{|\tilde{x}_i|}{|x_{0,i}|} + \sum_{x_{0,i}=0} \infty \cdot |\tilde{x}_i|. \tag{8.5}$$

To avoid the objective function diverging to infinity, the solution x must satisfy $\tilde{x}_i = 0$, for all i where $x_{0,i} = 0$. For indices i where $x_{0,i} \neq 0$, the weight $w_i = 1/|x_{0,i}|$. The objective function becomes $\sum_{x_{0,i}\neq 0} \frac{|\tilde{x}_i|}{|x_{0,i}|}$, which is equivalent to minimizing the relative deviation of $\tilde{x}_i$ from $x_{0,i}$. If $\tilde{x}$ matches x_0 on its support (i.e., $\tilde{x}_i = x_{0,i}$ for $x_{0,i} \neq 0$), the objective value equals S (the sparsity level), which is the smallest possible value.

Attention
Why other solutions cannot be better? Assume we adjust the coefficients only within the support set $\mathcal{S} = \{i | x_{0,i} \neq 0\}$, i.e., $\tilde{x}_i = \alpha_i x_{0,i}$ (with at least one $\alpha_i \neq 1$). The measurement constraint becomes

$$y = A\tilde{x} = \sum_{i\in\mathcal{S}} \alpha_i x_{0,i} a_i. \tag{8.6}$$

To satisfy $y = A\tilde{x}$, the coefficients $\{\alpha_i\}$ must satisfy the following linear system:

$$\sum_{i\in\mathcal{S}} (\alpha_i - 1) x_{0,i} a_i = 0. \tag{8.7}$$

Since the columns of A are linearly independent over the support set $\mathcal{S}$, the only solution is $\alpha_i = 1$ for all $i \in \mathcal{S}$. This means that any attempt to deviate from the true signal values, even within the support, would make the objective function worse (since the weighted ℓ_1 norm would increase if $|\alpha_i| \neq 1$). Thus, the ideal weighting not only suppresses false non-zero entries but also forces the recovered values on the support to match the true signal, ensuring exact recovery under minimal measurements.

It is of course impossible to construct the precise weights (8.4) without knowing the signal x_0 itself, but this suggests more generally that large weights could be used to discourage non-zero entries in the recovered signal, while small weights could be used to encourage non-zero entries.

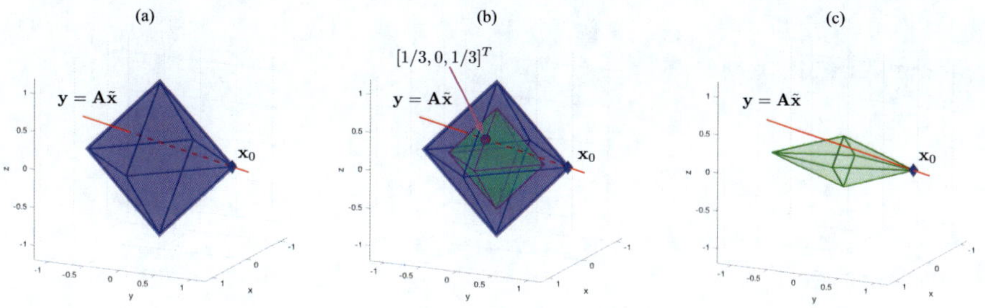

Fig. 8.1 (Adapted from Fig. 1 in [1]) An example of weighted ℓ_1-norm minimization for improving sparse signal recovery. (a) Sparse signal $\boldsymbol{x}_0$, feasible set $\boldsymbol{y} = \boldsymbol{A}\tilde{\boldsymbol{x}}$, and ℓ_1 ball of radius $\|\boldsymbol{x}_0\|_1$. (b) There exists an $\tilde{\boldsymbol{x}} = [1/3, 0, 1/3]^T \neq \boldsymbol{x}_0$ for which $\|\tilde{\boldsymbol{x}}\|_1 < \|\boldsymbol{x}_0\|_1$. (c) Weighted ℓ_1 ball, and there exists no $\tilde{\boldsymbol{x}} \neq \boldsymbol{x}_0$ for which $\|\boldsymbol{W}\tilde{\boldsymbol{x}}\|_1 \leq \|\boldsymbol{W}\boldsymbol{x}_0\|_1$

For ease of illustration, a simple 3-D example presented in [1] is given in Fig. 8.1, where $\boldsymbol{x}_0 = [0, 1, 0]^T$ and

$$\boldsymbol{A} = \begin{bmatrix} 2 & 1 & 1 \\ 1 & 1 & 2 \end{bmatrix}. \tag{8.8}$$

The observations are given by $\boldsymbol{y} = \boldsymbol{A}\tilde{\boldsymbol{x}} = [1, 1]^T$. As can be observed from Fig. 8.1a, the true signal $\boldsymbol{x}_0$ is marked as a blue diamond at coordinates $[0, 1, 0]^T$. The feasible set $\boldsymbol{y} = \boldsymbol{A}\tilde{\boldsymbol{x}}$ is represented by a red line parameterizing all solutions to the underdetermined system. This line is derived by solving the equations $2\tilde{x}_1 + \tilde{x}_2 + \tilde{x}_3 = 1$ and $\tilde{x}_1 + \tilde{x}_2 + 2\tilde{x}_3 = 1$, leading to $\tilde{x}_1 = \tilde{x}_3$ and $\tilde{x}_2 = 1 - 3\tilde{x}_1$. The unweighted ℓ_1 ball, defined by $\|\tilde{\boldsymbol{x}}\|_1 = |\tilde{x}_1| + |\tilde{x}_2| + |\tilde{x}_3| \leq 1$, is a symmetric convex polyhedron centered at the origin. From Fig. 8.1b, we can see that it intersects the feasible set at $\tilde{\boldsymbol{x}} = [1/3, 0, 1/3]^T$, which is non-sparse and differs from $\boldsymbol{x}_0$. This occurs because the symmetric geometry of the ℓ_1 ball prioritizes solutions with smaller total magnitudes, even if they violate sparsity. The ℓ_1 norm's uniform penalty on all coordinates fails to distinguish between true non-zero entries and spurious ones. In Fig. 8.1c, a hypothetical weighting matrix $\boldsymbol{W} = \text{diag}([3, 1, 3])$ is introduced. This assigns small weights to the true support ($w_2 = 1$) and large weights to zero positions ($w_1 = w_3 = 3$). The weighted ℓ_1 ball is defined by $\|\boldsymbol{W}\tilde{\boldsymbol{x}}\|_1 = 3|\tilde{x}_1| + |\tilde{x}_2| + 3|\tilde{x}_3| \leq 1$. The large weights on $\tilde{x}_1$ and $\tilde{x}_3$ drastically increase the penalty for non-zero values at these positions, effectively "pinching" the ℓ_1 ball at $\tilde{x}_0$. This reshaping ensures that the weighted ℓ_1 ball no longer intersects the feasible set $\boldsymbol{y} = \boldsymbol{A}\tilde{\boldsymbol{x}}$ at any point other than x_0. Consequently, the weighted ℓ_1 minimization (8.1) recovers the correct sparse solution $\boldsymbol{x}_0$.

This example implicitly requires weights to satisfy $w_2 < \frac{w_1 + w_3}{3}$, which ensures that the weighted ℓ_1 ball's geometry avoids intersections with the feasible set outside $\boldsymbol{x}_0$. More generally, weights should relate inversely to the magnitudes of the true signal's non-zero entries. Large weights suppress spurious non-zero terms, while small weights preserve true signal components.

Questions
In the above example, why the weights have to satisfy $w_2 < \frac{w_1 + w_3}{3}$ for exact sparse recovery?

8.1.2 Reweighted ℓ_1-Norm Minimization

Although the ideal weights $w_i = \frac{1}{|x_{0,i}|}$ (for non-zero entries) and $w_i = \infty$ (for zero entries) require prior knowledge of $\boldsymbol{x}_0$, adaptive strategies such as iterative reweighting can approximate this behavior. By iteratively adjusting weights based on intermediate solutions, one can enhance sparsity and improve recovery accuracy without exact prior information. The reweighted ℓ_1-norm minimization is a simple iterative algorithm that alternates between estimating $\boldsymbol{x}_0$ and redefining the weights. The algorithm is as follows:

1. Set the iteration index l to zero and $w_i^{(0)} = 1, i = 1, \ldots, N$.
2. Solve the weighted ℓ_1 minimization problem

$$\tilde{\boldsymbol{x}}^l = \arg\min_{\tilde{\boldsymbol{x}} \in \mathbb{R}^N} \|\boldsymbol{W}^{(l)}\tilde{\boldsymbol{x}}\|_1 \quad \text{s.t. } \boldsymbol{y} = \boldsymbol{A}\tilde{\boldsymbol{x}}. \tag{8.9}$$

3. Update the weights: for each $i = 1, ..., N$,

$$w_i^{(l+1)} = \frac{1}{|\tilde{x}_i^{(l)}| + \epsilon}. \tag{8.10}$$

4. Terminate on convergence or when l attains a specified maximum number of iterations l_{max}. Otherwise, set $l = l + 1$ and go to step 2.

Here, an important parameter, $\epsilon > 0$, is introduced during the weight update step to enhance stability and flexibility. This parameter prevents components temporarily estimated as zero in one iteration from being permanently excluded in subsequent steps. By adding ϵ to the denominator of the weight update formula, the algorithm avoids infinite weights that could rigidly suppress specific coefficients, even if they might later prove to be non-zero. Empirical studies suggest setting ϵ slightly smaller than the anticipated magnitudes of the true non-zero components in $\boldsymbol{x}_0$. While this guideline helps align the parameter with the signal's characteristics, the recovery process exhibits robustness to variations in ϵ, i.e., minor deviations from the ideal value do not significantly degrade performance. This tolerance is crucial in practical scenarios where prior knowledge of the signal's exact structure may be limited.

The iterative reweighting mechanism operates by progressively refining the estimation of non-zero coefficients. During initial iterations, the algorithm tends to identify the largest-magnitude components of the signal, as these dominate the energy of the solution under the ℓ_1 penalty. Once these prominent components are detected, their corresponding weights are reduced in subsequent iterations. This downweighting effectively diminishes their influence in the optimization objective, allowing the algorithm to shift focus toward smaller, previously overlooked coefficients. Over multiple iterations, this adaptive process mimics a "focusing" effect: early stages prioritize coarse recovery of dominant features, while later stages enhance sensitivity to subtler details. Even if early estimates are imperfect, for instance, missing small non-zero entries or misclassifying noise as weak signal components, the dynamic adjustment of weights enables self-correction. By iteratively rebalancing the penalties, the algorithm gradually disentangles true signal elements from artifacts, ultimately converging toward a more accurate reconstruction. This approach leverages the inherent sparsity structure without relying on exact prior knowledge, making it particularly effective in scenarios where signal characteristics are partially unknown or noisy.

8.1.3 Connection to Majorization-Minimization

The above iterative reweighted ℓ_1-norm minimization algorithm falls in the general class of MM algorithms, see [2] and references therein. In a nutshell, MM algorithms are more general than EM algorithms (as discussed in Chap. 4), and work by iteratively minimizing a simple surrogate function majorizing a given objective function. To establish this connection, consider the problem

$$\min_{\tilde{\boldsymbol{x}} \in \mathbb{R}^N} \sum_{i=1}^{N} \log(|\tilde{x}_i| + \epsilon) \quad \text{s.t. } \boldsymbol{y} = \boldsymbol{A}\tilde{\boldsymbol{x}}, \tag{8.11}$$

which is equivalent to

$$\min_{\tilde{\boldsymbol{x}}, \boldsymbol{u} \in \mathbb{R}^N} \sum_{i=1}^{N} \log(u_i + \epsilon) \quad \text{s.t. } \boldsymbol{y} = \boldsymbol{A}\tilde{\boldsymbol{x}}, |\tilde{x}_i| \le u_i, \ i = 1, \ldots, N. \tag{8.12}$$

Their equivalence hinges on the monotonicity of the logarithmic function and the structure of the constraints. By introducing auxiliary variables u_i in problem (8.12) to upper-bound $|\tilde{x}_i|$, the optimization task shifts to minimizing $\sum_{i=1}^{N} \log(u_i + \epsilon)$ under the constraints $|\tilde{x}_i| \le u_i$ and $\boldsymbol{y} = \boldsymbol{A}\tilde{\boldsymbol{x}}$. Since the logarithm is monotonically increasing, minimizing $\log(u_i + \epsilon)$ requires u_i to be as small as possible while satisfying $u_i \geq |\tilde{x}_i|$. This forces u_i to align exactly with $|\tilde{x}_i|$ at optimality, as any $u_i > |\tilde{x}_i|$ would unnecessarily increase the objective value. Consequently, if $\tilde{\boldsymbol{x}}^\star$ solves (8.11), setting $u_i^\star = |\tilde{x}_i^\star|$ satisfies all constraints in (8.12) and preserves the objective value, making $(\tilde{\boldsymbol{x}}^\star, \boldsymbol{u}^\star)$ a solution to (8.12). Conversely, if $(\tilde{\boldsymbol{x}}^\star, \boldsymbol{u}^\star)$ solves (8.12), optimality ensures $u_i^\star = |\tilde{x}_i^\star|$, reducing the objective to $\sum_{i=1}^N \log(|\tilde{x}_i^\star|+\epsilon)$, which matches problem (8.11). This bidirectional relationship confirms the equivalence, as both formulations yield identical solutions through this reparameterization.

In (8.12), the function $\sum_{i=1}^N \log(u_i + \epsilon)$ is concave with respect to $\{u_i\}$ and, therefore, below its tangent. Thus, one can improve on a guess $\{u_i\}$ at the solution by minimizing a linearization of $\sum_{i=1}^N \log(u_i + \epsilon)$ around $\{u_i\}$. This simple observation yields the following MM algorithm: starting with $\{\boldsymbol{u}^{(0)}, \tilde{\boldsymbol{x}}^{(0)}\}$ satisfying $\boldsymbol{y} = \boldsymbol{A}\tilde{\boldsymbol{x}}^{(0)}$, $|\tilde{x}_i^{(0)}| \leq u_i^{(0)}$, inductively define the following problem:

$$(\tilde{\boldsymbol{x}}^{(l+1)}, \boldsymbol{u}^{(l+1)}) = \arg\min_{\tilde{\boldsymbol{x}}, \boldsymbol{u} \in \mathbb{R}^N} \sum_{i=1}^N \frac{u_i}{u_i^{(l)} + \epsilon} \quad \text{s.t. } \boldsymbol{y} = \boldsymbol{A}\tilde{\boldsymbol{x}},\ |\tilde{x}_i| \leq u_i,\ i = 1, \ldots, N, \tag{8.13}$$

which is of course equivalent to

$$\tilde{\boldsymbol{x}}^{(l+1)} = \arg\min_{\tilde{\boldsymbol{x}} \in \mathbb{R}^N} \sum_{i=1}^N \frac{|\tilde{x}_i|}{|x_i^{(l)}| + \epsilon} \quad \text{s.t. } \boldsymbol{y} = \boldsymbol{A}\tilde{\boldsymbol{x}}. \tag{8.14}$$

Now, it is clear that the above MM algorithm is exactly the reweighted ℓ_1-norm minimization algorithm introduced in the previous subsection.

In [3–5], it is argued that the sequence $\log(|\tilde{x}_i^{(l)}| + \epsilon)$ converges to a local minimum of $\log(|\tilde{x}_i| + \epsilon)$, but this does not guarantee convergence of the sequence $\{\tilde{x}_i^{(l)}\}$ itself. Due to the concave nature of the log-sum penalty function, the algorithm inherently risks settling at local minima rather than global minima. To mitigate this limitation, careful selection of an appropriate starting point is critical. Following the approach in prior work [5], initializing the algorithm with the solution to the unweighted ℓ_1 minimization problem (7.11) is recommended, a strategy validated as effective in practical applications.

The connection between reweighted ℓ_1 minimization in Sect. 8.1.2 and the log-sum penalty function in (8.12) provides a basis for understanding why reweighted ℓ_1 minimization can improve the recovery of sparse signals. Consider, for example, the following three potential penalty functions for scalar magnitudes t:

$$f_0(t) = \mathbf{1}_{\{t \neq 0\}}, \quad f_1(t) = |t| \quad \text{and} \quad f_{\log,\epsilon}(t) \propto \log(1 + |t|/\epsilon), \tag{8.15}$$

where the constant of proportionality is set such that $f_{\log,\epsilon}(1) = 1 = f_0(1) = f_1(1)$ is satisfied (see Fig. 8.2). As shown, unlike the ℓ_1 norm, which imposes a linear penalty on signal magnitudes, the log-sum penalty $f_{\log,\epsilon}(t) \propto \log(1 + |t|/\epsilon)$ introduces a concave function that more aggressively promotes sparsity by penalizing small non-zero coefficients. This behavior stems from its steep slope near the origin, which scales approximately as $1/\epsilon$ when ϵ is small. As $\epsilon \to 0$, the log-sum penalty increasingly resembles the ideal ℓ_0 "norm" (a binary indicator of non-zero entries), which applies an infinite penalty to even infinitesimal deviations from zero. However, while reducing ϵ strengthens sparsity encouragement, it also heightens the risk of the iterative reweighted algorithm converging to suboptimal local minima, particularly when initial coefficient estimates are inaccurate. To balance these competing effects, ϵ is typically chosen to be slightly smaller than the expected magnitudes of the true non-zero components in the signal. This choice stabilizes the optimization process by preventing excessive sensitivity to noise or estimation errors while still enabling significant improvements over standard ℓ_1 minimization. The log-sum penalty thus bridges the gap between the intractable ℓ_0 objective and practical convex relaxations, leveraging its concave shape to mimic the sparsity-promoting behavior of ℓ_0 norm without fully inheriting its computational challenges. By adaptively adjusting weights based on intermediate estimates, the algorithm progressively suppresses insignificant coefficients while refining the recovery of meaningful signal components, ultimately achieving a more accurate reconstruction of sparse signals.

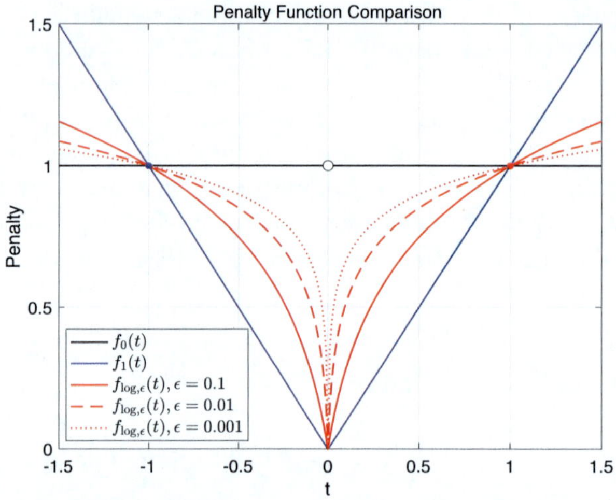

Fig. 8.2 Near the origin, the log-sum penalty function $f_{\log,\epsilon}(t)$ approximates the canonical ℓ_0 sparsity measure $f_0(t)$ more closely than the convex ℓ_1 norm $f_1(t)$. As $\epsilon \to 0$, $f_{\log,\epsilon}(t)$ increasingly resembles $f_0(t)$, mimicking its steep slope at $t = 0$ to more aggressively penalize small non-zero values and enhance sparsity

8.2 LASSO

This section is devoted to discussion of the LASSO, a method that combines the LS loss with an ℓ_1-constraint, or bound on the sum of the absolute values of the coefficients. Relative to the LS solution, this constraint has the effect of shrinking the coefficients, and even setting some to zero, for compressive sensing.

8.2.1 LASSO Estimator

With given $\boldsymbol{A}$ and $\boldsymbol{y}$, the LASSO problem is usually expressed as

$$\begin{aligned} &\min_{\tilde{\boldsymbol{x}} \in \mathbb{R}^N} \frac{1}{2M} \|\boldsymbol{y} - \boldsymbol{A}\tilde{\boldsymbol{x}}\|_2^2 \\ &\text{s.t. } \|\tilde{\boldsymbol{x}}\|_1 \leq t, \end{aligned} \tag{8.16}$$

where t is a prespecified free parameter that determines the degree of regularization, and it is also a kind of "budget": it limits the sum of the absolute values of the parameter estimates. Since a shrunken parameter estimate corresponds to a more heavily-constrained model, this budget limits how well we can fit the data. It must be specified by an external procedure such as cross-validation, and the detailed discussion is provided in [6].

It is often convenient to rewrite the LASSO problem in the so-called Lagrangian form, i.e.,

$$\min_{\tilde{\boldsymbol{x}} \in \mathbb{R}^N} \frac{1}{2M} \|\boldsymbol{y} - \boldsymbol{A}\tilde{\boldsymbol{x}}\|_2^2 + \lambda \|\tilde{\boldsymbol{x}}\|_1, \tag{8.17}$$

for some $\lambda \geq 0$. By Lagrangian duality, there is a one-to-one correspondence between the constrained problem (8.16) and the Lagrangian form (8.17): for each value of t in the range where the constraint $\|\tilde{\boldsymbol{x}}\|_1 \leq t$ is active, there is a corresponding value of λ that yields the same solution from the Lagrangian form (8.17). Conversely, the solution $\tilde{\boldsymbol{x}}_\lambda$ to problem (8.17) solves the bound problem with $t = \|\tilde{\boldsymbol{x}}_\lambda\|_1$.

We note that in many descriptions of the LASSO, the factor $\frac{1}{2M}$ appearing in (8.16) and (8.17) is replaced by 1/2 or simply 1. Although this makes no difference in (8.16), and corresponds to a simple reparametrization of λ in (8.17), this kind of standardization makes λ values comparable for different sample sizes (useful for cross-validation).

8.2.2 Solution to the LASSO Problem

The LASSO problem is a convex program, specifically a quadratic program (QP) with a convex constraint. As such, there are many sophisticated QP methods for solving the LASSO. However, there is a particularly simple and effective computational algorithm that gives insight into how the LASSO works. For convenience, we rewrite the LASSO problem in Lagrangian form as follows:

$$\min_{\tilde{x}\in\mathbb{R}^N} \frac{1}{2M}\sum_{i=1}^{M}\left(y_i - \sum_{j=1}^{N} A_{ij}\tilde{x}_j\right)^2 + \lambda\sum_{j=1}^{N}|\tilde{x}_j|. \tag{8.18}$$

We will assume that both y_i and A_{ij} have been standardized so that $\frac{1}{M}\sum_i y_i = 0$, $\frac{1}{M}\sum_i A_{ij} = 0$, and $\frac{1}{M}\sum_i A_{ij}^2 = 1$. The Lagrangian form in (8.18) is especially convenient for numerical computation of the solution by a simple procedure known as coordinate descent.

Let us first focus on a simple scenario where $N = 1$. The LASSO problem then is to solve

$$\min_{\tilde{x}} \frac{1}{2M}\sum_{i=1}^{M}(y_i - A_{i1}\tilde{x})^2 + \lambda|\tilde{x}|. \tag{8.19}$$

The standard approach to this univariate minimization problem would be to take the gradient (first derivative) with respect to $\tilde{x}$, and set it to zero. However, simple gradient-based optimization is a bit complicated here, because the squared error term $\sum_{i=1}^{M}(y_i - A_{i1}\tilde{x})^2$ is smooth, but the absolute value penalty $\lambda|\tilde{x}|$ introduces a non-differentiable point at $\tilde{x} = 0$. To resolve this, one should analyze the behavior of the objective function for three cases: $\tilde{x} > 0$, $\tilde{x} < 0$ and $\tilde{x} = 0$. Let $\tilde{x}_{\text{LS}} = \frac{1}{M}\langle \boldsymbol{a}, \boldsymbol{y}\rangle$ denote the unregularized LS estimate, where $\boldsymbol{a} \triangleq [A_{11}, A_{21}, \ldots, A_{M1}]^T$.

1. $\tilde{x} > 0$: For $\tilde{x} > 0$, the penalty term becomes $\lambda\tilde{x}$, and the objective function is

$$f(\tilde{x}) = \frac{1}{2M}\sum_{i=1}^{M}(y_i - A_{i1}\tilde{x})^2 + \lambda\tilde{x}. \tag{8.20}$$

Taking the derivative of $f(\tilde{x})$ with respect to $\tilde{x}$, we have $f'(\tilde{x}) = -\frac{1}{M}\sum_{i=1}^{M} A_{i1}(y_i - A_{i1}\tilde{x}) + \lambda$, and by setting $f'(\tilde{x}) = 0$ yields

$$-\tilde{x}_{\text{LS}} + \tilde{x}\frac{1}{M}\sum_{i=1}^{M} A_{i1}^2 + \lambda = 0. \tag{8.21}$$

Simplifying (8.21) using $\frac{1}{M}\sum_i A_{i1}^2 = 1$, we obtain

$$\tilde{x} = \tilde{x}_{\text{LS}} - \lambda, \tag{8.22}$$

which is valid only if $\tilde{x}_{\text{LS}} - \lambda > 0$.

2. $\tilde{x} < 0$: For $\tilde{x} < 0$, the penalty term becomes $-\lambda\tilde{x}$. By following a similar procedure as in the case of $\tilde{x} > 0$, we have

$$\tilde{x} = \tilde{x}_{\text{LS}} + \lambda, \tag{8.23}$$

and this is only valid when $\tilde{x}_{\text{LS}} + \lambda < 0$.

3. $\tilde{x} = 0$: At $\tilde{x} = 0$, the absolute value $|\tilde{x}|$ is non-differentiable. Thus, we use the concept of subgradients to determine optimality. Specifically, a subgradient at $\tilde{x} = 0$ must satisfy

$$0 \in -\tilde{x}_{\text{LS}} + \lambda \cdot \partial|\tilde{x}|\big|_{\tilde{x}=0}, \tag{8.24}$$

where $\partial|\tilde{x}|\big|_{\tilde{x}=0}$ is the subdifferential of $\tilde{x}$ at $\tilde{x} = 0$, and satisfies $\partial|\tilde{x}|\big|_{\tilde{x}=0} \in [-1, 1]$. Thus, we have

$$-\tilde{x}_{\text{LS}} + \lambda \cdot \partial|\tilde{x}|\big|_{\tilde{x}=0} = 0, \tag{8.25}$$

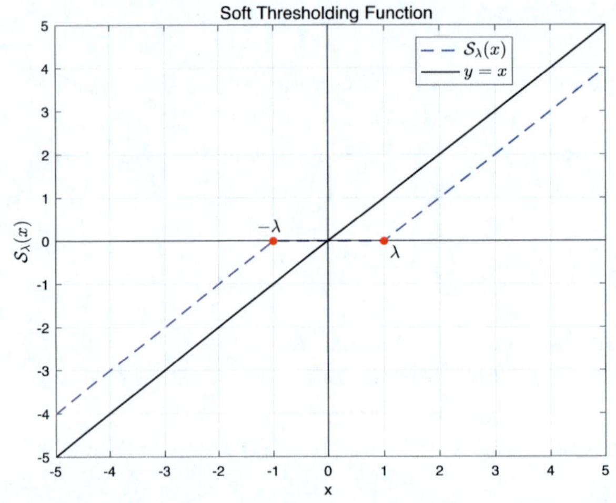

Fig. 8.3 (Adapted from Fig. 2.4 in [6]) Soft-thresholding function $\mathcal{S}_\lambda(x) = \text{sign}(x)(|x| - \lambda)_+$ (blue dashed line). The black solid line represents the 45° reference line (identity function). The soft-thresholding operator shrinks values toward zero for $|x| \leq \lambda$ and shifts them linearly beyond this threshold

for some $\partial|\tilde{x}|\big|_{\tilde{x}=0} \in [-1, 1]$. Since $|\partial|\tilde{x}|\big|_{\tilde{x}=0}| \leq 1$, $\tilde{x} = 0$ is optimal when $|\tilde{x}_{\text{LS}}| \leq \lambda$.

Combining the above three cases, the solution $\tilde{x}$ to problem (8.19) is determined by comparing $\tilde{x}_{\text{LS}}$ to λ as follows:

$$\tilde{x} = \begin{cases} \tilde{x}_{\text{LS}} - \lambda, & \text{if } \tilde{x}_{\text{LS}} > \lambda, \\ 0, & \text{if } |\tilde{x}_{\text{LS}}| \leq \lambda, \\ \tilde{x}_{\text{LS}} + \lambda, & \text{if } \tilde{x}_{\text{LS}} < -\lambda. \end{cases} \tag{8.26}$$

This result can be compactly expressed using the soft-thresholding operator:

$$\tilde{x} = \mathcal{S}_\lambda\left(\tilde{x}_{\text{LS}}\right), \tag{8.27}$$

where

$$\mathcal{S}_\lambda(x) = \text{sign}(x)(|x| - \lambda)_+, \tag{8.28}$$

and $(x)_+$ denotes the positive part of x, i.e., equal to x if $x > 0$ and 0 otherwise. The operator shifts $\tilde{x}_{\text{LS}}$ toward zero by λ, producing sparsity when $|\tilde{x}_{\text{LS}}| \leq \lambda$, as shown in Fig. 8.3.

For solving the full LASSO problem (8.18), one can extend the intuition gained from the univariate soft-thresholding case to the multivariate setting. The core idea is to iteratively update each coefficient $\tilde{x}_j$ while keeping all other coefficients $\{\tilde{x}_k, k \neq j\}$ fixed. This approach breaks the high-dimensional optimization problem into a sequence of simpler univariate subproblems, each solvable via the soft-thresholding operator derived earlier. Specifically, the algorithm cycles through predictors (i.e., the columns of the sensing matrix $\boldsymbol{A}$) in a fixed order, say $j = 1, 2, \ldots, N$. At each step j, only $\tilde{x}_j$ is updated, while all other coefficients remain unchanged. For the j-th predictor, the LASSO objective (8.18) is rewritten as

$$\frac{1}{2M}\sum_{i=1}^{M}\left(y_i - \sum_{k\neq j} A_{ik}\tilde{x}_k - A_{ij}\tilde{x}_j\right)^2 + \lambda\sum_{k\neq j}|\tilde{x}_k| + \lambda|\tilde{x}_j|, \tag{8.29}$$

where the terms involving $\{\tilde{x}_k, k \neq j\}$ are treated as constants, reducing the problem to a univariate optimization over $\tilde{x}_j$. Thus, the simplified objective for $\tilde{x}_j$ is

$$\frac{1}{2M}\sum_{i=1}^{M}\left(r_i^{(j)} - A_{ij}\tilde{x}_j\right)^2 + \lambda|\tilde{x}_j|, \tag{8.30}$$

where $r_i^{(j)} = y_i - \sum_{k \neq j} A_{ik} \tilde{x}_k$ represents the partial residual excluding the j-th predictor. The solution to this subproblem is given by the following soft-thresholding operator:

$$\tilde{x}_j = \mathcal{S}_\lambda \left(\frac{1}{M} \langle \boldsymbol{a}_j, \boldsymbol{r}^{(j)} \rangle \right). \tag{8.31}$$

Equivalently, the update can be expressed using the full residual $r_i = y_i - \sum_{j=1}^{N} A_{ij} \tilde{x}_j$, which accounts for all predictors, and the corresponding alternative formulation is given by

$$\tilde{x}_j \leftarrow \mathcal{S}_\lambda \left(\tilde{x}_j + \frac{1}{M} \langle \boldsymbol{a}_j, \boldsymbol{r} \rangle \right), \tag{8.32}$$

which combines the current coefficient estimate with a gradient-like term derived from the full residual. Note that from (8.31) to (8.32), we have used the following derivation:

$$\begin{aligned} \frac{1}{M} \langle \boldsymbol{a}_j, \boldsymbol{r}^{(j)} \rangle &= \frac{1}{M} \sum_{i=1}^{M} A_{ij} r_i^{(j)} = \frac{1}{M} \sum_{i=1}^{M} A_{ij} (r_i + A_{ij} \tilde{x}_j) \\ &= \frac{1}{M} \sum_{i=1}^{M} A_{ij} r_i + \tilde{x}_j \cdot \frac{1}{M} \sum_{i=1}^{M} A_{ij}^2. \end{aligned} \tag{8.33}$$

Then, using the normalization condition $\frac{1}{M} \sum_{i=1}^{M} A_{ij}^2 = 1$, we obtain

$$\frac{1}{M} \langle \boldsymbol{a}_j, \boldsymbol{r}^{(j)} \rangle = \frac{1}{M} \langle \boldsymbol{a}_j, \boldsymbol{r} \rangle + \tilde{x}_j, \tag{8.34}$$

which directly leads to (8.32).

The algorithm cyclically updates each coefficient in a fixed order, recalculating residuals after each adjustment. Since the LASSO objective is convex, this coordinate-wise minimization strategy, known as cyclical coordinate descent, guarantees convergence to the global minimum under mild conditions. The convexity ensures no local minima exist, and the separability of the LASSO penalty (i.e., the sum of absolute values) allows updates to proceed independently across coordinates without interference. However, note that non-separable penalties or ill-conditioned problems might disrupt convergence [6], though such issues do not arise in standard LASSO implementations.

By breaking the high-dimensional optimization into a sequence of univariate thresholding operations, the algorithm efficiently balances computational tractability with sparsity enforcement. The soft-thresholding step inherently promotes sparsity by zeroing out coefficients with weak correlations to the residuals, while larger coefficients are adaptively shrunk. This iterative process continues until updates stabilize, ensuring a sparse solution that minimizes both prediction error and model complexity.

In practice, one is often interested in generating the entire solution path of the LASSO regression across a range of regularization parameters λ, a process known as pathwise coordinate descent. The goal is to efficiently compute the sequence of LASSO solutions as λ decreases from its maximum value (λ_{max}) to near zero. The algorithm starts with λ_{max}, the smallest λ value large enough to force all coefficients to zero. This is determined by the maximum absolute correlation between any predictor $\boldsymbol{a}_j$ and the measurement $\boldsymbol{y}$, i.e.,

$$\lambda_{\text{max}} = \max_j \left| \frac{1}{M} \langle \boldsymbol{a}_j, \boldsymbol{y} \rangle \right|. \tag{8.35}$$

At this λ, the penalty dominates, yielding $\tilde{\boldsymbol{x}} = \boldsymbol{0}$. Starting from λ_{max}, the algorithm gradually decreases λ in small increments. For each new λ, the solution from the previous λ is used as a warm start, i.e., initializing the coefficients with values close to the new optimal solution. This warm-start strategy drastically accelerates convergence because small changes in λ typically result in minor adjustments to the coefficient estimates. At each λ, coordinate descent iteratively updates one coefficient $\tilde{x}_j$ at a time while holding others fixed, as in (8.31) or (8.32).

Furthermore, it is important to note that the soft-thresholding operator is crucial in both the LASSO method and signal denoising. Consider the coordinate minimization scheme: if the predictors are orthogonal (meaning $\frac{1}{M} \langle \boldsymbol{a}_j, \boldsymbol{a}_k \rangle = 0$ for $j \neq k$), this scheme takes a very simple form. In such a case, the update formula (8.31) simplifies a lot. Specifically, $\frac{1}{M} \langle \boldsymbol{a}_j, \boldsymbol{r}^{(j)} \rangle$

equals $\frac{1}{M}\langle \boldsymbol{a}_j, \boldsymbol{y}\rangle$. Consequently, $\tilde{x}_j$ becomes just the soft-thresholded version of the univariate LS estimate from regressing $\boldsymbol{y}$ on $\boldsymbol{a}_j$. Therefore, when the predictors are orthogonal, the LASSO has a clear, direct closed-form solution, eliminating the need for iterative computations.

The preceding derivation of the coordinate descent algorithm for LASSO assumed that the data $\boldsymbol{y}$ and the predictors (columns of $\boldsymbol{A}$) were standardized. Specifically, we assumed $\frac{1}{M}\sum_i y_i = 0$ (response centered), $\frac{1}{M}\sum_i A_{ij} = 0$ for all j (predictors centered), and $\frac{1}{M}\sum_i A_{ij}^2 = 1$ for all j (predictors scaled to unit variance). This standardization simplifies the mathematics significantly. It ensures that the univariate LS estimate is $\tilde{x}_{\mathrm{LS}} = \frac{1}{M}\langle \boldsymbol{a}_j, \boldsymbol{y}\rangle$ and, crucially, that the term $\frac{1}{M}\sum_i A_{ij}^2$, which appears in the derivative, equals 1.

In practical applications, however, data is rarely pre-standardized within the optimization algorithm itself. It is therefore essential to understand the general form of the solution. Without standardization, the univariate minimization problem for coordinate j becomes

$$\min_{\tilde{x}_j} \frac{1}{2M}\sum_{i=1}^{M}\left(r_i^{(j)} - A_{ij}\tilde{x}_j\right)^2 + \lambda|\tilde{x}_j|. \tag{8.36}$$

Following the same procedure of taking the subgradient and solving for the optimal $\tilde{x}_j$, the solution is given by the following generalized soft-thresholding operator:

$$\tilde{x}_j = \frac{1}{\gamma_j}\cdot \mathcal{S}_\lambda\left(\frac{1}{M}\langle \boldsymbol{a}_j, \boldsymbol{r}^{(j)}\rangle\right), \tag{8.37}$$

where $\gamma_j = \frac{1}{M}\sum_{i=1}^{M} A_{ij}^2$ is the average squared norm of the j-th predictor, and please see Appendix A for the detailed derivation. The key difference from the standardized case is the division by γ_j, the average squared norm of the j-th predictor. This factor accounts for the scale of the predictor. Specifically, a predictor with a large scale (large γ_j) will have its coefficient update shrunk more aggressively by the ℓ_1 penalty. The algorithm inherently penalizes coefficients for variables that have a larger scale more heavily. Conversely, a predictor with a small scale will see its coefficient update amplified.

8.2.3 Bayesian LASSO

The Bayesian paradigm treats the parameters as random quantities, along with a prior distribution that characterizes our belief in what their values might be. Here we adopt the approach of [7], involving a model of the form

$$\begin{aligned} \boldsymbol{y} \mid \tilde{\boldsymbol{x}}, \lambda, \sigma &\sim \mathcal{N}(\boldsymbol{A}\tilde{\boldsymbol{x}}, \sigma^2 \boldsymbol{I}_{M\times M}), \\ \tilde{\boldsymbol{x}} \mid \lambda, \sigma &\sim \prod_{j=1}^{N}\frac{\lambda}{2\sigma}e^{-\frac{\lambda}{\sigma}|\tilde{x}_j|}, \end{aligned} \tag{8.38}$$

i.e., the measurement vector $\boldsymbol{y}$ is modeled as a normal distribution conditional on $\tilde{\boldsymbol{x}}$, λ and σ, and a Laplacian (double exponential) prior is placed on the coefficients $\tilde{\boldsymbol{x}}$. Since the posterior density $p(\tilde{\boldsymbol{x}}|\boldsymbol{y}, \lambda, \sigma)$ is proportional to the product of the likelihood and the prior, we have

$$p(\tilde{\boldsymbol{x}}|\boldsymbol{y}, \lambda, \sigma) \propto \exp\left(-\frac{1}{2\sigma^2}\|\boldsymbol{y} - \boldsymbol{A}\tilde{\boldsymbol{x}}\|_2^2\right)\cdot \prod_{j=1}^{N}\frac{\lambda}{2\sigma}\exp\left(-\frac{\lambda}{\sigma}|\tilde{x}_j|\right). \tag{8.39}$$

Then, by taking the negative logarithm of $p(\tilde{\boldsymbol{x}}|\boldsymbol{y}, \lambda, \sigma)$ (and ignoring additive constants independent of $\tilde{\boldsymbol{x}}$), we obtain

$$-\log p(\tilde{\boldsymbol{x}}|\boldsymbol{y}, \lambda, \sigma) \propto \frac{1}{2\sigma^2}\|\boldsymbol{y} - \boldsymbol{A}\tilde{\boldsymbol{x}}\|_2^2 + \frac{\lambda}{\sigma}\|\tilde{\boldsymbol{x}}\|_1. \tag{8.40}$$

Consequently, for any fixed values of λ and σ, the posterior mode coincides with the LASSO estimate. Note that the posterior distribution provides more than point estimates, it provides an entire joint distribution.

As mentioned earlier, λ is a regularization parameter that controls the strength of sparsity. In practical applications, selecting an appropriate value of λ is crucial for model performance, but λ usually needs to be adjusted manually or determined via cross-validation, which may lead to the results being sensitive to the choice of λ and lacking robustness. Thus, one can improve robustness with respect to the uncertainty of λ by introducing a Bayesian framework that treats λ as a random variable with a gamma prior distribution, i.e.,

$$\lambda \sim \text{Gamma}(\alpha, \beta), \tag{8.41}$$

where α (shape) and β (rate) are hyperparameters. The gamma prior is chosen because it is conjugate to the Laplace (double-exponential) distribution, thus simplifies the posterior computations. In this setting, we can use the EM algorithm to estimate the hyperparameters α and β. Here, the latent variables are the coefficients $\tilde{\boldsymbol{x}}$ and the regularization parameter λ, while the hyperparameters α and β are the parameters to be estimated. The complete-data likelihood is

$$P(\boldsymbol{y}, \tilde{\boldsymbol{x}}, \lambda|\alpha, \beta) = P(\boldsymbol{y}|\tilde{\boldsymbol{x}}) \cdot P(\tilde{\boldsymbol{x}}|\lambda) \cdot P(\lambda|\alpha, \beta), \tag{8.42}$$

where

$$P(\boldsymbol{y}|\tilde{\boldsymbol{x}}) = \mathcal{N}(\boldsymbol{A}\tilde{\boldsymbol{x}}, \sigma^2\boldsymbol{I}),\ P(\tilde{\boldsymbol{x}}|\lambda) = \prod_{j=1}^{N} \frac{\lambda}{2\sigma} e^{-\frac{\lambda}{\sigma}|\tilde{x}_j|},\ P(\lambda|\alpha, \beta) = \frac{\beta^\alpha}{\Gamma(\alpha)} \lambda^{\alpha-1} e^{-\beta\lambda}.$$

The EM algorithm proceeds as follows. In the E-step, we compute the expected complete-data log-likelihood (the Q-function) with respect to the posterior distribution of the latent variables $\tilde{\boldsymbol{x}}$ and λ given the current hyperparameters $\alpha^{(k)}$, $\beta^{(k)}$, i.e.,

$$Q(\alpha, \beta \mid \alpha^{(k)}, \beta^{(k)}) = \mathbb{E}_{\tilde{\boldsymbol{x}}, \lambda|\boldsymbol{y}, \alpha^{(k)}, \beta^{(k)}} \{\log P(\boldsymbol{y}, \tilde{\boldsymbol{x}}, \lambda \mid \alpha, \beta)\}. \tag{8.43}$$

While in the M-step, we update the hyperparameters by maximizing the Q-function as follows:

$$(\alpha^{(k+1)}, \beta^{(k+1)}) = \arg\max_{\alpha, \beta} Q(\alpha, \beta \mid \alpha^{(k)}, \beta^{(k)}).$$

As shown, only the term $\log P(\lambda|\alpha, \beta)$ depends on α and β. Thus, we focus on

$$\mathbb{E}\{\log P(\lambda \mid \alpha, \beta)\} = \mathbb{E}\{\alpha \log \beta - \log \Gamma(\alpha) + (\alpha - 1) \log \lambda - \beta\lambda\}.$$

Let $A = \mathbb{E}\{\log \lambda\}$ and $B = \mathbb{E}\{\lambda\}$, where the expectations are taken with respect to the posterior $P(\tilde{\boldsymbol{x}}, \lambda \mid \boldsymbol{y}, \alpha^{(k)}, \beta^{(k)})$.[1] The Q-function (8.43) becomes

$$Q(\alpha, \beta) = \alpha \log \beta - \log \Gamma(\alpha) + (\alpha - 1)A - \beta B.$$

Taking the derivative of with $Q(\alpha, \beta)$ respect to β, we obtain

$$\frac{\partial Q(\alpha, \beta)}{\partial \beta} = \frac{\alpha}{\beta} - B \Rightarrow \beta = \frac{\alpha}{B}.$$

Substituting $\beta = \alpha/B$ into $Q(\alpha, \beta)$ leads to

$$Q(\alpha) = \alpha(\log \alpha - \log B) - \log \Gamma(\alpha) + (\alpha - 1)A - \alpha.$$

Taking the derivative of the above $Q(\alpha)$ with respect to α yields

$$\frac{dQ(\alpha)}{d\alpha} = \log \alpha - \log B + A - \psi(\alpha),$$

where $\psi(\alpha) = \frac{d}{d\alpha} \log \Gamma(\alpha)$ is the digamma function. Setting the derivative $\frac{dQ(\alpha)}{d\alpha}$ to zero, we have

$$\psi(\alpha) = \log \alpha + A - \log B,$$

which can be solved numerically for α using root-finding algorithms (e.g., Newton–Raphson). Once α is found, set $\beta = \alpha/B$.

This EM procedure respects the Bayesian hierarchy and properly accounts for uncertainty in both $\tilde{\boldsymbol{x}}$ and λ, leading to more robust hyperparameter estimates. Besides, it ensures that the Q-function increases monotonically, leading to convergence to

[1] These expectations can be approximated using samples from the posterior $P(\tilde{\boldsymbol{x}}, \lambda \mid \boldsymbol{y}, \alpha^{(k)}, \beta^{(k)})$ via Markov chain Monte Carlo (MCMC) methods or variational inference.

a stationary point of the marginal likelihood. The estimated hyperparameters can then be used to compute the posterior mean or mode of $\tilde{x}$ for final model inference.

8.2.4 Theoretical Insights

This subsection provides a brief overview of theoretical insights into the LASSO method, focusing on two key aspects: prediction consistency (how well the LASSO estimates perform in terms of prediction error) and sparsistency (the ability to correctly identify the non-zero coefficients in the true model). The LASSO is proven to achieve MSE consistency under certain conditions. Specifically, as the number of predictors N and sample size M grow to infinity, the prediction error of the LASSO estimator $\tilde{x}$ compared to the true parameter $\tilde{x}^\star$ is bounded by

$$\frac{\|A(\tilde{x} - \tilde{x}^\star)\|_2^2}{M} \leq C \cdot \|\tilde{x}^\star\|_1 \sqrt{\frac{\log(N)}{M}} \tag{8.44}$$

with high probability [8, 9]. Thus, for the LASSO to be consistent in prediction (i.e., prediction error approaches zero as M grows), the term $\|\tilde{x}^\star\|_1$ must grow slower than $\sqrt{M/\log(N)}$, i.e., $\|\tilde{x}^\star\|_1 = o\left(\sqrt{M/\log(N)}\right)$. This implies that the true parameter vector $\tilde{x}^\star$ must be sparse relative to the ratio $M/\log(N)$. In other words, the "complexity" of the true model (measured by $\|\tilde{x}^\star\|_1$) must not overwhelm the available data. The result only assumes that the design A is fixed and has no other conditions on A. Besides, achieving sparsistency, refers to the LASSO's ability to recover the correct (support) subset of non-zero coefficients in $\tilde{x}^\star$, requires stricter assumptions compared to prediction consistency, please see [6] for more details.

8.3 Generalized LASSO

The LASSO has revolutionized statistical modeling by unifying parameter estimation and variable selection through its ℓ_1-norm penalty. By imposing sparsity on regression coefficients, the LASSO efficiently balances bias and variance in high-dimensional settings where the number of predictors N may exceed the sample size M. However, while the classical LASSO penalizes the absolute magnitude of coefficients uniformly, real-world problems often demand more nuanced structural constraints. This limitation motivates the generalized LASSO, a flexible extension that adapts the regularization framework to accommodate domain-specific prior knowledge or complex dependencies among variables. Many applications require penalizing not the coefficients directly but transformations of them. For example, in wireless communications, due to the physical scattering structure, the significant elements in the angular domain massive MIMO channel [10] will appear in bursts (given in Definition 8.1 and Fig. 8.4); in signal processing, one might enforce smoothness in adjacent coefficients; in genomics, hierarchical structures among variables may need preservation.

Definition 8.1 (*Burst-Sparse Massive MIMO Channel*) A burst-sparse massive MIMO channel vector $\tilde{x}^\star$ with parameter $\{K, D\}$ is composed of K bursts: $\tilde{x}^\star = \sum_{k=1}^{K} \tilde{x}_k^\star$. Specifically, the support Ω_k of the k-th burst $\tilde{x}_k^\star$ satisfies $\Omega_k \subseteq I_{s_k^\star}$ for some $s_k^\star \in \{1, \ldots, N\}$, where $s_k^\star$ is termed the starting location of the k-th burst. I_i is defined as $I_i \triangleq \{i, i \oplus 1, \ldots, i \oplus (D-1)\}$

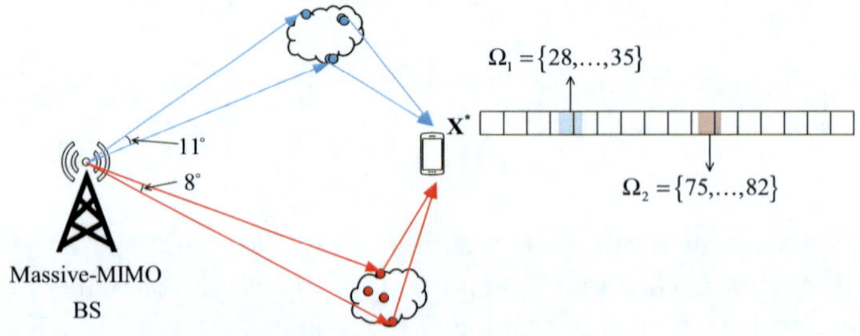

Fig. 8.4 (Adapted from Fig. 1 in [18]) Illustration of the system model and channel model for a massive MIMO system with the ULA of 128 antennas and half-wavelength inter-antenna distance at the base station (BS)

with the operation $i \oplus j$ given by

$$i \oplus j = \begin{cases} i + j, & \text{if } i + j \le N \\ i + j - N, & \text{if } i + j > N. \end{cases} \tag{8.45}$$

Moreover, D satisfies $D \ge \max_k |\Omega_k|$ and $D \ll N$, serving as an upper bound on the support size (sparsity) of all bursts.

Consider again the standard form of compressive sensing model as in

$$\boldsymbol{y} = \boldsymbol{A}\tilde{\boldsymbol{x}}^{\star} + \boldsymbol{n}, \tag{8.46}$$

where $\boldsymbol{n}$ is the noise vector with variance σ^2. In the generalized ℓ_2^2-LASSO approach, the recovery of the sparse signal $\tilde{\boldsymbol{x}}^{\star}$ can be formulated as the following convex optimization problem:

$$\min_{\tilde{\boldsymbol{x}}} \frac{1}{2} \|\boldsymbol{y} - \boldsymbol{A}\tilde{\boldsymbol{x}}\|^2 + \sigma\lambda M f(\tilde{\boldsymbol{x}}), \tag{8.47}$$

where $f(\tilde{\boldsymbol{x}})$ is a convex penalty function typically nonsmooth (e.g., ℓ_1-norm, nuclear-norm, ℓ_1/ℓ_2-norm), aiming to promote the structure of $\tilde{\boldsymbol{x}}^{\star}$ (e.g., sparse, low-rank, block sparse), and $\lambda > 0$ is a regularization parameter. By adopting this more general form, the specific sparse structure can be encoded in the penalty function $f(\tilde{\boldsymbol{x}})$. For example, if $\tilde{\boldsymbol{x}}^{\star}$ is a general sparse signal (i.e., the number of non-zero elements of $\tilde{\boldsymbol{x}}^{\star}$ is small compared to N) without any specific structure, we can choose $f(\tilde{\boldsymbol{x}}) = \|\tilde{\boldsymbol{x}}\|_1$ to impose the sparse constraint on $\tilde{\boldsymbol{x}}^{\star}$, as discussed before. On the other hand, if $\tilde{\boldsymbol{x}}^{\star}$ is a block sparse signal (i.e., $\tilde{\boldsymbol{x}}^{\star}$ can be divided into B non-overlapping blocks $\tilde{\boldsymbol{x}}^{\star} = [\tilde{\boldsymbol{x}}_1^{\star T}, \tilde{\boldsymbol{x}}_2^{\star T}, \ldots, \tilde{\boldsymbol{x}}_B^{\star T}]^T$ and there are only a few non-zero blocks), we can choose $f(\tilde{\boldsymbol{x}}) = \sum_{i=1}^{B} \|\tilde{\boldsymbol{x}}_i\|$ to impose the block-sparse constraint on $\tilde{\boldsymbol{x}}^{\star}$ [11, 12]. The design of the penalty function and the associated parameter λ are critical and require domain specific insight for different applications and different structural properties of the sparsity.

Attention
Note that from now on, we explicitly address sparse recovery for complex-valued signals, which are prevalent in wireless communications (e.g., massive MIMO channel estimation), radar sensing, and other signal processing applications where signals naturally exist in the complex domain. For the preceding algorithms, including the reweighted ℓ_1-minimization, standard LASSO, and Bayesian LASSO, extension to complex-valued signals and measurement matrices can be achieved through:

1. Magnitude Substitution: All real-domain absolute values $|\cdot|$ (e.g., in ℓ_1-norms, weight updates, or soft-thresholding) are replaced by the complex modulus $|\tilde{x}_i| = \sqrt{\Re(\tilde{x}_i)^2 + \Im(\tilde{x}_i)^2}$.
2. Complex Gaussian Models: For Bayesian frameworks, real Gaussians $\mathcal{N}(\cdot)$ are replaced with circularly symmetric complex Gaussians $\mathcal{CN}(\cdot)$.
3. Orthogonality Adjustments: Predictor orthogonality conditions use complex inner products $\langle \boldsymbol{a}_j, \boldsymbol{y}\rangle = \boldsymbol{a}_j^H \boldsymbol{y}$.
4. Convexity Preservation: All optimization problems (LASSO, reweighted ℓ_1) remain convex in $\mathbb{C}^N$ under modulus-based penalties.

8.3.1 Performance Analysis

The CGMT approach proposed in [13–15] is a useful tool to analyze the asymptotic normalized squared error of the generalized ℓ_2^2-LASSO with a general convex penalty function $f(\tilde{\boldsymbol{x}})$. In the following, we give an overview of the CGMT approach and the main results for the generalized LASSO method.

Definition 8.2 (*Normalized Squared Error*) Let $\tilde{\boldsymbol{x}}$ denote the estimate of $\tilde{\boldsymbol{x}}^{\star}$ obtained by the generalized ℓ_2^2-LASSO algorithm. The normalized squared error (NSE) of the generalized ℓ_2^2-LASSO with parameter λ is defined as

$$NSE(\lambda, \sigma) \triangleq \|\tilde{\boldsymbol{x}} - \tilde{\boldsymbol{x}}^{\star}\|^2/\sigma^2, \tag{8.48}$$

and the aNSE is defined as

$$aNSE(\lambda) \triangleq \lim_{\sigma \to 0} NSE(\lambda, \sigma). \tag{8.49}$$

The reasons for using aNSE as the performance metric are justified in [13]. It is conjectured in [13] that the aNSE is the worst-case NSE, i.e.,

$$aNSE(\lambda) = \sup_{\sigma > 0} NSE(\lambda, \sigma). \tag{8.50}$$

The CGMT approach in [13–15] for analyzing the aNSE works as follows. The key step is to calculate the following average Gaussian distance[2]

$$\bar{d}(\tau) \triangleq \mathbb{E}_{\boldsymbol{w}} \left\{ \left\| \boldsymbol{w} - \sqrt{M}\tau \partial f(\tilde{\boldsymbol{x}}^\star) \right\|^2 \right\} / M, \tag{8.51}$$

where $\boldsymbol{w} \sim \mathcal{CN}(\boldsymbol{0}, \boldsymbol{I})$, $\partial f(\tilde{\boldsymbol{x}}^\star)$ is the set of all subgradients of $f(\tilde{\boldsymbol{x}})$ at $\tilde{\boldsymbol{x}}^\star$, and $\left\| \boldsymbol{w} - \sqrt{M}\tau \partial f(\tilde{\boldsymbol{x}}^\star) \right\|^2$ is the Gaussian distance for fixed $\boldsymbol{w}$ defined as

$$\left\| \boldsymbol{w} - \sqrt{M}\tau \partial f(\tilde{\boldsymbol{x}}^\star) \right\|^2 \triangleq \min_{\boldsymbol{\alpha} \in \partial f(\tilde{\boldsymbol{x}}^\star)} \left\| \boldsymbol{w} - \sqrt{M}\tau \boldsymbol{\alpha} \right\|^2. \tag{8.52}$$

Then, the optimal λ and the corresponding aNSE are summarized in the following theorem.

Theorem 8.1 (aNSE for Generalized ℓ_2^2-LASSO) *Suppose the measurement matrix $\boldsymbol{A}$ has i.i.d. complex Gaussian entries with zero mean and unit variance and the penalty function $f(\tilde{\boldsymbol{x}})$ is convex. Let $\tau^\star = \arg\min_\tau \bar{d}(\tau)$. Let $\frac{M}{N} \to \delta \in (0, \infty)$ and $\bar{d}(\tau^\star) \to d^\star \in (0, 1)$. The optimal λ that minimizes the aNSE is given by*

$$\lambda^\star \triangleq \arg\min_\lambda aNSE(\lambda) = \tau^\star \sqrt{1 - d^\star}, \tag{8.53}$$

and the corresponding minimum aNSE is given by

$$aNSE(\lambda^\star) = \frac{d^\star}{1 - d^\star}. \tag{8.54}$$

Note that the results in Theorem 8.1 are based on the assumption that $\boldsymbol{A}$, $\boldsymbol{y}$, $\tilde{\boldsymbol{x}}^\star$, and $\boldsymbol{n}$ are complex matrix/vectors. Similar result as in Theorem 8.1 has been proved in [13] for the real case when $\boldsymbol{A}$, $\boldsymbol{y}$, $\tilde{\boldsymbol{x}}^\star$, and $\boldsymbol{n}$ are real matrix/vectors, the derivation of the average Gaussian distance in the real case is provided in Appendix B. The proof in [13] can be readily extended to the complex case in Theorem 8.1. The details are omitted here.

8.3.2 Group LASSO

Group LASSO is an extension of the classical LASSO method, designed to handle structured sparsity in regression problems. While standard LASSO promotes sparsity by shrinking individual coefficients to zero, group LASSO operates on predefined groups of variables, encouraging entire groups to be selected or excluded together. This is particularly useful when variables have inherent groupings, such as time-frequency components in signal processing, genes in biological pathways, or dummy variables for categorical features.

Suppose that the N predictors are divided into L groups, with p_l the number in group l. For ease of notation, we use a matrix $\boldsymbol{A}_l$ to represent the predictors corresponding to the l-th group, with corresponding coefficient vector $\tilde{\boldsymbol{x}}_l$. Then, the group LASSO, as proposed in [16], solves the following convex optimization problem:

$$\min_{\tilde{\boldsymbol{x}} \in \mathbb{R}^N} \left\| \boldsymbol{y} - \sum_{l=1}^{L} \boldsymbol{A}_l \tilde{\boldsymbol{x}}_l \right\|_2^2 + \lambda \sum_{l=1}^{L} \sqrt{p_l} \|\tilde{\boldsymbol{x}}_l\|_2, \tag{8.55}$$

[2] The scaling factor M in $\bar{d}(\tau)$ is used to ensure that $\bar{d}(\tau)$ is constant as $M, N \to \infty$.

where the $\sqrt{p_l}$ terms accounts for the varying group sizes. This procedure acts like the LASSO at the group level, i.e., depending on λ, an entire group of predictors may drop out of the model. In fact, if the group sizes are all one, it reduces to the LASSO.

Similar to the conventional LASSO as discussed in Sect. 8.2.2, we can obtain the solution of problem (8.55) via subgradient conditions. For each group l, the subgradient equations are

$$-\boldsymbol{A}_l^H\left(\boldsymbol{y}-\sum_k \boldsymbol{A}_k\tilde{\boldsymbol{x}}_k\right)+\lambda\sqrt{p_l}\boldsymbol{s}_l=\boldsymbol{0}, \tag{8.56}$$

where $\boldsymbol{s}_l$ is the subgradient vector defined as

$$\boldsymbol{s}_l=\begin{cases}\frac{\tilde{\boldsymbol{x}}_l}{\|\tilde{\boldsymbol{x}}_l\|_2}, & \text{if } \tilde{\boldsymbol{x}}_l\neq\boldsymbol{0},\\ \text{any vector with } \|\boldsymbol{s}_l\|_2<1, & \text{otherwise.}\end{cases} \tag{8.57}$$

Let the solutions be $\hat{\boldsymbol{x}}_1, \hat{\boldsymbol{x}}_2, \ldots, \hat{\boldsymbol{x}}_l$. Then, when $\tilde{\boldsymbol{x}}_l=\boldsymbol{0}$, by substituting $\tilde{\boldsymbol{x}}_l=\boldsymbol{0}$ into (8.56), we have

$$-\boldsymbol{A}_l^H\left(\boldsymbol{y}-\sum_{k\neq l} \boldsymbol{A}_k\hat{\boldsymbol{x}}_k\right)+\lambda\sqrt{p_l}\boldsymbol{s}_l=\boldsymbol{0}, \tag{8.58}$$

which can be further simplified to

$$\boldsymbol{A}_l^H\boldsymbol{r}_l=\lambda\sqrt{p_l}\boldsymbol{s}_l, \tag{8.59}$$

where $\boldsymbol{r}_l=\boldsymbol{y}-\sum_{k\neq l}\boldsymbol{A}_k\hat{\boldsymbol{x}}_k$ denotes the residual after excluding the l-th group. Since $\|\boldsymbol{s}\|_2<1$ (by definition when $\tilde{\boldsymbol{x}}_l=\boldsymbol{0}$), this implies

$$\left\|\boldsymbol{A}_l^H\boldsymbol{r}_l\right\|_2<\lambda\sqrt{p_l} \implies \hat{\boldsymbol{x}}_l=\boldsymbol{0}. \tag{8.60}$$

Similarly, when $\tilde{\boldsymbol{x}}_l\neq\boldsymbol{0}$, we obtain

$$\boldsymbol{A}_l^H\boldsymbol{A}_l\tilde{\boldsymbol{x}}_l-\boldsymbol{A}_l^H\boldsymbol{r}_l+\lambda\sqrt{p_l}\frac{\tilde{\boldsymbol{x}}_l}{\|\tilde{\boldsymbol{x}}_l\|_2}=\boldsymbol{0}, \tag{8.61}$$

which directly leads to

$$\hat{\boldsymbol{x}}_l=\left(\boldsymbol{A}_l^H\boldsymbol{A}_l+\frac{\lambda\sqrt{p_l}}{\|\hat{\boldsymbol{x}}_l\|_2}\boldsymbol{I}\right)^{-1}\boldsymbol{A}_l^H\boldsymbol{r}_l. \tag{8.62}$$

Now, if we assume $\boldsymbol{A}_l^H\boldsymbol{A}_l=\boldsymbol{I}$, the above solution simplifies significantly. Specifically, in this case, Eq. (8.62) becomes

$$\hat{\boldsymbol{x}}_l=\left(\boldsymbol{I}+\frac{\lambda\sqrt{p_l}}{\|\hat{\boldsymbol{x}}_l\|_2}\boldsymbol{I}\right)^{-1}\boldsymbol{A}_l^H\boldsymbol{r}_l=\left(1+\frac{\lambda\sqrt{p_l}}{\|\hat{\boldsymbol{x}}_l\|_2}\right)^{-1}\boldsymbol{A}_l^H\boldsymbol{r}_l. \tag{8.63}$$

Taking norms on both sides of (8.63), we have $\|\hat{\boldsymbol{x}}_l\|_2=\frac{\|\boldsymbol{A}_l^H\boldsymbol{r}_l\|_2}{1+\frac{\lambda\sqrt{p_l}}{\|\hat{\boldsymbol{x}}_l\|_2}}$, which yields

$$\|\hat{\boldsymbol{x}}_l\|_2^2+\lambda\sqrt{p_l}\|\hat{\boldsymbol{x}}_l\|_2-\|\boldsymbol{A}_l^H\boldsymbol{r}_l\|_2\|\hat{\boldsymbol{x}}_l\|_2=0 \implies \|\hat{\boldsymbol{x}}_l\|_2=\|\boldsymbol{A}_l^H\boldsymbol{r}_l\|_2-\lambda\sqrt{p_l}. \tag{8.64}$$

Thus, the solution in (8.60) and (8.62) becomes

$$\hat{\boldsymbol{x}}_l=\left(1-\frac{\lambda\sqrt{p_l}}{\|\boldsymbol{A}_l^H\boldsymbol{r}_l\|_2}\right)_+\boldsymbol{A}_l^H\boldsymbol{r}_l, \tag{8.65}$$

which is known as the group soft-thresholding operator. As can be seen in (8.65), the solution depends on the current residual $\boldsymbol{r}_l$, which is computed using fixed $\hat{\boldsymbol{x}}_k$ $(k\neq l)$, this leads to an algorithm that cycles through the groups, and is a block-wise coordinate descent procedure [16].

When the predictors within a group $\boldsymbol{A}_l$ are not orthonormal, one might attempt to orthonormalize them using SVD, i.e., let $\boldsymbol{A}_l=\boldsymbol{U}\boldsymbol{D}\boldsymbol{V}^H$, then the columns of $\boldsymbol{U}=\boldsymbol{A}_l\boldsymbol{V}\boldsymbol{D}^{-1}$ can be viewed as the orthonormalized predictors, and the original model

$A_l\tilde{x}_l$ can be rewritten as

$$A_l\tilde{x}_l = UDV^H\tilde{x}_l = U(DV^H\tilde{x}_l) = U\tilde{x}_l^*, \tag{8.66}$$

where $\tilde{x}_l^* = DV^H\tilde{x}_l$. In this case, the norm of the transformed coefficients $\tilde{x}_l^*$ and the original coefficients $\tilde{x}_l$ are related as

$$\|\tilde{x}_l^*\|_2 = \|DV^H\tilde{x}_l\|_2 \neq \|\tilde{x}_l\|_2 \quad \text{unless } D = I. \tag{8.67}$$

Thus, solving the group LASSO on U (orthonormalized predictors) alters the original penalty structure $\|\tilde{x}\|_2$, and this may lead to incorrect solutions.

For non-orthonormal A_l, the group LASSO solution resembles a ridge regression solution with a regularization parameter $\frac{\lambda\sqrt{p_l}}{\|\hat{x}_l\|_2}$. However, the parameter depends on $\|\hat{x}_l\|_2$, which creates a circular dependency, i.e., $\hat{x}_l$ depends on $\|\hat{x}_l\|_2$, while $\|\hat{x}_l\|_2$ also depends on $\hat{x}_l$. In this case, we can define $t_l = \|\hat{x}_l\|_2$, solve (8.62) for $\hat{x}_l$ in terms of t_l, derive a scalar equation for t_l by taking the norm of both sides of (8.62), i.e.,

$$t_l = \left\| \left(A_l^H A_l + \frac{\lambda\sqrt{p_l}}{t_l} I \right)^{-1} A_l^H r_l \right\|_2, \tag{8.68}$$

and iteratively solve for t_l and substitute back into (8.62). However, the above equation is a nonlinear equation in t_l, which cannot be solved analytically and needs to be approximated by numerical methods (such as Newton's iteration method). Such an approach can involve dividing by the norm of a vector that is very close to zero. It is also not guaranteed to converge. Please refer to [17] for a better solution.[3]

8.3.3 Burst LASSO

Consider the massive MIMO channel in Definition 8.1, the main challenge here is designing a proper penalty function $f(\tilde{x})$ to fully exploit the burst-sparse structure. Unlike the block-sparse signal where the support of each non-zero block is randomly generated from a set of non-overlapping candidate supports (as illustrated in Fig. 8.5a), the support of each non-zero burst in the burst-sparse channel (signal) in Definition 8.1 is randomly generated from a set of N overlapping candidate supports, as illustrated in Fig. 8.5b. If we directly apply existing block-sparse recovery algorithms such as group LASSO to recover the burst-sparse signal, the performance would be far from the optimal. As a result, both the design of the penalty function $f(\tilde{x})$ and the associated performance analysis for burst-sparse signals are much more difficult than those for block-sparse signals.

To address the above challenges, this subsection first presents a block-sparse lifting technique to transform a burst-sparse signal into a block-sparse signal with a higher dimension, from which a proper penalty function $f(\tilde{x})$ for burst-sparse signals is derived. Based on this, a burst LASSO algorithm is proposed for the recovery of the burst-sparse channel [18]. Note that the channel estimation performance depends on the available information at the user. In practice, the number of bursts K and the per-burst sparsity bound D changes slowly (over a very long timescale) compared to instantaneous channel support information such as the set of the starting locations of the bursts $\mathcal{S}^\star = \{s_1^\star, \ldots, s_K^\star\}$. As a result, the BS can obtain knowledge of $\{K, D\}$ using offline channel propagation measurement [19] or long term stochastic learning and estimation [20].

For a given burst-sparse channel $\tilde{x}^\star$ that is sparse but with non-zero values clustered in K contiguous blocks (bursts), it is expressed as

$$\tilde{x}^\star = \sum_{k=1}^{K} \tilde{x}_k^\star, \tag{8.69}$$

where each $\tilde{x}_k^\star \in \mathbb{C}^N$ represents a vector of length N that contains the k-th block. Define $z^\star = [z_1^{\star T}, \ldots, z_N^{\star T}]^T \in \mathbb{C}^{ND}$. Here, $z_{s_k^\star}^\star = \tilde{x}_k^\star[I_{s_k^\star}] \in \mathbb{C}^D$ for $k = 1, \ldots, K$, and $z_i^\star = \mathbf{0} \in \mathbb{C}^D$ for all $i \notin \mathcal{S}^\star$. This $z^\star$ represents a block-sparse signal with K non-zero blocks. The burst-sparse channel $\tilde{x}^\star$ is related to the block-sparse signal $z^\star$ through the following block-sparse lifting transformation:

$$\tilde{x}^\star = Lz^\star, \tag{8.70}$$

[3] Instead of tackling (8.68), one can also use convex solvers, such as CVX, to solve the full group LASSO problem (8.55). However, it may be inefficient for large-scale problems compared to block-wise coordinate descent.

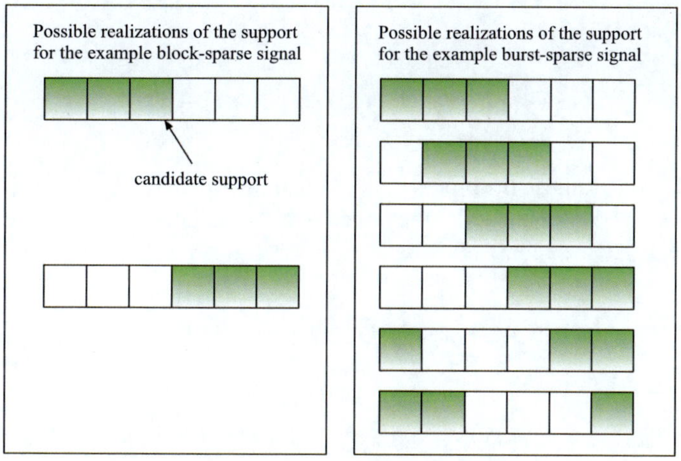

(a) In the example block-sparse signal, the support is selected from two non-overlapping candidate supports.

(b) For the example burst-sparse signal, the support is selected from six overlapping candidate supports.

Fig. 8.5 (Adapted from Fig. 2 in [18]) Comparison of the block-sparse signal and burst-sparse signal. The example block-sparse signal has the dimension $N = 6$ and a single non-zero block of size $D = 3$. The example burst-sparse signal has the dimension $N = 6$ and a single non-zero burst of size $D = 3$

where $\boldsymbol{L} \in \{0, 1\}^{N \times ND}$, and its $((i-1)D + j)$-th column ($\forall i \in \{1, \ldots, N\}, j \in \{1, \ldots, D\}$) has exactly one non-zero element 1 at the location $i \oplus (j-1)$.

Example 8.1 (*Block-Sparse Lifting Transformation*). Consider an example for $N = 8$, $D = 2$, $s_1^\star = 2$ and $s_2^\star = 8$. In this case, $\boldsymbol{z}^\star$ contains two non-zero blocks: block 2 (spanning positions 3 and 4, denoted as $\boldsymbol{z}_2^\star = [z_{2,1}^\star, z_{2,2}^\star]^T$) and block 8 (positions 15 and 16, denoted as $\boldsymbol{z}_8^\star = [z_{8,1}^\star, z_{8,2}^\star]^T$), while all other blocks are zero. The transformation is governed by the binary matrix

$$\boldsymbol{L} = \begin{bmatrix} 0,0,0,0,0,0,0,0,0,0,0,0,0,0,0,1 \\ 0,0,1,0,0,0,0,0,0,0,0,0,0,0,0,0 \\ 0,0,0,1,0,0,0,0,0,0,0,0,0,0,0,0 \\ 0,0,0,0,0,0,0,0,0,0,0,0,0,0,0,0 \\ 0,0,0,0,0,0,0,0,0,0,0,0,0,0,0,0 \\ 0,0,0,0,0,0,0,0,0,0,0,0,0,0,0,0 \\ 0,0,0,0,0,0,0,0,0,0,0,0,0,0,0,0 \\ 0,0,0,0,0,0,0,0,0,0,0,0,0,0,1,0 \end{bmatrix} \in \{0, 1\}^{8 \times 16}, \tag{8.71}$$

where each non-zero block in $\boldsymbol{z}^\star$ is mapped to specific rows in $\tilde{\boldsymbol{x}}^\star$ via cyclic shifts. For block 2, the first element $z_{2,1}^\star$ is placed at row $2 \oplus (1-1) = 2$ (column 3 of $\boldsymbol{L}$), and the second element $z_{2,2}^\star$ shifts to row $2 \oplus 1 = 3$ (column 4 of $\boldsymbol{L}$). Similarly, block 8's first element $z_{8,1}^\star$ maps to row $8 \oplus (1-1) = 8$ (column 15 of $\boldsymbol{L}$), while its second element $z_{8,2}^\star$ cyclically wraps to row $8 \oplus 1 = 1$ (column 16 of $\boldsymbol{L}$). Applying $\tilde{\boldsymbol{x}}^\star = \boldsymbol{L}\boldsymbol{z}^\star$, the resulting burst-sparse signal becomes

$$\tilde{\boldsymbol{x}}^\star = [z_{8,2}^\star, z_{2,1}^\star, z_{2,2}^\star, 0, 0, 0, 0, z_{8,1}^\star]^T, \tag{8.72}$$

where non-zero values occupy positions 1, 2, 3, and 8.

Using the block-sparse lifting transformation in (8.70), the burst-sparse channel estimation model can be equivalently formulated as

$$\boldsymbol{y} = \boldsymbol{A}\boldsymbol{L}\boldsymbol{z}^\star + \boldsymbol{n}. \tag{8.73}$$

Based on (8.73), a burst LASSO algorithm can be derived to recover the burst-sparse channel $\tilde{\boldsymbol{x}}^\star = \boldsymbol{L}\boldsymbol{z}^\star$. This is achieved by first solving a lifted-group LASSO problem and then applying the block-sparse lifting transformation. Specifically, the lifted-group LASSO problem is defined as

$$\min_{z} \frac{1}{2}\|y - ALz\|^2 + \sigma\lambda M \bar{f}(z), \tag{8.74}$$

where $\bar{f}(z) = \sum_{i=1}^{N} \|z_i\|$ is utilized to impose the block-sparse constraint on z. After solving an optimal solution $z^\star$ of (8.74), an estimation of $\tilde{x}^\star$ is given by $\hat{x} = Lz^\star$. It can be demonstrated that the output of the burst LASSO algorithm is the optimal solution of the following burst LASSO problem

$$\min_{\tilde{x}} \frac{1}{2}\|y - A\tilde{x}\|^2 + \sigma\lambda M f(\tilde{x}), \tag{8.75}$$

where

$$f(\tilde{x}) \triangleq \min_{z} \bar{f}(z), \quad \text{s.t. } \tilde{x} = Lz. \tag{8.76}$$

Hence, the burst-sparse structure of $\tilde{x}^\star$ is encoded in the above penalty function $f(\tilde{x})$, which is the optimal objective of the convex problem in (8.76). It can be shown that $f(\tilde{x})$ is also convex.

Theorem 8.1 cannot be applied to the lifted-group LASSO problem (8.74) because the measurement matrix AL is no longer an i.i.d. Gaussian matrix. Although Theorem 8.1 is valid for the burst LASSO problem in (8.75), it is difficult to calculate the average Gaussian distance $\bar{d}(\tau)$ and determine the algorithm parameter λ using Theorem 8.1 because the penalty function $f(\tilde{x})$ has no closed-form expression. In summary, due to the more complicated burst-sparse structure, both the algorithm parameter optimization and the associated performance analysis for burst-sparse signals are much more challenging than those for block-sparse signals.

According to Theorem 8.1, the aNSE performance of the burst LASSO with penalty function $f(\tilde{x})$ in (8.76) depends on the corresponding average Gaussian distance $\bar{d}(\tau)$, which further depends on the set of all subgradients $\partial f(\tilde{x}^\star)$. However, there is no closed-form expression for the penalty function $f(\tilde{x})$ given in (8.76), thus it is highly non-trivial to characterize $\partial f(\tilde{x}^\star)$. This challenge is addressed in the following theorem [18].

Theorem 8.2 (Property of the Subgradient of $f(\tilde{x})$) *The set of all subgradients of $f(\tilde{x})$ at $\tilde{x}^\star$ is given by*

$$\partial f(\tilde{x}^\star) = \left\{\alpha : L^T\alpha \in \partial \bar{f}(z^\star)\right\}, \tag{8.77}$$

where $\partial \bar{f}(z^\star) = \left[\partial_1^T \bar{f}(z^\star), \ldots, \partial_N^T \bar{f}(z^\star)\right]^T$ *with*

$$\partial_i \bar{f}(z^\star) = \begin{cases} \frac{z_i^\star}{\|z_i^\star\|}, & \|z_i^\star\| \neq 0 \\ \{g_i : \|g_i\| \leq 1\}, & \|z_i^\star\| = 0 \end{cases} \tag{8.78}$$

and it is the set of all subgradients of $\bar{f}(z)$ at $z^\star$, and $z^\star$ can be any optimal solution of problem (8.76) *with $\tilde{x} = \tilde{x}^\star$.*

Based on Theorem 8.2, the Gaussian distance and aNSE for the general burst-sparse channels with $K \geq 1$ bursts are given in the following theorem and corollary.

Theorem 8.3 (Gaussian Distance for General Burst-Sparse Channels) *Let $\tilde{x}^\star$ be a general burst-sparse channel. When $N, M, D \to \infty$ with constant ratios $\frac{M}{N} \to \delta \in (0, \infty)$ and $\frac{D}{M} \to \rho \in (0, 1)$, we almost surely have $\bar{d}(\tau) \geq \bar{d}^\circ(\tau)$ with*

$$\bar{d}^\circ(\tau) = K(\tau^2 + \rho) + \left(\frac{1}{\delta} - K\rho\right)\left[\left(1 - \sqrt{\frac{1}{\rho}}\tau\right)_+\right]^2. \tag{8.79}$$

Corollary 8.1 (aNSE for General Burst-Sparse Channels) *Consider the recovery of a general burst-sparse channel $\tilde{x}^\star$ with parameter $\{K, D\}$ using burst LASSO. When $N, M, D \to \infty$ with constant ratios $\frac{M}{N} \to \delta \in (0, \infty)$ and $\frac{D}{M} \to \rho \in (0, 1)$, and $2K\rho - K^2\rho^2\delta < 1$, the minimum aNSE is lower bounded as*

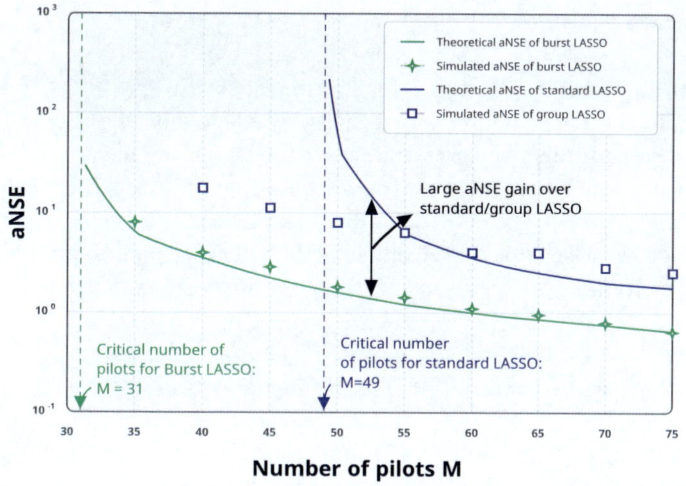

Fig. 8.6 (Adapted from Fig. 3 in [18]) Illustration of the advantages of exploiting the burst-sparsity predicted by Corollary 8.1. Consider a massive MIMO system with the ULA of $N = 128$ antennas at the BS. The burst-sparse channel parameters are given in Fig. 8.4. The per-burst sparsity bound information at the user is $D = 8$

$$aNSE\left(\lambda^{\star}\right) \geq \frac{2K\rho - K^2\rho^2\delta}{1-(2K\rho - K^2\rho^2\delta)}. \tag{8.80}$$

In practice, we can use a burst LASSO algorithm with parameter [18]

$$\lambda^{*} = (1 - K\rho\delta)\sqrt{\rho}\sqrt{1-(2K\rho - K^2\rho^2\delta)} \tag{8.81}$$

to approach the aNSE lower bound in (8.80). The rationale for the choice of such algorithm parameter is as follows. When $\bar{d}(\tau) = \bar{d}^{\circ}(\tau)$, λ^* is the optimal algorithm parameter and the burst LASSO with parameter λ^* achieves the aNSE lower bound in (8.80). Hence, if $|\bar{d}(\tau) - \bar{d}^{\circ}(\tau)|$ is smaller, the aNSE performance of the burst LASSO with parameter λ^* is expected to be close to the aNSE lower bound in (8.80). Corollary 8.1 quantifies the significant advantages of exploiting the burst-sparsity. Figure 8.6 illustrates the advantages predicted by Corollary 8.1 by comparing the theoretical aNSE of the burst LASSO (given in Corollary 8.1) and the standard ℓ_2^2-LASSO [22] (whose theoretical aNSE is given in [21]) for the massive MIMO channel estimation problem considered in [18]. The simulated aNSEs of burst LASSO with parameter λ^*, group LASSO[4] and standard LASSO are also plotted in Fig. 8.6 for comparison. It can be seen that the theoretical prediction obtained from the asymptotic analysis is still quite accurate for a system with a finite number of antennas (i.e., the length of the sparse signal). Moreover, from Fig. 8.6, we can make the following observation.

- Much Better Phase Transition Behavior: The theoretical aNSEs of the burst LASSO and standard LASSO exhibit a phase transition behavior. Specifically, when the number of pilots (i.e., the number of measurements) M is larger than a critical point, the aNSE is finite, which implies that robust channel recovery can be guaranteed. As M approaches the critical point, the aNSE grows to ∞ and robust channel recovery is impossible. However, the critical number of pilots required for robust channel recovery in the burst LASSO is much smaller than that in the standard LASSO, as shown in Fig. 8.6 (31 versus 49).
- Huge Saving in Pilot Overhead: For example, to achieve the same aNSE, say 3, the number of pilots required by the burst LASSO is only about 62% of that required by the standard/group LASSO (about 40 versus 65), as shown in Fig. 8.6.

Beyond burst sparsity, recent research has continued to explore more sophisticated structured sparsity patterns tailored to specific sensing and communication scenarios. For instance, in multi-target sensing systems operating in multipath environments, a cross-sparsity structure has been proposed to exploit the geometric relationships between direct and first-order reflection paths [23]. By modeling the dependencies between the angles of departure (AoDs) and AoAs of different targets, cross-sparsity enables more accurate target localization through a structured Bayesian inference framework.

[4] Since the burst-sparse signal is not a special case of block-sparse signal, the theoretical aNSE of using the group LASSO to recover the burst-sparse signal is unknown.

Summary

This chapter presents optimization-driven algorithms for sparse signal recovery in CS. We introduce reweighted ℓ_1-minimization, an iterative technique enhancing sparsity recovery by adaptively adjusting weights based on intermediate solutions, which significantly outperforms standard ℓ_1-minimization with fewer measurements. Besides, we extensively discuss the LASSO, detailing its formulation, Bayesian interpretation, coordinate descent solution via soft-thresholding, and theoretical guarantees (prediction consistency, sparsistency). Further, we explore structured sparsity extensions, including group LASSO for group-wise variable selection and burst LASSO tailored for burst-sparse signals (e.g., in massive MIMO channels), which employs a block-sparse lifting transformation. Performance is analyzed using metrics like aNSE via the CGMT.

Specifically, reweighted ℓ_1-minimization bridges the gap between ℓ_0 and ℓ_1 norms by iteratively refining weights to better approximate the ideal ℓ_0 sparsity count. While offering superior recovery accuracy, it increases computational cost compared to standard ℓ_1 methods. Besides, the reweighted ℓ_1 scheme is unified under the MM framework, linking it to minimizing a log-sum penalty. This ensures monotonic convergence and provides scalability to large-scale problems.

LASSO provides a convex relaxation for robust sparse regression and variable selection. Its Bayesian variant integrates prior distributions (e.g., Laplacian) for probabilistic interpretation and uncertainty quantification, trading off bias and variance via the regularization parameter λ.

Generalized LASSO is a flexible extension that adapts the regularization framework to accommodate domain-specific prior knowledge or complex dependencies among variables. Group LASSO and burst LASSO are two typical generalized LASSO variants that explicitly models dependencies (groups, contiguous bursts) to significantly improve recovery efficiency for structured signals. Burst LASSO, via its lifting transform, demonstrates substantial gains in pilot efficiency (e.g., about 40% reduction) and phase transition behavior over standard methods for the massive MIMO channel estimation problem. However, their performance critically depends on accurate parameter tuning (e.g., λ, structure parameters, etc.) matching the signal's inherent sparsity coherence, which may be difficult to determine in some practical application scenarios.

Appendix

Appendix 1: Derivation of the Generalized Soft-Thresholding Operator

Let us first define a key quantity that is constant with respect to $\tilde{x}_j$, i.e., $z_j = \frac{1}{M}\langle \boldsymbol{a}_j, \boldsymbol{r}^{(j)}\rangle = \frac{1}{M}\sum_{i=1}^{M} A_{ij} r_i^{(j)}$ (the unregularized LS estimate for $\tilde{x}_j$ based on the current residuals). Then, we can now rewrite the objective function in (8.36), denoted by $F(\tilde{x}_j)$, by expanding the quadratic term as follows:

$$
\begin{aligned}
F(\tilde{x}_j) &= \frac{1}{2M}\sum_{i=1}^{M}\left(\left(r_i^{(j)}\right)^2 - 2A_{ij} r_i^{(j)}\tilde{x}_j + A_{ij}^2\tilde{x}_j^2\right) + \lambda|\tilde{x}_j| \\
&= \text{Constant} - \left(\frac{1}{M}\sum_{i=1}^{M} A_{ij} r_i^{(j)}\right)\tilde{x}_j + \frac{1}{2}\left(\frac{1}{M}\sum_{i=1}^{M} A_{ij}^2\right)\tilde{x}_j^2 + \lambda|\tilde{x}_j| \\
&= C - z_j\tilde{x}_j + \frac{1}{2}\gamma_j\tilde{x}_j^2 + \lambda|\tilde{x}_j|.
\end{aligned}
$$

To find the minimizer of $F(\tilde{x}_j)$, we analyze its subgradient, which is given by

$$\partial F(\tilde{x}_j) = -z_j + \gamma_j\tilde{x}_j + \lambda\partial|\tilde{x}_j|,$$

where $\partial|\tilde{x}_j|$ is the subdifferential of the absolute value function, i.e.,

$$\partial|\tilde{x}_j| = \begin{cases} 1 & \text{if } \tilde{x}_j > 0, \\ [-1, +1] & \text{if } \tilde{x}_j = 0, \\ -1 & \text{if } \tilde{x}_j < 0. \end{cases}$$

A point $\tilde{x}_j$ is a minimizer if and only if $0 \in \partial F(\tilde{x}_j)$. We proceed by dividing the derivation into the following three cases.

1. $\tilde{x}_j > 0$: we have $\partial|\tilde{x}_j| = 1$, which leads to

$$0 = -z_j + \gamma_j \tilde{x}_j + \lambda \implies \gamma_j \tilde{x}_j = z_j - \lambda \implies \tilde{x}_j = \frac{z_j - \lambda}{\gamma_j}.$$

   This solution is valid only if $\tilde{x}_j > 0$, which requires $z_j > \lambda$.
2. $\tilde{x}_j < 0$: then, $\partial|\tilde{x}_j| = -1$ holds, and we can obtain

$$0 = -z_j + \gamma_j \tilde{x}_j - \lambda \implies \gamma_j \tilde{x}_j = z_j + \lambda \implies \tilde{x}_j = \frac{z_j + \lambda}{\gamma_j},$$

   which is valid only if $\tilde{x}_j < 0$, i.e., $z_j < -\lambda$.
3. $\tilde{x}_j = 0$: in this case, we obtain $\partial|\tilde{x}_j| = [-1, +1]$, and

$$0 \in -z_j + \lambda \cdot [-1, +1] \implies -z_j \in [-\lambda, +\lambda] \implies |z_j| \leq \lambda.$$

   Thus, $\tilde{x}_j = 0$ is optimal when $|z_j| \leq \lambda$.

By combining the above three cases, the solution can be expressed as

$$\tilde{x}_j = \begin{cases} \frac{z_j - \lambda}{\gamma_j} & \text{if } z_j > \lambda \\ 0 & \text{if } |z_j| \leq \lambda \\ \frac{z_j + \lambda}{\gamma_j} & \text{if } z_j < -\lambda, \end{cases}$$

which can be expressed compactly using the soft-thresholding operator $\mathcal{S}_\lambda(x) = \text{sign}(x)(|x| - \lambda)_+$, i.e.,

$$\tilde{x}_j = \frac{1}{\gamma_j} \mathcal{S}_\lambda(z_j),$$

which is exactly (8.37).

Appendix 2: Derivation of the Average Gaussian Distance for the Real Case

In the real case, according to [13], the normalized Gaussian squared distance to the scaled subdifferential is defined as

$$d(\tau) \triangleq d_{\partial f(\tilde{\boldsymbol{x}}^\star)}(\tau) \triangleq \frac{1}{N} \mathbb{E}_{\boldsymbol{w}} \left\{ \min_{\boldsymbol{\alpha} \in \partial f(\tilde{\boldsymbol{x}}^\star)} \|\boldsymbol{w} - \tau \boldsymbol{\alpha}\|^2 \right\}, \tag{8.82}$$

where $f : \mathbb{R}^N \to \mathbb{R}$ is convex, $\boldsymbol{w} \in \mathbb{R}^N$ have i.i.d. $\mathcal{N}(0, 1)$ entries. For $f(\cdot) = \|\cdot\|_1$, the derivation of (8.82) can be divided into the following four steps.

1. Calculate the Scaled Subdifferential Set $\tau \partial f(\tilde{\boldsymbol{x}}^\star)$: For $f(\cdot) = \|\cdot\|_1$, the subdifferential $\partial\|\tilde{\boldsymbol{x}}^\star\|_1$ depends on the support and signs of $\tilde{\boldsymbol{x}}^\star$. Since $\tilde{\boldsymbol{x}}^\star$ is S-sparse, let $\mathcal{S} \subset \{1, \ldots, N\}$ denote the support set (where $\tilde{\boldsymbol{x}}^\star$ has non-zero entries), with $|\mathcal{S}| = S$. For each component i, we have

   - If $i \in \mathcal{S}$ (i.e., $\tilde{x}_i^\star \neq 0$), the subdifferential is $\partial|\tilde{x}_i^\star| = \text{sign}(\tilde{x}_i^\star)$. Thus, $\tau \partial|\tilde{x}_i^\star| = \tau\, \text{sign}(\tilde{x}_i^\star)$.

- If $i \notin S$ (i.e., $\tilde{x}_i^\star = 0$), the subdifferential is $\partial|\tilde{x}_i^\star| = [-1, 1]$. Thus, $\tau\partial|\tilde{x}_i^\star| = [-\tau, \tau]$.

The scaled subdifferential set $\tau\partial f(\tilde{\boldsymbol{x}}^\star)$ is the Cartesian product of these component-wise sets. Therefore, we obtain

$$\tau\partial f(\tilde{\boldsymbol{x}}^\star) = \left\{\boldsymbol{s} \in \mathbb{R}^N \mid s_i = \tau \operatorname{sign}\left(\tilde{x}_i^\star\right) \text{ for } i \in \mathcal{S}, \text{ and } s_i \in [-\tau, \tau] \text{ for } i \notin \mathcal{S}\right\}. \tag{8.83}$$

2. Compute the Distance $\min_{\boldsymbol{\alpha}\in\partial f(\tilde{\boldsymbol{x}}^\star)} \|\boldsymbol{w} - \tau\boldsymbol{\alpha}\|^2$: The squared distance decomposes additively over the components because the set $\tau\partial f(\tilde{\boldsymbol{x}}^\star)$ is separable (Cartesian product), which yields

$$\min_{\boldsymbol{\alpha}\in\partial f(\tilde{\boldsymbol{x}}^\star)} \|\boldsymbol{w} - \tau\boldsymbol{\alpha}\|^2 = \sum_{i=1}^{N} \min_{\alpha_i\in\partial|(\tilde{\boldsymbol{x}}^\star)_i|} (w_i - \tau\alpha_i)^2. \tag{8.84}$$

For $i \in \mathcal{S}$, $\alpha_i = \tau \operatorname{sign}(\tilde{x}_i^\star)$, so we have

$$\min_{\alpha_i\in\partial|(\tilde{\boldsymbol{x}}^\star)_i|} (w_i - \tau\alpha_i)^2 = \left(w_i - \tau \operatorname{sign}(\tilde{x}_i^\star)\right)^2, \tag{8.85}$$

while for $i \notin \mathcal{S}$, $\alpha_i = [-\tau, \tau]$, we obtain

$$\min_{\alpha_i\in\partial|\tilde{x}_i^\star|} (w_i - \tau\alpha_i)^2 = (\max\{|w_i| - \tau, 0\})^2. \tag{8.86}$$

The sign $\operatorname{sign}(\tilde{x}_i^\star)$ is fixed for each $i \in \mathcal{S}$, but since the expectation is over $\boldsymbol{w}$ (which is symmetric and i.i.d. Gaussian), and the distribution of w_i is symmetric around 0, the value of $\operatorname{sign}(\tilde{x}_i^\star)$ does not affect the expectation. Without loss of generality, we can assume $\operatorname{sign}(\tilde{x}_i^\star) = 1$ for all $i \in \mathcal{S}$ (if some signs are -1, the expectation remains the same due to symmetry). Thus, we have

$$\mathbb{E}_{w_i}\left\{\min_{\alpha_i\in\partial|\tilde{x}_i^\star|} (w_i - \tau\alpha_i)^2\right\} = \begin{cases} \mathbb{E}_{w_i}\left\{(w_i - \tau)^2\right\}, \text{ for } i \in \mathcal{S}, \\ \mathbb{E}_{w_i}\left\{(\max\{|w_i| - \tau, 0\})^2\right\}, \text{ for } i \notin \mathcal{S}. \end{cases} \tag{8.87}$$

3. Compute the Expectation $\mathbb{E}_{\boldsymbol{w}}\left\{\min_{\boldsymbol{\alpha}\in\partial f(\tilde{\boldsymbol{x}}^\star)} \|\boldsymbol{w} - \tau\boldsymbol{\alpha}\|^2\right\}$: Since the w_i's are i.i.d. $\mathcal{N}(0, 1)$, this expectation simplifies to

$$\mathbb{E}_{\boldsymbol{w}}\left\{\min_{\boldsymbol{\alpha}\in\partial f(\tilde{\boldsymbol{x}}^\star)} \|\boldsymbol{w} - \tau\boldsymbol{\alpha}\|^2\right\} = \sum_{i\in\mathcal{S}} \mathbb{E}_{w_i}\left\{(w_i - \tau)^2\right\} + \sum_{i\notin\mathcal{S}} \mathbb{E}_{w_i}\left\{(\max\{|w_i| - \tau, 0\})^2\right\}. \tag{8.88}$$

It can be readily seen that the above sums depend only on the sizes of $\mathcal{S}$ and its complement, i.e., $|\mathcal{S}| = S = \rho N$, $|\mathcal{S}^c| = N - S = (1 - \rho)N$. As the w_i's are identically distributed, we compute the expectations for a single standard Gaussian component $h \sim \mathcal{N}(0, 1)$:

- Expectation for $i \in \mathcal{S}$: The expectation is computed using the linearity of expectation and properties of the standard Gaussian $w_i \sim \mathcal{N}(0, 1)$, i.e.,

$$\mathbb{E}\left\{(w_i - \tau)^2\right\} = \mathbb{E}\{w_i^2\} - 2\tau\mathbb{E}\{w_i\} + \tau^2. \tag{8.89}$$

Since $\mathbb{E}\{w_i\} = 0$ (symmetry) and $\mathbb{E}\{w_i^2\} = \operatorname{Var}(w_i) + (\mathbb{E}\{w_i\})^2 = 1$, substituting these into (8.89) gives

$$\mathbb{E}\left\{(w_i - \tau)^2\right\} = 1 + \tau^2. \tag{8.90}$$

- Expectation for $i \notin \mathcal{S}$: Since the squared distance involves the indicator that $|w_i| > \tau$, we rewrite the expectation as follows:

$$\mathbb{E}\left\{(\max\{|w_i| - \tau, 0\})^2\right\} = \mathbb{E}\left\{(|w_i| - \tau)^2\, \mathbf{1}_{\{|w_i|>\tau\}}\right\}. \tag{8.91}$$

Because w_i is symmetric (i.e., w_i and $-w_i$ have the same distribution), we have

$$\mathbb{E}\left\{(|w_i| - \tau)^2\, \mathbf{1}_{\{|w_i|>\tau\}}\right\} = 2\int_\tau^\infty (x - \tau)^2\phi(x)\, dx, \tag{8.92}$$

where $\phi(x) = \frac{1}{\sqrt{2\pi}} e^{-x^2/2}$ is the PDF of the standard Gaussian, and the integral is simplified by doubling the contribution from $x > \tau$. Define $I(\tau) = \int_\tau^\infty (x-\tau)^2 \phi(x)\,dx$, and expand $(x-\tau)^2 = x^2 - 2\tau x + \tau^2$, then the integral can be decomposed as

$$I(\tau) = \underbrace{\int_\tau^\infty x^2 \phi(x)\,dx}_{I_1(\tau)} - 2\tau \underbrace{\int_\tau^\infty x\phi(x)\,dx}_{I_2(\tau)} + \tau^2 \underbrace{\int_\tau^\infty \phi(x)\,dx}_{I_3(\tau)}. \tag{8.93}$$

For $I_1(\tau)$, use integration by parts and let $u = x$ and $dv = x\phi(x)dx$, then we have $du = dx$ and $v = -\phi(x)$ (since $\frac{d}{dx}\phi(x) = -x\phi(x)$), which leads to

$$\begin{aligned}\int_\tau^\infty x^2\phi(x)\,dx &= \int_\tau^\infty u dv = [-x\phi(x)]_\tau^\infty - \int_\tau^\infty (-\phi(x))dx \\ &= 0 - (-\tau\phi(\tau)) + \int_\tau^\infty \phi(x)dx = \tau\phi(\tau) + Q(\tau).\end{aligned} \tag{8.94}$$

For $I_2(\tau)$, use substitution $u = x^2/2$, we have $du = xdx$ and

$$\int_\tau^\infty x\phi(x)\,dx = \int_\tau^\infty x \cdot \frac{1}{\sqrt{2\pi}} e^{-x^2/2} dx = \frac{1}{\sqrt{2\pi}} \int_{\tau^2/2}^\infty e^{-u} du = \frac{1}{\sqrt{2\pi}} e^{-\tau^2/2}, \tag{8.95}$$

while for $I_3(\tau)$, we have $I_3(\tau) = \int_\tau^\infty \phi(x)\,dx = Q(\tau)$. Substituting $I_1(\tau)$, $I_2(\tau)$ and $I_3(\tau)$ into (8.93), we have

$$I(\tau) = Q(\tau)(1+\tau^2) - \tau\phi(\tau). \tag{8.96}$$

Now, by combining the above expectations, the following result can be readily obtained:

$$\mathbb{E}_{\boldsymbol{w}}\left\{ \min_{\boldsymbol{\alpha}\in\partial f(\tilde{\boldsymbol{x}}^\star)} \|\boldsymbol{w} - \tau\boldsymbol{\alpha}\|^2 \right\} = S(1+\tau^2) + (N-S)\left[2\left(1+\tau^2\right) Q(\tau) - \sqrt{\frac{2}{\pi}}\tau e^{-\tau^2/2} \right]. \tag{8.97}$$

4. Compute $d(\tau)$: Normalize by $\frac{1}{N}$ and substitute $\rho = \frac{k}{N}$, we have

$$d(\tau) = \rho(1+\tau^2) + (1-\rho)\left[2(1+\tau^2)Q(\tau) - \sqrt{\frac{2}{\pi}}\tau e^{-\tau^2/2} \right]. \tag{8.98}$$

Homework Problems

8.1 Implement the reweighted ℓ_1-minimization algorithm in MATLAB. Use a Gaussian sensing matrix $\boldsymbol{A} \in \mathbb{R}^{M\times N}$ with $M = 100$, $N = 256$, and generate a sparse signal $\boldsymbol{x}$ with S non-zero entries (random positions, the non-zero values are chosen randomly from a zero-mean unit-variance Gaussian distribution). Compare the recovery performance (NMSE and success rate over 100 trials) of standard ℓ_1-minimization and reweighted ℓ_1-minimization (4 iterations, $\epsilon = 0.1$).

8.2 Focus on the LASSO algorithm.

(a) Derive the soft-thresholding operator $\mathcal{S}_\lambda(x)$ for the following univariate LASSO problem:

$$\min_{\bar{x}} \frac{1}{2}(y - a\bar{x})^2 + \lambda|\bar{x}|.$$

Show the solution when $a \neq 0$.

(b) Consider the LASSO problem in the following Lagrangian form:

$$\min_{\tilde{x}} \frac{1}{2} \|y - A\tilde{x}\|_2^2 + \lambda \|\tilde{x}\|_1 .$$

Assume A has orthonormal columns. Show that the solution is given by

$$\bar{x}_j = \mathcal{S}_\lambda \left(a_j^T y \right) .$$

(c) Implement coordinate descent for LASSO in MATLAB. Use a design matrix $A \in \mathbb{R}^{M\times N}$ with $M = 50$, $N = 100$, and generate a sparse signal x with $S = 15$. Add Gaussian noise to measurements, i.e., $y = Ax + n$, $n \sim \mathcal{N}(0, 0.1)$. Plot the solution path for $\lambda \in \left[0, \max_j \left|\langle a_j, y\rangle\right|\right]$ and identify the value of λ that minimizes the MSE.

8.3 (a) Consider a burst-sparse signal $x_* \in \mathbb{C}^N$ with parameters $\{K = 2, D = 4\}$ and length $N = 8$. The non-zero bursts start at locations $s_1^* = 2$ and $s_2^* = 7$. Explicitly construct the binary lifting matrix $L \in \{0, 1\}^{8\times 32}$. Describe the structure of one of its columns in your own words. The corresponding block-sparse signal in the lifted space is $z_* \in \mathbb{C}^{32}$. Specify the indices of the only two non-zero blocks in z_* and explain why all other blocks must be zero. A key advantage of the burst LASSO is that it provides an estimate $\hat{x} = L\hat{z}$ in the original signal space. Explain why this is more desirable than directly interpreting the solution $\hat{z}$ in the high-dimensional lifted space.

(b) Figure 8.6 in the chapter compares the aNSE of standard LASSO, group LASSO, and burst LASSO for a massive MIMO channel estimation problem. Describe the two key performance advantages of burst LASSO over standard LASSO, as revealed by the phase transition behavior and pilot overhead in the figure. The group LASSO performance is also plotted. Even though a burst-sparse signal is a form of structured sparsity, why does the group LASSO (with non-overlapping blocks) fail to match the performance of the burst LASSO? Connect your answer to the fundamental difference in how the supports of non-zero blocks are defined in each model (see Fig. 8.5).

References

1. E.J. Candes, M.B. Wakin, S.P. Boyd, Enhancing sparsity by reweighted ℓ_1 minimization. J. Fourier Anal. Appl. **14**, 877–905 (2008)
2. K. Lange, Optimization, vol. 95 (Springer Science & Business Media, 2013)
3. M. Fazel, Matrix rank minimization with applications, Doctoral dissertation (Ph.D. thesis) (Stanford University, 2002)
4. M.S. Lobo, M. Fazel, S. Boyd, Portfolio optimization with linear and fixed transaction costs. Ann. Oper. Res. **152**, 341–365 (2007)
5. M. Fazel, H. Hindi, S.P. Boyd, Log-det heuristic for matrix rank minimization with applications to Hankel and Euclidean distance matrices, in *Proceedings of the 2003 American Control Conference*, vol. 3, (IEEE, 2003), pp. 2156–2162
6. T. Hastie, R. Tibshirani, M. Wainwright, Statistical Learning with Sparsity: The LASSO and Generalizations, 1st edn. (Chapman and Hall/CRC, 2015)
7. T. Park, G. Casella, The Bayesian LASSO. J. Am. Stat. Assoc. **103**(482), 681–686 (2008)
8. E. Greenshtein, Y.A. Ritov, Persistence in high-dimensional linear predictor selection and the virtue of over-parametrization. Bernoulli **10**(6), 971–988 (2004)
9. P. Bühlmann, S. Van De Geer, Statistics for high-dimensional data: methods, theory and applications (Springer Science & Business Media, 2011)
10. D. Tse, P. Viswanath, *Fundamentals of Wireless Communication* (Cambridge University Press, Cambridge, 2005)
11. M. Stojnic, F. Parvaresh, B. Hassibi, On the reconstruction of block-sparse signals with an optimal number of measurements. IEEE Trans. Signal Process. **57**(8), 3075–3085 (2009)
12. Y.C. Eldar, P. Kuppinger, H. Bolcskei, Block-sparse signals: uncertainty relations and efficient recovery. IEEE Trans. Signal Process. **58**(6), 3042–3054 (2010)
13. C. Thrampoulidis, A. Panahi, B. Hassibi, Asymptotically exact error analysis for the generalized equation-LASSO, in *2015 IEEE International Symposium on Information Theory (ISIT)*. (IEEE, 2015), pp. 2021–2025
14. C. Thrampoulidis, S. Oymak, B. Hassibi, Regularized linear regression: a precise analysis of the estimation error, in *Conference on Learning Theory* (PMLR, 2015), pp. 1683–1709
15. C. Thrampoulidis, E. Abbasi, B. Hassibi, Precise error analysis of regularized M-estimators in high dimensions. IEEE Trans. Inf. Theory **64**(8), 5592–5628 (2018)
16. M. Yuan, Y. Lin, Model selection and estimation in regression with grouped variables. J. R. Stat. Soc. Ser. B Stat. Methodol. **68**(1), 49–67 (2006)
17. J. Friedman, T. Hastie, R. Tibshirani, A note on the group LASSO and a sparse group LASSO (2010). arXiv:1001.0736
18. A. Liu, V.K. Lau, W. Dai, Exploiting burst-sparsity in massive MIMO with partial channel support information. IEEE Trans. Wireless Commun. **15**(11), 7820–7830 (2016)
19. Y. Barbotin, A. Hormati, S. Rangan, M. Vetterli, Estimation of sparse MIMO channels with common support. IEEE Trans. Commun. **60**(12), 3705–3716 (2012)
20. L. Bottou, N. Murata, Stochastic approximations and efficient learning, in *The Handbook of Brain Theory and Neural Networks*, 2nd edn. (The MIT Press, Cambridge, MA, 2002)

21. S. Oymak, C. Thrampoulidis, B. Hassibi, The squared-error of generalized LASSO: a precise analysis, in *51st Annual Allerton Conference on Communication, Control, and Computing*. (IEEE, Allerton, 2013), pp. 1002–1009
22. E.J. Candès, M.B. Wakin, An introduction to compressive sampling. IEEE Signal Process. Mag. **25**(2), 21–30 (2008)
23. X. Chen, M.M. Zhao, A. Liu, M. Li, Q. Shi, M.J. Zhao, Cross-sparsity-enabled multipath perception via structured Bayesian inference for multi-target estimation (2025). arXiv:2511.14051

Greedy CS Recovery Algorithms

9

This chapter presents greedy algorithms for sparse signal recovery. We introduce orthogonal matching pursuit (OMP), and detail its iterative atom selection via residual correlation, LS estimation, and computational complexity. Theoretical performance guarantees are established for admissible measurement matrices (e.g., Gaussian, Bernoulli ensembles). To address OMP's limitations, enhanced variants are explored, e.g., regularized OMP (ROMP) ensures uniform recovery under the RIP by selecting multiple correlated atoms with magnitude regularization, stagewise OMP (StOMP) accelerates convergence through staged thresholding of correlations, and compressive sampling matching pursuit (CoSaMP) combines iterative support estimation, LS projection, and pruning for robust recovery to deliver strong reconstruction guarantees under RIP. Theoretical analysis results including recovery conditions, noise resilience, and computational efficiency are provided.

9.1 Orthogonal Matching Pursuit

As studied in Chaps. 7 and 8, suppose that $\tilde{\boldsymbol{x}}$ is an arbitrary S-sparse signal in $\mathbb{R}^N$. Form an $M \times N$ matrix $\boldsymbol{A}$ whose columns are the measurement vectors, and observe that the M measurements of the signal can be collected in an M-dimensional data vector $\boldsymbol{y} = \boldsymbol{A}\tilde{\boldsymbol{x}} + \boldsymbol{n}$. To identify the ideal signal $\tilde{\boldsymbol{x}}$, we need to determine which columns of $\boldsymbol{A}$ participate in the measurement vector $\boldsymbol{y}$. The idea behind the algorithm is to pick columns in a greedy fashion. At each iteration, we choose the column of $\boldsymbol{A}$ that is most strongly correlated with the remaining part of $\boldsymbol{y}$. Then we subtract off its contribution to $\boldsymbol{y}$ and iterate on the residual. One hopes that, after S iterations, the algorithm will have identified the correct set of columns. The overall OMP algorithm is given as follows.

Input: 1. an $M \times N$ measurement matrix $\boldsymbol{A}$; 2. an M-dimensional data vector $\boldsymbol{y}$; 3. the sparsity level S of the ideal signal.
Output: 1. an estimate $\hat{\boldsymbol{x}} \in \mathbb{R}^N$ for the ideal signal; 2. a set Λ_S containing S elements from $\{1, \ldots, N\}$; 3. an M-dimensional approximation $\boldsymbol{y}_S$ of the data $\boldsymbol{y}$; 4. an M-dimensional residual $\boldsymbol{r}_M = \boldsymbol{y} - \boldsymbol{y}_S$.

1. Initialize the residual $\boldsymbol{r}_0 = \boldsymbol{y}$, the index set $\Lambda_0 = \emptyset$, and the iteration counter $t = 1$.
2. Find the index λ_t that solves the easy optimization problem

$$\lambda_t = \arg \max_{j=1,\ldots,N} |\langle \boldsymbol{r}_{t-1}, \boldsymbol{a}_j \rangle|. \tag{9.1}$$

   If the maximum occurs for multiple indices, break the tie deterministically.[1]
3. Augment the index set and the matrix of chosen atoms: $\Lambda_t = \Lambda_{t-1} \cup \{\lambda_t\}$ and $\boldsymbol{A}_t = [\boldsymbol{A}_{t-1}\ \boldsymbol{a}_{\lambda_t}]$. We use the convention that $\boldsymbol{A}_0$ is an empty matrix.
4. Solve the following LS problem to obtain a new signal estimate:

[1] Here, the phrase "break the tie deterministically" requires a predefined rule to select one index unambiguously without randomness. For example, one can select the smallest index, i.e., among all indices with the maximum inner product value, choose the smallest index (the column appearing first in the measurement matrix $\boldsymbol{A}$). Besides, one can also choose to use a predefined column order (use a secondary criterion based on a predefined order, e.g., column norms or prior importance) or lexicographic order (treat indices as strings and sort them lexicographically).

A. Liu and M.-M. Zhao, *Advanced Bayesian Signal Processing*, Textbooks in Telecommunication Engineering,
https://doi.org/10.1007/978-3-032-26656-9_9

$$\boldsymbol{x}_t = \arg\min_{\boldsymbol{x}} \|\boldsymbol{y} - \boldsymbol{A}_t\boldsymbol{x}\|_2, \tag{9.2}$$

which gives $\boldsymbol{x}_t = (\boldsymbol{A}_t^T\boldsymbol{A}_t)^{-1}\boldsymbol{A}_t^T\boldsymbol{y}$.

5. Calculate the new approximation of the data and the new residual:

$$\boldsymbol{y}_t = \boldsymbol{A}_t\boldsymbol{x}_t, \quad \boldsymbol{r}_t = \boldsymbol{y} - \boldsymbol{y}_t. \tag{9.3}$$

6. Increment t, and return to Step 2 if $t \leq S$.
7. The estimate $\hat{\boldsymbol{x}}$ for the ideal signal has non-zero indices at the components listed in Λ_S. The value of the estimate $\hat{\boldsymbol{x}}$ in component λ_j equals the jth component of $\boldsymbol{x}_t$.

In particular, step 4 computes the optimal coefficients for the selected atoms (columns of $\boldsymbol{A}_t$) to approximate the data vector $\boldsymbol{y}$. The solution to problem (9.2) satisfies the normal equation

$$\boldsymbol{A}_t^T(\boldsymbol{y} - \boldsymbol{A}_t\boldsymbol{x}_t) = \boldsymbol{0}, \tag{9.4}$$

which simplifies to $\boldsymbol{A}_t^T\boldsymbol{r}_t = \boldsymbol{0}$, i.e., the residual $\boldsymbol{r}_t$ is always orthogonal to the columns of $\boldsymbol{A}_t$. This property ensures that subsequent iterations select new atoms not yet correlated with the residual and the matrix $\boldsymbol{A}_t$ has full column rank, provided that the residual $\boldsymbol{r}_{t-1}$ is non-zero. In this case, the solution $\boldsymbol{x}_t$ to the LS problem in Step 4 is unique (because the matrix $\boldsymbol{A}_t$ has full column rank and satisfies that $\boldsymbol{A}_t^T\boldsymbol{A}_t$ is invertible). It should be noted that the approximation and residual calculated in Step 5 are always uniquely determined.

The computational complexity of the OMP algorithm is primarily dominated by two key components: atom selection (Step 2) and LS solution (Step 4). Specifically, in Step 2, OMP selects the atom (column of $\boldsymbol{A}$) most correlated with the current residual $\boldsymbol{r}_{t-1}$. This involves computing inner products between $\boldsymbol{r}_{t-1}$ and all N columns of $\boldsymbol{A}$, each of dimension M. Thus, the total cost over S iterations is O(SMN). This step dominates the runtime because N (dictionary size) is often much larger than S (sparsity level) or M (measurement dimension), especially in high-dimensional sparse recovery problems. At iteration t, the LS problem can be solved with marginal cost $O(tM)$. To do so, one should maintain a QR factorization of $\boldsymbol{A}_t$, i.e., $\boldsymbol{A}_t = \boldsymbol{Q}_t\boldsymbol{R}_t$, where $\boldsymbol{Q}_t$ contains the orthonormal columns spanning the selected atoms, and $\boldsymbol{R}_t$ is an upper-triangular matrix. In particular, at iteration t, the matrix $\boldsymbol{A}_t$ (composed of t selected atoms) grows by one column, and the corresponding QR factorization can be updated incrementally by orthogonalizing the new column against all previous columns in $\boldsymbol{Q}_{t-1}$. The detailed implementation involves the modified Gram–Schmidt (MGS) algorithm because the measurement matrix is unstructured and dense. The book [1] provides extensive details and a survey of alternate approaches. When the measurement matrix is structured, more efficient implementations of OMP are possible; see the paper [2] for one example.

9.1.1 Performance Analysis

Random measurement ensembles are critical in sparse signal recovery algorithms like OMP, as they ensure robust signal reconstruction from a small number of linear measurements. Two common distributions for constructing such measurement matrices $\boldsymbol{A}$ are Gaussian and Bernoulli, each with specific normalization to meet theoretical requirements. Below, we explain the properties of these matrices.

- Gaussian Ensemble: Each entry of the $M \times N$ matrix $\boldsymbol{A}$ is independently sampled from a normal distribution with mean 0 and variance M^{-1}, the corresponding PDF is

$$p(x) = \frac{1}{\sqrt{2\pi M^{-1}}}e^{-x^2M/2}, \quad x \in \mathbb{R}, \tag{9.5}$$

  where normalization ensures $\mathbb{E}\{\|\boldsymbol{a}_j\|_2^2\} = 1$ for each column $\boldsymbol{a}_j$.
- Bernoulli Ensemble: Each entry of $\boldsymbol{A}$ is independently set to $\pm 1/\sqrt{M}$ with equal probability. Note that binary entries are able to simplify hardware implementation (e.g., sensors with binary outputs), and normalization by $1/\sqrt{M}$ also ensures $\mathbb{E}\{\|\boldsymbol{a}_j\|_2^2\} = 1$ as in the Gaussian case.

Indeed, either one of these distributions can be used to collect measurements. More generally, the measurement ensemble can be chosen from any distribution that meets a few basic requirements, known as "admissible measurement matrix." These properties are briefly introduced even though we are primarily interested in the foregoing examples. Specifically, an admissible matrix $\boldsymbol{A}$ for recovering S-sparse signals in $\mathbb{R}^N$ must satisfy the following four properties:

- **Property (M0) Independence**. To meet this property, the columns of $\boldsymbol{A}$ are required to be statistically independent, which is supposed to prevent correlations between measurements, and thus ensure unbiased signal recovery. Note that the columns of $\boldsymbol{A}$ need not have the same distribution. Besides, property (M0) only requires independence of the columns of $\boldsymbol{A}$, the entries within each column may be correlated.
- **Property (M1) Normalization.** $\mathbb{E}\{\|\boldsymbol{a}_j\|_2^2\} = 1$ for all columns $\boldsymbol{a}_j$. This property standardizes the energy of measurements, and is supposed to avoid scaling issues during recovery, i.e., the signal recovery problem should not be dependent on the scale of the measurement matrix.
- **Property (M2) Joint Correlation**. For any sequence of S vectors $\{\boldsymbol{u}_t\}$ with $\|\boldsymbol{u}_t\|_2 \leq 1$, and let $\boldsymbol{a}$ denote a column of $\boldsymbol{A}$ that is independent of $\{\boldsymbol{u}_t\}$, then
$$\mathbb{P}\left(\max_t |\langle \boldsymbol{a}, \boldsymbol{u}_t\rangle| \leq \varepsilon\right) \geq 1 - 2Se^{-c\varepsilon^2 M}. \tag{9.6}$$
This property controls the maximum correlation between a random column and any S-sparse signal subspace. Here, c is a constant that measures the decay rate of the correlation between the column vectors of a random measurement matrix and sparse signals, and its value is determined by the distribution characteristics of the matrix.
- **Property (M3) Smallest Singular Value**. For any $M \times S$ submatrix $\boldsymbol{Z}$ of $\boldsymbol{A}$, the Sth largest singular value $\sigma_S(\boldsymbol{Z})$ satisfies
$$\mathbb{P}\left(\sigma_S(\boldsymbol{Z}) \geq 0.5\right) \geq 1 - e^{-cM}. \tag{9.7}$$
Intuitively, this property ensures that submatrices are well-conditioned, thus preventing rank deficiency. Besides, this property guarantees that, with high probability, the smallest singular value of $\boldsymbol{Z}$ is bounded below by 0.5. Thus, for a sparse vector $\boldsymbol{x}$ supported on the columns of $\boldsymbol{Z}$, the energy (squared norm) of the measurement $\boldsymbol{Zx}$ is bounded by
$$\|\boldsymbol{Zx}\|_2 \geq \sigma_S(\boldsymbol{Z})\|\boldsymbol{x}\|_2 \geq 0.5\|\boldsymbol{x}\|_2. \tag{9.8}$$
This ensures that the matrix $\boldsymbol{Z}$ does not reduce the energy of $\boldsymbol{x}$ by more than a factor of 0.5, thus preserving critical information for recovery, i.e., this prevents the matrix from collapsing sparse signals into a low-energy subspace. Otherwise, if $\sigma_S(\boldsymbol{Z})$ is too small, small measurement noise could disproportionately distort the sparse signal, leading to recovery errors. Therefore, property (M3) controls how much the matrix is likely to shrink a sparse vector.

It is worth mentioning that Gaussian and Bernoulli ensembles both yield admissible measurement matrices, please see [3] for the detailed proof. See also the technical report [4] for a detailed treatment of the Gaussian case, including explicit constants. Moreover, it is important to mention that the literature on CS introduces various types of measurement matrices (ensembles) to ensure reliable signal recovery. Below, we compare three prominent approaches and their underlying principles:

- Candès and Tao's Random Matrices: In their foundational work [5], Candès and Tao proposed matrices that satisfy two critical properties, i.e., (1) uniform uncertainty principle (UUP): ensures that the measurement matrix preserves the energy of sparse signals across all subsets of columns; (2) exact reconstruction principle (ERP): guarantees exact recovery of sparse signals via convex optimization (e.g., basis pursuit). Gaussian and Bernoulli matrices both meet these requirements. In another paper [6], as discussed in Chap. 7, they study a class of matrices whose "restricted isometry constants" are under control. Gaussian and Bernoulli matrices satisfy RIP with high probability when $M = O(S\log(N/S))$ (as given in (7.13)), making them ideal for sparse recovery.
- Donoho's Deterministic CS Matrices: Donoho introduced deterministic constructions of measurement matrices, which are different from probabilistic ensembles [7]. In this approach, matrices are explicitly constructed (e.g., using finite fields or structured codes) rather than sampled randomly. Donoho showed that Gaussian random matrices, despite being probabilistic, asymptotically behave like deterministic CS matrices with high probability.
- Rudelson and Vershynin's Geometric Approach: Rudelson and Vershynin analyzed CS through the geometry of random subspaces [8]. The core idea is that if the rows of the measurement matrix $\boldsymbol{A}$ span a random subspace in $\mathbb{R}^N$, then signal recovery succeeds if this subspace intersects the set of sparse signals "transversely." This method relies on leveraging

probabilistic geometry, particularly the behavior of random slices of high-dimensional cubes. As such, they describe measurement ensembles via the geometric properties of the subspace they span (e.g., random rotations of coordinate axes), instead of defining ensembles explicitly through entry-wise distributions (e.g., Gaussian/Bernoulli).

Based on the discussions, it was shown in [3] that if we take random measurements of a sparse signal using an admissible measurement matrix, then OMP can be used to recover the original signal with high probability, as given in the following theorem.

Theorem 9.1 (OMP With Admissible Measurements) *Fix* $\delta \in (0, 0.36)$, *and choose* $M \geq KS\log(N/\delta)$ *where* K *is an absolute constant. Suppose that* $\tilde{\boldsymbol{x}}$ *is an arbitrary* S*-sparse signal in* $\mathbb{R}^N$, *and draw a random* $M \times N$ *admissible measurement matrix* $\boldsymbol{A}$ *independent from the signal. Given the data* $\boldsymbol{y} = \boldsymbol{A}\tilde{\boldsymbol{x}}$, *OMP can reconstruct the signal with probability exceeding* $1 - \delta$.

For Gaussian measurements, more precise estimates for the constant K can be obtained [3]. In this case, the following theorem holds with $K \leq 20$. Moreover, when the number S of non-zero components approaches infinity, it is possible to take $K \leq 4 + \eta$ for any positive number η. See the technical report [4] for a detailed proof of these estimates.

Theorem 9.2 (OMP With Gaussian Measurements) *Fix* $\delta \in (0, 0.36)$, *and choose* $M \geq KS\ln(N/\delta)$. *Suppose that* $\tilde{\boldsymbol{x}}$ *is an arbitrary* S*-sparse signal in* $\mathbb{R}^N$. *Draw* M *measurement vectors independently from the standard Gaussian distribution on* $\mathbb{R}^N$ *and form a measurement matrix* $\boldsymbol{A}$. *Given the data* $\boldsymbol{y} = \boldsymbol{A}\tilde{\boldsymbol{x}}$, *OMP can reconstruct the signal with probability exceeding* $1 - 2\delta$. *The constant satisfies* $K \leq 20$. *For large values of* S, *it can be reduced to* $K \approx 4$.

As can be seen in Theorems 9.1 and 9.2, the required M scales linearly with S and logarithmically with N, which is aligned with CS principles.

Even though OMP may fail, the user can detect a success or failure in the present setting. A simple result for Gaussian measurements is provided below [3].

Proposition 9.1 *Choose an arbitrary* S*-sparse signal* $\tilde{\boldsymbol{x}}$ *from* $\mathbb{R}^N$, *and let* $M \geq 2S$. *Suppose that* $\boldsymbol{A}$ *is an* $M \times N$ *Gaussian measurement matrix, and execute OMP with the data* $\boldsymbol{y} = \boldsymbol{A}\tilde{\boldsymbol{x}}$. *If the residual* $\boldsymbol{r}_S$ *after* S *iterations is zero, then OMP has correctly identified* $\tilde{\boldsymbol{x}}$ *with probability one. Conversely, if the residual after* S *iterations is non-zero, then OMP has failed.*

For Bernoulli measurements, a similar proposition holds with probability exponentially close to one. This result follows from the fact that an exponentially small fraction of (square) sign matrices are singular [9].

9.1.2 Regularized OMP

This subsection presents a modified OMP algorithm, known as ROMP, which seeks to bridge the two major algorithmic approaches to sparse signal recovery from an incomplete set of linear measurements, i.e., ℓ_1-minimization methods and OMP. By comparing the ℓ_1-minimization algorithm and OMP, we can see that the former has two major strengths, i.e., strong uniform guarantees (if the measurement matrix $\boldsymbol{A}$ satisfies the RIP, ℓ_1-minimization guarantees exact recovery of all S-sparse signals) and algorithm-independent analysis (the RIP condition is a deterministic property of $\boldsymbol{A}$, decoupled from the recovery algorithm). However, its weakness is also obvious, i.e., it relies on LP, which lacks a strongly polynomial-time algorithm. LP solvers depend on problem-specific condition numbers, making runtime unpredictable, and they are usually computationally intensive and less intuitive compared to greedy methods like OMP.

On the other hand, OMP's greedy selection of atoms (columns of $\boldsymbol{A}$) is intuitive, i.e., at each step, it picks the atom most correlated with the residual, this yields a strongly polynomial runtime (polynomial in S, M, and N), making it faster than LP in practice. However, OMP's recovery guarantees are signal-specific, which means that for each fixed S-sparse signal $\tilde{\boldsymbol{x}}$, OMP succeeds with high probability over random $\boldsymbol{A}$. It cannot guarantee recovery for all sparse signals simultaneously. Moreover, OMP requires stricter conditions on $\boldsymbol{A}$ than RIP. For example, it is unknown whether OMP succeeds with partial Fourier matrices (subsampled DFT matrices), a critical class in applications like MRI.

The above content highlights two unresolved issues with OMP: (1) lack of uniform guarantees, and (2) limited applicability to structured matrices. These limitations motivated the ROMP algorithm, which perform correctly for all measurement matrices

A satisfying the restricted isometry condition, and for all sparse signals [10]. When we are trying to recover the signal $\tilde{x}$ from its measurements $y = A\tilde{x}$, the ROMP algorithm leverages the observation vector $u = A^H y$, which encodes correlations between the measurement vector y and the columns of A. Since $\tilde{x}$ is sparse, y has a sparse representation in the dictionary A. Under RIP, we can see that any subset of S columns in A forms an approximately orthonormal system. Consequently, the S-largest coordinates of u (in magnitude) closely approximate the corresponding non-zero coefficients of $\tilde{x}$.

Traditional OMP greedily selects the single column most correlated with the residual. However, this approach can be unstable under noise or matrix imperfections. The new ROMP algorithm instead selects the S-largest coordinates of u, and enforces regularity by ensuring these coordinates have comparable magnitudes. This regularization prevents dominance by a few large coordinates, thus distributing information more evenly and improving robustness. The overall ROMP algorithm is summarized as follows.

Input: 1. an $M \times N$ measurement matrix A; 2. an M-dimensional data vector y; 3. the sparsity level S of the ideal signal.
Output: Index set $I \subset \{1, \ldots, N\}$.

1. **Initialize**: Let the index set $I = \emptyset$, the residual $r_0 = y$ and the iteration counter $t = 1$.
2. **Identify**: Choose a set J of the S biggest coordinates in magnitude of the observation vector $u = A^H r_{t-1}$, or all of its non-zero coordinates, whichever set is smaller.
3. **Regularize**: Among all subsets $J_0 \subset J$ with comparable coordinates:
$$|u(i)| \leq 2|u(j)| \quad \text{for all } i, j \in J_0, \tag{9.9}$$
choose J_0 with the maximal energy $\|u_{J_0}\|_2$.
4. **Update**: Add the set J_0 to the index set $I \leftarrow I \cup J_0$, and update the residual:
$$z_t = \arg\min_{z \in \mathbb{R}^I} \|y - Az\|_2; \quad r_t = y - Az_t. \tag{9.10}$$
5. Increment t, and repeat the steps 2–4 S times or until $r_{t-1} = \mathbf{0}$, whichever occurs first.

It was shown in [10] that ROMP yields exact sparse recovery provided that the measurement matrix satisfies the RIP, which is given as follows.

Theorem 9.3 (Exact sparse recovery via ROMP) *Assume a measurement matrix A satisfies the RIP with parameters $(2S, \delta_{2S})$ for $\delta_{2S} = 0.03/\sqrt{\log S}$. Let $\tilde{x}$ be an S-sparse vector in $\mathbb{R}^N$ with measurements $y = A\tilde{x}$. Then, ROMP, in at most S iterations, outputs a set I such that*
$$supp(\tilde{x}) \subset I, \text{ and } |I| \leq 2S. \tag{9.11}$$

Theorem 9.3 tells us that

- Once I is identified, the original sparse vector $\tilde{x}$ can be reconstructed as $\tilde{x} = A_I^{-1} y$, where A_I is the submatrix of A with columns indexed by I, i.e., exact sparse recovery is guaranteed. The RIP ensures A_I has stable singular values, allowing reliable inversion.
- Unlike OMP, ROMP guarantees recovery for all S-sparse vectors $\tilde{x}$ if A satisfies RIP. This resolves a long-standing open problem, as OMP cannot provide such uniform guarantees.

Besides, ROMP gives sparse recovery for a broader class of measurement matrices, including random Gaussian, Bernoulli, and partial Fourier matrices, with number of measurements M almost linear in the sparsity level S. But for OMP, it has been unknown whether it gives sparse recovery for partial Fourier measurements (even with nonuniform guarantees).

Despite these advantages of ROMP, one can see that there is a logarithmic factor in δ_{2S}, which implies that as S grows, δ_{2S} becomes smaller, requiring A to behave more like a perfect isometry on larger subspaces (of dimension $2S$). This shrinking δ_{2S} reflects a trade-off, i.e., higher sparsity levels demand higher-quality measurement matrices. It is unclear if this factor is necessary for the theorem to hold, but it currently cannot be removed.

It may seem that the complexity of ROMP is higher than that of OMP, however, the running time of ROMP is comparable to that of OMP in theory. Specifically, the identification and regularization steps of ROMP can be performed efficiently. The identification step selects the subset J by identifying the largest S components of the correlation vector $\boldsymbol{u}$. This can be done by sorting the coordinates of $\boldsymbol{u}$ in non-increasing order. Sorting algorithms like Mergesort or Heapsort achieve this in $O(S \log S)$ time. The regularization step seeks a smaller subset $J_0 \subset J$, where the largest and smallest coefficients in J_0 differ by at most a factor of 2. Instead of exhaustively searching all possible subsets, ROMP leverages the observation that J_0 corresponds to an interval in the sorted coefficients, and it suffices to look for J_0 among $O(\log S)$ consecutive intervals [10]. In particular, one can partition the sorted coefficients into intervals where the magnitude halves at each step; for example, start with the largest coefficient u_{max}, we can define intervals where the smallest coefficient in each interval is at least $u_{max}/2$, $u_{max}/4$, $u_{max}/8$, etc. This creates $O(\log S)$ intervals, as the magnitudes decay geometrically, and therefore, the regularization step can be conducted in $O(S)$ time.

In addition, each iteration involves multiplying the matrix $\boldsymbol{A}^H$ (of size $N \times M$) with a residual vector, and solving the LS problem (9.10) with the $M \times |I|$ matrix $\boldsymbol{A}_I$. For unstructured matrices, the multiplication $\boldsymbol{A}^H \boldsymbol{r}_{t-1}$ costs $O(MN)$ time. Solving the LS is more computationally intensive, and for unstructured matrices, direct methods (e.g., QR factorization) require $O(S^2 M)$ time per iteration. However, iterative methods like the conjugate gradient method exploit the near-isometric property of $\boldsymbol{A}_I$, reducing the LS solver to $O(SM)$ time with a constant number of iterations. For structured matrices (e.g., Fourier), fast matrix–vector multiplication techniques further accelerate both the matrix multiplication and LS steps.

ROMP terminates in at most S iterations. Therefore, for unstructured matrices, using the methods mentioned above, the total running time of ROMP is

$$O(S)\,(O(MN) + O(SM) + O(N \log N) + O(S)). \tag{9.12}$$

In the interesting regime $M \geq \log N$, $O(N \log N)$ is dominated by $O(MN)$, and in the meantime $O(S)$ is dominated by $O(SM)$. Thus, the total running time of ROMP is $O(\mathrm{SMN})$, which is the same bound as for OMP, as discussed before.

9.1.3 Stagewise OMP

StOMP aims to achieve an approximate solution to $\boldsymbol{y} = \boldsymbol{A}\tilde{\boldsymbol{x}}$, where $\boldsymbol{A}$ comes from the uniform spherical ensemble (USE), i.e., its columns $\boldsymbol{a}_1, \boldsymbol{a}_2, \cdots, \boldsymbol{a}_N$ that are i.i.d. random vectors drawn uniformly from the unit sphere $\mathbb{S}^{M-1}$ and $\tilde{\boldsymbol{x}}$ is sparse. In this subsection, we describe its basic ingredients and some performance analysis results.

Unlike methods that select one component at a time, StOMP operates in P stages, progressively identifying clusters of significant coefficients to approximate the solution. A detailed breakdown of its procedure and logic is as follows, and Fig. 9.1 gives a schematic representation.

Input: 1. an $M \times N$ measurement matrix $\boldsymbol{A}$; 2. an M-dimensional data vector $\boldsymbol{y}$.
Output: The final estimate $\tilde{\boldsymbol{x}}_P$.

1. **Initialization**: The algorithm starts with an initial guess $\tilde{\boldsymbol{x}}_0$ and an initial residual $\boldsymbol{r}_0 = \boldsymbol{y}$ (the observed signal itself). A counter p tracks the current stage, beginning at $p = 1$. It also maintains a growing list I_p estimating the locations (indices) of non-zeros in $\tilde{\boldsymbol{x}}$.
2. **Matched Filtering**: At each stage p, the algorithm computes the correlation between the current residual $\boldsymbol{r}_{p-1}$ and the columns of $\boldsymbol{A}$, i.e.,

$$\boldsymbol{c}_p = \boldsymbol{A}^T \boldsymbol{r}_{p-1}. \tag{9.13}$$

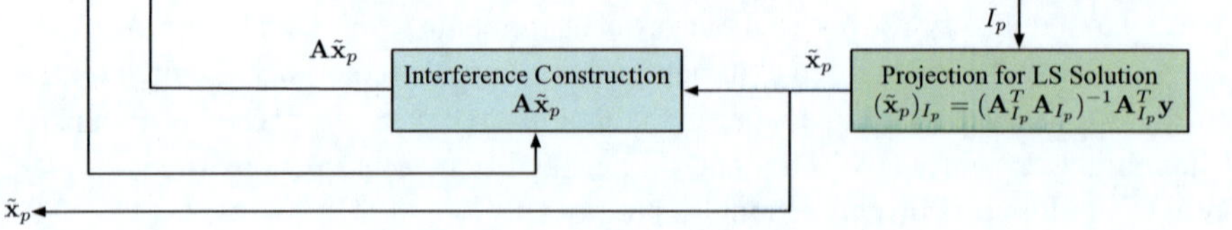

Fig. 9.1 Schematic representation of the StOMP algorithm (Adapted from Fig. 1 in [11])

This vector $\boldsymbol{c}_p$ represents how much each column of $\boldsymbol{A}$ aligns with the residual. Intuitively, large entries in $\boldsymbol{c}_p$ suggest columns (features) that might correspond to non-zero entries in $\tilde{\boldsymbol{x}}$.

3. **Hard Thresholding**: To distinguish meaningful correlations from noise, StOMP applies a hard threshold. Entries in $\boldsymbol{c}_p$ exceeding a threshold $t_p\sigma_p$ yields a small set J_p of "large" coordinates, are selected as candidates:
$$J_p = \{j : |\boldsymbol{c}_p(j)| > t_p\sigma_p\}. \tag{9.14}$$
Here, σ_p is an estimated noise level, and t_p is a thresholding parameter controlling selectivity. The thresholding step filters out weak correlations, assuming Gaussian noise in $\boldsymbol{c}_p$
4. **Support Set Update**: The newly identified indices J_p are merged with the previous support estimate I_{p-1}, i.e.,
$$I_p = I_{p-1} \cup J_p. \tag{9.15}$$
This expands the set of candidate locations for non-zeros in $\tilde{\boldsymbol{x}}$.
5. **Projection for LS Solution**: With the updated support I_p, StOMP solves an LS problem restricted to the columns of $\boldsymbol{A}$ indexed by I_p, i.e.,
$$(\tilde{\boldsymbol{x}}_p)_{I_p} = (\boldsymbol{A}_{I_p}^T \boldsymbol{A}_{I_p})^{-1} \boldsymbol{A}_{I_p}^T \boldsymbol{y}. \tag{9.16}$$
This step computes the best approximation of $\boldsymbol{y}$ using only the selected columns, effectively refining the coefficients of $\tilde{\boldsymbol{x}}_p$ on the support I_p.
6. **Residual Update**: The residual $\boldsymbol{r}_p$ is recalculated to reflect the remaining unexplained part of $\boldsymbol{y}$:
$$\boldsymbol{r}_p = \boldsymbol{y} - \boldsymbol{A}\tilde{\boldsymbol{x}}_p. \tag{9.17}$$
The residual shrinks as more structure (non-zeros) is identified and incorporated into $\tilde{\boldsymbol{x}}_p$.
7. **Stopping Condition**: The algorithm sets $p \leftarrow p+1$ and repeats stages until a stopping criterion is met (e.g., a maximum number of stages P, or when the residual becomes sufficiently small). The final estimate $\tilde{\boldsymbol{x}}_P = \tilde{\boldsymbol{x}}_p$ is returned.

StOMP can be seen as a generalized version of OMP. If StOMP were constrained to select only one new entry (index) per stage, i.e., mimicking OMP's greedy single-term selection, and if the number of stages P is equal to the measurement dimension M, the two algorithms would yield identical results.[2] However, StOMP diverges by intentionally allowing multiple selection of indices at each stage through its thresholding strategy. Instead of picking the single largest correlation, StOMP uses statistical thresholds to admit multiple candidates simultaneously. This accelerates convergence and reduces computational overhead, especially for large-scale problems. The trade-off is a need for careful threshold tuning to avoid false positives.

The threshold $t_p\sigma_p$ plays a central role in StOMP. The formal noise level σ_p is calculated as $\|\boldsymbol{r}_p\|/\sqrt{M}$. This normalization accounts for the expected energy of Gaussian noise distributed across M measurements. Besides, the parameter t_p typically lies between 2 and 3. Intuitively, this range aligns with statistical principles for Gaussian noise, i.e., values beyond 2–3 standard deviations are unlikely under a null hypothesis (i.e., pure noise), ensuring that only statistically significant correlations are selected. For example, in a Gaussian distribution, 95% of values lie within $\pm 2\sigma$, and 99.7% within $\pm 3\sigma$, making this range effective for controlling false discoveries.

The recommended settings for StOMP, such as limiting the number of stages to $P = 10$ and ensuring $|I_P| \leq M$—are rooted in theoretical and practical considerations. They are designed so that when the true $\tilde{\boldsymbol{x}}$ is sufficiently sparse, the following two conditions are likely to hold upon termination: (1) all non-zeros in $\tilde{\boldsymbol{x}}$ are selected in I_P; (2) I_P has no more than M entries. The following lemma motivates this design criterion.

Lemma 9.1 *Let the columns of $\boldsymbol{A}$ be in general position. Let I_P denote the support of $\tilde{\boldsymbol{x}}_P$. Suppose that the support I of the true signal $\tilde{\boldsymbol{x}}$ is a subset of I_P. Suppose in addition that the number of non-zero indices in I_P is less than or equal to M. Then we have perfect recovery:*
$$\tilde{\boldsymbol{x}}_P = \tilde{\boldsymbol{x}}. \tag{9.18}$$

[2] Typically, OMP stops after S iterations, as it aims to recover the true support Λ_S. However, it can theoretically run up to M iterations, selecting M linearly independent atoms, at which point the residual becomes zero (since $\boldsymbol{A}_\Lambda$ with $|\Lambda| = M$ spans $\mathbb{R}^M$). In practice, stopping criteria (e.g., residual norm or iteration limit) halt it earlier.

Please also refer to [11] for more detailed theoretical and empirical support for the large-system effectiveness of StOMP and several other ensembles (other than USE).

Attention
It is important to mention that OMP and ROMP both assume that the sparsity level S is known. While for StOMP, its general procedures are mechanical and driven by the data ($\boldsymbol{y}$ and $\boldsymbol{A}$) and the current state of the algorithm (residuals and support set). The stopping criterion, typically a fixed number of stages P (e.g., $P = 10$), does not depend on S. However, the effectiveness of StOMP in recovering $\tilde{\boldsymbol{x}}$ relies on the thresholding step, which is where S may or may not come into play. In [11], two strategies are proposed for setting the threshold, which are discussed below:

- False Alarm Control: this strategy aims to limit the number of false alarms, i.e., cases where a zero coefficient is incorrectly identified as non-zero, across all stages. The total false alarms should not exceed the "natural codimension" of the problem, defined as $M - S$. This strategy requires knowledge of the number of non-zeros S or some upper bound [11]. Without knowing S or an estimate, setting an appropriate threshold becomes infeasible, as the algorithm needs to balance the detection of true non-zeros against the risk of false alarms.
- False Discovery Control: this strategy seeks to control the false discovery rate, i.e., the proportion of false positives among all selected coefficients, typically aiming for a fixed fraction. The threshold t_p is determined based on statistical methods (e.g., Simes' rule [12, 13]) that assess the distribution of correlations. This strategy does not require the knowledge S, which makes it more convenient for applications, but also slightly more complex to implement and substantially more complex to analyze [12].

9.2 CoSaMP

This section introduces CoSaMP, from the acronym compressive sampling matching pursuit, a signal reconstruction algorithm designed to address limitations in existing approaches. In particular, combinatorial algorithms (e.g., Fourier sampling [14, 15], chaining pursuit [16], and HHS pursuit [17], as well as some algorithms of Cormode–Muthukrishnan [18] and Iwen [19]) are fast but require specialized sampling patterns, making them impractical for many real-world applications; convex relaxation methods (e.g., solving ℓ_1 minimization problem using interior-point methods [20, 21], projected gradient methods [22], and iterative thresholding [23]) work with fewer measurements but are computationally intensive; greedy methods like OMP [3], ROMP [10], StOMP [11], strike a balance but lack strong theoretical guarantees or robustness. CoSaMP belongs to the family of greedy pursuit algorithms, which iteratively approximate sparse signals from limited measurements. It builds on OMP but integrates insights from combinatorial algorithms and convex optimization to improve speed and reliability.

The core challenge in sparse signal recovery is identifying the locations (support) of the largest components in the target signal $\tilde{\boldsymbol{x}}$. CoSaMP leverages the RIP of the sampling matrix $\boldsymbol{A}$. If $\boldsymbol{A}$ satisfies RIP with a small constant $\delta_S \ll 1$, the proxy vector $\boldsymbol{u} = \boldsymbol{A}^H \boldsymbol{A}\tilde{\boldsymbol{x}}$ (derived from samples $\boldsymbol{y} = \boldsymbol{A}\tilde{\boldsymbol{x}}$ (noiseless case)) approximates the energy distribution of the original signal $\tilde{\boldsymbol{x}}$, similar to the ROMP algorithm introduced in Sect. 9.1.2. Specifically, the largest S entries of $\boldsymbol{u}$ correspond to the largest S entries of $\tilde{\boldsymbol{x}}$. This observation enables iterative refinement: at each step, the algorithm identifies significant components in the residual (unrecovered part of the signal) and updates the approximation accordingly. An overview of the CoSaMP algorithm is provided as follows.

Input: 1. an $M \times N$ measurement matrix $\boldsymbol{A}$; 2. an M-dimensional data vector $\boldsymbol{y}$; 3. the sparsity level S of the ideal signal.
Output: The final estimate $\hat{\boldsymbol{x}}$.

- **Initialization**: The CoSaMP algorithm can be initialized with a trivial approximation (all zeros), i.e., $\hat{\boldsymbol{x}}^0 = \boldsymbol{0}$. Let $t = 1$ denote the iteration index.
- **Identification**: Compute the proxy $\boldsymbol{u} = \boldsymbol{A}^H \boldsymbol{r}$ where $\boldsymbol{r}$ initially equals the input samples $\boldsymbol{y}$. Select the $2S$ largest magnitude components of the proxy $\boldsymbol{u}$. These indices form a tentative support set Ω that indicates the potential significant components in the residual, i.e., $\Omega \leftarrow \text{supp}(\boldsymbol{u}_{2S})$.[3]

[3] Suppose that $\boldsymbol{y}$ is a signal in $\mathbb{C}^N$ and S is a positive integer. We write $\boldsymbol{y}_S$ for the signal in $\mathbb{C}^N$ that is formed by restricting $\boldsymbol{y}$ to its S largest magnitude components. Ties are broken lexicographically.

- **Support Merger**: Merge Ω with the support of the current approximation $\hat{\boldsymbol{x}}^{t-1}$ (i.e., indices of non-zero entries), which is given by
$$T \leftarrow \Omega \cup \operatorname{supp}(\hat{\boldsymbol{x}}^{t-1}). \tag{9.19}$$
This step ensures that both newly identified components and previously retained components are considered. The merged support set T has size up to $3S$.
- **Estimation**: Solve an LS problem restricted to the merged support T and obtain
$$\boldsymbol{b}_T = \boldsymbol{A}_T^{\dagger}\boldsymbol{y}, \quad \boldsymbol{b}_{T^c} \leftarrow \boldsymbol{0}, \tag{9.20}$$
where $\boldsymbol{A}_T^{\dagger}$ is the pseudoinverse of the submatrix of $\boldsymbol{A}$ indexed by T. This minimizes the error between the current samples and the approximation on T. Entries outside T are set to zero.
- **Pruning**: Retain only the S largest components of the LS estimate $\boldsymbol{b}_T$, and discard weaker entries. This enforces sparsity and prevents overfitting. The pruned vector becomes the new approximation, i.e., $\hat{\boldsymbol{x}}^t = \boldsymbol{b}_S$.
- **Sample Update**: Update the residual samples $\boldsymbol{r} = \boldsymbol{y} - \boldsymbol{A}\hat{\boldsymbol{x}}^t$, which reflects the part of the signal that remains unrecovered. This residual drives the next iteration.
- Let $t \leftarrow t + 1$, and the above process repeats until the halting criterion (e.g., a fixed number of iterations) is met. The final output is an S-sparse approximation of the original signal.

The analysis on CoSaMP is inspired by the work on ROMP [10] and the work [24] on convex relaxation methods. In particular, the following result can be obtained.

Theorem 9.4 (CoSaMP [25]) *Suppose that* $\mathbf{A}$ *is an* $M \times N$ *sampling matrix with restricted isometry constant* $\delta_{2S} \leq c$. *Let* $\boldsymbol{y} = \mathbf{A}\tilde{\boldsymbol{x}} + \boldsymbol{n}$ *be a vector of samples of an arbitrary signal, contaminated with arbitrary noise. For a given precision parameter* η, *the algorithm CoSaMP produces an S-sparse approximation* $\hat{\boldsymbol{x}}$ *that satisfies*
$$\|\tilde{\boldsymbol{x}} - \hat{\boldsymbol{x}}\|_2 \leq C \cdot \max\left\{\eta, \frac{1}{\sqrt{S}}\|\tilde{\boldsymbol{x}} - \tilde{\boldsymbol{x}}_{S/2}\|_1 + \|\boldsymbol{n}\|_2\right\}, \tag{9.21}$$
where $\tilde{\boldsymbol{x}}_{S/2}$ *is a best* $(S/2)$*-sparse approximation to* $\tilde{\boldsymbol{x}}$. *The running time is* $O(\mathcal{L} \cdot \log(\|\tilde{\boldsymbol{x}}\|_2/\eta))$, *where* $\mathcal{L}$ *bounds the cost of a matrix–vector multiply with* $\mathbf{A}$ *or* $\mathbf{A}^*$. *Working storage is* $O(N)$.

From Theorem 9.4, we can see that CoSaMP works with any measurement matrix $\boldsymbol{A}$ satisfying the RIP, critically, random matrices (e.g., Gaussian, Bernoulli) and structured matrices (e.g., partial Fourier) meet RIP when $M = O(S \log^{\alpha} N)$, where α is a small integer.

Besides, CoSaMP delivers an S-sparse approximation where the ℓ_2 reconstruction error fundamentally reflects two unavoidable limitations: (1) it scales proportionally to the ℓ_1 approximation error of the best possible $S/2$-sparse representation of the original signal adjusted by a $1/\sqrt{S}$ factor; (2) it also directly inheriting the energy of any additive noise present in the measurements. This structure is not arbitrary but emerges as a consequence of computational efficiency: the disparity between the target sparsity S and the stricter $S/2$ benchmark in CoSaMP's error bound is a consequence of its specific algorithmic design, which is optimized for uniform guarantees and robustness. Unlike simpler greedy methods, CoSaMP employs an iterative process of support merger and pruning. This pruning step, while crucial for maintaining a fixed sparsity level S across iterations, can inadvertently discard components of the signal that are genuine but small. The error bound, therefore, compares the algorithm's output to the best possible approximation using only $S/2$ components, acknowledging that some fraction of the estimated support may be incorrect.

This design is a strategic trade-off that allows CoSaMP to provide strong uniform recovery guarantees (i.e., it works for all sparse signals under the RIP) and to be highly robust to noise and signal compressibility, all within polynomial time. It avoids the NP-hardness of universal ℓ_0 minimization not by being approximate for all algorithms, but by employing a strategy that is inherently robust at the cost of this slightly looser benchmark. This stands in contrast to algorithms like OMP, which can achieve perfect recovery for exactly S-sparse, noiseless signals. However, OMP's guarantees are often nonuniform (it works for a fixed signal with high probability over the measurement matrix) and typically require more stringent conditions on the measurement matrix (e.g., a smaller RIP constant). Furthermore, OMP lacks the same level of proven robustness for compressible signals or in high-noise scenarios as CoSaMP. Thus, the $S/2$ benchmark is not a necessity for all polynomial-time algorithms but is a feature of CoSaMP's particular approach to achieving uniform robustness and computational efficiency.

Simultaneously, the explicit noise term $\|\boldsymbol{n}\|_2$ acknowledges that no recovery process can eliminate noise already embedded in the input data; it can only control its amplification through the sensing matrix properties. Crucially, this error formulation is "structurally optimal," i.e., it represents the tightest bound achievable under information-theoretic and complexity-theoretic constraints for methods using $O(S \log N)$ samples [26].

Theorem 9.4 offers the CoSaMP's overarching performance bound; however, it does not reveal how the algorithm dynamically achieves this performance through its iterative process. The following theorem addresses this issue.

Theorem 9.5 (Iteration Invariant for CoSaMP [25]) *Let S be a fixed sparsity level, and let $\boldsymbol{A}$ be an $M \times N$ sampling operator with restricted isometry constant $\delta_{4s} \le 0.1$. Consider an arbitrary signal $\tilde{\boldsymbol{x}} \in \mathbb{C}^N$ and an arbitrary noise vector $\boldsymbol{n} \in \mathbb{C}^M$. Define the* unrecoverable energy ν *as*

$$\nu = \|\tilde{\boldsymbol{x}} - \tilde{\boldsymbol{x}}_S\|_2 + \frac{1}{\sqrt{S}}\|\tilde{\boldsymbol{x}} - \tilde{\boldsymbol{x}}_S\|_1 + \|\boldsymbol{n}\|_2, \tag{9.22}$$

where $\tilde{\boldsymbol{x}}_S$ is a best S-sparse approximation to $\tilde{\boldsymbol{x}}$. Then, for every iteration $t \ge 0$ of the CoSaMP algorithm, the S-sparse signal approximation $\hat{\boldsymbol{x}}^t$ satisfies

$$\|\tilde{\boldsymbol{x}} - \hat{\boldsymbol{x}}^{t+1}\|_2 \le \rho\|\tilde{\boldsymbol{x}} - \hat{\boldsymbol{x}}^t\|_2 + 10\nu, \tag{9.23}$$

and consequently,

$$\|\tilde{\boldsymbol{x}} - \hat{\boldsymbol{x}}^t\|_2 \le \rho^t\|\tilde{\boldsymbol{x}}\|_2 + 20\nu, \tag{9.24}$$

where $\rho = 0.5$.

This result holds uniformly for all signals and noise vectors under the stated conditions, demonstrating that CoSaMP reduces the approximation error by a constant factor in each iteration until it converges to an error floor determined by the unrecoverable energy ν. ν represents the theoretical baseline error that cannot be fully eliminated, even with ideal recovery methods. This quantity comprises three components: the ℓ_2-error $\|\tilde{\boldsymbol{x}} - \tilde{\boldsymbol{x}}_S\|_2$ between the signal $\tilde{\boldsymbol{x}}$ and its best S-sparse approximation $\tilde{\boldsymbol{x}}_S$; a scaled ℓ_1-error term $\frac{1}{\sqrt{S}}\|\tilde{\boldsymbol{x}} - \tilde{\boldsymbol{x}}_S\|_1$ intrinsically linked to the compressibility of the signal (i.e., its tail decay rate); and the energy $\|\boldsymbol{n}\|_2$ of the measurement noise. In particular, for the case where $\tilde{\boldsymbol{x}}$ is S-sparse and there is no noise ($\boldsymbol{n} = \boldsymbol{0}$), the best S-sparse approximation satisfies $\tilde{\boldsymbol{x}}_S = \tilde{\boldsymbol{x}}$. Consequently, the unrecoverable energy simplifies to $\nu = 0$, and the iteration error thus reduces to

$$\|\tilde{\boldsymbol{x}} - \tilde{\boldsymbol{x}}^t\|_2 \le 2^{-t}\|\tilde{\boldsymbol{x}}\|_2. \tag{9.25}$$

As $t \to \infty$, the error converges to 0. This result demonstrates that CoSaMP achieves exact signal recovery under these ideal conditions.

Attention

For S-sparse and noise-free signals, Theorem 9.5 demonstrates that the approximation error converges to 0 as the number of iterations increases, since the unrecoverable energy ν vanishes under these ideal conditions. While Theorem 9.4 provides an error bound involving the term $\frac{1}{\sqrt{S}}\|\tilde{\boldsymbol{x}} - \tilde{\boldsymbol{x}}_{S/2}\|_1$, this expression represents a conservative upper bound that may be significantly smaller in practice, or even zero if $\tilde{\boldsymbol{x}}$ itself is $S/2$-sparse. Crucially, no contradiction exists between these theorems as Theorem 9.5 makes a stronger assumption about the measurement matrix ($\delta_{4S} \le 0.1$) than Theorem 9.4 ($\delta_{2S} \le c$). A stronger assumption naturally leads to a stronger, more detailed conclusion. As clarified in [25, Remark 2.3], a simplified upper bound for ν can be derived as

$$\nu \le \frac{1.71}{\sqrt{S}}\|\tilde{\boldsymbol{x}} - \tilde{\boldsymbol{x}}_{S/2}\|_1 + \|\boldsymbol{n}\|_2. \tag{9.26}$$

This simplification underscores that ν is primarily governed by the ℓ_1-tail error of the signal at the $S/2$-sparsity level along with the noise energy, and it directly links Theorem 9.5 to 9.4.

Later work [27] makes a beneficial attempt to improve the theoretical guarantees for CoSaMP. It was shown that for S-sparse recovery, the restricted isometry constant is enlarged to $\delta_{4S} < 0.5$ with $\rho < 1$, which improve the known results significantly as stated in the following theorem.

Theorem 9.6 (Improved Results for CoSaMP [25]) *For the general CS model* $\mathbf{y} = \boldsymbol{A}\tilde{\boldsymbol{x}} + \boldsymbol{n}$, *if* $\delta_{4s} < 0.5$, *then the sequence of* $\hat{\boldsymbol{x}}^t$ *defined by CoSaMP satisfies*

$$\|\tilde{\boldsymbol{x}} - \hat{\boldsymbol{x}}^t\|_2 \leq \rho^t \|\tilde{\boldsymbol{x}}\|_2 + \tau \|\boldsymbol{n}\|_2, \tag{9.27}$$

where

$$\rho = \sqrt{\frac{2\delta_{4s}^2(1 + 2\delta_{4s}^2)}{1 - \delta_{4s}^2}} < 1, \tag{9.28}$$

$$(1 - \rho)\tau = \frac{(\sqrt{2} + 1)\delta_{4s}\left(\sqrt{2(1 - \delta_{4s})} + \sqrt{1 + \delta_{4s}}\right)}{1 - \delta_{4s}} + (2\sqrt{2} + 1)\sqrt{1 + \delta_{4s}}. \tag{9.29}$$

Besides the above version of CoSaMP, there are several adjustable parameters that may improve the performance of the CoSaMP algorithm discussed above, e.g., the number of components selected in the identification step and the number of components retained in the pruning step. Another variation appears in Appendix 1.

Example 9.1 (*Sparse Channel Estimation for Massive MIMO Communication System*). Consider a flat block-fading massive MIMO system operating in FDD mode (similar to that mentioned in Chap. 7), where the BS is equipped with a large-scale antenna array (typically with antenna number $N \gg 1$, transmitting signals to a single-antenna user. The core objective of downlink channel estimation is to enable the user to accurately acquire the channel state information (CSI) from the BS to itself. This is enabled by allowing the BS to transmit a sequence of M pilots for the user to estimate the downlink channel. Denote the aggregate transmitted pilot signals (pilot training matrix) as $\boldsymbol{\Psi} \in \mathbb{C}^{N \times M}$. The aggregate received pilot signal $\boldsymbol{y}^H \in \mathbb{C}^{1 \times M}$ (which is also called the compressed CSI at the transmitter (CSIT) measurements) at the user is

$$\boldsymbol{y}^H = \boldsymbol{h}^H \boldsymbol{\Psi} + \boldsymbol{n}^H, \tag{9.30}$$

where $\boldsymbol{h}^H \in \mathbb{C}^{1 \times N}$ is the channel vector between BS and the user, and $\boldsymbol{n}^H \in \mathbb{C}^{1 \times M}$ is the additive complex Gaussian noise with zero mean and covariance matrix $\sigma^2 \boldsymbol{I}$.

Using the virtual angular-domain representation [28, Chap. 7], the channel vector can be expressed as

$$\boldsymbol{h}^H = \tilde{\boldsymbol{x}}^H \boldsymbol{\Phi}, \tag{9.31}$$

where $\boldsymbol{\Phi} \in \mathbb{C}^{N \times N}$ denotes the unitary matrix for the angular-domain transformation and it is determined by the geometrical structure of the antenna array, $(\tilde{\boldsymbol{x}})^H \in \mathbb{C}^{1 \times N}$ is the angular-domain channel vector. For the considered ULA with half wavelength spacing, $\boldsymbol{\Phi}$ is given by the following normalized DFT matrix:

$$\boldsymbol{\Phi} = \frac{1}{\sqrt{N}} \begin{bmatrix} 1 & 1 & \cdots & 1 \\ 1 & e^{-j2\pi/N} & \cdots & e^{-j2\pi(N-1)/N} \\ 1 & e^{-j4\pi/N} & \cdots & e^{-j4\pi(N-1)/N} \\ \vdots & \vdots & \ddots & \vdots \\ 1 & e^{-j2\pi(N-1)/N} & \cdots & e^{-j2\pi(N-1)^2/N} \end{bmatrix}, \tag{9.32}$$

where

$$\boldsymbol{\Phi}_{i,k} = \frac{1}{\sqrt{N}} e^{-j\frac{2\pi ik}{N}}, \quad i, k = 0, 1, \ldots, N - 1, \tag{9.33}$$

The nth entry of $\tilde{\boldsymbol{x}}$ can be interpreted as the channel gain consisting of the aggregation of all the spatial paths whose transmit directions are within an angular window around the nth angular coordinate at the BS [28, 29].

In this example, the pilot training matrix $\boldsymbol{\Psi}$ is chosen as $\boldsymbol{\Psi} = \boldsymbol{\Phi}^H \boldsymbol{A}^H$, where $\boldsymbol{A} \in \mathbb{C}^{M \times N}$ has i.i.d. complex Gaussian entries with zero mean and unit variance. Substituting $\boldsymbol{\Psi} = \boldsymbol{\Phi}^H \boldsymbol{A}^H$ and (9.31) into (9.30) and taking the conjugate transpose of both sides, we can rewrite (9.30) into the standard form of the compressive sensing model:

$$\boldsymbol{y} = \boldsymbol{A}\tilde{\boldsymbol{x}} + \boldsymbol{n}, \tag{9.34}$$

where A is the measurement matrix and $\tilde{x}$ is the unknown sparse channel vector to be recovered. We consider a standard downlink narrowband massive MIMO model with $N = 128$ transmit antennas and $L_p = 8$ physical paths, and set the prior sparsity to $S = L_p$. The angular-domain dictionary is the DFT basis, the pilot (measurement) matrix is set to be random, and each operating point is averaged over 1000 independent Monte Carlo trials.

In the off-grid setting (true angles do not exactly align with the discrete dictionary), each physical path spreads into a small cluster of adjacent atoms in the angular domain, resulting in a coefficient vector that is not strictly S-sparse. In this case, if OMP/CoSaMP are forced to select exactly $S = 8$ atoms, a nonvanishing approximation error remains due to energy leakage, as shown in Fig. 9.2; as SNR increases, noise is reduced but this mismatch error does not diminish, which leads to performance saturation at the medium–high SNR regime. By contrast, LASSO does not constrain the number of non-zero elements to be exactly S, and the ℓ_1 penalty allows using more than S small-magnitude coefficients to better cover the leakage cluster, which yields lower NMSE. Besides, we can observe from Fig. 9.2 that LASSO exhibits a sharp, cliff-like performance transition at a critical measurement threshold $M \approx C \cdot S \log(N/S)$ in (7.13), where $N = 128$, $S = 8$, and C is a constant dependent on the measurement matrix distribution. As the number of measurements, i.e., M, increases beyond this threshold, the system transitions from an underdetermined regime with significant modeling error to a stable recovery regime where noise becomes the dominant error source, resulting in the observed abrupt NMSE drop. This reveals a fundamental phase transition behavior consistent with the CS theory. Furthermore, it would be interesting to note that in the small-sample regime (especially $M \approx 3S$), CoSaMP's update (i.e., identify candidates, merge with the current support, solve an LS problem on the merged set, then prune to S without a subsequent LS refinement) often yields LS problems that are close to square, i.e., with roughly as many unknowns as equations, rather than the comfortably overdetermined systems maintained by OMP. The resulting poorer conditioning modestly degrades its performance relative to OMP, which adds one atom per iteration and thus maintains better-conditioned, overdetermined LS systems. As M grows and the LS becomes clearly overdetermined, CoSaMP's pruning and residual correction are more effective and its performance catches up or slightly surpasses OMP. Nevertheless, in off-grid scenarios, both greedy methods are ultimately limited by mismatch-induced approximation error, whereas LASSO retains an advantage by fitting the leakage cluster with multiple atoms.

In the on-grid setting, as illustrated in Fig. 9.3, LASSO often outperforms OMP/CoSaMP at low SNR because ℓ_1-based thresholding simultaneously reduces noise-induced false support selections and suppresses the variance of the amplitude estimates. As SNR increases and the support is identified correctly with high probability, the dominant error source shifts from model selection to amplitude estimation, i.e., on a known support of size S, OMP/CoSaMP perform LS estimation, which is unbiased, whereas LASSO retains shrinkage bias and thus exhibits higher NMSE in the medium–high SNR regime. Moreover, CoSaMP is slightly inferior to OMP because, after merging the support and solving a one-shot LS, it prunes by magnitude but does not refit on the pruned support; the retained coefficients are therefore not the optimal LS solution on that support, and this effect is exacerbated when the merged support is large or ill-conditioned. To remove differences in prior information and estimator form, and enable a fair comparison, we fix the final supports of LASSO and CoSaMP to size S and perform an additional LS refinement; after refitting, their curves nearly coincide with OMP, indicating that in the medium–high SNR regime the observed gaps are explained by LASSO's shrinkage bias and CoSaMP's lack of post-pruning refinement, rather than by support errors or unequal priors.

Summary

This chapter explores greedy algorithms for sparse signal recovery, with an emphasis on their iterative, computationally efficient nature compared to optimization-based methods. Key insights reveal trade-offs between speed, robustness, and theoretical guarantees:

1. OMP selects atoms sequentially via residual correlation. While simple and fast, its recovery guarantees are signal-dependent and lack uniformity;besides it requires stricter matrix conditions (e.g., admissible matrices like Gaussian/Bernoulli ensembles).
2. ROMP addresses OMP's limitations by selecting multiple correlated atoms per iteration with magnitude regularization (ensuring comparable coefficients). This enables uniform recovery under RIP for broader matrices (e.g., partial Fourier), bridging greedy and convex optimization strengths.

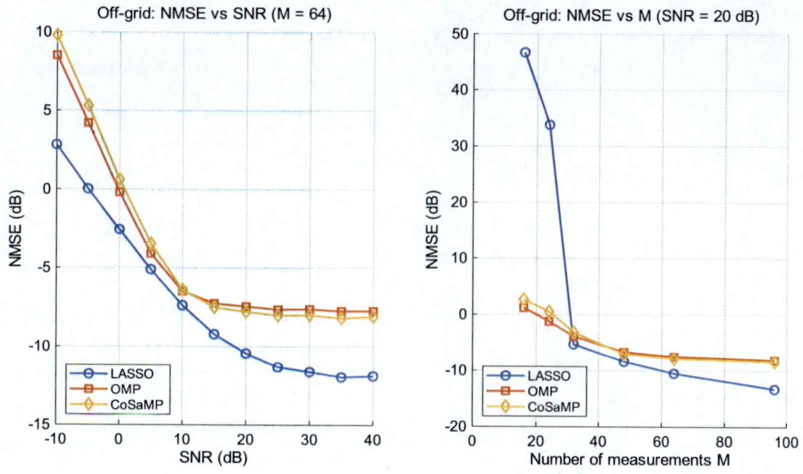

Fig. 9.2 Simulation results of Example 9.1 (off-grid case)

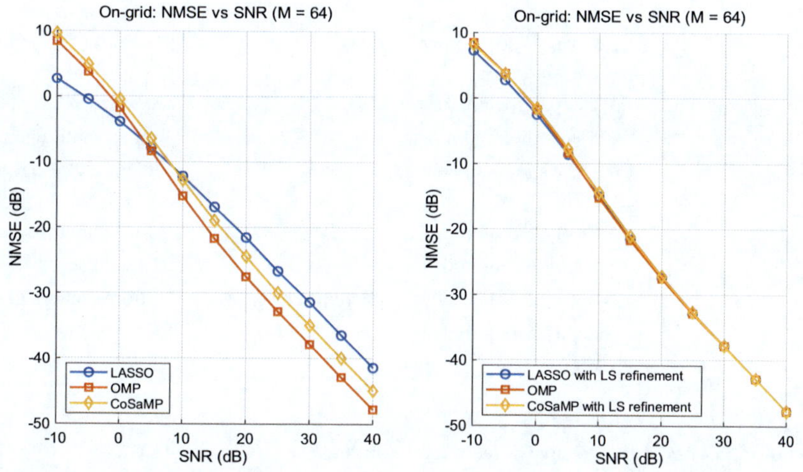

Fig. 9.3 Simulation results of Example 9.1 (on-grid case)

3. StOMP accelerates convergence via staged hard thresholding of correlations. Its statistical thresholding admits multiple coefficients per stage, trading parameter sensitivity for speed in large systems.
4. CoSaMP combines iterative support estimation, LS projection, and pruning. It achieves strong error bounds under RIP, with noise resilience proportional to the unrecoverable energy. Its error decays geometrically, ensuring robustness for compressible signals.

Besides, one may refer to Tables 9.1 and 9.2 to see the pros and cons of different optimization-based/greedy algorithms in terms of complexity, performance guarantee, required knowledge (e.g., sparsity levels S) and requirements on the sensing matrix.

Table 9.1 Comparison of optimization-based and greedy CS recovery algorithms

Algorithm	Complexity	Performance guarantees
Reweighted ℓ_1	High (iterative QP/LP)	Stronger than ℓ_1, log-sum penalty approximates ℓ_0
LASSO	Moderate (QP, coordinate descent)	Prediction consistency, sparsistency under conditions
Bayesian LASSO	High (Markov Chain Monte Carlo/EM)	Posterior inference, uncertainty quantification
Group LASSO	Moderate to High (block-wise coordinate descent)	Group-wise sparsity recovery
Burst LASSO	High (lifting + group LASSO)	aNSE bounds via CGMT; better phase transition
OMP	Low (O(SMN))	Signal-specific recovery; admissible matrices
ROMP	Moderate ($O(SNM + S^2M)$)	Uniform recovery under RIP
StOMP	Low to Moderate (O(PNM))	Large-system effectiveness (USE)
CoSaMP	Moderate ($O(\mathcal{L}\log(\frac{1}{\eta}))$)	Strong uniform guarantees under RIP
Algorithm	**Required prior knowledge**	**Requirements on sensing matrix**
Reweighted ℓ_1	No exact signal knowledge; ϵ must be chosen	Standard CS matrices (e.g., Gaussian, Bernoulli)
LASSO	Sparsity S not required; λ via CV	Standard CS matrices; RIP not strictly required
Bayesian LASSO	Hyperparameters for priors	Same as LASSO
Group LASSO	Group structure known	Groups should be orthonormal for simplicity
Burst LASSO	K, D (burst parameters)	Effective sensing matrix $\boldsymbol{AL}$ is non-i.i.d. (even if $\boldsymbol{A}$ is i.i.d.), complicating analysis.
OMP	Sparsity S required	Admissible matrices (Gaussian, Bernoulli)
ROMP	Sparsity S required	RIP with $\delta_{2S} \leq 0.03/\sqrt{\log S}$
StOMP	S not strictly needed; threshold tuning	Uniform spherical ensemble (USE)
CoSaMP	Sparsity S required	RIP with $\delta_{4S} \leq 0.1$ (or 0.5 in improved versions)

Table 9.2 Pros & Cons of optimization-based and greedy CS recovery algorithms

Algorithm	Pros & Cons
Reweighted ℓ_1	**Pros**: Better sparsity recovery, fewer measurements; **Cons**: Higher computation, sensitive to ϵ, may converge to local minima
LASSO	**Pros**: Convex, robust; **Cons**: λ tuning critical
Bayesian LASSO	**Pros**: Probabilistic interpretation, robust to λ uncertainty; **Cons**: Computationally intensive, prior specification needed
Group LASSO	**Pros**: Structured sparsity, group selection; **Cons**: Complex for non-orthonormal groups, circular dependency in updates
Burst LASSO	**Pros**: Exploits burst sparsity, significant measurement savings; **Cons**: Complex penalty, parameter tuning critical
OMP	**Pros**: Simple, fast, intuitive; **Cons**: No uniform guarantees, fails for some structured matrices
ROMP	**Pros**: Uniform guarantees, works with partial Fourier; **Cons**: Logarithmic factor in RIP constant
StOMP	**Pros**: Fast, multi-element selection per stage; **Cons**: Threshold sensitivity, less robust to noise
CoSaMP	**Pros**: Robust to noise, geometrically decaying error; **Cons**: Pruning may discard small true components

Appendices

Appendix 1: Another Variation of CoSaMP

This modified version of CoSaMP shifts focus from reconstructing the entire signal at each step to approximating the residual (unrecovered signal portion). This approach mirrors techniques like HHS pursuit and offers computational advantages when residuals decay over iterations. The modified workflow is shown as follows.

Input: 1. an $M \times N$ measurement matrix $\boldsymbol{A}$; 2. an M-dimensional data vector $\boldsymbol{y}$; 3. the sparsity level S of the ideal signal.
Output: The final estimate $\hat{\boldsymbol{x}}$.

- **Initialization**: Initialize with a trivial approximation (all zeros), i.e., $\hat{\boldsymbol{x}}^0 = \boldsymbol{0}$. Let $t = 1$ denote the iteration index.
- **Identification (unchanged)**: Compute the proxy $\boldsymbol{u} = \boldsymbol{A}^*\boldsymbol{r}$, where $\boldsymbol{r} = \boldsymbol{y}$ and obtain $\Omega \leftarrow \text{supp}(\boldsymbol{u}_{2S})$.
- **Estimation (key change)**: Solve a residual-focused LS problem, and obtain $\boldsymbol{b} = \boldsymbol{A}_\Omega^\dagger \boldsymbol{r}$. This approximates the residual signal (not the full signal) restricted to Ω. In this case, one initializes the iterative LS algorithm with the zero vector to take advantage of the fact that the residual is becoming small.
- **Approximation Merger (new step)**: Combine the previous signal approximation $\hat{\boldsymbol{x}}^{t-1}$ with the residual estimate $\boldsymbol{b}$:

$$\boldsymbol{c} = \hat{\boldsymbol{x}}^{t-1} + \boldsymbol{b}. \tag{9.35}$$

  Unlike standard CoSaMP (which merges support sets), this directly merges signal estimates.
- **Pruning**: Retain only the S largest components of $\boldsymbol{c}$:

$$\hat{\boldsymbol{x}}^t = \boldsymbol{c}_S. \tag{9.36}$$

  This enforces sparsity in the updated approximation.
- **Sample Update**: Recompute the residual sample vector: $\boldsymbol{r} = \boldsymbol{y} - \boldsymbol{A}\hat{\boldsymbol{x}}^t$.
- Let $t \leftarrow t + 1$, and the above process repeats until the halting criterion (e.g., a fixed number of iterations) is met. The final output is an S-sparse approximation of the original signal.

This algorithm also satisfies Theorem 9.4.

Appendix 2: Complex-Valued LASSO

For the complex-valued LASSO problem in Example 9.1, we minimize the following objective:

$$f(\tilde{x}_j) = \underbrace{\frac{1}{2M}\sum_{i=1}^{M} |r_i^{(j)} - A_{ij}\tilde{x}_j|^2}_{\text{Quadratic loss}} + \underbrace{\lambda|\tilde{x}_j|}_{\ell_1 \text{ penalty}}, \tag{9.37}$$

where $r_i^{(j)} = y_i - \sum_{k \neq j} A_{ik}\tilde{x}_k$ (complex partial residual), $A_{ij} \in \mathbb{C}$ (complex sensing matrix entries), $\tilde{x}_j \in \mathbb{C}$ (complex signal coefficient), and $|\cdot|$ denotes complex modulus, i.e., $|z| = \sqrt{\Re(z)^2 + \Im(z)^2}$. Then, the quadratic loss in (9.37) can be expanded as follows using Hermitian inner products:

$$\frac{1}{2M}\sum_{i=1}^{M}|r_i^{(j)} - A_{ij}\tilde{x}_j|^2 = \frac{1}{2M}\sum_{i=1}^{M}(r_i^{(j)} - A_{ij}\tilde{x}_j)^H(r_i^{(j)} - A_{ij}\tilde{x}_j)$$
$$= \frac{1}{2M}\left[\underbrace{\sum_i |r_i^{(j)}|^2}_{\text{const}} - 2\Re\left(\sum_i \overline{r_i^{(j)}} A_{ij}\tilde{x}_j\right) + |\tilde{x}_j|^2 \sum_i |A_{ij}|^2\right]. \tag{9.38}$$

Define $v_j = \sum_{i=1}^{M} |A_{ij}|^2$ and $c_j = \sum_{i=1}^{M} \overline{A_{ij}} r_i^{(j)}$ as the aggregate energy and correlation respectively, the objective simplifies to

$$f(\tilde{x}_j) = \frac{v_j}{2M}|\tilde{x}_j|^2 - \frac{1}{M}\Re\left(\overline{c_j}\tilde{x}_j\right) + \lambda|\tilde{x}_j| + \text{const}. \tag{9.39}$$

Similar to that in Sect. 8.2.2, the minimum of $f(\tilde{x}_j)$ occurs where 0 is in the Wirtinger subdifferential, i.e.,

$$0 \in \frac{v_j}{2M}\tilde{x}_j - \frac{c_j}{2M} + \lambda\frac{\partial|\tilde{x}_j|}{\partial\tilde{x}_j^*}, \tag{9.40}$$

where the subdifferential of the modulus is defined as

$$\frac{\partial|z|}{\partial z^*} = \begin{cases} \frac{z}{2|z|} & z \neq 0, \\ \{u \in \mathbb{C} : |u| \leq 1\} & z = 0. \end{cases} \tag{9.41}$$

Then, we proceed to analyze the behavior of (9.40) for two cases, i.e., $\tilde{x}_j \neq 0$ and $\tilde{x}_j = 0$.

1. $\tilde{x}_j \neq 0$: in this case, by substituting $\frac{\partial|\tilde{x}_j|}{\partial\tilde{x}_j^*} = \frac{\tilde{x}_j}{2|\tilde{x}_j|}$ into (9.40), we obtain

$$\frac{v_j}{2M}\tilde{x}_j - \frac{c_j}{2M} + \lambda\frac{\tilde{x}_j}{2|\tilde{x}_j|} = 0. \tag{9.42}$$

   Rearranging the terms in (9.42) leads to

$$\tilde{x}_j\left(v_j + \lambda M\frac{1}{|\tilde{x}_j|}\right) = c_j. \tag{9.43}$$

   Express $\tilde{x}_j = re^{j\phi}$ (polar form) and substitute it into (9.43), we have

$$re^{j\phi}v_j + \lambda M e^{j\phi} = c_j. \tag{9.44}$$

   Multiply both sides of (9.44) by $e^{-j\phi}$, we arrive at

$$rv_j + \lambda M = c_j e^{-j\phi}. \tag{9.45}$$

   Since left side of (9.45) is real and positive, the corresponding right side must also be real and positive. Thus, we have

$$e^{j\phi} = \frac{c_j}{|c_j|}, \quad r = \frac{|c_j| - \lambda M}{v_j} \quad (\text{require } |c_j| > \lambda M). \tag{9.46}$$

   Therefore, the LASSO solution in this case can be expressed as

$$\tilde{x}_j = \frac{c_j}{|c_j|}\frac{|c_j| - \lambda M}{v_j} = \frac{c_j}{v_j}\left(1 - \frac{\lambda M}{|c_j|}\right). \tag{9.47}$$

2. $\tilde{x}_j = 0$: the subdifferential condition in this case requires

$$0 \in -\frac{c_j}{M} + \lambda\{u : |u| \leq 1\}, \tag{9.48}$$

which is equivalent to

$$\left|\frac{c_j}{M}\right| \leq \lambda \implies |c_j| \leq \lambda M. \tag{9.49}$$

By combining the results in both cases, we can obtain the complex-valued LASSO solution as follows:

$$\tilde{x}_j = \begin{cases} \frac{c_j}{v_j}\left(1 - \frac{\lambda M}{|c_j|}\right) & |c_j| > \lambda M \\ 0 & |c_j| \leq \lambda M. \end{cases} \tag{9.50}$$

Define the complex soft-thresholding operator as follows:

$$\mathcal{S}_\kappa^{\mathbb{C}}(z) = \frac{z}{|z|}(|z| - \kappa)_+ = e^{j\arg(z)}\max(|z| - \kappa, 0), \tag{9.51}$$

then the solution in (9.50) can be more compactly written as

$$\tilde{x}_j = \frac{1}{v_j}\mathcal{S}_{\lambda M}^{\mathbb{C}}(c_j). \tag{9.52}$$

Besides, please see Appendix 3 for the detailed derivation on how to derive the average Gaussian distance in the complex case, based on which and by resorting to Theorem 8.1, we can find a value of λ that achieves fairly good performance in practice.

Appendix 3: Derivation of Average Gaussian Distance for Complex Case

In the complex case, the subgradient set $\partial\|\tilde{\boldsymbol{x}}^\star\|_1$ is defined as

$$\partial\|\tilde{\boldsymbol{x}}^\star\|_1 = \left\{\boldsymbol{\alpha} \in \mathbb{C}^N : \alpha_i = \begin{cases} \text{sign}(\tilde{x}_i^\star) & \text{if } \tilde{x}_i^\star \neq 0, \\ \{z : |z| \leq 1\} & \text{if } \tilde{x}_i^\star = 0 \end{cases}\right\}, \tag{9.53}$$

where sign(z) for a complex number z is defined as $\text{sign}(z) = \frac{z}{|z|}$ for $z \neq 0$. Recall that the average Gaussian distance $\bar{d}(\tau)$ is defined as

$$\bar{d}(\tau) = \frac{1}{M}\mathbb{E}_{\boldsymbol{w}}\left\{\min_{\boldsymbol{\alpha}\in\partial\|\tilde{\boldsymbol{x}}^\star\|_1} \|\boldsymbol{w} - \sqrt{M}\tau\boldsymbol{\alpha}\|^2\right\}, \quad \boldsymbol{w} \sim \mathcal{CN}(0, \boldsymbol{I}_N). \tag{9.54}$$

Next, similar to the real case, we decompose the inner minimization problem in $\bar{d}(\tau)$ into support sets. Specifically, let $\mathcal{S} = \{i : \tilde{x}_i^\star \neq 0\}$ (support of $\boldsymbol{x}^\star$) and $\mathcal{S}^c = \{1, \ldots, N\}\backslash\mathcal{S}$. Accordingly, the minimization in (9.54) separates as

$$\min_{\boldsymbol{\alpha}\in\partial\|\tilde{\boldsymbol{x}}^\star\|_1} \|\boldsymbol{w} - \sqrt{M}\tau\boldsymbol{\alpha}\|^2 = \underbrace{\sum_{i\in\mathcal{S}}\left|w_i - \sqrt{M}\tau\,\text{sign}(\tilde{x}_i^\star)\right|^2}_{\text{Fixed for } \boldsymbol{\alpha}_\mathcal{S}} + \underbrace{\sum_{i\in\mathcal{S}^c}\min_{|\alpha_i|\leq 1}\left|w_i - \sqrt{M}\tau\alpha_i\right|^2}_{\text{Optimized over } \boldsymbol{\alpha}_{\mathcal{S}^c}}. \tag{9.55}$$

Focus on the first part on the right-hand side of (9.55), i.e., the non-zero support $\mathcal{S}$ part, since $\alpha_i = \text{sign}(\tilde{x}_i^\star)$ is fixed for $i \in \mathcal{S}$, we have

$$\sum_{i\in\mathcal{S}}\left|w_i - \sqrt{M}\tau\,\text{sign}(\tilde{x}_i^\star)\right|^2 = \sum_{i\in\mathcal{S}}|w_i|^2 + M\tau^2 S - 2\sqrt{M}\tau\,\text{Re}\left(\sum_{i\in\mathcal{S}} w_i\,\text{sign}(\tilde{x}_i^\star)^*\right). \tag{9.56}$$

In the asymptotic analysis, the cross-term vanishes because $\boldsymbol{w}$ is independent of $\boldsymbol{x}^\star$, i.e., $\mathbb{E}_{w_i}\{w_i\,\text{sign}(\tilde{x}_i^\star)^*\} = 0$. Moreover, we have

$$\mathbb{E}\left\{\sum_{i\in\mathcal{S}} |w_i|^2\right\} = S, \tag{9.57}$$

and the first part can be written as $S + M\tau^2 S$.

For the second part on the right-hand side of (9.55), i.e., for $i \in \mathcal{S}^c$, we solve

$$\min_{|\alpha_i|\le 1} |w_i - \sqrt{M}\tau\alpha_i|^2 = \begin{cases} 0 & \text{if } |w_i| \le \sqrt{M}\tau, \\ \left(|w_i| - \sqrt{M}\tau\right)^2 & \text{otherwise.} \end{cases} \tag{9.58}$$

The expectation of (9.58) over $\boldsymbol{w}$ is

$$\mathbb{E}_{w_i}\left\{\left(|w_i| - \sqrt{M}\tau\right)_+^2\right\}, \quad w_i \sim \mathcal{CN}(0, 1). \tag{9.59}$$

For complex Gaussian w_i, $|w_i|$ follows a Rayleigh distribution with PDF $f_R(r) = 2re^{-r^2}$, i.e., $\mathbb{P}(|w_i| \le r) = 1 - e^{-r^2}$. Thus, we can obtain

$$\mathbb{E}\left\{\left(|w_i| - \sqrt{M}\tau\right)_+^2\right\} = \int_{\sqrt{M}\tau}^{\infty} (r - \sqrt{M}\tau)^2 \cdot 2re^{-r^2} dr. \tag{9.60}$$

To solve the integral in (9.60), let $c = \sqrt{M}\tau$ and $I(c) = \int_c^\infty (r - c)^2 \cdot 2re^{-r^2} dr$. Then, the integrand can be expanded as

$$(r - c)^2 \cdot 2re^{-r^2} = 2r(r^2 - 2cr + c^2)e^{-r^2} = 2r^3 e^{-r^2} - 4cr^2 e^{-r^2} + 2c^2 re^{-r^2}. \tag{9.61}$$

Accordingly, by defining $u = r^2$, $du = 2rdr$, the integral of the first term in (9.61) can be obtained as

$$vc^{2\infty} ue^{-u} du = \left[-ue^{-u}\right]_{c^2}^{\infty} + \int_{c^2}^{\infty} e^{-u} du = c^2 e^{-c^2} + e^{-c^2}.$$

For the second term $\int_c^\infty -4cr^2 e^{-r^2} dr$, by using integration by parts, we obtain

$$\int_c^\infty -4cr^2 e^{-r^2} dr = -2c\left(\frac{\sqrt{\pi}}{2}\operatorname{erfc}(c) + ce^{-c^2}\right), \tag{9.62}$$

please see Appendix 4 for the detailed derivation. Similarly, for the third term, we have

$$\int_c^\infty 2c^2 re^{-r^2} dr = c^2 \int_{c^2}^\infty e^{-u} du = c^2 e^{-c^2}.$$

Combining the above results leads to

$$\begin{aligned} I(c) &= \underbrace{(c^2 e^{-c^2} + e^{-c^2})}_{\text{Term 1}} \underbrace{- 2c\left(\frac{\sqrt{\pi}}{2}\operatorname{erfc}(c) + ce^{-c^2}\right)}_{\text{Term 2}} + \underbrace{c^2 e^{-c^2}}_{\text{Term 3}} \\ &= e^{-c^2} - c\sqrt{\pi}\operatorname{erfc}(c). \end{aligned}$$

Based on the above derivation, the average Gaussian distance is

$$\bar{d}(\tau) = \frac{S}{M} + \tau^2 S + \frac{N - S}{M} I(\sqrt{M}\tau), \tag{9.63}$$

and by substituting $I(\sqrt{M}\tau) = e^{-M\tau^2} - \sqrt{M}\tau\sqrt{\pi}\,\mathrm{erfc}(\sqrt{M}\tau)$ into (9.63), we have

$$\bar{d}(\tau) = \frac{S}{M} + \tau^2 S + \frac{N-S}{M}\left[e^{-M\tau^2} - \sqrt{M}\tau\sqrt{\pi}\,\mathrm{erfc}(\sqrt{M}\tau)\right]. \tag{9.64}$$

As can be seen, in order to avoid divergence, the term $\tau^2 S$ must be $O(1)$ as $N \to \infty$. To achieve this, we redefine $\tau = \frac{t}{\sqrt{N}}$, now we have $\sqrt{M}\tau = \sqrt{\delta N} \cdot \frac{t}{\sqrt{N}} = \sqrt{\delta}\,t$, $M\tau^2 = \delta N \cdot \frac{t^2}{N} = \delta t^2$. Substituting them back into $\bar{d}(\tau)$ yields (sparsity ratio $\rho = \frac{S}{N}$, measurement ratio $\delta = \frac{M}{N}$)

$$\lim_{N\to\infty} \bar{d}(t) = \frac{\rho}{\delta} + \rho t^2 + \frac{1-\rho}{\delta}\left(e^{-\delta t^2} - \sqrt{\delta}\,t\sqrt{\pi}\,\mathrm{erfc}(\sqrt{\delta}\,t)\right). \tag{9.65}$$

Appendix 4: Derivation of (9.62)

First use integration by parts on $\int_c^\infty r^2 e^{-r^2} dr$, and set $v = r$, $dw = re^{-r^2} dr$, then we have

$$dv = dr, \quad w = \int re^{-r^2} dr = -\frac{1}{2}e^{-r^2}.$$

Accordingly, the integration-by-parts formula $\int v\,dw = vw - \int w\,dv$ gives

$$\int_c^\infty r^2 e^{-r^2} dr = r \cdot \left(-\frac{1}{2}e^{-r^2}\right)\Big|_c^\infty - \int_c^\infty \left(-\frac{1}{2}e^{-r^2}\right) dr.$$

In the above equation, the boundary term can be calculated as

$$-\frac{r}{2}e^{-r^2}\Big|_c^\infty = \lim_{r\to\infty}\left(-\frac{r}{2}e^{-r^2}\right) - \left(-\frac{c}{2}e^{-c^2}\right) = 0 + \frac{c}{2}e^{-c^2} = \frac{c}{2}e^{-c^2},$$

where the limit is 0 because e^{-r^2} decays faster than r grows, and the remaining integral is

$$-\int_c^\infty \left(-\frac{1}{2}e^{-r^2}\right) dr = \frac{1}{2}\int_c^\infty e^{-r^2} dr.$$

Given the definition of the complementary error function below:

$$\mathrm{erfc}(z) = \frac{2}{\sqrt{\pi}}\int_z^\infty e^{-t^2} dt,$$

we can directly obtain (9.62).

Homework Problems

9.1 (a) Consider the measurement matrix

$$\boldsymbol{A} = \begin{bmatrix} 1 & 0 & -\frac{1}{\sqrt{2}} \\ 0 & 1 & -\frac{1}{\sqrt{2}} \end{bmatrix}.$$

Prove that any 1-sparse vector $\boldsymbol{x} \in \mathbb{R}^3$ is recovered with one iteration of the OMP algorithm using $\boldsymbol{A}$.

(b) Now, add a measurement by appending the row [3, 0, 1] to $\boldsymbol{A}$, forming the matrix

$$\hat{\boldsymbol{A}} = \begin{bmatrix} 1 & 0 & -\frac{1}{\sqrt{2}} \\ 0 & 1 & -\frac{1}{\sqrt{2}} \\ 3 & 0 & 1 \end{bmatrix}.$$

Prove that the 1-sparse vector $\boldsymbol{x} = [0, 0, 1]^T$ cannot be recovered via the OMP algorithm with the measurement matrix $\hat{\boldsymbol{A}}$.

9.2 Using MATLAB, implement the OMP algorithm. Run it on a CS model $\boldsymbol{y} = \boldsymbol{A}\boldsymbol{x} + \boldsymbol{n}$, where $\boldsymbol{A}$ is a 32×64 sensing matrix generated randomly with i.i.d. standard normal distributed entries, $\boldsymbol{n}$ is the AWGN with different noise variances, $\boldsymbol{x}$ is a 64×1 sparse signal with i.i.d. standard normally distributed non-zero entries, and the number of non-zero elements is set as 2, 4, 8, 16, 32 in different experiments.

Plot the average normalized mean squared error (NMSE) versus the number of iterations for different numbers of non-zero elements. Explain the simulation results.

9.3 Prove that in the noiseless case ($\boldsymbol{n} = \boldsymbol{0}$) with an exactly S-sparse signal $\tilde{\boldsymbol{x}}$, the OMP algorithm will always select a correct atom (one from the true support) in each iteration, provided that the sensing matrix $\boldsymbol{A}$ satisfies the following exact recovery condition (ERC):

$$\max_{j \notin \Lambda} \|\boldsymbol{A}_\Lambda^\dagger \boldsymbol{a}_j\|_1 < 1,$$

where Λ is the true support set of $\tilde{\boldsymbol{x}}$. Discuss the implications of this result for the exact recovery guarantee of OMP.

References

1. Å. Björck, *Numerical Methods for Least Squares Problems*. Society for Industrial and Applied Mathematics (2024)
2. S. Kunis, H. Rauhut, Random sampling of sparse trigonometric polynomials, II. Orthogonal matching pursuit versus basis pursuit. Found. Comput. Math. **8**, 737–763 (2008)
3. J.A. Tropp, A.C. Gilbert, Signal recovery from random measurements via orthogonal matching pursuit. IEEE Trans. Inf. Theory **53**(12), 4655–4666 (2007)
4. J.A. Tropp, A.C. Gilbert, Signal recovery from random measurements via orthogonal matching pursuit: the Gaussian case, Caltech, Pasadena, CA, 2007, ACM Technical Report 2007-01 (2007), http://www.acm.caltech.edu/~jtropp/reports/TG07-Signal-Recovery-TR.pdf
5. E.J. Candès, T. Tao, Near-optimal signal recovery from random projections: Universal encoding strategies? IEEE Trans. Inf. Theory **52**(12), 5406–5425 (2006)
6. E.J. Candès, T. Tao, Decoding by linear programming. IEEE Trans. Inf. Theory **51**(12), 4203–4215 (2005)
7. D.L. Donoho, Compressed sensing. IEEE Trans. Inf. Theory **52**(4), 1289–1306 (2006)
8. M. Rudelson, R. Vershynin, Geometric approach to error-correcting codes and reconstruction of signals. Int. Math. Res. Not. **2005**(64), 4019–4041 (2005)
9. J. Kahn, J. Komlós, E. Szemerédi, On the probability that a random ±1-matrix is singular. J. Am. Math. Soc. **8**(1), 223–240 (1995)
10. D. Needell, R. Vershynin, Uniform uncertainty principle and signal recovery via regularized orthogonal matching pursuit. Found. Comput. Math. **9**, 317–334 (2009)
11. D.L. Donoho, Y. Tsaig, I. Drori, J.L. Starck, Sparse solution of underdetermined systems of linear equations by stagewise orthogonal matching pursuit. IEEE Trans. Inf. Theory **58**(2), 1094–1121 (2012)
12. F. Abramovich, Y. Benjamini, D.L. Donoho, I.M. Johnstone, Adapting to unknown sparsity by controlling the false discovery rate. Ann. Stat. **34**(2), 584–653 (2006)
13. Y. Benjamini, Y. Hochberg, Controlling the false discovery rate: a practical and powerful approach to multiple testing. J. Royal Stat. Soc. Ser. B (Methodol.) **57**, 289–300 (1995)
14. A.C. Gilbert, S. Guha, P. Indyk, S. Muthukrishnan, M. Strauss, Near-optimal sparse Fourier representations via sampling, in *Proceedings of the Thirty-Fourth Annual ACM Symposium on Theory of Computing* (2002), pp. 152–161
15. A.C. Gilbert, S. Muthukrishnan, M. Strauss, Improved time bounds for near-optimal sparse Fourier representations, in *Wavelets XI*, vol. 5914 (2005), pp. 398–412 (SPIE)
16. A.C. Gilbert, M.J. Strauss, J.A. Tropp, R. Vershynin, Algorithmic linear dimension reduction in the ℓ_1 norm for sparse vectors (2006). arXiv preprint arXiv:cs/0608079
17. A.C. Gilbert, M.J. Strauss, J.A. Tropp, R. Vershynin, One sketch for all: fast algorithms for compressed sensing, in *Proceedings of the Thirty-Ninth Annual ACM Symposium on Theory of Computing* (2007), pp. 237–246
18. G. Cormode, S. Muthukrishnan, *Combinatorial algorithms for compressed sensing, in Structural Information and Communication Complexity: 13th International Colloquium, SIROCCO 2006, Chester, UK, July 2–5, Proceedings 13* (Springer, Berlin Heidelberg, 2006), pp. 280–294
19. M.A. Iwen, A deterministic sub-linear time sparse Fourier algorithm via non-adaptive compressed sensing methods (2007), arXiv preprint arXiv:0708.1211
20. E.J. Candés, J. Romberg, T. Tao, Robust uncertainty principles: exact signal reconstruction from highly incomplete frequency information. IEEE Trans. Inf. Theory **52**(2), 489–509 (2006)

21. S.J. Kim, K. Koh, M. Lustig, S. Boyd, D. Gorinevsky, An interior-point method for large-scale ℓ_1-regularized least squares. IEEE J. Sel. Top. Signal Process. **1**(4), 606–617 (2008)
22. M.A. Figueiredo, R.D. Nowak, S.J. Wright, Gradient projection for sparse reconstruction: application to compressed sensing and other inverse problems. IEEE J. Sel. Top. Signal Process. **1**(4), 586–597 (2007)
23. I. Daubechies, M. Defrise, C. De Mol, An iterative thresholding algorithm for linear inverse problems with a sparsity constraint. Commun. Pure Appl. Math.: J. Issued Courant Inst. Math. Sci. **57**(11), 1413–1457 (2004)
24. E.J. Candès, J.K. Romberg, T. Tao, Stable signal recovery from incomplete and inaccurate measurements. Commun. Pure Appl. Math. **59**, 1207–1223 (2006)
25. D. Needell, J.A. Tropp, CoSaMP: iterative signal recovery from incomplete and inaccurate samples. Appl. Comput. Harmon. Anal. **26**(3), 301–321 (2009)
26. G. Reeves, M. Gastpar, Sampling bounds for sparse support recovery in the presence of noise, in *2008 IEEE International Symposium on Information Theory*. (IEEE, 2008), pp. 2187–2191
27. C. Song, S. Xia, X. Liu, Improved analysis for SP and CoSaMP algorithms in terms of restricted isometry constants (2013), arXiv preprint arXiv:1309.6073
28. D. Tse, P. Viswanath, *Fundamentals of Wireless Communication* (Cambridge University Press, 2005)
29. A.M. Sayeed, Deconstructing multiantenna fading channels. IEEE Trans. Signal Process. **50**(10), 2563–2579 (2002). https://doi.org/10.1109/TSP.2002.803337

10 Approximate Message Passing

Building upon the optimization-based and greedy CS recovery algorithms studied in previous chapters, this chapter introduces Approximate Message Passing (AMP) as an efficient alternative that achieves an improved performance-complexity tradeoff, particularly for sensing matrices with i.i.d. entries. We first discuss the motivation for AMP development, then systematically present its algorithmic derivation and provide an intuitive interpretation of the underlying message passing mechanism, including practical considerations for denoising function selection. Furthermore, we introduce the state evolution (SE) framework that enables accurate performance prediction of AMP algorithms. The chapter concludes with a comprehensive discussion of various AMP generalizations and extensions that enhance its applicability to broader problem settings.

10.1 Motivation for Approximate Message Passing

Prior to the development of AMP, numerous optimization-based and greedy CS recovery algorithms had been successfully applied to various wireless communication problems, including massive MIMO channel estimation [1, 2] and spectrum sensing [3]. However, these approaches presented significant limitations: optimization-based methods typically incurred high computational complexity, while greedy algorithms often sacrificed recovery performance. Furthermore, accurately analyzing or predicting the performance of these algorithms proved challenging. Existing performance analyses for many optimization-based and greedy CS recovery methods typically provided only loose bounds that were too conservative to offer practical insights. Although more precise performance characterizations existed for LASSO under i.i.d. sensing matrices [4], these results lacked generalizability to broader settings.

This landscape created a pressing need for low-complexity, scalable algorithms with theoretical performance guarantees capable of solving high-dimensional linear regression problems in CS. Against this backdrop, the AMP algorithm emerged around 2009 [5] as a promising solution.

Compared to conventional optimization-based and greedy CS recovery algorithms, AMP offers several distinct advantages: low computational complexity, rapid convergence, asymptotically exact performance characterization, and testable conditions for optimality. These beneficial properties have led to AMP's widespread adoption across numerous applications, particularly in solving large-scale CS recovery problems such as ultra-massive MIMO channel estimation (involving thousands of base station antennas), wireless scattering environment sensing, and magnetic resonance imaging (MRI).

10.2 Derivation of AMP

10.2.1 Problem Setup

Consider the following sparse signal recovery problem of a standard linear model:

$$\boldsymbol{y} = \boldsymbol{A}\boldsymbol{x} + \boldsymbol{w}, \tag{10.1}$$

where $\boldsymbol{x} \in \mathbb{R}^N$ is an unknown sparse signal vector, $\boldsymbol{A} \in \mathbb{R}^{M \times N}$ is a known sensing matrix, $\boldsymbol{w} \in \mathbb{R}^M$ is the AWGN with zero mean and covariance $\sigma_w^2 \boldsymbol{I}_M$, and $\boldsymbol{y} \in \mathbb{R}^M$ is the received observation. This chapter addresses the standard CS problem with $M < N$.

A. Liu and M.-M. Zhao, *Advanced Bayesian Signal Processing*, Textbooks in Telecommunication Engineering,
https://doi.org/10.1007/978-3-032-26656-9_10

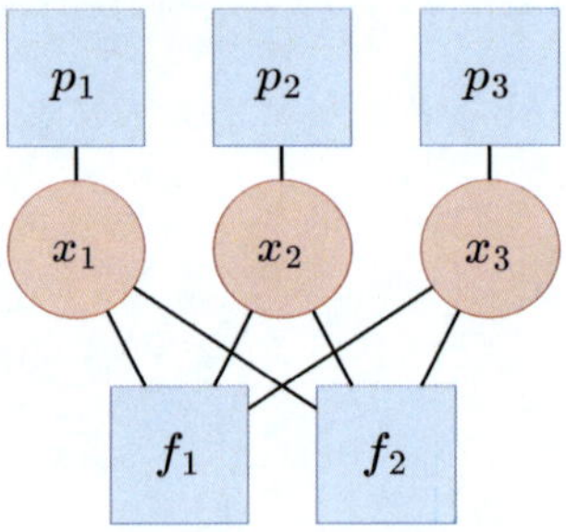

Fig. 10.1 Factor graph corresponding to (10.3), where $N = 3$, $M = 2$, p_n and f_m represent $p(x_n)$ and $\mathcal{N}\left(y_m; \boldsymbol{a}_m^T \boldsymbol{x}, \sigma_w^2\right)$, respectively

To solve the problem, one possible method is the well-known LASSO, which estimates $\boldsymbol{x}$ by solving the following optimization problem:

$$\hat{\boldsymbol{x}} = \arg\min_{\boldsymbol{x}} \frac{1}{2\sigma_w^2} \|\boldsymbol{y} - \boldsymbol{A}\boldsymbol{x}\|_2^2 + \beta \|\boldsymbol{x}\|_1. \tag{10.2}$$

Another potential method is Bayesian inference that aims at calculating the posterior distribution. In particular, assuming the elements $x_n, n = 1, \ldots, N$ in $\boldsymbol{x}$ are i.i.d. with a sparse prior $p(x_n)$, the posterior distribution of $\boldsymbol{x}$ is given by

$$p(\boldsymbol{x}|\boldsymbol{y}) \propto \prod_{n=1}^{N} p(x_n) \prod_{m=1}^{M} \mathcal{N}\left(y_m; \boldsymbol{a}_m^T \boldsymbol{x}, \sigma_w^2\right), \tag{10.3}$$

where $\boldsymbol{a}_m^T$ denotes the m-th row in $\boldsymbol{A}$. When the sparse prior is set as Laplace prior with Laplace parameter $\beta > 0$, i.e., $p(x_n) = \frac{\beta}{2} \mathrm{e}^{-\beta|x_n|}$, the MAP estimator $\hat{\boldsymbol{x}} = \arg\max_{\boldsymbol{x}} p(\boldsymbol{x}|\boldsymbol{y})$ becomes equivalent to the solution of the LASSO problem.

Although this Bayesian interpretation is theoretically appealing, exact inference is often computationally intractable. To this end, graphical models and message passing algorithms have been developed to approximate the MAP/MMSE estimator by computing the marginal posterior distributions via a sequence of iterations.

10.2.2 Factor Graph and Message Passing

The factorization in (10.3) corresponds to the factor graph illustrated in Fig. 10.1. In this graph, message passing is performed iteratively through two types of messages:

- **Factor-to-Variable Messages**: Let $\Delta_{m\to n}^t$ denote the message from factor node f_m to variable node x_n at the t-th iteration. This message is proportional to

$$\Delta_{m\to n}^t \propto \int_{x_{n'}, n' \neq n} \mathcal{N}\left(y_m; \boldsymbol{a}_m^T \boldsymbol{x}, \sigma_w^2\right) \prod_{n' \neq n} \Delta_{n'\to m}^t. \tag{10.4}$$

  It effectively aggregates information from all other variables to update the belief about x_n based on the measurement y_m.
- **Variable-to-Factor Messages**: Let $\Delta_{n\to m}^{t+1}$ denote the message from variable node x_n to factor node f_m at the $(t + 1)$-th iteration. This message is proportional to

$$\Delta_{n\to m}^{t+1} \propto \mathrm{e}^{-\beta|x_n|} \prod_{m' \neq m} \Delta_{m'\to n}^t, \tag{10.5}$$

  combining the local prior $\mathrm{e}^{-\beta|x_n|}$ with incoming messages from all other factors to form an updated belief about x_n.

After each iteration t, the marginal posterior distribution of each variable x_n is approximated by combining the local prior with all incoming factor-to-variable messages:

$$p^{t+1}(x_n|\mathbf{y}) \propto e^{-\beta|x_n|} \prod_{m=1}^{M} \Delta_{m\to n}^{t}. \tag{10.6}$$

Based on these marginal posteriors, the MAP or MMSE estimator can be obtained via one-dimensional search or integration, respectively.

However, directly implementing this exact message passing is computationally infeasible in practice, as it requires computing and storing $O(MN)$ message functions per iteration. To overcome this, the AMP algorithm introduces efficient approximations to the message updates. The AMP framework branches into two approximation strategies, each tailored to a specific estimation objective: one for the MAP estimator and another for the MMSE estimator.

- For the **MAP estimator**, derived from **min-sum message passing (MSMP)**, AMP operates in the log-domain. It applies a **quadratic Legendre transform** to the log-messages. This technique approximates a function by a quadratic form that matches both its extremum point and the second derivative at that point. In essence, this preserves the key properties of the original function while making the updates tractable.
- For the **MMSE estimator**, based on **sum-product message passing (SPMP)**, AMP performs a **moment-matching Gaussian approximation** directly on the messages in the probability domain. This ensures that the approximated message shares the same mean and variance as the original, thereby capturing the essential statistical characteristics while simplifying the computations.

Although the two AMP variants employ different approximation strategies, their final update rules converge to a similar form [6]. This unified behavior underscores the robustness and generality of the AMP approach. In the following, we provide a concise derivation of the AMP algorithm based on SPMP for MMSE estimation. The detailed derivation of the MAP-oriented AMP via MSMP can be found in [7].

10.2.3 From SPMP to AMP

The AMP framework achieves computational tractability from the exact but intractable SPMP rules in (10.4), (10.5) through two foundational approximations:

1. **Gaussian Approximation of Messages**: All messages are approximated as Gaussian distributions, completely characterized by their mean and variance. This converts the message passing procedure from propagating full distributions to updating a finite set of parameters. Formally, the mean and variance of the message $\Delta_{a\to b}^{t}$ in the t-th iteration are denoted by $\mu_{a\to b}^{t}$ and $v_{a\to b}^{t}$, respectively.
2. **Large-System Limit**: In the large-system limit, i.e., $M, N \to \infty$ with $\delta = M/N$ fixed, under i.i.d. sub-Gaussian $\boldsymbol{A}$ (elements a_{mn}, $\forall m, n$ are i.i.d. with the tail of the distribution decays at least as fast as the tail of a Gaussian [8]), by defining aggregated parameters, the iterations of $O(MN)$ parameters of messages on edges (i.e., $\Delta_{m\to n}^{t}$, $\Delta_{n\to m}^{t+1}$, $\forall m, n$) reduce to iterations of $O(N)$ aggregated parameters by neglecting vanishing terms [7, 8].

The subsequent subsections detail the application of these approximations to derive the core AMP updates.

10.2.3.1 Gaussian Approximation of Factor-to-Variable Messages

We begin by deriving the approximate factor-to-variable message $\Delta_{m\to n}^{t}$. A key conceptual step is to introduce the hidden variable $z_m = \boldsymbol{a}_m^T \boldsymbol{x}$, which represents the noiseless linear measurement of the signal $\boldsymbol{x}$ via the m-th sensing vector $\boldsymbol{a}_m$.

When the problem dimension N is large, the central limit theorem (CLT) provides a theoretical justification for approximating the distribution of z_m. As a weighted sum of many random variables, z_m tends toward a Gaussian distribution under mild conditions. This approximation significantly simplifies the computation. Specifically, the updating rule in (10.4) can be approximated as

$$\Delta_{m\to n}^{t} \propto \int_{z_m} p\left(y_m|z_m\right) \int_{x_{n'},n'\neq n} p\left(z_m|\mathbf{x}\right) \prod_{n'\neq n} \Delta_{n'\to m}^{t}$$
$$\approx \int_{z_m} \mathcal{N}\left(y_m; z_m, \sigma_w^2\right) \mathcal{N}\left(z_m; \mu_{z,m\backslash n}^{t} + a_{mn}x_n, v_{z,m\backslash n}^{t}\right), \tag{10.7}$$

where the prior distribution of z_m, derived from the input messages $\prod_{n'\neq n} \Delta_{n'\to m}^{t}$, is approximated by a Gaussian with parameters

$$\mu_{z,m\backslash n}^{t} = \sum_{n'\neq n} a_{mn'}\mu_{n'\to m}^{t}, \qquad v_{z,m\backslash n}^{t} = \sum_{n'\neq n} a_{mn'}^{2} v_{n'\to m}^{t}. \tag{10.8}$$

These parameters have intuitive interpretations:

- $\mu_{z,m\backslash n}^{t}$ is the expected value of z_m excluding the contribution from x_n.
- $v_{z,m\backslash n}^{t}$ quantifies the uncertainty in z_m arising from all variables except x_n.

The notation $m\backslash n$ emphasizes the exclusion of variable n at factor m, a fundamental aspect of message passing that prevents double-counting of information.

A direct consequence of this Gaussian approximation is that the entire message $\Delta_{m\to n}^{t}$ itself becomes a Gaussian distribution, i.e.,

$$\Delta_{m\to n}^{t} = \mathcal{N}\left(x_n; \mu_{m\to n}^{t}, v_{m\to n}^{t}\right), \tag{10.9}$$

with its parameters given by

$$\mu_{m\to n}^{t} = \frac{y_m - \mu_{z,m\backslash n}^{t}}{a_{mn}}, \qquad v_{m\to n}^{t} = \frac{\sigma_w^2 + v_{z,m\backslash n}^{t}}{a_{mn}^{2}}. \tag{10.10}$$

These parameters are interpretable:

- The mean $\mu_{m\to n}^{t}$ suggests the value x_n should take to make the observation y_m consistent with the current estimates of all other variables.
- The variance $v_{m\to n}^{t}$ accumulates uncertainty from both the measurement noise σ_w^2 and the estimates of other variables $v_{z,m\backslash n}^{t}$.

This Gaussian approximation transforms the computationally challenging message passing into a tractable algorithm that only requires maintaining and updating means and variances, thereby achieving the efficiency needed for practical applications.

10.2.3.2 Gaussian Approximation of Variable-to-Factor Messages

Building upon the Gaussian approximation of factor-to-variable messages, we now examine the variable-to-factor message $\Delta_{n\to m}^{t+1}$. This step completes the iterative cycle of message passing by propagating updated beliefs back to the factor nodes.

Substituting the Gaussian approximations for $\Delta_{m\to n}^{t}$ into the original sum-product update equation yields

$$\Delta_{n\to m}^{t+1} \propto \mathrm{e}^{-\beta|x_n|} \prod_{m'\neq m} \mathcal{N}\left(x_n; \mu_{m'\to n}^{t}, v_{m'\to n}^{t}\right). \tag{10.11}$$

Exploiting the mathematical property that the product of multiple Gaussian distributions is itself proportional to a Gaussian, we consolidate the product term as

$$\Delta_{n\to m}^{t+1} \propto \mathrm{e}^{-\beta|x_n|} \mathcal{N}\left(x_n; \mu_{r,n\backslash m}^{t}, v_{r,n\backslash m}^{t}\right), \tag{10.12}$$

where the parameters of the consolidated Gaussian are

$$\mu_{r,n\backslash m}^{t} = v_{r,n\backslash m}^{t} \left(\sum_{m'\neq m} \frac{\mu_{m'\to n}^{t}}{v_{m'\to n}^{t}}\right), \qquad v_{r,n\backslash m}^{t} = \left(\sum_{m'\neq m} \frac{1}{v_{m'\to n}^{t}}\right)^{-1}. \tag{10.13}$$

These parameters have meaningful interpretations:

- $\mu^t_{r,n\backslash m}$ is the precision-weighted average of incoming messages, excluding the m-th factor.
- $v^t_{r,n\backslash m}$ is the total precision (inverse variance) from all incoming messages excluding the m-th factor.

The notation $n\backslash m$ indicates that we're combining information from all factors except m, preventing over-confidence through the "exclusion principle" of message passing.

The core challenge now lies in handling the product of this Gaussian with the sparse prior $\frac{\beta}{2}\mathrm{e}^{-\beta|x_n|}$. While the product of Gaussians remains Gaussian, the incorporation of the non-Gaussian prior breaks this convenient property. To maintain computational tractability, AMP employs a moment-matching strategy, i.e., projecting the resulting distribution back to a Gaussian form as

$$\Delta^{t+1}_{n\to m} \approx \mathcal{N}\left(x_n; \mu^{t+1}_{n\to m}, v^{t+1}_{n\to m}\right), \tag{10.14}$$

where the mean and variance are determined by the following specific functions:

$$\mu^{t+1}_{n\to m} = F_\beta\left(\mu^t_{r,n\backslash m}, v^t_{r,n\backslash m}\right), \tag{10.15}$$

$$v^{t+1}_{n\to m} = G_\beta\left(\mu^t_{r,n\backslash m}, v^t_{r,n\backslash m}\right). \tag{10.16}$$

The functions F_β and G_β play a pivotal role—they effectively "project" the non-Gaussian posterior back to Gaussian form while preserving the first two moments (mean and variance). This projection ensures consistency with the original distribution while maintaining the Gaussian message structure essential for algorithm efficiency. The specific forms of F_β and G_β depend on the choice of sparse prior $p(x_n)$. For common priors such as Laplace, Bernoulli–Gaussian, or spike-and-slab distributions, these functions can be derived analytically. Comprehensive examples and detailed derivations are available in [8].

This Gaussian approximation, though seemingly drastic, preserves the essential statistical information while reducing computational complexity from managing full distributions to tracking only means and variances. This is the key innovation that makes AMP scalable to high-dimensional problems.

10.2.3.3 Gaussian Approximation of Marginal Posterior Distributions

Following the SPMP framework, we now approximate the marginal posterior distribution $p^{t+1}(x_n|\mathbf{y})$ for each variable x_n after t iterations, which represents the complete belief about x_n given all observations.

The exact marginal posterior combines the sparse prior with all incoming Gaussian messages from factor nodes as

$$p^{t+1}(x_n|\mathbf{y}) \propto \mathrm{e}^{-\beta|x_n|}\prod_{m=1}^{M}\mathcal{N}(x_n; \mu^t_{m\to n}, v^t_{m\to n}). \tag{10.17}$$

Leveraging the conjugate properties of Gaussian distributions, the product of multiple Gaussians can be consolidated into a single Gaussian, yielding the compact form:

$$p^{t+1}(x_n|\mathbf{y}) \propto \mathrm{e}^{-\beta|x_n|}\mathcal{N}(x_n; \mu^t_{r,n}, v^t_{r,n}), \tag{10.18}$$

where the parameters of this consolidated Gaussian are given by

$$\mu^t_{r,n} = v^t_{r,n}\left(\sum_{m=1}^{M}\frac{\mu^t_{m\to n}}{v^t_{m\to n}}\right) = v^t_{r,n}\left(\sum_{m=1}^{M}\frac{a_{mn}\left(y_m - \mu^t_{z,m\backslash n}\right)}{\sigma_w^2 + v^t_{z,m\backslash n}}\right), \tag{10.19}$$

$$v^t_{r,n} = \left(\sum_{m=1}^{M}\frac{1}{v^t_{m\to n}}\right)^{-1} = \left(\sum_{m=1}^{M}\frac{a^2_{mn}}{\sigma_w^2 + v^t_{z,m\backslash n}}\right)^{-1}. \tag{10.20}$$

These parameters have clear interpretations:

Table 10.1 AMP variable scalings in the large-system limit

μ_n^t	$O(1)$	$\mu_{n\to m}^t$	$O(1)$	$\mu_n^t - \mu_{n\to m}^t$	$O\left(\frac{1}{\sqrt{N}}\right)$
v_n^t	$O(1)$	$v_{n\to m}^t$	$O(1)$	$v_n^t - v_{n\to m}^t$	$O\left(\frac{1}{\sqrt{N}}\right)$
$\mu_{r,n}^t$	$O(1)$	$\mu_{r,n\backslash m}^t$	$O(1)$	$\mu_{r,n}^t - \mu_{r,n\backslash m}^t$	$O\left(\frac{1}{\sqrt{N}}\right)$
$v_{r,n}^t$	$O(1)$	$v_{r,n\backslash m}^t$	$O(1)$	$v_{r,n}^t - v_{r,n\backslash m}^t$	$O\left(\frac{1}{N}\right)$
$\mu_{z,m}^t$	$O(1)$	$\mu_{z,m\backslash n}^t$	$O(1)$	$\mu_{z,m}^t - \mu_{z,m\backslash n}^t$	$O\left(\frac{1}{\sqrt{N}}\right)$
$v_{z,m}^t$	$O(1)$	$v_{z,m\backslash n}^t$	$O(1)$	$v_{z,m}^t - v_{z,m\backslash n}^t$	$O\left(\frac{1}{N}\right)$

- The mean $\mu_{r,n}^t$ represents a precision-weighted average of suggestions from all measurements, where each measurement y_m contributes an estimate of x_n adjusted by the reliability of other variables (i.e., $\mu_{z,m\backslash n}^t$) and overall uncertainty.
- The variance $v_{r,n}^t$ captures the total confidence from all measurements, with each term $a_{mn}^2/(\sigma_w^2 + v_{z,m\backslash n}^t)$ representing the effective precision contributed by the m-th measurement.

To maintain computational tractability with the non-Gaussian sparse prior, we apply a Gaussian approximation via moment matching:

$$p^{t+1}(x_n|\mathbf{y}) \approx \mathcal{N}(x_n; \mu_n^{t+1}, v_n^{t+1}), \tag{10.21}$$

where the final parameters are determined by the same projection functions encountered earlier as

$$\mu_n^{t+1} = F_\beta\left(\mu_{r,n}^t, v_{r,n}^t\right), \qquad v_n^{t+1} = G_\beta\left(\mu_{r,n}^t, v_{r,n}^t\right). \tag{10.22}$$

This approximation strikes a balance between statistical accuracy and computational efficiency. The functions F_β and G_β ensure that while the distribution form is simplified to Gaussian, the first two moments (mean and variance) match those of the true posterior as closely as possible. The resulting Gaussian approximation provides not only point estimates (through the mean μ_n^{t+1}) but also uncertainty quantification (through the variance v_n^{t+1}), enabling comprehensive statistical inference for each variable in the sparse signal recovery problem.

10.2.3.4 Taylor Series Expansion for Message Aggregation

Despite the Gaussian approximation, directly computing all $O(MN)$ edge messages remains computationally expensive. To further reduce complexity, we introduce Taylor series expansion to approximate edge message parameters as functions of aggregated quantities.

1. **System Scaling and Key Assumptions**

In the large-system limit, we make the following crucial assumptions:

- $a_{mn}^2 \sim O(1/N)$, $\forall m, n$—each measurement coefficient squared scales inversely with system dimension.
- $\mathbb{E}(x_n^2) \sim O(1)$, $\forall n$—signal elements have finite variance.

These scaling properties ensure that as the system grows large, individual contributions become small, making approximations increasingly accurate. Under these assumptions, the systematic scaling of various quantities in the large-system limit is summarized in Table 10.1, which provides the foundation for subsequent Taylor series approximations.

2. **Taylor Series Expansion and Approximation for Mean and Variance Updates**

The simplification of both the mean and variance parameters leverages the scaling properties from Table 10.1 through a first-order Taylor expansion of the functions F_β and G_β:

$$
\begin{aligned}
\mu_{n\to m}^{t+1} &= F_\beta\left(\mu_{r,n\setminus m}^t, v_{r,n\setminus m}^t\right) \\
&\approx F_\beta\left(\mu_{r,n}^t, v_{r,n}^t\right) + \frac{\partial F_\beta\left(\mu, v_{r,n}^t\right)}{\partial \mu}\Big|_{\mu=\mu_{r,n}^t} \left(\mu_{r,n\setminus m}^t - \mu_{r,n}^t\right) \\
&+ \frac{\partial F_\beta\left(\mu_{r,n}^t, v\right)}{\partial v}\Big|_{v=v_{r,n}^t} \left(v_{r,n\setminus m}^t - v_{r,n}^t\right),
\end{aligned} \tag{10.23}
$$

$$
\begin{aligned}
v_{n\to m}^{t+1} &= G_\beta\left(\mu_{r,n\setminus m}^t, v_{r,n\setminus m}^t\right) \\
&\approx G_\beta\left(\mu_{r,n}^t, v_{r,n}^t\right) + \frac{\partial G_\beta\left(\mu, v_{r,n}^t\right)}{\partial \mu}\Big|_{\mu=\mu_{r,n}^t} \left(\mu_{r,n\setminus m}^t - \mu_{r,n}^t\right) \\
&+ \frac{\partial G_\beta\left(\mu_{r,n}^t, v\right)}{\partial v}\Big|_{v=v_{r,n}^t} \left(v_{r,n\setminus m}^t - v_{r,n}^t\right).
\end{aligned} \tag{10.24}
$$

The key insight here is that the differences between the "leave-one-out" quantities and the full aggregates are small, specifically,

$$
v_{r,n\setminus m}^t - v_{r,n}^t = \frac{v_{r,n\setminus m}^t v_{r,n}^t}{v_{m\to n}^t} \sim O\left(\frac{1}{N}\right), \tag{10.25}
$$

$$
\begin{aligned}
\mu_{r,n\setminus m}^t - \mu_{r,n}^t &= v_{r,n\setminus m}^t \left(\sum_{m'\neq m} \frac{\mu_{m'\to n}^t}{v_{m'\to n}^t}\right) - v_{r,n}^t \left(\sum_{m=1}^{M} \frac{\mu_{m\to n}^t}{v_{m\to n}^t}\right) \\
&= -v_{r,n}^t \frac{\mu_{m\to n}^t}{v_{m\to n}^t} + O\left(\frac{1}{N}\right) = -v_{r,n}^t \frac{a_{mn}\left(y_m - \mu_{z,m}^t\right)}{\sigma_w^2 + v_{z,m}^t} + O\left(\frac{1}{N}\right).
\end{aligned} \tag{10.26}
$$

These scaling relationships justify neglecting the vanishing terms in both Taylor expansions while maintaining approximation accuracy in the large-system limit.

3. **Simplified Update Rules**

Discarding the vanishing terms yields the simplified update rules. For the mean:

$$
\mu_{n\to m}^{t+1} \approx \mu_n^{t+1} - v_n^{t+1} \frac{a_{mn}(y_m - \mu_{z,m}^t)}{\sigma_w^2 + v_{z,m}^t}, \tag{10.27}
$$

and for the variance:

$$
v_{n\to m}^{t+1} = v_n^{t+1} + O\left(\frac{1}{\sqrt{N}}\right) \approx v_n^{t+1},
$$

where the aggregated quantities are defined as

$$
\mu_{z,m}^t = \sum_{n=1}^{N} a_{mn} \mu_{n\to m}^t, \qquad v_{z,m}^t = \sum_{n=1}^{N} a_{mn}^2 v_{n\to m}^t. \tag{10.28}
$$

The above approximation yields significant benefits:

- **Complexity Reduction:** The updates now depend only on aggregated quantities $\mu_{z,m}^t$ and $v_{z,m}^t$, reducing per-iteration complexity from $O(MN)$ to $O(N)$.
- **Intuitive Structure**: The mean update has a clear corrective form—the second term adjusts the estimate based on the residual $(y_m - \mu_{z,m}^t)$, weighted by reliability measures.
- **Theoretical Foundation:** The $O(1/N)$ scaling ensures approximation errors vanish in the large-system limit, providing mathematical rigor.
- **Practical Efficiency**: Variance updates become particularly simple, as individual edge variances are well approximated by their node-level counterparts.

The above Taylor series approach is the final key insight that transforms approximate message passing from a theoretical concept into a practically feasible algorithm for high-dimensional inference.

10.2.3.5 Final Approximations and Algorithm Summary

After applying the Taylor series expansion, the final step is to neglect vanishing terms of order $O(1/\sqrt{N})$ and $O(1/N)$ in the large-system limit. This yields a set of simplified, computationally efficient update rules for the key aggregated quantities:

- **Variance of the refined estimate**:

$$v_{r,n}^t = \left(\sum_{m=1}^{M} \frac{a_{mn}^2}{\sigma_w^2 + v_{z,m}^t}\right)^{-1} + O\left(\frac{1}{N}\right) \approx \left(\sum_{m=1}^{M} \frac{a_{mn}^2}{\sigma_w^2 + v_{z,m}^t}\right)^{-1}. \tag{10.29}$$

  Here, $\left(v_{r,n}^t\right)^{-1}$ represents the total precision (inverse variance) contributed by all measurements, each weighted by the reliability of the corresponding linear observation.
- **Mean of the refined estimate**:

$$\mu_{r,n}^t = \mu_n^t + v_{r,n}^t \sum_{m=1}^{M} \frac{a_{mn}\left(y_m - \mu_{z,m}^t\right)}{\sigma_w^2 + v_{z,m}^t} + O\left(\frac{1}{N}\right) \approx \mu_n^t + v_{r,n}^t \sum_{m=1}^{M} \frac{a_{mn}\left(y_m - \mu_{z,m}^t\right)}{\sigma_w^2 + v_{z,m}^t}. \tag{10.30}$$

  This update adjusts the current estimate μ_n^t by adding a correction term proportional to the weighted sum of residuals across all measurements.
- **Predicted measurement variance**:

$$v_{z,m}^t = \sum_{n=1}^{N} a_{mn}^2 v_n^t + O\left(\frac{1}{\sqrt{N}}\right) \approx \sum_{n=1}^{N} a_{mn}^2 v_n^t. \tag{10.31}$$

  This aggregates the uncertainty from all variable nodes to quantify the overall reliability of the m-th predicted measurement.
- **Predicted measurement mean**:

$$\begin{aligned} \mu_{z,m}^t &= \sum_{n=1}^{N} a_{mn}\left(\mu_n^t - v_n^t \frac{a_{mn}\left(y_m - \mu_{z,m}^{t-1}\right)}{\sigma_w^2 + v_{z,m}^{t-1}}\right) + O\left(\frac{1}{N^{3/2}}\right) \\ &\approx \sum_{n=1}^{N} a_{mn}\mu_n^t - \frac{v_{z,m}^t\left(y_m - \mu_{z,m}^{t-1}\right)}{\sigma_w^2 + v_{z,m}^{t-1}}. \end{aligned} \tag{10.32}$$

  This expression has an important interpretation: the first term is the standard linear prediction, while the second term acts as an **Onsager correction**—a memory term that accounts for dependencies between iterations and ensures correct error evolution.

Based on these approximations, the complete AMP algorithm is summarized in Algorithm 10.1. The algorithm elegantly decomposes into two complementary phases:

1. **Linear steps (Steps 2–6)**: Process the observation model by propagating beliefs through the linear measurement system, effectively implementing a matched filter in the measurement domain.
2. **Nonlinear steps (Steps 8–9)**: Incorporate the independent sparse prior through component-wise denoising functions F_β and G_β, which shrink each variable estimate toward zero according to the sparsity-inducing prior.

This separation reflects the underlying statistical structure: the linear steps handle the Gaussian likelihood, while the nonlinear steps handle the non-Gaussian prior. The result is an iterative algorithm that achieves near-optimal performance for high-dimensional sparse recovery while maintaining computational tractability.

Algorithm 10.1 AMP under Laplace Prior

Input: Sensing matrix $\boldsymbol{A}$, observation $\boldsymbol{y}$.

Output: $p^{T+1}(\boldsymbol{x}|\boldsymbol{y}) = \prod_{n=1}^{N} N\left(x_n; \mu_n^{T+1}, v_n^{T+1}\right)$

Initialization: $\forall m, n : \mu_n^1 = 0, v_n^1 = 1, \mu_{z,m}^0 = y_m, v_{z,m}^0 = 1$

1: **for** $t = 1, 2, \ldots, T$ **do**

2: %% Linear Estimation Steps

3: $v_{z,m}^t = \sum_{n=1}^{N} a_{mn}^2 v_n^t, \forall m.$

4: $\mu_{z,m}^t = \sum_{n=1}^{N} a_{mn}\mu_n^t - \frac{v_{z,m}^t\left(y_m - \mu_{z,m}^{t-1}\right)}{\sigma_w^2 + v_{z,m}^{t-1}}, \forall m.$

5: $v_{r,n}^t = \left(\sum_{m=1}^{M} \frac{a_{mn}^2}{\sigma_w^2 + v_{z,m}^t}\right)^{-1}, \forall n.$

6: $\mu_{r,n}^t = \mu_n^t + v_{r,n}^t \sum_{m=1}^{M} \frac{a_{mn}\left(y_m - \mu_{z,m}^t\right)}{\sigma_w^2 + v_{z,m}^t}, \forall n.$

7: %% Nonlinear Estimation Steps

8: $v_n^{t+1} = G_\beta\left(\mu_{r,n}^t, v_{r,n}^t\right), \forall n.$

9: $\mu_n^{t+1} = F_\beta\left(\mu_{r,n}^t, v_{r,n}^t\right), \forall n.$

10: **end for**

10.2.3.6 Min-Sum Formulation and Thresholding Function

The AMP algorithm above estimates the marginal posterior distributions $p\left(x_n|\boldsymbol{y}\right), n = 1, \ldots, N$ based on the SPMP, yielding the MMSE estimator for x_n in (10.22) via F_β. AMP can also be derived based on MSMP framework [6]. Remarkably, the resulting iterative updating rules are identical to that of Algorithm 10.1, with the only difference residing in the specific forms of F_β and G_β. For a Laplace prior $p(x_n) \propto \mathrm{e}^{-\beta|x_n|}$, the posterior distribution in Eq. (10.18) yields closed-form expressions for μ_n^{t+1} and v_n^{t+1} through an explicit thresholding function [7]:

$$\mu_n^{t+1} = \arg\max_{x_n} p^{t+1}(x_n|\boldsymbol{y}) = \eta(\mu_{r,n}^t, \beta v_{r,n}^t), \tag{10.33}$$

where the thresholding function $\eta(\cdot, \cdot)$ is defined as

$$\eta(\mu_{r,n}^t, \beta v_{r,n}^t) = \begin{cases} \mu_{r,n}^t - \beta v_{r,n}^t & \mu_{r,n}^t > \beta v_{r,n}^t, \\ 0 & -\beta v_{r,n}^t \le \mu_{r,n}^t \le \beta v_{r,n}^t, \\ \mu_{r,n}^t + \beta v_{r,n}^t & \mu_{r,n}^t < -\beta v_{r,n}^t. \end{cases} \tag{10.34}$$

This thresholding operation has a clear interpretation:

- When the refined estimate $|\mu_{r,n}^t|$ is sufficiently large compared to the uncertainty-scaled threshold $\beta v_{r,n}^t$, the estimate is shrunk toward zero by $\beta v_{r,n}^t$.
- When $|\mu_{r,n}^t|$ falls below this threshold, the estimate is set to exactly zero, promoting sparsity in the solution.

The corresponding variance update is given by

$$v_n^{t+1} = -\left(\frac{\partial^2 \log p^{t+1}(x_n|\boldsymbol{y})}{\partial x_n^2}\bigg|_{\mu_n^{t+1}}\right)^{-1} = v_{r,n}^t \cdot \eta'(\mu_{r,n}^t, \beta v_{r,n}^t), \tag{10.35}$$

where the derivative of the thresholding function is

$$\eta'(\mu_{r,n}^t, \beta v_{r,n}^t) = \begin{cases} 1 & |\mu_{r,n}^t| > \beta v_{r,n}^t, \\ 0 & |\mu_{r,n}^t| \le \beta v_{r,n}^t. \end{cases} \tag{10.36}$$

This variance update reflects the certainty of our estimate:

- For non-zero estimates ($|\mu^t_{r,n}| > \beta v^t_{r,n}$), the variance remains unchanged as $v^t_{r,n}$.
- For zero estimates ($|\mu^t_{r,n}| \leq \beta v^t_{r,n}$), the variance becomes zero, indicating high confidence in the sparse solution.

The threshold $\beta v^t_{r,n}$ elegantly combines the sparsity parameter β with the estimation uncertainty $v^t_{r,n}$, automatically adapting the threshold level based on the reliability of the current estimate. This adaptive thresholding mechanism is fundamental to AMP's success in sparse signal recovery, providing both computational efficiency and statistical accuracy.

10.2.3.7 Simplification Under Homogeneous Sensing Matrix

The AMP algorithm admits significant computational simplifications when the sensing matrix exhibits homogeneous properties—specifically, when its entries satisfy $a^2_{mn} \approx \frac{1}{M}$ for all m, n. This condition arises naturally in large-scale systems with i.i.d. Gaussian measurement matrices and leads to a more efficient and interpretable algorithm.

Under this homogeneity assumption, several quantities become uniform across indices:

- The **predicted measurement variance** becomes uniform

$$v^t_{z,m} = \frac{1}{M}\sum_{n=1}^{N} v^t_n = v^t_z, \quad \forall m. \tag{10.37}$$

  This indicates that the uncertainty in the linear observation model is identical across all measurements.
- The **residual term** $\mu^t_{s,m} = y_m - \mu^t_{z,m}$ is updated as

$$\mu^t_{s,m} = y_m - \sum_{n=1}^{N} a_{mn}\mu^t_n + \mu^{t-1}_{s,m}\frac{v^t_z}{\sigma^2_w + v^{t-1}_z}, \quad \forall m. \tag{10.38}$$

  Here, the last term acts as a damping or memory component that stabilizes the iteration.
- The **refined estimate parameters** also become uniform

$$v^t_{r,n} = \sigma^2_w + v^t_z = v^t_r, \qquad \mu^t_{r,n} = \mu^t_n + \sum_{m=1}^{M} a_{mn}\mu^t_{s,m}, \quad \forall n. \tag{10.39}$$

Substituting the thresholding expressions for μ^t_n and v^t_n into Eqs. (10.37), (10.38), and (10.39) leads to a compact vector-matrix form. Define

$$\begin{aligned}
\boldsymbol{r}^t &= [\mu^t_{s,1}, \ldots, \mu^t_{s,M}]^T, \\
\boldsymbol{x}^t &= [\mu^t_1, \ldots, \mu^t_N]^T, \\
b_t &= \frac{1}{M}\sum_{n=1}^{N}\eta'\left(\mu^t_n + \sum_{m=1}^{M} a_{mn}\mu^t_{s,m}, \beta\left(\sigma^2_w + v^t_z\right)\right) = \frac{1}{M}\|\boldsymbol{x}^{t+1}\|_0, \\
\theta^t &= \beta\left(\sigma^2_w + v^t_z\right) = \beta\sigma^2_w + b_{t-1}\theta^{t-1}.
\end{aligned} \tag{10.40}$$

Then the AMP iteration reduces to the canonical form originally proposed in [7], i.e.,

$$\boldsymbol{x}^{t+1} = \eta\left(\boldsymbol{x}^t + \boldsymbol{A}^T\boldsymbol{r}^t; \theta^t\right), \tag{10.41}$$

$$\boldsymbol{r}^t = \boldsymbol{y} - \boldsymbol{A}\boldsymbol{x}^t + b_t\boldsymbol{r}^{t-1}, \tag{10.42}$$

where $\eta(\cdot; \theta^t)$ is referred to as the **denoising function**, as it suppresses noise and promotes sparsity. A common choice is the soft-thresholding function derived from the Laplace prior, though its exact form depends on the sparse prior $p(x_n)$, as will be discussed in Sect. 10.4.

The resulting AMP iteration resembles **iterative soft thresholding (IST)**. However, a crucial distinction lies in the term $b_t\boldsymbol{r}^{t-1}$ in the residual update. This can be interpreted as a momentum term with a rigorously derived coefficient b_t, which

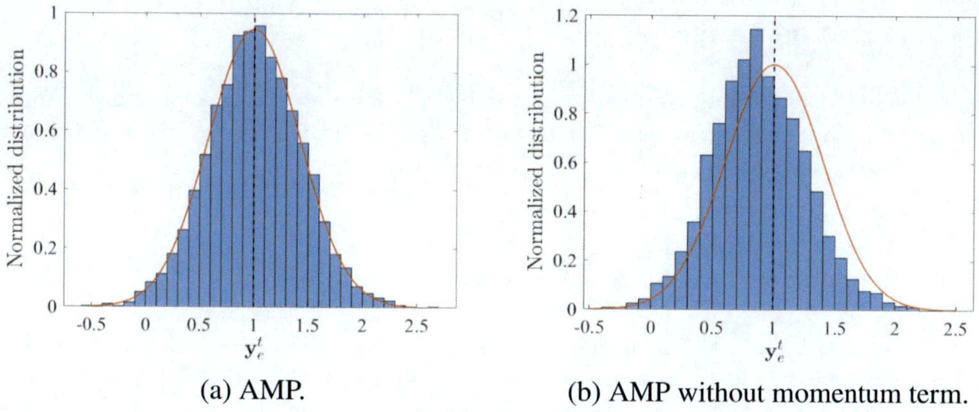

Fig. 10.2 Histogram of $\boldsymbol{y}_e^{10}$. The red curve is the distribution of the effective AWGN observation $\boldsymbol{x} + \tau_t \boldsymbol{Z}$

not only accelerates convergence compared to standard IST, but also improves statistical performance and stability, ensuring correct state evolution in the large-system limit. This term embodies the "memory" of previous residuals and is key to AMP's superior performance over conventional sparse recovery algorithms.

10.3 Intuitive Interpretation of AMP

Despite the empirical success of the AMP algorithm, its mathematical derivation is complex and can obscure the physical intuition underlying the final iterative steps. This section provides an intuitive interpretation of the AMP update rules in (10.41) and (10.42). Assume that the entries of the sensing matrix $\boldsymbol{A} \in \mathbb{R}^{M\times N}$ are i.i.d. and have zero mean and variance $1/M$. In the large-system limit where $M, N \to \infty$ with $\frac{M}{N} \to \delta$ for some constant $\delta \in (0, 1)$, the following asymptotic relationship holds [7]:

$$\boldsymbol{A}^T \boldsymbol{r}^t + \boldsymbol{x}^t \approx \boldsymbol{x} + \tau_t \boldsymbol{Z}, \tag{10.43}$$

where $\boldsymbol{Z} \sim \mathcal{N}(\boldsymbol{0}, \boldsymbol{I})$ has i.i.d. Gaussian distributed entries with zero mean and variance 1, and τ_t can be predicted by the state evolution (SE) equation as will be explained later. In other words, the term $\boldsymbol{y}_e^t \triangleq \boldsymbol{A}^T \boldsymbol{r}^t + \boldsymbol{x}^t$ generated by the AMP iterations can be interpreted as the true signal $\boldsymbol{x} = [x_1, x_2, \ldots, x_N]^T$ observed through an effective AWGN channel with noise variance determined by SE. Note that $\boldsymbol{y}_e^t$ is equivalent to $[\mu_{r,1}^t, \ldots, \mu_{r,N}^t]^T$ in (10.39). The following example illustrates this phenomenon and demonstrates the importance of the momentum term in AMP updates.

Example 10.1 Consider a noiseless observation model $\boldsymbol{y} = \boldsymbol{A}\boldsymbol{x}$ where $\boldsymbol{A} \in \mathbb{R}^{M\times N}$ has i.i.d. normal distributed entries with $M = 1000$ and $N = 4000$, $\boldsymbol{x}$ has 200 non-zero entries which are generated independently from uniform distribution over $\{-1, 1\}$. Figure 10.2 shows the histogram of $\boldsymbol{y}_e^t$ at iteration $t = 10$ for indices where x_n is $+1$. In the left subplot, the momentum term $b_t \boldsymbol{r}^{t-1}$ is retained in the update equation in (10.42). In the right subplot, the momentum term is removed. It can be seen from the left subplot in Fig. 10.2 that $\boldsymbol{y}_e^t$ is indeed close to the effective AWGN observation $\boldsymbol{x} + \tau_t \boldsymbol{Z}$ when $b_t \boldsymbol{r}^{t-1}$ is kept in (10.42). However, if $b_t \boldsymbol{r}^{t-1}$ is removed, it becomes increasingly inaccurate to model $\boldsymbol{y}_e^t$ as an AWGN observation, as shown in the right subplot of Fig. 10.2. This demonstrates the importance of the momentum term $b_t \boldsymbol{r}^{t-1}$.

As demonstrated above, the AMP iterations essentially convert the original coupled observation model $\boldsymbol{y} = \boldsymbol{A}\boldsymbol{x} + \boldsymbol{w}$ into a decoupled AWGN observation model $\boldsymbol{y}_e^t = \boldsymbol{x} + \tau_t \boldsymbol{Z}$ at each iteration. This insight is fundamental to understanding the design of the denoising function $\eta(\cdot)$ in the next section.

10.4 Denoising Function Design

The design of the denoising function is intrinsically determined by the choice of the sparse prior distribution. In this section, we illustrate this dependency by considering several commonly used priors and deriving their corresponding denoisers, based on the intuition that $\boldsymbol{y}_e^t$ can be viewed as an effective AWGN observation.

10.4.1 Denoising Function for I.i.d. Bernoulli–Gaussian Distribution

The i.i.d. Bernoulli–Gaussian (BG) distribution models sparse signals with i.i.d. elements. Under this prior, a support vector $\boldsymbol{s}$ indicates the active components of the signal $\boldsymbol{x}$, where $s_n = 1$ denotes an active (non-zero) coefficient and $s_n = 0$ denotes an inactive (zero) coefficient. The support vector $\boldsymbol{s}$ has i.i.d. Bernoulli entries, i.e.,

$$p(\boldsymbol{s}) = \prod_{n=1}^{N} p(s_n), \qquad p(s_n) = (1-\lambda)^{1-s_n}\lambda^{s_n}, \tag{10.44}$$

where λ denotes the probability that $s_n = 1$ (i.e., the sparsity ratio). Given $\boldsymbol{s}$, $\boldsymbol{x}$ is also assumed to have i.i.d. entries with conditional distribution:

$$p(\boldsymbol{x}|\boldsymbol{s}) = \prod_{n=1}^{N} p(x_n|s_n), \qquad p(x_n|s_n) = s_n \mathcal{N}\left(x_n; 0, \sigma_x^2\right) + (1-s_n)\,\delta(x_n). \tag{10.45}$$

Based on the effective AWGN observation model $y_{e,n}^t = x_n + \tau_t Z_n$, the original MAP/MMSE estimator decomposes into N independent scalar estimators. For each x_n, the optimal denoising function minimizing the MSE is the MMSE estimator given by

$$\eta^t\left(y_{e,n}^t\right) = \mathbb{E}\left[x_n|y_{e,n}^t\right] = \frac{\lambda\mathcal{N}\left(0; y_{e,n}^t, \sigma_x^2+\tau_t^2\right)}{\lambda\mathcal{N}\left(0; y_{e,n}^t, \sigma_x^2+\tau_t^2\right) + (1-\lambda)\,\mathcal{N}\left(0; y_{e,n}^t, \tau_t^2\right)} \frac{\sigma_x^2 y_{e,n}^t}{\sigma_x^2+\tau_t^2}. \tag{10.46}$$

10.4.2 Denoising Function for a General Independent Sparse Prior Distribution

The BG prior serves as a concrete example for denoiser design. To generalize this approach, we now consider a broader class of independent sparse priors, where the support vector $\boldsymbol{s}$ still has i.i.d. Bernoulli entries. But conditioned on the support vector $\boldsymbol{s}$, $\boldsymbol{x}$ is assumed to have independent entries, i.e., $p(\boldsymbol{x}|\boldsymbol{s}) = \prod_{n=1}^{N} p(x_n|s_n)$ with

$$p(x_n|s_n) = s_n g_n(x_n) + (1-s_n)\,\delta(x_n), \tag{10.47}$$

where $g_n(x_n)$ denotes the PDF of x_n conditioned on $s_n = 1$, which can be a general PDF instead of the Gaussian PDF.

The effective AWGN observation model again enables decomposition into independent scalar estimators. The optimal denoising function for MMSE estimation remains the posterior mean, i.e.,

$$\eta_n^t\left(y_{e,n}^t\right) = \mathbb{E}\left[x_n|y_{e,n}^t\right], \tag{10.48}$$

where the exact form depends on $g_n(x_n)$. Note that denoising functions $\eta_n^t(\cdot)$ may vary across components if the prior is not identical.

For discrete-valued non-zero elements, $g_n(x_n)$ becomes a probability mass function (PMF). In such cases, the MAP estimator may be preferred to maximize the probability of correct detection, i.e.,

$$\eta_n^t\left(y_{e,n}^t\right) = \underset{x_n}{\operatorname{argmax}}\, p\left(x_n|y_{e,n}^t\right).$$

Example 10.2 Consider a sparse prior where non-zero elements are binary: $g_n(x_n) = 0.5^{0.5(1-x_n)} 0.5^{0.5(1+x_n)}, \quad \forall x_n \in \{-1, 1\}$. In other words, each non-zero element x_n is a binary variable that is equally likely to be -1 or 1. The posterior distribution is

$$p\left(x_n | y_{e,n}^t\right) \propto \left(\lambda g_n(x_n) + (1-\lambda)\,\delta(x_n)\right) \mathcal{N}\left(y_{e,n}^t; x_n, \tau_t^2\right)$$
$$= \begin{cases} \frac{\lambda}{2}\mathcal{N}\left(y_{e,n}^t; 1, \tau_t^2\right), & x_n = 1, \\ \frac{\lambda}{2}\mathcal{N}\left(y_{e,n}^t; -1, \tau_t^2\right), & x_n = -1, \\ (1-\lambda)\,\mathcal{N}\left(y_{e,n}^t; 0, \tau_t^2\right), & x_n = 0. \end{cases} \tag{10.49}$$

The MAP denoising function $\eta_n^t\left(y_{e,n}^t\right)$ is

$$\eta_n^t\left(y_{e,n}^t\right) = \underset{x_n}{\operatorname{argmax}}\; p\left(x_n | y_{e,n}^t\right)$$
$$= \begin{cases} 1, & \lambda > \frac{\mathcal{N}(0; y_{e,n}^t, \tau_t^2)}{\mathcal{N}(0; y_{e,n}^t, \tau_t^2) + \frac{1}{2}\mathcal{N}(1; y_{e,n}^t, \tau_t^2)},\; y_{e,n}^t > 0, \\ -1, & \lambda > \frac{\mathcal{N}(0; y_{e,n}^t, \tau_t^2)}{\mathcal{N}(0; y_{e,n}^t, \tau_t^2) + \frac{1}{2}\mathcal{N}(-1; y_{e,n}^t, \tau_t^2)},\; y_{e,n}^t < 0, \\ 0, & \text{else.} \end{cases} \tag{10.50}$$

10.4.3 Denoising Function for Unknown Independent Sparse Prior Distribution

In practice, the exact sparsity ratio λ and distribution of non-zero elements may be unknown. In this case, we can assume instead that the sparsity ratio is bounded by $\lambda_0 < 1$, i.e., the prior belongs to a family $\mathcal{F}_{\lambda_0}$ of distributions with sparsity ratio at most λ_0. Since the variables $x_n, n = 1, \ldots, N$ share identical prior distributions and observation models, we denote a generic variable by X with prior p_X. A robust denoising function for an unknown prior p_X is the MAP estimator with a Laplace prior, equivalent to the scalar LASSO estimator:

$$\eta^t\left(y_{e,n}^t; \beta\right) = \underset{x}{\operatorname{argmin}} \left\{ \frac{\left(y_{e,n}^t - x\right)^2}{2\tau_t^2} + \beta\,|x| \right\}, \tag{10.51}$$

where $\beta > 0$ is the LASSO parameter set according to the sparsity ratio bound λ_0, with the specific selection criterion discussed later.

By checking the optimality condition of the above LASSO problem (10.51), it can be shown that the optimal solution is the soft-thresholding function:

$$\eta^t\left(y_{e,n}^t; \beta\right) = \begin{cases} y_{e,n}^t - \beta\tau_t^2, & y_{e,n}^t > \beta\tau_t^2, \\ 0, & \left|y_{e,n}^t\right| < \beta\tau_t^2, \\ y_{e,n}^t + \beta\tau_t^2, & y_{e,n}^t < -\beta\tau_t^2, \end{cases} \tag{10.52}$$

which is visually illustrated in Fig. 10.3.

The MSE of this LASSO estimator is

$$\operatorname{mse}\left(\tau_t; p_X, \beta\right) = \mathbb{E}\left[\eta^t\left(X_n + \tau_t Z; \beta\right) - X_n\right]^2, \tag{10.53}$$

where $Z \sim \mathcal{N}(0, 1)$ and $X_n \sim p_X$. If p_X is known, the optimal β^* can be found by minimizing this MSE, i.e.,

$$\beta^* = \underset{\beta}{\operatorname{argmin}}\; \operatorname{mse}\left(\tau_t; p_X, \beta\right). \tag{10.54}$$

For unknown p_X, a robust approach is to minimize the worst-case MSE over $\mathcal{F}_{\lambda_0}$:

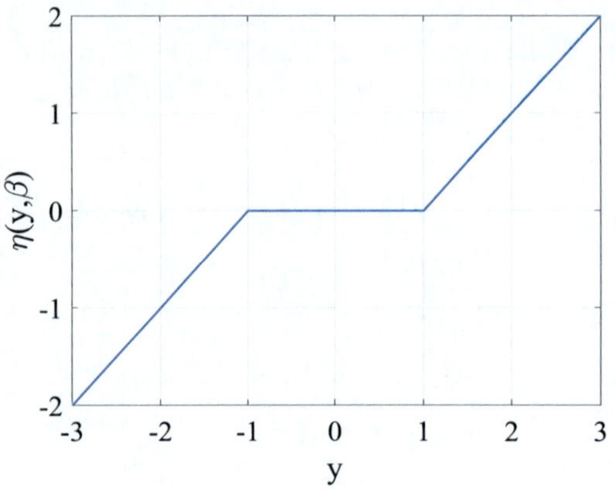

Fig. 10.3 Illustration of the soft-thresholding function $\eta^t(\cdot;\beta)$, where we set $\beta\tau_t^2 = 1$

$$\beta^* = \underset{\beta}{\operatorname{argmin}} \max_{p_X \in \mathcal{F}_{\lambda_0}} \operatorname{mse}(\tau_t; p_X, \beta).$$

It is shown in [7] that the worst-case distribution is

$$p_X = \frac{\lambda_0}{2}\delta_{+\infty} + \frac{\lambda_0}{2}\delta_{-\infty} + (1-\lambda_0)\,\delta_0,$$

and the corresponding optimal LASSO parameter and MSE are given by

$$\beta^* = \alpha^*(\lambda_0)\,\tau_t,\ \alpha^*(\lambda_0) \approx \sqrt{2\log(1/\lambda_0)}, \tag{10.55}$$

$$\operatorname{mse}^*(\tau_t; p_X, \beta) = M^*(\lambda_0)\,\tau_t^2,\ M^*(\lambda_0) \approx 2\lambda_0 \log(1/\lambda_0). \tag{10.56}$$

10.5 State Evolution for AMP

In all of the above discussions, there is still one piece missing, that is how to determine the effective noise variance τ_t^2 in the effective AWGN observation model $y_{e,n}^t = x_n + \tau_t Z_n$. Our objective in this section is to derive the SE equation, which is essential for determining τ_t^2. This derivation proceeds under the standard modeling assumption that the variables x_n's are i.i.d distributed. The resulting SE equation is presented as follows:

$$\tau_t^2 = \sigma_w^2 + \frac{1}{\delta}\mathbb{E}\left[\|X - \eta_{t-1}(X + \tau_{t-1}Z)\|^2\right], \tag{10.57}$$

where $\tau_0^2 = \sigma_w^2 + \frac{1}{\delta}\mathbb{E}\left[\|X\|^2\right]$, $Z \sim \mathcal{N}(0,1)$ is independent of $X \sim p_X$.

This equation has an intuitive interpretation: The first term σ_w^2 represents the actual physical noise power in the system. The second term $\frac{1}{\delta}\mathbb{E}[\|X - \eta_{t-1}(X + \tau_{t-1}Z)\|^2]$ represents the **estimation error** from the previous iteration, scaled by the measurement compression factor $1/\delta$. Therefore, at each iteration, the "effective noise" comes from two sources: the original measurement noise plus the error from our imperfect estimate in the previous iteration. The denoising function η_{t-1} attempts to recover X from its noisy version $X + \tau_{t-1}Z$, and the expected squared error of this denoising operation precisely determines the additional noise in the next iteration.

In the AMP algorithm, the message associated with each variable x_n is essentially approximated as Gaussian distributed. The mean of these messages is updated via the AMP recursion in (10.41) and (10.42), while its variance is governed by the SE equation in (10.57). In the following, we first provide an intuitive interpretation of the SE, followed by a rigorous mathematical presentation.

10.5.1 A Concise Derivation of SE

Recall that $\tau_t \boldsymbol{Z} = \boldsymbol{x} - \boldsymbol{x}^t - \boldsymbol{A}^T \boldsymbol{r}^t$ according to (10.43). Expanding this expression, we have

$$\begin{aligned}\tau_t \boldsymbol{Z} &= \boldsymbol{x} - \boldsymbol{x}^t - \boldsymbol{A}^T \boldsymbol{r}^t \\ &= \boldsymbol{x} - \boldsymbol{x}^t - \boldsymbol{A}^T \left(\boldsymbol{y} - \boldsymbol{A}\boldsymbol{x}^t\right) \\ &= \boldsymbol{x} - \boldsymbol{x}^t - \boldsymbol{A}^T \left(\boldsymbol{A}\boldsymbol{x} + \boldsymbol{w} - \boldsymbol{A}\boldsymbol{x}^t\right) \\ &= \left(\boldsymbol{I}_N - \boldsymbol{A}^T\boldsymbol{A}\right)\left(\boldsymbol{x} - \boldsymbol{x}^t\right) - \boldsymbol{A}^T \boldsymbol{w}.\end{aligned} \tag{10.58}$$

Assuming independence between $\boldsymbol{x}^t$ and $\boldsymbol{w}$, the expected squared norm is

$$\underbrace{\mathbb{E}\left(\|\tau_t \boldsymbol{Z}\|^2\right)}_{\text{Term1}} = \underbrace{\mathbb{E}\left(\left\|\left(\boldsymbol{I}_N - \boldsymbol{A}^T\boldsymbol{A}\right)\left(\boldsymbol{x} - \boldsymbol{x}^t\right)\right\|^2\right)}_{\text{Term2}} + \underbrace{\mathbb{E}\left(\left\|\boldsymbol{A}^T\boldsymbol{w}\right\|^2\right)}_{\text{Term3}}. \tag{10.59}$$

Breaking down Eq. (10.59), we begin with its first component, which is calculated as

$$\text{Term1} = N\tau_t^2. \tag{10.60}$$

Since the elements of $\boldsymbol{A}$ are i.i.d. and $a_{mn} \sim \mathcal{N}(0, 1/M)$, $\forall m, n$, the elements of $\boldsymbol{I}_N - \boldsymbol{A}^T\boldsymbol{A} \triangleq \boldsymbol{B}$ are also i.i.d. and $B_{mn} \sim \mathcal{N}(0, 1/M)$, $\forall m, n$ in the large-system limit. Thus, we have $\mathbb{E}\left[\boldsymbol{B}\boldsymbol{B}^T\right]_{ii} = \frac{N}{M}$, $\forall i$, and the second term in (10.59) is computed as

$$\begin{aligned}\text{Term2} &= \mathbb{E}\left(\left(\boldsymbol{x} - \boldsymbol{x}^t\right)^T \boldsymbol{B}^T\boldsymbol{B}\left(\boldsymbol{x} - \boldsymbol{x}^t\right)\right) \\ &= \mathbb{E}\left(\text{Tr}\left(\left(\boldsymbol{x} - \boldsymbol{x}^t\right)^T \boldsymbol{B}^T\boldsymbol{B}\left(\boldsymbol{x} - \boldsymbol{x}^t\right)\right)\right) \\ &= \mathbb{E}\left(\text{Tr}\left(\boldsymbol{B}\left(\boldsymbol{x} - \boldsymbol{x}^t\right)\left(\boldsymbol{x} - \boldsymbol{x}^t\right)^T \boldsymbol{B}^T\right)\right) \\ &= \mathbb{E}\left(\text{Tr}\left(\boldsymbol{B}\boldsymbol{B}^T\right)\right)\mathbb{E}\left(\left\|\left(\boldsymbol{x} - \boldsymbol{x}^t\right)\right\|^2\right) \\ &= \frac{N^2}{M}\mathbb{E}\left(\left\|\left(\boldsymbol{x} - \boldsymbol{x}^t\right)\right\|^2\right).\end{aligned} \tag{10.61}$$

Similarly, the third term in (10.59) can be obtained by

$$\text{Term3} = \mathbb{E}\left(\left\|\boldsymbol{A}^T\boldsymbol{w}\right\|^2\right) = N\sigma_w^2. \tag{10.62}$$

Submitting (10.60)–(10.62) to (10.59) yields

$$\tau_t^2 = \frac{1}{\delta}\mathbb{E}\left[\left\|\boldsymbol{x} - \boldsymbol{x}^t\right\|^2\right] + \sigma_w^2, \tag{10.63}$$

$$\approx \sigma_w^2 + \frac{1}{\delta}\mathbb{E}\left[X - \eta_{t-1}\left(X + \tau_{t-1}Z\right)\right]^2, \tag{10.64}$$

where $\delta = \frac{M}{N}$, which is consistent with the SE in (10.57). In particular, in the first iteration, i.e., when $t = 0$, we set $x_n^0 = 0$, $\forall n$, thus $\tau_0^2 = \frac{1}{\delta}\mathbb{E}\left[\|\boldsymbol{x}\|^2\right] + \sigma_w^2$.

Note that in the above derivation, we have assumed that both $\boldsymbol{A}$ and $\boldsymbol{w}$ are independent of $\boldsymbol{x}^t$, which is usually not true since $\boldsymbol{x}^t$ is calculated from $\boldsymbol{y}$ that further depends on $\boldsymbol{A}$ and $\boldsymbol{w}$. However, it turns out that the SE result derived from the above simplified assumptions coincides with the rigorous result presented in the next subsection.

10.5.2 Rigorous Results for SE

To present the rigorous results for SE, we must first establish a set of standard technical assumptions and define a specific class of functions. This framework is crucial for conducting precise high-dimensional probability analysis of the AMP algorithm.

Assumption 10.1 (*Technical Assumptions for SE*)

1. **Sensing Matrix Distribution**: The sensing matrix $\boldsymbol{A}$ has i.i.d. normal distributed entries $a_{mn} \sim \mathcal{N}\left(0, \frac{1}{M}\right)$, $\forall m, n$.
2. **Signal Prior**: The sparse signal $\boldsymbol{x}$ has i.i.d. entries $x_n \sim p_X$, $\forall n$, where p_X is a sub-Gaussian distribution.
3. **Noise Statistics**: The noise vector $\boldsymbol{w}$ has i.i.d. entries $w_m \sim p_W$, $\forall m$, where p_W is a sub-Gaussian distribution with variance $E\left[\|w_m\|^2\right] = \sigma_w^2$.
4. **Denoiser Regularity**: The denoising functions η_t, $\forall t$ are Lipschitz continuous. Consequently, it is differentiable almost everywhere, and its derivative η_t' is bounded wherever it exists.

These four conditions create a mathematically tractable yet practically relevant framework for analyzing AMP. In particular, the assumption on the Lipschitz continuity of the denoiser is one of the most critical assumptions. This assumption ensures that small changes in its input lead to only small changes in its output. This stability property is essential for preventing the propagation of errors and guaranteeing that the AMP iteration remains well behaved. Common denoisers, including the soft-thresholding operator, satisfy this condition.

To state the main theorem, we need a broad class of functions that allow us to measure the distance between the true signal and the AMP estimate. Standard Lipschitz functions prove too restrictive for this high-dimensional setting, necessitating the introduction of pseudo-Lipschitz functions.

Definition 10.1 (*Pseudo-Lipschitz (PL) Loss Functions*) A function ϕ: $\mathbb{R}^M \to \mathbb{R}$ is PL if there exists a constant $L > 0$ such that for all $\boldsymbol{x}, \boldsymbol{y} \in \mathbb{R}^M$:

$$|\phi(\boldsymbol{x}) - \phi(\boldsymbol{y})| \leq L\left(1 + \|\boldsymbol{x}\| + \|\boldsymbol{y}\|\right)\|\boldsymbol{x} - \boldsymbol{y}\| . \tag{10.65}$$

The key difference from a standard Lipschitz condition is the multiplicative factor $(1+\|\boldsymbol{x}\|+\|\boldsymbol{y}\|)$, which permits moderate growth in function sensitivity with input norms. This factor allows the function's sensitivity to grow modestly with the norm of its inputs. This is vital in high dimensions because the Euclidean norm $\|\boldsymbol{x}\|$ typically grows with $\sqrt{N}$. Many important loss functions are PL, including the squared-error loss $\phi(\boldsymbol{x}) = \|\boldsymbol{x}\|_2^2$ and the l_1 loss $\phi(\boldsymbol{x}) = \|\boldsymbol{x}\|_1$, which makes this class highly applicable.

We now present the cornerstone theoretical result for SE, which provides finite-sample guarantees for AMP performance.

Theorem 10.1 (Theoretical Convergence Result for SE *[9]) Under Assumption 10.1, for any PL function ϕ: $\mathbb{R}^M \to \mathbb{R}$, any tolerance level $\triangle \in (0, 1)$ and iteration $t \geq 0$, there exist positive constants $K_t > 0$, $\kappa_t > 0$, $\forall t \geq 0$ such that*

$$\Pr\left(\left|\frac{1}{N}\sum_{n=1}^{N}\phi\left(x_n^{t+1}, x_n\right) - \delta\mathbb{E}\left[\phi\left(\eta_t\left(X + \tau_t Z\right), X\right)\right]\right| \geq \triangle\right) \leq K_t e^{-\kappa_t N \triangle^2}, \tag{10.66}$$

where $Z \sim \mathcal{N}(0, 1)$, $X \sim p_X$, and τ_t evolves according to the state evolution equations.

This theorem is profound. It states that for any large but finite system size N and at any iteration t, the empirical average of a PL loss function (e.g., the MSE calculated from the algorithm's outputs $\boldsymbol{x}^{t+1}$ and the true signal $\boldsymbol{x}$) concentrates exponentially around a **deterministic** value predicted by the SE equations. The deterministic value is an expectation involving the **effective scalar channel**: $X + \tau_t Z$. This means that from the perspective of any individual coordinate n, the estimate x_n^{t+1} behaves **as if** it were produced by applying the denoiser η_t to a single scalar value $X + \tau_t Z$, where X is the true signal component and Z is an independent Gaussian noise of variance τ_t^2. The SE equations precisely track how this effective noise variance τ_t^2 evolves over time. Therefore, Theorem 10.1 rigorously justifies the i.i.d. Gaussianity of the AMP's effective observation noise, which is a key heuristic in the derivation of the algorithm.

The generality of Theorem 10.1 enables direct performance guarantees for AMP through appropriate choice of PL function (e.g., the squared error).

Theorem 10.2 (Performance Guarantee of AMP [9]) *Under Assumption 10.1, for any $\Delta \in (0, 1)$ and iteration $t \geq 0$, there exist positive constants K_t, $\kappa_t > 0$ such that*

$$\Pr\left(\left|\frac{1}{N}\left\|\boldsymbol{x}^{t+1}-\boldsymbol{x}\right\|^2-\delta\left(\tau_{t+1}^2-\sigma_w^2\right)\right|\geq\Delta\right)\leq K_t e^{-\kappa_t N\Delta^2}. \tag{10.67}$$

This non-asymptotic result establishes that the true MSE of the AMP estimate at iteration $t+1$ is, with high probability, exponentially close to the value $\delta(\tau_{t+1}^2-\sigma_w^2)$, which depends solely on the SE state τ_{t+1} and noise variance. This is a powerful **non-asymptotic** result: it gives a precise performance certificate for AMP even for finite-dimensional problems.

The finite-sample concentration results in Theorems 10.1 and 10.2 [9] immediately imply almost sure convergence in the large-system limit:

$$\lim_{N\to\infty}\frac{1}{N}\left\|\boldsymbol{x}^{t+1}-\boldsymbol{x}\right\|^2=\delta\left(\tau_{t+1}^2-\sigma_w^2\right)\quad\text{almost surely.} \tag{10.68}$$

The asymptotic convergence and performance guarantee results in (10.68) for $N\to\infty$ were firstly established in [10]. In other words, for large-scale problems with large M, N, the MSE performance achieved by AMP can be well predicted by $\delta\left(\tau_{t+1}^2-\sigma_w^2\right)$ with τ_{t+1}^2 obtained by the SE equation in (10.57). This decouples a complex high-dimensional inference problem into a simple, trackable scalar dynamic system.

10.5.3 Connection Between the AMP Algorithm and State Evolution

The connection between the AMP algorithm and SE is profound yet elegant: the iterative AMP algorithm implicitly and asymptotically constructs the very same scalar Gaussian channel that SE explicitly describes.

In the derived AMP algorithm, the belief about variable x_n is summarized by a Gaussian message $\mathcal{N}(x_n;\mu_{r,n}^t,v_{r,n}^t)$. This representation is statistically equivalent to a scalar observation model:

$$\mu_{r,n}^t=x_n+\sqrt{v_{r,n}^t}Z,\quad Z\sim\mathcal{N}(0,1). \tag{10.69}$$

The denoiser η_t then processes this effective observation $\mu_{r,n}^t$.

The critical insight emerges in the large-system limit: the message parameters $\mu_{r,n}^t$ and $v_{r,n}^t$ become statistically equivalent to their SE counterparts.

- The quantity $\mu_{r,n}^t$ becomes statistically equivalent to the SE's effective observation $y_{e,n}^t=x_n+\tau_t Z$.
- More significantly, the variance $v_{r,n}^t$ converges to a deterministic quantity that evolves identically to the SE variance τ_t^2:

$$v_{r,n}^t\to\sigma_w^2+\frac{1}{\delta}\cdot\frac{1}{N}\sum_{n=1}^{N}\mathbb{E}|x_n-x_n^t|^2=\tau_t^2. \tag{10.70}$$

This profound equivalence is what permits the dramatic simplification of the AMP updating rules. Specifically, the node-specific parameters $\mu_{r,n}^t$ and $v_{r,n}^t$ can be simplified into the forms $\mu_n^t+\sum_{m=1}^{M}a_{mn}\mu_{s,m}^t$ and $\sigma_w^2+v_z^t$, respectively, leading to the final, computationally efficient AMP rules (10.41) and (10.42).

Thus, SE is not merely an external analysis tool—it precisely describes the internal state evolution of the AMP algorithm itself. The practical AMP updates (10.41) and (10.42) automatically generate the parameters $(\mu_{r,n}^t, v_{r,n}^t)$ that track the theoretical SE process, creating a self-consistent loop between algorithm execution and theoretical prediction.

10.6 Generalizations and Extensions of AMP

The preceding sections have presented the fundamental AMP framework, primarily focusing on real-valued signals with linear observation models and independent sparse priors. However, the versatility of the AMP methodology has inspired numerous extensions that broaden its applicability to more complex scenarios. This section provides a systematic overview of key generalizations developed in the literature.

10.6.1 Complex-Valued AMP

In many practical applications such as communications and radar systems [11], both the sensing matrix $\boldsymbol{A}$ and sparse signal $\boldsymbol{x}$ are complex-valued. The straightforward approach of treating N-dimensional complex signals and M-dimensional complex measurements as their $2N$- and $2M$-dimensional real-valued counterparts, while mathematically valid, fails to exploit the joint sparsity structure between real and imaginary components that often exists in practice. This naive separation can lead to suboptimal recovery performance. To address this limitation, the Complex AMP (CAMP) algorithm was further proposed in [12], which explicitly accounts for the grouping of real and imaginary parts. This approach fully leverages the statistical dependencies between components, leading to substantial performance improvements. Furthermore, the state evolution framework has been extended to the complex domain [12], providing theoretical performance guarantees analogous to the real-valued case.

10.6.2 Extensions to Different Sensing Matrices

Although the theoretical results in Sect. 10.5.2 pertain specifically to i.i.d. Gaussian sensing matrices, the AMP algorithm exhibits robust performance in simulations with a wider class of i.i.d. non-Gaussian random matrices. However, the performance of the AMP algorithm often degrades a lot if the sensing matrix has strongly correlated entries. Several algorithmic innovations have emerged to address this challenge:

- For i.i.d. sub-Gaussian sensing matrices, the state evolution results have been rigorously extended [13], maintaining theoretical guarantees under broader distributional assumptions.
- For right orthogonally invariant matrices, the Vector AMP (VAMP) algorithm [14] achieves high-performance CS recovery by exploiting the rotational invariance properties of such matrices.
- Spatially coupled sensing matrices have been shown to improve MSE performance [15, 16], with corresponding AMP variants developed to leverage this structure.
- Memory AMP (MAMP) [17] introduces an orthogonalization procedure against previous messages, ensuring the asymptotic i.i.d. Gaussianity assumption remains valid throughout iterations by guaranteeing orthogonality between current output estimation errors and all previous input estimation errors.

10.6.3 Dependent Signal Structures and Non-separable Denoisers

In many practical scenarios, the signal $\boldsymbol{x}$ exhibits structured sparsity patterns that cannot be captured by independent priors. For example, in the wireless-channel-estimation-related literature, the AoD/AoA may concentrate on a few clusters, making Markov priors particularly appropriate [18]. In this case, the denoiser $\eta(\cdot)$ in (10.34) becomes a sliding window instead of an element-wise function, i.e., the updating rule for x_n^{t+1} in (10.41) may be related to $\eta_n(\cdot)$ and the $(n-k)$-th to $(n+k)$-th elements in vector $\boldsymbol{x}^t + \boldsymbol{A}^T\boldsymbol{r}^t$. The SE mentioned above that predicts the AMP performance is not applicable in this case. Instead, [19] introduced a new form of SE and proved its accuracy. The work [20] generalizes SE to Lipschitz continuous non-separable denoisers.

10.6.4 Multiple Measurement Vector Problems

Multiple measurement vector (MMV) problems come from dynamic compressed sensing problems, where the temporal correlation may help improve the estimation performance [21]. Specifically, when multiple measurement vectors share a common, time-invariant sensing matrix, the observation model generalizes to $\boldsymbol{Y} = \boldsymbol{A}\boldsymbol{X} + \boldsymbol{W}$, where $\boldsymbol{Y} = [\boldsymbol{y}_1, \boldsymbol{y}_2, \ldots, \boldsymbol{y}_I] \in \mathbb{R}^{M\times I}$, $\boldsymbol{X} = [\boldsymbol{x}_1, \boldsymbol{x}_2, \ldots, \boldsymbol{x}_I] \in \mathbb{R}^{N\times I}$, and $\boldsymbol{W} = [\boldsymbol{w}_1, \boldsymbol{w}_2, \ldots, \boldsymbol{w}_I] \in \mathbb{R}^{M\times I}$, with I denoting the number of temporal observations. When the time-varying signals $\boldsymbol{x}_i, i = 1, \ldots, I$ share a common support pattern $\boldsymbol{s}$ due to temporal correlation, the Dynamic Compressive Sensing AMP (DCS-AMP) algorithm [22] provides an effective solution. This approach applies AMP to compute messages with respect to the observation $\boldsymbol{Y}$, while employing standard sum-product rules for messages pertaining to the support $\boldsymbol{s}$.

10.6.5 Bilinear Models

In numerous applications, both the sensing matrix $\boldsymbol{A}$ and signal $\boldsymbol{x}$ constitute unknown random variables that must be jointly estimated from the coupled observations $\boldsymbol{Y}$. In this case, the factor graph contains more loops and the message passing procedure is more complex. To address these issues, bilinear generalized approximate message passing (BiG-AMP) algorithm was proposed in [23], where the messages w.r.t. $\boldsymbol{A}$ and $\boldsymbol{x}$ are symmetric, and the derivation is similar to that mentioned in Sect. 10.2. The BiG-AMP has demonstrated excellent performance across diverse applications, such as matrix factorization and blind deconvolution.[1]

Summary

AMP represents an efficient Bayesian inference algorithm that has been extensively studied for solving high-dimensional linear regression problems in CS. This chapter derives AMP as an approximation of loopy belief propagation (BP) for dense factor graphs. Although the detailed derivation of AMP involves considerable complexity, the resulting update equations in (10.41) and (10.42) possess remarkable simplicity and admit an intuitive interpretation that enhances understanding of the algorithm's underlying mechanics.

Specifically, the term $\boldsymbol{y}_e^t = \boldsymbol{A}^T\boldsymbol{r}^t + \boldsymbol{x}^t$ generated through AMP iterations can be interpreted as an effective AWGN observation $\boldsymbol{y}_e^t = \boldsymbol{x} + \tau_t\boldsymbol{Z}$. This interpretation provides a foundation for designing denoising functions $\eta^t(\cdot)$ in the AMP update rules based on either MMSE or MAP criteria to optimize CS recovery performance.

The chapter further introduces SE as a powerful analytical tool for predicting AMP performance across iterations. The original AMP update rules in (10.41) and (10.42), along with the SE formulation in (10.57), can be understood as governing the mean and variance updates of Gaussian messages within the AMP framework, respectively.

Finally, we discuss various generalizations and extensions of AMP, including algorithmic variants (such as complex-valued AMP), relaxed matrix assumptions (accommodating non-i.i.d. and non-Gaussian sensing matrices), generalized signal priors (incorporating dependent entries and non-separable denoisers), and extended problem frameworks (notably bilinear models), significantly broadening the applicability of the AMP methodology to diverse signal processing and machine learning applications.

Homework Problems

10.1 Derive the denoising function given by Eqs. (10.46) and (10.50) in the chapter. Clearly state all assumptions and key steps in your derivation.

10.2 Consider the scalar AWGN observation model:

$$y = x + w, \quad w \sim \mathcal{N}(0, \sigma^2),$$

where the goal is to estimate x from y. We examine two distinct scenarios:

---

[1] Matrix factorization [24, 25] includes matrix completion (some entries in $\boldsymbol{Y}$ are missing), dictionary learning ($\boldsymbol{x}$ has a sparse prior) and robust PCA ($\boldsymbol{Y} = \boldsymbol{Z} + \boldsymbol{E}$, where $\boldsymbol{Z} = \boldsymbol{A}\boldsymbol{X}$ is the low-rank component, and $\boldsymbol{E}$ is the sparse corruption), as special cases [26]. The blind deconvolution problem estimates the kernel $\boldsymbol{h}$ and the input signal $\boldsymbol{u}$ from convolutional observation model $\boldsymbol{z} = \boldsymbol{h} * \boldsymbol{u} + \boldsymbol{e}$ [27, 28]. By applying Fourier transformation $\mathcal{F}$, the problem is a bilinear recovery problem with $\boldsymbol{y} = \mathcal{F}(\boldsymbol{z})$, $\boldsymbol{A} = diag(\mathcal{F}(\boldsymbol{h}))$, $\boldsymbol{x} = \mathcal{F}(\boldsymbol{u})$, and $\boldsymbol{w} = \mathcal{F}(\boldsymbol{e})$.

We can distinguish two cases:

- I. **Known Prior**: The signal distribution $p_X(x)$ is known. To model sparsity, we consider a distribution with a point mass at zero: $P(x = 0) = 1 - \lambda_0$.
- II. **Unknown Sparse Prior**: The exact distribution $p_X(x)$ is unknown, but it is known to belong to the class $\mathcal{F}_{\lambda_0}$ of sparse distributions satisfying $P(X = 0) \geq 1 - \lambda_0$.

(a) For Case I, assume the specific sparse prior $p_X(x) = \lambda_0 \mathcal{N}(x; 0, \sigma_x^2) + (1 - \lambda_0)\delta(x)$. Derive the closed-form expression for the MMSE estimator, $\mathbb{E}[X \mid Y = y]$.

(b) For Case II, we employ the soft-thresholding estimator $\eta(y; \beta) = \text{sign}(y)(|y| - \beta\sigma)_+$, where the threshold is $\beta\sigma$. Let $\text{mse}(\sigma; p_X, \beta)$ denote the MSE achieved by this estimator under the true (but unknown) distribution p_X. Since p_X is unknown, the parameter β should be chosen to minimize the worst-case MSE over the class $\mathcal{F}_{\lambda_0}$:

$$\beta^* = \underset{\beta}{\text{argmin}} \max_{p_X \in \mathcal{F}_{\lambda_0}} \text{mse}(\sigma; p_X, \beta).$$

Derive the optimal threshold β^*.

(c) Now, suppose the *true* signal prior is $p_X(x) = \lambda_0 \mathcal{N}(x; 0, 1) + (1 - \lambda_0)\delta(x)$, with $\lambda_0 = 0.1$ and noise variance $\sigma^2 = 1$.

- For the Case I (MMSE) estimator, calculate the average MSE when it is derived using a *potentially mismatched* prior: $p_X(x) = \lambda_0 \mathcal{N}(x; 0, \sigma_x^2) + (1 - \lambda_0)\delta(x)$. Perform this calculation for $\sigma_x^2 = 0.1$, $\sigma_x^2 = 1$, and $\sigma_x^2 = 10$.
- For the Case II (robust) estimator, use the optimal threshold β^* derived in part (b) and calculate its average MSE.

Summarize the results of this experiment. What conclusion can you draw regarding the performance of the mismatched MMSE estimator versus the robust estimator?

10.3 Consider the AMP algorithm described in Example 10.1 of this chapter.

(a) Write down the specific State Evolution (SE) equations for this example. Plot the theoretical SE curve, i.e., the predicted MSE $\delta(\tau_{t+1}^2 - \sigma_w^2)$ versus the iteration t.

(b) Implement the AMP algorithm for this example (you may use the MATLAB code from the repository at https://github.com/biwcs-book/codebase as a starting point). Simulate the algorithm and plot the empirical MSE curve, i.e., $\frac{1}{N}\|\boldsymbol{x}^{t+1} - \boldsymbol{x}\|_2^2$ versus t. Compare this empirical curve with the theoretical SE curve from part (a). Do they match? Discuss your observations.

10.4 Given the provided MATLAB code for the AMP algorithm with an unknown prior (which uses the robust soft-thresholding denoiser from Eq. (10.52)), modify the code to leverage the *known* prior $p_X(x_n) = \lambda \mathcal{N}(x_n; 0, 1) + (1 - \lambda)\delta(x_n)$ (GitHub link: https://github.com/biwcs-book/codebase). This involves replacing the robust denoiser with the corresponding MMSE denoiser you derived in Problem 10.1. Conduct numerical simulations to compare the performance (e.g., in terms of final MSE and convergence rate) of your modified algorithm against the original robust version. Present and discuss your results.

References

1. C. Qi, Y. Huang, S. Jin, L. Wu, Sparse channel estimation based on compressed sensing for massive MIMO systems, in *Proceedings of the IEEE International Conference on Communications (ICC)* (London, UK, 2015), pp. 4558–4563
2. P.N. Jayanthi, S. Ravishankar, Sparse channel estimation for MIMO-OFDM systems using compressed sensing, in *Proceedings of the IEEE International Conference on Recent Trends in Electronics, Information and Communication Technology (RTEICT)* (Bangalore, India, 2016), pp. 1060–1064
3. W. Nasar, A.R. Maud, H. Wang, H.-N. Dai, An efficient greedy algorithm for wide band spectrum sensing in cognitive radio networks, in *Proceedings of the IEEE iThings/GreenCom/CPSCom/SmartData* (Halifax, NS, Canada, 2018), pp. 257–262

4. M. Bayati, A. Montanari, The LASSO risk for Gaussian matrices. IEEE Trans. Inf. Theory **58**(4), 1997–2017 (2012)
5. D.L. Donoho, A. Maleki, A. Montanari, Message-passing algorithms for compressed sensing. Proc. Natl. Acad. Sci. U.S.A. **106**(45), 18914–18919 (2009)
6. S. Rangan, Generalized approximate message passing for estimation with random linear mixing, in *Proceedings of the IEEE International Symposium on Information Theory* (St. Petersburg, Russia, 2011), pp. 2168–2172
7. A. Montanari, Graphical models concepts in compressed sensing, in *Compressed Sensing: Theory and Applications*, ed. by Y.C. Eldar, G. Kutyniok. (Cambridge University Press, Cambridge, U.K., 2012), pp. 394–438 (Chapter 9)
8. Q. Zou, H. Yang, A concise tutorial on approximate message passing (2022) [Online]. Available: arXiv:2201.07487
9. C. Rush, R. Venkataramanan, Finite sample analysis of approximate message passing algorithms. IEEE Trans. Inf. Theory **64**(11), 7264–7286 (2018)
10. C. Rush, R. Venkataramanan, The error probability of sparse superposition codes with approximate message passing decoding. IEEE Trans. Inf. Theory **65**(5), 3278–3303 (2019)
11. M.A. Herman, T. Strohmer, High-resolution radar via compressed sensing. IEEE Trans. Signal Process. **57**(6), 2275–2284 (2009)
12. A. Maleki, L. Anitori, Z. Yang, R.G. Baraniuk, Asymptotic analysis of complex LASSO via complex approximate message passing (CAMP). IEEE Trans. Inf. Theory **59**(7), 4290–4308 (2013)
13. M. Bayati, M. Lelarge, A. Montanari, Universality in polytope phase transitions and message passing algorithms. Ann. Appl. Probab. **25**(2), 753–822 (2015)
14. S. Rangan, P. Schniter, A.K. Fletcher, Vector approximate message passing. IEEE Trans. Inf. Theory **65**(10), 6664–6684 (2019)
15. D.L. Donoho, A. Javanmard, A. Montanari, Information-theoretically optimal compressed sensing via spatial coupling and approximate message passing. IEEE Trans. Inf. Theory **59**(11), 7434–7464 (2013)
16. C. Rush, K. Hsieh, R. Venkataramanan, Capacity-achieving spatially coupled sparse superposition codes with AMP decoding. IEEE Trans. Inf. Theory **67**(7), 4446–4484 (2021)
17. L. Liu, S. Huang, B.M. Kurkoski, Memory AMP. IEEE Trans. Inf. Theory **68**(12), 8015–8039 (2022)
18. J. Dai, A. Liu, V.K.N. Lau, FDD massive MIMO channel estimation with arbitrary 2D-array geometry. IEEE Trans. Signal Process. **66**(10), 2584–2599 (2018)
19. Y. Ma, C. Rush, D. Baron, Analysis of approximate message passing with non-separable denoisers and Markov random field priors. IEEE Trans. Inf. Theory **65**(11), 7367–7389 (2019)
20. R. Berthier, A. Montanari, P.-M. Nguyen, State evolution for approximate message passing with non-separable functions. Inf. Inference **9**(1), 33–79 (2020)
21. J. Ziniel, P. Schniter, Efficient high-dimensional inference in the multiple measurement vector problem. IEEE Trans. Signal Process. **61**(2), 340–354 (2013)
22. J. Ziniel, P. Schniter, Dynamic compressive sensing of time-varying signals via approximate message passing. IEEE Trans. Signal Process. **61**(21), 5270–5284 (2013)
23. J.T. Parker, P. Schniter, V. Cevher, Bilinear generalized approximate message passing—part I: derivation. IEEE Trans. Signal Process. **62**(22), 5839–5853 (2014)
24. K.-L. Du, M.N.S. Swamy, Z.-Q. Wang, W.H. Mow, Matrix factorization techniques in machine learning, signal processing, and statistics. Mathematics **11**(12), 2674 (2023)
25. Y. Kabashima, F. Krzakala, M. Mézard, A. Sakata, L. Zdeborová, Phase transitions and sample complexity in Bayes-optimal matrix factorization. IEEE Trans. Inf. Theory **62**(7), 4228–4265 (2016)
26. J.T. Parker, P. Schniter, V. Cevher, Bilinear generalized approximate message passing—part II: applications. IEEE Trans. Signal Process. **62**(22), 5854–5867 (2014)
27. A. Ahmed, B. Recht, J. Romberg, Blind deconvolution using convex programming. IEEE Trans. Inf. Theory **60**(3), 1711–1732 (2014)
28. S. Sarkar, A.K. Fletcher, S. Rangan, P. Schniter, Bilinear recovery using adaptive vector-AMP. IEEE Trans. Signal Process. **67**(13), 3383–3396 (2019)

Sparse Bayesian Learning 11

This chapter presents sparse Bayesian learning (SBL) as a comprehensive Bayesian framework for compressive sensing that overcomes key limitations of previously discussed recovery algorithms. While Chaps. 8 and 9 focused on optimization-based and greedy methods that yield only point estimates, and AMP requires restrictive assumptions about sensing matrices, SBL provides complete Bayesian inference, including uncertainty quantification, for general sensing matrices through an elegant EM formulation. The method automatically learns uncertain parameters in both observation and sparse prior models, enhancing robustness in practical applications. We systematically develop the SBL approach beginning with independent sparse priors, extend it to group-sparse variants that capture more structured sparsity patterns, and demonstrate its practical utility through an FDD massive MIMO channel estimation application where uplink-downlink angular reciprocity is effectively exploited as prior information.

11.1 Observation and Sparse Prior Models in SBL

In numerous wireless applications, the introduction of a grid parameter $\boldsymbol{\theta}$ proves essential for obtaining sparse representations of channels and signals. For instance, in the massive MIMO channel estimation problem examined in Chap. 9, we assume that the angles of active propagation paths are constrained to a discrete set of grid points. The corresponding array responses at these grid points form a dictionary of basis vectors that enable the transformation of antenna-domain channels into angular-domain sparse representations.

Traditional CS formulations, as discussed in previous chapters, typically employ predetermined fixed grid parameters. However, practical scenarios frequently exhibit mismatches between true parameter values and these fixed grid points. In massive MIMO systems, where $\boldsymbol{\theta}$ represents the angles of arriving paths, the use of a fixed angular grid inherently limits estimation accuracy by the grid resolution. This limitation introduces energy leakage effects that ultimately degrade channel estimation performance [1].

The emergence of 6G wireless systems, with their integrated sensing capabilities, further emphasizes the need for super-resolution parameter estimation. High-accuracy sensing and localization applications demand precise estimation of angle and delay parameters beyond conventional grid resolution limits [2, 3]. This requirement motivates the development of dynamic grid approaches that can adaptively adjust grid positions to improve parameter estimation accuracy while mitigating energy leakage.

Furthermore, practical implementations must address uncertainties in both sparse prior models and observation noise characteristics. Key parameters such as the variance of non-zero signal components and noise power are typically unknown a priori and must be estimated from data. Consequently, modern CS frameworks must simultaneously address two critical challenges: the integration of dynamic grid parameters and the handling of uncertain model parameters to maintain robustness in real-world applications.

The SBL method studied in this chapter is able to automatically learn both the uncertain grids and model parameters [4, 5]. Specifically, the goal of SBL is to recover a sparse signal $\boldsymbol{x} = [x_1, \ldots, x_N]^T \in \mathbb{C}^{N\times 1}$ from significantly fewer measurements $\boldsymbol{y} \in \mathbb{C}^{M\times 1}$ (i.e., $M \ll N$) under the following linear observation model:

$$\boldsymbol{y} = \boldsymbol{A}(\boldsymbol{\theta})\,\boldsymbol{x} + \boldsymbol{w}, \tag{11.1}$$

A. Liu and M.-M. Zhao, *Advanced Bayesian Signal Processing*, Textbooks in Telecommunication Engineering,
https://doi.org/10.1007/978-3-032-26656-9_11

where $\boldsymbol{A}(\boldsymbol{\theta}) \in \mathbb{C}^{M\times N}$ is the sensing matrix with dynamic grid parameters $\boldsymbol{\theta}$, and $\boldsymbol{w} \in \mathbb{C}^{M\times 1}$ is the noise vector with independent Gaussian entries.

In SBL, a core idea is to not just assume the signal $\boldsymbol{x}$ is sparse, but to learn the specifics of that sparsity from the data. To do this, a precision vector $\boldsymbol{\rho} = [\rho_1, \ldots, \rho_N]^T$ is introduced to control the sparsity of $\boldsymbol{x}$. It is helpful to remember that precision is the inverse of variance: $\rho_n = 1/\mathrm{var}(x_n)$. Therefore, a large ρ_n forces the variance of x_n to be small, effectively pulling it toward zero and promoting sparsity.

This precision vector $\boldsymbol{\rho}$ is itself modeled as a random variable with a Gamma distribution, making this a two-layer hierarchical model, i.e.,

$$p(\boldsymbol{\rho}) = \prod_{n=1}^{N} \Gamma(\rho_n; a_n, b_n), \tag{11.2}$$

where

$$\Gamma(\rho_n; a_n, b_n) = \frac{b_n^{a_n}}{\Gamma(a_n)} \rho_n^{a_n-1} \mathrm{e}^{-b_n\rho_n} \tag{11.3}$$

is a Gamma hyperprior with shape parameter a_n and rate parameter b_n. The parameters a_n, b_n are chosen such that the expected value $\mathbb{E}[\rho_n] = a_n/b_n \gg 1$. This high expected precision is what encourages most x_n to be zero. A typical choice that sets a very uninformative prior favoring high precision is $a_n = 1$ and $b_n = 0.001$.

The conditional probability for the sparse signal, $p(\boldsymbol{x} \mid \boldsymbol{\rho})$, is assumed to factorize and each component is a complex Gaussian whose variance is controlled by the corresponding precision as

$$p(\boldsymbol{x} \mid \boldsymbol{\rho}) = \prod_{n=1}^{N} p(x_n \mid \rho_n), \quad p(x_n \mid \rho_n) = \mathcal{CN}\left(x_n; 0, \rho_n^{-1}\right), \quad \forall n = 1, \ldots, N. \tag{11.4}$$

This two-layer hierarchical prior, a Gaussian conditional prior for $\boldsymbol{x}$ and a Gamma hyperprior for $\boldsymbol{\rho}$, integrates to give the following overall marginal prior for $\boldsymbol{x}$

$$p(\boldsymbol{x}) = \prod_{n=1}^{N} \int \mathcal{CN}\left(x_n; 0, \rho_n^{-1}\right) \Gamma(\rho_n; a_n, b_n)\, \mathrm{d}\rho_n \propto \prod_{n=1}^{N} \left(b_n + |x_n|^2\right)^{-(a_n+1)}. \tag{11.5}$$

This resulting prior is not a Gaussian! It has two key properties that enforce sparsity: a sharp peak at zero (which pulls small values strongly toward zero) and heavy tails (which allow large values to survive without excessive shrinkage). The connection to sparsity is profound; it can be shown that finding the MAP estimate of $\boldsymbol{x}$ with this prior is equivalent to finding the minimum ℓ_p-norm solution using FOCUSS with $p \to 0$ [6], which is a well-known sparse recovery method. This explains why SBL often outperforms basic ℓ_1-norm minimization, as it approximates a much more aggressive sparsity-promoting penalty.

The distribution of the noise also affects the results. The noise vector $\boldsymbol{w} \in \mathbb{C}^{M\times 1}$ is assumed to have independent Gaussian entries $w_m \sim \mathcal{CN}\left(w_m; 0, \kappa^{-1}\right)$. Similar to the signal precision, we also employ a Gamma distribution for the noise precision, i.e.,

$$p(\kappa) = \Gamma(\kappa; c, d).$$

Setting $c, d \to 0$ creates a very broad, non-informative hyperprior for κ, allowing its value to be learned entirely from the data. The Gamma distribution is chosen primarily for mathematical convenience: it is the conjugate prior for the precision of a Gaussian distribution, which simplifies subsequent Bayesian inference [7, 8].

Finally, given the noise precision κ, the signal precision $\boldsymbol{\rho}$, and the grid parameters $\boldsymbol{\theta}$, the likelihood function that connects the observed data $\boldsymbol{y}$ to the sparse signal $\boldsymbol{x}$ is

$$p(\boldsymbol{y}|\boldsymbol{x}, \boldsymbol{\xi}) = \mathcal{CN}\left(\boldsymbol{y}; \boldsymbol{A}(\boldsymbol{\theta})\boldsymbol{x}, \kappa^{-1}\boldsymbol{I}\right),$$

where $\boldsymbol{\xi} = \{\kappa, \boldsymbol{\rho}, \boldsymbol{\theta}\}$ denotes all the unknown parameters.

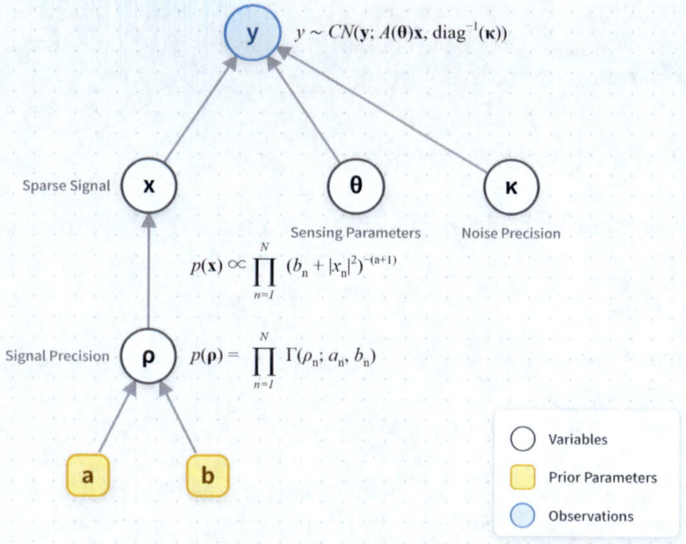

Fig. 11.1 An illustration of sparse prior and observation models in SBL

In summary, the sparse prior and observation models in SBL form a complete probabilistic graphical model, as illustrated in Fig. 11.1. The key takeaway is that by modeling the signal and noise precisions with Gamma distributions, SBL creates a framework where these crucial parameters are not fixed but are automatically learned from the data, typically using an EM framework, which will be detailed in the next section.

11.2 Derivation of SBL Based on the EM Framework

The SBL method is derived from the EM framework by treating the sparse signal $\boldsymbol{x}$ as the hidden variable and the other variables, including the grid parameters $\boldsymbol{\theta}$, the signal precision $\boldsymbol{\rho}$ and the noise precision κ, as parameters to be estimated based on the MAP criteria. Specifically, the SBL algorithm performs iterations between the following two major steps until convergence.

- **SBL-E Step**: With given grid parameters $\boldsymbol{\theta}$, signal precision and noise precision $\dot{\boldsymbol{\xi}} = \left\{\dot{\kappa}, \dot{\boldsymbol{\rho}}, \dot{\boldsymbol{\theta}}\right\}$ obtained in the previous iteration, calculate the posterior of the sparse signal $\boldsymbol{x}$

$$\begin{aligned} p\left(\boldsymbol{x}|\boldsymbol{y}, \dot{\boldsymbol{\xi}}\right) &\propto p\left(\boldsymbol{x} \mid \dot{\boldsymbol{\rho}}\right) p\left(\boldsymbol{y}|\boldsymbol{x}, \dot{\boldsymbol{\xi}}\right) \\ &= \mathcal{CN}\left(\boldsymbol{x}; \boldsymbol{\mu}\left(\dot{\boldsymbol{\xi}}\right), \boldsymbol{\Sigma}\left(\dot{\boldsymbol{\xi}}\right)\right), \end{aligned} \tag{11.6}$$

  where the posterior mean and covariance matrix of $\boldsymbol{x}$ are given by

$$\begin{aligned} \boldsymbol{\mu}\left(\dot{\boldsymbol{\xi}}\right) &= \dot{\kappa}\boldsymbol{\Sigma}\left(\dot{\boldsymbol{\xi}}\right) \boldsymbol{A}^H\left(\dot{\boldsymbol{\theta}}\right) \boldsymbol{y}, \\ \boldsymbol{\Sigma}\left(\dot{\boldsymbol{\xi}}\right) &= \left(\dot{\kappa}\boldsymbol{A}^H\left(\dot{\boldsymbol{\theta}}\right) \boldsymbol{A}\left(\dot{\boldsymbol{\theta}}\right) + \operatorname{diag}\left(\dot{\boldsymbol{\rho}}\right)\right)^{-1}, \end{aligned} \tag{11.7}$$

  respectively.
- **SBL-M Step**: Given $p\left(\boldsymbol{x}|\boldsymbol{y}, \dot{\boldsymbol{\xi}}\right) = \mathcal{CN}\left(\boldsymbol{x}; \boldsymbol{\mu}\left(\dot{\boldsymbol{\xi}}\right), \boldsymbol{\Sigma}\left(\dot{\boldsymbol{\xi}}\right)\right)$, first construct a surrogate function $u\left(\boldsymbol{\xi}; \dot{\boldsymbol{\xi}}\right)$ for the log joint PDF function, i.e., $\ln p\left(\boldsymbol{y}, \boldsymbol{\xi}\right)$, around the previous iterate $\dot{\boldsymbol{\xi}}$ based on the EM as

$$u\left(\boldsymbol{\xi}; \dot{\boldsymbol{\xi}}\right) = \int p\left(\boldsymbol{x}|\boldsymbol{y}, \dot{\boldsymbol{\xi}}\right) \ln \frac{p\left(\boldsymbol{x}, \boldsymbol{y}, \boldsymbol{\xi}\right)}{p\left(\boldsymbol{x}|\boldsymbol{y}, \dot{\boldsymbol{\xi}}\right)} \mathrm{d}\boldsymbol{x}, \tag{11.8}$$

  then alternatively maximize the surrogate function w.r.t. κ, $\boldsymbol{\rho}$, $\boldsymbol{\theta}$.

In the following, we shall elaborate the SBL-M step, which is based on the in-exact block majorization-minimization (MM) method [9]. In particular, an overview of the in-exact block MM method will be introduced first, followed by its detailed implementation, and finally the convergence of SBL.

11.2.1 Overview of the In-Exact Block MM Method

The SBL-M step aims to find the most probable values of the parameters $\boldsymbol{\xi}$ by maximizing the posterior, equivalent to maximizing $\ln p(\boldsymbol{y}, \boldsymbol{\xi})$, i.e.,

$$\boldsymbol{\xi}^* = \arg\max_{\boldsymbol{\xi}} \ln p\left(\boldsymbol{y}, \boldsymbol{\xi}\right). \tag{11.9}$$

However, this objective function is difficult to handle directly because it involves a computationally intractable multi-dimensional integral over the latent variable $\boldsymbol{x}$.

To overcome this, the Block MM method is employed. The core idea is to iteratively replace the complex objective function with a simpler surrogate function $u(\boldsymbol{\xi}; \dot{\boldsymbol{\xi}})$ that acts as a tractable lower bound. This surrogate is designed to be easier to optimize while ensuring the algorithm progresses toward the maximum of the original objective.

A key strategy to simplify the optimization further is to partition the parameters into blocks: $\boldsymbol{\xi} = (\boldsymbol{\xi}_1, \boldsymbol{\xi}_2, \boldsymbol{\xi}_3)$, where $\boldsymbol{\xi}_1 = \boldsymbol{\kappa}$, $\boldsymbol{\xi}_2 = \boldsymbol{\rho}$, and $\boldsymbol{\xi}_3 = \boldsymbol{\theta}$. The algorithm then proceeds by alternately maximizing the surrogate function w.r.t. one block of parameters at a time, keeping the others fixed. This block-wise approach often leads to simple, closed-form updates for each block.

The convergence of the Block MM method is guaranteed by the specific properties of the surrogate function, as outlined in the following lemma. Essentially, the surrogate must be a tight lower bound to the original function at the current estimate, ensuring that improving the surrogate also improves the true objective.

Lemma 11.1 (Properties of Surrogate Function) *The surrogate function $u\left(\boldsymbol{\xi}; \dot{\boldsymbol{\xi}}\right)$ in (11.8) satisfies the following properties:*

1. *It is a global lower bound, i.e.,*

$$u\left(\boldsymbol{\xi}; \dot{\boldsymbol{\xi}}\right) \leq \ln p(\boldsymbol{y}, \boldsymbol{\xi}),\ \forall \boldsymbol{\xi}. \tag{11.10}$$

2. *It is tight at the current estimate, i.e.,*

$$u\left(\dot{\boldsymbol{\xi}}; \dot{\boldsymbol{\xi}}\right) = \ln p(\boldsymbol{y}, \dot{\boldsymbol{\xi}}). \tag{11.11}$$

3. *Their gradients agree at the current estimate, i.e.,*

$$\frac{\partial u\left(\boldsymbol{\xi}; \dot{\boldsymbol{\xi}}\right)}{\partial \boldsymbol{\xi}}|_{\boldsymbol{\xi}=\dot{\boldsymbol{\xi}}} = \frac{\partial \ln p\left(\boldsymbol{y}, \boldsymbol{\xi}\right)}{\partial \boldsymbol{\xi}}|_{\boldsymbol{\xi}=\dot{\boldsymbol{\xi}}}. \tag{11.12}$$

Proof For any distribution $q(\boldsymbol{x})$, we can construct a lower bound using Jensen's inequality:

$$\ln p(\boldsymbol{y}, \boldsymbol{\xi}) = \ln \int p(\boldsymbol{x}, \boldsymbol{y}, \boldsymbol{\xi}) \mathrm{d}\boldsymbol{x} \geq \int q(\boldsymbol{x}) \ln \frac{p(\boldsymbol{x}, \boldsymbol{y}, \boldsymbol{\xi})}{q(\boldsymbol{x})} \mathrm{d}\boldsymbol{x}. \tag{11.13}$$

The inequality becomes an equality when $q(\boldsymbol{x})$ is chosen to be the posterior distribution $p(\boldsymbol{x} \mid \boldsymbol{y}, \boldsymbol{\xi})$. Setting $q(\boldsymbol{x}) = p(\boldsymbol{x} \mid \boldsymbol{y}, \dot{\boldsymbol{\xi}})$ for a fixed $\dot{\boldsymbol{\xi}}$ yields our surrogate function $u(\boldsymbol{\xi}; \dot{\boldsymbol{\xi}})$. This construction automatically satisfies properties (11.10) and (11.11).

To establish property (11.12), we examine the gradient of the surrogate function. Starting from the definition in (11.8), its derivative with respect to $\boldsymbol{\xi}$ evaluated at $\boldsymbol{\xi} = \dot{\boldsymbol{\xi}}$ can be expressed as

$$\begin{aligned}
&\frac{\partial u\left(\boldsymbol{\xi}; \dot{\boldsymbol{\xi}}\right)}{\partial \boldsymbol{\xi}}|_{\boldsymbol{\xi}=\dot{\boldsymbol{\xi}}} \\
&= \int p\left(\boldsymbol{x}|\boldsymbol{y}, \dot{\boldsymbol{\xi}}\right) \frac{\partial \ln p\left(\boldsymbol{x}, \boldsymbol{y}, \boldsymbol{\xi}\right)}{\partial \boldsymbol{\xi}} \mathrm{d}\boldsymbol{x}|_{\boldsymbol{\xi}=\dot{\boldsymbol{\xi}}} \\
&= \int \frac{p\left(\boldsymbol{x}|\boldsymbol{y}, \dot{\boldsymbol{\xi}}\right)}{p\left(\boldsymbol{x}, \boldsymbol{y}, \boldsymbol{\xi}\right)} \cdot \frac{\partial p\left(\boldsymbol{x}, \boldsymbol{y}, \boldsymbol{\xi}\right)}{\partial \boldsymbol{\xi}} \mathrm{d}\boldsymbol{x}|_{\boldsymbol{\xi}=\dot{\boldsymbol{\xi}}}.
\end{aligned} \tag{11.14}$$

Recognizing that $p(\boldsymbol{x}|\boldsymbol{y}, \dot{\boldsymbol{\xi}})/p(\boldsymbol{x}, \boldsymbol{y}, \dot{\boldsymbol{\xi}}) = 1/p(\boldsymbol{y}, \dot{\boldsymbol{\xi}})$, which is independent of $\boldsymbol{x}$, we can simplify the above expression to

$$\begin{aligned} &\frac{\partial u\left(\boldsymbol{\xi}; \dot{\boldsymbol{\xi}}\right)}{\partial \boldsymbol{\xi}}\Big|_{\boldsymbol{\xi}=\dot{\boldsymbol{\xi}}} \\ &= \frac{1}{p\left(\boldsymbol{y}, \dot{\boldsymbol{\xi}}\right)} \int \left.\frac{\partial p(\boldsymbol{x}, \boldsymbol{y}, \boldsymbol{\xi})}{\partial \boldsymbol{\xi}}\right|_{\boldsymbol{\xi}=\dot{\boldsymbol{\xi}}} \mathrm{d}\boldsymbol{x} \\ &= \frac{1}{p(\boldsymbol{y}, \dot{\boldsymbol{\xi}})} \cdot \frac{\partial}{\partial \boldsymbol{\xi}} \int p(\boldsymbol{x}, \boldsymbol{y}, \boldsymbol{\xi}) \mathrm{d}\boldsymbol{x} \Big|_{\boldsymbol{\xi}=\dot{\boldsymbol{\xi}}} \\ &= \frac{1}{p(\boldsymbol{y}, \dot{\boldsymbol{\xi}})} \cdot \left.\frac{\partial p(\boldsymbol{y}, \boldsymbol{\xi})}{\partial \boldsymbol{\xi}}\right|_{\boldsymbol{\xi}=\dot{\boldsymbol{\xi}}}. \end{aligned} \tag{11.15}$$

The interchange of integration and differentiation in the second line is justified by the dominated convergence theorem under mild regularity conditions. Meanwhile, the gradient of the log-marginal likelihood on the right-hand side of (11.12) is

$$\left.\frac{\partial \ln p(\boldsymbol{y}, \boldsymbol{\xi})}{\partial \boldsymbol{\xi}}\right|_{\boldsymbol{\xi}=\dot{\boldsymbol{\xi}}} = \frac{1}{p(\boldsymbol{y}, \dot{\boldsymbol{\xi}})} \cdot \left.\frac{\partial p(\boldsymbol{y}, \boldsymbol{\xi})}{\partial \boldsymbol{\xi}}\right|_{\boldsymbol{\xi}=\dot{\boldsymbol{\xi}}}. \tag{11.16}$$

Comparing the two expressions above, we immediately obtain

$$\left.\frac{\partial u(\boldsymbol{\xi}; \dot{\boldsymbol{\xi}})}{\partial \boldsymbol{\xi}}\right|_{\boldsymbol{\xi}=\dot{\boldsymbol{\xi}}} = \left.\frac{\partial \ln p(\boldsymbol{y}, \boldsymbol{\xi})}{\partial \boldsymbol{\xi}}\right|_{\boldsymbol{\xi}=\dot{\boldsymbol{\xi}}}, \tag{11.17}$$

which completes the proof of property (11.12). □

In the i-th iteration, each parameter block is updated sequentially by maximizing the surrogate as

$$\boldsymbol{\xi}_j^{(i+1)} = \underset{\boldsymbol{\xi}_j}{\operatorname{argmax}}\, u\left(\boldsymbol{\xi}_j, \boldsymbol{\xi}_{-j}^{(i)}; \boldsymbol{\xi}_j^{(i)}, \boldsymbol{\xi}_{-j}^{(i)}\right), \tag{11.18}$$

where $\boldsymbol{\xi}_{-j}^{(i)} \triangleq \left(\boldsymbol{\xi}_1^{(i+1)}, \ldots, \boldsymbol{\xi}_{j-1}^{(i+1)}, \boldsymbol{\xi}_{j+1}^{(i)}, \ldots, \boldsymbol{\xi}_3^{(i)}\right)$ contains the most recent values of the other blocks. This update rule guarantees that the algorithm converges to a stationary point of the original problem.

A practical challenge arises when maximizing w.r.t. a particular block, say $\boldsymbol{\xi}_3 = \boldsymbol{\theta}$, is itself a non-convex and difficult problem. Finding its global maximum in every iteration can be computationally prohibitive.

In such cases, we can use an **in-exact MM** algorithm. Here, instead of fully maximizing the surrogate for $\boldsymbol{\theta}$, we perform a simple one-step gradient ascent to obtain $\boldsymbol{\theta}^{(i+1)}$. Although this is an approximate (in-exact) update, it can be proven that the algorithm still converges to a stationary solution, making it a highly flexible and practical approach. The subsequent sections will detail the updates for κ, $\boldsymbol{\rho}$, and $\boldsymbol{\theta}$, and formally establish this convergence guarantee.

11.2.2 Detailed Implementations of the SBL-M Step

We now detail the updates for the three parameter blocks κ, $\boldsymbol{\rho}$, and $\boldsymbol{\theta}$. The core idea of the block MM algorithm is to optimize one block at a time while keeping the others fixed at their most recent values.

We begin by updating the noise precision κ. The surrogate function is simplified by retaining only terms relevant to κ, leading to

$$
\begin{aligned}
& u\left(\kappa, \boldsymbol{\rho}^{(i)}, \boldsymbol{\theta}^{(i)}; \kappa^{(i)}, \boldsymbol{\rho}^{(i)}, \boldsymbol{\theta}^{(i)}\right) \\
&= \int p\left(\boldsymbol{x} \mid \boldsymbol{y}, \kappa^{(i)}, \boldsymbol{\rho}^{(i)}, \boldsymbol{\theta}^{(i)}\right) \ln p\left(\boldsymbol{y} \mid \boldsymbol{x}, \kappa, \boldsymbol{\theta}^{(i)}\right) \mathrm{d}\boldsymbol{x} \\
&\quad + \int p\left(\boldsymbol{x} \mid \boldsymbol{y}, \kappa^{(i)}, \boldsymbol{\rho}^{(i)}, \boldsymbol{\theta}^{(i)}\right) \ln p\left(\kappa\right) \mathrm{d}\boldsymbol{x} \\
&= -\kappa \int p\left(\boldsymbol{x} \mid \boldsymbol{y}, \kappa^{(i)}, \boldsymbol{\rho}^{(i)}, \boldsymbol{\theta}^{(i)}\right) \left\| \boldsymbol{y} - \boldsymbol{A}\left(\boldsymbol{\theta}^{(i)}\right) \boldsymbol{x} \right\|_2^2 \mathrm{d}\boldsymbol{x} + M \ln \kappa + (c-1) \ln \kappa - \mathrm{d}\kappa \\
&= (M + c - 1) \ln \kappa - \kappa \left(d + \eta\left(\kappa^{(i)}, \boldsymbol{\rho}^{(i)}, \boldsymbol{\theta}^{(i)}\right)\right),
\end{aligned} \tag{11.19}
$$

where $\eta\left(\kappa, \boldsymbol{\rho}, \boldsymbol{\theta}\right) = \mathrm{tr}\left(\boldsymbol{A}\left(\boldsymbol{\theta}\right) \boldsymbol{\Sigma}\left(\boldsymbol{\xi}\right) \boldsymbol{A}^H\left(\boldsymbol{\theta}\right)\right) + \left\| \boldsymbol{y} - \boldsymbol{A}\left(\boldsymbol{\theta}\right) \boldsymbol{\mu}\left(\boldsymbol{\xi}\right) \right\|_2^2$.

This expression has an intuitive structure: the term η represents the **total expected model fitting error**, combining the uncertainty in the signal estimate ($\boldsymbol{\Sigma}$) and the mean squared error of the reconstruction ($\boldsymbol{\mu}$). The function itself is strictly concave in κ. Maximizing it is straightforward: we take the derivative w.r.t. κ, set it to zero, and solve, yielding the unique update given by

$$
\kappa^{(i+1)} = \frac{M + c - 1}{d + \eta\left(\kappa^{(i)}, \boldsymbol{\rho}^{(i)}, \boldsymbol{\theta}^{(i)}\right)}. \tag{11.20}
$$

This result is intuitive: the noise precision is updated as (**number of measurements** + **prior shape** − **1**)/(**prior rate** + **total expected error**). A larger model misfit leads to a lower noise precision (higher noise variance), which is a logical outcome.

Next, we update the signal precision vector $\boldsymbol{\rho}$, which controls the sparsity of $\boldsymbol{x}$. Again, we isolate terms depending on $\boldsymbol{\rho}$ in the surrogate function and obtain

$$
\begin{aligned}
& u\left(\kappa^{(i+1)}, \boldsymbol{\rho}, \boldsymbol{\theta}^{(i)}; \kappa^{(i+1)}, \boldsymbol{\rho}^{(i)}, \boldsymbol{\theta}^{(i)}\right) \\
&= \int p\left(\boldsymbol{x} \mid \boldsymbol{y}, \kappa^{(i+1)}, \boldsymbol{\rho}^{(i)}, \boldsymbol{\theta}^{(i)}\right) \ln p\left(\boldsymbol{x} \mid \boldsymbol{\rho}\right) \mathrm{d}\boldsymbol{x} \\
&\quad + \int p\left(\boldsymbol{x} \mid \boldsymbol{y}, \kappa^{(i+1)}, \boldsymbol{\rho}^{(i)}, \boldsymbol{\theta}^{(i)}\right) \ln p\left(\boldsymbol{\rho}\right) \mathrm{d}\boldsymbol{x} \\
&= -\ln \left|\mathrm{diag}\left(\boldsymbol{\rho}^{-1}\right)\right| + (a-1) \sum_{n=1}^{N} \ln \rho_n - b \sum_{n=1}^{N} \rho_n \\
&\quad - \int p\left(\boldsymbol{x} \mid \boldsymbol{y}, \kappa^{(i+1)}, \boldsymbol{\rho}^{(i)}, \boldsymbol{\theta}^{(i)}\right) \left(\boldsymbol{x}^H \mathrm{diag}\left(\boldsymbol{\rho}\right) \boldsymbol{x}\right) \mathrm{d}\boldsymbol{x} \\
&= a \sum_{n=1}^{N} \ln \rho_n - b \sum_{n=1}^{N} \rho_n - \mathrm{tr}\left(\Xi\left(\kappa^{(i+1)}, \boldsymbol{\rho}^{(i)}, \boldsymbol{\theta}^{(i)}\right) \mathrm{diag}\left(\boldsymbol{\rho}\right)\right),
\end{aligned} \tag{11.21}
$$

where $\Xi\left(\kappa, \boldsymbol{\rho}, \boldsymbol{\theta}\right) = \boldsymbol{\Sigma}\left(\boldsymbol{\xi}\right) + \boldsymbol{\mu}\left(\boldsymbol{\xi}\right) \boldsymbol{\mu}^H\left(\boldsymbol{\xi}\right)$ is the **posterior second moment** of the signal $\boldsymbol{x}$.

The key term here is $[\Xi]_{nn}$, which is the expected power of the n-th signal element. Taking the derivative w.r.t. a single ρ_n and setting it to zero gives the following update:

$$
\rho_n^{(i+1)} = \frac{a}{b + \left[\Xi\left(\kappa^{(i+1)}, \boldsymbol{\rho}^{(i)}, \boldsymbol{\theta}^{(i)}\right)\right]_{nn}}. \tag{11.22}
$$

This update has a clear Bayesian interpretation: the posterior precision for x_n is the prior 'rate' parameter b plus the signal's expected power, scaled by the prior shape a. If the expected power $[\Xi]_{nn}$ is small, the precision ρ_n becomes large, forcing the corresponding x_n to be closer to zero and thus promoting sparsity.

Finally, we address the grid parameters $\boldsymbol{\theta}$. Unlike the previous parameters, maximizing the surrogate for $\boldsymbol{\theta}$ is often a non-convex problem without a closed-form solution. Therefore, we employ an inexact update. Similar to the updates of $\boldsymbol{\rho}$ and κ, the objective function in (11.18) becomes

$$
\begin{aligned}
& u\left(\kappa^{(i+1)}, \boldsymbol{\rho}^{(i+1)}, \boldsymbol{\theta}; \kappa^{(i+1)}, \boldsymbol{\rho}^{(i+1)}, \boldsymbol{\theta}^{(i)}\right) \\
& = \int p\left(\boldsymbol{x} \mid \boldsymbol{y}, \kappa^{(i+1)}, \boldsymbol{\rho}^{(i+1)}, \boldsymbol{\theta}^{(i)}\right) \ln p\left(\boldsymbol{y} \mid \boldsymbol{x}, \kappa^{(i+1)}, \boldsymbol{\theta}\right) \mathrm{d}\boldsymbol{x} \\
& = -\kappa^{(i+1)} \int p\left(\boldsymbol{x} \mid \boldsymbol{y}, \kappa^{(i+1)}, \boldsymbol{\rho}^{(i+1)}, \boldsymbol{\theta}^{(i)}\right) \|\boldsymbol{y} - \boldsymbol{A}(\boldsymbol{\theta})\boldsymbol{x}\|_2^2 \, \mathrm{d}\boldsymbol{x} \\
& = -\kappa^{(i+1)} \mathrm{tr}\left(\boldsymbol{A}(\boldsymbol{\theta}) \boldsymbol{\Sigma}\left(\boldsymbol{\xi}_{-3}^{(i)}, \boldsymbol{\theta}^{(i)}\right) \boldsymbol{A}^H(\boldsymbol{\theta})\right) + \left\|\mathbf{y} - \boldsymbol{A}(\boldsymbol{\theta}) \boldsymbol{\mu}\left(\boldsymbol{\xi}_{-3}^{(i)}, \boldsymbol{\theta}^{(i)}\right)\right\|_2^2 .
\end{aligned} \tag{11.23}
$$

This shows that maximizing the surrogate for $\boldsymbol{\theta}$ is equivalent to minimizing the total expected model error η. Since a full optimization is costly, we take a single gradient ascent step to increase u as

$$
\boldsymbol{\theta}^{(i+1)} = \boldsymbol{\theta}^{(i)} + \Delta_\theta^{(i)} \frac{\partial u\left(\kappa^{(i+1)}, \boldsymbol{\rho}^{(i+1)}, \boldsymbol{\theta}; \kappa^{(i+1)}, \boldsymbol{\rho}^{(i+1)}, \boldsymbol{\theta}^{(i)}\right)}{\partial \boldsymbol{\theta}} \Big|_{\boldsymbol{\theta}=\boldsymbol{\theta}^{(i)}}, \tag{11.24}
$$

where the step size $\Delta_\theta^{(i)}$ is determined via backtracking line search to ensure stable convergence [10]. This process is often called grid refining, as it adjusts the dictionary matrix $\boldsymbol{A}(\boldsymbol{\theta})$ to better fit the data, thereby compensating for the off-grid gap between the assumed and true parameters.

11.2.3 Convergence Analysis of SBL

Clearly, the update rule of SBL ensures that the surrogate function $u(\boldsymbol{\xi}; \dot{\boldsymbol{\xi}})$ is non-decreasing. Then, it follows from the properties of the surrogate function in Lemma 11.1 that the sequence $\ln p\left(\boldsymbol{y}, \boldsymbol{\xi}^{(i)}\right)$, $i = 1, 2, 3, \ldots$ generated by the SBL algorithm is also non-decreasing and it converges to a limit because the objective function has an upper bound. In the following, we further prove that the sequence of iterates generated by the SBL algorithm converges to a stationary point of problem (11.9).

Theorem 11.1 (Convergenece of SBL) *For the surrogate function defined in (11.8), if variables are iteratively updated by (11.20), (11.22) and (11.24), the iterates generated by the in-exact block MM algorithm converge to a stationary solution of problem (11.9). Apparently, if the inexact block MM algorithm converges, the overall SBL algorithm will also converge.*

Proof According to Theorem 2-b in [11] , the block MM algorithm will converge to a stationary solution if the following additional conditions are satisfied:

1. All the properties in (11.10)–(11.12) hold true with the surrogate function.
2. At least two of the problems in (11.18) have a unique solution.

Lemma 11.1 and the unique solutions in (11.20) and (11.22) guarantee that the above two conditions hold true, respectively. This completes the proof. □

11.3 Extension to Handle Group Sparsity

Consider a group sparse signal $\boldsymbol{x}$, where the index set $\{1, \ldots, N\}$ of $\boldsymbol{x}$ can be partitioned into Q non-overlapping subsets $\mathcal{I}_1, \ldots, \mathcal{I}_Q$ such that all the elements of $\boldsymbol{x}\left[\mathcal{I}_i\right]$, $\forall i \in \{1, \ldots, Q\}$ are simultaneously active or inactive. Correspondingly, we introduce a common precision control vector $\boldsymbol{\rho} = \left[\rho_1, \ldots, \rho_Q\right]^T$ to control the group sparsity, where each ρ_i determine whether the i-th group signal $\boldsymbol{x}\left[\mathcal{I}_i\right]$ is zero or non-zero. The individual precision for each element in the i-th group is given by $\rho_i \gamma_n$, $\forall n \in \mathcal{I}_i$, where $\gamma_n > 0$ is a positive number to control the variance of x_n since different elements in the same group may still have different variances.

As before, the common precision control vector $\boldsymbol{\rho}$ is still modeled as i.i.d. Gamma distributed, i.e.,

$$p(\boldsymbol{\rho}) = \prod_{i=1}^{Q} \Gamma(\rho_i; a_i, b_i), \tag{11.25}$$

where the shape parameter a_i and rate parameter b_i are chosen such that $\frac{a_i}{b_i} = \mathbb{E}[\rho_i] \gg 1$ to promote sparsity. The individual precision control vector $\boldsymbol{\gamma} = [\gamma_1, \ldots, \gamma_N]^T$ is assumed to have an i.i.d. non-informative uniform prior. Such a modeling method allows the elements in the same group to share the same sparse structure (i.e., simultaneously zero or non-zero) but with different varainces/amplitudes.

Based on the above group sparse prior model, we can derive an SBL algorithm to recover the group sparse signal by following similar steps as in Sect. 11.2, with only a few minor modifications. In particular, let $\dot{\boldsymbol{\xi}} = \{\dot{\kappa}, \dot{\boldsymbol{\rho}}, \dot{\boldsymbol{\gamma}}, \dot{\boldsymbol{\theta}}\}$ denote the set of noise precision $\dot{\kappa}$, common/individual precision control vector $\dot{\boldsymbol{\rho}}/\dot{\boldsymbol{\gamma}}$ and grid parameter $\dot{\boldsymbol{\theta}}$ obtained in the previous iteration. For given $\dot{\boldsymbol{\xi}}$, the posterior mean and covariance matrix of $\boldsymbol{x}$ can be calculated as

$$\begin{aligned}\boldsymbol{\mu}(\dot{\boldsymbol{\xi}}) &= \dot{\kappa}\boldsymbol{\Sigma}(\dot{\boldsymbol{\xi}})\boldsymbol{A}^H(\dot{\boldsymbol{\theta}})\boldsymbol{y},\\ \boldsymbol{\Sigma}(\dot{\boldsymbol{\xi}}) &= \left(\dot{\kappa}\boldsymbol{A}^H(\dot{\boldsymbol{\theta}})\boldsymbol{A}(\dot{\boldsymbol{\theta}}) + \operatorname{diag}(\dot{\boldsymbol{\rho}} \odot \dot{\boldsymbol{\gamma}})\right)^{-1},\end{aligned} \tag{11.26}$$

where $\odot$ denotes the Hadamard product.

Then, we can construct the surrogate function for the SBL-M step as in (11.8), where the joint PDF $p(\boldsymbol{x}, \boldsymbol{y}, \boldsymbol{\xi})$ in (11.8) is now given by

$$\begin{aligned}p(\boldsymbol{x}, \boldsymbol{y}, \boldsymbol{\xi}) = &\prod_{i=1}^{Q} \Gamma(\rho_i; a_i, b_i) \prod_{n\in\mathcal{I}_i} \mathcal{CN}\left(x_n; 0, (\gamma_n\rho_i)^{-1}\right)\\ &\times \mathcal{CN}\left(\boldsymbol{y}; \boldsymbol{A}(\boldsymbol{\theta})\boldsymbol{x}, \kappa^{-1}\boldsymbol{I}\right)\Gamma(\kappa; c, d).\end{aligned} \tag{11.27}$$

Partition $\boldsymbol{\xi}$ into four blocks $\boldsymbol{\xi} = (\boldsymbol{\xi}_1, \boldsymbol{\xi}_2, \boldsymbol{\xi}_3, \boldsymbol{\xi}_4)$ with $\boldsymbol{\xi}_1 = \kappa$, $\boldsymbol{\xi}_2 = \boldsymbol{\rho}$, $\boldsymbol{\xi}_3 = \boldsymbol{\gamma}$ and $\boldsymbol{\xi}_4 = \boldsymbol{\theta}$. Accordingly, in the SBL-M step of the i-th iteration, we update $\boldsymbol{\xi}_j$ for $j = 1, 2, 3$ as

$$\boldsymbol{\xi}_j^{(i+1)} = \underset{\boldsymbol{\xi}_j}{\operatorname{argmax}}\, u\left(\boldsymbol{\xi}_j, \boldsymbol{\xi}_{-j}^{(i)}; \boldsymbol{\xi}_j^{(i)}, \boldsymbol{\xi}_{-j}^{(i)}\right), \tag{11.28}$$

while for $j = 4$, we do the following gradient update for $\boldsymbol{\theta}$:

$$\boldsymbol{\theta}^{(i+1)} = \boldsymbol{\theta}^{(i)} + \tau^{(i)} \frac{\partial u\left(\boldsymbol{\theta}, \kappa^{(i)}, \boldsymbol{\rho}^{(i)}, \boldsymbol{\gamma}^{(i)}; \boldsymbol{\theta}^{(i)}, \kappa^{(i)}, \boldsymbol{\rho}^{(i)}, \boldsymbol{\gamma}^{(i)}\right)}{\partial \boldsymbol{\theta}}\Big|_{\boldsymbol{\theta}=\boldsymbol{\theta}^{(i)}}, \tag{11.29}$$

where $\tau^{(i)}$ is the step size determined by the Armijo rule, which is a type of backtracking line search [10]. In particular, following a similar analysis as before, the update equations for κ, $\boldsymbol{\rho}$, and $\boldsymbol{\gamma}$ can be obtained as follows:

$$\begin{aligned}\kappa^{(i+1)} &= \frac{M + c - 1}{d + \eta\left(\boldsymbol{\xi}^{(i)}\right)},\\ \rho_n^{(i+1)} &= \frac{a + |\mathcal{I}_g| - 1}{b + \sum_{k\in\mathcal{I}_g} \gamma_k \left[\boldsymbol{\Xi}\left(\kappa^{(i+1)}, \boldsymbol{\rho}^{(i)}, \boldsymbol{\gamma}^{(i)}, \boldsymbol{\theta}^{(i)}\right)\right]_{kk}}, \quad \forall n \in \mathcal{I}_g,\\ \gamma_n^{(i+1)} &= \frac{1}{\left[\rho_n^{(i+1)}\boldsymbol{\Xi}\left(\kappa^{(i+1)}, \boldsymbol{\rho}^{(i+1)}, \boldsymbol{\gamma}^{(i)}, \boldsymbol{\theta}^{(i)}\right)\right]_{nn}}, \quad \forall n,\end{aligned} \tag{11.30}$$

where $\boldsymbol{\Xi}(\cdot)$ denotes the posterior second moment matrix as defined after (11.21), evaluated at the specified parameters.

11.4 Application: FDD Massive MIMO Channel Estimation with Angular Reciprocity

This example, a simplification of [9], addresses the critical challenge of high feedback overhead in FDD massive MIMO. By exploiting angular reciprocity, where the propagation paths are similar between uplink and downlink, we formulate the joint downlink-uplink channel estimation as a sparse signal recovery problem. The SBL framework is then applied to accurately estimate the channel with significantly reduced overhead.

11.4.1 Angular Reciprocity in FDD Systems

Consider an FDD massive MIMO system as depicted in Fig. 11.2. The BS employs a ULA with $N \gg 1$ antenna elements, while K mobile users (MUs) are each equipped with a single antenna. Under flat fading conditions, the downlink channel vector from the BS to the k-th user follows the geometric channel model:

$$\boldsymbol{h}_k = \sum_{c=1}^{N_c} \sum_{s=1}^{N_s} \xi_{c,s}^k \boldsymbol{a}(\theta_{c,s}^k), \tag{11.31}$$

where N_c denotes the number of scattering clusters, N_s represents the number of sub-paths within each cluster, $\xi_{c,s}^k$ is the downlink complex gain of the s-th sub-path in the c-th cluster for user k, and $\theta_{c,s}^k$ is the corresponding angle of departure (AoD).

The downlink steering vector $\boldsymbol{a}(\theta) \in \mathbb{C}^{N \times 1}$ is defined as

$$\boldsymbol{a}(\theta) = \left[1, \mathrm{e}^{-j2\pi \frac{d}{\lambda_d} \sin\theta}, \ldots, \mathrm{e}^{-j2\pi \frac{(N-1)d}{\lambda_d} \sin\theta}\right]^T, \tag{11.32}$$

where λ_d denotes the downlink wavelength and d represents the inter-antenna spacing.

Similarly, the uplink channel vector from user k to the BS is modeled as

$$\bar{\boldsymbol{h}}_k = \sum_{c=1}^{N_c} \sum_{s=1}^{N_s} \bar{\xi}_{c,s}^k \bar{\boldsymbol{a}}(\bar{\theta}_{c,s}^k), \tag{11.33}$$

where $\bar{\xi}_{c,s}^k$ is the uplink complex gain and $\bar{\theta}_{c,s}^k$ is the corresponding angle of arrival (AoA).

The uplink steering vector takes the form:

$$\bar{\boldsymbol{a}}(\theta) = \left[1, \mathrm{e}^{-j2\pi \frac{d}{\lambda_u} \sin\theta}, \ldots, \mathrm{e}^{-j2\pi \frac{(N-1)d}{\lambda_u} \sin\theta}\right]^T, \tag{11.34}$$

with λ_u representing the uplink wavelength.

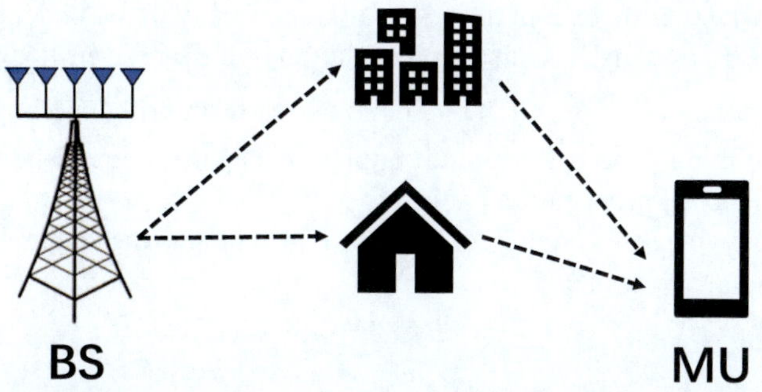

Fig. 11.2 An illustration of the massive MIMO system

Although complete channel reciprocity does not hold in FDD systems due to different operating frequencies, when uplink and downlink transmissions occur within a sufficiently short time interval, the underlying physical propagation paths remain essentially unchanged. This gives rise to the angular reciprocity principle, which establishes the key relationship:

$$\theta_{c,s}^{k} \approx \bar{\theta}_{c,s}^{k}, \quad \forall k, c, s. \tag{11.35}$$

This angular correspondence enables the transfer of uplink-derived angle information to downlink channel estimation. In the following subsection, we will develop a joint off-grid model to effectively exploit this angular reciprocity for enhanced channel estimation performance.

11.4.2 Joint Off-Grid Model for FDD Channel Estimation

11.4.2.1 Training Protocol for FDD Channel Estimation

Assume that the BS broadcasts a sequence of T training pilot symbols, denoted by $\boldsymbol{X} \in \mathbb{C}^{T\times N}$, for each MU to estimate the downlink channel. In massive MIMO systems with $N \gg 1$, we often have $T < N$ due to the limited pilot resource. Then, the downlink received signal $\boldsymbol{y}_k \in \mathbb{C}^{T\times 1}$ at the k-th MU is given by

$$\boldsymbol{y}_k = \boldsymbol{X}\boldsymbol{h}_k + \boldsymbol{n}_k, \tag{11.36}$$

where $\boldsymbol{n}_k \in \mathbb{C}^{T\times 1}$ is additive complex Gaussian noise with zero-mean and variance σ^2 per element in the downlink, and $\mathrm{tr}(\boldsymbol{X}\boldsymbol{X}^H) = PTN$ with P/σ^2 measuring the training SNR.

For uplink channel estimation, assume that each MU broadcasts a sequence of $\bar{T}$ training pilot symbols, denoted by $\boldsymbol{s}_k \in \mathbb{C}^{\bar{T}\times 1}$, $k = 1, 2, \ldots, K$. Then, the received signal $\bar{\boldsymbol{Y}} \in \mathbb{C}^{N\times \bar{T}}$ at the BS is given by

$$\bar{\boldsymbol{Y}} = \bar{\boldsymbol{H}}\boldsymbol{S} + \bar{\boldsymbol{N}},$$

where $\bar{\boldsymbol{H}} = [\bar{\boldsymbol{h}}_1, \bar{\boldsymbol{h}}_2, \ldots, \bar{\boldsymbol{h}}_K] \in \mathbb{C}^{N\times K}$ with $\bar{\boldsymbol{h}}_k$ being the channel vector associated with the k-th MU, $\boldsymbol{S} = [\boldsymbol{s}_1, \boldsymbol{s}_2, \ldots, \boldsymbol{s}_K]^T \in \mathbb{C}^{K\times \bar{T}}$, and $\bar{\boldsymbol{N}} \in \mathbb{C}^{N\times \bar{T}}$ is additive complex Gaussian noise with zero-mean and variance $\bar{\sigma}^2$ per element in the uplink. If the number of MUs is small, i.e., $K \le \bar{T}$, the uplink channel matrix $\boldsymbol{H}_u$ can be easily obtained by the conventional LS estimate, i.e.,

$$[\bar{\boldsymbol{h}}_1^{ls}, \bar{\boldsymbol{h}}_2^{ls}, \ldots, \bar{\boldsymbol{h}}_K^{ls}] \triangleq \bar{\boldsymbol{Y}}\boldsymbol{S}^{\dagger} = \bar{\boldsymbol{H}} + \boldsymbol{E}, \tag{11.37}$$

where $\bar{\boldsymbol{h}}_k^{ls}$ stands for the LS estimate of $\bar{\boldsymbol{h}}_k$, $\boldsymbol{S}^{\dagger} = \boldsymbol{S}^H(\boldsymbol{S}\boldsymbol{S}^H)^{-1}$ denotes the Moore-Penrose pseudoinverse of $\boldsymbol{S}$ and $\boldsymbol{E} \triangleq [\boldsymbol{e}_1, \boldsymbol{e}_2, \ldots, \boldsymbol{e}_K]$ stands for the estimation error. In particular, if $\boldsymbol{S}$ consists of an orthogonal pilot sequence, then $\boldsymbol{E}$ is i.i.d. Gaussian distributed. Besides, it is noteworthy that compared with the requirement $T \ge N$ for downlink channel estimation, it is much easier to meet the requirement $\bar{T} \ge K$ for uplink channel estimation.

11.4.2.2 Sparse Angular-Domain Channel Representation

For notational simplicity, we omit the user index k in the rest of this section, with the understanding that the same processing applies independently to each user. To establish a sparse representation of the massive MIMO channel, we consider a discrete angular grid spanning the spatial domain. Let $\hat{\boldsymbol{\vartheta}} = \left\{\hat{\vartheta}_l\right\}_{l=1}^{Q}$ denote a uniform grid covering the angular range $[-\pi/2, \pi/2]$, where Q represents the grid resolution. The true azimuth angles of departure are denoted by $\{\theta_l, l = 1, 2, \ldots, L\}$ with $L = N_c N_s$ indicating the total number of propagation paths.

In the idealized on-grid scenario where true angles coincide with grid points, the channel vector admits a sparse representation

$$\boldsymbol{h} = \boldsymbol{A}\boldsymbol{w}, \tag{11.38}$$

where the dictionary matrix $\boldsymbol{A} = \left[\boldsymbol{a}(\hat{\vartheta}_1), \boldsymbol{a}(\hat{\vartheta}_2), \ldots, \boldsymbol{a}(\hat{\vartheta}_Q)\right] \in \mathbb{C}^{N\times Q}$ contains array response vectors evaluated at grid points, and $\boldsymbol{w} \in \mathbb{C}^{Q\times 1}$ is a sparse vector whose non-zero entries identify active propagation paths.

Practical systems inevitably experience grid mismatches since true angles rarely align perfectly with predefined grid points. To address this limitation, we introduce an off-grid model that compensates for discretization errors, as illustrated in Fig. 11.3.

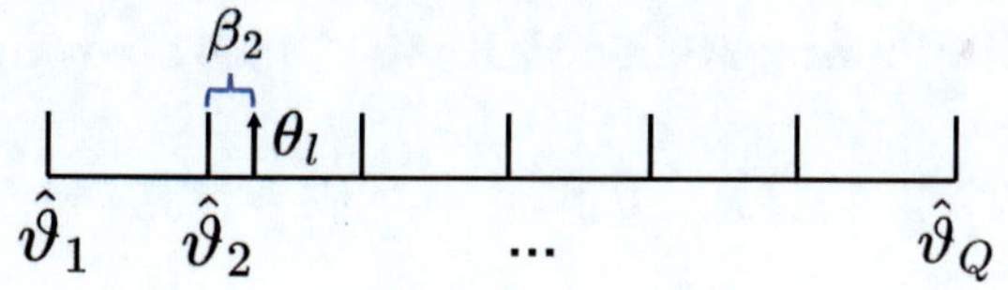

Fig. 11.3 An illustration of the off-grid model

For each true angle θ_l with nearest grid point $\hat{\vartheta}_{n_l}$, we define

$$\theta_l = \hat{\vartheta}_{n_l} + \beta_{n_l}, \tag{11.39}$$

where β_{n_l} represents the angular offset. This formulation leads to the off-grid dictionary matrix

$$\boldsymbol{A}(\boldsymbol{\beta}) = \left[\boldsymbol{a}\left(\hat{\vartheta}_1 + \beta_1\right), \ldots, \boldsymbol{a}\left(\hat{\vartheta}_Q + \beta_Q\right)\right], \tag{11.40}$$

with the offset vector $\boldsymbol{\beta} = \left[\beta_1, \beta_2, \ldots, \beta_Q\right]^T$ satisfying

$$\beta_{n_l} = \begin{cases} \theta_l - \hat{\vartheta}_{n_l}, & l = 1, 2, \ldots, L \\ 0, & \text{otherwise} \end{cases}. \tag{11.41}$$

The resulting channel model becomes

$$\boldsymbol{h} = \boldsymbol{A}(\boldsymbol{\beta})\,\boldsymbol{w}, \tag{11.42}$$

This off-grid formulation provides exact representation of arbitrary angles while maintaining the computational advantages of sparse recovery algorithms, effectively bridging the gap between discrete grid assumptions and continuous parameter spaces encountered in practical massive MIMO systems.

11.4.2.3 CS Observation Model with Angular Reciprocity

The received downlink signal in (11.36) can be reformulated using the off-grid sparse representation as

$$\boldsymbol{y} = \boldsymbol{X}\boldsymbol{A}(\boldsymbol{\beta})\,\boldsymbol{w} + \boldsymbol{n} = \boldsymbol{\Phi}(\boldsymbol{\beta})\,\boldsymbol{w} + \boldsymbol{n}, \tag{11.43}$$

where we define the effective sensing matrix $\boldsymbol{\Phi}(\boldsymbol{\beta}) \triangleq \boldsymbol{X}\boldsymbol{A}(\boldsymbol{\beta})$.

Building upon the uplink channel estimate from Eq. (11.37) and the downlink off-grid model (11.43), we obtain a pair of sparse measurement equations that can be processed independently for each mobile user

$$\begin{aligned} \boldsymbol{y} &= \boldsymbol{\Phi}(\boldsymbol{\beta})\boldsymbol{w} + \boldsymbol{n}, \\ \bar{\boldsymbol{h}}^{ls} &= \bar{\boldsymbol{\Phi}}(\boldsymbol{\beta})\bar{\boldsymbol{w}} + \boldsymbol{e}, \end{aligned} \tag{11.44}$$

where $\bar{\boldsymbol{\Phi}}(\boldsymbol{\beta}) = [\bar{\boldsymbol{a}}(\hat{\vartheta}_1 + \beta_1), \bar{\boldsymbol{a}}(\hat{\vartheta}_2 + \beta_2), \ldots, \bar{\boldsymbol{a}}(\hat{\vartheta}_Q + \beta_Q)]$ constitutes the uplink dictionary matrix, and $\bar{\boldsymbol{w}}$ represents the sparse representation of the uplink channel $\bar{\boldsymbol{h}}$.

A crucial observation emerges from the physical propagation characteristics: under the angular reciprocity assumption, the support sets of the sparse vectors satisfy $\text{supp}(\boldsymbol{w}) = \text{supp}(\bar{\boldsymbol{w}})$. This common support structure reflects the shared physical scattering environment between uplink and downlink transmissions, despite their operation at different carrier frequencies.

The following subsections develop a joint recovery framework based on SBL that systematically exploits this support equivalence through an inexact MM algorithm, enabling enhanced estimation accuracy by leveraging the complementary information from both uplink and downlink measurements.

11.4.3 Statistical Modeling with Angular Reciprocity for SBL Framework

Under the assumption of circular symmetric complex Gaussian noises, we have

$$p(\boldsymbol{y}|\boldsymbol{w}, \kappa, \boldsymbol{\beta}) = \mathcal{CN}(\boldsymbol{y}|\boldsymbol{\Phi}(\boldsymbol{\beta})\boldsymbol{w}, \kappa^{-1}\boldsymbol{I}),$$
$$p(\bar{\boldsymbol{h}}^{ls}|\bar{\boldsymbol{w}}, \bar{\kappa}, \boldsymbol{\beta}) = \mathcal{CN}(\bar{\boldsymbol{h}}^{ls}|\bar{\boldsymbol{\Phi}}(\boldsymbol{\beta})\bar{\boldsymbol{w}}, \bar{\kappa}^{-1}\boldsymbol{I}), \tag{11.45}$$

where κ and $\bar{\kappa}$ denote the noise precision for downlink and uplink, respectively. Both are modeled as Gamma hyperpriors

$$p(\kappa) = \Gamma(\kappa; c, d), \tag{11.46}$$
$$p(\bar{\kappa}) = \Gamma(\bar{\kappa}; \bar{c}, \bar{d}), \tag{11.47}$$

with distinct hyperparameters (c, d) for downlink and $(\bar{c}, \bar{d})$ for uplink. Furthermore, $\boldsymbol{\rho}$ is used to control the sparsity of $\boldsymbol{w}$ as follows

$$p(\boldsymbol{w}|\boldsymbol{\rho}) = \mathcal{CN}(\boldsymbol{w}; \boldsymbol{0}, \text{diag}(\boldsymbol{\rho}^{-1})). \tag{11.48}$$

If we let $\boldsymbol{\gamma} = [\gamma_1, \gamma_2, \ldots, \gamma_Q]^T$ be a non-negative vector and let

$$p(\bar{\boldsymbol{w}}|\boldsymbol{\rho}) = \mathcal{CN}(\bar{\boldsymbol{w}}; \boldsymbol{0}, \text{diag}((\boldsymbol{\gamma} \odot \boldsymbol{\rho})^{-1})), \tag{11.49}$$

then $\boldsymbol{w}$ and $\bar{\boldsymbol{w}}$ will share a joint sparse structure. For example, if ρ_l^{-1} tends to zero, so is $(\rho_l\gamma_l)^{-1}$. Therefore, (11.48) and (11.49) provide a mathematical representation of the angular reciprocity. Note that the estimated uplink channel $\bar{\boldsymbol{h}}^{ls}$ contains both angular and small-scale fading information. While the azimuth AoA information can be exploited to identify the downlink azimuth AoD (i.e., the downlink channel support) via the angular reciprocity in (11.48) and (11.49), the small-scale fading information is not reusable due to the lack of channel reciprocity in FDD systems.

11.4.4 SBL Algorithm for FDD Channel Estimation

11.4.4.1 Notation and Surrogate Function Construction

For notational compactness, we define the aggregated variables:

- $\boldsymbol{w}^a \triangleq \{\boldsymbol{w}, \bar{\boldsymbol{w}}\}$ (sparse vectors)
- $\boldsymbol{y}^a \triangleq \left\{\boldsymbol{y}, \bar{\boldsymbol{h}}^{ls}\right\}$ (measurements)
- $\Theta \triangleq \{\kappa, \bar{\kappa}, \boldsymbol{\rho}, \boldsymbol{\gamma}, \boldsymbol{\beta}\}$ (parameters).

Assuming a non-informative uniform prior for $\boldsymbol{\gamma}$, we construct the surrogate function at fixed point $\dot{\Theta} = \{\dot{\kappa}, \dot{\bar{\kappa}}, \dot{\boldsymbol{\rho}}, \dot{\boldsymbol{\gamma}}, \dot{\boldsymbol{\beta}}\}$ as

$$u(\Theta|\dot{\Theta}) = \int p(\boldsymbol{w}^a|\boldsymbol{y}^a, \dot{\Theta}) \ln \frac{p(\boldsymbol{w}^a, \boldsymbol{y}^a, \Theta)}{p(\boldsymbol{w}^a|\boldsymbol{y}^a, \dot{\Theta})} \mathrm{d}\boldsymbol{w}^a, \tag{11.50}$$

where the complete joint distribution factorizes as

$$p(\boldsymbol{w}^a, \boldsymbol{y}^a, \Theta) = p(\boldsymbol{y}|\boldsymbol{w}, \kappa, \boldsymbol{\beta})p(\boldsymbol{w}|\boldsymbol{\rho})p(\bar{\boldsymbol{h}}^{ls}|\bar{\boldsymbol{w}}, \bar{\kappa}, \boldsymbol{\beta})$$
$$\times\, p(\bar{\boldsymbol{w}}|\boldsymbol{\rho}, \boldsymbol{\gamma})p(\kappa)p(\bar{\kappa})p(\boldsymbol{\rho})p(\boldsymbol{\gamma})p(\boldsymbol{\beta}), \tag{11.51}$$

and the posterior distribution admits the factorized Gaussian form

$$p(\boldsymbol{w}^a|\boldsymbol{y}^a, \Theta) = \mathcal{CN}(\boldsymbol{w}; \boldsymbol{\mu}(\kappa, \boldsymbol{\rho}, \boldsymbol{\beta}), \boldsymbol{\Sigma}(\kappa, \boldsymbol{\rho}, \boldsymbol{\beta})),$$
$$\times\, \mathcal{CN}(\bar{\boldsymbol{w}}; \bar{\boldsymbol{\mu}}(\bar{\kappa}, \boldsymbol{\rho}, \boldsymbol{\gamma}, \boldsymbol{\beta}), \bar{\boldsymbol{\Sigma}}(\bar{\kappa}, \boldsymbol{\rho}, \boldsymbol{\gamma}, \boldsymbol{\beta})), \tag{11.52}$$

with the posterior moments given by

$$\begin{aligned}\boldsymbol{\mu}(\kappa,\boldsymbol{\rho},\boldsymbol{\beta}) &= \kappa\boldsymbol{\Sigma}(\kappa,\boldsymbol{\rho},\boldsymbol{\beta})\boldsymbol{\Phi}^H(\boldsymbol{\beta})\mathbf{y},\\ \boldsymbol{\Sigma}(\kappa,\boldsymbol{\rho},\boldsymbol{\beta}) &= \left(\kappa\boldsymbol{\Phi}^H(\boldsymbol{\beta})\boldsymbol{\Phi}(\boldsymbol{\beta}) + \mathrm{diag}(\boldsymbol{\rho})\right)^{-1}.\end{aligned} \tag{11.53}$$

$$\begin{aligned}\bar{\boldsymbol{\mu}}(\bar{\kappa},\boldsymbol{\rho},\boldsymbol{\gamma},\boldsymbol{\beta}) &= \bar{\kappa}\bar{\boldsymbol{\Sigma}}(\bar{\kappa},\boldsymbol{\rho},\boldsymbol{\gamma},\boldsymbol{\beta})\bar{\boldsymbol{\Phi}}^H(\boldsymbol{\beta})\bar{\mathbf{h}}^{ls},\\ \bar{\boldsymbol{\Sigma}}(\bar{\kappa},\boldsymbol{\rho},\boldsymbol{\gamma},\boldsymbol{\beta}) &= \left(\bar{\kappa}\bar{\boldsymbol{\Phi}}^H(\boldsymbol{\beta})\bar{\boldsymbol{\Phi}}(\boldsymbol{\beta}) + \mathrm{diag}(\boldsymbol{\rho}\odot\boldsymbol{\gamma})\right)^{-1}.\end{aligned} \tag{11.54}$$

11.4.4.2 Iterative Parameter Updates

The maximization step at iteration $(i+1)$ updates parameters sequentially as

$$\begin{aligned}\kappa^{(i+1)} &= \arg\max_{\kappa} u(\kappa,\bar{\kappa}^{(i)},\boldsymbol{\rho}^{(i)},\boldsymbol{\gamma}^{(i)},\boldsymbol{\beta}^{(i)}|\Theta^{(i)}),\\ \bar{\kappa}^{(i+1)} &= \arg\max_{\bar{\kappa}} u(\kappa^{(i+1)},\bar{\kappa},\boldsymbol{\rho}^{(i)},\boldsymbol{\gamma}^{(i)},\boldsymbol{\beta}^{(i)}|\Theta_1^{(i)}),\\ \boldsymbol{\rho}^{(i+1)} &= \arg\max_{\boldsymbol{\rho}} u(\kappa^{(i+1)},\bar{\kappa}^{(i+1)},\boldsymbol{\rho},\boldsymbol{\gamma}^{(i)},\boldsymbol{\beta}^{(i)}|\Theta_2^{(i)}),\\ \boldsymbol{\gamma}^{(i+1)} &= \arg\max_{\boldsymbol{\gamma}} u(\kappa^{(i+1)},\bar{\kappa}^{(i+1)},\boldsymbol{\rho}^{(i+1)},\boldsymbol{\gamma},\boldsymbol{\beta}^{(i)}|\Theta_3^{(i)}),\\ \boldsymbol{\beta}^{(i+1)} &= \arg\max_{\boldsymbol{\beta}} u(\kappa^{(i+1)},\bar{\kappa}^{(i+1)},\boldsymbol{\rho}^{(i+1)},\boldsymbol{\gamma}^{(i+1)},\boldsymbol{\beta}|\Theta_4^{(i)}),\end{aligned} \tag{11.55}$$

where $\Theta_j^{(i)}$ denotes the parameter set with the first j elements from iteration $(i+1)$ and the remainder from iteration i.

The closed-form updates for noise and sparsity parameters are

$$\begin{aligned}\kappa^{(i+1)} &= \frac{T+c}{d+\eta_d(\Theta^{(i)})},\\ \bar{\kappa}^{(i+1)} &= \frac{N+\bar{c}}{\bar{d}+\eta_u(\Theta_1^{(i)})},\\ \rho_l^{(i+1)} &= \frac{a+1}{b+\left[\Xi_d(\Theta_2^{(i)})+\gamma_l\Xi_u(\Theta_2^{(i)})\right]_{ll}}, \quad \forall l,\end{aligned} \tag{11.56}$$

with the auxiliary quantities

$$\begin{aligned}\eta_d(\Theta) &= \mathrm{tr}\left(\boldsymbol{\Phi}(\boldsymbol{\beta})\boldsymbol{\Sigma}(\kappa,\boldsymbol{\rho},\boldsymbol{\beta})\boldsymbol{\Phi}^H(\boldsymbol{\beta})\right) + \|\mathbf{y}-\boldsymbol{\Phi}(\boldsymbol{\beta})\boldsymbol{\mu}(\kappa,\boldsymbol{\rho},\boldsymbol{\beta})\|_2^2,\\ \eta_u(\Theta) &= \mathrm{tr}\left(\bar{\boldsymbol{\Phi}}(\boldsymbol{\beta})\bar{\boldsymbol{\Sigma}}(\bar{\kappa},\boldsymbol{\rho},\boldsymbol{\gamma},\boldsymbol{\beta})\bar{\boldsymbol{\Phi}}^H(\boldsymbol{\beta})\right) + \left\|\bar{\mathbf{h}}^{ls}-\bar{\boldsymbol{\Phi}}(\boldsymbol{\beta})\bar{\boldsymbol{\mu}}(\bar{\kappa},\boldsymbol{\rho},\boldsymbol{\gamma},\boldsymbol{\beta})\right\|_2^2,\\ \Xi_d(\Theta) &= \boldsymbol{\Sigma}(\kappa,\boldsymbol{\rho},\boldsymbol{\beta}) + \boldsymbol{\mu}(\kappa,\boldsymbol{\rho},\boldsymbol{\beta})\boldsymbol{\mu}^H(\kappa,\boldsymbol{\rho},\boldsymbol{\beta}),\\ \Xi_u(\Theta) &= \bar{\boldsymbol{\Sigma}}(\bar{\kappa},\boldsymbol{\rho},\boldsymbol{\gamma},\boldsymbol{\beta}) + \bar{\boldsymbol{\mu}}(\bar{\kappa},\boldsymbol{\rho},\boldsymbol{\gamma},\boldsymbol{\beta})\bar{\boldsymbol{\mu}}^H(\bar{\kappa},\boldsymbol{\rho},\boldsymbol{\gamma},\boldsymbol{\beta}).\end{aligned} \tag{11.57}$$

11.4.4.3 Update for Grid Parameters

Finding the exact global maximizer for $\boldsymbol{\beta}$ is computationally intractable, as the objective function is non-convex and does not admit a closed-form solution. Nevertheless, thanks to the convergence properties established in Theorem 11.1, it suffices to find an update direction that guarantees a monotonic increase in the objective value at each iteration. A natural and efficient choice is to perform a gradient ascent step. To this end, we compute the gradient of the surrogate function with respect to $\boldsymbol{\beta}$, which receives contributions from both the downlink and uplink observations. The gradient vector is given by

$$\boldsymbol{\zeta}_{\boldsymbol{\beta}}^{(i)} = [\zeta^{(i)}(\beta_1), \zeta^{(i)}(\beta_2), \ldots, \zeta^{(i)}(\beta_Q)]^T, \tag{11.58}$$

where each component $\zeta^{(i)}(\beta_l)$ can be expressed as the sum of downlink and uplink contributions

$$\zeta^{(i)}(\beta_l) = \zeta_d^{(i)}(\beta_l) + \zeta_u^{(i)}(\beta_l). \tag{11.59}$$

The downlink contribution $\zeta_d^{(i)}(\beta_l)$ is given by

$$\zeta_d^{(i)}(\beta_l) = 2\mathrm{Re}\left((\boldsymbol{a}'(\hat{\vartheta}_l + \beta_l))^H \boldsymbol{X}^H \boldsymbol{X}\boldsymbol{a}(\hat{\vartheta}_l + \beta_l)\right) \cdot c_{d1}^{(i)} + 2\mathrm{Re}\left((\boldsymbol{a}'(\hat{\vartheta}_l + \beta_l))^H \boldsymbol{X}^H \boldsymbol{c}_{d2}^{(i)}\right), \tag{11.60}$$

with coefficients

$$c_{d1}^{(i)} = -\kappa^{(i+1)}\left(\chi_{d,ll}^{(i)} + |\mu_{d,l}^{(i)}|^2\right), \tag{11.61}$$

$$\boldsymbol{c}_{d2}^{(i)} = \kappa^{(i+1)}\left((\mu_{d,l}^{(i)})^* \boldsymbol{y}_{-l}^{(i)} - \sum_{j\neq l} \chi_{d,jl}^{(i)} \boldsymbol{X}\boldsymbol{a}(\hat{\vartheta}_j + \beta_j)\right), \tag{11.62}$$

$$\boldsymbol{y}_{-l}^{(i)} = \boldsymbol{y} - \sum_{j\neq l} \mu_{d,j}^{(i)} \boldsymbol{X}\boldsymbol{a}(\hat{\vartheta}_j + \beta_j). \tag{11.63}$$

Similarly, the uplink contribution $\zeta_u^{(i)}(\beta_l)$ is given by

$$\zeta_u^{(i)}(\beta_l) = 2\mathrm{Re}\left((\bar{\boldsymbol{a}}'(\hat{\vartheta}_l + \beta_l))^H \bar{\boldsymbol{a}}(\hat{\vartheta}_l + \beta_l)\right) \cdot c_{u1}^{(i)} + 2\mathrm{Re}\left((\bar{\boldsymbol{a}}'(\hat{\vartheta}_l + \beta_l))^H \boldsymbol{c}_{u2}^{(i)}\right), \tag{11.64}$$

with coefficients

$$c_{u1}^{(i)} = -\bar{\kappa}^{(i+1)}\left(\chi_{u,ll}^{(i)} + |\mu_{u,l}^{(i)}|^2\right), \tag{11.65}$$

$$\boldsymbol{c}_{u2}^{(i)} = \bar{\kappa}^{(i+1)}\left((\mu_{u,l}^{(i)})^* \bar{\boldsymbol{y}}_{-l}^{(i)} - \sum_{j\neq l} \chi_{u,jl}^{(i)} \bar{\boldsymbol{a}}(\hat{\vartheta}_j + \beta_j)\right), \tag{11.66}$$

$$\bar{\boldsymbol{y}}_{-l}^{(i)} = \bar{\boldsymbol{h}}^{ls} - \sum_{j\neq l} \mu_{u,j}^{(i)} \bar{\boldsymbol{a}}(\hat{\vartheta}_j + \beta_j). \tag{11.67}$$

Here, $\boldsymbol{a}'(\hat{\vartheta}_j + \beta_l) = \mathrm{d}\boldsymbol{a}(\hat{\vartheta}_j + \beta_l)/\mathrm{d}\beta_l$ and $\bar{\boldsymbol{a}}'(\hat{\vartheta}_j + \beta_l) = \mathrm{d}\bar{\boldsymbol{a}}(\hat{\vartheta}_j + \beta_l)/\mathrm{d}\beta_l$ denote the derivatives of the downlink and uplink steering vectors, respectively. The quantities $\mu_{d,l}^{(i)}$, $\mu_{u,l}^{(i)}$ are elements of the downlink and uplink channel posterior means, while $\chi_{d,jl}^{(i)}$, $\chi_{u,jl}^{(i)}$ denote the corresponding elements of the posterior covariance matrices. The noise precisions $\kappa^{(i+1)}$ and $\bar{\kappa}^{(i+1)}$ are the current estimates for downlink and uplink, respectively.

Then the grid parameter $\boldsymbol{\beta}$ is updated via gradient ascent as

$$\boldsymbol{\beta}^{(i+1)} = \boldsymbol{\beta}^{(i)} + \Delta_\beta \cdot \boldsymbol{\zeta}_\beta^{(i)}, \tag{11.68}$$

where Δ_β represents the step size.

A backtracking line search strategy can be employed to determine the maximum step size Δ_β that ensures strict objective function improvement before reaching a stationary point. However, this approach incurs significant computational overhead, as each line search requires multiple evaluations of the objective function. Specifically, if R_b represents the number of function evaluations per backtracking iteration, the computational complexity scales as $O(R_b T Q^2)$ per main algorithm iteration.

To reduce computational complexity while preserving estimation accuracy, we adopt a simplified fixed step size update

$$\boldsymbol{\beta}^{(i+1)} = \boldsymbol{\beta}^{(i)} + \frac{r_\beta}{100} \cdot \mathrm{sign}\left(\boldsymbol{\zeta}_\beta^{(i)}\right), \tag{11.69}$$

where $r_\beta = \pi/Q$ is the grid interval and $\mathrm{sign}(\cdot)$ denotes the element-wise signum function. The step size is set to a small fraction (1%) of the grid resolution. This choice is motivated by practical considerations: once the angular estimates are within this small neighborhood of the true values, further refinement yields diminishing returns in channel estimation accuracy. Moreover, this step size guarantees that under worst-case initial offsets, the estimates converge to within 1% of the true grid spacing within at most 100 iterations. Compared to backtracking line search, this fixed-step approach significantly reduces per-iteration complexity while still achieving satisfactory performance in practice.

11.4.5 Simulation Results and Performance Analysis

This subsection presents a comprehensive performance evaluation of the proposed uplink-aided off-grid SBL method through comparative simulations against five established baseline approaches:

- **Baseline 1 (Off-grid)**: Standalone off-grid SBL without uplink information.
- **Baseline 2 (SBL)**: Standard SBL without off-grid adjustment or uplink assistance [4].
- **Baseline 3 (DFT)**: l_1-norm minimization with DFT basis [12].
- **Baseline 4 (Overcomplete DFT)**: l_1-norm minimization with overcomplete DFT basis [12].
- **Baseline 5 (Dictionary Learning)**: l_1-norm minimization with learned dictionary [13].

The simulation environment employs a ULA configuration with channel realizations generated according to the 3GPP Spatial Channel Model (SCM) for urban microcell scenarios [14]. System parameters include uplink frequency of 1980 MHz, downlink frequency of 2170 MHz, and inter-antenna spacing $d = c/(2f_0)$ where c denotes the speed of light and $f_0 = 2000$ MHz. All 3D channel parameters adhere to the 3D-UMa-NLoS specification [15].

Performance is quantified using NMSE defined as

$$\text{NMSE} = \frac{1}{M_c} \sum_{m=1}^{M_c} \frac{\|\boldsymbol{h}_m^e - \boldsymbol{h}_m\|_2^2}{\|\boldsymbol{h}_m\|_2^2}, \tag{11.70}$$

where $\boldsymbol{h}_m^e$ represents the channel estimate in the m-th Monte Carlo trial, with $M_c = 200$ total trials.

Figure 11.4 examines the relationship between pilot symbol quantity and downlink channel estimation performance for a system with 150 BS antennas serving 10 mobile users. Each channel realization incorporates $N_c = 3$ randomly distributed scattering clusters within $[-40°, 40°]$, with each cluster containing $N_s = 10$ sub-paths concentrated within a 20° angular spread. Random training pilots are employed with SNR levels of 0 and 10 dB, while maintaining 200 grid points for all methods except the standard DFT approach.

Key observations reveal

1. All methods exhibit decreasing NMSE with increasing pilot symbols (T), with the standard DFT method demonstrating the poorest overall performance.
2. While overcomplete DFT and dictionary learning approaches provide modest improvements over standard DFT, their performance remains inferior to standard SBL.
3. Both off-grid methods consistently outperform alternative approaches, with the proposed uplink-aided off-grid SBL achieving superior performance by effectively leveraging angular reciprocity information from uplink measurements.

These results underscore the critical importance of both off-grid modeling and uplink-downlink information transfer for achieving high-precision channel estimation in FDD massive MIMO systems.

Summary
SBL has emerged as a powerful methodology for sparse signal recovery, distinguished by its robust performance in CS applications and inherent resilience to model parameter uncertainties. This chapter has established a comprehensive derivation of the SBL framework through the EM principle, which iterates between two fundamental steps:

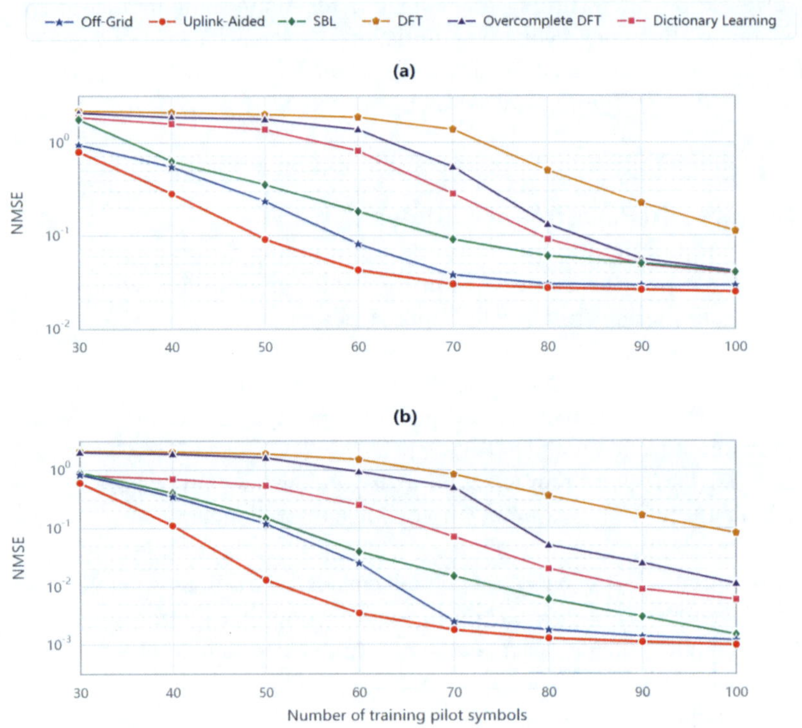

Fig. 11.4 (Adapted from Fig. 6 in [9]) NMSE of downlink channel estimate versus the number of training pilot symbols T. **a** SNR = 0 dB; **b** SNR = 10 dB

- **SBL-E Step**: Computes the posterior distribution of the sparse signal conditioned on current estimates of model parameters and grid configurations, yielding a tractable Gaussian distribution.
- **SBL-M Step**: Updates the uncertain parameters through maximization of the expected complete-data log-likelihood.

A significant theoretical contribution presented herein is the convergence guarantee: despite employing approximate updates in practical implementations, the SBL algorithm provably converges to a stationary point of the MAP estimation problem.

The chapter further extends the basic SBL framework to incorporate structured sparsity patterns through a group-sparse variant that effectively captures dependencies among coefficient groups. This extension enhances the method's applicability to problems with inherent structured sparsity.

Finally, we demonstrate the practical efficacy of SBL through a sophisticated application to FDD massive MIMO channel estimation. By exploiting angular reciprocity between uplink and downlink transmissions, the proposed approach achieves significant performance gains over conventional methods, validating SBL's effectiveness in real-world wireless communication scenarios with practical constraints.

Homework Problems

11.1 The SBL hierarchical prior is defined as

$$p(x_n|\rho_n) = \mathcal{CN}(x_n; 0, \rho_n^{-1}), \qquad p(\rho_n) = \Gamma(\rho_n; a_n, b_n).$$

The marginal sparse prior for x_n is obtained by integrating out the precision ρ_n as

$$p(x_n) = \int_0^\infty \mathcal{CN}(x_n; 0, \rho_n^{-1}) \cdot \Gamma(\rho_n; a_n, b_n)\, \mathrm{d}\rho_n.$$

It can be shown that this integral yields

$$p(x_n) \propto \left(b_n + |x_n|^2\right)^{-(a_n+1)}.$$

Write down the detailed derivation for the marginal prior $p(x_n)$ as shown in Eq. (11.5) of the textbook. Plot the resulting marginal prior $p(x_n)$ for $a_n = 1$ and $b_n = 0.01$. Discuss its shape: Does it have a sharp peak at zero and heavy tails? Explain how this particular shape encourages sparsity in the estimated signal $\boldsymbol{x}$.

11.2 In the SBL-E step, we compute the posterior distribution $p(\boldsymbol{x}|\boldsymbol{y}, \boldsymbol{\xi})$, which is a complex Gaussian. Its mean $\boldsymbol{\mu}$ and covariance matrix $\boldsymbol{\Sigma}$ are given by

$$\boldsymbol{\mu} = \kappa \boldsymbol{\Sigma} \boldsymbol{A}^H(\boldsymbol{\theta})\boldsymbol{y}, \qquad \boldsymbol{\Sigma} = \left(\kappa \boldsymbol{A}^H(\boldsymbol{\theta})\boldsymbol{A}(\boldsymbol{\theta}) + \mathrm{diag}(\boldsymbol{\rho})\right)^{-1}.$$

Starting from Bayes' theorem and the prior/likelihood definitions, derive the closed-form expressions for the posterior mean $\boldsymbol{\mu}$ and covariance $\boldsymbol{\Sigma}$ (Eq. (11.7)).

11.3 Derive the update rule for the precisions κ, $\boldsymbol{\rho}$, and $\boldsymbol{\gamma}$ in (11.30).

11.4 In this problem, you will experiment with different step-size strategies for the off-grid parameter update in the provided MATLAB code (GitHub link: https://github.com/biwcs-book/codebase) for the FDD massive MIMO channel estimation example.

(a) Implement three different step-size rules Δ_β for the gradient update of $\boldsymbol{\beta}$:

- **Strategy A (Fixed)**: A constant step size as used in the original code.
- **Strategy B (Backtracking Line Search)**: Implement the backtracking line search algorithm to adaptively choose the step size, ensuring a sufficient increase in the objective function.
- **Strategy C (Decreasing)**: Use a decreasing step size rule: $\Delta_\beta^{(i)} = \frac{c_1}{(c_2+i)^\kappa}$. You need to choose suitable hyperparameters $c_1, c_2 > 0$ and $0.5 < \kappa \leq 1$.

(b) Compare the performance of the three strategies. Create plots showing the convergence of the negative log-posterior (or the surrogate function) versus iteration number. Also, compare the final NMSE of the channel estimate.

(c) Write a brief conclusion summarizing your findings. Which strategy converges fastest? Which one gives the most stable or accurate final result? Discuss the trade-offs between computational cost and performance for each method.

References

1. Z. Yang, L. Xie, C. Zhang, Off-grid direction of arrival estimation using sparse bayesian inference. IEEE Trans. Signal Process. **61**(1), 38–43 (2013)
2. F. Liu, Y. Cui, C. Masouros, J. Xu, T.X. Han, Y.C. Eldar, S. Buzzi, Integrated sensing and communications: toward dual-functional wireless networks for 6g and beyond. IEEE J. Sel. Areas Commun. **40**(6), 1728–1767 (2022)
3. C. De Lima, D. Belot, R. Berkvens, A. Bourdoux, D. Dardari, M. Guillaud, M. Isomursu, E.-S. Lohan, Y. Miao, A.N. Barreto, M.R.K. Aziz, J. Saloranta, T. Sanguanpuak, H. Sarieddeen, G. Seco-Granados, J. Suutala, T. Svensson, M. Valkama, B. Van Liempd, H. Wymeersch, Convergent communication, sensing and localization in 6g systems: an overview of technologies, opportunities and challenges. IEEE Access **9**, 26902–26925 (2021)
4. M.E. Tipping, Sparse bayesian learning and the relevance vector machine. J. Mach. Learn. Res. **1**, 211–244 (2001)
5. S. Ji, Y. Xue, L. Carin, Bayesian compressive sensing. IEEE Trans. Signal Process. **56**(6), 2346–2356 (2008)
6. B.D. Rao, K. Kreutz-Delgado, An affine scaling methodology for best basis selection. IEEE Trans. Signal Process. **47**(1), 187–200 (1999)
7. H. Duan, L. Yang, J. Fang, H. Li, Fast inverse-free sparse bayesian learning via relaxed evidence lower bound maximization. IEEE Signal Process. Lett. **24**(6), 774–778 (2017)
8. A. Liu, L. Lian, V. Lau, G. Liu, M.-J. Zhao, Cloud-assisted cooperative localization for vehicle platoons: a turbo approach. IEEE Trans. Signal Process. **68**, 605–620 (2020)

9. J. Dai, A. Liu, V.K. Lau, FDD massive MIMO channel estimation with arbitrary 2D-array geometry. IEEE Trans. Signal Process. **66**(10), 2584–2599 (2018)
10. D.P. Bertsekas, *Nonlinear Programming*, 2nd edn. (Athena Scientific, Belmont, MA, 1999)
11. M. Razaviyayn (2014) Successive convex approximation: analysis and applications. Ph.D. dissertation, Faculty Graduate School, University of Minnesota, Minneapolis, MN, USA
12. D.L. Donoho, Y. Tsaig, Fast solution of ℓ_1-norm minimization problems when the solution may be sparse. IEEE Trans. Inf. Theory **54**(11), 4789–4812 (2008). (Nov.)
13. Y. Ding, B.D. Rao, Dictionary learning-based sparse channel representation and estimation for FDD massive MIMO systems. IEEE Trans. Wireless Commun. **17**(8), 5437–5451 (2018)
14. 3GPP, Universal mobile telecommunications system (UMTS); spatial channel model for multiple input multiple output (MIMO) simulations, in *3GPP TR 25.996 Version 11.0.0 Release 11* (2012)
15. 3GPP, 3rd generation partnership project; technical specification group radio access network; study on 3D channel model for LTE, in *3GPP TR 36.873 Version 12.2.0 Release 12* (2015)

Variational Bayesian Inference 12

This chapter revisits the central theme of Bayesian inference, which treats unknown parameters as random variables and computes their posterior distributions to obtain both point estimators (e.g., MMSE or MAP estimators) and uncertainty measures. We present Variational Bayesian Inference (VBI) as a powerful computational framework that transforms the intractable problem of exact posterior inference into a tractable optimization task: minimizing the Kullback-Leibler divergence (KLD) between an approximating variational distribution and the true posterior. We systematically develop the mathematical foundation through the Evidence Lower Bound (ELBO) and mean-field factorization, deriving the general Mean-Field VBI (MFVBI) algorithm for conjugate models. To overcome limitations in non-conjugate and nonlinear settings, we introduce two advanced variants: Particle VBI (PVBI), which employs non-parametric particle representations to capture complex multi-modal posteriors, and Successive Linear Approximation VBI (SLAVBI), which enables efficient closed-form updates through iterative linearization of nonlinear models. We further specialize the framework to compressive sensing problems, deriving a Sparse VBI method with hierarchical sparsity-promoting priors and clarifying its relationship to Sparse Bayesian Learning. The practical effectiveness of these VBI approaches is demonstrated through detailed wireless localization applications, showcasing their capability to handle challenging high-dimensional estimation tasks while providing a unifying Bayesian perspective that encompasses many classical algorithms as special cases.

12.1 Motivation of VBI

Future multi-functional wireless networks envisioned by next-generation communication standards (e.g., 6G, Wi-Fi 7, and Wi-Fi 8) are expected to provide not only enhanced communications but also various high-accuracy sensing services, such as indoor positioning, activity recognition, radar sensing, and integrated sensing and communications (ISAC) [1]. These applications require accurate estimation of multiple target parameters (e.g., distance, angle, and speed) in complex wireless environments, which often leads to multi-dimensional non-convex estimation problems.

Conventional parameter estimation methods such as ML and MAP estimation are prone to converge to poor local optima in such challenging settings. Bayesian inference provides a powerful alternative by incorporating prior information and modeling parameter uncertainty, thereby offering robustness against local optima and returning the full posterior distribution rather than a single point estimate. However, a long-standing challenge in Bayesian methods is the intractability of the posterior distribution, as the required multi-dimensional integrals are rarely available in closed form for practical problems, necessitating the use of approximate inference techniques.

Approaches for posterior approximation fall mainly into two categories: Monte Carlo methods [2, 5, 6] and VBI [3, 4]. Monte Carlo techniques (e.g., MCMC [5], SMC [6], and importance sampling [2, 7]) approximate the posterior using random samples. While highly flexible and capable of handling non-standard distributions, such methods are generally computationally expensive and slow to converge, particularly in high-dimensional spaces [8].

In contrast, VBI is a deterministic alternative that approximates the posterior by optimizing a tractable variational distribution to minimize the KLD from the true posterior. This formulation often leads to efficient, scalable algorithms suitable for large-scale data and complex models [8], making VBI a widely adopted tool for Bayesian inference in areas such as channel estimation, signal detection, and deep learning.

This chapter develops the VBI methodology from its general formulation to specialized algorithms and concrete applications. Section 12.2 establishes the mathematical foundation, defining the KLD and the crucial mean-field factorization assumption that makes optimization tractable. Section 12.3 derives the general MFVBI framework, yielding a coordinate descent

A. Liu and M.-M. Zhao, *Advanced Bayesian Signal Processing*, Textbooks in Telecommunication Engineering,
https://doi.org/10.1007/978-3-032-26656-9_12

algorithm for variational updates. We also discuss the limitations of the basic framework when models are non-conjugate, setting the stage for more advanced methods. Section 12.4 introduces PVBI, which uses a discrete set of particles to approximate complex, multi-modal posteriors, offering flexibility at a higher computational cost. Section 12.5 presents SLAVBI, an efficient algorithm that iteratively linearizes nonlinear models to enable closed-form Gaussian variational updates. Section 12.6 specializes the framework to CS, deriving a Sparse VBI algorithm. We explicitly compare it to the SBL method from Chap. 11, highlighting how Sparse VBI performs full Bayesian inference over all parameters (including noise variance), often leading to more robust performance. The effectiveness of these VBI algorithms is validated through wireless localization applications in Sect. 12.7.

12.2 Mathematical Foundations: The KLD and Mean-Field Approximation

12.2.1 The Optimization Goal: Minimizing the Reverse KLD

We begin with a general observation model for non-convex parameter estimation

$$y_m = f_m(\boldsymbol{\theta}) + w_m, \quad \forall m = 1, 2, \ldots, M, \tag{12.1}$$

where y_m is the m-th observation, $w_m \sim \mathcal{CN}(0, \sigma_w^2)$ is AWGN, and $\boldsymbol{\theta} = [\theta_1, \ldots, \theta_N]^T$ represents the unknown parameters to be estimated. The function $f_m(\boldsymbol{\theta})$ describes how the parameters generate the m-th measurement. Stacking all observations gives $\mathbf{y} = [y_1, \ldots, y_M]^T$, and the corresponding likelihood is

$$p(\mathbf{y}|\boldsymbol{\theta}) = \prod_{m=1}^{M} p(y_m|\boldsymbol{\theta}) = \prod_{m=1}^{M} \frac{1}{\pi\sigma_w^2} \exp\left(\frac{-|y_m - f_m(\boldsymbol{\theta})|^2}{\sigma_w^2}\right). \tag{12.2}$$

When $f_m(\boldsymbol{\theta})$ is a nonlinear function of $\boldsymbol{\theta}$, this likelihood function typically becomes non-convex, meaning it has multiple peaks and valleys. This makes finding the global optimum challenging for conventional estimators.

In Bayesian inference, our goal is to compute the posterior distribution $p(\boldsymbol{\theta}|\mathbf{y})$, which combines the likelihood with our prior knowledge $p(\boldsymbol{\theta})$ using Bayes' theorem: $p(\boldsymbol{\theta}|\mathbf{y}) \propto p(\mathbf{y}|\boldsymbol{\theta})p(\boldsymbol{\theta})$. VBI formulates posterior approximation as an optimization problem. Given the true posterior $p(\boldsymbol{\theta}|\mathbf{y})$ and a family of tractable variational distributions $\mathcal{Q}$, we seek the distribution $q(\boldsymbol{\theta}) \in \mathcal{Q}$ that is most similar to $p(\boldsymbol{\theta}|\mathbf{y})$. The standard measure of similarity (or rather, dissimilarity) is the KLD.

For two probability distributions $q(\boldsymbol{\theta})$ and $p(\boldsymbol{\theta})$, the KLD from q to p is defined as

$$D_{\text{KL}}(q||p) = \int q(\boldsymbol{\theta}) \ln \frac{q(\boldsymbol{\theta})}{p(\boldsymbol{\theta})} d\boldsymbol{\theta}. \tag{12.3}$$

It is non-negative ($D_{\text{KL}}(q||p) \geq 0$) and zero if and only if $q = p$ almost everywhere. However, it is not symmetric ($D_{\text{KL}}(q||p) \neq D_{\text{KL}}(p||q)$), making the choice of direction important.

In VBI, we minimize the reverse KL divergence $D_{\text{KL}}(q||p)$, where $p = p(\boldsymbol{\theta}|\mathbf{y})$. Substituting the posterior, we have

$$D_{\text{KL}}\big(q(\boldsymbol{\theta}) \,||\, p(\boldsymbol{\theta}|\mathbf{y})\big) = \int q(\boldsymbol{\theta}) \ln \frac{q(\boldsymbol{\theta})}{p(\boldsymbol{\theta}|\mathbf{y})} d\boldsymbol{\theta}. \tag{12.4}$$

The reverse KLD has a mode-seeking property: it tends to find a variational distribution q that concentrates on a single mode of the posterior p, often providing a good local approximation. Expanding $p(\boldsymbol{\theta}|\mathbf{y}) = p(\mathbf{y}|\boldsymbol{\theta})p(\boldsymbol{\theta})/p(\mathbf{y})$ yields

$$D_{\text{KL}}\big(q||p\big) = \int q(\boldsymbol{\theta}) \ln \frac{q(\boldsymbol{\theta})}{p(\mathbf{y}|\boldsymbol{\theta})p(\boldsymbol{\theta})} d\boldsymbol{\theta} + \ln p(\mathbf{y}). \tag{12.5}$$

Since the model evidence $\ln p(\mathbf{y})$ is constant with respect to q, minimizing $D_{\text{KL}}(q||p)$ is equivalent to minimizing

$$\mathcal{L}(q) = \int q(\boldsymbol{\theta}) \ln \frac{q(\boldsymbol{\theta})}{p(\mathbf{y}|\boldsymbol{\theta})p(\boldsymbol{\theta})} d\boldsymbol{\theta}. \tag{12.6}$$

This quantity $\mathcal{L}(q)$ is known as the negative ELBO. Minimizing $\mathcal{L}(q)$ is equivalent to maximizing the ELBO, which serves as a lower bound on the log-evidence $\ln p(\boldsymbol{y})$.

12.2.2 The Mean-Field Variational Family: The Factorization Assumption

The choice of the variational family Q is crucial. It must be flexible enough to provide a good approximation, yet simple enough to make optimization tractable. The mean-field variational family is the most common and foundational choice. It assumes that the variational distribution factorizes into independent partitions over groups of parameters [4]

$$q(\boldsymbol{\theta}) = \prod_{j=1}^{J} q(\boldsymbol{\theta}_j), \tag{12.7}$$

where $\boldsymbol{\theta}$ is partitioned into J disjoint blocks $\{\boldsymbol{\theta}_1, \ldots, \boldsymbol{\theta}_J\}$. This is the mean-field assumption. It implies that under the variational approximation, the parameters in different blocks are statistically independent.

While this independence assumption is a strong simplification that ignores posterior correlations between blocks, it brings immense computational benefits. It breaks down a high-dimensional optimization over $q(\boldsymbol{\theta})$ into a set of lower-dimensional optimizations over each factor $q(\boldsymbol{\theta}_j)$. The factorization turns coupled high-dimensional integrals into products of simpler, lower-dimensional integrals.

12.2.3 Formalizing the Variational Optimization Problem

Combining the reverse KLD objective with the mean-field constraint yields the core optimization problem of MFVBI

$$\begin{aligned} \min_{q_1,\ldots,q_J} \quad & \mathcal{L}(q) = \int \left(\prod_{j=1}^{J} q(\boldsymbol{\theta}_j) \right) \ln \frac{\prod_{j=1}^{J} q(\boldsymbol{\theta}_j)}{p(\boldsymbol{y}, \boldsymbol{\theta})} \, d\boldsymbol{\theta} \\ \text{s.t.} \quad & \int q(\boldsymbol{\theta}_j)\, d\boldsymbol{\theta}_j = 1, \quad q(\boldsymbol{\theta}_j) \geq 0, \quad \forall j = 1, \ldots, J. \end{aligned} \tag{12.8}$$

Here, $p(\boldsymbol{y}, \boldsymbol{\theta}) = p(\boldsymbol{y} \mid \boldsymbol{\theta})\, p(\boldsymbol{\theta})$ denotes the joint distribution of the observed data $\boldsymbol{y}$ and the unknown parameters $\boldsymbol{\theta}$, while $q(\boldsymbol{\theta}_j)$ is written compactly as q_j. Notice that we are performing functional optimization: optimizing over probability density functions q_j rather than parameter values. Although this appears daunting, the structure of the problem under the mean-field assumption leads to a remarkably simple and intuitive solution via coordinate descent, as will be derived in the next section.

To build intuition, consider a simple case with two blocks, $\boldsymbol{\theta} = \{\boldsymbol{\theta}_1, \boldsymbol{\theta}_2\}$. The objective becomes

$$\mathcal{L}(q_1, q_2) = \iint q(\boldsymbol{\theta}_1) q(\boldsymbol{\theta}_2) \left[\ln q(\boldsymbol{\theta}_1) + \ln q(\boldsymbol{\theta}_2) - \ln p(\boldsymbol{y}, \boldsymbol{\theta}_1, \boldsymbol{\theta}_2) \right] d\boldsymbol{\theta}_1 d\boldsymbol{\theta}_2. \tag{12.9}$$

If we fix q_2 and optimize over q_1, all terms involving only q_2 become constants. The problem reduces to minimizing a KLD involving an expectation over q_2, i.e.,

$$\min_{q_1} \int q(\boldsymbol{\theta}_1) \ln \frac{q(\boldsymbol{\theta}_1)}{\exp\left(\mathbb{E}_{q_2}[\ln p(\boldsymbol{y}, \boldsymbol{\theta}_1, \boldsymbol{\theta}_2)] \right)} \, d\boldsymbol{\theta}_1 + \text{const.} \tag{12.10}$$

This reveals the general principle: the optimal variational factor for one block is proportional to the exponential of the expected log-joint probability, where the expectation is taken over all other blocks. This principle will guide the derivation of the explicit update equations in Sect. 12.3.

12.3 MFVBI: The General Framework

This section derives the general MFVBI algorithm, which provides a principled and often tractable methodology for approximate Bayesian inference. We begin with the optimization problem formalized in (12.8) and show how the mean-field assumption leads to a simple, iterative coordinate descent procedure.

12.3.1 Derivation of the ELBO

Recall our goal: to minimize the negative ELBO, $\mathcal{L}(q)$, subject to the mean-field factorization $q(\boldsymbol{\theta}) = \prod_{j=1}^{J} q(\boldsymbol{\theta}_j)$. Expanding $\mathcal{L}(q)$, we obtain two distinct components as

$$\mathcal{L}(q) = \underbrace{\int q(\boldsymbol{\theta}) \ln q(\boldsymbol{\theta})\, d\boldsymbol{\theta}}_{\text{(A) Negative Entropy}} - \underbrace{\int q(\boldsymbol{\theta}) \ln p(\boldsymbol{y}, \boldsymbol{\theta})\, d\boldsymbol{\theta}}_{\text{(B) Expected Log-Joint}} . \tag{12.11}$$

Component (A) is the negative entropy of the variational distribution, $-\mathbb{H}[q(\boldsymbol{\theta})]$. Under the mean-field factorization, it simplifies significantly because the entropy of a product of independent distributions is the sum of their individual entropies, i.e.,

$$\begin{aligned}\int q(\boldsymbol{\theta}) \ln q(\boldsymbol{\theta})\, d\boldsymbol{\theta} &= \int \left(\prod_{j=1}^{J} q(\boldsymbol{\theta}_j)\right) \sum_{j=1}^{J} \ln q(\boldsymbol{\theta}_j)\, d\boldsymbol{\theta} \\ &= \sum_{j=1}^{J} \int q(\boldsymbol{\theta}_j) \ln q(\boldsymbol{\theta}_j)\, d\boldsymbol{\theta}_j .\end{aligned} \tag{12.12}$$

Component (B) is the negative expected log-joint distribution. This term couples all variational factors through the joint distribution $p(\boldsymbol{y}, \boldsymbol{\theta}) = p(\boldsymbol{y}|\boldsymbol{\theta})p(\boldsymbol{\theta})$.

12.3.2 The Closed-Form Solution for Optimal Variational Factors

The key to solving the coupled optimization is to employ a coordinate descent strategy. We optimize one variational factor $q(\boldsymbol{\theta}_j)$ at a time while keeping all others $\{q(\boldsymbol{\theta}_i)\}_{i\neq j}$ fixed.

Let us isolate the terms in $\mathcal{L}(q)$ that depend on a specific factor q_j. All terms involving other factors $\{q_i\}_{i\neq j}$ become additive constant C with respect to q_j. Therefore,

$$\begin{aligned}\mathcal{L}(q) = &\int q(\boldsymbol{\theta}_j) \ln q(\boldsymbol{\theta}_j)\, d\boldsymbol{\theta}_j - \\ &\int q(\boldsymbol{\theta}_j) \underbrace{\left[\int \ln p(\boldsymbol{y}, \boldsymbol{\theta}) \prod_{i\neq j} q(\boldsymbol{\theta}_i)\, d\boldsymbol{\theta}_i\right]}_{\mathbb{E}_{q_{\sim j}}[\ln p(\boldsymbol{y}, \boldsymbol{\theta})]} d\boldsymbol{\theta}_j + C.\end{aligned} \tag{12.13}$$

We recognize the inner integral as the expectation of the log-joint distribution with respect to all variational factors except q_j, denoted as $\mathbb{E}_{q_{\sim j}}[\ln p(\boldsymbol{y}, \boldsymbol{\theta})]$. Defining a new distribution

$$\widetilde{q}(\boldsymbol{\theta}_j) \triangleq \frac{1}{Z_j} \exp\left(\mathbb{E}_{q_{\sim j}}[\ln p(\boldsymbol{y}, \boldsymbol{\theta})]\right), \tag{12.14}$$

where Z_j is the normalization factor, allows us to rewrite the q_j-dependent part of $\mathcal{L}(q)$ succinctly. Observe that

$$\mathcal{L}(q) = \int q(\boldsymbol{\theta}_j) \ln \frac{q(\boldsymbol{\theta}_j)}{\widetilde{q}(\boldsymbol{\theta}_j)} d\boldsymbol{\theta}_j - \ln Z_j + C. \tag{12.15}$$

Since $-\ln Z_j + C$ is constant in q_j, minimizing $\mathcal{L}(q)$ with respect to q_j is equivalent to minimizing the first term, which is precisely the KLD $D_{\mathrm{KL}}(q_j \,||\, \widetilde{q}_j)$. This divergence is minimized (to zero) if and only if $q_j = \widetilde{q}_j$.

Thus, we arrive at the fundamental MFVBI update equation

$$q^*(\boldsymbol{\theta}_j) = \frac{\exp\left(\mathbb{E}_{q_{\sim j}}[\ln p(\mathbf{y}, \boldsymbol{\theta})]\right)}{\int \exp\left(\mathbb{E}_{q_{\sim j}}[\ln p(\mathbf{y}, \boldsymbol{\theta})]\right) d\boldsymbol{\theta}_j}. \tag{12.16}$$

Interpretation: The optimal variational factor for block j is proportional to the exponential of the expected log-joint distribution, where the expectation is taken over the current estimates of all other blocks. The denominator ensures normalization. This result has an intuitive Bayesian flavor: the prior for updating q_j is effectively provided by the expectations over the other variables.

12.3.3 The Coordinate Descent Algorithm

Equation (12.16) suggests a natural iterative algorithm, as the update for each q_j depends on the current estimates of all others.

Step 1 (Initialization): Initialize all variational factors $q^{(0)}(\boldsymbol{\theta}_j)$, for $j = 1, \ldots, J$. A common choice is to set them to their respective prior distributions $p(\boldsymbol{\theta}_j)$.

Step 2 (Iterative Update): For iteration $t = 0, 1, 2, \ldots$, update each factor sequentially ($j = 1$ to J) using

$$q^{(t+1)}(\boldsymbol{\theta}_j) \propto \exp\left(\mathbb{E}_{q_{\sim j}^{(t+1,t)}}[\ln p(\mathbf{y}, \boldsymbol{\theta})]\right), \tag{12.17}$$

where the expectation uses the most recent estimates $q_{\sim j}^{(t+1,t)}$: $q_i^{(t+1)}$ for blocks $i < j$ already updated in the current cycle, and $q_i^{(t)}$ for blocks $i > j$ not yet updated.

Step 3 (Convergence Check): Monitor the decrease in the negative ELBO $\mathcal{L}(q^{(t)})$. The algorithm stops when the change in $\mathcal{L}(q^{(t)})$ falls below a predefined threshold, indicating convergence to a local optimum.

This procedure is guaranteed to monotonically increase the ELBO (and thus decrease $\mathcal{L}(q)$) with each update, ensuring stable convergence.

12.3.4 Properties, Convergence and Practical Challenges

The MFVBI algorithm possesses well-defined theoretical properties that simultaneously establish its scope and reveal its limitations. The coordinate descent procedure guarantees monotonic decrease of the negative ELBO, ensuring that $\mathcal{L}(q^{(t+1)}) \leq \mathcal{L}(q^{(t)})$ at each iteration and convergence to a stationary point. Computationally, the algorithm's efficiency is tightly linked to model structure: it achieves closed-form, efficient updates specifically within conjugate-exponential family models, such as Gaussian-Gaussian or Gamma-Poisson pairs, where variational factors share the same distributional family as the corresponding priors. This conjugacy ensures that expectations of the form $\mathbb{E}_{q_{\sim j}}[\ln p(\mathbf{y}, \boldsymbol{\theta})]$ and the required normalization integrals remain tractable.

However, the practical deployment of MFVBI is constrained by several interrelated challenges. The framework fundamentally relies on conjugacy; outside this class, especially in models with nonlinear likelihoods $p(\mathbf{y}|\boldsymbol{\theta})$, the requisite expectations become analytically intractable. Moreover, the mean-field factorization imposes a structural simplification by neglecting posterior dependencies between parameter blocks, often resulting in underestimated uncertainties and potentially biased estimates. Finally, the ELBO itself is generally non-convex, rendering convergence sensitive to initialization and frequently leading to suboptimal local solutions. These limitations motivate the development of advanced VBI techniques that extend inference beyond the conjugate-exponential setting while better capturing complex posterior geometries.

12.4 Particle Variational Bayesian Inference

12.4.1 Motivation: Capturing Complex Posteriors with Particles

PVBI emerges as a direct response to the aforementioned constraints of MFVBI, particularly in applications involving highly non-convex objectives or multi-modal posteriors. While standard parametric variational families impose a restrictive form (often unimodal) on the approximate posterior, PVBI adopts a non-parametric, particle-based representation. Concretely, the variational distribution for each parameter block is expressed as a discrete measure [3, 9]

$$q(\boldsymbol{\theta}_j; \boldsymbol{w}_j) = \sum_{n=1}^{N_p} w_{j,n}\, \delta(\boldsymbol{\theta}_j - \boldsymbol{p}_{j,n}), \tag{12.18}$$

where $\{\boldsymbol{p}_{j,1}, \ldots, \boldsymbol{p}_{j,N_p}\}$ denotes a fixed set of particle positions, and $\boldsymbol{w}_j = [w_{j,1}, \ldots, w_{j,N_p}]^T$ represents the corresponding non-negative weights satisfying $\sum_n w_{j,n} = 1$. Particles are typically initialized either by sampling from the prior $p(\boldsymbol{\theta}_j)$ or via a deterministic grid covering the plausible parameter region, with initial weights set uniformly as $w_{j,n}^{(0)} = 1/N_p$.

This formulation delivers notable representational flexibility: as $N_p \to \infty$, the particle ensemble can in principle approximate any posterior density, including those exhibiting multiple modes, skewness, or intricate correlation patterns. Consequently, the variational inference problem is transformed from a functional optimization over distributional forms into a finite-dimensional numerical optimization over particle weights, enabling inference in models where traditional mean-field assumptions would fail.

12.4.2 Problem Reformulation: Variational Optimization with Discrete Measures

Substituting the particle-based representation $q(\boldsymbol{\theta}; \boldsymbol{w}) = \prod_{j=1}^{J} q(\boldsymbol{\theta}_j; \boldsymbol{w}_j)$ into the negative ELBO $\mathcal{L}(q)$ in (12.11) yields a finite-dimensional optimization problem over the aggregated weight vector $\boldsymbol{w} = [\boldsymbol{w}_1^T, \ldots, \boldsymbol{w}_J^T]^T$. Specifically, the negative entropy term simplifies dramatically because the entropy of a discrete distribution is the Shannon entropy of its weight vector, i.e.,

$$\int q(\boldsymbol{\theta}; \boldsymbol{w}) \ln q(\boldsymbol{\theta}; \boldsymbol{w}) d\boldsymbol{\theta} = \sum_{j=1}^{J} \sum_{n=1}^{N_p} w_{j,n} \ln w_{j,n} = \sum_{j=1}^{J} \boldsymbol{w}_j^T \ln \boldsymbol{w}_j, \tag{12.19}$$

where $\ln \boldsymbol{w}_j$ denotes the element-wise logarithm. The expected log-joint term becomes a weighted sum over all possible combinations of particles from different blocks, i.e.,

$$\int q(\boldsymbol{\theta}; \boldsymbol{w}) \ln p(\boldsymbol{y}, \boldsymbol{\theta}) d\boldsymbol{\theta} = \sum_{n_1=1}^{N_p} \cdots \sum_{n_J=1}^{N_p} \left(\prod_{j=1}^{J} w_{j,n_j} \right) \ln p\,(\boldsymbol{y}, \boldsymbol{\theta}) \Big|_{\boldsymbol{\theta}=[\boldsymbol{p}_{1,n_1}, \ldots, \boldsymbol{p}_{J,n_J}]^T}. \tag{12.20}$$

Combining these, the PVBI optimization problem is given by

$$\begin{aligned} \min_{\boldsymbol{w}} \quad & \mathcal{L}(\boldsymbol{w}) = \sum_{j=1}^{J} \boldsymbol{w}_j^T \ln \boldsymbol{w}_j - \sum_{n_1=1}^{N_p} \cdots \sum_{n_J=1}^{N_p} \left(\prod_{j=1}^{J} w_{j,n_j} \right) \ln p \left(\boldsymbol{y}, \boldsymbol{\theta}^{(n_1, \ldots, n_J)} \right) \\ \text{s.t.} \quad & \boldsymbol{1}^T \boldsymbol{w}_j = 1, \quad \epsilon \le w_{j,n} \le 1, \quad \forall j, n. \end{aligned} \tag{12.21}$$

Here, $\boldsymbol{\theta}^{(n_1, \ldots, n_J)} = [\boldsymbol{p}_{1,n_1}^T, \ldots, \boldsymbol{p}_{J,n_J}^T]^T$, and a small $\epsilon > 0$ is used for numerical stability to prevent weights from becoming exactly zero.

Algorithm 12.1 PVBI Algorithm

Input: Measurement data $\boldsymbol{y}$, prior $p(\boldsymbol{\theta})$.

Initialization: $\left\{\boldsymbol{p}_{j,n} \sim p(\boldsymbol{\theta}_j), w_{j,n} = \frac{1}{N_p}\right\}_{n=1}^{N_p}, \forall j.$

While not converged, do

  For $j = 1 : J$

    Solve sub-problem $\mathcal{P}_{w_j}$ in (12.22);

    Update the particle weight $\boldsymbol{w}_j$ of variable $\boldsymbol{\theta}_j$ according to (12.26);

  end

end

Output: Find the position $\boldsymbol{p}_j^*$, $j = 1, 2, \ldots J$ of the particle with the maximum weight.

12.4.3 Derivation of the Particle Weight Update Rule

Problem (12.21) is optimized using a block coordinate descent approach, iteratively updating the weight vector $\boldsymbol{w}_j$ for one block while keeping others fixed. Isolating terms that depend on $\boldsymbol{w}_j$, the subproblem for block j is

$$\begin{aligned} \mathcal{P}_{w_j} : \min_{\boldsymbol{w}_j} \quad & \mathcal{L}_j(\boldsymbol{w}_j) = \boldsymbol{w}_j^T \ln \boldsymbol{w}_j - \boldsymbol{g}_j^T \boldsymbol{w}_j \\ \text{s.t.} \quad & \mathbf{1}^T \boldsymbol{w}_j = 1, \quad \epsilon \leq w_{j,n} \leq 1, \quad \forall n, \end{aligned} \tag{12.22}$$

where the n_j-th element of the vector $\boldsymbol{g}_j$ is defined as

$$g_{j,n_j} = \sum_{n_1=1}^{N_p} \cdots \sum_{n_{j-1}=1}^{N_p} \sum_{n_{j+1}=1}^{N_p} \cdots \sum_{n_J=1}^{N_p} \left(\prod_{i \neq j} w_{i,n_i} \right) \ln p\left(\boldsymbol{y}, \boldsymbol{\theta}^{(n_1,\ldots,n_J)}\right). \tag{12.23}$$

The constant g_{j,n_j} can be interpreted as the expected log-joint probability evaluated at the particle $\boldsymbol{p}_{j,n_j}$, where the expectation is taken over all combinations of particles from the other blocks using their current weights.

The objective in (12.22) is convex in $\boldsymbol{w}_j$ (sum of entropy and a linear term). Ignoring the box constraints $\epsilon \leq w_{j,n} \leq 1$ for a moment (they are typically inactive at the optimum if ϵ is very small), we form the Lagrangian

$$\mathcal{J}(\boldsymbol{w}_j, \lambda) = \boldsymbol{w}_j^T \ln \boldsymbol{w}_j - \boldsymbol{g}_j^T \boldsymbol{w}_j + \lambda(\mathbf{1}^T \boldsymbol{w}_j - 1). \tag{12.24}$$

Taking the gradient with respect to $w_{j,n}$ and setting it to zero gives

$$\frac{\partial \mathcal{J}}{\partial w_{j,n}} = \ln w_{j,n} + 1 - g_{j,n} + \lambda = 0 \quad \Rightarrow \quad w_{j,n} = \exp\left(g_{j,n} - \lambda - 1\right). \tag{12.25}$$

Applying the normalization constraint $\sum_n w_{j,n} = 1$ solves for the Lagrange multiplier, yielding the elegant closed-form update

$$w_{j,n_j}^* = \frac{\exp\left(g_{j,n_j}\right)}{\sum_{n_j=1}^{N_p} \exp\left(g_{j,n_j}\right)}. \tag{12.26}$$

Interpretation: The new weight for particle $\boldsymbol{p}_{j,n}$ is proportional to the exponential of the expected log-joint probability, where particles that contribute to higher-probability joint configurations receive larger weights. This mirrors the MFVBI update (12.16) but in a discrete, non-parametric setting.

12.4.4 Algorithm Summary, Complexity and Trade-Offs

The PVBI algorithm proceeds as outlined in Algorithm 12.1. It iteratively cycles through parameter blocks, computing $\boldsymbol{g}_j$ via (12.23) and updating weights via (12.26) until convergence of the negative ELBO $\mathcal{L}(\boldsymbol{w})$.

The computational profile and practical characteristics of PVBI are defined by its inherent trade-offs between flexibility and efficiency. The primary computational bottleneck arises from the evaluation of g_{j,n_j} in Eq. (12.23), which requires summing over N_p^{J-1} terms, resulting in an exponential complexity in the number of parameter blocks J. This complexity fundamentally limits the basic PVBI algorithm to problems with a small number of blocks. For larger-scale problems, advanced techniques such as Stochastic Successive Convex Approximation (SSCA) can be employed to obtain unbiased gradient estimates with substantially reduced computational burden [12]. Under standard regularity conditions, the block-coordinate descent procedure ensures convergence to a stationary point of $\mathcal{L}(\boldsymbol{w})$.

The algorithm's principal advantage lies in its representational flexibility: the non-parametric particle-based formulation can approximate arbitrary posterior distributions, providing robustness against multi-modality and enabling inference in models that fall outside the conjugate-exponential family. This generality comes at the cost of significant computational intensity, making PVBI particularly suitable for problems where posterior geometry is complex and the parameter structure naturally decomposes into a modest number of blocks. Upon convergence, the weighted particle set $\{\boldsymbol{p}_{j,n}, w_{j,n}\}$ furnishes a direct discrete approximation to the marginal posterior $p(\boldsymbol{\theta}_j|\boldsymbol{y})$, from which point estimates such as the MMSE estimate $\hat{\boldsymbol{\theta}}_j^{\text{MMSE}} = \sum_n w_{j,n}\boldsymbol{p}_{j,n}$ or the MAP estimate (the particle with highest weight) can be immediately extracted, along with empirical uncertainty metrics like sample variance.

In summary, PVBI sacrifices computational efficiency for flexibility and generality, making it a powerful tool for complex, low-to-moderate dimensional inference problems where the posterior may have a challenging structure.

12.5 Successive Linear Approximation VBI (SLAVBI)

12.5.1 Motivation: Enabling Closed-Form Updates in Nonlinear Models

The Particle VBI algorithm provides flexibility for complex posteriors at the cost of exponential computational complexity. For high-dimensional nonlinear problems, we often seek a more efficient deterministic alternative that retains some of the advantages of conjugate updates. The primary obstacle to using the standard MFVBI framework in such settings is non-conjugacy: the nonlinear likelihood $p(\boldsymbol{y}|\boldsymbol{\theta})$ prevents the expectations $\mathbb{E}_{q_{\sim j}}[\ln p(\boldsymbol{y}, \boldsymbol{\theta})]$ from having a tractable, closed form.

SLAVBI addresses this by a strategy of iterative local linearization. At each iteration, it constructs a Gaussian approximation to the likelihood around the current parameter estimate. Within this locally linearized model, the standard conjugate Gaussian-Gaussian model relations hold, enabling closed-form variational updates reminiscent of the Kalman filter. This approach trades off global flexibility for local computational efficiency and stability.

12.5.2 Core Idea: Iterative Linearization and Gaussian Variational Posteriors

Consider the general nonlinear observation model with additive Gaussian noise

$$\boldsymbol{y} = \boldsymbol{f}(\boldsymbol{\theta}) + \boldsymbol{w}, \quad \boldsymbol{w} \sim \mathcal{CN}(\boldsymbol{0}, \sigma_w^2 \boldsymbol{I}), \tag{12.27}$$

where $\boldsymbol{f}(\boldsymbol{\theta}) = [f_1(\boldsymbol{\theta}), \ldots, f_M(\boldsymbol{\theta})]^T$. The SLAVBI core procedure consists of two key operations.

The first involves **linearization**: at iteration t, using a current linearization point $\boldsymbol{\theta}^{(t)}$ (typically taken as the posterior mean from the preceding iteration), the nonlinear function is approximated through a first-order Taylor expansion

$$\boldsymbol{f}(\boldsymbol{\theta}) \approx \boldsymbol{f}(\boldsymbol{\theta}^{(t)}) + \sum_{j=1}^{J} \boldsymbol{G}_j^{(t)}(\boldsymbol{\theta}_j - \boldsymbol{\theta}_j^{(t)}), \tag{12.28}$$

where $\boldsymbol{G}_j^{(t)} = \nabla_{\boldsymbol{\theta}_j} \boldsymbol{f}(\boldsymbol{\theta})\big|_{\boldsymbol{\theta}=\boldsymbol{\theta}^{(t)}}$ denotes the Jacobian matrix of $\boldsymbol{f}$ with respect to the j-th parameter block, evaluated at $\boldsymbol{\theta}^{(t)}$.

The second operation entails the adoption of a **Gaussian variational family** for each block, specified as $q(\boldsymbol{\theta}_j) = \mathcal{CN}(\boldsymbol{\theta}_j; \boldsymbol{\mu}_j, \boldsymbol{\Sigma}_j)$, alongside Gaussian priors $p(\boldsymbol{\theta}_j) = \mathcal{CN}(\boldsymbol{\theta}_j; \boldsymbol{\vartheta}_j, \boldsymbol{\Lambda}_j)$. This parametric choice ensures conjugacy with the resulting linearized Gaussian likelihood, thereby enabling closed-form variational updates.

12.5.3 Derivation: Linearization and Variational Update

Substituting the linear approximation into the log-likelihood yields a conditionally Gaussian form

$$\ln p(\mathbf{y}|\boldsymbol{\theta}) \approx -\frac{1}{\sigma_w^2}\left\| \mathbf{y} - \boldsymbol{f}(\boldsymbol{\theta}^{(t)}) - \sum_{j=1}^{J} \mathbf{G}_j^{(t)}(\boldsymbol{\theta}_j - \boldsymbol{\theta}_j^{(t)}) \right\|^2 + \text{const.} \tag{12.29}$$

We now apply the MFVBI update rule from Eq. (12.16) to find $q^{(t)}(\boldsymbol{\theta}_j)$. The log of the optimal variational factor is proportional to the expected log-joint under $q_{\sim j}^{(t)}$, i.e.,

$$\ln q^{(t)}(\boldsymbol{\theta}_j) \propto \mathbb{E}_{q_{\sim j}^{(t)}}[\ln p(\mathbf{y}|\boldsymbol{\theta})] + \ln p(\boldsymbol{\theta}_j). \tag{12.30}$$

Crucially, because the approximated log-likelihood is quadratic in $\boldsymbol{\theta}_j$, and the prior is quadratic in $\boldsymbol{\theta}_j$, the expectation results in another quadratic form. Performing this expectation involves replacing the other blocks $\boldsymbol{\theta}_i (i \neq j)$ in the linearized model with their current variational means $\boldsymbol{\mu}_i^{(t)}$.

After algebraic manipulation and completing the square, we find that $\ln q^{(t)}(\boldsymbol{\theta}_j)$ is a quadratic function, confirming that $q^{(t)}(\boldsymbol{\theta}_j)$ is Gaussian. The parameters of this Gaussian are given by the following **SLAVBI update equations**

$$\begin{aligned} \boldsymbol{\Sigma}_j^{(t)} &= \left(\frac{1}{\sigma_w^2} (\mathbf{G}_j^{(t)})^H \mathbf{G}_j^{(t)} + \boldsymbol{\Lambda}_j^{-1} \right)^{-1}, \\ \boldsymbol{\mu}_j^{(t)} &= \boldsymbol{\Sigma}_j^{(t)} \left(\frac{1}{\sigma_w^2} (\mathbf{G}_j^{(t)})^H \boldsymbol{b}_j^{(t)} + \boldsymbol{\Lambda}_j^{-1} \boldsymbol{\vartheta}_j \right), \end{aligned} \tag{12.31}$$

where the effective observation vector $\boldsymbol{b}_j^{(t)}$ is defined as

$$\boldsymbol{b}_j^{(t)} = \mathbf{y} - \boldsymbol{f}(\boldsymbol{\theta}^{(t)}) - \sum_{i \neq j} \mathbf{G}_i^{(t)}(\boldsymbol{\mu}_i^{(t)} - \boldsymbol{\theta}_i^{(t)}) + \mathbf{G}_j^{(t)} \boldsymbol{\theta}_j^{(t)}. \tag{12.32}$$

Interpretation: These updates have an intuitive Bayesian structure. The posterior precision matrix $(\boldsymbol{\Sigma}_j^{(t)})^{-1}$ is the sum of the prior precision $\boldsymbol{\Lambda}_j^{-1}$ and the information from the linearized model $\frac{1}{\sigma_w^2}(\mathbf{G}_j^{(t)})^H \mathbf{G}_j^{(t)}$. The posterior mean $\boldsymbol{\mu}_j^{(t)}$ is a precision-weighted average of the prior mean $\boldsymbol{\vartheta}_j$ and the data-dependent term $\frac{1}{\sigma_w^2}(\mathbf{G}_j^{(t)})^H \boldsymbol{b}_j^{(t)}$, which represents the portion of the residual not explained by other blocks, projected onto the subspace of block j.

12.5.4 Algorithm Summary, Complexity and Connection to Extended Kalman Filtering

The overall SLAVBI algorithm is summarized in Algorithm 12.2. It iterates between linearization and variational Gaussian updates. After updating all blocks, the linearization point is refreshed to the new posterior means, $\boldsymbol{\theta}^{(t+1)} = \{\boldsymbol{\mu}_1^{(t)}, \ldots, \boldsymbol{\mu}_J^{(t)}\}$, for the next iteration.

Computational Complexity: The dominant cost per iteration is the matrix inversion in Eq. (12.31) for each block, which is $O(d_j^3)$ for a block of dimension d_j. This is polynomial and far more scalable than the exponential cost of PVBI for large J. Complexity can be further reduced by exploiting structure (e.g., diagonal $\boldsymbol{\Lambda}_j$) or using iterative solvers and MM techniques to avoid explicit inversion [15].

Connection to Extended Kalman Filtering (EKF): SLAVBI bears a strong resemblance to the EKF. The EKF performs a single linearization and update. SLAVBI can be viewed as a multi-block, variational generalization of the EKF: it iteratively re-linearizes and performs a full Bayesian update (providing a covariance $\boldsymbol{\Sigma}_j$) for each parameter block within a coordinated variational framework, rather than a single joint state update.

Trade-offs: The SLAVBI algorithm presents a distinct set of trade-offs between efficiency and approximation quality. On one hand, it offers computational efficiency for high-dimensional problems while providing closed-form Gaussian uncertainty estimates through covariance matrices, making it particularly well-suited for inference tasks where the posterior is expected

Algorithm 12.2 SLAVBI algorithm

Input: $\boldsymbol{y}$, prior $p(\boldsymbol{\theta})$.
Initialization: $\boldsymbol{\theta}_j^{(1)} = \boldsymbol{\vartheta}_j, \forall j$ and $t = 1$.
While not converged, do
  Perform linear approximation around $\boldsymbol{\theta} = \boldsymbol{\theta}^{(t)}$, using (12.28).
  For $j = 1 : J$
    Update $\boldsymbol{\mu}_j$ and $\boldsymbol{\Sigma}_j$, using (12.31).
  end
  Let $\boldsymbol{\theta}_j^{(t+1)} = \boldsymbol{\mu}_j^{(t)}, \forall j$ and $t = t + 1$.
end
Output: the MMSE estimate $\boldsymbol{\theta}_j^{(t)}, \forall j$.

to be unimodal or where a local MAP approximation suffices. On the other hand, its reliance on local linearization introduces sensitivity to initialization; when the initial point is poorly chosen or the likelihood landscape is severely non-convex, the algorithm may converge to suboptimal local modes. Furthermore, the Gaussian variational family inherently lacks the flexibility to represent multi-modal distributions or capture posterior skewness, limiting its applicability in problems exhibiting more complex posterior geometries.

In summary, SLAVBI is a powerful and efficient deterministic algorithm for approximate Bayesian inference in smooth nonlinear models. It fills the gap between the flexible but expensive PVBI and the restricted but fast conjugate MFVBI, making it a preferred choice for many large-scale nonlinear estimation problems in signal processing and communications.

12.6 Sparse VBI for Compressive Sensing

12.6.1 Problem Setup: Linear Models with Uncertain Parameters

We now apply the VBI framework to a cornerstone problem in modern signal processing: CS recovery with uncertain model parameters. This problem bridges the general methodology of this chapter with the specific SBL algorithm from Chap. 11. We consider the linear observation model with potential parameter uncertainty in the sensing matrix

$$\boldsymbol{y} = \boldsymbol{A}(\boldsymbol{\theta})\boldsymbol{x} + \boldsymbol{w}, \tag{12.33}$$

where $\boldsymbol{y} \in \mathbb{C}^M$ is the measurement vector, $\boldsymbol{A}(\boldsymbol{\theta}) \in \mathbb{C}^{M\times N}$ is the sensing matrix that may depend on parameters $\boldsymbol{\theta}$ (e.g., grid offsets in direction-of-arrival estimation), $\boldsymbol{x} \in \mathbb{C}^N$ is the sparse signal to recover, and $\boldsymbol{w} \sim \mathcal{CN}(\boldsymbol{0}, \kappa^{-1}\boldsymbol{I})$ is complex Gaussian noise with precision κ.

To focus on the core VBI mechanism for sparsity, we initially assume the sensing matrix is fixed, i.e., $\boldsymbol{A}(\boldsymbol{\theta}) = \boldsymbol{A}$. We will later comment on extending the algorithm to estimate $\boldsymbol{\theta}$.

12.6.2 Revisiting the Hierarchical Sparse Prior

The challenge is to infer $\boldsymbol{x}$ from $\boldsymbol{y}$ under the prior belief that $\boldsymbol{x}$ is sparse. Following the SBL framework, we employ a two-layer hierarchical prior that is conjugate to the Gaussian likelihood, enabling tractable VBI updates.

First Layer (Signal Prior): Conditional on a precision parameter ρ_n, each element x_n of the signal is assumed independent and Gaussian

$$p(x_n|\rho_n) = \mathcal{CN}(x_n; 0, \rho_n^{-1}), \quad \forall n. \tag{12.34}$$

Thus, $p(\boldsymbol{x}|\boldsymbol{\rho}) = \prod_{n=1}^{N} \mathcal{CN}(x_n; 0, \rho_n^{-1})$.

Second Layer (Sparsity-Promoting Hyperprior): To promote sparsity, we assign an independent Gamma prior to each precision ρ_n

$$p(\rho_n) = \Gamma(\rho_n; a_n, b_n), \quad \forall n. \tag{12.35}$$

The hyperparameters a_n, b_n are chosen so that the prior mean $\mathbb{E}[\rho_n] = a_n/b_n \gg 1$. This places most prior mass on large ρ_n, which corresponds to small variances ρ_n^{-1}, effectively forcing x_n toward zero. Typically, we use $a_n = 1$ and $b_n = \epsilon$, a small positive constant (e.g., 10^{-4}), to create a broad, non-informative prior that allows the data to drive ρ_n to large values for irrelevant coefficients.

Noise Precision Prior: Similarly, we place a Gamma prior on the noise precision κ

$$p(\kappa) = \Gamma(\kappa; c, d), \tag{12.36}$$

with $c, d \to 0$ (e.g., 10^{-5}) for a non-informative hyperprior.

Let $\boldsymbol{v} = \{\boldsymbol{x}, \boldsymbol{\rho}, \kappa\}$ denote the complete set of unknown variables. The joint distribution of data and parameters is

$$\begin{aligned} p(\boldsymbol{y}, \boldsymbol{v}) &= p(\boldsymbol{y}|\boldsymbol{x}, \kappa)\, p(\boldsymbol{x}|\boldsymbol{\rho})\, p(\boldsymbol{\rho})\, p(\kappa) \\ &= \mathcal{CN}(\boldsymbol{y}; \boldsymbol{A}\boldsymbol{x}, \kappa^{-1}\boldsymbol{I}) \prod_{n=1}^{N} \left[\mathcal{CN}(x_n; 0, \rho_n^{-1})\Gamma(\rho_n; a_n, b_n)\right] \Gamma(\kappa; c, d). \end{aligned} \tag{12.37}$$

12.6.3 Derivation of Exact Variational Updates Under Mean-Field

We apply the MFVBI framework with $J = 3$ blocks: $\boldsymbol{v}_1 = \boldsymbol{x}$, $\boldsymbol{v}_2 = \boldsymbol{\rho}$, and $\boldsymbol{v}_3 = \kappa$. The mean-field assumption is

$$q(\boldsymbol{v}) = q(\boldsymbol{x})q(\boldsymbol{\rho})q(\kappa). \tag{12.38}$$

The optimal variational factors are given by Eq. (12.16): $q(\boldsymbol{v}_j) \propto \exp\left(\mathbb{E}_{q_{\sim j}}[\ln p(\boldsymbol{y}, \boldsymbol{v})]\right)$. We initialize $q(\boldsymbol{\rho})$ and $q(\kappa)$ to their prior distributions and proceed with cyclic updates.

Update for $q(\boldsymbol{x})$ (Variational E-step for Signal):

$$\begin{aligned} \ln q(\boldsymbol{x}) &\propto \mathbb{E}_{q(\boldsymbol{\rho})q(\kappa)}[\ln p(\boldsymbol{y}|\boldsymbol{x}, \kappa) + \ln p(\boldsymbol{x}|\boldsymbol{\rho})] \\ &\propto \mathbb{E}_{q(\kappa)}[-\kappa\|\boldsymbol{y} - \boldsymbol{A}\boldsymbol{x}\|^2] + \mathbb{E}_{q(\boldsymbol{\rho})}[-\boldsymbol{x}^H \mathrm{diag}(\boldsymbol{\rho})\boldsymbol{x}] \\ &\propto -\langle\kappa\rangle(\boldsymbol{y} - \boldsymbol{A}\boldsymbol{x})^H(\boldsymbol{y} - \boldsymbol{A}\boldsymbol{x}) - \boldsymbol{x}^H \mathrm{diag}(\langle\boldsymbol{\rho}\rangle)\boldsymbol{x}. \end{aligned} \tag{12.39}$$

This is a quadratic form in $\boldsymbol{x}$, confirming that $q(\boldsymbol{x})$ is complex Gaussian: $q(\boldsymbol{x}) = \mathcal{CN}(\boldsymbol{x}; \boldsymbol{\mu}, \boldsymbol{\Sigma})$. Completing the square yields the sparse VBI signal update

$$\boldsymbol{\Sigma} = \left(\mathrm{diag}(\langle\boldsymbol{\rho}\rangle) + \langle\kappa\rangle\boldsymbol{A}^H\boldsymbol{A}\right)^{-1}, \tag{12.40}$$

$$\boldsymbol{\mu} = \langle\kappa\rangle\boldsymbol{\Sigma}\boldsymbol{A}^H\boldsymbol{y}. \tag{12.41}$$

The posterior mean $\boldsymbol{\mu}$ is a regularized least-squares solution, where the diagonal matrix $\mathrm{diag}(\langle\boldsymbol{\rho}\rangle)$ applies a weighted ℓ_2 regularization, with larger expected precisions forcing corresponding coefficients toward zero.

Update for $q(\boldsymbol{\rho})$ (Variational E-step for Precision):

$$\begin{aligned} \ln q(\boldsymbol{\rho}) &\propto \mathbb{E}_{q(\boldsymbol{x})}[\ln p(\boldsymbol{x}|\boldsymbol{\rho})] + \ln p(\boldsymbol{\rho}) \\ &\propto \sum_{n=1}^{N} \left(a_n \ln \rho_n - \left(b_n + \langle|x_n|^2\rangle\right)\rho_n\right). \end{aligned} \tag{12.42}$$

This is a sum of terms each proportional to the log of a Gamma distribution. Therefore, $q(\boldsymbol{\rho}) = \prod_{n=1}^{N} \Gamma(\rho_n; \tilde{a}_n, \tilde{b}_n)$, with parameters updated as

$$\tilde{a}_n = a_n + 1, \tag{12.43}$$

$$\tilde{b}_n = b_n + \langle|x_n|^2\rangle = b_n + |\mu_n|^2 + \Sigma_{nn}, \forall n. \tag{12.44}$$

The updated rate $\tilde{b}_n$ increases with the posterior second moment $\langle |x_n|^2 \rangle$. A large $\langle |x_n|^2 \rangle$ (significant coefficient) leads to a smaller $\mathbb{E}[\rho_n] = \tilde{a}_n/\tilde{b}_n$, relaxing the prior for that coefficient. A small $\langle |x_n|^2 \rangle$ leads to a large $\mathbb{E}[\rho_n]$, further shrinking the coefficient toward zero.

Update for $q(\kappa)$ (Variational E-step for Noise):

$$\begin{aligned}\ln q(\kappa) &\propto \mathbb{E}_{q(\boldsymbol{x})}[\ln p(\boldsymbol{y}|\boldsymbol{x}, \kappa)] + \ln p(\kappa) \\ &\propto (M + c - 1) \ln \kappa - \left(d + \mathbb{E}_{q(\boldsymbol{x})}[\|\boldsymbol{y} - \boldsymbol{A}\boldsymbol{x}\|^2]\right) \kappa.\end{aligned} \tag{12.45}$$

Thus, $q^*(\kappa) = \Gamma(\kappa; \tilde{c}, \tilde{d})$, where

$$\begin{aligned}\tilde{c} &= c + M, \\ \tilde{d} &= d + \left\langle \|\boldsymbol{y} - \boldsymbol{A}\boldsymbol{x}\|^2 \right\rangle_{q(\boldsymbol{x})} = d + \mathrm{tr}(\boldsymbol{A}\boldsymbol{\Sigma}\boldsymbol{A}^H) + \|\boldsymbol{y} - \boldsymbol{A}\boldsymbol{\mu}\|^2\end{aligned} \tag{12.46}$$

The expected residual power includes both the fitting error $\|\boldsymbol{y} - \boldsymbol{A}\boldsymbol{\mu}\|^2$ and the trace term representing the uncertainty in $\boldsymbol{x}$.

The required expectations are

$$\langle \rho_n \rangle = \frac{\tilde{a}_n}{\tilde{b}_n}, \quad \langle \kappa \rangle = \frac{\tilde{c}}{\tilde{d}}, \quad \langle |x_n|^2 \rangle = |\mu_n|^2 + \Sigma_{nn}. \tag{12.47}$$

The algorithm iterates Eqs. (12.40)–(12.46) until convergence. The final sparse signal estimate is the posterior mean $\boldsymbol{\mu}$. The diagonal of $\boldsymbol{\Sigma}$ provides uncertainty estimates for each coefficient.

12.6.4 Relationship to SBL

Sparse VBI is deeply connected to the SBL algorithm from Chap. 11, as both employ the same two-layer hierarchical prior to promote sparsity. The fundamental distinction lies in their respective approximation strategies within the iterative inference framework. In the SBL approach, which follows an EM scheme, the E-step calculates the exact Gaussian posterior distribution of the signal, $p(\boldsymbol{x}|\boldsymbol{y}, \boldsymbol{\rho}^{(t)}, \kappa^{(t)})$. However, SBL treats the hyperparameters $\boldsymbol{\rho}$ and κ as deterministic point estimates. This is formally equivalent to assuming their variational posteriors collapse to Dirac delta functions, i.e., $q(\boldsymbol{\rho}) = \delta(\boldsymbol{\rho} - \bar{\boldsymbol{\rho}})$ and $q(\kappa) = \delta(\kappa - \bar{\kappa})$, within a variational inference perspective. In contrast, the sparse VBI framework maintains a full mean-field variational distribution over all latent variables. While it also represents $q(\boldsymbol{x})$ as a Gaussian, it concurrently models $q(\boldsymbol{\rho})$ and $q(\kappa)$ as full Gamma distributions. The primary consequence of this difference is that sparse VBI generally provides a more accurate approximation to the full joint posterior $p(\boldsymbol{x}, \boldsymbol{\rho}, \kappa|\boldsymbol{y})$ compared to SBL. By accounting for uncertainty in all unknown quantities throughout the iterative process, sparse VBI tends to demonstrate enhanced robustness and slightly superior recovery performance. This advantage is particularly noticeable in challenging scenarios such as low SNRs or when dealing with highly coherent sensing matrices, as the method's inherent uncertainty quantification helps prevent premature overconfidence in estimated values.

This probabilistic treatment extends naturally to scenarios involving additional model parameters. Although our derivation has assumed a fixed grid parameter $\boldsymbol{\theta}$ for simplicity, the framework readily accommodates its estimation. In particular, the grid parameter $\boldsymbol{\theta}$ can be refined based on the EM framework using a similar in-exact gradient update as discussed in the SBL-M step of Chap. 11. The main difference is that the Bayesian estimators of all the other variables, including the signal precision $\boldsymbol{\rho}$ and noise precision κ, are now calculated in the E-step via sparse VBI.

12.7 Application Examples: VBI in Wireless Localization

This section presents two concrete applications of VBI in wireless localization systems. Through these carefully selected examples, we demonstrate how the theoretical VBI framework can be effectively implemented to solve challenging estimation problems in modern wireless systems.

The first application focuses on **wireless cooperative localization**, where we formulate the problem as a general non-convex parameter estimation task and develop a PVBI solution. As a simplified case from [12], this example illustrates VBI's capability to handle complex, high-dimensional estimation problems with non-standard probability distributions.

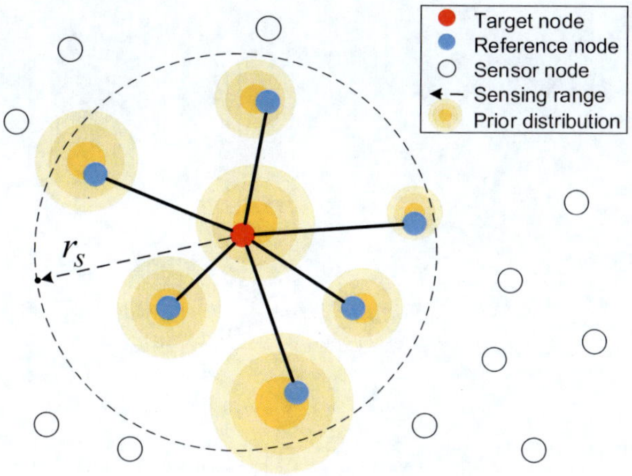

Fig. 12.1 Illustration of network node deployment in wireless cooperative localization

The second application addresses **massive MIMO based localization**, which we formulate as a CS problem and solve using sparse VBI. This case study, adapted from [2], showcases VBI's adaptability to group sparsity pattern and its efficiency in large-scale antenna systems.

12.7.1 PVBI for Wireless Cooperative Localization

Consider a two-dimensional wireless network with uniformly distributed location-aware nodes, as illustrated in Fig. 12.1. A target node lies within measurement range r_s of M reference nodes, all possessing imperfect initial location estimates. In the absence of anchor nodes with known positions, these devices must cooperatively estimate their locations through shared information and relative measurements.

Let $\boldsymbol{s}_0 = [p_{x_0}, p_{y_0}]^T$ and $\boldsymbol{s}_m = [p_{x_m}, p_{y_m}]^T$ $(m = 1, \ldots, M)$ denote the true positions of the target and reference nodes, respectively. Each node has a coarse location estimate $\boldsymbol{\mu}_m$ with associated precision matrix $\boldsymbol{U}_m$. The true positions are modeled as independent Gaussian random variables:

$$\boldsymbol{s}_m \sim \mathcal{N}\left(\boldsymbol{\mu}_m, \boldsymbol{U}_m^{-1}\right), \quad \forall m = 0, 1, \ldots, M. \tag{12.48}$$

The target node collects prior information $\{\boldsymbol{\mu}_m, \boldsymbol{U}_m\}$ from all reference nodes along with distance measurements y_m between itself and each reference node.

The measurement model follows:

$$y_m = f(\boldsymbol{s}_m, \boldsymbol{s}_0) + w_m, \quad \forall m = 1, \ldots, M, \tag{12.49}$$

where $w_m \sim \mathcal{N}\left(0, \kappa_m^{-1}\right)$ is measurement noise with precision κ_m. Using RSS-based ranging, the observation function becomes:

$$f(\boldsymbol{s}_m, \boldsymbol{s}_0) = \phi_{ref} - 10\lambda \log_{10} \|\boldsymbol{s}_m - \boldsymbol{s}_0\|_2, \tag{12.50}$$

where ϕ_{ref} denotes reference power loss and λ is the path-loss exponent [2].

Let $\boldsymbol{c} = \text{vec}[\boldsymbol{s}_m]_{m=1}^M$ aggregate reference node positions and $\boldsymbol{y} = \text{vec}[y_m]_{m=1}^M$ stack all measurements. The likelihood function is:

$$p(\boldsymbol{y}|\boldsymbol{s}_0, \boldsymbol{c}) = \prod_{m=1}^{M} \frac{|\kappa_m|^{\frac{1}{2}}}{\sqrt{2\pi}} \exp\left[-\frac{\kappa_m}{2}(y_m - f(\boldsymbol{s}_m, \boldsymbol{s}_0))^2\right]. \tag{12.51}$$

The posterior distribution of the target position is proportional to:

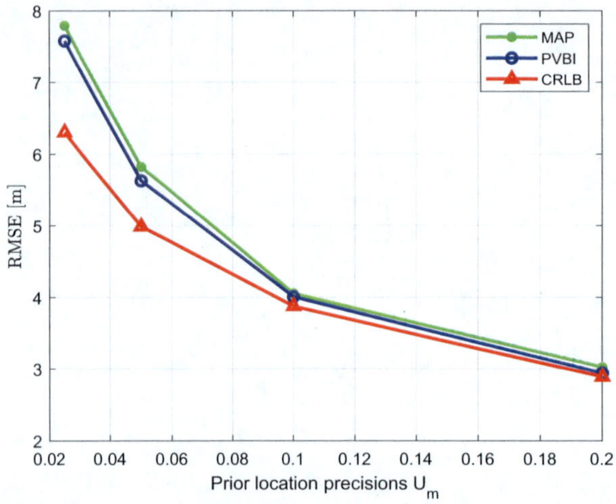

Fig. 12.2 Localization error over various prior location precisions

$$p(\boldsymbol{s}_0 \mid \boldsymbol{y}) \propto \mathcal{N}\left(\boldsymbol{s}_0; \boldsymbol{\mu}_0, \boldsymbol{U}_0^{-1}\right) \cdot \prod_{m=1}^{M} \int \mathcal{N}\left(\boldsymbol{s}_m; \boldsymbol{\mu}_m, \boldsymbol{U}_m^{-1}\right) \frac{|\kappa_m|^{1/2}}{\sqrt{2\pi}} \exp\left[-\frac{\kappa_m}{2}\left(y_m - f\left(\boldsymbol{s}_m, \boldsymbol{s}_0\right)\right)^2\right] d\boldsymbol{s}_m. \tag{12.52}$$

The nonlinear observation function, when combined with uncertainties in reference node positions, gives rise to a non-convex posterior distribution characterized by multiple poor local optima. This poses considerable challenges for conventional optimization methods [2].

To address this difficult inference problem, the PVBI approach is applied through the following procedure. First, particle representations for all node positions are initialized by placing particles on grid points within $\pm$ 3 standard deviations of the prior means, with weights assigned according to normalized 2D Gaussian distributions. Next, an iterative weight optimization process is carried out by constructing subproblems $\mathcal{P}_{w_j}$. These subproblems incorporate the log-prior distribution

$$\ln p\left(\boldsymbol{s}_m\right) = \frac{1}{2}\ln\frac{|\boldsymbol{U}_m|}{4\pi^2} + \left[-\frac{1}{2}\left(\boldsymbol{s}_m - \boldsymbol{\mu}_m\right)^T \boldsymbol{U}_m\left(\boldsymbol{s}_m - \boldsymbol{\mu}_m\right)\right] \tag{12.53}$$

and the log-likelihood function

$$\ln p\left(\boldsymbol{y}|\boldsymbol{s}_0, \boldsymbol{c}\right) = \frac{1}{2}\sum_{m=1}^{M}\ln\frac{|\kappa_m|}{2\pi} + \sum_{m=1}^{M}\left[-\frac{\kappa_m}{2}\left(y_m - f\left(\boldsymbol{s}_m, \boldsymbol{s}_0\right)\right)^2\right]. \tag{12.54}$$

Each subproblem is iteratively solved using $g_j(\boldsymbol{w}_{\sim j}^{(t)})$ as defined in Eq. (12.23) until convergence is achieved. Final estimates are then extracted via MAP or MMSE criteria from the optimized particle distribution.

To evaluate performance under the simulation setup described in [2], an RSS-based scenario is considered with one target node and six reference nodes ($M = 6$). Key parameters include a path-loss exponent $\lambda = 3$, reference path loss $\phi_{\text{ref}} = -5$ dBm, detection range $r_s = 50$ m, location precision $\boldsymbol{U}_m = U_m \boldsymbol{I}$ with varying U_m, and measurement precision $\kappa_m = 4/75$.

As shown in Fig. 12.2, which summarizes localization error under varying prior location precisions over 500 Monte Carlo trials, two main observations emerge. First, more accurate prior location information consistently enhances localization performance. Second, PVBI outperforms gradient-based MAP estimation, largely due to its non-parametric nature. This flexibility allows PVBI to better capture complex posterior structures and avoid premature convergence to suboptimal solutions.

Overall, this example illustrates the effectiveness of PVBI in addressing the challenges of wireless cooperative localization, particularly in scenarios involving significant nonlinearities and positional uncertainties where traditional optimization methods often fall short.

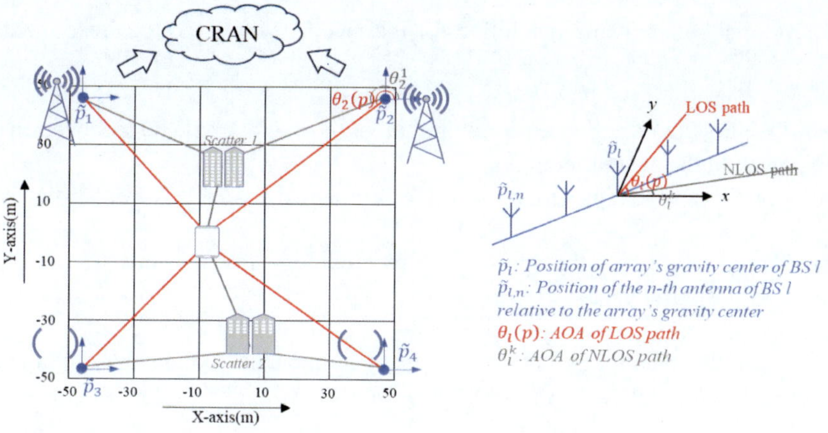

Fig. 12.3 A localization model in 5G massive MIMO system

12.7.2 Sparse VBI for Massive MIMO Based Localization

A fundamental requirement for autonomous driving systems is the ability to achieve precise and dependable vehicle localization. Current solutions typically rely on a fusion of different technologies, such as Global Navigation Satellite Systems (GNSS), visual sensors from cameras, and other on-board sensors [17]. Recent studies have highlighted the substantial potential of massive MIMO technology to enhance localization accuracy in this context, outperforming traditional methods [18].

We consider a two-dimensional geographical region $\mathcal{R}$, as depicted in Fig. 12.3. An autonomous vehicle, outfitted with a single-antenna wireless transceiver, operates within the coverage area of L massive MIMO BSs. Each BS is equipped with a large antenna array of size $M \gg 1$. The positions of the BSs are known and fixed, and they are interconnected via a cloud network.

At a given time slot, the vehicle's true position is $\boldsymbol{p} = [p^x, p^y]^T \in \mathcal{R}$, while the center of the l-th BS's array is at $\tilde{\boldsymbol{p}}_l = [\tilde{p}_l^x, \tilde{p}_l^y]^T$, for $l = 1, \ldots, L$. The response vector for the array at BS l, corresponding to an angle-of-arrival (AoA) θ, is denoted by $\boldsymbol{a}_l(\theta) \in \mathbb{C}^{M\times 1}$.

For position estimation, the vehicle transmits T pilot symbols $u(t)$, with $t = 1, ..., T$, to all BSs. Assuming a flat fading channel with a total of $K + 1$ propagation paths, the signal received at BS l can be modeled as [18]

$$\boldsymbol{y}_l(t) = \underbrace{\boldsymbol{a}_l\left(\theta_l(\boldsymbol{p})\right)\alpha_l u(t)}_{\text{LoS path}} + \underbrace{\sum_{k=1}^{K} \boldsymbol{a}_l\left(\theta_l^k\right)\alpha_l^k u(t)}_{\text{NLoS paths}} + \boldsymbol{w}_l(t), \tag{12.55}$$

where $\boldsymbol{w}_l(t) \sim \mathcal{CN}\left(\boldsymbol{0}, \sigma_l^2 \boldsymbol{I}\right)$ is the additive complex Gaussian noise. In this model, α_l and $\theta_l(\boldsymbol{p})$ are the complex gain and AoA for the line-of-sight (LoS) path, respectively, while α_l^k and θ_l^k are the gain and AoA for the k-th non-line-of-sight (NLoS) path.

The AoA of the LoS path is geometrically linked to the vehicle's position [18] as

$$\theta_l(\boldsymbol{p}) = \arctan\left(\frac{p^y - \tilde{p}_l^y}{p^x - \tilde{p}_l^x}\right) + \pi \cdot \mathbf{1}\left(p^x < \tilde{p}_l^x\right), \tag{12.56}$$

where the angle is measured anticlockwise from the positive x-axis. A Gamma prior is assigned to the noise precision $\kappa_l = \sigma_l^{-2}$ for each BS, i.e.,

$$p(\kappa_l) = \Gamma(\kappa_l; a_\kappa, b_\kappa). \tag{12.57}$$

As suggested in [19], we set $a_\kappa, b_\kappa \to 0$ to create a non-informative hyperprior, allowing the data to drive the noise precision estimate.

In the following, we develop a comprehensive solution for massive MIMO-based localization, structured into three key parts.

1. **Sparse Representation**: We first derive a sparse representation of the massive MIMO localization problem that efficiently captures the inherent sparsity in the angular domain.
2. **CS Formulation**: Based on this sparse representation, we reformulate the vehicle localization task as a CS problem with group sparsity.
3. **Sparse VBI Implementation**: We apply the sparse VBI methodology to solve this CS formulation and evaluate its localization performance against baseline algorithms.

This methodology capitalizes on the inherent sparsity of multipath components in the angular domain, while also providing probabilistic uncertainty estimates through its Bayesian foundation.

12.7.2.1 Sparse Representation for Massive MIMO Based Localization

To harness the inherent sparsity of the wireless channel, the geographical area is discretized into a uniform grid of $\widetilde{Q}$ candidate locations [18]

$$\mathcal{P} = \left\{ \boldsymbol{r}_1, ..., \boldsymbol{r}_{\widetilde{Q}} \right\} \subset \mathcal{R}, \tag{12.58}$$

Simultaneously, the angular space $[0, 2\pi)$ is discretized into a uniform grid of $\widetilde{M}$ candidate AoAs

$$\mathcal{A} = \left\{ \vartheta_1, ..., \vartheta_{\widetilde{M}} \right\} \subset [0, 2\pi). \tag{12.59}$$

As true positions and AoAs rarely fall exactly on these predefined grid points, off-grid correction terms are introduced to minimize modeling errors. Let $i^* = \arg\min_i \|\boldsymbol{p} - \boldsymbol{r}_i\|$ be the index of the location grid point nearest to the true vehicle position. A position offset vector $\Delta \boldsymbol{r} \in \mathbb{R}^{2\widetilde{Q}}$ is defined, where its constituent vectors are

$$\Delta \boldsymbol{r}_i = \begin{cases} \boldsymbol{p} - \boldsymbol{r}_{i^*}, & i = i^*, \\ \boldsymbol{0}, & i \neq i^*. \end{cases} \tag{12.60}$$

Similarly, for the k-th NLoS path at BS l, let $m_l^k = \arg\min_m \|\theta_l^k - \vartheta_m\|$ denote the index of the AoA grid point nearest to the true AoA. The corresponding AoA offset vector $\Delta \boldsymbol{\vartheta}_l \in \mathbb{R}^{\widetilde{M}}$ for BS l is defined with elements

$$\Delta \vartheta_l^m = \begin{cases} \theta_l^k - \vartheta_{m_l^k}, & m \in \left\{ m_l^1, \ldots, m_l^K \right\}, \\ 0, & \text{otherwise}. \end{cases} \tag{12.61}$$

With these offset definitions, we can construct off-grid array response matrices. For the LoS path at BS l, this matrix is

$$\overline{\boldsymbol{A}}_l (\Delta \boldsymbol{r}) = \left[\boldsymbol{a}_l (\theta_l (\boldsymbol{r}_1 + \Delta \boldsymbol{r}_1)), \ldots, \boldsymbol{a}_l \left(\theta_l \left(\boldsymbol{r}_{\widetilde{Q}} + \Delta \boldsymbol{r}_{\widetilde{Q}}\right)\right) \right], \tag{12.62}$$

and for the NLoS paths

$$\overline{\boldsymbol{A}}_l^{\text{NL}} (\Delta \boldsymbol{\vartheta}_l) = \left[\boldsymbol{a}_l \left(\vartheta_1 + \Delta \vartheta_l^1\right), \ldots, \boldsymbol{a}_l \left(\vartheta_{\widetilde{M}} + \Delta \vartheta_l^{\widetilde{M}}\right) \right]. \tag{12.63}$$

The received signal at BS l can now be expressed in a sparse form

$$\boldsymbol{y}_l(t) = u(t) \overline{\boldsymbol{A}}_l(\Delta \boldsymbol{r}) \boldsymbol{x}_l + u(t) \overline{\boldsymbol{A}}_l^{\text{NL}}(\Delta \boldsymbol{\vartheta}_l) \boldsymbol{z}_l + \boldsymbol{w}_l(t), \tag{12.64}$$

where $\boldsymbol{x}_l \in \mathbb{C}^{\widetilde{Q}}$ and $\boldsymbol{z}_l \in \mathbb{C}^{\widetilde{M}}$ are sparse vectors representing the LoS and NLoS channel components at BS l, respectively. These sparse vectors have key structural properties:

- Each $\boldsymbol{x}_l$ contains exactly one non-zero element corresponding to the true vehicle position
- All $\boldsymbol{x}_l$ share a common support across different BSs
- Each $\boldsymbol{z}_l$ contains $K_l \ll M$ non-zero elements corresponding to NLoS paths.

This structure motivates a dual-precision Bayesian model. A shared precision vector $\boldsymbol{\rho} = [\rho_1, \ldots, \rho_{\widetilde{Q}}]^T$ is introduced to enforce group sparsity across all $\boldsymbol{x}_l$ vectors. Concurrently, individual precision vectors $\boldsymbol{v}_l = [v_{l,1}, \ldots, v_{l,\widetilde{M}}]^T$ promote element-wise sparsity within each z_l.

The coarse position estimate is obtained from the grid point $\boldsymbol{r}_{i^*}$, where i^* is found by

$$i^* = \arg\max_i \left[\sum_{l=1}^{L} \boldsymbol{x}_l \odot \boldsymbol{x}_l^* \right]_i . \tag{12.65}$$

The final, refined position estimate is then $\boldsymbol{p} \approx \boldsymbol{r}_{i^*} + \Delta \boldsymbol{r}_{i^*}$. This sparse representation effectively transforms the localization task into a structured sparse recovery problem that inherently accounts for off-grid errors.

12.7.2.2 CS Formulation for Massive MIMO Based Localization

The signal model in (12.64) can be rewritten more compactly by stacking the time-domain samples. Defining $\boldsymbol{y}_l \triangleq [\boldsymbol{y}_l(1); \ldots; \boldsymbol{y}_l(T)]$ and $\boldsymbol{w}_l \triangleq [\boldsymbol{w}_l(1); \ldots; \boldsymbol{w}_l(T)]$, we obtain

$$\boldsymbol{y}_l = \boldsymbol{A}_l \left(\Delta \boldsymbol{r}, \Delta \boldsymbol{\vartheta}_l\right) \begin{bmatrix} \boldsymbol{x}_l \\ z_l \end{bmatrix} + \boldsymbol{w}_l, \quad \forall l, \tag{12.66}$$

where the combined sensing matrix $\boldsymbol{A}_l(\Delta \boldsymbol{r}, \Delta \boldsymbol{\vartheta}_l) \in \mathbb{C}^{MT \times (\widetilde{Q}+\widetilde{M})}$ has the following block structure

$$\boldsymbol{A}_l \left(\Delta \boldsymbol{r}, \Delta \boldsymbol{\vartheta}_l\right) = \begin{bmatrix} u(1)\overline{\boldsymbol{A}}_l(\Delta \boldsymbol{r}) & u(1)\overline{\boldsymbol{A}}_l^{\mathrm{NL}}(\Delta \boldsymbol{\vartheta}_l) \\ \vdots & \vdots \\ u(T)\overline{\boldsymbol{A}}_l(\Delta \boldsymbol{r}) & u(T)\overline{\boldsymbol{A}}_l^{\mathrm{NL}}(\Delta \boldsymbol{\vartheta}_l) \end{bmatrix} . \tag{12.67}$$

The primary goal is to recover the set of LoS channel vectors $\{\boldsymbol{x}_l\}_{l=1}^{L}$ from the aggregate measurements $\boldsymbol{y} = [\boldsymbol{y}_1, ..., \boldsymbol{y}_L]$. This recovery enables two key steps for localization:

- **Coarse Position Estimation**: Compute $\sum_{l=1}^{L} \boldsymbol{x}_l \odot \boldsymbol{x}_l^*$, where the index of the largest element i^* indicates the coarse vehicle position on the grid
- **Position Refinement**: Determine the position offset $\Delta \boldsymbol{r}_{i^*}$ for precise localization.

This problem naturally exhibits a group sparsity structure with $Q = \widetilde{Q} + \widetilde{M}L$ groups and dynamic grid parameters $\Delta \boldsymbol{r}$ and $\Delta \boldsymbol{\vartheta}$. The group organization is as follows. The first $\widetilde{Q}$ groups, defined by $\mathcal{I}_i = \{(l-1)\widetilde{Q}+i : l = 1, \ldots, L\}$ for $i = 1, \ldots, \widetilde{Q}$, couple the coefficients across all BSs that correspond to the same candidate location $\boldsymbol{r}_i$, enforcing a common support for the LoS path. The remaining $\widetilde{M}L$ groups (indexed by $i = \widetilde{Q}+1, \ldots, \widetilde{Q}+\widetilde{M}L$) each contain a single element, corresponding to the unstructured sparsity of the NLoS paths at each BS.

This formulation elegantly captures the physical constraints of the problem: the single vehicle position creates a joint sparsity pattern across BSs for the LoS component, while the multipath environment at each BS contributes its own independent sparse structure. The final result is a well-defined group sparse signal recovery task with dynamic grid adaptation, providing a powerful framework for high-accuracy vehicle positioning.

12.7.2.3 Implementation Details

E-step: Extension to Group Sparsity

The E-step generalizes the sparse VBI framework from Sect. 12.6 from independent sparsity to group sparsity. This extension follows established methodological principles and is omitted here for brevity; it is recommended as an exercise for readers seeking deeper understanding of group sparse variational inference.

Denote the approximate posteriors calculated via the E-step as

$$\begin{aligned} q\left(\boldsymbol{x}_l; \dot{\boldsymbol{\xi}}\right) &= \mathcal{CN}\left(\boldsymbol{x}_l; \boldsymbol{\mu}_l, \boldsymbol{\Sigma}_l\right), \\ q\left(z_l; \dot{\boldsymbol{\xi}}\right) &= \mathcal{CN}\left(z_l; \boldsymbol{\mu}_l^z, \boldsymbol{\Sigma}_l^z\right). \end{aligned} \tag{12.68}$$

where $\dot{\boldsymbol{\xi}} \triangleq \{\triangle \dot{\boldsymbol{r}}, \triangle \dot{\boldsymbol{\vartheta}}\}$ denotes previous estimates of the dynamic grid parameters.

M-step: Surrogate Function and Gradient Updates
The surrogate function $u(\boldsymbol{\xi}; \dot{\boldsymbol{\xi}})$ for the log joint PDF under the EM framework is:

$$u(\boldsymbol{\xi}; \dot{\boldsymbol{\xi}}) = \int p(\boldsymbol{x}|\boldsymbol{y}, \dot{\boldsymbol{\xi}}) \ln \frac{p(\boldsymbol{x}, \boldsymbol{y}, \boldsymbol{\xi})}{p(\boldsymbol{x}|\boldsymbol{y}, \dot{\boldsymbol{\xi}})} d\boldsymbol{x}. \tag{12.69}$$

In the M-step, parameters $\triangle \boldsymbol{r}$ and $\triangle \boldsymbol{\vartheta}$ are updated via gradient ascent on $u(\boldsymbol{\xi}; \dot{\boldsymbol{\xi}})$, where the gradients are derived as follows.

For $i = 1, \ldots, \widetilde{Q}$, with $\triangle \boldsymbol{r}_i = [\triangle r_i^x, \triangle r_i^y]$ and $\boldsymbol{r}_i = [r_i^x, r_i^y]$, the partial gradient w.r.t. $\triangle r_i^x$ is:

$$\frac{\partial u(\boldsymbol{\xi}; \dot{\boldsymbol{\xi}})}{\partial \triangle r_i^x} = \sum_{l=1}^{L} 2\mathrm{Re}\left(\boldsymbol{a}_{l,i}^{'H} \left(\boldsymbol{a}_l(\theta_l(\triangle \boldsymbol{r}_i)) c_{l,i} + \bar{\boldsymbol{c}}_{l,i}\right) c_{l,i}^x\right), \tag{12.70}$$

where the constituent terms are:

$$\begin{aligned}
c_{l,i} &= -\frac{1}{\sigma^2}\left(|\mu_{l,i}|^2 + \Sigma_{l,i,i}\right),\\
c_{l,i}^x &= -\left(r_i^y + \triangle r_i^y - \tilde{p}_l^y\right) / \left\|\boldsymbol{r}_i + \triangle \boldsymbol{r}_i - \tilde{\boldsymbol{p}}_l\right\|^2,\\
\bar{\boldsymbol{c}}_{l,i} &= \frac{1}{\sigma^2}\left(\mu_{l,i}^* \boldsymbol{y}_{-l,i} - \sum_{j\neq i} \Sigma_{l,j,i} \boldsymbol{a}_l(\theta_l(\triangle \boldsymbol{r}_j))\right),\\
\boldsymbol{y}_{-l,i} &= \boldsymbol{y}_l - \sum_{j\neq i}\left(\mu_{l,j} \boldsymbol{a}_l(\theta_l(\triangle \boldsymbol{r}_j))\right),\\
\boldsymbol{a}_{l,i}^{'} &= \frac{\partial \boldsymbol{a}_l(\theta)}{\partial \theta}|_{\theta=\theta_l(\triangle \boldsymbol{r}_i)}.
\end{aligned} \tag{12.71}$$

Here, $\mu_{l,i}$ is the i-th element of $\boldsymbol{\mu}_l$, $\Sigma_{l,j,i}$ is the (j, i)-th element of $\boldsymbol{\Sigma}_l$, and $\theta_l(\triangle \boldsymbol{r}_i)$ abbreviates $\theta_l(\boldsymbol{r}_i + \triangle \boldsymbol{r}_i)$ as defined in (12.56).

The partial gradient w.r.t. $\triangle r_i^y$ follows similarly, with $c_{l,i}^x$ replaced by:

$$c_{l,i}^y = \frac{r_i^x + \triangle r_i^x - \tilde{p}_l^x}{\|\boldsymbol{r}_i + \triangle \boldsymbol{r}_i - \tilde{\boldsymbol{p}}_l\|^2}. \tag{12.72}$$

For $m = 1, \ldots, \widetilde{M}$, the partial gradient w.r.t. $\triangle \vartheta_l^m$ is:

$$\frac{\partial u(\boldsymbol{\xi}; \dot{\boldsymbol{\xi}})}{\partial \triangle \vartheta_l^m} = 2\mathrm{Re}\left(\boldsymbol{a}_l^{'}(\vartheta_l^m)^H \left(\boldsymbol{a}_l(\vartheta_l^m) c_{l,m}^z + \bar{\boldsymbol{c}}_{l,m}^z\right)\right), \tag{12.73}$$

where:

$$\begin{aligned}
\vartheta_l^m &= \vartheta_m + \triangle \vartheta_l^m,\\
c_{l,m}^z &= -\frac{1}{\sigma^2}\left(|\mu_{l,m}^z|^2 + \Sigma_{l,m,m}^z\right),\\
\bar{\boldsymbol{c}}_{l,m}^z &= \frac{1}{\sigma^2}\left(\mu_{l,m}^{z*} \boldsymbol{y}_{-l,m}^z - \sum_{n\neq m} \Sigma_{l,n,n}^z \boldsymbol{a}_l(\vartheta_l^n)\right),\\
\boldsymbol{y}_{-l,m}^z &= \boldsymbol{y}_l - \sum_{n\neq m}\left(\mu_{l,n} \boldsymbol{a}_l(\vartheta_l^n)\right).
\end{aligned} \tag{12.74}$$

Here, $\mu_{l,m}^z$ is the m-th element of $\boldsymbol{\mu}_l^z$, and $\Sigma_{l,n,m}^z$ is the (n, m)-th element of $\boldsymbol{\Sigma}_l^z$.

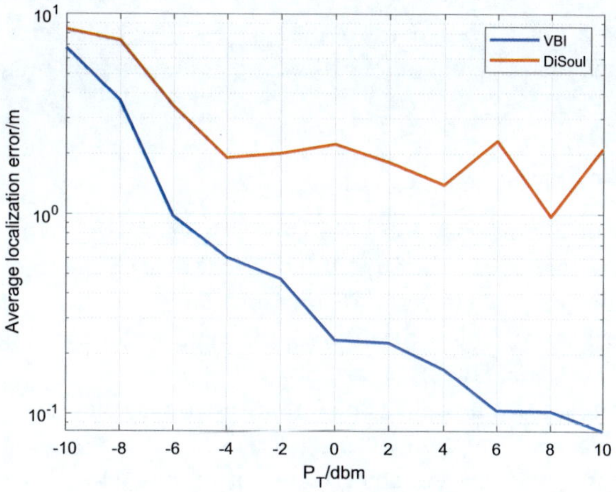

Fig. 12.4 Average localization error versus the transmit power

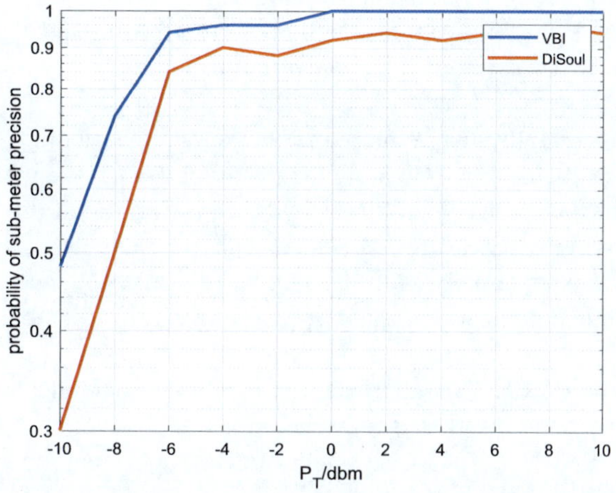

Fig. 12.5 Probability of sub-meter precision versus the transmit power

12.7.2.4 Simulation Results

The simulation environment consists of four BSs positioned at the corners of a 100 m × 100 m square area, with the target source randomly located within the central 50 m × 50 m region. Each BS employs a ULA of 32 antennas with half-wavelength spacing. The massive MIMO channel, including both LoS and NLoS components, is generated according to the Clustered Delay Line (CDL-B) channel model specified in 3GPP TR 38.901 Release 15 (Table 7.7.1-2) [20].

For performance evaluation, we employ the Direct Source Localization (DiSouL) method [18] as a benchmark. This approach iteratively estimates LoS path complex gains, vehicle position, and grid parameters. The AMP algorithm is excluded from comparison due to its known performance degradation in scenarios with highly correlated sensing matrix columns, which characterizes the current setup.

Figure 12.4 presents the average localization error as a function of transmit power. The sparse VBI algorithm demonstrates consistently superior performance compared to DiSouL, particularly at higher transmit power levels. This performance advantage stems from VBI's enhanced numerical stability, which enables more precise localization as signal power increases. In contrast, DiSouL exhibits instability that can lead to substantial localization errors even under favorable high-power conditions.

Figure 12.5 illustrates the probability of achieving sub-meter precision (defined as the likelihood of localization errors below one meter) across varying transmit power levels. The sparse VBI method again demonstrates clear superiority in this critical performance metric, consistently achieving higher probabilities of sub-meter accuracy across the evaluated power range.

These results collectively demonstrate the effectiveness of the sparse VBI approach for massive MIMO based localization, particularly in demanding applications such as autonomous driving that require both high accuracy and reliability.

Summary

VBI serves as a powerful and versatile framework for addressing a wide range of Bayesian inference problems. It is applicable to multi-dimensional parameter estimation with nonlinear observation models, as well as CS problems involving dynamic grids and uncertain model parameters. Within this framework, parameter estimation is reformulated as a variational optimization problem, where the objective is to approximate the true posterior distribution by minimizing the KLD between a variational distribution and the true posterior.

The MFVBI approach is commonly employed to solve this variational problem. It assumes a factorized form for the variational distribution and iteratively optimizes each variable block. While variational updates in MFVBI generally involve high-dimensional integrals and are analytically intractable, closed-form solutions become feasible under specific conditions. For instance, when the likelihood is Gaussian (such as in linear observation models with Gaussian noise), selecting conjugate priors (e.g., Gaussian or Gamma distributions) enables tractable variational updates. A representative example is the CS problem with uncertain parameters, where a two-layer hierarchical prior, combining Gamma-distributed precision and Gaussian sparse signals, facilitates closed-form updates while promoting sparsity.

In more general settings, two primary methods facilitate variational updates.

1. **Particle-Based VBI**: This method employs a parameterized variational distribution, where the parameters correspond to the weights of discrete particles approximating the posterior. By applying MFVBI, PVBI transforms the KLD minimization into an optimization over particle weights, converting multidimensional integrals into finite summations. Although PVBI excels at capturing multi-modal distributions and avoiding poor local optima, its computational complexity grows exponentially with the number of particles and variables. Therefore, PVBI is best suited to problems with a limited number of variables, requiring careful balancing between particle count and performance.
2. **Successive Linear Approximation VBI**: This approach iteratively approximates nonlinear observation models using a sequence of linear models, enabling closed-form variational updates at each step. While SLAVBI is computationally more efficient than PVBI, its reliance on local linear approximations may lead to convergence to suboptimal solutions, particularly in highly non-convex scenarios.

It is noteworthy that many established CS recovery algorithms can be interpreted as special cases or approximations of VBI. For example:

- **SBL** can be viewed as a sparse VBI variant where the variational posteriors for the signal and noise precision collapse to single-particle approximations (i.e., delta distributions). Specifically, the noise precision distribution becomes $q(\kappa) = \delta(\kappa - \bar{\kappa})$, and its optimal value corresponds to the point estimate derived in the SBL M-step.
- Similarly, **LASSO** is equivalent to a VBI implementation where the sparse signal prior is Laplacian and its variational posterior is restricted to a delta distribution $q(\boldsymbol{x}) = \delta(\boldsymbol{x} - \bar{\boldsymbol{x}})$, with the optimal value $\bar{\boldsymbol{x}}$ recovering the LASSO MAP estimate.

In summary, Bayesian inference offers a unifying perspective for understanding a broad class of parameter estimation and CS algorithms, with VBI serving as a flexible and generalized methodology that encompasses many existing techniques as specific instances.

Homework Problems

12.1 Using the Lagrange multiplier method, show that the optimal solution of Problem $\mathcal{P}_{w_j}$ in Eq. (12.22) is given by the closed-form expression in (12.26). In your derivation, clearly state the Lagrangian function, derive the Karush-Kuhn-Tucker (KKT) conditions, and show how solving these conditions leads to the final weight update formula.

12.2 Discuss the advantages and disadvantages of the PVBI and SLAVBI algorithms. Your comparison should consider the following aspects:

- Computational complexity and scalability
- Flexibility in modeling posterior distributions
- Convergence behavior and stability
- Applicability to different types of problems (e.g., high-dimensional vs. low-dimensional, linear vs. nonlinear models)
- Ease of implementation.

12.3 Consider the wireless cooperative localization example described in Sect. 12.7.1.
(a) **Algorithm Derivation**: Derive an SLAVBI algorithm specifically for this application. Your derivation should include:

- The linear approximation of the observation model
- The Gaussian prior assumptions for all parameters
- The complete update rules for all variational distributions.

(b) **Implementation and Comparison**:

- Implement your derived SLAVBI algorithm using MATLAB
- Compare its performance with the provided PVBI algorithm in terms of: localization accuracy, computational time and convergence speed
- Use the MATLAB code for the PVBI algorithm available at: https://github.com/biwcs-book/codebase.

12.4 Consider the Massive MIMO based localization example in Sect. 12.7.2.
(a) **Gradient Verification**:

- Write MATLAB code to numerically verify the gradient expressions in Eqs. (12.70) and (12.73)
- Use finite difference methods to compute numerical gradients and compare them with your analytical gradient computations
- Provide evidence (e.g., plot the error norms) that confirms the correctness of your gradient implementation.

(b) **Algorithm Implementation and Comparison**:

- Implement the SBL algorithm using MATLAB for this application
- Compare the performance of your implemented VBI and SBL algorithms in terms of localization precision, computational efficiency, robustness to noise and sparsity recovery capability
- The MATLAB code for the VBI algorithm available at https://github.com/biwcs-book/codebase provides a good starting point for implementing the SBL algorithm.

References

1. F. Liu, Y. Cui, C. Masouros, J. Xu, T.X. Han, Y.C. Eldar, S. Buzzi, Integrated sensing and communications: toward dual-functional wireless networks for 6G and Beyond. IEEE J. Sel. Areas Commun. **40**(6), 1728–1767 (2022)
2. B. Zhou, Q. Chen, On the particle-assisted stochastic search mechanism in wireless cooperative localization. IEEE Trans. Wirel. Commun. **15**(7), 4765–4777 (2016)
3. B. Zhou, Q. Chen, H. Wymeersch, P. Xiao, L. Zhao, Variational inference-based positioning with nondeterministic measurement accuracies and reference location errors. IEEE Trans. Mobile Comput. **16**(10), 2955–2969 (2017)
4. M.J. Beal, Variational algorithms for approximate Bayesian inference. Ph.D. thesis, University of London, 2003
5. R.M. Neal, Probabilistic inference using Markov chain Monte Carlo methods. Tech. Rep. CRG-TR-93-1 (Department of Computer Science, University of Toronto Toronto, ON, Canada, 1993)
6. A. Doucet, N. De Freitas, N.J. Gordon et al., *Sequential Monte Carlo Methods in Practice*, vol. 1, no. 2 (Springer, 2001)

7. M.S. Arulampalam, S. Maskell, N. Gordon, T. Clapp, A tutorial on particle filters for online nonlinear/non-Gaussian Bayesian tracking. IEEE Trans. Signal Process. **50**(2), 174–188 (2002)
8. Q. Liu, D. Wang, Stein variational gradient descent: a general purpose Bayesian inference algorithm, in *Proceedings of Advances in Neural Information Processing Systems (NIPS)*, vol. 29 (2016)
9. B. Dai, N. He, H. Dai, L. Song, Provable Bayesian inference via particle mirror descent, in *Proceedings of the International Conference on Artificial Intelligence and Statistics (AISTATS)* (PMLR, 2016), pp. 985–994
10. Y. Yang, G. Scutari, D.P. Palomar, M. Pesavento, A parallel decomposition method for nonconvex stochastic multi-agent optimization problems. IEEE Trans. Signal Process. **64**(11), 2949–2964 (2016)
11. A. Liu, V.K.N. Lau, B. Kananian, Stochastic successive convex approximation for non-convex constrained stochastic optimization. IEEE Trans. Signal Process. **67**(16), 4189–4203 (2019)
12. Z. Hu, A. Liu, Y. Wan, T.X. Han, M. Zhao, A two-stage multiband delay estimation scheme via stochastic particle-based variational Bayesian inference. IEEE Internet Things J. **11**(11), 19632–19645 (2024)
13. Z. Hu, A. Liu, Y. Wan, T.Q.S. Quek, M. Zhao, Two-stage multiband Wi-Fi sensing for ISAC via stochastic particle-based variational Bayesian inference, in *Proceedings of the IEEE Global Communications Conference (GLOBECOM)*, pp. 5617–5622 (2023)
14. Y. Sun, P. Babu, D.P. Palomar, Majorization-minimization algorithms in signal processing, communications, and machine learning. IEEE Trans. Signal Process. **65**(3), 794–816 (2017)
15. W. Xu, A. Liu, B. Zhou, M.-J. Zhao, Successive linear approximation VBI for joint sparse signal recovery and dynamic grid parameters estimation. [Online]. https://arxiv.org/pdf/2307.09149
16. A. Liu, G. Liu, L. Lian, V.K. Lau, M.-J. Zhao, Robust recovery of structured sparse signals with uncertain sensing matrix: a turbo-VBI approach. IEEE Trans. Wirel. Commun. **19**(5), 3185–3198 (2020)
17. J.A. del Peral-Rosado, J. A. López-Salcedo, S. Kim, G. Seco-Granados, Feasibility study of 5G-based localization for assisted driving, in *2016 International conference on localization and GNSS (ICL-GNSS)* (IEEE, 2016), pp. 1–6
18. N. Garcia, H. Wymeersch, E.G. Larsson, A.M. Haimovich, M. Coulon, Direct localization for massive MIMO. IEEE Trans. Signal Process. **65**(10), 2475–2487 (2017)
19. S. Ji, Y. Xue, L. Carin, Bayesian compressive sensing. IEEE Trans. Signal Process. **56**(6), 2346–2356 (2008)
20. Study on Channel Model for Frequencies From 0.5 to 100 GHz, document TR 38.901 V15.0.0, 3GPP (2018)

Part III
Advanced Topics in Future Wireless Networks

Structured Compressive Sensing with a Dynamic Grid 13

In Part II of this book, we addressed standard CS problems under an independent sparse prior. This chapter extends that foundation to a more general class of sparse Bayesian inference problems, focusing on the recovery of structured sparse signals from linear observations where the sensing matrix incorporates an uncertain dynamic grid. We begin by motivating the need for structured CS models, which arise when sparse signals exhibit complex dependencies beyond simple sparsity. To capture these rich structures, we introduce several multi-layer sparse priors and formally formulate the corresponding inference problem. The practical utility of this framework is demonstrated through two key 6G wireless applications: channel estimation in massive MIMO-OFDM systems, encompassing both frequency division duplex (FDD) and time division duplex hybrid beamforming (TDD-HBF) scenarios, and integrated sensing and communication (ISAC). These case studies illustrate how structured sparse Bayesian inference provides a powerful methodology for tackling challenging real-world problems in advanced wireless communications and sensing.

13.1 Motivation

The sparse structure inherent in wireless channels and signals is fundamentally governed by the physical propagation environment. For instance, the limited number of scatterers observed by a mobile station (MS) leads to channel sparsity. More importantly, these scatterers are not uniformly distributed but tend to form structured clusters, giving rise to *structured sparsity* in wireless channels and signals—a property that can be leveraged to design more efficient CS recovery algorithms. Several representative forms of structured sparsity arising from typical propagation scenarios are detailed below.

A common form is *burst sparsity* caused by clustered scatterers. As shown in Fig. 13.1, scatterers in real environments often concentrate in a few clusters. This results in a small number of dominant channel paths, each comprising several subpaths spanning a narrow range of delay and angle. Consequently, significant components in the delay-angular domain channel manifest as contiguous bursts, each corresponding to a scattering cluster. This structure has been utilized in designing the burst-LASSO channel estimator [1].

When multiple users coexist in a shared environment, *joint burst sparsity* may arise due to partially overlapping scattering clusters. As illustrated in Fig. 13.2, users in proximity often observe common scatterers, leading to angular-domain channel supports that partially overlap across users, while each channel still exhibits burst sparsity internally. Such joint sparsity has been exploited to improve compressive channel estimation efficiency in [2, 3].

Another important structure is *common sparsity*, typically caused by collocated antennas or multi-carrier transmission. Due to the small physical size of an MS, different antennas of the same user are effectively co-located and likely to observe the same set of scatterers, leading to a common support in the angular domain. Similarly, in massive MIMO-OFDM systems, channels across subcarriers share a common sparsity pattern, as shown in Fig. 13.3. This property has been utilized in [4] to enhance channel estimation.

In all the above cases, the underlying structured sparsity cannot be adequately captured by independent priors. It is therefore essential to develop structured CS methods that use more sophisticated sparse priors to explicitly model such structures, thereby improving recovery performance and reducing measurement requirements. Furthermore, incorporating a *dynamic grid* in the sensing model can achieve super-resolution estimation of channel parameters like angle and delay, while mitigating energy leakage effects—a crucial capability for high-accuracy channel estimation and wireless sensing in future 6G systems.

A. Liu and M.-M. Zhao, *Advanced Bayesian Signal Processing*, Textbooks in Telecommunication Engineering,
https://doi.org/10.1007/978-3-032-26656-9_13

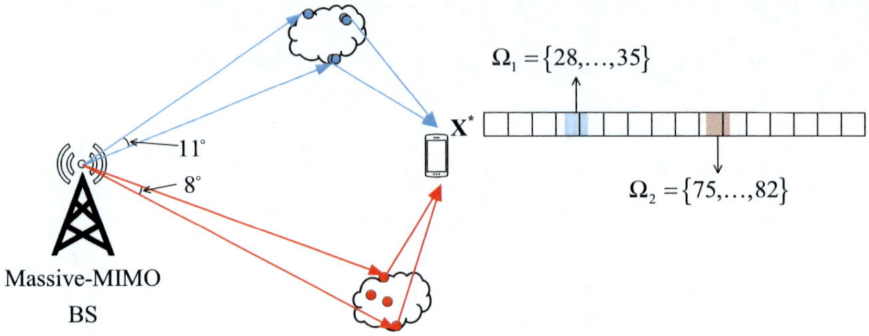

Fig. 13.1 Illustration of the burst sparsity for a massive MIMO system with a ULA of 128 antennas and inter-antenna distance $\lambda/2$ at the BS, where λ is the wavelength of the carrier frequency, Ω_1 and Ω_2 are the index sets of the non-zero elements in the angular-domain channel (adapted from Fig. 1 in [1])

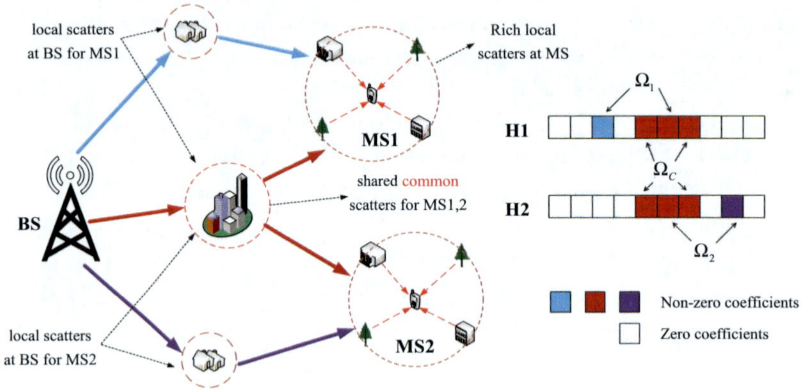

Fig. 13.2 Illustration of the joint burst sparsity (adapted from Fig. 2 in [3])

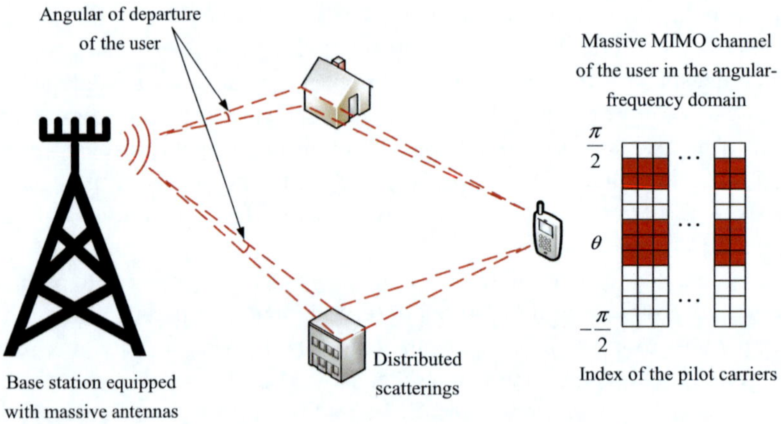

Fig. 13.3 Illustration of the common sparsity among the massive MIMO channels on different subcarriers

13.2 Structured Sparsity Model and Structured CS Problem Formulation

13.2.1 Structured Sparsity Model

A well-defined probabilistic model for structured sparsity is fundamental to leveraging specific sparse patterns effectively. An ideal sparse probability model should satisfy three key criteria: *flexibility* to accurately capture the underlying signal structure, *robustness* to imperfect prior information in practical scenarios, and *tractability* to facilitate the design of high-performance, low-complexity inference algorithms. In the following, we introduce three sparse prior models, each suited to different classes of sparse Bayesian inference algorithms.

13.2.1.1 Precision-Based Two-Layer Sparse Prior Model

The precision-based hierarchical prior serves as the foundational framework for many sparse Bayesian learning and variational Bayesian inference methods, employing an elegant two-layer structure that effectively promotes sparsity while maintaining analytical tractability. At the core of this approach lies a precision vector $\boldsymbol{\rho} = [\rho_1, \ldots, \rho_N]^T$, where each element ρ_n controls the inverse variance of the corresponding signal component x_n. This precision parameter functions as an effective activity indicator: values of order $\rho_n = \Theta(1)$ suggest active components with significant amplitude, while $\rho_n \gg 1$ effectively drives the corresponding x_n toward zero, enforcing sparsity through variance suppression.

The complete hierarchical specification is given below:

1. **Conditional signal distribution**:

$$p(\boldsymbol{x}|\boldsymbol{\rho}) = \prod_{n=1}^{N} \mathcal{CN}\left(x_n; 0, \rho_n^{-1}\right). \tag{13.1}$$

2. **Precision prior distribution**:

$$p(\boldsymbol{\rho}) = \prod_{n=1}^{N} \Gamma\left(\rho_n; a_n, b_n\right). \tag{13.2}$$

This construction, known as the independent Gamma-Gaussian (GG) prior, typically employs small values for the hyperparameters a_n, b_n to enforce sparsity by favoring precision values that strongly shrink insignificant coefficients toward zero.

A significant advantage of this precision-based model is its inherent robustness to uncertainties in the sensing matrix, coupled with natural extensibility to group sparsity patterns through shared precision parameters across elements within the same group. This flexibility makes the framework particularly suitable for applications where signal components exhibit block-sparse characteristics, as the shared precision mechanism automatically induces correlated sparsity patterns within predefined groups.

Despite its mathematical elegance and computational advantages, the precision-based model faces fundamental theoretical limitations that constrain its applicability to complex real-world scenarios. The requirement for independent Gamma-distributed precisions, while crucial for maintaining tractability in SBL and VBI algorithms, intrinsically limits its capacity to capture sophisticated dependency structures. Specifically, the model cannot represent Markovian dependencies between adjacent components [5], hidden Markov model structures [6], or arbitrary correlation patterns among precision parameters that often arise in practical applications. Furthermore, the framework lacks a natural mechanism to incorporate prior support information—a significant limitation in scenarios where partial knowledge of component activation probabilities is available from domain expertise or previous measurements [7]. These theoretical constraints motivate the development of more sophisticated prior models that can accommodate structured dependencies while preserving computational feasibility, which we will explore in subsequent sections.

13.2.1.2 Support-Based Two-Layer Sparse Prior Model

In contrast to precision-based approaches, support-based sparse modeling provides an alternative framework particularly suited for capturing structured sparsity patterns. This methodology forms the theoretical foundation for many AMP algorithms and employs a discrete-continuous mixture distribution to explicitly model coefficient activity through a fundamentally different mechanism.

The core of this approach centers on a binary support vector $\boldsymbol{s} = [s_1, \ldots, s_N]^T$, where $s_n = 1$ indicates an active (non-zero) coefficient x_n, while $s_n = 0$ corresponds to an inactive element that is explicitly set to zero. The conditional distribution of the sparse signal $\boldsymbol{x}$ given the support follows a discrete-continuous mixture structure defined as

$$p(\boldsymbol{x}|\boldsymbol{s}) = \prod_{n=1}^{N} p\left(x_n \mid s_n\right), \tag{13.3}$$

with the elemental distribution taking the form:

$$p\left(x_n|s_n\right) = s_n g_n\left(x_n\right) + \left(1 - s_n\right)\delta\left(x_n\right). \tag{13.4}$$

Here, $g_n(x_n)$ represents the probability density function of active coefficients, typically modeled as Gaussian for analytical tractability. When $g_n(x_n) = \mathcal{CN}(x_n; \mu_n, \sigma_n^2)$, this formulation specializes to the well-known Bernoulli–Gaussian (BG) prior, which provides a balanced compromise between model flexibility and computational feasibility.

The principal advantage of this support-based framework lies in its enhanced capacity to encode complex structured sparsity patterns through deliberate design of the support prior $p(\boldsymbol{s})$. By incorporating appropriate correlation structures into this prior distribution, the model can capture various sophisticated sparsity configurations including Markov chain sparsity [9], Markov tree sparsity [5], and other spatially correlated support patterns that commonly arise in practical applications such as image processing and communications. This structural flexibility represents a significant advantage over precision-based methods, which struggle to explicitly model such dependencies due to their continuous nature.

However, the discrete nature of the support vector $\boldsymbol{s}$ introduces significant computational challenges that cannot be overlooked. Direct optimization and inference become combinatorially complex and computationally intractable, as the number of possible support configurations grows exponentially with the signal dimension. Consequently, support-based models are predominantly implemented within AMP frameworks, which provide efficient approximate Bayesian inference through iterative message passing algorithms that circumvent the combinatorial explosion.

This inherent dependency on AMP algorithms introduces an important theoretical constraint: performance guarantees and convergence properties typically require the sensing matrix to satisfy specific properties, such as being i.i.d. Gaussian or possessing other well-behaved random matrix characteristics. Consequently, the applicability of support-based models becomes restricted to scenarios where these matrix conditions are satisfied—a notable limitation in practical applications with structured sensing operators like partial Fourier matrices or spatially correlated measurement systems encountered in radar and medical imaging.

13.2.1.3 Three-Layer Sparse Prior Model

To effectively capture the diverse and complex structured sparsity patterns encountered in practical wireless applications, this chapter introduces a three-layer hierarchical sparse prior model. Originally proposed in [10], this hybrid framework strategically combines the strengths of two complementary approaches: the explicit group-level activity modeling characteristic of support-based methods and the fine-grained variance control mechanism inherent to precision-based techniques.

Model Structure and Rationale

The construction begins by partitioning the index set $\{1, \ldots, N\}$ of the sparse signal $\boldsymbol{x}$ into Q non-overlapping subsets $\mathcal{I}_1, \ldots, \mathcal{I}_Q$. Within each subvector $\boldsymbol{x}[\mathcal{I}_i]$, all elements share a common activity state, reflecting structural dependencies often present in real-world data, for instance, wavelet coefficients in image processing naturally form clusters, and multipath components in wireless communications tend to arrive in groups.

To capture this group-level structure, we introduce a binary support vector $\boldsymbol{s} = [s_1, \ldots, s_Q]^T \in \{0, 1\}^Q$, where $s_i = 1$ indicates that the i-th group is active (containing non-zero elements), and $s_i = 0$ signifies an inactive group. This macroscopic representation encodes block-sparse patterns at the group level.

Complementing this coarse-grained structure, a precision vector $\boldsymbol{\rho} = [\rho_1, \ldots, \rho_N]^T$ provides fine-grained control at the individual coefficient level, with $1/\rho_n$ representing the variance of x_n. For inactive groups where $s_i = 0$, the hyperparameters are selected to satisfy $\mathbb{E}[\rho_n] \gg 1$, effectively driving the variance toward zero and ensuring comprehensive suppression of irrelevant coefficients. This dual-layer control mechanism enables the model to simultaneously capture structured sparsity at the group level while preserving the robustness benefits of precision-based shrinkage at the element level.

To accommodate imperfect prior knowledge and enable adaptive learning from data, the support distribution $p(\boldsymbol{s}|\boldsymbol{\phi})$ is parameterized by uncertain parameters $\boldsymbol{\phi}$ with a known hyperprior $p(\boldsymbol{\phi})$. This creates a hierarchical Bayesian structure that can automatically adjust to observed sparsity patterns. The complete three-layer prior is formally expressed as

$$p(\boldsymbol{x}, \boldsymbol{\rho}, \boldsymbol{s}|\boldsymbol{\phi}) = \underbrace{p(\boldsymbol{s}|\boldsymbol{\phi})}_{\text{Support}} \underbrace{p(\boldsymbol{\rho}|\boldsymbol{s})}_{\text{Precision}} \underbrace{p(\boldsymbol{x}|\boldsymbol{\rho})}_{\text{Sparse signal}}, \tag{13.5}$$

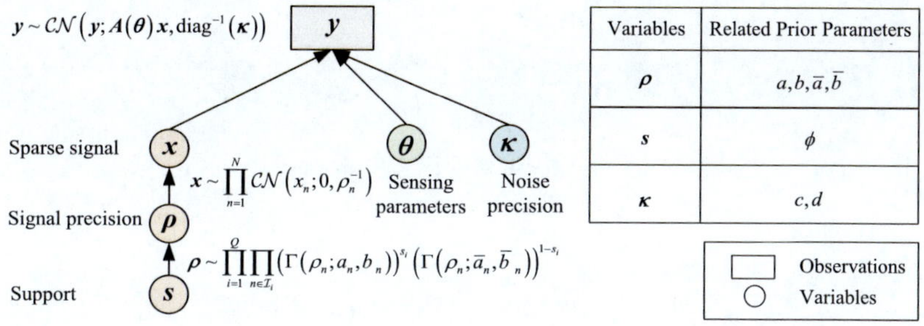

Fig. 13.4 Three-layer sparse prior model (adapted from Fig. 1 in [10])

which we refer to as the Bernoulli–Gamma–Gaussian (BGG) prior, illustrated in Fig. 13.4. This integrated framework leverages the complementary strengths of both foundational approaches: the support layer efficiently encodes structured sparsity patterns, the precision layer provides robust coefficient-level regularization, and the full hierarchical specification facilitates adaptive learning of model parameters from data. In the following, we elaborate the statistical specifications of each layer.

Layer 1: Support Modeling

The prior $p(\boldsymbol{s}|\boldsymbol{\phi})$ is designed to capture application-specific structured sparsity patterns. For example, in massive MIMO channel estimation where scatterers tend to cluster [5], a Markov chain prior can effectively model sequential dependencies in support activities. In such cases, $\boldsymbol{\phi}$ represents transition probabilities that can be learned from data during inference.

Layer 2: Precision Modeling

Conditioned on the support vector, the precision parameters follow a Gamma distribution with support-dependent hyperparameters

$$p(\boldsymbol{\rho}|\boldsymbol{s}) = \prod_{i=1}^{Q}\prod_{n\in\mathcal{I}_i} (\Gamma(\rho_n; a_n, b_n))^{s_i} \left(\Gamma(\rho_n; \bar{a}_n, \bar{b}_n)\right)^{1-s_i}, \tag{13.6}$$

where $\Gamma(\rho; a, b)$ denotes the Gamma distribution. For active groups where $s_i = 1$, the hyperparameters are chosen such that $\mathbb{E}[\rho_n] = a_n/b_n = \Theta(1)$, allowing moderate variance for non-zero coefficients. Conversely, for inactive groups, we enforce $\mathbb{E}[\rho_n] = \bar{a}_n/\bar{b}_n \gg 1$, forcing the variance toward zero. This design offers two crucial advantages: (i) the Gamma prior is conjugate to the Gaussian signal layer, enabling closed-form Bayesian updates; (ii) the support-dependent parameterization directly embeds sparsity structure into the precision specification.

Layer 3: Sparse Signal Modeling

At the most fundamental level, the signal layer follows a conditional Gaussian distribution:

$$p(\boldsymbol{x}|\boldsymbol{\rho}) = \prod_{n=1}^{N} \mathcal{CN}(x_n; 0, \rho_n^{-1}). \tag{13.7}$$

This choice is motivated by several theoretical and practical considerations that have established Gaussian conditionals as a standard building block in sparse Bayesian learning. First, the central limit theorem provides a theoretical foundation for Gaussian modeling, ensuring robust performance across diverse application domains where exact signal statistics may be unknown. Second, the conditional Gaussian structure affords exceptional tractability in variational Bayesian inference [11], enabling efficient computation of posterior distributions through closed-form updates, a critical advantage over alternative distributional assumptions that would render inference intractable. Third, extensive empirical evidence suggests that CS recovery performance depends primarily on first-order sparsity structure rather than exact signal distributions [12], making the Gaussian approximation both sufficient and near optimal for most practical scenarios while avoiding unnecessary model complexity.

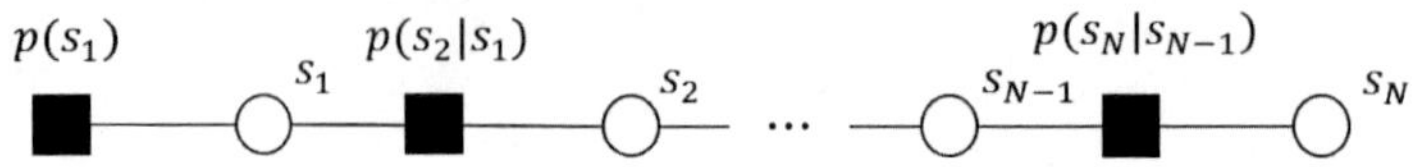

Fig. 13.5 Factor graph representation of the Markov chain $p(\boldsymbol{s})$ for modeling burst sparsity

Example 13.1 (*Markov Chain Model for Burst Sparsity*) To demonstrate how the support prior $p(\boldsymbol{s})$ can be tailored to specific sparsity patterns, consider the case of burst sparsity arising from clustered scatterers. This structure can be effectively modeled using a first-order Markov chain as

$$p(\boldsymbol{s}) = p(s_1)\prod_{n=2}^{N} p(s_n|s_{n-1}), \tag{13.8}$$

with transition probabilities

$$p(s_n|s_{n-1}) = \begin{cases} (1-p_{01})^{1-s_n} p_{01}^{s_n}, & s_{n-1}=0, \\ p_{10}^{1-s_n}(1-p_{10})^{s_n}, & s_{n-1}=1. \end{cases} \tag{13.9}$$

The parameters p_{10} and p_{01} govern the cluster characteristics: smaller p_{10} values promote larger contiguous clusters by reducing the probability of transitioning from active to inactive states, while smaller p_{01} values increase the spacing between clusters by making transitions from inactive to active states less likely. The initial distribution is set to the stationary distribution

$$p(s_1 = 1) = \lambda \triangleq \frac{p_{01}}{p_{01}+p_{10}}, \tag{13.10}$$

ensuring homogeneous marginal distributions $p(s_n) = \lambda^{s_n}(1-\lambda)^{1-s_n}$ for all n. The factor graph representation of this Markov chain is illustrated in Fig. 13.5.

13.2.2 Structured CS Problem Formulation with a Dynamic Grid

We begin by recalling the CS model with dynamic grid parameterization of the sensing matrix:

$$\boldsymbol{y} = \boldsymbol{A}(\boldsymbol{\theta})\boldsymbol{x} + \boldsymbol{w}, \tag{13.11}$$

where $\boldsymbol{w} \in \mathbb{C}^{M\times 1}$ represents additive noise with independent circularly symmetric Gaussian components $w_m \sim \mathcal{CN}(0, \kappa_m^{-1})$. The noise precision vector $\boldsymbol{\kappa} = [\kappa_1, \ldots, \kappa_M]^T$ is assigned a Gamma prior as

$$p(\kappa_m) = \Gamma(\kappa_m; c, d), \quad \forall m, \tag{13.12}$$

thereby modeling uncertainty in noise levels within the Bayesian framework. The following development focuses on the three-layer BGG prior model for structured sparsity, though the formulation naturally extends to support-based two-layer BG priors.

Let $p(\boldsymbol{\theta})$ and $p(\boldsymbol{\phi})$ denote the prior distributions for the dynamic grid parameters and prior hyperparameters, respectively. Our primary objectives are to estimate:

1. the sparse signal $\boldsymbol{x}$,
2. the support vector $\boldsymbol{s}$, and
3. the uncertain parameter vector $\boldsymbol{\xi} = [\boldsymbol{\theta}; \boldsymbol{\phi}; \boldsymbol{\kappa}]$

given the observations $\boldsymbol{y}$.

For fixed $\boldsymbol{\xi}$, we seek the conditional marginal posteriors for the sparse signal:

$$\begin{aligned} p(\boldsymbol{x}|\boldsymbol{y},\boldsymbol{\xi}) &\propto \sum_{\boldsymbol{s}} \int p(\boldsymbol{y},\boldsymbol{x},\boldsymbol{\rho},\boldsymbol{s}|\boldsymbol{\xi})\, d\boldsymbol{\rho} \\ &= \sum_{\boldsymbol{s}} \int p(\boldsymbol{x},\boldsymbol{\rho},\boldsymbol{s}|\boldsymbol{\phi})\, p(\boldsymbol{y}|\boldsymbol{x},\boldsymbol{\xi})\, d\boldsymbol{\rho}, \end{aligned} \tag{13.13}$$

and for each support element

$$p(s_i|\boldsymbol{y},\boldsymbol{\xi}) \propto \sum_{\boldsymbol{s}_{-i}} \iint p(\boldsymbol{x},\boldsymbol{\rho},\boldsymbol{s}|\boldsymbol{\phi})\, p(\boldsymbol{y}|\boldsymbol{x},\boldsymbol{\xi})\, d\boldsymbol{\rho} d\boldsymbol{x}, \tag{13.14}$$

where the likelihood function is

$$p(\boldsymbol{y}|\boldsymbol{x},\boldsymbol{\xi}) = \mathcal{CN}\left(\boldsymbol{y}; \boldsymbol{A}(\boldsymbol{\theta})\boldsymbol{x}, \operatorname{diag}(\boldsymbol{\kappa})^{-1}\right). \tag{13.15}$$

Here, $\boldsymbol{s}_{-i}$ denotes all support variables excluding s_i, and $\propto$ indicates proportionality up to a normalizing constant.

For parameter estimation, the uncertain parameters are determined via MAP inference as

$$\boldsymbol{\xi}^* = \arg\max_{\boldsymbol{\xi}} \ln p(\boldsymbol{\xi}|\boldsymbol{y}) = \arg\max_{\boldsymbol{\xi}} \ln \sum_{\boldsymbol{s}} \iint p(\boldsymbol{y},\boldsymbol{v},\boldsymbol{\xi})\, d\boldsymbol{\rho} d\boldsymbol{x}, \tag{13.16}$$

where $\boldsymbol{v} = \{\boldsymbol{x},\boldsymbol{\rho},\boldsymbol{s}\}$ comprises all latent variables.

Given the optimal parameters $\boldsymbol{\xi}^*$, we obtain the conditional MAP estimators as

$$\boldsymbol{x}^* = \arg\max_{\boldsymbol{x}} p\left(\boldsymbol{x} \mid \boldsymbol{y},\boldsymbol{\xi}^*\right), \quad s_i^* = \arg\max_{s_i} p\left(s_i \mid \boldsymbol{y},\boldsymbol{\xi}^*\right). \tag{13.17}$$

When the posterior of $\boldsymbol{x}$ is approximately Gaussian, the MAP estimator coincides with the MMSE estimator.

From a computational perspective, exact computation of these posteriors is generally intractable due to the loopy factor graph structure of the underlying model. This fundamental challenge motivates the approximate inference methods developed in subsequent chapters, where we introduce Turbo sparse Bayesian inference algorithms to efficiently approximate the required marginal distributions. Depending on specific algorithm implementations, the noise precision $\boldsymbol{\kappa}$ can be estimated either via EM (point estimate) or variational Bayesian inference (full Bayesian estimate), offering flexibility in balancing computational efficiency with statistical accuracy.

This structured CS formulation with dynamic grid parameters provides a comprehensive framework applicable to numerous practical scenarios. Its versatility will be demonstrated through two important 6G wireless communication applications in subsequent sections, showcasing how the theoretical framework translates to solving real-world signal processing challenges in advanced communication systems.

13.3 Channel Estimation for Massive MIMO-OFDM Systems

Massive MIMO technology stands as a cornerstone of 5G and emerging 6G wireless systems, offering substantial spatial multiplexing and array gains that significantly enhance both channel capacity and energy efficiency. When combined with orthogonal frequency division multiplexing (OFDM)—a waveform widely adopted in 4G and 5G systems for its high spectral efficiency, robustness to multipath fading, and efficient FFT implementation—massive MIMO-OFDM systems unlock unprecedented spectral and spatial gains. However, realizing these benefits requires accurate channel state information (CSI) acquisition in dynamically varying wireless environments, a challenging task given the high dimensionality of massive MIMO-OFDM channels.

The difficulties in CSI estimation vary significantly between FDD and TDD system configurations:

- **In FDD systems**, CSI is typically acquired through downlink pilot transmission to user equipment, followed by CSI feedback to the BS. Conventional approaches require pilot symbols and feedback overhead that scale linearly with the

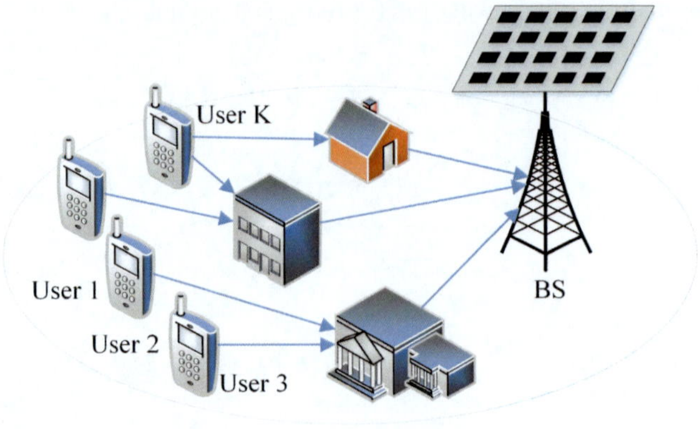

Fig. 13.6 An illustration of massive MIMO-OFDM system

number of BS antennas, becoming prohibitively expensive in massive MIMO deployments. Additional complications arise from real-time processing requirements, irregular array geometries [13], and beam squint effects [14].

- **In TDD systems**, channel reciprocity enables the BS to estimate downlink channels directly from uplink pilots (known as sounding reference signals in 5G NR [15]). While this significantly reduces pilot overhead compared to FDD systems, practical implementations often employ hybrid beamforming (HBF) architectures where the number of radio frequency (RF) chains is substantially smaller than the number of antennas due to cost and power constraints, introducing additional complexity to the channel estimation problem.

This section proceeds as follows: we first establish the channel model for massive MIMO-OFDM systems, then formulate channel estimation problems for both FDD and TDD HBF scenarios. Importantly, we demonstrate how both problems can be naturally cast as special cases of the structured CS framework with a dynamic grid introduced earlier.

13.3.1 Channel Model for Massive MIMO-OFDM Systems

We consider a massive MIMO-OFDM system where the BS employs a uniform planar array (UPA) with $N_b = N_x \times N_y$ antenna elements arranged in N_x horizontal and N_y vertical positions, while the user equipment utilizes a ULA with N_u antenna elements, as illustrated in Fig. 13.6.

The uplink channel frequency response (CFR) matrix $\boldsymbol{H}[n] \in \mathbb{C}^{N_b \times N_u}$ at subcarrier n $(0 \leq n \leq N-1)$ is characterized by the following geometric channel model with K propagation paths:

$$\boldsymbol{H}[n] = \sum_{k=1}^{K} \alpha_k e^{-j2\pi n f_s \tau_k} \boldsymbol{a}_R(\theta_k, \phi_k) \boldsymbol{a}_T^T(\bar{\theta}_k), \tag{13.18}$$

where $\alpha_k \in \mathbb{C}$ represents the complex gain of the k-th path, τ_k denotes the time delay, f_s is the subcarrier spacing, θ_k and ϕ_k are the azimuth and elevation angles of arrival at the BS, and $\bar{\theta}_k$ is the angle of departure at the user equipment. This multipath representation captures the essential spatial and frequency characteristics of the wireless propagation environment.

Using the coordinate system in Fig. 13.7, the BS's UPA steering vector can be decomposed as

$$\boldsymbol{a}_R(\theta, \phi) = \boldsymbol{a}_x(\theta, \phi) \otimes \boldsymbol{a}_y(\phi) \in \mathbb{C}^{N_b \times 1}, \tag{13.19}$$

where the horizontal and vertical components are given by [16]

$$[\boldsymbol{a}_x(\theta, \phi)]_{n_x} = \frac{1}{\sqrt{N_x}} e^{j\pi(n_x-1)\sin(\theta)\cos(\phi)}, \quad n_x = 1, \ldots, N_x, \tag{13.20}$$

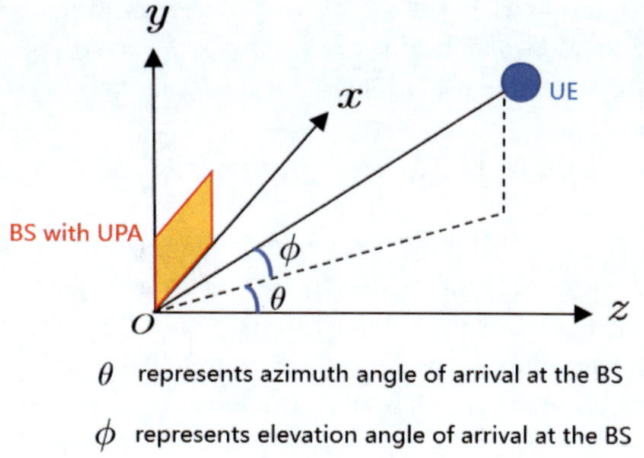

Fig. 13.7 Illustration of the BS's UPA steering vector

$$\left[\boldsymbol{a}_y(\phi)\right]_{n_y} = \frac{1}{\sqrt{N_y}} e^{j\pi(n_y-1)\sin(\phi)}, \quad n_y = 1, \ldots, N_y, \tag{13.21}$$

and $\otimes$ denotes the Kronecker product. Complementing this, the user equipment's ULA steering vector is given by

$$\left[\boldsymbol{a}_T(\bar{\theta})\right]_n = \frac{1}{\sqrt{N_u}} e^{-j(n-1)\pi\sin(\bar{\theta})}, \quad n = 1, \ldots, N_u. \tag{13.22}$$

These array response formulations enable precise modeling of the spatial characteristics essential for massive MIMO beamforming and spatial multiplexing.

For the downlink channel, the CFR matrix exhibits a similar structure as

$$\bar{\boldsymbol{H}}[n] = \sum_{k=1}^{K} \bar{\alpha}_k e^{-j2\pi n f_s \tau_k} \boldsymbol{a}_T(\bar{\theta}_k)\, \boldsymbol{a}_R^T(\theta_k, \phi_k), \tag{13.23}$$

with an important distinction between duplexing modes: in TDD systems, channel reciprocity typically yields $\bar{\alpha}_k = \alpha_k^*$, whereas in FDD systems, no such reciprocity exists between α_k and $\bar{\alpha}_k$, necessitating separate estimation procedures for uplink and downlink channels.

For single-antenna users ($N_u = 1$), the channel representations simplify to

$$\boldsymbol{h}[n] = \sum_{k=1}^{K} \alpha_k e^{-j2\pi n f_s \tau_k} \boldsymbol{a}_R(\theta_k, \phi_k), \tag{13.24}$$

$$\bar{\boldsymbol{h}}[n] = \sum_{k=1}^{K} \bar{\alpha}_k e^{-j2\pi n f_s \tau_k} \boldsymbol{a}_R^T(\theta_k, \phi_k). \tag{13.25}$$

This parametric channel model effectively captures the joint spatial-frequency characteristics of massive MIMO-OFDM systems and provides the mathematical foundation for developing advanced channel estimation techniques in subsequent sections, enabling efficient exploitation of the abundant spatial degrees of freedom available in massive antenna arrays.

13.3.2 Channel Estimation for FDD Systems

13.3.2.1 Downlink Pilot Signal Model

We consider an FDD massive MIMO-OFDM system where a BS serves multiple single-antenna users. Focusing on a reference user, we examine the downlink channel estimation problem during the pilot transmission phase.

The BS transmits M training OFDM symbols $\boldsymbol{x}_m[n] \in \mathbb{C}^{N_b \times 1}$ over successive time slots, where $n \in \mathcal{N}_p$ denote the pilot subcarrier indices, and $m = 1, \ldots, M$. The pilot subcarrier set $\mathcal{N}_p$ contains $P = |\mathcal{N}_p|$ elements.

The received signal $\boldsymbol{y}[n] \in \mathbb{C}^{M \times 1}$ at the reference user on pilot subcarrier n is given by

$$\boldsymbol{y}[n] = \boldsymbol{X}\bar{\boldsymbol{h}}^T[n] + \boldsymbol{w}[n], \quad n \in \mathcal{N}_p, \tag{13.26}$$

where $\boldsymbol{X} = [\boldsymbol{x}_1, \ldots, \boldsymbol{x}_M]^T \in \mathbb{C}^{M \times N_b}$ denotes the pilot matrix, which is assumed to be identical across all pilot subcarriers for implementation simplicity; $\bar{\boldsymbol{h}}[n]$ is the downlink channel vector at subcarrier n, as previously defined in Eq. (13.25); and $\boldsymbol{w}[n] \sim \mathcal{CN}(\boldsymbol{0}, \sigma^2\boldsymbol{I})$ represents AWGN that corrupts the received signal.

This signal model forms the basis for developing structured CS-based channel estimation techniques that exploit the inherent sparsity of massive MIMO channels in the angular-delay domain.

13.3.2.2 CS Model Formulation with a Dynamic Grid

To formulate the channel estimation problem within a CS framework, we first construct a sparse representation of the signal model in the delay-angular domain by defining discrete grids over the potential parameter ranges. This discretization approach transforms the continuous parameter estimation problem into a sparse recovery task that can be efficiently solved using CS techniques. Specifically, we establish a uniform delay grid $\bar{\boldsymbol{\tau}} = \{\bar{\tau}_1, \ldots, \bar{\tau}_L\}$ with $L \gg K$ points spanning $[-\frac{1}{4}T_d, T_d]$, where T_d denotes the upper bound of maximum delay spread, along with uniform angle grids $\bar{\boldsymbol{\theta}} = \{\bar{\theta}_1, \ldots, \bar{\theta}_{N_1}\}$ and $\bar{\boldsymbol{\phi}} = \{\bar{\phi}_1, \ldots, \bar{\phi}_{N_2}\}$ with N_1 and N_2 points covering $[-\frac{\pi}{2}, \frac{\pi}{2}]$. The inclusion of negative delays in the grid design accounts for potential time synchronization errors at the receiver, while maintaining the practical assumption that such errors remain small relative to $\frac{1}{4}T_d$ to ensure model validity.

In the ideal case where the true delay and angle parameters align perfectly with the discrete grids, an assumption that provides a theoretical foundation though rarely holds exactly in practice, the received signal model can be reformulated as

$$\boldsymbol{Y} = \boldsymbol{X}\boldsymbol{A}_R\left(\bar{\boldsymbol{\theta}}, \bar{\boldsymbol{\phi}}\right)\boldsymbol{H}_d\boldsymbol{F}_{\mathcal{N}_p}^T(\bar{\boldsymbol{\tau}}) + \boldsymbol{W}, \tag{13.27}$$

where $\boldsymbol{Y} = [\boldsymbol{y}[n]]_{n \in \mathcal{N}_p} \in \mathbb{C}^{M \times P}$ and $\boldsymbol{W} = [\boldsymbol{w}[n]]_{n \in \mathcal{N}_p} \in \mathbb{C}^{M \times P}$ are the concatenated received signal and noise matrices across all pilot subcarriers. The delay dictionary matrix

$$\boldsymbol{F}(\bar{\boldsymbol{\tau}}) = [\boldsymbol{a}(\bar{\tau}_1), \boldsymbol{a}(\bar{\tau}_2), \ldots, \boldsymbol{a}(\bar{\tau}_L)] \tag{13.28}$$

is constructed with steering vectors $\boldsymbol{a}(\tau) = [1, e^{-j2\pi f_s \tau}, \ldots, e^{-j2\pi(N-1)f_s\tau}]^T \in \mathbb{C}^{N \times 1}$, while $\boldsymbol{F}_{\mathcal{N}_p}(\bar{\boldsymbol{\tau}}) \in \mathbb{C}^{P \times L}$ contains the P rows of $\boldsymbol{F}(\bar{\boldsymbol{\tau}})$ corresponding specifically to pilot subcarriers. The angular dictionary matrix

$$\boldsymbol{A}_R\left(\bar{\boldsymbol{\theta}}, \bar{\boldsymbol{\phi}}\right) = \left[\boldsymbol{a}_R\left(\bar{\theta}_1, \bar{\phi}_1\right), \ldots, \boldsymbol{a}_R\left(\bar{\theta}_{N_1}, \bar{\phi}_{N_2}\right)\right] \in \mathbb{C}^{N_b \times Q}$$

with $Q = N_1 \times N_2$ captures the spatial response of the BS array across the discretized angle space. The matrix $\boldsymbol{H}_d \in \mathbb{C}^{Q \times L}$ represents the sparse delay-angular-domain (DAD) channel matrix, where non-zero elements correspond to active path complex gains at specific grid points, effectively encoding the channel's inherent sparsity in this transformed domain.

Example 13.2 (*Uniform Grids Construction for ULA and UPA*) **For a ULA with 32 antennas**, a uniform grid of Q AoA points can be defined such that $[\sin\bar{\theta}_1, \ldots, \sin\bar{\theta}_Q]$ is uniformly spaced over $[-1, 1]$, i.e.,

$$\sin\bar{\theta}_i = -\frac{Q-1}{Q} + \frac{2(i-1)}{Q}, i = 1, \ldots, Q. \tag{13.29}$$

When $Q = N_b = 32$, the array response matrix $\boldsymbol{A}_R(\bar{\boldsymbol{\theta}})$ becomes unitary and equivalent to a DFT dictionary.

For a UPA with 8 × 8 antennas, we define $Q \triangleq N_1 \times N_2$ angle grid points $(\bar{\boldsymbol{\theta}}, \bar{\boldsymbol{\phi}}) = \{(\bar{\theta}_1, \bar{\phi}_1), \ldots, (\bar{\theta}_Q, \bar{\phi}_Q)\}$ satisfying

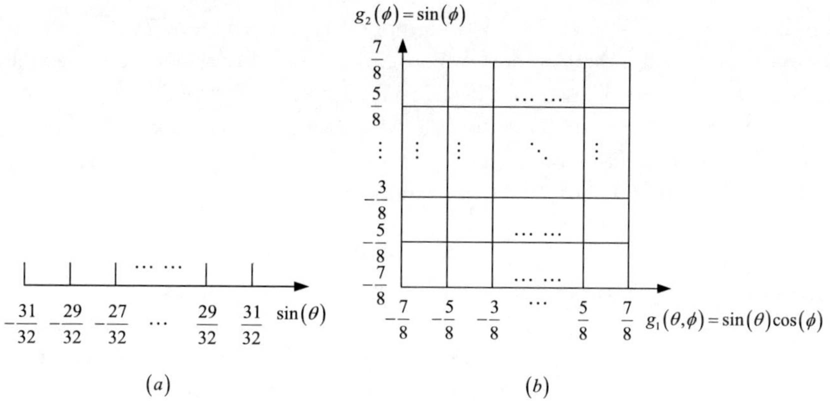

Fig. 13.8 An illustration of (**a**) the one-dimensional uniform grid for ULA when $Q = 32$ and (**b**) two-dimensional uniform grids for UPA when $N_1 = 8$ and $N_2 = 8$

$$g_1\left(\bar{\theta}_{i\times j}, \bar{\phi}_{i\times j}\right) = -\frac{N_1 - 1}{N_1} + \frac{2(i-1)}{N_1},$$
$$g_2\left(\bar{\phi}_{i\times j}\right) = -\frac{N_2 - 1}{N_2} + \frac{2(j-1)}{N_2}, \tag{13.30}$$

for $i = 1, \ldots, N_1$ and $j = 1, \ldots, N_2$, where $g_1(\theta, \phi) = \sin(\theta)\cos(\phi)$, $g_2(\phi) = \sin(\phi)$ (Fig. 13.8). This creates a uniform grid in the transformed angular domain (g_1, g_2). When $N_1 = N_x$ and $N_2 = N_y$, $\boldsymbol{A}_R(\bar{\boldsymbol{\theta}}, \bar{\boldsymbol{\phi}})$ becomes the Kronecker product of two DFT matrices.

Oversampling considerations: Practical implementations often employ oversampled grids (e.g., $N_1 = N_o N_x$ and $N_2 = N_o N_y$ with oversampling rate N_o between 1 and 4) to enhance angular resolution, though this increases computational complexity and basis correlation.

Delay grid construction: For an OFDM system with $N = 256$ subcarriers and CP length $L_{CP} = 32$, a uniform delay grid $\bar{\boldsymbol{\tau}} = [-\frac{L_{CP}}{4B}, -\frac{L_{CP}}{4B} + \frac{1}{B}, -\frac{L_{CP}}{4B} + \frac{2}{B}, \ldots, \frac{L_{CP}}{B}]$ safely covers the maximum delay spread, where B is the system bandwidth.

In practical scenarios where true channel parameters rarely align perfectly with predefined discrete grids, energy leakage occurs across adjacent grid points, significantly degrading estimation performance. To mitigate this fundamental limitation, we introduce dynamic grids $\boldsymbol{\tau}, \boldsymbol{\theta}, \boldsymbol{\phi}$ that are initialized from uniform baseline grids but are subsequently adaptively adjusted to better match the true underlying channel parameters through an iterative refinement process. This adaptive approach leads to the final compressed sensing formulation expressed as

$$\boldsymbol{y} = \boldsymbol{A}(\boldsymbol{\tau}, \boldsymbol{\theta}, \boldsymbol{\phi})\, \boldsymbol{h}_d + \boldsymbol{w}, \tag{13.31}$$

where the sensing matrix

$$\boldsymbol{A}(\boldsymbol{\tau}, \boldsymbol{\theta}, \boldsymbol{\phi}) = \boldsymbol{F}_{N_p}(\boldsymbol{\tau}) \otimes \boldsymbol{X}\boldsymbol{A}_R(\boldsymbol{\theta}, \boldsymbol{\phi}) \in \mathbb{C}^{PM\times LQ} \tag{13.32}$$

combines the delay and angular dictionaries through Kronecker product structure, $\boldsymbol{h}_d \triangleq \mathrm{vec}(\boldsymbol{H}_d) \in \mathbb{C}^{LQ\times 1}$ represents the vectorized sparse DAD channel coefficients, $\boldsymbol{y} \triangleq \mathrm{vec}(\boldsymbol{Y}) \in \mathbb{C}^{MP\times 1}$ is the vectorized received signal across all pilot subcarriers, and $\boldsymbol{w} \triangleq \mathrm{vec}(\boldsymbol{W}) \in \mathbb{C}^{MP\times 1}$ denotes the concatenated noise vector. This dynamic grid approach enables super-resolution parameter estimation by continuously refining the dictionary matrices to reduce model mismatch, while simultaneously maintaining the sparse structural representation essential for efficient CS recovery algorithms to operate effectively.

It is important to mention that the structured sparsity essential for efficient channel estimation is sometimes hidden and not directly exploitable in the original signal domain. In such cases, sparsity-aware methods must first construct a transformed representation to reveal the underlying low-dimensional structure. For instance, one representative approach [17] formulates the problem within a tensor framework, where the channel is decomposed into a polar-angle-delay domain representation.

This transformation is key to exposing a hierarchical spatio-temporal sparsity pattern, encompassing 2D block sparsity, 1D clustered sparsity, and temporal correlation that is otherwise obscured. The subsequent recovery algorithm, which integrates coarse tensor matching pursuit with a refined stochastic PVBI (will be introduced in Chap. 17), demonstrates a methodology for converting a high-dimensional estimation problem into a tractable sparse recovery one. This highlights a broader principle: effectively leveraging sparsity in complex systems frequently requires careful domain transformation and structured prior modeling, rather than direct application of standard sparse techniques.

13.3.2.3 Support-Based Two-Layer Sparse Prior Model for the DAD Channel

The DAD channel matrix $\boldsymbol{H}_d$ possesses a rich structural sparsity that can be effectively modeled using a two-layer hierarchical framework. This model introduces two distinct levels of binary latent variables to capture the complex dependencies inherent in the channel, specifically addressing sparsity patterns across both delay taps and angular bins.

Let $h^l_{d,q}$ denote the (q, l)-th element in $\boldsymbol{H}_d$, the channel coefficient corresponding to the l-th delay tap and q-th angular bin. At the most granular level, each individual channel coefficient $h^l_{d,q}$ is associated with a binary latent state $s^l_{d,q} \in \{0, 1\}$ that indicates its support. The conditional distribution of the coefficient given this state follows a spike-and-slab formulation:

$$p\left(h^l_{d,q} | s^l_{d,q}\right) = \left(1 - s^l_{d,q}\right) \delta\left(h^l_{d,q}\right) + s^l_{d,q} \mathcal{CN}\left(h^l_{d,q}; 0, \left(\sigma^l_d\right)^2\right), \tag{13.33}$$

where $\delta(\cdot)$ denotes the Dirac delta function. This construction ensures that inactive coefficients ($s^l_{d,q} = 0$) are deterministically zero, while active coefficients ($s^l_{d,q} = 1$) follow a complex Gaussian distribution, thereby modeling sparsity at the elemental level.

To capture the fact that entire delay taps may be inactive, we introduce a second-level binary state $t^l_d \in \{0, 1\}$ associated with each column of $\boldsymbol{H}_d$. When $t^l_d = 0$, the entire l-th column $\boldsymbol{h}^l_d$ is constrained to be a zero vector, representing an inactive delay tap. Conversely, $t^l_d = 1$ indicates that the column contains at least one non-zero element. The column states $\boldsymbol{t}_d = [t^1_d, \ldots, t^L_d]$ follow independent Bernoulli distributions:

$$p\left(t^l_d\right) = \left(1 - \gamma^l_d\right)^{1-t^l_d} \left(\gamma^l_d\right)^{t^l_d}, \tag{13.34}$$

where γ^l_d represents the prior probability that the l-th delay tap is active. This formulation reflects the physical reality that only a limited number of delay taps typically contain significant energy due to the finite delay spread of the channel.

For each active column where $t^l_d = 1$, the vector of element-level states $\boldsymbol{s}^l_d = [s^l_{d,1}, \ldots, s^l_{d,Q}]^T$ must capture the clustered nature of non-zero elements in the angular domain. This clustering phenomenon, where active angular bins tend to appear in contiguous groups rather than in isolation, is naturally modeled using a first-order Markov chain

$$p\left(\boldsymbol{s}^l_d | t^l_d\right) = p\left(s^l_{d,1} | t^l_d\right) \prod_{q=2}^{Q} p\left(s^l_{d,q} | s^l_{d,q-1}, t^l_d\right), \tag{13.35}$$

with transition probabilities defined conditionally on the column state as

$$p\left(s^l_{d,q} | s^l_{d,q-1}, t^l_d\right) = \begin{cases} \left(1 - \rho^l_{01}\right)^{1-s^l_{d,q}} \left(\rho^l_{01}\right)^{s^l_{d,q}}, & s^l_{d,q-1} = 0,\ t^l_d = 1, \\ \left(\rho^l_{10}\right)^{1-s^l_{d,q}} \left(1 - \rho^l_{10}\right)^{s^l_{d,q}}, & s^l_{d,q-1} = 1,\ t^l_d = 1, \\ \mathbb{I}(s^l_{d,q} = 0), & t^l_d = 0. \end{cases} \tag{13.36}$$

Here, $\mathbb{I}(\cdot)$ denotes the indicator function, which enforces that when $t^l_d = 0$, all element-level states are deterministically zero regardless of the previous state. The parameters ρ^l_{01} and ρ^l_{10} govern the dynamics of the Markov chain for active columns: ρ^l_{01} is the probability of transitioning from an inactive to an active state, while ρ^l_{10} is the probability of transitioning from active to inactive. Smaller values of ρ^l_{10} encourage longer contiguous blocks of active elements, effectively modeling larger scattering clusters.

The initial state distribution for each active column is given by

$$p\left(s^l_{d,1} = 1 | t^l_d = 1\right) = \lambda^l_d, \quad p\left(s^l_{d,1} = 0 | t^l_d = 1\right) = 1 - \lambda^l_d, \tag{13.37}$$

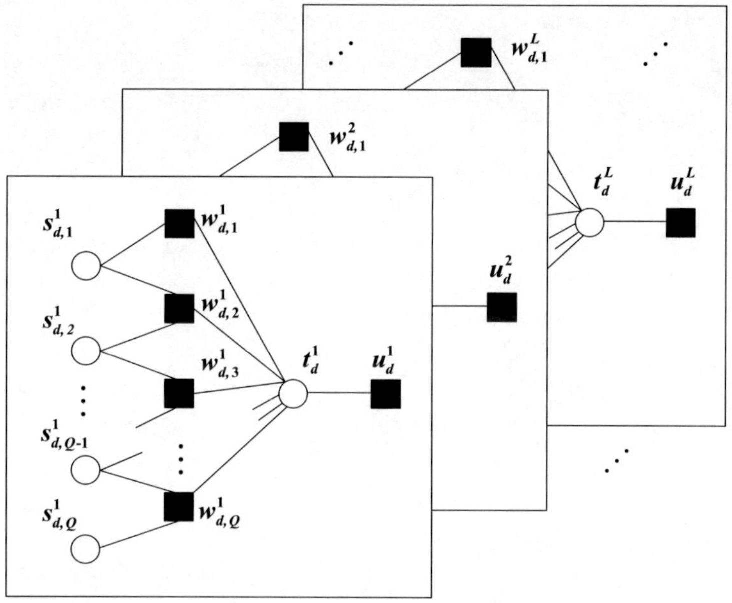

Fig. 13.9 Factor graph for the support-based two-layer sparse prior model of the DAD channel $\boldsymbol{H}_d$. The factor nodes u_d^l and $w_{d,q}^l$ correspond to the probability distributions $p\left(t_d^l\right)$ and $p\left(s_{d,q}^l | s_{d,q-1}^l, t_d^l\right)$, respectively

and when $t_d^l = 0$, we have $p(s_{d,1}^l = 0 | t_d^l = 0) = 1$. The parameter λ_d^l represents the marginal probability of activation for the first angular bin in an active column and is related to the steady-state distribution of the Markov chain through $\lambda_d^l = \rho_{01}^l / (\rho_{01}^l + \rho_{10}^l)$.

This hierarchical construction offers several desirable properties for modeling DAD channels. The column-level states t_d^l provide a mechanism to enforce global sparsity across delay taps, ensuring that entire columns corresponding to insignificant delays are forced to zero. For columns deemed active, the Markov structure on s_d^l naturally captures the tendency of physical scatterers to create contiguous angular support rather than isolated non-zero coefficients. This clustering behavior reflects the underlying geometry of the propagation environment, where scatterers with similar angles of arrival/departure contribute to neighboring angular bins.

The structured sparse prior described above can be represented as a factor graph, as illustrated in Fig. 13.9. When combined with the CS observation model developed earlier, this prior forms a complete probabilistic model that casts the FDD MIMO-OFDM channel estimation task as a structured CS problem with dynamic grid adaptation. This unified formulation enables the simultaneous exploitation of sparsity in both the delay and angular domains. By leveraging these inherent structural constraints, the resulting inference framework can achieve superior estimation accuracy compared to conventional unstructured sparse recovery methods.

13.3.3 Channel Estimation for TDD-HBF Systems

13.3.3.1 System and Signal Models

We consider a TDD massive MIMO system employing a partially connected hybrid beamforming (PC-HBF) architecture. The BS is equipped with a UPA comprising $N_b = N_x \times N_y$ antenna elements and N_{RF} RF chains, where $N_{RF} < N_b$, as illustrated in Fig. 13.10.

In TDD systems, orthogonal pilot resources allocated across time or frequency domains eliminate inter-user antenna interference during uplink channel estimation. Given the potentially large number of BS antennas (e.g., 2048) and to facilitate efficient parallel processing across subcarriers, channel estimation is typically performed independently on each subcarrier. Consequently, we focus on the channel estimation problem for a single user antenna at a reference subcarrier and will omit the subcarrier index n in subsequent formulations for notational simplicity.

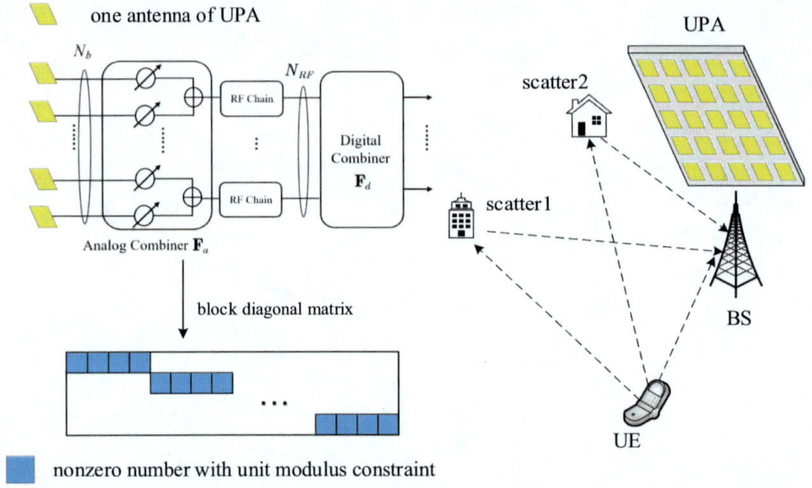

Fig. 13.10 Architecture of the PC-HBF massive MIMO systems

During the channel training phase, each user antenna transmits an uplink pilot signal. The received signal $\boldsymbol{y} \in \mathbb{C}^{N_{RF}\times 1}$ at the BS on the reference subcarrier is modeled as

$$\boldsymbol{y} = \boldsymbol{F}\boldsymbol{h}u + \boldsymbol{w}, \tag{13.38}$$

where u represents the transmitted pilot symbol, $\boldsymbol{h} \in \mathbb{C}^{N_b\times 1}$ denotes the channel vector between the user antenna and BS array, and $\boldsymbol{F} = \boldsymbol{F}_d\boldsymbol{F}_a \in \mathbb{C}^{N_{RF}\times N_b}$ constitutes the overall combining matrix formed by the cascade of analog combining matrix $\boldsymbol{F}_a \in \mathbb{C}^{N_{RF}\times N_b}$ and digital combining matrix $\boldsymbol{F}_d \in \mathbb{C}^{N_{RF}\times N_{RF}}$. The term $\boldsymbol{w} \sim \mathcal{CN}(\boldsymbol{0}, \sigma^2\boldsymbol{I})$ represents AWGN that corrupts the received signal. This hybrid combining architecture allows the massive MIMO system to efficiently process signals from N_b antenna elements through a limited number of N_{RF} RF chains, balancing performance requirements with hardware complexity and power consumption constraints.

In hybrid beamforming architectures, the analog combining matrix $\boldsymbol{F}_a$ is physically implemented using phase shifters, which constrains its non-zero elements to have unit modulus. Under the PC-HBF configuration, each RF chain is connected exclusively to a dedicated subarray of antennas, resulting in $\boldsymbol{F}_a$ having a block-diagonal structure where each block is a row vector of N_b/N_{RF} elements, as illustrated in Fig. 13.10. To ensure that the overall combining matrix $\boldsymbol{F}$ satisfies the Restricted Isometry Property with high probability—a crucial condition for reliable channel estimation—the system employs a carefully designed randomization strategy: the phases of non-zero elements in $\boldsymbol{F}_a$ are independently drawn from a uniform distribution over $[0, 2\pi)$, while the digital combining matrix $\boldsymbol{F}_d$ is constructed as a random orthogonal matrix. This structured yet randomized approach enables efficient channel estimation while rigorously respecting the practical implementation constraints inherent in HBF architectures.

13.3.3.2 CS Model Formulation with a Dynamic Grid

Based on the physical channel model established in Eq. (13.24), the channel vector $\boldsymbol{h}$ at a reference subcarrier admits a structured representation:

$$\boldsymbol{h} = \sum_{k=1}^{K} \alpha_k \boldsymbol{a}_R(\theta_k, \phi_k), \tag{13.39}$$

where α_k denotes the complex gain of the k-th propagation path, θ_k and ϕ_k represent the azimuth and elevation angles of arrival, respectively, and $\boldsymbol{a}_R(\theta_k, \phi_k)$ is the array response vector of the UPA. This parametric formulation captures the essential physical characteristics of the massive MIMO channel while providing a foundation for sparse signal recovery.

To enable effective sparse representation of the massive MIMO channel for high-accuracy estimation, we employ a grid-based approach with dynamic adjustment capabilities. Specifically, we construct a two-dimensional angular grid comprising $Q = N_1 \times N_2$ discrete points, where the q-th grid point corresponds to the angular coordinates (θ_q, ϕ_q). The dynamic grid vectors $\boldsymbol{\theta} \triangleq [\theta_1, \ldots, \theta_Q]^T$ and $\boldsymbol{\phi} \triangleq [\phi_1, \ldots, \phi_Q]^T$ are typically initialized using a uniform angular sampling pattern across the spatial domain, providing comprehensive coverage of potential arrival directions while maintaining computational efficiency.

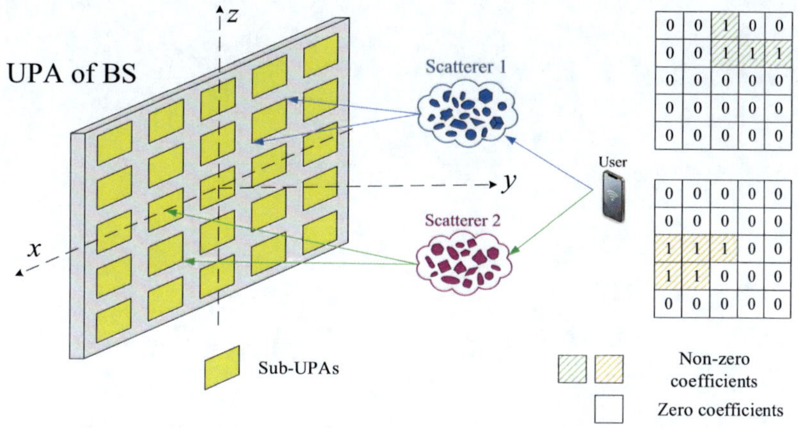

Fig. 13.11 Illustration of the 2D burst sparsity for a massive MIMO system with the UPA

Using this dynamic angular grid framework, we reformulate the received signal model from Eq. (13.38) as a CS problem with adaptive grid parameters:

$$y = uFA(\theta, \phi)\, x + w, \tag{13.40}$$

where

$$A(\theta, \phi) \triangleq \left[a(\theta_1, \phi_1), \ldots, a\left(\theta_Q, \phi_Q\right)\right] \in \mathbb{C}^{N_b \times Q} \tag{13.41}$$

constitutes the dictionary matrix of array response vectors evaluated at the grid points, and $x \in \mathbb{C}^{Q\times 1}$ represents the sparse angular-domain channel vector. This formulation effectively transforms the channel estimation problem into a structured sparse recovery task that can leverage advanced CS algorithms.

The angular-domain channel vector x exhibits strong sparsity characteristics, with only $K \ll Q$ non-zero elements corresponding to the K dominant propagation paths present in the physical environment. Specifically, the q-th element x_q represents the complex gain of the channel path whose angles are closest to the q-th grid point (θ_q, ϕ_q), creating a sparse representation where the non-zero coefficients directly correspond to actual physical propagation paths. This formulation, combined with dynamic grid adjustment, enables super-resolution estimation of the angle-of-arrival parameters while maintaining the computational advantages of compressed sensing techniques, effectively bridging the gap between discrete grid methods and continuous parameter estimation.

13.3.3.3 Three-Layer Sparse Prior Model to Capture the 2D Clustered Sparsity

The angular-domain sparse channel vector x exhibits a distinctive two-dimensional clustered sparsity pattern. When reshaped into an $N_1 \times N_2$ matrix, the non-zero elements of x concentrate within a few localized 2D clusters, as illustrated in Fig. 13.11. To effectively capture this structural sparsity, we employ the three-layer BGG prior model from Equation (13.5) and incorporate a specialized 2D Markov random field model [18] for the support vector s.

Let s_{n_1,n_2} represent the element of the support matrix s at spatial position (n_1, n_2), corresponding to the $((n_2-1)N_1+n_1)$-th element in vectorized form. The joint distribution of the support vector follows a two-dimensional Markov random field structure that captures spatial dependencies through the factorization:

$$\begin{aligned} p(s) = & p\left(s_{1,1}\right) \prod_{n_1=1}^{N_1} \prod_{n_2=2}^{N_2} p\left(s_{n_1,n_2} \mid s_{n_1,n_2-1}\right) \\ & \times \prod_{n_1=2}^{N_1} \prod_{n_2=1}^{N_2} p\left(s_{n_1,n_2} \mid s_{n_1-1,n_2}\right), \end{aligned} \tag{13.42}$$

with the transition probabilities governing state evolution along both spatial dimensions defined as

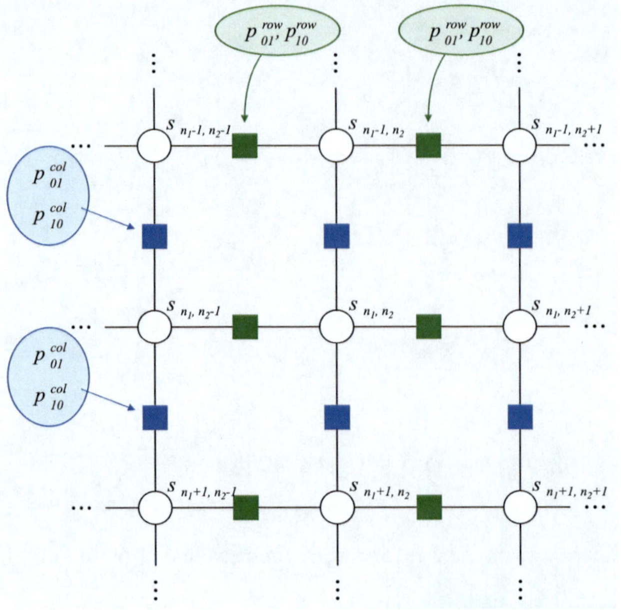

Fig. 13.12 Factor graph of the 2D Markov model

$$
\begin{aligned}
p\left(s_{n_1,n_2}=1 \mid s_{n_1,n_2-1}=0\right) &= p_{01}^{\text{row}},\\
p\left(s_{n_1,n_2}=0 \mid s_{n_1,n_2-1}=1\right) &= p_{10}^{\text{row}},\\
p\left(s_{n_1,n_2}=1 \mid s_{n_1-1,n_2}=0\right) &= p_{01}^{\text{col}},\\
p\left(s_{n_1,n_2}=0 \mid s_{n_1-1,n_2}=1\right) &= p_{10}^{\text{col}}.
\end{aligned}
\tag{13.43}
$$

As established in [18], there exists a consistent set of transition parameters $\{p_{01}^{\text{row}}, p_{10}^{\text{row}}, p_{01}^{\text{col}}, p_{10}^{\text{col}}\}$ that satisfies the steady-state condition of this 2D Markov model, ensuring uniform marginal activation probability $p(s_{n_1,n_2}=1)=\lambda$ for all positions (n_1, n_2), where λ represents the global sparsity ratio of $\boldsymbol{s}$. Consequently, the initial distribution $p(s_{1,1})$ naturally assumes this steady-state distribution with $p(s_{1,1}=1)=\lambda$, ensuring spatial homogeneity throughout the support pattern.

The transition parameters directly govern the morphological properties of the resulting sparsity patterns, providing explicit control over cluster characteristics. Smaller values of p_{10}^{row} and p_{10}^{col} promote larger average cluster sizes by reducing the probability of transitioning from active to inactive states, thereby encouraging spatial continuity. Conversely, smaller values of p_{01}^{row} and p_{01}^{col} increase the average separation between adjacent clusters by making transitions from inactive to active states less probable, effectively enforcing stronger sparsity constraints. The factor graph representation of this 2D Markov model, shown in Fig. 13.12, visually illustrates the conditional dependencies between neighboring support variables and the message passing structure required for efficient inference.

By combining the observation model from Eq. (13.40) with the structured sparse prior model defined above, the channel estimation problem for TDD hybrid beamforming systems is formally cast as a structured compressed sensing problem with dynamic grid adaptation. This integrated formulation simultaneously exploits the inherent two-dimensional clustered sparsity in the angular domain while enabling super-resolution parameter estimation through iterative grid refinement, creating a powerful framework that bridges discrete Markov random field modeling with continuous parameter estimation for enhanced massive MIMO channel recovery.

13.4 Integrated Sensing and Communication

Future 6G wireless systems are envisioned to seamlessly integrate radar sensing and communication functionalities, enabling transformative applications such as autonomous driving and smart city infrastructure [19]. This integration represents a significant departure from traditional approaches where radar and communication systems operated in separate frequency

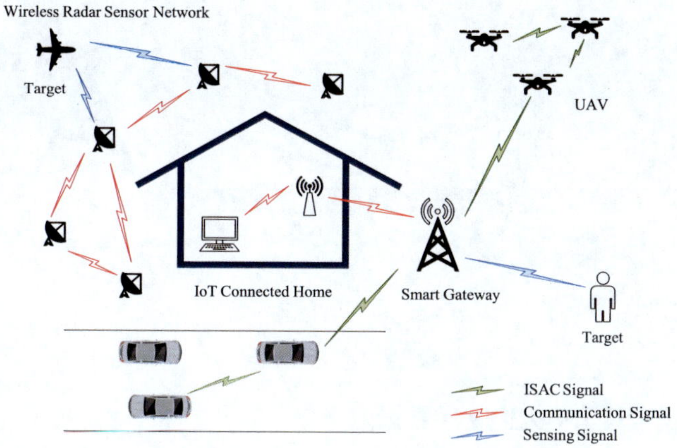

Fig. 13.13 Illustration for the applications of ISAC (adapted from Fig. 1 in [19])

bands to minimize mutual interference. The technological evolution toward integration has been catalyzed by recent advances in millimeter-wave technology and massive MIMO architectures, which have endowed modern communication signals with unprecedented resolution in both temporal and spatial domains, thereby creating new opportunities for high-accuracy sensing capabilities using communication infrastructure.

This technological convergence has given rise to ISAC, which utilizes shared hardware platforms and spectral resources to simultaneously deliver communication and sensing services [20–23]. This unified approach represents a cornerstone technology for future wireless systems, particularly in applications demanding both high-data-rate connectivity and precise environmental awareness. The potential applications of ISAC span diverse domains with distinct requirements: autonomous vehicle networks require both high-resolution environmental perception for centimeter-level obstacle detection [25] and reliable high-speed connectivity for navigation and traffic information exchange [24]; indoor localization systems enable simultaneous data transmission and precise positioning; aerial networks support UAV communication and coordinated environmental monitoring; extended reality applications demand high-speed data delivery with precise motion tracking; and joint radar-communication systems provide unified infrastructure for target tracking and data communications, as illustrated in Fig. 13.13.

Our focus centers on scattering environment characterization in massive MIMO-OFDM ISAC systems, which reveals fundamental interconnections between radar sensing and communication channel properties. The scattering environment comprises two complementary subsets: radar targets detected through echo signals for localization and tracking, and communication scatterers that facilitate multipath propagation for communication links. In practical ISAC scenarios, significant overlap exists between these two subsets. For instance, in vehicular networks (illustrated in Fig. 13.14), the BS simultaneously performs vehicle localization via radar echo processing and uplink channel estimation from vehicle transmissions. In this context, neighboring vehicles frequently serve dual roles as radar targets for detection and as communication scatterers for nearby vehicles. Moreover, both radar targets and communication scatterers tend to form spatial clusters, e.g., a vehicle can be modeled as a cluster of point targets/scatterers.

This structural overlap creates joint burst sparsity patterns in appropriate sparse domains, enabling the development of unified algorithms for joint scattering environment sensing and channel estimation. By exploiting these shared sparsity structures, radar sensing and communication functions can mutually enhance each other, leading to improved performance in both sensing accuracy and channel estimation quality. The inherent synergy between these traditionally separate functionalities allows for more efficient spectrum utilization and hardware sharing while providing comprehensive environmental awareness essential for advanced 6G applications. One may refer to [26] for another example of leveraging this structural overlap to enhance both location sensing and channel estimation performance in millimeter-wave ISAC systems.

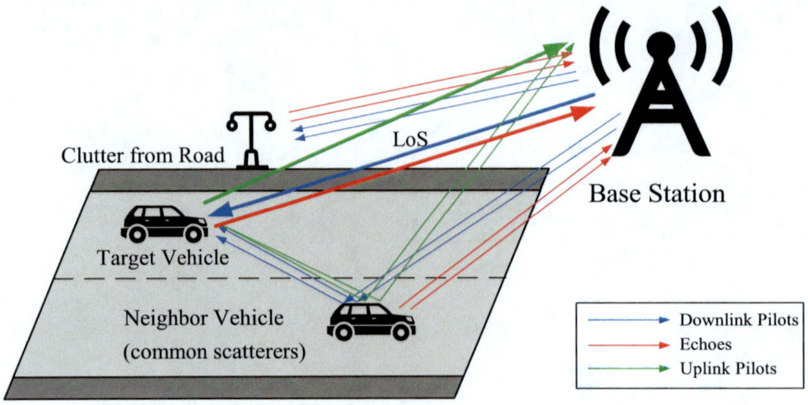

Fig. 13.14 An example of vehicle communication and sensing in ISAC scenarios. The BS sends downlink pilots to detect the target via the received echo signals, and the user sends uplink pilots for the BS to estimate the uplink communication channel

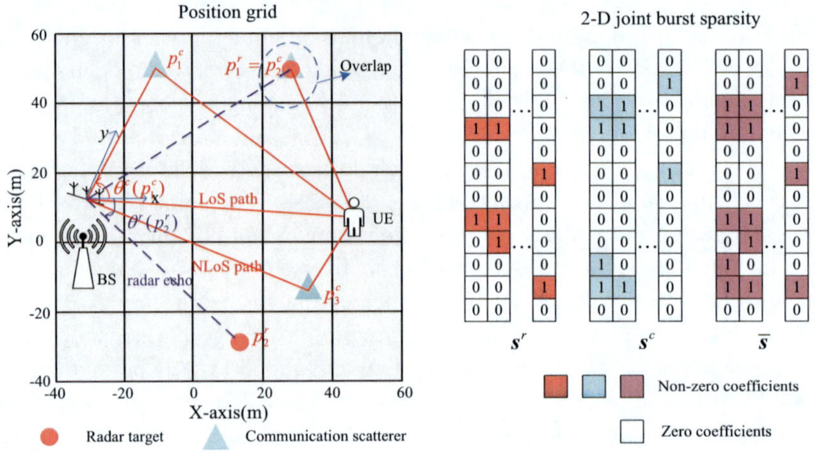

Fig. 13.15 Illustration of location domain radar and communication channels and their non-zero coefficients (Adapted from Fig. 1 in [29])

13.4.1 System Model

13.4.1.1 System Architecture and Frame Structure

We consider a TDD massive MIMO-OFDM ISAC system where a BS equipped with $N_b \gg 1$ antennas simultaneously serves a single-antenna user while performing environmental sensing, as illustrated in Fig. 13.15. This integrated architecture operates through coordinated transmission of downlink pilots for target sensing and uplink pilots for communication channel estimation, creating a unified framework that efficiently shares hardware and spectral resources between these traditionally separate functions.

The scattering environment contains K radar targets and L communication scatterers, with significant potential overlap between these sets that forms the foundation for joint processing. The system employs a comprehensive spatial coordinate system where the user location $\boldsymbol{p}_u = [p^x, p^y]^T \in \mathcal{R}$ exists within a two-dimensional coverage area $\mathcal{R}$, while the BS location $\boldsymbol{p}_b = [\tilde{p}^x, \tilde{p}^y]^T$ is precisely known. The target and scatterer coordinates are denoted as $\boldsymbol{p}_k^r = [p_k^{r,x}, p_k^{r,y}]^T$ for $k = 1, \ldots, K$ and scatterer coordinates $\boldsymbol{p}_l^c = [p_l^{c,x}, p_l^{c,y}]^T$ for $l = 1, \ldots, L$, respectively. We assume that the BS maintains approximate user location information through GPS or previous localization results to facilitate initial beam alignment and sector selection.

As shown in Fig. 13.16, the system employs a frame-based structure with two distinct operational phases designed to maximize efficiency. During the pilot transmission phase, the system combines environmental sensing and channel estimation through coordinated procedures: the BS performs angular sector scanning with downlink pilots for target detection while the user transmits uplink pilots for communication channel estimation. This approach leverages the significant overlap between radar targets and communication scatterers within the same angular sector containing the user, creating natural synergies

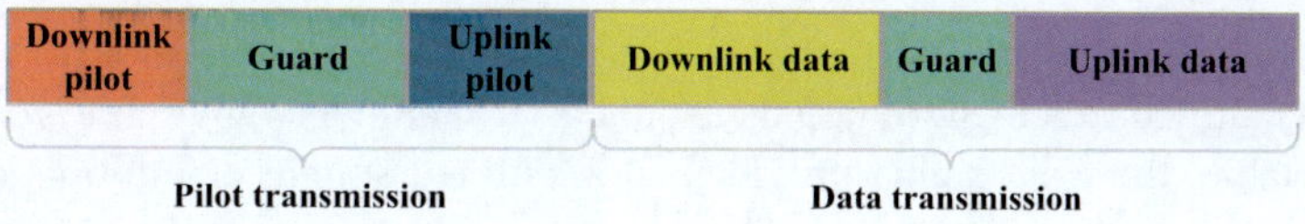

Fig. 13.16 Frame structure of the ISAC system (adapted from Fig. 2 in [29])

between the two functions. The subsequent data transmission phase builds upon the channel parameters acquired during pilot transmission, enabling efficient communication while maintaining environmental awareness.

The joint sensing and estimation procedure begins with the BS determining user angular sector affiliation through GPS-based prior location information. For each angular sector, the BS executes integrated scattering environment sensing and channel estimation using both reflected downlink pilot signals from targets and received uplink pilot signals from the user. A carefully designed guard interval between downlink and uplink transmissions prevents interference while maintaining temporal efficiency. The remainder of this section focuses specifically on the joint estimation problem within a single angular sector, developing the mathematical framework that enables simultaneous environmental characterization and communication channel estimation through shared signal processing techniques.

13.4.1.2 Reflected Downlink Pilot Signal

The objective of target sensing is twofold: to determine the presence of targets within the surveillance area and to estimate their spatial locations. This is accomplished by analyzing the echoes of downlink pilot signals reflected off various objects in the environment. On each subcarrier designated for sensing, the BS transmits a known pilot signal and processes the returned echo.

Specifically, on the n-th subcarrier, where n belongs to the set $\mathcal{N}_b$ of subcarriers allocated for target sensing within a given angular sector, the BS transmits a downlink pilot vector $\boldsymbol{v}_n^r \in \mathbb{C}^{N_b}$. The reflected signal received back at the BS is modeled as

$$\boldsymbol{y}_n^r = \boldsymbol{H}_n^r \boldsymbol{v}_n^r + \boldsymbol{z}_n^r, \quad \forall n \in \mathcal{N}_b, \tag{13.44}$$

where $\boldsymbol{H}_n^r \in \mathbb{C}^{N_b \times N_b}$ denotes the radar channel matrix characterizing the scattering environment, and $\boldsymbol{z}_n^r \in \mathbb{C}^{N_b \times 1}$ is AWGN with variance $(\sigma_z^r)^2$.

The parameters of interest, namely, the AoA and propagation delay, are fundamentally determined by the geometric relationship between each target and the BS. For the k-th target located at position $\boldsymbol{p}_k^r$, its AoA $\theta^r(\boldsymbol{p}_k^r)$ and delay $\tau^r(\boldsymbol{p}_k^r)$ are given by

$$\theta^r(\boldsymbol{p}_k^r) = \arctan\left(\frac{p_k^{r,y} - \tilde{p}^y}{p_k^{r,x} - \tilde{p}^x}\right) + \pi \cdot \mathbb{I}\left(p_k^{r,x} < \tilde{p}^x\right), \tag{13.45}$$

$$\tau^r(\boldsymbol{p}_k^r) = \frac{2\left\|\boldsymbol{p}_b - \boldsymbol{p}_k^r\right\|}{c}, \tag{13.46}$$

where $\mathbb{I}(E)$ is the indicator function (evaluating to 1 when the logical expression E is true and 0 otherwise), c represents the speed of light, and $\boldsymbol{p}_b = [\tilde{p}^x, \tilde{p}^y]^T$ denotes the known BS position. The AoA is measured anticlockwise with respect to the positive x-axis, ensuring consistent angular disambiguation across the coverage area.

With these geometric relationships established, the radar channel matrix can be expressed as a superposition of contributions from all detectable targets

$$\boldsymbol{H}_n^r = \sum_{k=0}^{K} x_k^r e^{-j2\pi n f_0 \tau^r(\boldsymbol{p}_k^r)} \boldsymbol{a}(\theta^r(\boldsymbol{p}_k^r)) \boldsymbol{a}^T(\theta^r(\boldsymbol{p}_k^r)), \tag{13.47}$$

where x_k^r represents the radar cross section (RCS) of the k-th target, f_0 is the subcarrier spacing, and $\boldsymbol{a}(\theta) \in \mathbb{C}^{N_b \times 1}$ denotes the array response vector of the BS. This formulation captures both the angular and delay characteristics of each scattering contribution through the steering vector and the exponential phase term, respectively.

A notable feature of this model is the inclusion of the mobile user as a potential target. Specifically, we define the user as the 0-th target with position $\boldsymbol{p}_0^r \triangleq \boldsymbol{p}_u$. The corresponding coefficient x_0^r indicates whether the BS can directly detect the user via radar echo: $x_0^r > 0$ signifies that the user is radar-detectable, while $x_0^r = 0$ indicates that the user is not directly observable through radar. This unified treatment of targets and the user within a single framework offers several advantages. When the user is radar-detectable, the echo signal provides an additional source of information for user localization that complements conventional communication-based estimation, thereby enhancing overall system awareness and positioning reliability. Moreover, this unified formulation facilitates joint processing of sensing and communication functionalities, enabling mutual assistance between these two system objectives.

13.4.1.3 Received Uplink Pilot Signal

The uplink pilot transmission fulfills two complementary objectives: estimating the communication channel and sensing the surrounding communication scatterers. On the n-th subcarrier, where n belongs to the set $\mathcal{N}_u$ of subcarriers designated for uplink channel estimation, the user transmits a pilot symbol $u_n^c \in \mathbb{C}$. The signal received at the BS is modeled as

$$\boldsymbol{y}_n^c = \boldsymbol{h}_n^c u_n^c + \boldsymbol{z}_n^c, \quad \forall n \in \mathcal{N}_u, \tag{13.48}$$

where $\boldsymbol{h}_n^c \in \mathbb{C}^{N_b \times 1}$ represents the communication channel vector, and $\boldsymbol{z}_n^c \in \mathbb{C}^{N_b \times 1}$ is AWGN with variance $(\sigma_z^c)^2$.

The communication channel consists of a LoS component and L single-bounce NLoS components arising from interactions with communication scatterers. The AoAs for these paths are determined by the spatial configuration of the BS, the user, and the scatterers as

$$\theta^c(\boldsymbol{p}_u) = \arctan\left(\frac{p^y - \tilde{p}^y}{p^x - \tilde{p}^x}\right) + \pi \cdot \mathbb{I}\left(p^x < \tilde{p}^x\right), \tag{13.49}$$

$$\theta^c(\boldsymbol{p}_l^c) = \arctan\left(\frac{p_l^{c,y} - \tilde{p}^y}{p_l^{c,x} - \tilde{p}^x}\right) + \pi \cdot \mathbb{I}\left(p_l^{c,x} < \tilde{p}^x\right), \tag{13.50}$$

where $\theta^c(\boldsymbol{p}_u)$ denotes the AoA of the LoS path, and $\theta^c(\boldsymbol{p}_l^c)$ denotes the AoA corresponding to the l-th scatterer. As before, all angles are measured anticlockwise from the positive x-axis.

For each NLoS path, the propagation delay relative to the LoS path carries important information about the scatterer location. This relative delay $\tau^c(\boldsymbol{p}_l^c, \boldsymbol{p}_u)$ is given by

$$\tau^c(\boldsymbol{p}_l^c, \boldsymbol{p}_u) = \frac{\|\boldsymbol{p}_b - \boldsymbol{p}_l^c\|}{c} + \frac{\|\boldsymbol{p}_u - \boldsymbol{p}_l^c\|}{c} - \frac{\|\boldsymbol{p}_b - \boldsymbol{p}_u\|}{c}, \tag{13.51}$$

which quantifies the excess distance traveled by the signal when it reflects off a scatterer compared to the direct LoS path.

Based on this geometric characterization, the communication channel vector can be decomposed into its LoS and NLoS components:

$$\boldsymbol{h}_n^c = \boldsymbol{h}_n^{c,L} + \boldsymbol{h}_n^{c,NL}, \tag{13.52}$$

with the individual components expressed as

$$\boldsymbol{h}_n^{c,L} = x_0^c e^{-j2\pi n f_0 \tau_o} \boldsymbol{a}\left(\theta^c\left(\boldsymbol{p}_u\right)\right), \tag{13.53}$$

$$\boldsymbol{h}_n^{c,NL} = \sum_{l=1}^{L} x_l^c e^{-j2\pi n f_0 \left(\tau^c(\boldsymbol{p}_l^c, \boldsymbol{p}_u) + \tau_o\right)} \boldsymbol{a}\left(\theta^c\left(\boldsymbol{p}_l^c\right)\right). \tag{13.54}$$

Here, x_0^c and x_l^c denote the complex gains of the LoS path and the l-th NLoS path, respectively. The term τ_o represents a common time offset introduced by timing synchronization errors at the BS, which affects all paths uniformly. This offset, if unaccounted for, can lead to biased range estimates and must therefore be estimated and compensated during processing.

Several modeling assumptions merit comment. First, we consider only single-bounce NLoS paths, as multiple-bounce reflections typically experience significantly higher path loss and contribute negligibly to the received signal power. Second, the model assumes low mobility for all entities (targets, scatterers, and users) relative to the coherence time of the wideband ISAC system. In high-mobility scenarios, the channel models would need to be augmented with Doppler shift terms to capture time-varying effects. Finally, practical imperfections such as timing offset τ_o and user localization errors can introduce ambiguity in scatterer localization and degrade estimation performance. However, these non-ideal factors can be treated as unknown parameters within the Bayesian framework and estimated jointly with the channel and scatterer parameters using techniques such as EM, thereby mitigating their impact on overall system performance.

13.4.2 Formulation as a Structured CS Problem with a Dynamic Grid

13.4.2.1 Location-Domain Sparse Representation

To improve the accuracy of both sensing and channel estimation, we adopt a grid-based methodology that reformulates the channel representation in a sparse location domain. The first step involves constructing a two-dimensional uniform grid of candidate positions:

$$\mathcal{P} = \{\bar{\boldsymbol{r}}_1, \ldots, \bar{\boldsymbol{r}}_Q\} \subset \mathcal{R}, \tag{13.55}$$

which covers the surveillance area with $Q = H \times W$ grid points, arranged in H rows and W columns, as illustrated in Fig. 13.15. Here, Q is chosen to be much larger than the total number of targets and scatterers, i.e., $Q \gg (K + L)$, to ensure an overcomplete representation.

In practice, the true positions of targets and scatterers rarely align exactly with the predefined discrete grid points. Such misalignment can lead to energy leakage and degraded estimation performance. To address this issue, we introduce a dynamic position grid $\boldsymbol{r} = [\boldsymbol{r}_1; \ldots; \boldsymbol{r}_Q]$, which is initialized using the uniform grid $\mathcal{P}$ and then iteratively refined during the inference procedure to better match the actual physical locations.

Based on this dynamic grid, we define a sparse basis dictionary that maps grid points to their corresponding array responses:

$$\boldsymbol{A}(\boldsymbol{r}) \triangleq \left[\boldsymbol{a}\left(\theta^r(\boldsymbol{r}_1)\right), \ldots, \boldsymbol{a}\left(\theta^r(\boldsymbol{r}_Q)\right)\right] \in \mathbb{C}^{N_b \times Q}, \tag{13.56}$$

where $\theta^r(\boldsymbol{r}_q)$ denotes the AoA associated with the grid point $\boldsymbol{r}_q$. This dictionary serves as the fundamental building block for representing both radar and communication channels in the location domain.

Using this dynamic grid framework, we obtain sparse representations for the radar channel matrix and the NLoS communication channel vector. On the n-th subcarrier, the radar channel matrix is expressed as

$$\boldsymbol{H}_n^r = \boldsymbol{A}(\boldsymbol{r})\boldsymbol{D}_n^r \mathrm{diag}(\boldsymbol{x}^r)\boldsymbol{A}^T(\boldsymbol{r}) + x_0^r e^{-j2\pi n f_0 \tau^r(\boldsymbol{p}_u)} \boldsymbol{a}\left(\theta^r(\boldsymbol{p}_u)\right) \boldsymbol{a}^T\left(\theta^r(\boldsymbol{p}_u)\right), \tag{13.57}$$

and the NLoS component of the communication channel vector is given by

$$\boldsymbol{h}_n^{c,\mathrm{NL}} = \boldsymbol{A}(\boldsymbol{r})\boldsymbol{D}_n^c \boldsymbol{x}^c. \tag{13.58}$$

In these expressions, $\boldsymbol{x}^r \in \mathbb{C}^{Q\times 1}$ and $\boldsymbol{x}^c \in \mathbb{C}^{Q\times 1}$ are the location-domain sparse channel vectors for radar and communication, respectively. The matrices $\boldsymbol{D}_n^r$ and $\boldsymbol{D}_n^c$ are diagonal delay matrices, whose diagonal entries encode the phase shifts due to propagation delays. Specifically, the q-th diagonal element of $\boldsymbol{D}_n^r$ is $e^{-j2\pi n f_0 \tau^r(\boldsymbol{r}_q)}$, capturing the delay associated with a target at grid point $\boldsymbol{r}_q$. Similarly, the q-th diagonal element of $\boldsymbol{D}_n^c$ is $e^{-j2\pi n f_0 (\tau^c(\boldsymbol{r}_q, \boldsymbol{p}_u) + \tau_o)}$, accounting for both the scatterer delay and the timing offset τ_o. Here, f_0 denotes the subcarrier spacing.

The user's own echo signal is modeled separately in (13.57) rather than being absorbed into the sparse vector $\boldsymbol{x}^r$. This treatment reflects the practical consideration that the BS typically has more accurate prior information about the user's location compared to arbitrary targets, and separating this component can improve estimation reliability.

The location-domain sparse vectors exhibit structural sparsity patterns that directly mirror the physical environment. The vector $\boldsymbol{x}^r$ contains exactly K non-zero entries, each corresponding to the position of a distinct radar target. Likewise, $\boldsymbol{x}^c$ contains exactly L non-zero entries corresponding to the positions of communication scatterers. The values of these non-zero elements carry clear physical interpretations: x_q^r represents the complex reflection coefficient of a target located near $\boldsymbol{r}_q$, while x_q^c denotes the complex channel gain associated with a scatterer at the same location.

By casting the joint sensing and communication problem in this location-domain sparse framework with dynamic grid adaptation, we transform a challenging continuous parameter estimation task into a structured compressed sensing problem. This formulation not only exploits the inherent spatial sparsity of the environment but also enables super-resolution parameter estimation through iterative grid refinement, providing a unified foundation for jointly optimizing sensing and communication performance.

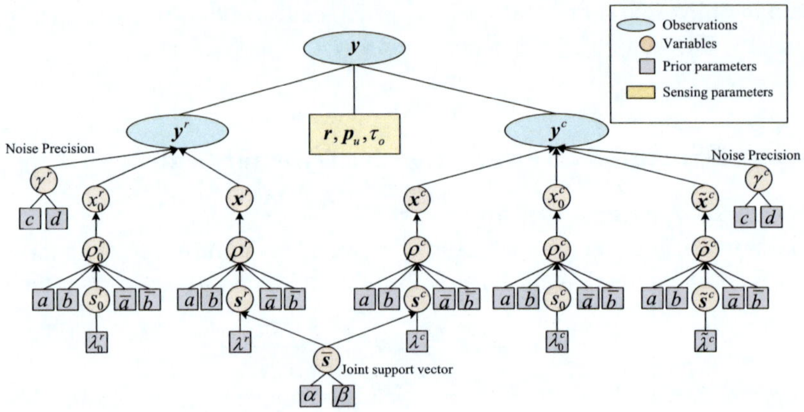

Fig. 13.17 Illustration of the three-layer sparse prior model for 2D joint burst sparsity (adapted from Fig. 3 in [29])

13.4.2.2 Markov Random Field for 2D Joint Burst Sparsity

To adequately represent the rich structural sparsity present in location-domain channels, we develop a three-layer hierarchical prior model that utilizes a Markov Random Field (MRF) to characterize the two-dimensional joint burst sparsity pattern. This model introduces several layers of latent variables, each capturing different aspects of the sparsity structure.

- **Precision vectors**: $\boldsymbol{\rho}^r \triangleq [\rho_1^r, \ldots, \rho_Q^r]^T$ and $\boldsymbol{\rho}^c \triangleq [\rho_1^c, \ldots, \rho_Q^c]^T$, which control the variance of the channel coefficients at each grid point.
- **Support vectors**: $\boldsymbol{s}^r \triangleq [s_1^r, \ldots, s_Q^r]^T \in \{-1, 1\}^Q$ and $\boldsymbol{s}^c \triangleq [s_1^c, \ldots, s_Q^c]^T \in \{-1, 1\}^Q$, indicating the presence of radar targets and communication scatterers, respectively.
- **Joint support vector**: $\bar{\boldsymbol{s}} \triangleq [\bar{s}_1, \ldots, \bar{s}_Q]^T \in \{-1, 1\}^Q$, representing the union of active positions across both sensing and communication functionalities.

The binary support variables take values in $\{-1, 1\}$, a convention that aligns naturally with standard MRF formulations. In this representation, $s_q^r = 1$ signifies the existence of a radar target in the vicinity of grid point $\boldsymbol{r}_q$, while $s_q^c = 1$ indicates a communication scatterer near the same location. To capture the combined occupancy of the spatial grid, we define the joint support variable $\bar{s}_q = 1$ whenever either $s_q^r = 1$ or $s_q^c = 1$, effectively representing the union of all target and scatterer positions.

As illustrated in Fig. 13.17, the complete joint distribution factorizes into three interconnected layers as

$$
\begin{aligned}
p(\boldsymbol{x}^r, \boldsymbol{x}^c, \boldsymbol{\rho}^r, \boldsymbol{\rho}^c, \boldsymbol{s}^r, \boldsymbol{s}^c, \bar{\boldsymbol{s}}) = & \underbrace{p(\boldsymbol{s}^r, \boldsymbol{s}^c, \bar{\boldsymbol{s}})}_{\text{Support}} \\
& \times \underbrace{p(\boldsymbol{\rho}^r \mid \boldsymbol{s}^r) p(\boldsymbol{\rho}^c \mid \boldsymbol{s}^c)}_{\text{Precision}} \\
& \times \underbrace{p(\boldsymbol{x}^r \mid \boldsymbol{\rho}^r) p(\boldsymbol{x}^c \mid \boldsymbol{\rho}^c)}_{\text{Sparse signal}} .
\end{aligned} \tag{13.59}
$$

This hierarchical construction enables the model to capture dependencies both within and across the radar and communication channels.

Signal and Precision Layer Specifications

At the signal layer, the sparse channel vectors follow zero-mean complex Gaussian distributions, where the variance of each coefficient is inversely proportional to its corresponding precision parameter, i.e.,

$$p(\boldsymbol{x}^r \mid \boldsymbol{\rho}^r) = \prod_{q=1}^{Q} \mathcal{CN}\left(x_q^r; 0, 1/\rho_q^r\right), \tag{13.60}$$

$$p(\boldsymbol{x}^c \mid \boldsymbol{\rho}^c) = \prod_{q=1}^{Q} \mathcal{CN}\left(x_q^c; 0, 1/\rho_q^c\right). \tag{13.61}$$

The precision parameters themselves are governed by Gamma distributions, conditioned on the support states

$$p(\boldsymbol{\rho}^r \mid \boldsymbol{s}^r) = \prod_{q=1}^{Q} \left[\delta(s_q^r - 1)\Gamma(\rho_q^r; a, b) + \delta(s_q^r + 1)\Gamma(\rho_q^r; \bar{a}, \bar{b})\right], \tag{13.62}$$

$$p(\boldsymbol{\rho}^c \mid \boldsymbol{s}^c) = \prod_{q=1}^{Q} \left[\delta(s_q^c - 1)\Gamma(\rho_q^c; a, b) + \delta(s_q^c + 1)\Gamma(\rho_q^c; \bar{a}, \bar{b})\right]. \tag{13.63}$$

The hyperparameters (a, b) and $(\bar{a}, \bar{b})$ are carefully chosen to create distinct statistical profiles for active and inactive elements. For active states where $s_q = 1$, we maintain $a/b = \Theta(1)$, yielding a moderate variance that allows coefficients to take on meaningful non-zero values. Conversely, for inactive states where $s_q = -1$, we enforce $\bar{a}/\bar{b} \gg 1$, resulting in extremely small variances that effectively drive the corresponding coefficients toward zero. A typical parameter configuration that achieves this contrast is $a = 1$, $b = 1$, $\bar{a} = 1$, and $\bar{b} = 10^{-5}$.

Support Layer Conditional Relationships

The support layer captures dependencies between the radar and communication support vectors through the joint support variable. The factorization proceeds as

$$\begin{aligned} p(\boldsymbol{s}^r, \boldsymbol{s}^c, \bar{\boldsymbol{s}}) &= p(\boldsymbol{s}^r \mid \bar{\boldsymbol{s}}) p(\boldsymbol{s}^c \mid \bar{\boldsymbol{s}}) p(\bar{\boldsymbol{s}}) \\ &= \prod_{q=1}^{Q} p(s_q^r \mid \bar{s}_q) \prod_{q=1}^{Q} p(s_q^c \mid \bar{s}_q) p(\bar{\boldsymbol{s}}), \end{aligned} \tag{13.64}$$

with the conditional distributions given by

$$\begin{aligned} p(s_q^r \mid \bar{s}_q) &= \delta(\bar{s}_q + 1)\delta(s_q^r + 1) \\ &\quad + \delta(\bar{s}_q - 1)\left[\delta(s_q^r - 1)\lambda_q^r + \delta(s_q^r + 1)(1 - \lambda_q^r)\right], \end{aligned} \tag{13.65}$$

$$\begin{aligned} p(s_q^c \mid \bar{s}_q) &= \delta(\bar{s}_q + 1)\delta(s_q^c + 1) \\ &\quad + \delta(\bar{s}_q - 1)\left[\delta(s_q^c - 1)\lambda_q^c + \delta(s_q^c + 1)(1 - \lambda_q^c)\right]. \end{aligned} \tag{13.66}$$

Here, λ_q^r and λ_q^c represent the conditional probabilities of radar target and communication scatterer presence, respectively, given that the joint support indicates activity at that grid point.

These conditional distributions have an intuitive interpretation. When the joint support indicates no activity ($\bar{s}_q = -1$), both s_q^r and s_q^c are deterministically set to -1, reflecting the absence of any target or scatterer. Conversely, when the joint support indicates activity ($\bar{s}_q = 1$), the individual supports may be either active or inactive according to probabilities λ_q^r and λ_q^c, allowing for scenarios where a grid point hosts only a radar target, only a communication scatterer, or both simultaneously.

To model the spatial clustering of active positions, a key characteristic of realistic scattering environments, we employ the Ising model, a well-established framework from statistical physics originally developed to describe ferromagnetic spin systems. This model is particularly well suited for capturing the tendency of radar targets and communication scatterers to form contiguous spatial clusters rather than being independently distributed. Specifically, the joint support vector follows a spatially non-stationary MRF

$$p(\bar{\boldsymbol{s}}) = \frac{1}{Z(\boldsymbol{\zeta})} \exp\left(\sum_{q=1}^{Q}\left(\frac{1}{2}\sum_{i \in \mathcal{N}_q} \beta_{iq}\bar{s}_i - \alpha_q\right)\bar{s}_q\right), \tag{13.67}$$

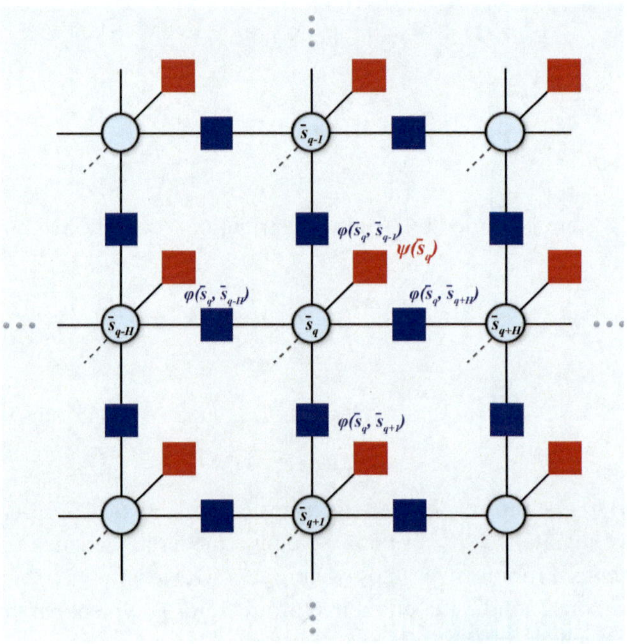

Fig. 13.18 Factor graph of the 4-connected MRF model (adapted from Fig. 4 in [29])

where $\mathcal{N}_q$ denotes the set of neighboring indices for grid point q, α_q controls the local sparsity bias, β_{iq} governs the interaction strength between neighboring sites, and $Z(\boldsymbol{\zeta})$ is the partition function ensuring proper normalization. This formulation can be equivalently expressed in a factorized form that reveals its structure more explicitly

$$p(\bar{\mathbf{s}}) = \frac{1}{Z(\boldsymbol{\zeta})} \left(\prod_{q=1}^{Q} \prod_{i \in \mathcal{N}_q} \varphi(\bar{s}_q, \bar{s}_i) \right)^{\frac{1}{2}} \prod_{q=1}^{Q} \psi(\bar{s}_q), \tag{13.68}$$

where $\varphi(\bar{s}_q, \bar{s}_i) = \exp(\beta_{iq}\bar{s}_i\bar{s}_q)$ represents the pairwise interaction potential that encourages neighboring sites to share the same state, and $\psi(\bar{s}_q) = \exp(-\alpha_q\bar{s}_q)$ is the unary potential encoding the local tendency toward activity or inactivity. The collection of all MRF parameters is denoted by $\boldsymbol{\zeta} \triangleq \{\alpha_q, \beta_{iq} \mid i \in \mathcal{N}_q, \forall q\}$.

For the two-dimensional grid structure under consideration, we adopt a 4-connected (first-order) neighborhood system, which strikes an effective balance between capturing essential spatial correlations and maintaining computational tractability. In this configuration, each position $\bar{s}_q$ interacts with its four immediate neighbors: left, right, top, and bottom. For a grid with $Q = H \times W$ points arranged in H rows and W columns, the neighborhood relationships are defined as

- **Left neighbor**: $\bar{s}_{q-H}$ (when $q > H$).
- **Right neighbor**: $\bar{s}_{q+H}$ (when $q \leq Q - H$).
- **Top neighbor**: $\bar{s}_{q-1}$ (when q mod $H \neq 1$).
- **Bottom neighbor**: $\bar{s}_{q+1}$ (when q mod $H \neq 0$).

Nodes at the grid boundaries naturally have reduced connectivity, reflecting the physical constraints of the finite surveillance area.

The factor graph of this 4-connected MRF, shown in Fig. 13.18, visually represents these interactions. The pairwise factors $\varphi(\bar{s}_q, \bar{s}_i)$ encourage neighboring positions to share the same state, with the interaction strength controlled by β_{iq}: larger values promote the formation of larger, more coherent clusters. Meanwhile, the unary factors $\psi(\bar{s}_q)$ encode local sparsity tendencies through α_q: higher values of α_q favor inactive states ($\bar{s}_q = -1$), while lower values favor activity.

The spatially varying parameters in this non-stationary formulation provide substantial flexibility to accommodate diverse scattering environments. The interaction parameters β_{iq} can vary across the grid to model regions with different cluster sizes, while the bias parameters α_q can adapt to spatial variations in target/scatterer density. This adaptability makes the model particularly suitable for real-world scenarios where target and scatterer distributions are rarely uniform or stationary.

Finally, to complete the Bayesian hierarchy, we assign Gamma priors to the noise precision parameters:

$$p(\gamma^r) = \Gamma(\gamma^r; c, d), \quad p(\gamma^c) = \Gamma(\gamma^c; c, d). \tag{13.69}$$

For the special user-related components x_0^r and x_0^c, we employ simplified priors that follow a similar Bernoulli–Gamma–Gaussian structure but without the joint support constraints, using individual support parameters λ_0^r and λ_0^c.

13.4.2.3 Structured CS Problem Formulation with a Dynamic Grid

Building upon the location-domain sparse representations developed in (13.57) and (13.58), we now construct a unified observation model by aggregating measurements across all available subcarriers. This concatenation yields compact expressions for the reflected downlink pilot signal and the received uplink pilot signal as

$$\boldsymbol{y}^r = \boldsymbol{\Phi}^r\left(\boldsymbol{r}, \boldsymbol{p}_u\right)\begin{bmatrix} x_0^r \\ \boldsymbol{x}^r \end{bmatrix} + \boldsymbol{z}^r, \tag{13.70}$$

$$\boldsymbol{y}^c = \boldsymbol{\Phi}^c\left(\boldsymbol{r}, \boldsymbol{p}_u, \tau_o\right)\begin{bmatrix} x_0^c \\ \boldsymbol{x}^c \end{bmatrix} + \boldsymbol{z}^c, \tag{13.71}$$

where the stacked vectors are defined by collecting the corresponding quantities across subcarriers: $\boldsymbol{y}^r \triangleq [\boldsymbol{y}_n^r]_{n\in N_b} \in \mathbb{C}^{M|N_b|\times 1}$, $\boldsymbol{y}^c \triangleq [\boldsymbol{y}_n^c]_{n\in N_u} \in \mathbb{C}^{M|N_u|\times 1}$, with analogous definitions for the noise vectors $\boldsymbol{z}^r$ and $\boldsymbol{z}^c$.

The measurement matrices in (13.70) and (13.71) admit a natural block structure that separates the contribution of the user-specific components from those of the general targets and scatterers, i.e.,

$$\boldsymbol{\Phi}^r\left(\boldsymbol{r}, \boldsymbol{p}_u\right) \triangleq \left[\boldsymbol{\Phi}^{r,0}, \boldsymbol{\Phi}^{r,1}\right] \in \mathbb{C}^{M|N_b|\times(Q+1)}, \tag{13.72}$$

$$\boldsymbol{\Phi}^c\left(\boldsymbol{r}, \boldsymbol{p}_u, \tau_o\right) \triangleq \left[\boldsymbol{\Phi}^{c,0}, \boldsymbol{\Phi}^{c,1}\right] \in \mathbb{C}^{M|N_u|\times(Q+1)}. \tag{13.73}$$

The constituent submatrices are defined as follows. For the radar channel, the user-specific component is

$$\boldsymbol{\Phi}^{r,0} = \left[e^{-j2\pi n f_0 \tau^r(\boldsymbol{p}_u)}\boldsymbol{a}(\theta^r(\boldsymbol{p}_u))\boldsymbol{a}^T(\theta^r(\boldsymbol{p}_u))\boldsymbol{v}_n^r\right]_{n\in N_b}, \tag{13.74}$$

while the target-related component is constructed from an auxiliary matrix:

$$\bar{\boldsymbol{\Phi}}^{r,1} = \left[\left((\boldsymbol{v}_n^r)^T\boldsymbol{A}(\boldsymbol{r})\right) \otimes \left(\boldsymbol{A}(\boldsymbol{r})\boldsymbol{D}_n^r\right)\right]_{n\in N_b}, \tag{13.75}$$

from which $\boldsymbol{\Phi}^{r,1} \in \mathbb{C}^{M|N_b|\times Q}$ is obtained by selecting the $((q-1)Q+q)$-th column for each $q = 1, \ldots, Q$. For the communication channel, the LoS and NLoS components are, respectively, captured by

$$\boldsymbol{\Phi}^{c,0} = \left[u_n^c e^{-j2\pi n f_0 \tau_o}\boldsymbol{a}(\theta^c(\boldsymbol{p}_u))\right]_{n\in N_u}, \tag{13.76}$$

$$\boldsymbol{\Phi}^{c,1} = \left[u_n^c \boldsymbol{A}(\boldsymbol{r})\boldsymbol{D}_n^c\right]_{n\in N_u}. \tag{13.77}$$

With these building blocks in place, we consolidate the radar and communication observations into a single linear model as

$$\boldsymbol{y} = \boldsymbol{A}(\boldsymbol{\vartheta})\boldsymbol{x} + \boldsymbol{z}, \tag{13.78}$$

where the consolidated variables are defined as

- **Sensing parameters:** $\boldsymbol{\vartheta} \triangleq \{\boldsymbol{r}, \boldsymbol{p}_u, \tau_o\}$, encompassing the dynamic position grid, user location, and timing offset.
- **Aggregated observation**: $\boldsymbol{y} \triangleq [(\boldsymbol{y}^r)^T, (\boldsymbol{y}^c)^T]^T$, stacking radar and communication measurements.
- **Composite noise**: $\boldsymbol{z} \triangleq [(\boldsymbol{z}^r)^T, (\boldsymbol{z}^c)^T]^T$, combining the noise from both subsystems.
- **Augmented sparse signal**: $\boldsymbol{x} \triangleq [x_0^r, (\boldsymbol{x}^r)^T, x_0^c, (\boldsymbol{x}^c)^T]^T$, collecting all unknown channel coefficients.
- **Block-diagonal sensing matrix**: $\boldsymbol{A}(\boldsymbol{\vartheta}) \triangleq \mathrm{BlockDiag}\left(\boldsymbol{\Phi}^r(\boldsymbol{r}, \boldsymbol{p}_u), \boldsymbol{\Phi}^c(\boldsymbol{r}, \boldsymbol{p}_u, \tau_o)\right)$.

The unified formulation in (13.78) integrates the radar and communication subsystems within a common mathematical framework. When combined with the MRF-based sparse prior model illustrated in Fig. 13.17, this formulation formally casts the joint scattering environment sensing and channel estimation problem as a structured compressed sensing task with dynamic grid adaptation. The MRF prior captures the inherent spatial correlations among targets and scatterers, while the dynamic grid $\boldsymbol{r}$ enables super-resolution parameter estimation through iterative refinement. This integrated approach allows the sensing and communication functionalities to mutually benefit from shared structural information, leading to enhanced performance in both tasks.

Summary

In this chapter, we have demonstrated that many important wireless communication and sensing problems, such as massive MIMO-OFDM channel estimation, wireless localization, and joint scattering environment sensing and channel estimation in ISAC systems, can be formulated as *structured CS problems with a dynamic grid*. Formulating such problems generally involves two key steps: first, obtaining a sparse representation of the wireless channel based on a dynamic grid, and second, designing a sparse prior model that captures the specific sparse structure of the channel.

Step 1: Sparse Representation via Dynamic Grid

Depending on the application scenario, the wireless channel can be transformed into different domains (such as the angular domain, delay-angular domain, or location domain) by discretizing the angle/delay/location space into a large number of dynamic grid points. The dynamic grid is typically initialized as a uniform grid with an appropriately chosen grid size:

- If the **grid size is too small**, the number of columns in the sensing matrix increases significantly, leading to strong inter-column correlation, high computational complexity, and degraded CS recovery performance.
- If the **grid size is too large**, the energy leakage effect becomes more severe, and gradient-based grid adjustment is more likely to converge to poor local optima.

In practice, the grid size is often determined by the underlying signal resolution. For example, the delay grid size is usually set to the delay resolution, i.e., the inverse of the signal bandwidth. The angular grid size is typically chosen as the angular resolution, e.g., the inverse of the number of antennas in a half-wavelength-spaced ULA.

Step 2: Structured Sparse Prior Modeling

The random clustering of scatterers in the wireless propagation environment can often be characterized using a Markov sparse prior or an MRF sparse prior:

- A **2D Markov sparse prior** (Sect. 13.3.3) involves fewer parameters and is more suitable for scenarios where cluster density and size do not vary significantly over space.
- An **MRF-based sparse prior** (Sect. 13.4.2) includes more parameters and offers greater flexibility for modeling spatially non-stationary environments where cluster characteristics change dramatically.

While the MRF-based model can better adapt to complex and non-stationary propagation conditions, it also requires learning a larger number of unknown parameters, which increases estimation complexity. Furthermore, conventional EM methods become computationally prohibitive for such models due to exponential complexity. To address this issue, a low-complexity approach based on pseudo-likelihood approximation was developed in [29]; interested readers are referred there for details.

We have provided three concrete examples showing how structured CS formulations with dynamic grids can be applied to wireless channel estimation and sensing. In the following chapter, we will introduce a unified algorithmic framework for solving general structured CS problems, based on which detailed solutions for the examples discussed here will be presented in the subsequent chapters.

Homework Problems

13.1 Read the related literature to understand the structural differences between group-sparse (block-sparse) signals and burst-sparse signals.

(a) Clearly define each type of structured sparsity with illustrative examples.

(b) Considering the same compressive observation model $\boldsymbol{y} = \boldsymbol{A}\boldsymbol{x} + \boldsymbol{w}$, analyze which type of signal is generally easier to recover.

(c) Provide a theoretical justification for your answer, explaining the underlying reasons based on the structural properties of each sparsity type.

13.2 Calculate the marginal prior $p\left(x_n\right)$ for:

(a) The independent Bernoulli–Gaussian (BG) prior.

(b) The independent Bernoulli–Gamma–Gaussian (BGG) prior.

For both cases:

- Plot the probability density function (PDF) of $p\left(x_n\right)$.
- Analyze whether it indeed constitutes a sparse prior when $p\left(s_n = 1\right) = \lambda$ is small and the precision parameters are properly chosen.
- Discuss how the choice of prior parameters affects the sparsity-inducing properties.

13.3 For the initial uniform delay-angular-domain grid defined in Sect. 13.3.2.2:

(a) Analyze how the choice of grid resolution parameters N_1, N_2, and L affects

- The accuracy of channel parameter estimation.
- The computational complexity of the CS recovery algorithm.
- The sensitivity to off-grid effects.

(b) Discuss the trade-offs involved in selecting these parameters and provide guidelines for choosing appropriate values in practical implementations.

13.4 The uniform location-domain grid in Sect. 13.4.2.1 uses a Cartesian coordinate system. Consider defining an alternative grid using a spherical coordinate system.

(a) Compare and contrast the two coordinate systems for constructing location-domain grids.

(b) Analyze how the choice of coordinate system might affect

- The sparsity pattern of the channel representation.
- The performance of sparse recovery algorithms.
- The computational complexity of the estimation process.
- The sensitivity to measurement geometry.

(c) Discuss scenarios where one coordinate system might be preferred over the other, providing theoretical justification for your recommendations.

References

1. A. Liu, V.K.N. Lau, W. Dai, Exploiting burst-sparsity in massive MIMO with partial channel support information. IEEE Trans. Wirel. Commun. **15**(11), 7820–7830 (2016)
2. M. Masood, L.H. Afify, T.Y. Al-Naffouri, Efficient coordinated recovery of sparse channels in massive MIMO. IEEE Trans. Signal Process. **63**(1), 104–118 (2015)
3. X. Rao, V.K.N. Lau, X. Kong, CSIT estimation and feedback for FDD multi-user massive MIMO systems, in *Proceedings of the 2014 IEEE International Conference on Acoustics, Speech, and Signal Processing (ICASSP)*, pp. 3157–3161

4. X. Kuai, L. Chen, X. Yuan, A. Liu, Structured turbo compressed sensing for downlink massive MIMO-OFDM channel estimation. IEEE Trans. Wirel. Commun. **18**(8), 3813–3826 (2019)
5. L. Chen, A. Liu, X. Yuan, Structured turbo compressed sensing for massive MIMO channel estimation using a Markov prior. IEEE Trans. Veh. Technol. **67**(5), 4635–4639 (2018)
6. M. Crouse, R. Nowak, R. Baraniuk, Wavelet-based statistical signal processing using hidden Markov models. IEEE Trans. Signal Process. **46**(4), 886–902 (1998)
7. L. Lian, A. Liu, V.K.N. Lau, Weighted lasso for sparse recovery with statistical prior support information. IEEE Trans. Signal Process. **66**(6), 1607–1618 (2018)
8. P. Schniter, Turbo reconstruction of structured sparse signals, in *Proceedings of 2010 Annual Conference on Information Sciences and Systems (CISS)*, pp. 1–6
9. S. Som, P. Schniter, Compressive imaging using approximate message passing and a Markov-tree prior. IEEE Trans. Signal Process. **60**(7), 3439–3448 (2012)
10. A. Liu, G. Liu, L. Lian, V.K.N. Lau, M.-J. Zhao, Robust recovery of structured sparse signals with uncertain sensing matrix: a turbo-VBI approach. IEEE Trans. Wirel. Commun. **19**(5), 3185–3198 (2020)
11. D.G. Tzikas, A.C. Likas, N.P. Galatsanos, The variational approximation for Bayesian inference. IEEE Signal Process. Mag. **25**(6), 131–146 (2008)
12. D.L. Donoho, A. Maleki, A. Montanari, The noise-sensitivity phase transition in compressed sensing. IEEE Trans. Inf. Theory **57**(10), 6920–6941 (2011)
13. L. Cheng, C. Xing, Y.C. Wu, Irregular array manifold aided channel estimation in massive MIMO communications. IEEE J. Sel. Top. Signal Process. **13**(5), 974–988 (2019)
14. L. Xu, L. Cheng, N. Wong, Y.C. Wu, H.V. Poor, Overcoming beam squint in dual-wideband mmWave MIMO channel estimation: a Bayesian multi-band sparsity approach. IEEE Trans. Signal Process. **72**, 1219–1234 (2024)
15. T.V. R. 3GPP, Technical specification group radio access network; NR; physical channels and modulation (2021)
16. X. Wu, X. Yang, S. Ma, B. Zhou, G. Yang, Hybrid channel estimation for UPA-assisted millimeter-wave massive MIMO IoT systems. IEEE Internet Things J. **9**(4), 2829–2842 (2022)
17. X. Tuo, Z. Chen, M.M. Zhao, C. You, M.J. Zhao, Near-field sparse Bayesian channel estimation and tracking for XL-IRS-aided wideband mmWave systems (2025). arXiv:2511.18752
18. E. Fornasini, 2D Markov chains. Linear Algebra Appl. **140**, 101–127 (1990)
19. A. Liu, Z. Huang, M. Li, Y. Wan, W. Li, T.X. Han, C. Liu, R. Du, D.K.P. Tan, J. Lu, Y. Shen, F. Colone, K. Chetty, A survey on fundamental limits of integrated sensing and communication. IEEE Commun. Surv. Tutor. **24**(2), 994–1034 (2022)
20. B. Paul, A.R. Chiriyath, D.W. Bliss, Survey of rf communications and sensing convergence research. IEEE Access **5**, 252–270 (2017)
21. O.B. Akan, M. Arik, Internet of radars: sensing versus sending with joint radar-communications. IEEE Commun. Mag. **58**(9), 13–19 (2020)
22. F. Liu, C. Masouros, A.P. Petropulu, H. Griffiths, L. Hanzo, Joint radar and communication design: applications, state-of-the-art, and the road ahead. IEEE Trans. Commun. **68**(6), 3834–3862 (2020)
23. Z. Feng, Z. Fang, Z. Wei, X. Chen, Z. Quan, D. Ji, Joint radar and communication: a survey. China Commun. **17**(1), 1–27 (2020)
24. K. Zheng, Q. Zheng, H. Yang, L. Zhao, L. Hou, P. Chatzimisios, Reliable and efficient autonomous driving: the need for heterogeneous vehicular networks. IEEE Commun. Mag. **53**(12), 72–79 (2015)
25. I. Bilik, O. Longman, S. Villeval, J. Tabrikian, The rise of radar for autonomous vehicles: signal processing solutions and future research directions. IEEE Signal Process. Mag. **36**(5), 20–31 (2019)
26. Z. Chen, M.M. Zhao, M. Li, F. Xu, Q. Wu, M.J. Zhao, Joint location sensing and channel estimation for IRS-aided mmWave ISAC systems. IEEE Trans. Wirel. Commun. **23**(9), 11985–12002 (2024)
27. S. Som, P. Schniter, Approximate message passing for recovery of sparse signals with Markov-random-field support structure, in *Proceedings of the International Conference on Machine Learning*, pp. 1–15 (2011)
28. M. Zhang, X. Yuan, Z.-Q. He, Variance state propagation for structured sparse Bayesian learning. IEEE Trans. Signal Process. **68**, 2386–2400 (2020)
29. W. Xu, Y. Xiao, A. Liu, M. Lei, M.-J. Zhao, Joint scattering environment sensing and channel estimation based on non-stationary Markov random field. IEEE Trans. Wirel. Commun. **23**(5), 3903–3917 (2024)

14 Expectation Maximization-Based Turbo Sparse Bayesian Inference

In the previous chapter, we demonstrated that numerous problems in wireless communications and sensing can be cast as structured CS with a dynamic grid, where the objective is to recover a structured sparse signal from linear observations, while the sensing matrix itself depends on unknown grid parameters. This chapter introduces EM-based Turbo sparse Bayesian inference (EM-Turbo-SBI)—a unified and efficient algorithmic framework for solving such problems. It operates by alternating between two key steps until convergence: In the E-step, a Bayesian estimate of the sparse signal is computed using a Turbo-SBI algorithm. In the M-step, a point estimate of the grid parameter is updated via maximum likelihood or MAP estimation. We present three principal Turbo-SBI variants for the E-step: Turbo compressive sensing (Turbo-CS), Turbo approximate message passing (Turbo-AMP), and Turbo variational Bayesian inference (Turbo-VBI). We further analyze the strengths and limitations of each method and offer practical guidance on selecting an appropriate Turbo-SBI algorithm based on specific problem structure, sensing matrix properties, and computational constraints.

14.1 Outline of EM-Turbo-SBI Framework

14.1.1 Problem Formulation

We begin by recalling the following CS observation model with dynamic grid parameters:

$$\boldsymbol{y} = \boldsymbol{A}(\boldsymbol{\theta})\boldsymbol{x} + \boldsymbol{w}, \tag{14.1}$$

where $\boldsymbol{A}(\boldsymbol{\theta}) \in \mathbb{C}^{M\times N}$ is the sensing matrix parameterized by grid vector $\boldsymbol{\theta}$, $\boldsymbol{x} \in \mathbb{C}^{N\times 1}$ is a structured sparse signal, and $\boldsymbol{w} \in \mathbb{C}^{M\times 1}$ represents additive noise with precision $\boldsymbol{\kappa}$.

The structured sparsity of $\boldsymbol{x}$ follows either the two-layer or three-layer hierarchical priors introduced in Chap. 13. To facilitate our discussion, we define two main variable groups. The first group comprises sparse signal variables $\boldsymbol{v} = \left\{\boldsymbol{v}^k, \forall k \in \mathcal{H}\right\}$, where for the two-layer prior we have $\boldsymbol{v} = \{\boldsymbol{x}, \boldsymbol{s}\}$ with $\mathcal{H} = \{1, 2\}$, and for the three-layer prior, $\boldsymbol{v} = \{\boldsymbol{x}, \boldsymbol{\rho}, \boldsymbol{s}\}$ with $\mathcal{H} = \{1, 2, 3\}$. The second group consists of system parameters $\boldsymbol{\xi} = \{\boldsymbol{\theta}, \boldsymbol{\phi}, \boldsymbol{\kappa}, \ldots\}$, which include grid parameters $\boldsymbol{\theta}$ in the sensing matrix, prior parameters $\boldsymbol{\phi}$ such as Markov transition probabilities, noise precision parameters $\boldsymbol{\kappa}$, and additional imperfection factors like timing offsets and phase noise.

The EM-Turbo-SBI framework addresses the following two coupled estimation problems:

1. **Bayesian Signal Recovery**: For fixed parameters $\boldsymbol{\xi}$, compute the following conditional marginal posteriors:

$$p\left(\boldsymbol{v}^k|\boldsymbol{y},\boldsymbol{\xi}\right) \propto \int p\left(\boldsymbol{y},\boldsymbol{v}|\boldsymbol{\xi}\right) \mathrm{d}\boldsymbol{v}^{-k} = \int p\left(\boldsymbol{v}\right) p\left(\boldsymbol{y}|\boldsymbol{v},\boldsymbol{\xi}\right) \mathrm{d}\boldsymbol{v}^{-k}, \tag{14.2}$$

where $\propto$ denotes equality after scaling, $\boldsymbol{v}^{-k}$ denotes the set of all variables in $\boldsymbol{v}$ except $\boldsymbol{v}^k$, and the likelihood is Gaussian, i.e.,

$$p\left(\boldsymbol{y}|\boldsymbol{x},\boldsymbol{\xi}\right) = \mathcal{CN}\left(\boldsymbol{y}; \boldsymbol{A}(\boldsymbol{\theta})\boldsymbol{x}, \operatorname{diag}\left(\boldsymbol{\kappa}\right)^{-1}\right).$$

2. **Parameter Estimation**: Obtain MAP estimates of the system parameters $\boldsymbol{\xi}$ as

A. Liu and M.-M. Zhao, *Advanced Bayesian Signal Processing*, Textbooks in Telecommunication Engineering,
https://doi.org/10.1007/978-3-032-26656-9_14

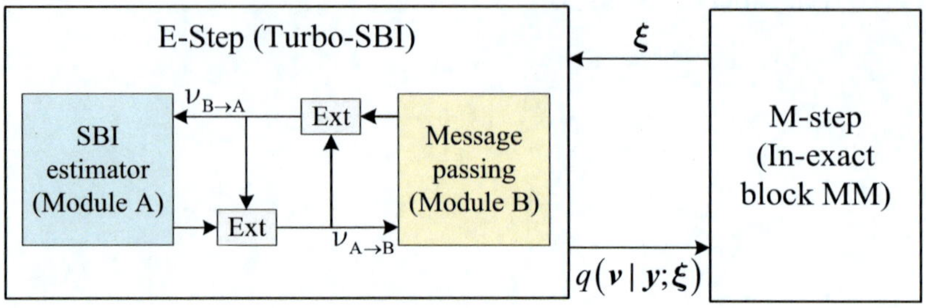

Fig. 14.1 Framework of the EM-Turbo-SBI algorithm

$$\begin{aligned}\boldsymbol{\xi}^* &= \underset{\boldsymbol{\xi}}{\operatorname{argmax}} \ln p(\boldsymbol{\xi} \mid \boldsymbol{y}) \\ &= \underset{\boldsymbol{\xi}}{\operatorname{argmax}} \ln p(\boldsymbol{y}, \boldsymbol{\xi}) \triangleq \ln \int p(\boldsymbol{y}, \boldsymbol{v}, \boldsymbol{\xi})\, \mathrm{d}\boldsymbol{v}.\end{aligned} \tag{14.3}$$

From (14.2) and (14.3), we can derive MAP estimates $\hat{\boldsymbol{v}}_{\mathrm{MAP}}^k = \arg\max p\left(\boldsymbol{v}^k \mid \boldsymbol{y}, \boldsymbol{\xi}^*\right)$ and MMSE estimates $\hat{\boldsymbol{v}}_{\mathrm{MMSE}}^k = \mathbb{E}\left[\boldsymbol{v}^k \mid \boldsymbol{y}, \boldsymbol{\xi}^*\right]$. However, direct computation of the conditional marginal posteriors $p\left(\boldsymbol{v}^k \mid \boldsymbol{y}, \boldsymbol{\xi}\right)$ and the MAP estimates of the system parameters $\boldsymbol{\xi}^*$ faces significant obstacles: the underlying factor graph contains cycles preventing exact inference, high-dimensional integrals lack closed-form solutions, and strong coupling exists between signal and parameter estimates. The EM-Turbo-SBI framework, developed in subsequent subsections, provides efficient approximate solutions that overcome these limitations while maintaining strong empirical performance.

14.1.2 EM-Turbo-SBI Framework

The EM-Turbo-SBI framework provides a systematic approach to solving joint estimation problems through an iterative procedure that combines EM with turbo processing. As illustrated in Fig. 14.1, the algorithm alternates between two complementary phases until convergence is achieved.

- **E-step (Sparse Signal Estimation)**: Given the current parameter estimate $\boldsymbol{\xi}$, this step computes an approximate posterior distribution for the sparse signal variables given by

$$q(\boldsymbol{v}; \boldsymbol{\xi}) \approx p(\boldsymbol{v} \mid \boldsymbol{y}, \boldsymbol{\xi}). \tag{14.4}$$

  The estimation combines standard sparse Bayesian inference methods (such as AMP or VBI) with message passing through the turbo approach. This integration enables the algorithm to handle both independent sparse priors and complex structured sparsity patterns, leveraging the strengths of both methodologies.
- **M-step (Block Coordinate Parameter Optimization)**: Using the variational distribution $q(\boldsymbol{v}; \boldsymbol{\xi})$ from the E-step, this phase constructs a surrogate function that minorizes the log-joint probability density $\ln p(\boldsymbol{y}, \boldsymbol{\xi})$. The parameters $\boldsymbol{\xi}$ are partitioned into B manageable blocks $\boldsymbol{\xi} = \left(\boldsymbol{\xi}_1, \ldots, \boldsymbol{\xi}_B\right)$, and the algorithm alternately maximizes the surrogate function w.r.t. each block. This block coordinate approach enables efficient optimization in high-dimensional parameter spaces.

The M-step builds upon the inexact block MM framework developed in Chap. 12. This foundation provides theoretical guarantees of convergence to a stationary point, offers flexibility in handling non-convex objective functions, and achieves computational efficiency through block-wise optimization. The construction of effective surrogate functions relies critically on the quality of the approximate posterior $q(\boldsymbol{v}; \boldsymbol{\xi})$ obtained in the E-step, creating a symbiotic relationship between the two steps.

14.1.3 M-Step: Inexact Block MM

The M-step of the EM-Turbo-SBI framework addresses the optimization of the intractable objective function $\ln p(\boldsymbol{y}, \boldsymbol{\xi})$ through a sophisticated MM approach. This approach combines block coordinate decomposition with surrogate function optimization to achieve efficient and stable parameter estimation.

A fundamental aspect of the M-step is the systematic decomposition of the parameter vector into manageable blocks

$$\boldsymbol{\xi} = (\boldsymbol{\xi}_1, \boldsymbol{\xi}_2, \ldots, \boldsymbol{\xi}_B). \tag{14.5}$$

This partitioning is guided by two key considerations: physical interpretation and computational tractability. Parameters that share similar physical roles or mathematical properties are naturally grouped together. For example, in array processing applications, angle parameters and delay parameters typically form separate blocks because they exhibit distinct characteristics, angles influence the steering vectors nonlinearly through sinusoidal functions, while delays appear as phase shifts linear in frequency. Such grouping facilitates the design of efficient update strategies tailored to each block's specific structure.

At each iteration, we construct a surrogate function $u(\boldsymbol{\xi}; \dot{\boldsymbol{\xi}})$ that serves as a local approximation to the original objective $\ln p(\boldsymbol{y}, \boldsymbol{\xi})$. For this surrogate to enable provable convergence, it must satisfy three essential properties:

1. **Global Lower Bound**:
$$u(\boldsymbol{\xi}; \dot{\boldsymbol{\xi}}) \leq \ln p(\boldsymbol{y}, \dot{\boldsymbol{\xi}}), \quad \forall \boldsymbol{\xi}. \tag{14.6}$$
This property ensures that maximizing the surrogate function guarantees improvement in the original objective, as any increase in the surrogate translates to an increase in the true objective.
2. **Current Point Matching**:
$$u(\dot{\boldsymbol{\xi}}; \dot{\boldsymbol{\xi}}) = \ln p(\boldsymbol{y}, \dot{\boldsymbol{\xi}}). \tag{14.7}$$
The surrogate must exactly match the original objective at the current iterate $\dot{\boldsymbol{\xi}}$, ensuring that the approximation is tight at the point around which it is constructed.
3. **Gradient Consistency**:
$$\left.\frac{\partial u(\boldsymbol{\xi}; \dot{\boldsymbol{\xi}})}{\partial \boldsymbol{\xi}}\right|_{\boldsymbol{\xi}=\dot{\boldsymbol{\xi}}} = \left.\frac{\partial \ln p(\boldsymbol{y}, \boldsymbol{\xi})}{\partial \boldsymbol{\xi}}\right|_{\boldsymbol{\xi}=\dot{\boldsymbol{\xi}}}. \tag{14.8}$$
This first-order consistency condition preserves the local gradient information, which is crucial for proper convergence behavior, particularly when the surrogate is used in gradient-based updates.

With an appropriate surrogate function constructed, the M-step proceeds to optimize it through a Gauss–Seidel-type alternating maximization scheme, i.e.,

$$\boldsymbol{\xi}_j^{(i+1)} = \underset{\boldsymbol{\xi}_j}{\operatorname{argmax}}\, u\left(\boldsymbol{\xi}_j, \boldsymbol{\xi}_{-j}^{(i)}; \boldsymbol{\xi}_j^{(i)}, \boldsymbol{\xi}_{-j}^{(i)}\right), \quad j = 1, \ldots, B, \tag{14.9}$$

where the superscript (i) denotes the i-th iteration, and

$$\boldsymbol{\xi}_{-j}^{(i)} = \left(\boldsymbol{\xi}_1^{(i+1)}, \ldots, \boldsymbol{\xi}_{j-1}^{(i+1)}, \boldsymbol{\xi}_{j+1}^{(i)}, \ldots, \boldsymbol{\xi}_B^{(i)}\right) \tag{14.10}$$

collects the most recent estimates of all blocks other than j.

This sequential update strategy, where each block is updated using the latest available information from previously updated blocks, often yields more stable convergence compared to simultaneous (Jacobi-type) updates, particularly when strong interactions exist between parameters. In addition, the algorithm adapts its update strategy to the convexity properties of each block. Let $\mathcal{J}_c$ denote the set of blocks for which $u(\boldsymbol{\xi}; \dot{\boldsymbol{\xi}})$ is concave with respect to $\boldsymbol{\xi}_j$, and $\bar{\mathcal{J}}_c = \{1, \ldots, B\} \setminus \mathcal{J}_c$ the set for which it is non-concave. The index sets are determined by examining the structure of the surrogate function for each block.

For blocks in $\mathcal{J}_c$ (concave blocks), we perform exact maximization. This can be achieved through: closed-form analytical solutions when available, or efficient convex optimization methods that leverage the concavity to guarantee finding the global maximum for that block.

For blocks in $\bar{\mathcal{J}}_c$ (non-concave blocks), where exact maximization is computationally prohibitive or analytically intractable, we employ a gradient ascent update with adaptive step size selection:

$$\boldsymbol{\xi}_j^{(i+1)} = \boldsymbol{\xi}_j^{(i)} + \gamma^{(i)} \left. \frac{\partial u\left(\boldsymbol{\xi}_j, \boldsymbol{\xi}_{-j}^{(i)}; \boldsymbol{\xi}_j^{(i)}, \boldsymbol{\xi}_{-j}^{(i)}\right)}{\partial \boldsymbol{\xi}_j} \right|_{\boldsymbol{\xi}_j = \boldsymbol{\xi}_j^{(i)}}. \tag{14.11}$$

The step size $\gamma^{(i)}$ is determined using the Armijo rule, a backtracking line search procedure that ensures sufficient increase in the surrogate function while maintaining numerical stability [16]. This inexact update for non-concave blocks strikes a practical balance between computational efficiency and convergence guarantees.

The following theorem establishes the convergence properties of this inexact block MM approach.

Theorem 14.1 (Convergence of Inexact MM) *Suppose the surrogate function* $u\left(\boldsymbol{\xi}; \dot{\boldsymbol{\xi}}\right)$ *satisfies the three key properties in (14.6)–(14.8) and is strongly concave w.r.t. each block* $\boldsymbol{\xi}_j$, $\forall j \in \mathcal{J}_c$. *If at each iteration we perform exact updates for concave blocks* ($j \in \mathcal{J}_c$) *and gradient updates for non-concave blocks* ($j \in \bar{\mathcal{J}}_c$), *then the sequence of iterates converges to a stationary point of the original optimization problem in (14.3).*

The complete proof, available in Appendix A of [1], extends classical MM theory to handle problems with multiple non-concave blocks. The key insight is that the exact updates for concave blocks ensure progress in directions where the problem structure is favorable, while the gradient updates for non-concave blocks provide a reliable mechanism for exploring directions where the landscape is more complex.

14.1.4 EM-Based Surrogate Function Construction

The construction of an effective surrogate function is central to the M-step optimization procedure. This subsection presents a practical surrogate function that balances two competing objectives: maintaining the theoretical convergence guarantees of the exact EM approach while remaining computationally tractable for real-world problems.

14.1.4.1 Ideal Surrogate Function from EM Theory

Building upon the classical EM formulation, we first define an ideal surrogate function that would be used if exact posterior inference were possible

$$u(\boldsymbol{\xi}; \dot{\boldsymbol{\xi}}) = u^{\mathrm{EM}}(\boldsymbol{\xi}; \dot{\boldsymbol{\xi}}) - \sum_{j \in \mathcal{J}_c^1} \tau_j \|\boldsymbol{\xi}_j - \dot{\boldsymbol{\xi}}_j\|^2, \tag{14.12}$$

where the standard EM surrogate is given by

$$u^{\mathrm{EM}}(\boldsymbol{\xi}; \dot{\boldsymbol{\xi}}) = \int p(\boldsymbol{v}|\boldsymbol{y}, \dot{\boldsymbol{\xi}}) \ln \frac{p(\boldsymbol{v}, \boldsymbol{y}, \boldsymbol{\xi})}{p(\boldsymbol{v}|\boldsymbol{y}, \dot{\boldsymbol{\xi}})} \mathrm{d}\boldsymbol{v}. \tag{14.13}$$

Here, $\mathcal{J}_c^1 \subseteq \{1, \ldots, B\}$ denotes the index set of blocks for which $u^{\mathrm{EM}}(\boldsymbol{\xi}; \dot{\boldsymbol{\xi}})$ is concave, but not strongly concave, with respect to $\boldsymbol{\xi}_j$. The parameters $\tau_j > 0$ are positive constants that can be chosen arbitrarily.

The quadratic regularization terms $-\tau_j \|\boldsymbol{\xi}_j - \dot{\boldsymbol{\xi}}_j\|^2$ serve a crucial theoretical purpose: they transform mere concavity into strong concavity for the blocks in $\mathcal{J}_c^1$. This strengthening of the curvature properties is essential for satisfying the conditions of Theorem 14.1, which requires strong concavity with respect to each block in $\mathcal{J}_c$ (the set of blocks where exact updates are performed). By ensuring strong concavity, these terms guarantee that each block subproblem has a unique maximizer and that the overall algorithm converges to a stationary point.

14.1.4.2 Practical Surrogate Using Variational Approximation

In practice, exact computation of the posterior $p(\boldsymbol{v}|\boldsymbol{y}, \dot{\boldsymbol{\xi}})$ is typically intractable. To overcome this obstacle, we leverage the variational approximation $q(\boldsymbol{v}; \dot{\boldsymbol{\xi}}) \approx p(\boldsymbol{v}|\boldsymbol{y}, \dot{\boldsymbol{\xi}})$ obtained from the E-step. Substituting this approximation yields a practical surrogate function:

$$\hat{u}(\boldsymbol{\xi}; \dot{\boldsymbol{\xi}}) = \hat{u}^{\mathrm{EM}}(\boldsymbol{\xi}; \dot{\boldsymbol{\xi}}) - \sum_{j \in \hat{\mathcal{J}}_c^1} \tau_j \|\boldsymbol{\xi}_j - \dot{\boldsymbol{\xi}}_j\|^2, \tag{14.14}$$

where the approximated EM surrogate is

$$\hat{u}^{\mathrm{EM}}(\boldsymbol{\xi}; \dot{\boldsymbol{\xi}}) = \int q(\boldsymbol{v}; \dot{\boldsymbol{\xi}}) \ln \frac{p(\boldsymbol{v}, \boldsymbol{y}, \boldsymbol{\xi})}{q(\boldsymbol{v}; \dot{\boldsymbol{\xi}})} \mathrm{d}\boldsymbol{v}. \tag{14.15}$$

The index set $\hat{\mathcal{J}}_c^1$ is defined analogously to $\mathcal{J}_c^1$, but based on the concavity properties of $\hat{u}^{\mathrm{EM}}$ rather than the exact u^{EM}. In practice, these sets often coincide or can be determined through analysis of the problem structure.

This practical surrogate function $\hat{u}(\boldsymbol{\xi}; \dot{\boldsymbol{\xi}})$ possesses several important characteristics that make it suitable for optimization. First, the accuracy of the surrogate function approximation depends directly on the quality of the variational posterior $q(\boldsymbol{v}; \dot{\boldsymbol{\xi}})$, and the Turbo-SBI framework is specifically designed to provide high-fidelity approximations for this purpose. Second, although $\hat{u}(\boldsymbol{\xi}; \dot{\boldsymbol{\xi}})$ only approximately satisfies the theoretical conditions in (14.6)–(14.8), extensive simulations demonstrate that this approximation has negligible impact on convergence behavior, ensuring robustness. Third, the introduced regularization terms add minimal computational overhead while maintaining the strong concavity required for convergence guarantees, thus preserving efficiency. Finally, the parameters $\{\tau_j\}$ can be adaptively tuned to balance convergence speed and solution accuracy based on problem-specific characteristics, providing flexibility across different applications.

It is worth emphasizing the symbiotic relationship between the E-step and the surrogate construction described here. The E-step provides the variational approximation $q(\boldsymbol{v}; \dot{\boldsymbol{\xi}})$ that makes the surrogate tractable; in return, the M-step uses this surrogate to refine parameter estimates, which then improve the posterior approximations in subsequent E-step iterations. This mutual reinforcement, i.e., accurate posteriors enabling better surrogate functions, which in turn yield better parameter estimates for even more accurate posteriors, lies at the core of the EM-Turbo-SBI framework's effectiveness.

14.1.5 Outline of the E-Step

The E-step of the EM-Turbo-SBI framework addresses the challenge of joint inference through a strategic decomposition of the underlying factor graph. By partitioning the graph into complementary subgraphs, the framework transforms an intractable global inference problem into a set of coordinated local inference tasks that can be solved efficiently.

14.1.5.1 Factor Graph Structure and Decomposition

Consider the factor graph $\mathcal{G}$ representing the joint distribution $p(\boldsymbol{y}, \boldsymbol{v}|\boldsymbol{\xi})$, where $\boldsymbol{v}$ encompasses all latent variables associated with the sparse signal. This graph exhibits an inherent structural dichotomy that motivates its decomposition into two distinct subgraphs with complementary characteristics:

- **Subgraph** $\mathcal{G}_A$ encodes the linear observation model $p(\boldsymbol{y}|\boldsymbol{v}, \boldsymbol{\xi})$ together with independent sparse priors. This subgraph is characterized by dense connectivity and numerous loops, a direct consequence of the coupling introduced by the sensing matrix $\boldsymbol{A}(\boldsymbol{\theta})$. The presence of these loops makes direct application of sum-product message passing [2] computationally prohibitive and numerically unstable.
- **Subgraph** $\mathcal{G}_B$ captures the structured sparse prior model $p(\boldsymbol{s})$, where $\boldsymbol{s}$ denotes the support vector. This subgraph features simpler variable types (notably binary variables) and, depending on the specific prior structure, may be loop-free or have only limited local cycles. These characteristics make it well suited for efficient message-passing algorithms.

This structural dichotomy, where one subgraph handles the complex, densely connected observation model while the other manages the structured, potentially loop-free prior, provides the foundation for the turbo approach. By strategically combining standard sparse Bayesian inference (SBI) methods for $\mathcal{G}_A$ with structured message passing for $\mathcal{G}_B$, the framework enables effective approximate inference across the entire graph $\mathcal{G}$.

14.1.5.2 Dual-Module Architecture

Building on this graph decomposition, the E-step implements a dual-module architecture that leverages the distinct characteristics of each subgraph.

- **Module A (Observation Processing)**: This module employs standard SBI methods, such as AMP, LMMSE estimation, or sparse VBI, to compute approximate posterior distributions. It integrates two sources of information: the linear observation model $p(\boldsymbol{y}|\boldsymbol{v}, \boldsymbol{\xi})$ and independent sparse prior information received from Module B. Module A's role is to extract information from the measurements while respecting the prior constraints provided by Module B.
- **Module B (Structured Prior Processing)**: This module utilizes sum-product message passing to compute marginal posteriors by combining the structured sparse prior $p(\boldsymbol{s})$ with decoupled observation information received from Module A. Module B's role is to enforce the structural dependencies encoded in $p(\boldsymbol{s})$, such as Markov chain correlations or MRF neighborhood constraints, while incorporating the observational evidence provided by Module A.

This division of computational labor allows each module to focus on what it does best: Module A handles the challenging linear inverse problem using robust SBI techniques, while Module B enforces structural constraints using efficient message passing on well-structured graphs.

14.1.5.3 Information Exchange via Extrinsic Messages

The two modules engage in an iterative information exchange governed by a carefully designed protocol. The exchange proceeds in two directions.

- **Prior information flow (B→ A)**: Module B provides Module A with an effective independent sparse prior, capturing the structural knowledge encoded in $p(\boldsymbol{s})$ in a form that Module A can readily incorporate.
- **Observation information flow (A→ B)**: Module A provides Module B with decoupled observation information, summarizing what has been learned from the measurements in a way that isolates each variable's contribution.

Central to this exchange is the concept of extrinsic message computation. For each module, the procedure follows three steps:

1. **Belief computation**: The module computes an updated belief (posterior distribution) using all available information, its own model (observation model for Module A, structured prior for Module B) plus the incoming message from the other module.
2. **Information subtraction**: The module removes from this belief the contribution of the incoming message. Mathematically, this corresponds to division in the probability domain: the extrinsic message is proportional to the ratio of the posterior to the incoming message.
3. **Message transmission**: The resulting extrinsic message, representing only the new information the module has gleaned, is sent to the other module.

This subtraction step is critical: It prevents circular reasoning by ensuring that information already possessed by the receiving module is not redundantly fed back. Each exchanged message contains fresh, actionable evidence that drives the inference forward.

This strategic decomposition yields several important benefits. First, the original problem, structured sparse signal recovery with coupled linear observations, is transformed into two more manageable subproblems: independent sparse signal recovery with coupled observations (handled by Module A) and structured sparse signal recovery with decoupled observations (handled by Module B). Second, despite the decomposition, the coordinated information exchange ensures that the approximate posterior distributions closely track those that would be obtained from exact inference on the full graph. Finally, different SBI methods can be plugged into Module A depending on problem characteristics. The framework supports three primary instantiations:

- **Turbo-CS Algorithm** (Sect. 14.2): Combines turbo processing with LMMSE principles, particularly effective for partially orthogonal sensing matrices.
- **Turbo-AMP Algorithm** (Sect. 14.3): Leverages AMP for efficient inference, well suited for i.i.d. or weakly correlated sensing matrices.
- **Turbo-VBI Algorithm** (Sect. 14.4): Employs VBI methods for flexible prior modeling, offering robustness for general, ill-conditioned sensing matrices.

The modular design of the EM-Turbo-SBI framework offers practitioners significant flexibility in algorithm selection. By choosing the most appropriate SBI method for Module A based on specific problem characteristics, sensing matrix structure, sparsity pattern, and computational constraints, the framework can be tailored to achieve an optimal balance between estimation accuracy and computational efficiency. This adaptability, combined with the theoretical foundations of the turbo approach, makes the framework suitable for a wide range of applications in wireless communications, signal processing, and beyond.

14.2 Turbo-CS Algorithm

The Turbo-CS algorithm represents a significant advancement in sparse signal recovery, first introduced in [3] for the standard CS model with partially orthogonal sensing matrices and independent sparse priors. Since its initial development, the algorithm has undergone substantial extensions to address more complex and realistic scenarios. The evolution of Turbo-CS has progressed along two primary dimensions:

1. **Structured Sparsity Extension**: Building upon the original formulation, researchers have extended Turbo-CS to accommodate structured sparse prior models including Markov random fields and hidden Markov models [4–6], enabling the algorithm to leverage sophisticated statistical dependencies in sparse signals.
2. **Sensing Matrix Generalization**: The framework has been further generalized to handle arbitrary sensing matrices [7], significantly expanding its applicability beyond the original partially orthogonal constraint.

This section presents a unified Turbo-CS algorithm that simultaneously addresses both structured sparse priors and arbitrary sensing matrices. This generalized formulation maintains the computational efficiency of the original approach while substantially broadening its theoretical foundations and practical utility.

The algorithm preserves the modular architecture described in Sect. 14.1.5, with Module A implementing LMMSE-based inference and Module B handling structured prior processing through appropriate message-passing techniques. The iterative information exchange between modules enables effective recovery of structured sparse signals even under challenging sensing conditions.

14.2.1 Outline of Turbo-CS Algorithm

The Turbo-CS algorithm is built upon a two-layer hierarchical Bayesian framework that effectively captures structured sparsity while maintaining computational efficiency. This framework comprises three essential components that together define the statistical relationships among the sparse signal, its support structure, and the observations.

14.2.1.1 Hierarchical Prior Model

The first layer models the sparse signal conditioned on its support. Given the support vector $\boldsymbol{s} = [s_1, \ldots, s_N]^T$ with $s_n \in \{0, 1\}$, the signal elements are assumed to be independent and follow a Bernoulli–Gaussian distribution:

$$p(\boldsymbol{x}|\boldsymbol{s}) = \prod_{n=1}^{N} p(x_n|s_n), \tag{14.16}$$

where each conditional distribution takes the form of a spike-and-slab mixture:

$$p(x_n|s_n) = s_n \mathcal{CN}(x_n; 0, \sigma_{x,n}^2) + (1 - s_n)\delta(x_n). \tag{14.17}$$

Here, $\delta(\cdot)$ denotes the Dirac delta function, ensuring that inactive components ($s_n = 0$) are exactly zero, while active components follow a complex Gaussian distribution.

The second layer captures structured sparsity patterns through a prior distribution $p(\boldsymbol{s})$ on the support vector. This distribution can be tailored to specific application requirements, for instance, a Markov chain prior for burst sparsity or a Markov random field for two-dimensional clustered patterns, providing the flexibility to encode domain-specific structural knowledge.

The observation model follows the standard compressive sensing formulation

$$\boldsymbol{y} = \boldsymbol{A}\boldsymbol{x} + \boldsymbol{w}, \tag{14.18}$$

with independent and identically distributed Gaussian noise $w_m \sim \mathcal{CN}(0, \sigma_w^2)$ for $m = 1, \ldots, M$.

14.2.1.2 Factor Graph Decomposition

The joint probability density function $p(\boldsymbol{s})p(\boldsymbol{x}|\boldsymbol{s})p(\boldsymbol{y}|\boldsymbol{x})$ can be represented by a factor graph $\mathcal{G}$, as illustrated in Fig. 14.2a. A key insight underlying the Turbo-CS algorithm is the strategic decomposition of this graph into two complementary subgraphs, shown in Fig. 14.2b. This partitioning creates a natural division of computational labor:

- **Subgraph** $\mathcal{G}_A$ encompasses the linear observation model together with effectively independent Gaussian priors. This subgraph is responsible for integrating the measurement information.
- **Subgraph** $\mathcal{G}_B$ contains the structured sparse prior model $p(\boldsymbol{s})$ along with decoupled AWGN observations. This subgraph enforces the structural constraints on the support.

The specific functional forms of each factor node in $\mathcal{G}$ are provided in Table 14.1, which serves as the mathematical foundation for the subsequent message-passing operations.

14.2.1.3 Modular Architecture and Information Exchange

The above graph decomposition naturally gives rise to a modular architecture comprising two specialized processing units that exchange extrinsic information iteratively until convergence.

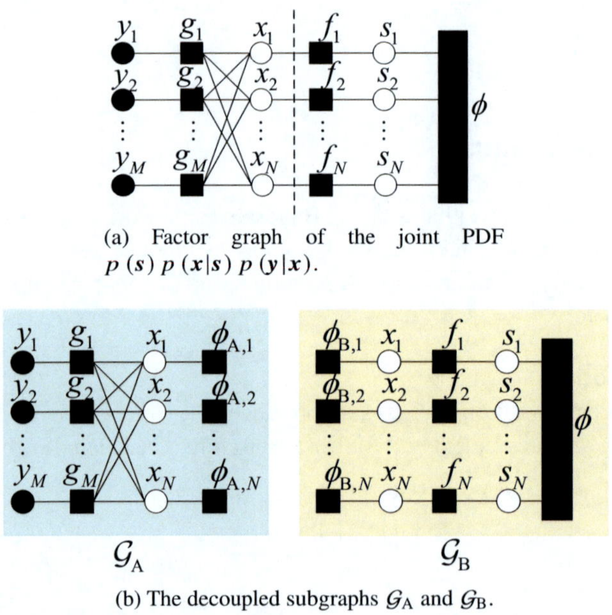

(a) Factor graph of the joint PDF $p(\boldsymbol{s})\, p(\boldsymbol{x}|\boldsymbol{s})\, p(\boldsymbol{y}|\boldsymbol{x})$.

(b) The decoupled subgraphs $\mathcal{G}_A$ and $\mathcal{G}_B$.

Fig. 14.2 Partitioning of the factor graph $\mathcal{G}$ in the Turbo-CS algorithm

Table 14.1 Factors, distributions, and functional forms in the factor graph $\mathcal{G}$ in Fig. 14.2a

Factor	Distribution	Functional form
$g_m(y_m, \boldsymbol{x})$	$p(y_m\|\boldsymbol{x}, \boldsymbol{\xi})$	$\mathcal{CN}(y_m; \boldsymbol{A}_m(\boldsymbol{\theta})\boldsymbol{x}, \sigma_w^2)$
$f_n(x_n, s_n)$	$p(x_n\|s_n)$	$s_n\mathcal{CN}(x_n; 0, \sigma_{x,n}^2) + (1 - s_n)\delta(x_n)$
ϕ	$p(\boldsymbol{s})$	Depends on application

$\boldsymbol{A}_m(\boldsymbol{\theta})$ denotes the m-th row of $\boldsymbol{A}(\boldsymbol{\theta})$

Module A functions as an LMMSE estimator. It receives from Module B extrinsic messages

$$\phi_{\mathrm{A},n}(x_n) \triangleq \nu_{f_n \to x_n}(x_n) = \mathcal{CN}(x_n; x_{A,n}^{\mathrm{pri}}, v_{A,n}^{\mathrm{pri}}), \quad \forall n, \tag{14.19}$$

that serve as independent Gaussian priors. Using these priors together with the linear observation model $\boldsymbol{y} = \boldsymbol{Ax} + \boldsymbol{w}$, Module A computes the posterior distribution of $\boldsymbol{x}$.

Module B acts as a structured sparsity combiner. It interprets the extrinsic messages received from Module A

$$\phi_{\mathrm{B},n}(x_n) \triangleq \nu_{x_n \to f_n}(x_n) = \mathcal{CN}(x_n; x_{B,n}^{\mathrm{ob}}, v_{B,n}^{\mathrm{ob}}), \quad \forall n, \tag{14.20}$$

as decoupled observations from an equivalent AWGN channel, i.e.,

$$x_{B,n}^{\mathrm{ob}} = x_n + z_n, \tag{14.21}$$

with independent noise $z_n \sim \mathcal{CN}(0, v_{B,n}^{\mathrm{ob}})$. Module B then performs sum-product message passing over this effective observation model combined with the structured prior $p(\boldsymbol{s})$, yielding refined posterior distributions for both $\boldsymbol{x}$ and $\boldsymbol{s}$.

14.2.1.4 Iterative Information Exchange

The two modules engage in a carefully orchestrated exchange of extrinsic information. In the forward direction (from Module A to Module B), the extrinsic messages are computed by excluding the prior information from Module A's posterior as

$$\phi_{\mathrm{B},n}(x_n) = \nu_{x_n \to f_n}(x_n) \propto \frac{\mathcal{CN}(x_n; x_{A,n}^{\mathrm{post}}, v_{A,n}^{\mathrm{post}})}{\phi_{\mathrm{A},n}(x_n)}, \tag{14.22}$$

where $\mathcal{CN}(x_n; x_{A,n}^{\mathrm{post}}, v_{A,n}^{\mathrm{post}})$ denotes Module A's posterior distribution for the n-th signal component.

In the reverse direction (from Module B to Module A), the extrinsic messages are similarly obtained by removing the observation information from Module B's posterior as

$$\phi_{\mathrm{A},n}(x_n) \propto \frac{\mathcal{CN}(x_n; x_{B,n}^{\mathrm{post}}, v_{B,n}^{\mathrm{post}})}{\phi_{\mathrm{B},n}(x_n)}, \tag{14.23}$$

with $\mathcal{CN}(x_n; x_{B,n}^{\mathrm{post}}, v_{B,n}^{\mathrm{post}})$ representing Module B's posterior.

This iterative exchange of extrinsic information enables progressive refinement of the estimates. Each iteration allows Module A to incorporate fresh measurement information while Module B enforces structural constraints, with the extrinsic message calculation ensuring that information already possessed by the other module is not redundantly fed back. Through this mechanism, the algorithm progressively hones in on a solution that simultaneously satisfies both the linear observation constraints and the structured sparsity patterns.

14.2.2 LMMSE in Module A

Module A implements a linear minimum mean square error (LMMSE) estimator that integrates information from two sources: the linear observation model and the prior knowledge provided by Module B. This section details the mathematical formulation of this estimator, and the computation of extrinsic messages.

The prior information received from Module B takes the form of independent Gaussian distributions for each signal component $\phi_{\mathrm{A},n}(x_n) = \mathcal{CN}(x_n; x_{A,n}^{\mathrm{pri}}, v_{A,n}^{\mathrm{pri}}), \forall n$. Collecting these component-wise priors, we obtain the prior mean vector $\boldsymbol{x}_A^{\mathrm{pri}} = [x_{A,1}^{\mathrm{pri}}, \ldots, x_{A,N}^{\mathrm{pri}}]^T$ and a diagonal prior covariance matrix $\boldsymbol{V}_A^{\mathrm{pri}} = \mathrm{diag}(v_{A,1}^{\mathrm{pri}}, \ldots, v_{A,N}^{\mathrm{pri}})$. The diagonal structure reflects the independence assumption embedded in the messages from Module B, a design choice that simplifies subsequent computations while preserving the essential statistical information needed for effective inference.

Given this Gaussian prior and the linear observation model $\boldsymbol{y} = \boldsymbol{Ax} + \boldsymbol{w}$ with i.i.d. Gaussian noise of variance σ_w^2, the posterior distribution of $\boldsymbol{x}$ remains Gaussian. The posterior mean and covariance matrix are given by the standard LMMSE formulas:

$$\boldsymbol{x}_A^{\text{post}} = \boldsymbol{V}_A^{\text{post}} \left((\boldsymbol{V}_A^{\text{pri}})^{-1} \boldsymbol{x}_A^{\text{pri}} + \frac{\boldsymbol{A}^H \boldsymbol{y}}{\sigma_w^2} \right), \tag{14.24}$$

$$\boldsymbol{V}_A^{\text{post}} = \left(\frac{\boldsymbol{A}^H \boldsymbol{A}}{\sigma_w^2} + (\boldsymbol{V}_A^{\text{pri}})^{-1} \right)^{-1}. \tag{14.25}$$

From these, the marginal posterior for each coefficient is $\mathcal{CN}(x_n; x_{A,n}^{\text{post}}, v_{A,n}^{\text{post}})$, where $x_{A,n}^{\text{post}}$ denotes the n-th element of $\boldsymbol{x}_A^{\text{post}}$ and $v_{A,n}^{\text{post}}$ is the corresponding diagonal entry of $\boldsymbol{V}_A^{\text{post}}$.

A fundamental principle of the turbo architecture is that each module should transmit to the other only the new information it has gleaned, information not already available to the receiving module. This is achieved through extrinsic message computation, which effectively removes the contribution of the incoming prior from the posterior. In the probability domain, this operation corresponds to division. The extrinsic variance is obtained by

$$\boldsymbol{V}_B^{\text{ob}} = \left((\bar{\boldsymbol{V}}_A^{\text{post}})^{-1} - (\boldsymbol{V}_A^{\text{pri}})^{-1} \right)^{-1}, \tag{14.26}$$

where $\bar{\boldsymbol{V}}_A^{\text{post}} = \text{diag}(v_{A,1}^{\text{post}}, \ldots, v_{A,N}^{\text{post}})$ is the diagonalized posterior covariance matrix. The use of the diagonalized version, rather than the full covariance matrix, is a deliberate approximation that significantly reduces computational complexity while empirically maintaining good performance. This approximation effectively ignores off-diagonal correlations in the posterior when computing extrinsic information for Module B. The extrinsic mean is computed similarly:

$$\boldsymbol{x}_B^{\text{ob}} = \boldsymbol{V}_B^{\text{ob}} \left((\bar{\boldsymbol{V}}_A^{\text{post}})^{-1} \boldsymbol{x}_A^{\text{post}} - (\boldsymbol{V}_A^{\text{pri}})^{-1} \boldsymbol{x}_A^{\text{pri}} \right). \tag{14.27}$$

These extrinsic quantities, $\boldsymbol{x}_B^{\text{ob}}$ and $\boldsymbol{V}_B^{\text{ob}}$, are then passed to Module B, where they are interpreted as observations from an equivalent AWGN channel, as described in the next subsection.

The implementation of Module A requires careful attention to several computational aspects. The computation of $\boldsymbol{V}_A^{\text{post}}$ involves inversion of an $N \times N$ matrix, which can be computationally intensive for large N, necessitating efficient implementations that often exploit matrix inversion lemmas or iterative methods. Numerical stability is another important consideration, particularly the subtraction in the extrinsic variance calculation that requires careful handling to maintain positive definiteness. Additionally, the use of $\bar{\boldsymbol{V}}_A^{\text{post}}$ (diagonalized posterior covariance) rather than the full $\boldsymbol{V}_A^{\text{post}}$ represents a computational approximation that significantly reduces complexity while maintaining good empirical performance by ignoring off-diagonal correlations in the posterior.

From a theoretical perspective, the LMMSE formulation in Module A can be interpreted in multiple ways: as a Bayesian estimator that optimally combines prior information with observational evidence, as a Wiener filter operating within the Bayesian framework, or as the solution to a regularized least squares problem with prior covariance serving as regularization. This mathematically rigorous foundation ensures that Module A provides statistically optimal estimates under the Gaussian assumptions, forming a crucial component of the overall Turbo-CS algorithm's effectiveness by delivering reliable estimates that enable productive information exchange with Module B.

14.2.3 Message Passing in Module B

Module B performs structured Bayesian inference by combining two sources of information: the structured sparse prior $p(\boldsymbol{s})$ and the decoupled observation information received from Module A. This section details the message-passing procedures that enable this inference, organized as a forward–backward schedule over the factor graph $\mathcal{G}_B$.

14.2.3.1 Effective Observation Model

A key conceptual step is interpreting the extrinsic messages from Module A as observations from an equivalent AWGN channel. Specifically, for each component n, the message $\phi_{\text{B},n}(x_n) = \mathcal{CN}(x_n; x_{B,n}^{\text{ob}}, v_{B,n}^{\text{ob}})$ is treated as a noisy observation of

Table 14.2 Factors, distributions, and functional forms in $\mathcal{G}_B$ of Fig. 14.2b

Factor	Distribution	Functional form
$\phi_{B,n}(x_n)$	$p\left(x^{ob}_{B,n}\|x_n\right)$	$\mathcal{CN}\left(x_n; x^{ob}_{B,n}, v^{ob}_{B,n}\right)$
$f_n(x_n, s_n)$	$p(x_n\|s_n)$	$s_n\mathcal{CN}\left(x_n; 0, \sigma^2_{x,n}\right) + (1 - s_n)\,\delta(x_n)$
ϕ	$p(\boldsymbol{s})$	Depends on application

the true signal x_n: $x^{ob}_{B,n} = x_n + z_n$, $z_n \sim \mathcal{CN}(0, v^{ob}_{B,n})$, or in vector form $\boldsymbol{x}^{ob}_B = \boldsymbol{x} + \boldsymbol{z}$ with $\boldsymbol{z} \sim \mathcal{CN}(\boldsymbol{0}, \boldsymbol{V}^{ob}_B)$ and diagonal covariance $\boldsymbol{V}^{ob}_B = \text{diag}(v^{ob}_{B,1}, \ldots, v^{ob}_{B,N})$. This interpretation is natural because the Gaussian form of the extrinsic messages exactly matches the likelihood function that would arise from such an AWGN channel. By adopting this perspective, the structured inference problem is transformed into a more tractable form while preserving all essential statistical relationships.

The factor graph corresponding to this effective observation model combined with the structured prior is exactly $\mathcal{G}_B$, shown in Fig. 14.2b. Table 14.2 provides the complete functional expressions for each factor node in this graph. The super factor node ϕ encapsulates the structured sparse prior $p(\boldsymbol{s})$; depending on the application, this node can be further expanded into constituent factor nodes representing, for example, Markov chains for burst sparsity or Markov random fields for two-dimensional clustered patterns.

14.2.3.2 Forward Message Passing

Message passing in Module B follows a systematic forward–backward schedule. In the forward direction, information flows along the path $x_n \to f_n \to s_n$ for each n. The process begins with the message from the observation node to the factor node f_n:

$$\nu_{x_n \to f_n}(x_n) = \mathcal{CN}(x_n; x^{ob}_{B,n}, v^{ob}_{B,n}). \tag{14.28}$$

Integrating over x_n, the message from factor node f_n to the support node s_n is obtained as

$$\begin{aligned} \nu_{f_n \to s_n}(s_n) &\propto \int f_n(x_n, s_n)\, \nu_{x_n \to f_n}(x_n)\, \mathrm{d}x_n \\ &= \pi_n \delta(s_n - 1) + (1 - \pi_n)\delta(s_n), \end{aligned} \tag{14.29}$$

where the activation probability π_n is given by

$$\pi_n = \left(1 + \frac{\mathcal{CN}(0; x^{ob}_{B,n}, v^{ob}_{B,n})}{\mathcal{CN}(0; x^{ob}_{B,n}, v^{ob}_{B,n} + \sigma^2_{x,n})}\right)^{-1}. \tag{14.30}$$

This expression has an intuitive interpretation: it compares the likelihood of the observation under the hypothesis that $x_n = 0$ (the numerator) against the hypothesis that x_n is drawn from the active component's Gaussian distribution (the denominator).

The collection of messages $\{\nu_{f_n \to s_n}(s_n)\}_{n=1}^N$ is then processed through the structured prior $p(\boldsymbol{s})$ represented by the super factor node ϕ. The specific implementation of this processing depends on the structure of $p(\boldsymbol{s})$:

- For independent priors, the processing reduces to component-wise multiplication.
- For a Markov chain prior, forward–backward algorithm on a chain graph yields exact inference.
- For a Markov random field, loopy belief propagation or approximate methods may be employed.

This step produces refined beliefs for each support variable

$$\nu_{s_n \to f_n}(s_n) \propto \nu_{\phi \to s_n}(s_n) \triangleq \hat{\pi}_n s_n + (1 - \hat{\pi}_n)(1 - s_n), \tag{14.31}$$

where $\hat{\pi}_n$ integrates both the local observation evidence (captured in the incoming messages $\{\nu_{f_n \to s_n}(s_n)\}_{n=1}^N$) and the global structural constraints encoded in $p(\boldsymbol{s})$.

14.2.3.3 Backward Message Passing

In the backward direction, information flows along the reverse path $s_n \rightarrow f_n \rightarrow x_n$. Using the refined belief $\nu_{s_n \rightarrow f_n}(s_n)$, the message from factor node f_n back to the signal node is

$$\begin{aligned}\nu_{f_n \rightarrow x_n}(x_n) &\propto \sum_{s_n \in \{0,1\}} f_n(x_n, s_n)\, \nu_{s_n \rightarrow f_n}(s_n) \\ &= \hat{\pi}_n \mathcal{CN}(x_n; 0, \sigma_{x,n}^2) + (1 - \hat{\pi}_n)\delta(x_n). \end{aligned} \tag{14.32}$$

The approximate posterior distribution for x_n combines this message with the original observation message

$$p(x_n \mid \boldsymbol{x}_B^{\text{ob}}) \propto \nu_{f_n \rightarrow x_n}(x_n)\, \phi_{\text{B},n}(x_n), \quad \forall n. \tag{14.33}$$

This yields a Bernoulli–Gaussian mixture distribution. From this posterior, we can compute several key quantities.

The posterior activation probability (the probability that x_n is non-zero given all observations) is

$$\gamma_n = \left(1 + \frac{(1 - \hat{\pi}_n)\mathcal{CN}(x_{B,n}^{\text{ob}}; 0, v_{B,n}^{\text{ob}})}{\hat{\pi}_n \mathcal{CN}(x_{B,n}^{\text{ob}}; 0, v_{B,n}^{\text{ob}} + \sigma_{x,n}^2)}\right)^{-1}. \tag{14.34}$$

The posterior mean, which serves as the MMSE estimate, is

$$x_{B,n}^{\text{post}} = \int x_n\, p(x_n \mid \boldsymbol{x}_B^{\text{ob}})\, \mathrm{d}x_n = \gamma_n \frac{\sigma_{x,n}^2}{v_{B,n}^{\text{ob}} + \sigma_{x,n}^2} x_{B,n}^{\text{ob}}. \tag{14.35}$$

The posterior variance, quantifying estimation uncertainty, is

$$\begin{aligned} v_{B,n}^{\text{post}} &= \int |x_n - x_{B,n}^{\text{post}}|^2\, p(x_n \mid \boldsymbol{x}_B^{\text{ob}})\, \mathrm{d}x_n \\ v_{B,n}^{\text{post}} &= \gamma_n \frac{v_{B,n}^{\text{ob}} \sigma_{x,n}^2}{v_{B,n}^{\text{ob}} + \sigma_{x,n}^2} + \gamma_n(1 - \gamma_n) \left| \frac{\sigma_{x,n}^2}{v_{B,n}^{\text{ob}} + \sigma_{x,n}^2} x_{B,n}^{\text{ob}} \right|^2. \end{aligned} \tag{14.36}$$

14.2.3.4 Extrinsic Message to Module A

Following the turbo principle, Module B must transmit to Module A only the new information it has derived, information not already present in the incoming messages. This is achieved by excluding the contribution of the observation $\phi_{\text{B},n}(x_n)$ from the posterior when computing the outgoing message.

The extrinsic mean is computed as

$$x_{A,n}^{\text{pri}} = v_{A,n}^{\text{pri}} \left(\frac{x_{B,n}^{\text{post}}}{v_{B,n}^{\text{post}}} - \frac{x_{B,n}^{\text{ob}}}{v_{B,n}^{\text{ob}}} \right), \tag{14.37}$$

and the extrinsic variance as

$$v_{A,n}^{\text{pri}} = \left(\frac{1}{v_{B,n}^{\text{post}}} - \frac{1}{v_{B,n}^{\text{ob}}} \right)^{-1}. \tag{14.38}$$

These extrinsic quantities, $\boldsymbol{x}_A^{\text{pri}}$ and $\boldsymbol{V}_A^{\text{pri}} = \text{diag}(v_{A,1}^{\text{pri}}, \ldots, v_{A,N}^{\text{pri}})$, are then passed back to Module A, completing one full iteration of the turbo loop.

The above message-passing scheme exhibits several desirable characteristics including computational efficiency through leveraging algorithms tailored to specific graph structures, theoretical consistency with sum-product algorithm foundations, numerical stability via careful handling of probability ratios, and modular design that enables adaptation to different structured prior models. The complete Turbo-CS algorithm, integrating both Module A and Module B, is summarized in Algorithm 14.1, providing an efficient solution for structured sparse signal recovery with theoretical guarantees and strong empirical performance.

Algorithm 14.1 Turbo-CS algorithm

Input: y, A, iteration number I.
Output: x^* and s^*.

```
Initialize x_A^pri = 0 and V_A^pri.
for i = 1, ···, I do
    %Module A: LMMSE estimator
    Update x_A^post and V_A^post, using (14.24) and (14.25).
    Calculate the extrinsic messages x_B^ob and V_B^ob, using(14.27) and (14.26).
    %Module B: Message Passing
    Perform message passing, using (14.28) - (14.32).
    Calculate the approximate marginal posterior distribution, using (14.33).
    Update x_B^post and V_B^post, using (14.35) and (14.36).
    Calculate the extrinsic messages x_A^pri and V_A^pri, using (14.37) and (14.38).
end for
Output x* = x_A^post.
```

14.2.4 Implementation Issues

The practical implementation of the Turbo-CS algorithm requires careful consideration of several key aspects, including initialization strategies, numerical stability enhancements, and computational complexity reduction techniques. Proper initialization of model parameters and message values is particularly crucial as it significantly influences both algorithm convergence and final performance.

14.2.4.1 Initialization

The choice of prior parameters $\boldsymbol{\phi}$ depends fundamentally on the specific structured sparse prior model employed in the application. For an independent sparse prior where $p(\boldsymbol{s}) = \prod_{n=1}^{N} p(s_n)$ with $p(s_n = 1) = \lambda_n$, the relevant parameters are $\boldsymbol{\phi} = \{\lambda_n, \sigma_{x,n}^2\}$. In contrast, for a Markov sparse prior with $p(\boldsymbol{s}) = p(s_1)\prod_{n=2}^{N} p(s_n|s_{n-1})$ and transition probabilities defined by

$$p(s_n|s_{n-1}) = \begin{cases} (1-p_{01})^{1-s_n} p_{01}^{s_n} & s_{n-1} = 0, \\ p_{10}^{1-s_n}(1-p_{10})^{s_n} & s_{n-1} = 1, \end{cases} \tag{14.39}$$

the parameter set becomes $\boldsymbol{\phi} = \{\lambda, p_{01}, \sigma_{x,n}^2\}$ where $\lambda = \frac{p_{01}}{p_{01}+p_{10}}$. This parameterization ensures the Markov chain reaches a stationary distribution consistent with the expected sparsity level.

When limited prior information is available, conservative initialization strategies provide a robust starting point for the algorithm. Specifically:

1. For **signal parameters**, a common approach sets
   - $x_{A,n}^{(0)} = 0$ for all n.
   - $v_{A,n}^{(0)} = \sigma_{x,n}^2 = E_{\max}$, where $E_{\max}$ is an upper bound on signal energy.
2. For **sparsity parameters**, the initialization strategy depends on the specific structured sparse prior model, for example:
   - For independent prior, set $\lambda_n = \lambda_{\max}$ for all n, where $\lambda_{\max}$ is an upper bound for the sparsity ratio.
   - For Markov prior, set $\lambda = \lambda_{\max}$, $p_{01} = \lambda_{\max}$, which effectively reduces the model to an i.i.d. case initially. This conservative approach prevents premature convergence to incorrect solutions while allowing the algorithm to refine its estimates through iterative processing.

When more detailed prior information is available, such as from target tracking predictions or domain-specific knowledge, these values can be initialized accordingly to accelerate convergence. Importantly, if the initial value for $\boldsymbol{\phi}$ proves inaccurate, the framework provides a built-in correction mechanism: by treating $\boldsymbol{\phi}$ as one block within the collection of uncertain parameters $\boldsymbol{\xi}$, the inexact MM method in the M-step can automatically learn more accurate prior parameter information during execution. This adaptive capability enhances the algorithm's robustness to imperfect initialization while maintaining the theoretical convergence guarantees of the overall framework.

14.2.4.2 Numerical Stability

The iterative message passing in Turbo-CS, particularly the inexact posterior calculations within Module B, can occasionally produce numerical instabilities that threaten algorithm convergence. These instabilities most commonly manifest as unrealistic variance estimates that violate theoretical constraints, typically arising from approximations made in the factor graph decomposition or from accumulated numerical errors in successive iterations. To maintain robust operation under practical conditions, we present two complementary stabilization methods that address different manifestations of this challenge.

The first approach, variance averaging, addresses distributed instability patterns by computing global average posterior variances across all components, i.e.:

$$v_B^{\text{post}} = \frac{1}{N} \sum_{n=1}^{N} v_{B,n}^{\text{post}}, \qquad v_A^{\text{post}} = \frac{1}{N} \sum_{n=1}^{N} v_{A,n}^{\text{post}}. \tag{14.40}$$

These averaged values then establish uniform extrinsic variances throughout the system, with

$$\begin{aligned} v_{A,n}^{\text{pri}} &= v_A^{\text{pri}} \triangleq \left(\frac{1}{v_B^{\text{post}}} - \frac{1}{v_B^{\text{ob}}} \right)^{-1}, \quad \forall n, \\ v_{B,n}^{\text{ob}} &= v_B^{\text{ob}} \triangleq \left(\frac{1}{v_A^{\text{post}}} - \frac{1}{v_A^{\text{pri}}} \right)^{-1}, \quad \forall n. \end{aligned} \tag{14.41}$$

This method effectively smooths out localized numerical anomalies while preserving the overall statistical relationships between modules, though it does sacrifice some component-specific information in exchange for enhanced stability.

The second method, selective resetting, targets more concentrated instability by identifying specific elements with problematic variance estimates through the set $\mathcal{N}^- \triangleq \left\{ n : v_{B,n}^{\text{post}} > v_{B,n}^{\text{ob}} \right\}$, a condition that violates the theoretical relationship between posterior and observation variances. For these identified components, the algorithm resets the priors to their initial values as

$$x_{A,n}^{\text{pri}} = x_{A,n}^{(0)}, \qquad v_{A,n}^{\text{pri}} = v_{A,n}^{(0)}, \quad \forall n \in \mathcal{N}^-. \tag{14.42}$$

This approach preserves detailed information for stable components while intervening only where necessary, making it particularly suitable for handling isolated numerical outliers.

The selection between these stabilization strategies should be guided by the characteristics of the sensing matrix $\boldsymbol{A}$. Variance averaging generally performs better with well-conditioned, nearly partially orthogonal sensing matrices where instability tends to be uniformly distributed rather than concentrated in specific components. In contrast, selective resetting proves more effective for irregular or ill-conditioned sensing matrices where instability often manifests in isolated elements corresponding to particular matrix columns. In practice, the conditioning of $\boldsymbol{A}$ can be assessed through its coherence properties or condition number, with high coherence or condition number suggesting the potential need for selective resetting. For severely ill-conditioned systems, a hybrid approach that applies selective resetting followed by mild variance averaging may provide the most robust performance.

14.2.4.3 Complexity Reduction

In practical applications of the Turbo-CS algorithm, computational complexity emerges as a critical consideration, with the main bottleneck occurring during the posterior covariance calculation in Module A. This computation originally involves inversion of an $N \times N$ matrix, whose $O(N^3)$ complexity becomes prohibitive for large-scale problems. To address this challenge, we have developed several acceleration techniques that leverage different properties of the sensing matrix $\boldsymbol{A}$.

When the sensing matrix $\boldsymbol{A}$ has a general structure, SVD-based acceleration effectively reduces computational burden. Using the SVD decomposition $\boldsymbol{A} = \boldsymbol{U}\boldsymbol{\Lambda}\boldsymbol{V}^H$, the original posterior covariance calculation transforms into an equivalent but more efficient form as

$$\begin{aligned} \boldsymbol{V}_A^{\text{post}} &= \left(\frac{\boldsymbol{V}\boldsymbol{\Lambda}^2\boldsymbol{V}^H}{\sigma_w^2} + \left(\boldsymbol{V}_A^{\text{pri}}\right)^{-1}\right)^{-1} \\ &= \boldsymbol{V}_A^{\text{pri}} - \boldsymbol{V}_A^{\text{pri}}\boldsymbol{V}\left(\boldsymbol{V}^H\boldsymbol{V}_A^{\text{pri}}\boldsymbol{V} + \sigma_w^2\boldsymbol{\Lambda}^{-2}\right)^{-1}\boldsymbol{V}^H\boldsymbol{V}_A^{\text{pri}}. \end{aligned} \tag{14.43}$$

The key advantage of this transformation lies in reducing the dimensionality of the large-scale matrix inversion from N to M (typically $M \ll N$), thereby significantly lowering computational complexity from $O(N^3)$ to $O(N^2M)$. From a computational perspective, the efficiency gain of this method stems from exploiting the low-dimensional representation of $\boldsymbol{A}$ while maintaining numerical stability.

For the important special case of partially orthogonal matrices, when variance averaging yields uniform prior variance $\boldsymbol{V}_A^{\text{pri}} = v_A^{\text{pri}}\boldsymbol{I}$, the computation simplifies further. The posterior covariance reduces to

$$\boldsymbol{V}_A^{\text{post}} = v_A^{\text{pri}}\boldsymbol{I} - \frac{\left(v_A^{\text{pri}}\right)^2}{v_A^{\text{pri}} + \sigma_w^2}\boldsymbol{A}^H\boldsymbol{A}. \tag{14.44}$$

When $\boldsymbol{A}$ has entries of equal magnitude (such as partial DFT matrices) or when the matrix dimension is large, the diagonal elements of $\boldsymbol{V}_A^{\text{post}}$, representing the marginal posterior variances, become identical or approximately identical. In this scenario, the posterior covariance v_A^{post} and mean $\boldsymbol{x}_A^{\text{post}}$ simplify to

$$v_A^{\text{post}} = v_A^{\text{pri}} - \frac{M}{N}\frac{\left(v_A^{\text{pri}}\right)^2}{v_A^{\text{pri}} + \sigma_w^2}, \tag{14.45}$$

$$\boldsymbol{x}_A^{\text{post}} = v_A^{\text{post}}\left(\frac{\boldsymbol{x}_A^{\text{pri}}}{v_A^{\text{pri}}} + \frac{\boldsymbol{A}^H\boldsymbol{y}}{\sigma_w^2}\right). \tag{14.46}$$

This remarkable simplification reduces complexity to $O(MN)$, or even to $O(N \log N)$ for DFT matrices when employing FFT algorithms. The underlying reason for this efficiency gain is that the matrix structure enables us to bypass explicit matrix inversion and instead work directly with scalar variance parameters.

These implementation strategies ensure the Turbo-CS algorithm remains computationally feasible while maintaining robust performance across diverse application scenarios. The choice between acceleration methods depends on the specific properties of $\boldsymbol{A}$: SVD-based methods offer general applicability, while the partial orthogonal simplification provides exceptional efficiency for structured matrices commonly encountered in CS applications such as medical imaging and wireless communications.

14.3 Turbo-AMP Algorithm

The Turbo-AMP algorithm represents a powerful fusion of turbo processing principles with AMP, first introduced in [8] for systems employing two-layer Bernoulli–Gaussian sparse priors. This algorithm achieves an exceptional balance between estimation performance and computational efficiency, particularly when the sensing matrix possesses i.i.d. or weakly correlated entries. The following subsections will detail the specific update equations and practical implementation considerations for Turbo-AMP, highlighting both its theoretical foundations and empirical performance advantages in suitable application scenarios.

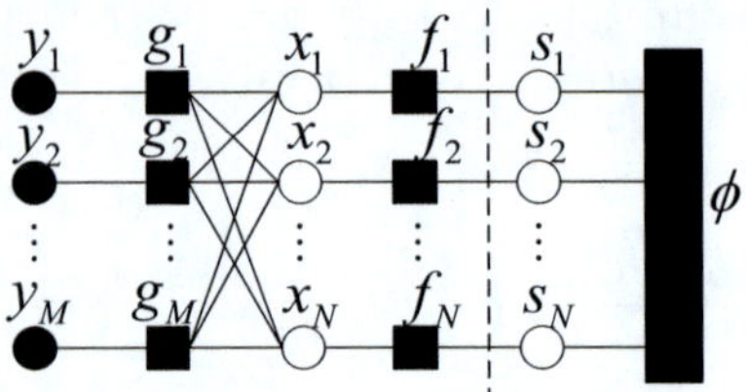

(a) Factor graph of the joint PDF $p(\boldsymbol{s})\,p(\boldsymbol{x}|\boldsymbol{s})\,p(\boldsymbol{y}|\boldsymbol{x})$.

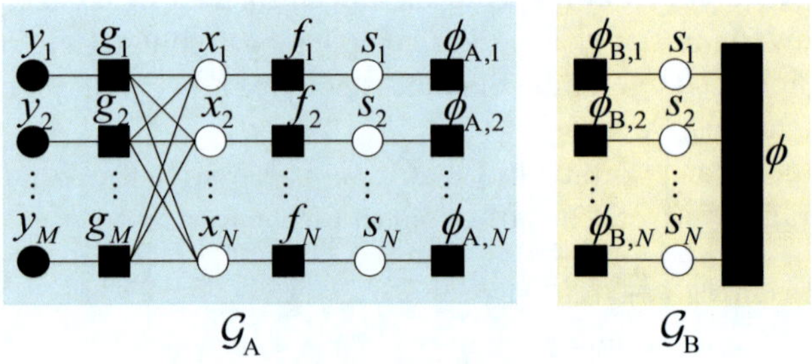

(b) The decoupled subgraphs $\mathcal{G}_A$ and $\mathcal{G}_B$.

Fig. 14.3 Partitioning of the factor graph $\mathcal{G}$ in the Turbo-AMP algorithm

14.3.1 Outline of Turbo-AMP Algorithm

The Turbo-AMP algorithm employs a sophisticated factor graph decomposition that enables efficient inference for structured sparse signal recovery problems, building upon the same support-based two-layer sparse prior and linear observation model used in Turbo-CS but with a distinct partitioning strategy that yields complementary advantages. The algorithm operates on the factor graph $\mathcal{G}$ representing the joint distribution $p(\boldsymbol{s})p(\boldsymbol{x}|\boldsymbol{s})p(\boldsymbol{y}|\boldsymbol{x})$, which undergoes a strategic decomposition into two interconnected subgraphs (Fig. 14.3).

Unlike Turbo-CS, the Turbo-AMP algorithm partitions the factor graph along the edges between variable nodes s_n and factor nodes f_n, creating two specialized subgraphs with clearly separated responsibilities. Subgraph $\mathcal{G}_A$ encodes the linear observation model with an effective two-layer independent BG prior, while Subgraph $\mathcal{G}_B$ captures the structured sparsity pattern through the prior $p(\boldsymbol{s})$ with decoupled binary observations. This particular partitioning strategy is crucial as it allows Module A to leverage the computational efficiency of AMP while enabling Module B to focus exclusively on exploiting structural dependencies within the support pattern.

The algorithm's modular processing framework consists of two specialized units that engage in iterative extrinsic message exchange until convergence. Module A implements AMP by incorporating effective independent prior information $\phi_{A,n}(s_n) \triangleq \nu_{\phi\to s_n}(s_n), \forall n$ (which are extrinsic messages from Module B) and processes the linear observation model $\boldsymbol{y} = \boldsymbol{A}\boldsymbol{x} + \boldsymbol{w}$ using the standard AMP algorithm. This module computes approximate marginal posteriors for both $\boldsymbol{x}$ and $\boldsymbol{s}$ and outputs updated binary observation information $\phi_{B,n}(s_n)$ for Module B.

Complementing this, Module B performs structured sparsity processing by incorporating binary observation information $\phi_{B,n}(s_n) \triangleq \nu_{f_n\to s_n}(s_n), \forall n$ (which are extrinsic messages from Module A) and executes sum-product message passing over the structured prior $p(\boldsymbol{s})$ to refine support estimates by leveraging structural constraints before outputting updated prior information $\phi_{A,n}(s_n)$ for Module A.

The communication between modules occurs through carefully designed extrinsic messages that ensure each module contributes unique information. In the forward direction, Module B sends to Module A the messages

$$\phi_{A,n}(s_n) \triangleq \nu_{\phi\to s_n}(s_n), \tag{14.47}$$

which represent an effective independent BG prior characterized by

$$\hat{p}(\boldsymbol{x},\boldsymbol{s}) = \hat{p}(\boldsymbol{s})\, p(\boldsymbol{x}|\boldsymbol{s}),$$
$$\hat{p}(\boldsymbol{s}) = \prod_n (\pi_n)^{s_n} (1-\pi_n)^{1-s_n}, \tag{14.48}$$

with activation probabilities

$$\pi_n = \frac{\phi_{\mathrm{A},n}(1)}{\phi_{\mathrm{A},n}(1)+\phi_{\mathrm{A},n}(0)}. \tag{14.49}$$

The key distinction from the original prior is that the structured distribution $p(\boldsymbol{s})$ is replaced by an independent approximation $\hat{p}(\boldsymbol{s})$, enabling efficient AMP implementation while preserving the essential sparsity characteristics. This approximation is justified because Module B has already incorporated the structural information when generating the extrinsic messages $\phi_{\mathrm{A},n}(s_n), \forall n$.

In the reverse direction, Module A sends to Module B the messages

$$\phi_{\mathrm{B},n}(s_n) \triangleq \nu_{f_n \to s_n}(s_n) \propto \frac{q(s_n)}{\phi_{\mathrm{A},n}(s_n)}, \tag{14.50}$$

which encapsulate the binary observation information computed by excluding the input prior from Module A's posterior $q(s_n)$. This exclusion operation ensures that Module B receives only new information derived from the linear observations, preventing redundant feedback of prior knowledge.

This architectural design provides several important benefits, including computational efficiency through AMP's near-linear complexity for suitable sensing matrices, effective structural exploitation via specialized message passing in Module B, theoretical foundations through state evolution guarantees under appropriate conditions, and robust empirical performance across diverse sparse recovery scenarios. Since Module B implements conventional sum-product message passing over a binary factor graph, the subsequent subsections will focus primarily on elaborating the specific implementations of Module A (AMP) along with practical implementation considerations that ensure the algorithm's effectiveness in real-world applications.

14.3.2 AMP Algorithm for Independent BG Prior (Module A)

The AMP algorithm for the independent BG prior described in Eq. (14.48) represents a specialized instance of the general AMP framework discussed in Chap. 10, serving as the computational core of Module A within the Turbo-AMP architecture. For completeness and reader convenience, we provide a concise summary of this algorithm in Algorithm 14.2, highlighting its role in enabling efficient Bayesian inference under the effective independent BG prior while maintaining the theoretical guarantees and computational advantages characteristic of AMP algorithms.

The algorithm operates through an iterative procedure that maintains and updates key statistical quantities at each iteration t, tracking the evolving beliefs about the signal and its support structure. Specifically, it tracks the posterior mean μ_n^t and variance v_n^t for each signal component x_n, along with the posterior probability α_n^t that the support indicator $s_n = 1$. These quantities collectively capture the algorithm's current state of knowledge about both the amplitude and sparsity pattern of the unknown signal, with the variances providing crucial uncertainty measures that guide the iterative refinement process.

After executing the algorithm for a maximum of T iterations, it yields approximate posterior distributions that reflect the synthesized information from both the linear observations and the prior constraints. The signal posterior is given by

$$q(\boldsymbol{x}) = \prod_{n=1}^{N} \mathcal{CN}\left(x_n; \mu_n^{T+1}, v_n^{T+1}\right), \tag{14.51}$$

representing the final estimate of the sparse signal with associated uncertainty quantification. Simultaneously, the support posterior takes the form:

$$q(\boldsymbol{s}) = \prod_{n=1}^{N} \left(\alpha_n^{T+1}\delta(s_n-1) + \left(1-\alpha_n^{T+1}\right)\delta(s_n)\right), \tag{14.52}$$

providing probabilistic estimates of the sparsity pattern that are essential for structured recovery.

Algorithm 14.2 AMP for the independent BG prior

Input: Sensing matrix $\boldsymbol{A}$, observation $\boldsymbol{y}$ and π_n, $\forall n$.

Output:

$$q(\boldsymbol{x}) = \prod_{n=1}^{N} \mathcal{CN}\left(x_n; \mu_n^{T+1}, v_n^{T+1}\right),$$

$$q(\boldsymbol{s}) = \prod_{n=1}^{N} \left(\alpha_n^{T+1} \delta(s_n - 1) + \left(1 - \alpha_n^{T+1}\right) \delta(s_n)\right).$$

Initialization: $\forall m, n : \mu_n^1 = 0, v_n^1 = 1, \mu_{z,m}^0 = y_m, v_{z,m}^0 = 1.$

for $t = 1, \ldots, T$ **do**

  %% Linear Estimation Steps

  $v_{z,m}^t = \sum_{n=1}^N |a_{mn}|^2 v_n^t, \forall m.$

  $\mu_{z,m}^t = \sum_{n=1}^N a_{mn} \mu_n^t - \frac{v_{z,m}^t \left(y_m - \mu_{z,m}^{t-1}\right)}{\sigma_w^2 + v_{z,m}^t}, \forall m.$

  $v_{r,n}^t = \left(\sum_{m=1}^M \frac{|a_{mn}|^2}{\sigma_w^2 + v_{z,m}^t}\right)^{-1}, \forall n.$

  $\mu_{r,n}^t = \mu_n^t + v_{r,n}^t \sum_{m=1}^M \frac{a_{mn}^* \left(y_m - \mu_{z,m}^t\right)}{\sigma_w^2 + v_{z,m}^t}, \forall n.$

  %% Nonlinear Estimation Steps

  $\alpha_n^{t+1} = \frac{1}{1 + \left(\frac{\pi_n}{1-\pi_n} \frac{\mathcal{CN}\left(\mu_{r,n}^t; 0, v_{r,n}^t + \sigma_{x,n}^2\right)}{\mathcal{CN}\left(\mu_{r,n}^t; 0, v_{r,n}^t\right)}\right)^{-1}}, \forall n.$

  $\gamma_n^{t+1} = \frac{\mu_{r,n}^t / v_{r,n}^t}{1/v_{r,n}^t + 1/\sigma_{x,n}^2}, \forall n.$

  $\eta_n^{t+1} = \frac{1}{1/v_{r,n}^t + 1/\sigma_{x,n}^2}, \forall n.$

  $\mu_n^{t+1} = \alpha_n^{t+1} \gamma_n^{t+1}, \forall n.$

  $v_n^{t+1} = \alpha_n^{t+1} \left(|\gamma_n^{t+1}|^2 + \eta_n^{t+1}\right) - |\mu_n^{t+1}|^2, \forall n.$

end for

The effectiveness of this AMP implementation stems from its careful balance between computational efficiency and statistical accuracy. By leveraging the independent prior assumption within each iteration while incorporating the Onsager correction to account for statistical dependencies across iterations, the algorithm achieves near-optimal performance for a broad class of sensing matrices while maintaining computational scalability that makes it practical for high-dimensional problems. This combination of attributes makes it particularly well suited for its role as Module A within the broader Turbo-AMP framework, where it must efficiently process the linear observations while seamlessly integrating with the structured processing occurring in Module B.

14.3.3 Implementation Issues

This subsection addresses two critical practical considerations for the Turbo-AMP algorithm: automatic parameter learning through EM and ensuring numerical stability for general sensing matrices.

14.3.3.1 EM Update for BG Prior and Noise Variance σ_w^2

In real-world applications, the parameters governing signal and noise distributions are typically unknown a priori, creating a need for methods that can jointly estimate both the sparse signal and these underlying statistical parameters.

To address this challenge, the EM algorithm can be seamlessly integrated with AMP within a unified framework. Considering the independent BG prior model

$$p(\boldsymbol{x}) = \prod_{n=1}^{N} \left(\lambda \mathcal{CN}\left(x_n; 0, \sigma_x^2\right) + (1-\lambda)\,\delta(x_n)\right), \tag{14.53}$$

let $\boldsymbol{\phi} = \{\lambda, \sigma_x^2, \sigma_w^2\}$ represent the complete set of unknown parameters. The EM algorithm alternates between two complementary phases: in the E-step, posterior distribution $q^i(\boldsymbol{x})$ of $\boldsymbol{x}$ is computed using AMP/Turbo-AMP with current parameter estimates $\boldsymbol{\phi}^i$; in the M-step, parameters are updated by maximizing the expected complete-data log-likelihood according to [17]

$$\begin{aligned}\boldsymbol{\phi}^{i+1} &= \underset{\boldsymbol{\phi}}{\operatorname{argmax}}\, \mathbb{E}\left[\ln p\left(\boldsymbol{x}, \boldsymbol{y}; \boldsymbol{\phi}\right) | \boldsymbol{y}, \boldsymbol{\phi}^i\right] \\ &= \sum_{n=1}^{N} \underset{\lambda, \sigma_x^2}{\operatorname{argmax}} \int_{x_n} q^i\left(x_n\right) \ln p\left(x_n\right) dx_n \\ &+ \sum_{m=1}^{M} \underset{\sigma_w^2}{\arg\max} \int_{\boldsymbol{x}} q^i\left(\boldsymbol{x}\right) \ln \mathcal{CN}\left(y_m; \boldsymbol{A}_m \boldsymbol{x}, \sigma_w^2\right) d\boldsymbol{x}.\end{aligned}$$

This optimization yields closed-form update equations that possess intuitive statistical interpretations. The sparsity rate updates as

$$\lambda^{i+1} = \frac{1}{N}\sum_{n=1}^{N} \alpha_n^i, \tag{14.54}$$

effectively becoming the average posterior probability of activation across all components. The signal variance updates according to

$$\left(\sigma_x^2\right)^{i+1} = \frac{1}{N\lambda^{i+1}} \sum_{n=1}^{N} \alpha_n^i \left(|\gamma_n^i|^2 + \eta_n^i\right), \tag{14.55}$$

which represents the average energy of components deemed active, properly normalized by the estimated sparsity rate. The noise variance updates via

$$\left(\sigma_w^2\right)^{i+1} = \frac{1}{M} \sum_{m=1}^{M} \left(|y_m - \mu_{z,m}^i|^2 + v_{z,m}^i\right), \tag{14.56}$$

capturing the residual between observations and their predicted values while accounting for estimation uncertainty. Here, α_n^i, γ_n^i, η_n^i, $\mu_{z,m}^i$ and $v_{z,m}^i$ are calculated in the final inner iteration of the E-step, ensuring consistency between the parameter estimates and the current signal beliefs.

This EM framework extends naturally beyond the BG prior to various structured priors including non-negative priors and Gaussian mixture priors [9, 10]. While the specific update equations vary according to the prior structure, they all follow the same underlying EM principle of iteratively refining parameter estimates based on current posterior beliefs. This adaptability makes the approach particularly valuable in practical applications where the true signal statistics may be unknown or time-varying, as the algorithm can automatically learn and adapt to the prevailing statistical characteristics without requiring manual parameter tuning.

14.3.3.2 Numerical Stability

While AMP demonstrates near-optimal performance for i.i.d. Gaussian sensing matrices with strong theoretical guarantees through state evolution, its behavior can become unstable when applied to more general matrix structures encountered in practical applications. Standard AMP implementations may diverge or exhibit oscillatory behavior, fail to converge to meaningful solutions, or display sensitivity to matrix condition number. Those limitations stem from the algorithm's reliance on statistical properties that may not hold for arbitrary sensing matrices.

To address these stability challenges, several specialized techniques have been developed to adapt AMP for broader applicability. One prominent approach is the damping method [18], which introduces memory into the iterative updates to smooth convergence behavior. This technique modifies the standard AMP updates through weighted averaging between current and previous estimates, as expressed by

$$\mu_{z,m}^{t} = \beta_z \left(\sum_{n=1}^{N} a_{mn}\mu_n^t - \frac{v_{z,m}^t \left(y_m - \mu_{z,m}^{t-1} \right)}{\sigma_w^2 + v_{z,m}^{t-1}} \right) + (1-\beta_z)\,\mu_{z,m}^{t-1}, \tag{14.57}$$

$$\mu_n^{t+1} = \alpha_n^{t+1}\gamma_n^{t+1} + (1-\beta_x)\,\mu_n^t. \tag{14.58}$$

The damping factors β_z and β_x are typically selected based on the peak-to-average ratio of squared singular values of the sensing matrix:

$$\kappa\left(\boldsymbol{A}\right) = \frac{||\boldsymbol{A}||_2^2}{||\boldsymbol{A}|_F^2/\min\left(M,N\right)}, \tag{14.59}$$

which quantifies the matrix's deviation from ideal i.i.d. Gaussian characteristics. While this damping approach guarantees convergence for Gaussian priors, it may significantly slow convergence rates and lacks theoretical guarantees for general priors, representing a practical trade-off between stability and efficiency.

A more fundamentally robust approach is provided by Unitary Transformation AMP (UT-AMP) [20], which employs SVD preprocessing to transform the original problem into a more stable domain. By decomposing $\boldsymbol{A} = \boldsymbol{U}\boldsymbol{\Lambda}\boldsymbol{V}$ and transforming the observation model to

$$\boldsymbol{r} = \boldsymbol{U}^H\boldsymbol{y} = (\boldsymbol{\Lambda}\boldsymbol{V})\,\boldsymbol{x} + \boldsymbol{U}^H\boldsymbol{w}, \tag{14.60}$$

UT-AMP applies standard AMP to the transformed system with sensing matrix $\boldsymbol{\Lambda}\boldsymbol{V}$. This transformation ensures convergence for any sensing matrix under Gaussian priors [19], provides enhanced numerical stability by normalizing the effective sensing operator, and maintains computational efficiency through the unitary properties that preserve statistical characteristics while improving conditioning.

For practical implementation, several strategic considerations can guide algorithm selection and parameter tuning. First, assessing sensing matrix properties through metrics like condition number or eigenvalue distribution before algorithm selection can prevent stability issues. Second, employing an adaptive strategy that uses UT-AMP for ill-conditioned matrices while reserving standard AMP for well-behaved ones optimizes the trade-off between robustness and efficiency. Third, implementing robust convergence monitoring with appropriate stopping criteria and divergence detection mechanisms allows for early intervention when numerical issues arise. Finally, considering adaptive damping factors that evolve based on iteration progress rather than using fixed values can provide dynamic stability control throughout the optimization process.

These implementation strategies collectively enhance the robustness and applicability of the Turbo-AMP algorithm across diverse practical scenarios, ensuring reliable performance even when dealing with challenging sensing matrices that violate the ideal assumptions underlying standard AMP theory. The choice between stabilization methods ultimately depends on the specific application requirements, with damping offering simplicity for mildly non-ideal matrices and UT-AMP providing stronger guarantees for severely ill-conditioned systems.

14.4 Turbo-VBI Algorithm

While the Turbo-AMP algorithm is effective for i.i.d. sensing matrices, its performance can deteriorate with more general matrices, particularly those that are ill-conditioned or possess strongly correlated entries. Although the general Turbo-CS framework, previously introduced in this chapter, is applicable to such matrices, it is subject to several limitations. First, its Module A utilizes a conventional LMMSE estimator, which has a limited capacity to promote sparsity. Second, when the sensing matrix deviates from a partially orthogonal structure, the method can become numerically unstable, thereby compromising its robustness.

To address these shortcomings in both Turbo-AMP and Turbo-CS, the Turbo-VBI algorithm was proposed in [1] for the recovery of structured sparse signals under uncertain model parameters and general sensing matrices. The original Turbo-VBI framework has since inspired the development of several variants. These include extensions that enable Bayesian estimation of grid parameters [11], as well as techniques to reduce computational complexity through inverse-free VBI [12] and subspace-constrained VBI [13]. This section will focus on the foundational Turbo-VBI framework in [1]; its subsequent variations will be detailed in later chapters.

14.4.1 Outline of Turbo-VBI Algorithm

The Turbo-VBI algorithm distinguishes itself from Turbo-CS and Turbo-AMP through its use of a three-layer Bernoulli–Gamma–Gaussian (BGG) sparse prior and its fully Bayesian treatment of both signal and noise parameters. This section outlines the probabilistic model, factor graph decomposition, and modular architecture that form the foundation of the algorithm.

14.4.1.1 Three-Layer BGG Sparse Prior and Observation Model

The BGG prior introduces an additional layer of latent variables compared to the two-layer models used in Turbo-CS and Turbo-AMP. In addition to the support vector $\boldsymbol{s}$, we introduce a precision vector $\boldsymbol{\rho} = [\rho_1, \ldots, \rho_N]^T$, where $\mathrm{Var}(x_n) = 1/\rho_n$ controls the variance of each element in the sparse signal $\boldsymbol{x}$. This precision-based parameterization provides fine-grained control over coefficient shrinkage.

To capture structured sparsity patterns where coefficients exhibit group-level dependencies, the index set $\{1, \ldots, N\}$ is partitioned into Q non-overlapping subsets $\mathcal{I}_1, \ldots, \mathcal{I}_Q$. Within each subvector $\boldsymbol{x}[\mathcal{I}_i]$, all elements share the same activity state, they are either all active or all inactive. A support vector $\boldsymbol{s} = [s_1, \ldots, s_Q]^T \in \{0, 1\}^Q$ indicates the state of each group, with $s_i = 1$ denoting that the i-th group is active and $s_i = 0$ indicating inactivity.

The joint distribution of all latent variables factorizes across three hierarchical layers as

$$p(\boldsymbol{x}, \boldsymbol{\rho}, \boldsymbol{s}) = \underbrace{p(\boldsymbol{s})}_{\text{Support}} \underbrace{p(\boldsymbol{\rho}|\boldsymbol{s})}_{\text{Precision}} \underbrace{p(\boldsymbol{x}|\boldsymbol{\rho})}_{\text{Sparse signal}} . \tag{14.61}$$

Layer 1: Support Distribution. The support vector $\boldsymbol{s}$ encodes the activity pattern of the sparse signal. For example, for the independent BGG prior, the support variables are modeled as independent Bernoulli random variables

$$p(\boldsymbol{s}) = \prod_{i=1}^{Q} \lambda_i^{s_i} (1 - \lambda_i)^{1-s_i}, \tag{14.62}$$

where λ_i denotes the prior probability that the i-th group is active. More complex structured sparsity patterns can be incorporated by replacing this independent prior with a more sophisticated distribution, such as a Markov chain or Markov random field, as discussed in Chap. 13.

Layer 2: Precision Distribution. Conditioned on the support vector, the precision parameters follow a Gamma distribution whose parameters depend on the group activity state as

$$p(\boldsymbol{\rho}|\boldsymbol{s}) = \prod_{i=1}^{Q} \prod_{n \in \mathcal{I}_i} (\Gamma(\rho_n; a_n, b_n))^{s_i} \left(\Gamma(\rho_n; \bar{a}_n, \bar{b}_n)\right)^{1-s_i}, \tag{14.63}$$

where $\Gamma(\rho; a, b)$ denotes the Gamma distribution with shape a and rate b. The hyperparameters are chosen to create distinct statistical profiles for active and inactive components. For active groups ($s_i = 1$), we set $a_n/b_n = \mathbb{E}[\rho_n] = O(1)$, allowing moderate variance for non-zero coefficients. For inactive groups ($s_i = 0$), we enforce $\bar{a}_n/\bar{b}_n = \mathbb{E}[\rho_n] \gg 1$, driving the variance toward zero and effectively suppressing irrelevant coefficients.

Layer 3: Signal Distribution. At the lowest layer, the sparse signal follows a factorized Gaussian distribution conditioned on the precisions

$$p(\boldsymbol{x}|\boldsymbol{\rho}) = \prod_{n=1}^{N} \mathcal{CN}(x_n; 0, \rho_n^{-1}), \quad \forall n = 1, \ldots, N. \tag{14.64}$$

This conditional Gaussian form is chosen for its conjugacy with the Gamma prior, which enables closed-form variational updates as will be shown in subsequent subsections.

The observation model follows the standard compressed sensing formulation with a dynamically gridded sensing matrix

$$\boldsymbol{y} = \boldsymbol{A}(\boldsymbol{\theta})\boldsymbol{x} + \boldsymbol{w}, \tag{14.65}$$

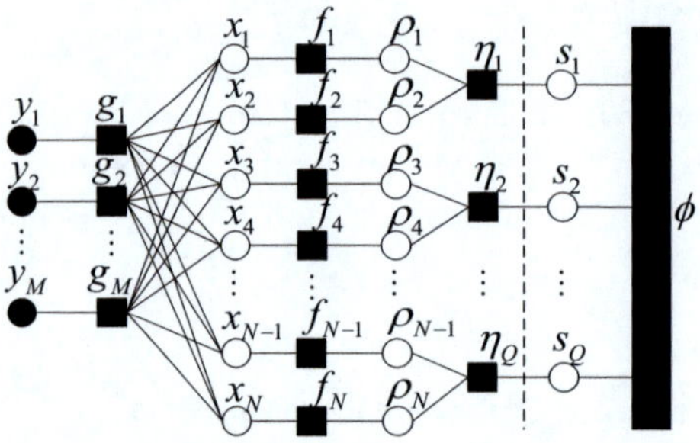

(a) Factor graph of the joint PDF. For easy illustration, we assume every two adjacent elements of the sparse signal $\boldsymbol{x}$ form a group, i.e., $Q = N/2$ and $\mathcal{I}_i = \{2i-1, 2i\}$.

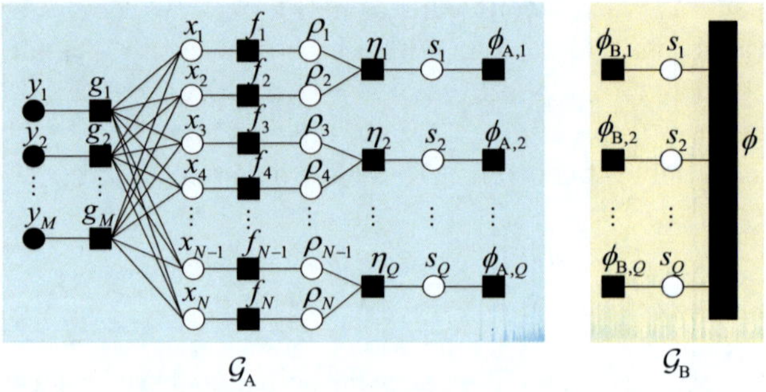

(b) The decoupled subgraphs $\mathcal{G}_A$ and $\mathcal{G}_B$.

Fig. 14.4 Partitioning of the factor graph $\mathcal{G}$ in the Turbo-VBI algorithm

Table 14.3 Factors, distributions, and functional forms in the factor graph $\mathcal{G}$ in Fig. 14.4a

Factor	Distribution	Functional form
$g_m(y_m, \boldsymbol{x})$	$p(y_m \vert \boldsymbol{x}, \boldsymbol{\xi})$	$\mathcal{CN}(y_m; \boldsymbol{A}_m(\boldsymbol{\theta})\boldsymbol{x}, \kappa^{-1})$
$f_n(x_n, \rho_n)$	$p(x_n \vert \rho_n)$	$\mathcal{CN}(x_n; 0, \rho_n^{-1})$
$\eta_i(\boldsymbol{\rho}[\mathcal{I}_i], s_i)$	$p(\boldsymbol{\rho}[\mathcal{I}_i] \vert s_i)$	$\prod_{n\in\mathcal{I}_i} (\Gamma(\rho_n; a_n, b_n))^{s_i} (\Gamma(\rho_n; \bar{a}_n, \bar{b}_n))^{1-s_i}$
ϕ	$p(\boldsymbol{s})$	Depends on application

$\boldsymbol{A}_m(\boldsymbol{\theta})$ denotes the m-th row of $\boldsymbol{A}(\boldsymbol{\theta})$

where $\boldsymbol{w} \in \mathbb{C}^{M\times 1}$ is additive Gaussian noise with independent entries $w_m \sim \mathcal{CN}(0, \kappa^{-1})$. A key distinction from Turbo-CS and Turbo-AMP is that Turbo-VBI treats the noise precision κ as an unknown random variable to be inferred within the variational framework, rather than as a fixed parameter. This is achieved by placing a Gamma hyperprior on κ

$$p(\kappa) = \Gamma(\kappa; c, d), \tag{14.66}$$

with $c, d \to 0$ yielding a non-informative prior that allows the data to dominate the inference. For simplicity, we assume a common noise precision across all observations; extension to heterogeneous precisions is straightforward.

14.4.1.2 Factor Graph Representation and Decomposition

The joint probability distribution $p(\boldsymbol{s})p(\boldsymbol{\rho}|\boldsymbol{s})p(\boldsymbol{x}|\boldsymbol{\rho})p(\boldsymbol{y}|\boldsymbol{x})$ can be represented by the factor graph $\mathcal{G}$ shown in Fig. 14.4a. Table 14.3 provides the complete functional expressions for each factor node.

Following the turbo framework, $\mathcal{G}$ is strategically decomposed into two complementary subgraphs along the edges connecting the group support nodes $\{s_i\}$ and the factor nodes $\{\eta_i\}$ (which represent $p(\boldsymbol{\rho}[\mathcal{I}_i]|s_i)$):

- **Subgraph** $\mathcal{G}_\text{A}$ encompasses the linear observation model together with an independent BGG prior. This subgraph is responsible for extracting information from the measurements while respecting the prior constraints.
- **Subgraph** $\mathcal{G}_\text{B}$ captures the structured sparsity constraints embodied in $p(\boldsymbol{s})$, with the incoming messages from $\mathcal{G}_\text{A}$ interpreted as binary observations about each support variable.

This graph decomposition naturally gives rise to a dual-module architecture, where two specialized computational units engage in iterative message exchange:

- **Module A** implements sparse variational Bayesian inference over $\mathcal{G}_\text{A}$, processing the observations $\boldsymbol{y}$ together with prior messages received from Module B.
- **Module B** performs sum-product message passing over $\mathcal{G}_\text{B}$, enforcing the structured sparsity constraints encoded in $p(\boldsymbol{s})$ while incorporating observation information from Module A.

The two modules communicate via extrinsic messages, defined as

$$\phi_{\text{A},i}(s_i) \triangleq \nu_{\phi\to s_i}(s_i), \quad i = 1, \ldots, Q, \tag{14.67}$$

$$\phi_{\text{B},i}(s_i) \triangleq \nu_{\eta_i\to s_i}(s_i), \quad i = 1, \ldots, Q. \tag{14.68}$$

In this framework, $\{\phi_{\text{A},i}(s_i)\}$ serve as effective prior probabilities for Module A, summarizing the structural information extracted by Module B. Conversely, $\{\phi_{\text{B},i}(s_i)\}$ act as effective binary likelihoods for Module B, conveying the observational evidence gathered by Module A in a form that can be processed by the sum-product algorithm.

14.4.1.3 Module A: Sparse VBI with Independent Prior

Building on the messages received from Module B, Module A performs inference under an effective independent prior that approximates the true structured prior

$$\hat{p}(\boldsymbol{x}, \boldsymbol{\rho}, \boldsymbol{s}) = \hat{p}(\boldsymbol{s})\, p(\boldsymbol{\rho}|\boldsymbol{s})\, p(\boldsymbol{x}|\boldsymbol{\rho}), \tag{14.69}$$

$$\hat{p}(\boldsymbol{s}) = \prod_{i=1}^{Q} (\pi_i)^{s_i}(1-\pi_i)^{1-s_i}, \tag{14.70}$$

where the activation probabilities are derived from the incoming extrinsic messages

$$\pi_i \triangleq \hat{p}(s_i = 1) = \frac{\phi_{\text{A},i}(1)}{\phi_{\text{A},i}(1) + \phi_{\text{A},i}(0)}. \tag{14.71}$$

The sparse VBI procedure applied to this independent prior yields a Gaussian posterior $q(\boldsymbol{x})$ for the sparse signal and a Bernoulli posterior $q(s_i)$ for each support element.

Following the turbo principle, Module A then computes extrinsic messages to send back to Module B by excluding the contribution of the incoming prior as

$$\phi_{\text{B},i}(s_i) \propto \frac{q(s_i)}{\phi_{\text{A},i}(s_i)}. \tag{14.72}$$

This subtraction ensures that Module B receives only the new information gleaned from the observations, preventing redundant feedback of previously exchanged knowledge.

14.4.1.4 Module B: Structured Message Passing

Module B receives the extrinsic messages $\{\phi_{\text{B},i}(s_i)\}$ from Module A and interprets them as binary observations in an equivalent probabilistic model. It then performs sum-product message passing over $\mathcal{G}_\text{B}$, combining these observations with the structured prior $p(\boldsymbol{s})$ to compute refined beliefs for each support variable. The updated beliefs are then converted into extrinsic messages $\{\phi_{\text{A},i}(s_i)\}$ and sent back to Module A, completing one full turbo iteration.

This iterative exchange continues until convergence, with each module progressively refining its estimates based on the other's specialized processing. The following subsections provide a detailed exposition of the sparse VBI implementation in Module A, which constitutes the computational core of Turbo-VBI and distinguishes it from other turbo algorithms through its fully Bayesian treatment of both signal and noise parameters.

14.4.2 Sparse VBI Algorithm (Module A)

14.4.2.1 Framework of Sparse VBI

Let $\boldsymbol{v} = \{\boldsymbol{v}^k, \forall k \in \mathcal{H}\}$ denote the collection of all latent variables associated with the sparse signal, where for the three-layer BGG prior we have $\boldsymbol{v} = \{\boldsymbol{x}, \boldsymbol{\rho}, \boldsymbol{s}, \kappa\}$ and $\mathcal{H} = \{1, 2, 3, 4\}$. The noise precision κ can be handled in two ways within the EM-Turbo-VBI framework: either estimated as part of the sparse VBI procedure during the E-step (the approach presented here), or treated as a fixed parameter and updated via point estimation in the M-step.

The core idea of VBI is to approximate the intractable true posterior $\hat{p}(\boldsymbol{v}|\boldsymbol{y}, \boldsymbol{\xi})$ with a simpler, factorized distribution $q(\boldsymbol{v}; \boldsymbol{\xi})$ by minimizing the KLD between them. This leads to the following optimization problem:

$$\mathcal{A}_{\text{VBI}} : q^*(\boldsymbol{v}; \boldsymbol{\xi}) = \arg\min_{q(\boldsymbol{v};\boldsymbol{\xi})} \int q(\boldsymbol{v}; \boldsymbol{\xi}) \ln \frac{q(\boldsymbol{v}; \boldsymbol{\xi})}{\hat{p}(\boldsymbol{v}|\boldsymbol{y}, \boldsymbol{\xi})} d\boldsymbol{v} \tag{14.73}$$

$$\text{s.t.}\, q(\boldsymbol{v}; \boldsymbol{\xi}) = \prod_{k\in\mathcal{H}} q(\boldsymbol{v}^k; \boldsymbol{\xi}), \quad \int q(\boldsymbol{v}^k; \boldsymbol{\xi})\mathrm{d}\boldsymbol{v}^k = 1, \quad \forall k \in \mathcal{H}. \tag{14.74}$$

Here, $\hat{p}(\boldsymbol{v}|\boldsymbol{y}, \boldsymbol{\xi})$ denotes the posterior distribution under the effective independent prior $\hat{p}(\boldsymbol{x}, \boldsymbol{\rho}, \boldsymbol{s})$ defined in (14.69). For discrete variables such as the support vector $\boldsymbol{s}$, the integral in the normalization constraint is understood as summation over all possible discrete values. To simplify notation, we will omit the explicit dependence on $\boldsymbol{\xi}$ in $q(\boldsymbol{v}; \boldsymbol{\xi})$ throughout the remainder of this subsection.

Solving $\mathcal{A}_{\text{VBI}}$ yields a tractable approximation to the true posterior. The solution methodology closely follows the sparse VBI framework introduced in Chap. 12, with one key distinction: Chap. 12 employed a precision-based two-layer sparse prior, whereas the current chapter adopts the more expressive three-layer BGG prior. This difference manifests in the specific update equations derived below, but the overall variational inference principle remains unchanged.

Problem $\mathcal{A}_{\text{VBI}}$ is non-convex due to the factorization constraint and the form of the posterior. Consequently, uniqueness of the optimal solution cannot be guaranteed. However, the existence of at least one optimal solution has been established in the literature [14]. In practice, we seek a stationary solution $q^*(\boldsymbol{v})$ of $\mathcal{A}_{\text{VBI}}$, defined as follows.

Definition 14.1 (*Stationary Solution of VBI Problem*) A factorized distribution $q^*(\boldsymbol{v}) = \prod_{k\in\mathcal{H}} q^*\left(\boldsymbol{v}^k\right)$ is called a stationary solution of Problem $\mathscr{A}_{\text{VBI}}$ if it satisfies all constraints in $\mathscr{A}_{\text{VBI}}$, and for all $k \in \mathcal{H}$,

$$q^*\left(\boldsymbol{v}^k\right) = \arg\min_{q\left(\boldsymbol{v}^k\right)} \int \prod_{l\neq k} q^*\left(\boldsymbol{v}^l\right) q\left(\boldsymbol{v}^k\right) \ln \frac{\prod_{l\neq k} q^*\left(\boldsymbol{v}^l\right) q\left(\boldsymbol{v}^k\right)}{\hat{p}\left(\boldsymbol{v}|\boldsymbol{y}, \boldsymbol{\xi}\right)} \mathrm{d}\boldsymbol{v}^{-k}. \tag{14.75}$$

Intuitively, a stationary solution is one where each factor $q(\boldsymbol{v}^k)$ is optimal given all other factors held fixed. Such solutions can be obtained via AO, where we cyclically update each factor while keeping the others fixed. The optimal update for a single factor, given fixed distributions for all other variables, follows a standard result in variational inference [14]

$$q(\boldsymbol{v}^k) \propto \exp\left(\langle \ln p(\boldsymbol{v}, \boldsymbol{y}|\boldsymbol{\xi})\rangle_{\prod_{l\neq k} q(\boldsymbol{v}^l)}\right), \tag{14.76}$$

where $\langle f(x)\rangle_{q(x)} = \int f(x)q(x)\mathrm{d}x$ denotes expectation with respect to the distribution $q(x)$. Equation (14.76) serves as the foundation for deriving all subsequent update equations. For notational convenience, we will abbreviate $\langle f(\boldsymbol{v}^k)\rangle_{q(\boldsymbol{v}^k)}$ as $\langle f(\boldsymbol{v}^k)\rangle$ when the distribution is clear from context.

14.4.2.2 Update for x

Applying (14.76) to $\boldsymbol{x}$, we obtain

$$\begin{aligned} \ln q(\boldsymbol{x}) &\propto \langle \ln p(\boldsymbol{x}|\boldsymbol{\rho})\rangle_{q(\boldsymbol{\rho})} + \ln p(\boldsymbol{y}|\boldsymbol{x}, \kappa) \\ &\propto -\boldsymbol{x}^H \mathrm{diag}\left(\langle \boldsymbol{\rho}\rangle\right)\boldsymbol{x} - \|\langle\kappa\rangle(\boldsymbol{y} - \boldsymbol{A}\boldsymbol{x})\|^2 \\ &\propto -(\boldsymbol{x} - \boldsymbol{\mu})^H \boldsymbol{\Sigma}^{-1}(\boldsymbol{x} - \boldsymbol{\mu}). \end{aligned} \tag{14.77}$$

The quadratic form reveals that $q(\boldsymbol{x})$ is complex Gaussian. Given the current variational distributions $q(\boldsymbol{\rho}) = \prod_{n=1}^{N} \Gamma(\rho_n; \tilde{a}_n, \tilde{b}_n)$ and $q(\kappa) = \Gamma(\kappa; \tilde{c}, \tilde{d})$, the posterior parameters are

$$q(\boldsymbol{x}) = \mathcal{CN}(\boldsymbol{x}; \boldsymbol{\mu}, \boldsymbol{\Sigma}), \tag{14.78}$$

with

$$\boldsymbol{\Sigma} = \left(\mathrm{diag}\left(\frac{\tilde{a}_1}{\tilde{b}_1}, \ldots, \frac{\tilde{a}_N}{\tilde{b}_N}\right) + \frac{\tilde{c}}{\tilde{d}}\boldsymbol{A}^H\boldsymbol{A}\right)^{-1}, \tag{14.79}$$

$$\boldsymbol{\mu} = \frac{\tilde{c}}{\tilde{d}}\boldsymbol{\Sigma}\boldsymbol{A}^H\boldsymbol{y}. \tag{14.80}$$

The posterior mean $\boldsymbol{\mu}$ represents the current MMSE estimate of the sparse signal, while the covariance $\boldsymbol{\Sigma}$ quantifies the associated uncertainty.

14.4.2.3 Update for ρ

For the precision vector $\boldsymbol{\rho}$, we again apply (14.76) using the current distributions $q(\boldsymbol{x})$ and $q(\boldsymbol{s}) = \prod_{i=1}^{Q}(\tilde{\pi}_i)^{s_i}(1-\tilde{\pi}_i)^{1-s_i}$

$$\begin{aligned} \ln q(\boldsymbol{\rho}) &\propto \langle \ln p(\boldsymbol{x}|\boldsymbol{\rho})\rangle_{q(\boldsymbol{x})} + \langle \ln p(\boldsymbol{\rho}|\boldsymbol{s})\rangle_{q(\boldsymbol{s})} \\ &\propto \sum_{i=1}^{Q}\sum_{n\in\mathcal{I}_i}\left[\left(\langle s_i\rangle a_n + \langle 1-s_i\rangle \bar{a}_n\right)\ln\rho_n - \left(\langle|x_n|^2\rangle + \langle s_i\rangle b_n + \langle 1-s_i\rangle\bar{b}_n\right)\rho_n\right]. \end{aligned} \tag{14.81}$$

This expression reveals that $q(\boldsymbol{\rho})$ factorizes as a product of independent Gamma distributions:

$$q(\boldsymbol{\rho}) = \prod_{n=1}^{N}\Gamma(\rho_n; \tilde{a}_n, \tilde{b}_n), \tag{14.82}$$

with parameters updated as

$$\tilde{a}_n = \langle s_i\rangle a_n + \langle 1-s_i\rangle\bar{a}_n + 1 = \tilde{\pi}_i a_n + (1-\tilde{\pi}_i)\bar{a}_n + 1, \tag{14.83}$$

$$\begin{aligned} \tilde{b}_n &= \langle|x_n|^2\rangle + \langle s_i\rangle b_n + \langle 1-s_i\rangle\bar{b}_n \\ &= |\mu_n|^2 + \Sigma_n + \tilde{\pi}_i b_n + (1-\tilde{\pi}_i)\bar{b}_n, \quad \forall n\in\mathcal{I}_i, \forall i, \end{aligned} \tag{14.84}$$

where μ_n is the n-th element of $\boldsymbol{\mu}$, and Σ_n is the n-th diagonal element of $\boldsymbol{\Sigma}$. The expectations $\langle s_i\rangle = \tilde{\pi}_i$ and $\langle|x_n|^2\rangle = |\mu_n|^2 + \Sigma_n$ follow from the current variational distributions.

14.4.2.4 Update for s

Applying (14.76) to the support vector $\boldsymbol{s}$ yields

$$\begin{aligned}\ln q(\boldsymbol{s}) &\propto \langle \ln p(\boldsymbol{\rho}|\boldsymbol{s})\rangle_{q(\boldsymbol{\rho})} + \ln \hat{p}(\boldsymbol{s}) \\ &\propto \sum_{i=1}^{Q}\Bigg[s_i \sum_{n\in\mathcal{I}_i}\left(\ln\frac{\pi_i b_n^{a_n}}{\Gamma(a_n)} + (a_n-1)\langle \ln\rho_n\rangle - b_n\langle\rho_n\rangle\right) \\ &\quad + (1-s_i)\sum_{n\in\mathcal{I}_i}\left(\ln\frac{(1-\pi_i)\bar{b}_n^{\bar{a}_n}}{\Gamma(\bar{a}_n)} + (\bar{a}_n-1)\langle\ln\rho_n\rangle - \bar{b}_n\langle\rho_n\rangle\right)\Bigg].\end{aligned} \tag{14.85}$$

This expression factorizes across groups, leading to a product of independent Bernoulli distributions:

$$q(\boldsymbol{s}) = \prod_{i=1}^{Q} (\tilde{\pi}_i)^{s_i}(1-\tilde{\pi}_i)^{1-s_i}, \tag{14.86}$$

where the posterior activation probability for group i is

$$\tilde{\pi}_i = \frac{1}{C}\prod_{n\in\mathcal{I}_i}\frac{\pi_i b_n^{a_n}}{\Gamma(a_n)} \mathrm{e}^{(a_n-1)\langle\ln\rho_n\rangle - b_n\langle\rho_n\rangle}, \tag{14.87}$$

and C is the normalization constant ensuring $\tilde{\pi}_i \in [0, 1]$

$$\begin{aligned}C = &\prod_{n\in\mathcal{I}_i}\frac{\pi_i b_n^{a_n}}{\Gamma(a_n)} \mathrm{e}^{(a_n-1)\langle\ln\rho_n\rangle - b_n\langle\rho_n\rangle} \\ &+ \prod_{n\in\mathcal{I}_i}\frac{(1-\pi_i)\bar{b}_n^{\bar{a}_n}}{\Gamma(\bar{a}_n)} \mathrm{e}^{(\bar{a}_n-1)\langle\ln\rho_n\rangle - \bar{b}_n\langle\rho_n\rangle}.\end{aligned} \tag{14.88}$$

The required expectations involve $\langle\rho_n\rangle = \tilde{a}_n/\tilde{b}_n$ (from the Gamma distribution's mean) and $\langle\ln\rho_n\rangle$, which is given by

$$\langle\ln\rho_n\rangle = \psi(\tilde{a}_n) - \ln(\tilde{b}_n), \tag{14.89}$$

where $\psi(x) = \frac{\mathrm{d}}{\mathrm{d}x}\ln\Gamma(x)$ is the digamma function.

14.4.2.5 Update for $q(\kappa)$

Finally, applying (14.76) to the noise precision κ yields

$$\begin{aligned}\ln q(\kappa) &\propto \langle \ln p(\boldsymbol{y}|\boldsymbol{x},\kappa)\rangle_{q(\boldsymbol{x})} + \ln p(\kappa) \\ &\propto (M+c-1)\ln\kappa - \left(d + \left\langle\|\boldsymbol{y}-\boldsymbol{A}\boldsymbol{x}\|^2\right\rangle_{q(\boldsymbol{x})}\right)\kappa.\end{aligned} \tag{14.90}$$

Thus, $q(\kappa)$ follows a Gamma distribution

$$q(\kappa) = \Gamma(\kappa; \tilde{c}, \tilde{d}), \tag{14.91}$$

with parameters

$$\tilde{c} = c + M, \tag{14.92}$$

$$\begin{aligned}\tilde{d} &= d + \left\langle\|\boldsymbol{y}-\boldsymbol{A}\boldsymbol{x}\|^2\right\rangle_{q(\boldsymbol{x})} \\ &= d + \mathrm{tr}(\boldsymbol{A}\boldsymbol{\Sigma}\boldsymbol{A}^H) + \|\boldsymbol{y}-\boldsymbol{A}\boldsymbol{\mu}\|^2.\end{aligned} \tag{14.93}$$

The expected residual power decomposes naturally into two terms: the squared error of the mean estimate $\|\boldsymbol{y}-\boldsymbol{A}\boldsymbol{\mu}\|^2$, and a trace term $\mathrm{tr}(\boldsymbol{A}\boldsymbol{\Sigma}\boldsymbol{A}^H)$ that accounts for the uncertainty captured by the posterior covariance.

14.4.2.6 Convergence Analysis of Sparse VBI

The sparse VBI algorithm can be viewed as an instance of AO applied to Problem $\mathscr{A}_{\mathrm{VBI}}$ [15]. This perspective provides a rigorous foundation for analyzing its convergence behavior. The algorithm possesses two key properties that together ensure convergence to a meaningful solution.

1. **Monotonic Decrease**: Each update of any factor $q(\boldsymbol{v}^k)$ strictly decreases (or at least does not increase) the KLD objective function. Consequently, the sequence of objective values generated by the algorithm is non-increasing.
2. **Unique Subproblem Solutions**: For any fixed set of distributions $\{q(\boldsymbol{v}^l)\}_{l\neq k}$, the optimization subproblem with respect to $q(\boldsymbol{v}^k)$ has a unique minimizer, given explicitly by (14.76).

These properties align precisely with the sufficient conditions for convergence of block coordinate descent methods established in the optimization literature [15]. They lead to the following rigorous convergence guarantee.

Theorem 14.2 (Convergence of Sparse VBI) *Let $\left\{q^{(t)}(\boldsymbol{v})\right\}_{t=1}^{\infty}$ be the sequence of distributions generated by the sparse VBI algorithm. Then any accumulation point $q^*(\boldsymbol{v}) = \prod_{k\in\mathcal{H}} q^*\left(\boldsymbol{v}^k\right)$ of this sequence is a stationary solution of Problem $\mathscr{A}_{\mathrm{VBI}}$, satisfying the optimality conditions defined in Definition 14.1.*

This theorem ensures that regardless of initialization, the sparse VBI algorithm will converge (in the sense of accumulation points) to a point where no further improvement can be made by updating any individual factor, a stationary point of the variational optimization problem.

Within the broader EM-Turbo-VBI framework, the sparse VBI algorithm serves as the computational engine for Module A. Its convergence guarantee, provided by Theorem 14.2, ensures that this core component reliably produces stable posterior approximations, contributing to the overall algorithmic stability of the proposed approach. The complete EM-Turbo-VBI procedure, integrating sparse VBI within an EM architecture for joint estimation of latent variables and model parameters, is systematically summarized in Algorithm 14.3.

14.4.3 Implementation Considerations

14.4.3.1 Initialization Strategy

Proper initialization is crucial for ensuring stable convergence of the Turbo-VBI algorithm, as it establishes the starting point for the iterative refinement process and significantly influences both convergence speed and final solution quality. The initialization procedures are strategically differentiated between the first outer iteration, where no prior information from previous cycles exists, and subsequent iterations that can leverage results from earlier processing.

During the first outer iteration, the algorithm must establish initial beliefs without the benefit of previous inference results, as detailed below.

- The support prior is initialized as $\hat{p}(\boldsymbol{s}) = \prod_{i=1}^{Q} \hat{p}(s_i)$ with
$$\hat{p}(s_i) = \left(\pi_i^0\right)^{s_i}\left(1-\pi_i^0\right)^{1-s_i}, \forall i, \tag{14.94}$$
where π_i^0 is derived from available prior knowledge. For instance, given an upper bound $\lambda_{\max}$ on the ratio of non-zero groups, a conservative approach sets $\pi_i^0 = \lambda_{\max}$, effectively assuming the sparsest plausible signal configuration initially.
- Correspondingly, the support posterior is initialized as $q(\boldsymbol{s}) = \prod_{i=1}^{Q}(\tilde{\pi}_i)^{s_i}(1-\tilde{\pi}_i)^{1-s_i}$ with $\tilde{\pi}_i = \pi_i^0$, creating consistent starting beliefs across prior and posterior distributions.
- The precision distribution follows a similar conservative strategy, initialized as $q(\boldsymbol{\rho}) = \prod_{n=1}^{N}\Gamma(\rho_n; \tilde{a}_n, \tilde{b}_n)$, where for all $n \in \mathcal{I}_i$ and all i,
$$\tilde{a}_n = \pi_i^0 a_n + \left(1-\pi_i^0\right)\bar{a}_n, \quad \tilde{b}_n = \pi_i^0 b_n + \left(1-\pi_i^0\right)\bar{b}_n. \tag{14.95}$$
This initialization effectively averages the active and inactive precision parameters according to the initial sparsity belief, preventing premature commitment to either state.

Algorithm 14.3 EM-Turbo-VBI algorithm

Input: $\boldsymbol{y}$, prior distributions $p(\boldsymbol{\theta})$, $p(\boldsymbol{\phi})$ and $p(\kappa)$, $\boldsymbol{A}(\boldsymbol{\theta})$, maximum iteration number I_m, threshold ε.
Output: $\boldsymbol{\xi}^*$, $\boldsymbol{x}^*$, s_i^*, $\forall i$.

Initialize the parameters $\boldsymbol{\xi}$, and the message π_i.
for $k = 1, \cdots, I_m$ **do**
  Turbo-VBI-E Step:
  %Module A: Sparse VBI Estimator
  Initialize the distribution functions $q(\boldsymbol{s})$ and $q(\boldsymbol{\rho})$.
  while not converge **do**
    Update $q^k(\boldsymbol{x}; \boldsymbol{\xi}), q^k(\boldsymbol{\rho}; \boldsymbol{\xi}), q^k(\boldsymbol{s}; \boldsymbol{\xi}), q^k(\kappa; \boldsymbol{\xi})$.
  end while
  Calculate the extrinsic information of s_i based on (14.72), send $\nu_{\eta_i \to s_i}(s_i)$ to Module B.
  % Module B:
  Perform the SPMP over the support subgraph $\mathcal{G}_{\mathrm{B}}$, send $\nu_{\phi \to s_i}(s_i)$ to Module A.
  Turbo-VBI-M Step:
  Construct the surrogate function $\hat{u}$ in (14.14) using the output of Module A, i.e., $q(\boldsymbol{v}; \boldsymbol{\xi})$.
  Update $\boldsymbol{\xi}_j$ alternatively for $j = 1, ..., B$ using (14.9) for $j \in \mathcal{J}_c$ and (14.11) for $j \in \overline{\mathcal{J}}_c$.
  if $\left\|\boldsymbol{\mu}^{(k-1)} - \boldsymbol{\mu}^{(k)}\right\| \leq \epsilon$ and $\left\|\boldsymbol{\Sigma}^{(k-1)} - \boldsymbol{\Sigma}^{(k)}\right\| \leq \varepsilon$ **then**
    break
  end if
  $k = k + 1$.
end for
Output $\boldsymbol{\xi}^*$, $\boldsymbol{x}^* = \arg\max_{\boldsymbol{x}} q^k(\boldsymbol{x}; \boldsymbol{\xi}) = \boldsymbol{\mu}^k$ and $s_i^* = \arg\max_{s_i} q^k(s_i; \boldsymbol{\xi})$.

For subsequent outer iterations, the algorithm transitions to a knowledge-reusing strategy that leverages information gained from previous processing cycles, as detailed below:

- The support prior $\hat{p}(\boldsymbol{s})$ is updated using extrinsic messages from Module B, with π_i obtained from Eq. (14.71), allowing the algorithm to build upon the structured information extracted in prior iterations.
- Similarly, the support posterior $q(\boldsymbol{s})$ is initialized using posterior probabilities $\tilde{\pi}_i$ computed by Module A in the previous iteration, creating continuity in the belief evolution.
- The precision distribution $q(\boldsymbol{\rho})$ follows the same principle, initialized with posterior parameters $\tilde{a}_n$ and $\tilde{b}_n$ from Module A in the previous iteration, which preserves the refined uncertainty estimates developed through variational inference.

Beyond the signal parameters, noise precision initialization requires separate consideration as it significantly impacts the algorithm's ability to distinguish signal from noise. In practical implementations where exact noise statistics are unknown, heuristic methods such as median absolute deviation or pilot symbol-based estimation can provide rough initial estimates of noise variance. The prior parameters c and d for noise precision κ should be chosen such that $\mathbb{E}[\kappa] = c/d$ matches the estimated noise precision, with the ratio c/d controlling the mean and the individual values of c and d influencing the concentration around this mean. Setting larger values for both parameters creates a more informative prior that strongly pulls estimates toward the initial guess, while smaller values yield a less influential prior that allows greater adaptation to evidence from observations.

This stratified initialization approach distinguishes between cold starts and warm starts and provides sensible defaults for noise parameters, ensuring robust algorithm performance across diverse operating conditions, from completely unknown scenarios to those with varying degrees of prior knowledge.

14.4.3.2 Implementation Details

Several key implementation aspects require careful attention to ensure optimal performance of the Turbo-VBI algorithm, spanning hyperparameter selection, computational complexity management, and convergence optimization. These practical considerations often determine the algorithm's effectiveness in real-world applications where theoretical assumptions may not perfectly hold.

The selection of hyperparameters for the Gamma priors plays a critical role in balancing the algorithm's sensitivity to active versus inactive components. Typical parameter values are $a_n = 1$, $b_n = 1$, $\bar{a}_n = 1$, and $\bar{b}_n = 10^{-4}$, which establish a clear distinction between active components with moderate precision ($\mathbb{E}[\rho_n] = 1$) and inactive components with high precision ($\mathbb{E}[\rho_n] = 10^4$). While algorithm performance is generally robust to variations in $\bar{b}_n$, adapting this parameter to specific SNR conditions can yield significant performance improvements. In low-SNR regimes where noise may mask weak signal components, larger $\bar{b}_n$ values (e.g., 10^{-3}) typically yield better estimation performance by reducing the penalty on potentially active components. Conversely, in high SNR regimes where distinguishing true zeros from noise is more reliable, smaller $\bar{b}_n$ values (e.g., 10^{-5}) are preferred as they enforce stronger sparsity constraints and produce cleaner reconstructions.

Beyond hyperparameter tuning, managing computational complexity represents another crucial implementation challenge, particularly for high-dimensional problems. The dominant computational cost arises from the matrix inversion in Eq. (14.79), which exhibits $O(N^3)$ complexity that becomes prohibitive for large N. Two complementary approaches effectively mitigate this computational burden while maintaining estimation accuracy. First, applying the matrix inverse lemma allows complexity reduction from $O(N^3)$ to $O(NM^2) + O(M^3)$ when $M < N$ by exploiting the relationship between matrix dimensions—this approach is particularly beneficial in CS applications where $M \ll N$. Second, optimizing the iteration strategy provides substantial efficiency gains: since $q(s)$ is intelligently initialized using results from the previous EM iteration rather than starting from scratch each time, the E-step typically requires only 1–2 inner iterations to achieve convergence, significantly reducing the computational overhead associated with repeated matrix operations.

These implementation considerations collectively ensure both computational efficiency and estimation accuracy across various operating conditions. The adaptive hyperparameter selection allows the algorithm to maintain robustness across different SNR environments, while the computational optimizations make the method more practical for realistic problem sizes without sacrificing the theoretical benefits of the variational Bayesian approach.

14.5 Discussions and Comparative Analysis

14.5.1 Insight into Turbo-SBI Algorithms

The Turbo-SBI family of algorithms, in their E-step, can be fundamentally understood as a variant of the sum-product message-passing algorithm employing a two-timescale message scheduling strategy. This scheduling proceeds as follows:

1. **Module A Update (Fast Timescale)**: Module A first executes its internal inference procedure over subgraph $\mathcal{G}_{\mathrm{A}}$ until convergence. During this phase, all messages originating from Module B, which serve as effective prior information, are held fixed.
2. **Extrinsic Message Passing (Slow Timescale)**: Upon convergence of Module A, the updated (extrinsic) messages along the edges from $\mathcal{G}_{\mathrm{A}}$ to $\mathcal{G}_{\mathrm{B}}$ are computed and passed to Module B.
3. **Module B Update (Fast Timescale)**: Module B then performs standard sum-product message passing over the structured sparsity subgraph $\mathcal{G}_{\mathrm{B}}$ until convergence. In this phase, the incoming messages from Module A are treated as fixed effective observation information.
4. **Closing the Loop (Slow Timescale)**: Finally, the updated extrinsic messages from Module B are passed back to Module A, completing one global iteration. This cyclic process, adhering to the same two-timescale schedule, repeats until global convergence is achieved.

Table 14.4 Comparison of the Turbo-CS, Turbo-AMP, and Turbo-VBI, where PO stands for partially orthogonal

	Turbo-CS	Turbo-AMP	Turbo-VBI
Sparse prior	Two-layer BG	Two-layer BG	Three-layer BGG
Passed messages	Messages of $\{x_n\}$	Messages of $\{s_n\}$	Messages of $\{s_n\}$
Module A	LMMSE	AMP	Sparse VBI
Complexity	High (low for PO)	Low	High
Stable situation	PO sensing matrix	I.i.d. sensing matrix	General cases
Learn noise precision	No	No	Yes

In this framework, intra-module messages (within $\mathcal{G}_\text{A}$ or $\mathcal{G}_\text{B}$) are updated at a faster timescale, while extrinsic messages exchanged between the modules are updated at a slower timescale.

A critical distinction arises from the structure of the subgraphs. Since $\mathcal{G}_\text{A}$ typically contains dense loops, exact sum-product message passing is often intractable. Consequently, Module A does not implement the standard sum-product algorithm directly. Instead, it employs specialized, computationally efficient sparse Bayesian inference algorithms (such as LMMSE estimation, AMP, or sparse VBI) to first approximate the marginal posteriors. The extrinsic messages are then derived according to the sum-product rule: by effectively subtracting (in the log-probability domain) the incoming extrinsic prior from the computed approximate posterior.

14.5.2 Comparative Analysis of Turbo-SBI Algorithms

The distinguishing features among different Turbo-SBI algorithms primarily stem from three key design choices: the sparse prior model, the factor graph partitioning strategy, and the Bayesian inference algorithm employed in Module A. These choices lead to distinct performance characteristics and computational trade-offs, as summarized in Table 14.4 and elaborated below.

Turbo-CS Algorithm:

- **Prior Model and Partitioning**: Employs a support-based two-layer sparse prior. The factor graph $\mathcal{G}$ is partitioned along the edges between variable nodes x_n and factor nodes f_n (associated with $p(x_n|s_n)$). Consequently, the extrinsic messages convey Gaussian-distributed information about x_n.
- **Module A Inference**: The resulting inference problem in Module A involves a linear observation model with a Gaussian prior, which is solved efficiently using LMMSE estimation.
- **Strengths and Limitations**: Turbo-CS is particularly effective for CS problems with partially orthogonal sensing matrices (e.g., randomly selected rows of a DFT matrix). In such cases, the matrix inversion in LMMSE can be efficiently approximated, reducing per-iteration complexity to $O(MN)$ or even $O(N \log N)$. Empirical evidence suggests that under these conditions, Turbo-CS can outperform Turbo-AMP and occasionally even the more complex Turbo-VBI. However, for general sensing matrices, it requires a full matrix inversion. Its primary limitations include: (1) the limited sparsity-promoting capability of conventional LMMSE, leading to potential performance degradation or slow convergence with limited measurements or high sparsity ratios and (2) numerical instability and reduced robustness with irregular, non-orthogonal sensing matrices.

Turbo-AMP Algorithm:

- **Prior Model and Partitioning**: Also uses a support-based two-layer sparse prior, but partitions $\mathcal{G}$ along the edges between variable nodes s_n and factor nodes f_n. This leads to extrinsic messages that are binary distributions over s_n.
- **Module A Inference**: Module A tackles a linear observation model with an independent BG prior, approximated via the AMP algorithm.
- **Strengths and Limitations**: Turbo-AMP excels with i.i.d. or weakly correlated sensing matrices, achieving an excellent balance between computational complexity ($O(MN)$ per iteration) and recovery performance. However, its performance can significantly degrade, and the algorithm may even diverge, when applied to more general sensing matrix structures.

Turbo-VBI Algorithm:

- **Prior Model and Partitioning**: Utilizes a more expressive three-layer BGG sparse prior. The graph is partitioned along the edges between group support nodes s_i and factor nodes η_i (associated with $\prod_{n \in \mathcal{I}_i} p(\rho_n | s_i)$), resulting in binary extrinsic messages for s_i.
- **Module A Inference**: Module A handles a linear model with an independent BGG prior, solved approximately using the sparse VBI method.
- **Strengths and Limitations**: Turbo-VBI offers superior robustness for general sensing matrices and can automatically estimate the noise precision. Theoretically, its sparse VBI core guarantees convergence to a stationary point of the KLD minimization problem. The stronger sparsity promotion of the BGG prior often makes it more effective than Turbo-CS in challenging scenarios with limited measurements or high sparsity. The main drawback is its higher computational complexity: even after applying the matrix inversion lemma, the per-iteration cost is $O\left(NM^2\right) + O\left(M^3\right)$. Subsequent chapters will explore variants like inverse-free VBI and subspace-constrained VBI designed to alleviate this complexity burden.

14.5.3 Guidance on Algorithm Selection

In practical applications, the choice of an appropriate EM-Turbo-SBI algorithm should be guided by the specific structure of the sensing matrix and the available computational resources. The following guidelines are recommended:

- For partially orthogonal sensing matrices (or those approximating this structure), Turbo-CS generally provides the optimal balance between estimation performance and computational efficiency.
- When dealing with i.i.d. sensing matrices (or those with weak correlations), Turbo-AMP emerges as an excellent choice due to its competitive performance and low complexity.
- For general sensing matrices that exhibit irregular structure or ill-conditioning, Turbo-VBI is preferred when robust recovery performance is prioritized over computational cost.

To illustrate these selection principles, we present several representative scenarios from wireless communications and sensing applications.

Example 14.1 (Sensing Matrix Structures in Wireless Systems)

Case 1: Uplink Massive MIMO-OFDM Channel Estimation (TDD)
Consider the uplink massive MIMO-OFDM system described in Chap. 13, where the BS employs a ULA or UPA and users have single antennas. When angle and delay grids are uniformly discretized with appropriate resolutions (e.g., when the azimuth grid spacing satisfies $\sin \Delta\theta = \frac{1}{N_b}$ for an N_b-antenna ULA, and the delay grid spacing is $\Delta\tau = \frac{1}{B}$, where B is the system bandwidth), the resulting sensing matrix exhibits a partially orthogonal structure, thereby making it well suited for implementation with the Turbo-CS algorithm.

Case 2: Downlink Massive MIMO-OFDM Channel Estimation (FDD)
In FDD downlink massive MIMO-OFDM channel estimation, the sensing matrix structure depends critically on the pilot design. With partially orthogonal pilots, the matrix maintains partial orthogonality, enabling computational simplification and performance enhancement for Turbo-CS. Alternatively, i.i.d. complex Gaussian pilots yield a sensing matrix with i.i.d. entries, making Turbo-AMP a suitable choice for this scenario.

Case 3: Dynamic Grid Scenarios
When dynamic grids $\boldsymbol{\theta}$ are employed in the M-step, the sensing matrix $\boldsymbol{A}(\boldsymbol{\theta})$ typically loses exact i.i.d. or partial orthogonality. However, with sufficiently large antenna array/bandwidth and sparse channels, $\boldsymbol{A}(\boldsymbol{\theta})$ often approximates these ideal structures, allowing Turbo-AMP or Turbo-CS to maintain satisfactory performance.

Case 4: Oversampled Grid Scenarios
When oversampled angle/delay grids are used to enhance resolution (e.g., with $\sin\Delta\theta = \frac{1}{4N_b}$ and $\Delta\tau = \frac{1}{4B}$), the resulting sensing matrix exhibits strongly correlated entries. In such cases, Turbo-VBI typically delivers more robust performance compared to Turbo-AMP and Turbo-CS.

Case 5: ISAC Environment Sensing
In the joint scattering environment sensing and channel estimation problem for ISAC systems (as detailed in Chap. 13), the location-domain channel matrix typically exhibits strong correlations among its entries, especially when dense location grids are employed. Under such challenging conditions, the relative robustness of the Turbo algorithms varies significantly. Turbo-AMP often struggles with convergence or delivers unsatisfactory performance due to its sensitivity to matrix structure. Turbo-CS, utilizing full matrix inversion, may remain functional but with limited robustness. In contrast, Turbo-VBI demonstrates the strongest resilience against ill-conditioned matrices, followed by Turbo-CS, while Turbo-AMP proves most vulnerable to deviations from ideal matrix properties.

14.6 Applications

This section demonstrates the practical utility of the EM-Turbo-SBI framework by applying it to two representative problems in wireless communications and sensing, as introduced in Chap. 13:

1. **Channel Estimation in FDD Massive MIMO-OFDM Systems** (a simplified case from [6]; see Sect. 13.3.2 for the structured CS formulation with a dynamic grid).
2. **Joint Scattering Environment Sensing and Channel Estimation in ISAC Systems** (adapted from [12]; see Sect. 13.4 for the structured CS formulation with a dynamic grid).

Through these case studies, we will illustrate how the theoretical framework can be adapted to address specific system models and performance requirements, highlighting its versatility and effectiveness in solving complex real-world inference problems.

14.6.1 Turbo-CS for FDD Massive MIMO-OFDM Channel Estimation

14.6.1.1 Problem Setup and System Parameters

We consider a downlink massive MIMO-OFDM system operating in FDD mode. The BS is equipped with a ULA consisting of $N_b = 256$ antennas. To facilitate channel estimation, pilot symbols are placed on $P \triangleq |\mathcal{N}_p| = 32$ uniformly spaced subcarriers across the frequency band.

The delay-domain channel taps $\boldsymbol{H}_d$ are generated according to a structured sparsity model that reflects realistic propagation conditions. Specifically, the state transition probabilities for the support variables $\{s^l_{d,q}\}$ are set to $\rho^l_{01} = 1/240$ and $\rho^l_{10} = 1/16$, which control the average cluster size and gap between clusters in the angular domain. The maximum delay spread is limited to 16 taps, and the training sequence length is $M = 103$ (approximately $0.4N_b$). Channel estimation accuracy is evaluated using the NMSE metric.

14.6.1.2 Application of the Turbo-CS Framework

To apply the Turbo-CS algorithm to this FDD channel estimation problem, we follow the modular architecture described in Sect. 14.2. The algorithm operates by iteratively exchanging extrinsic information between two specialized modules until convergence.

Module A: LMMSE Estimation. In Module A, the sparse DAD channel vector $\boldsymbol{h}_d$ is estimated using an LMMSE estimator. This module incorporates the current observation $\boldsymbol{y}$ together with a Gaussian prior received from Module B. The prior is characterized by mean $\boldsymbol{h}^{\text{pri}}_A$ and diagonal covariance $\boldsymbol{V}^{\text{pri}}_A$, which encapsulate the structured sparsity information processed by Module B. The posterior mean and covariance are computed via standard LMMSE updates, and the extrinsic information passed back to Module B is obtained by subtracting the contribution of the input prior from the posterior.

Module B: Structured Sparsity Processing. Module B exploits the structured sparsity of the DAD channel using the support-based two-layer sparse prior model introduced in Sect. 13.3.2. In this module, the extrinsic mean $\boldsymbol{h}_B^{\text{pri}}$ from Module A is interpreted as a virtual AWGN observation of the true DAD channel. The module then performs sum-product message passing over the factor graph that encodes the structured prior, which includes: a Bernoulli distribution for the column-level support variables t_d^l, a Markov chain for the element-level support variables $s_{d,q}^l$ within each active column, and the deterministic relationship linking the channel coefficients to the support and amplitude variables.

The detailed message-passing rules for this structured prior, including the forward–backward algorithm over the Markov chain and the coupling between column-level and element-level states, are provided in [6]. Interested readers are referred there for the complete derivation. The output of Module B includes refined posterior means and variances for the DAD channel coefficients, which are then converted into extrinsic messages and fed back to Module A.

This iterative exchange between the two modules enables the algorithm to jointly exploit both the linear observation model (via Module A) and the structured sparsity of the DAD channel (via Module B), leading to accurate channel estimation with fast convergence. We refer to this algorithm as Structured Turbo-CS with Delay Support (STCS-DS).

14.6.1.3 Sensing Matrix Design

Following the framework developed in Chap. 13, we employ a uniform angular grid for the ULA to obtain a sparse representation of the channel in the angular domain. With this discretization, the array response matrix $\boldsymbol{A}_R(\bar{\boldsymbol{\theta}})$ becomes a DFT dictionary, and consequently the resulting sensing matrix exhibits a partially orthogonal structure, a property that makes the Turbo-CS algorithm particularly well suited for this application.

However, the channel support in massive MIMO systems typically exhibits clustered sparsity rather than an independent structure. Direct application of Turbo-CS with a standard partial DFT sensing matrix can lead to degraded performance, as the algorithm's assumptions about signal structure are violated. To address this issue, we introduce a random permutation (RP) operation that effectively decorrelates the sparse signal. The resulting sensing matrix, referred to as a partial DFT-RP matrix, is constructed as

$$\boldsymbol{A} \triangleq \boldsymbol{X}\boldsymbol{A}_R(\bar{\boldsymbol{\theta}}) = \boldsymbol{S}\boldsymbol{F}\boldsymbol{R}, \tag{14.96}$$

where $\boldsymbol{S}$ is a selection matrix comprising M randomly selected and reordered rows from an $N_b \times N_b$ identity matrix, $\boldsymbol{F}$ is an $N_b \times N_b$ unitary DFT matrix, and $\boldsymbol{R}$ is a random permutation matrix.

The corresponding pilot matrix is given by $\boldsymbol{X} = \boldsymbol{S}\boldsymbol{F}\boldsymbol{R}\boldsymbol{A}_R^H$. This design offers two significant practical advantages. First, the storage burden at the user equipment is substantially reduced, as only the permutation indices defined by $\boldsymbol{S}$ and $\boldsymbol{R}$ need to be stored rather than the entire sensing matrix. Second, matrix-vector multiplications involving $\boldsymbol{A}$ can be implemented efficiently using fast FFT algorithms, leveraging the underlying DFT structure to achieve $O(N_b \log N_b)$ complexity.

14.6.1.4 Simulation Results

To assess the effectiveness of STCS-DS, we compare its performance against two baseline schemes.

1. Turbo-CS with i.i.d. Support (TCS-IID): This variant assumes an independent Bernoulli–Gaussian prior on the channel support, ignoring any structural dependencies between adjacent angular bins or delay taps.
2. Structured Turbo-CS with Frequency Support (STCS-FS): This approach exploits structured sparsity in the angle-frequency domain rather than the angle-delay domain. Details of this algorithm can be found in [6].

The performance of the Turbo-CS algorithms for FDD massive MIMO-OFDM channel estimation is illustrated in Figs. 14.5 and 14.6. These results highlight both convergence behavior and final estimation accuracy.

Convergence Behavior: Figure 14.5 depicts the NMSE as a function of iteration number for the different algorithms at an SNR of 10 dB. The STCS-DS algorithm demonstrates remarkably rapid convergence, reaching a stable solution within approximately five iterations. This fast convergence is attributable to the effective exploitation of the angle-delay domain sparsity structure, which provides strong prior information that accelerates the inference process. In contrast, the baseline algorithms require more iterations to achieve their final performance levels.

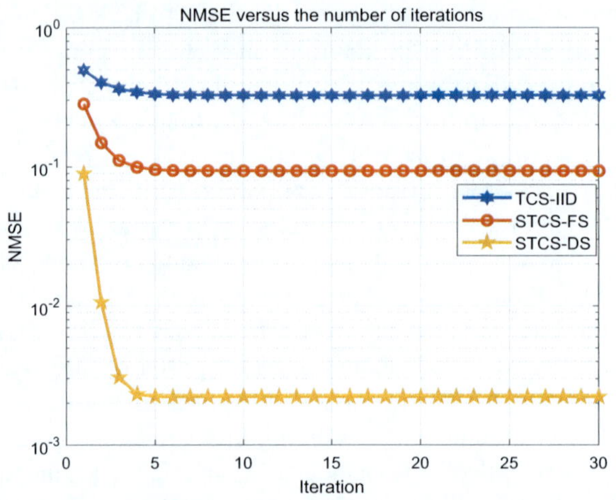

Fig. 14.5 NMSE of channel estimation versus the number of iterations under SNR = 10 dB

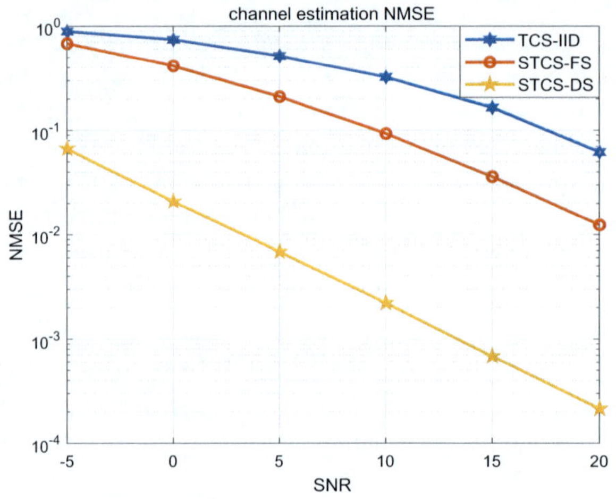

Fig. 14.6 NMSE of channel estimation versus SNR

Estimation Accuracy: Figure 14.6 presents the NMSE performance versus SNR for all algorithms. As expected, the estimation accuracy of all methods improves with increasing SNR. However, the STCS-DS algorithm consistently outperforms both baseline schemes across the entire SNR range. The performance gain is particularly pronounced at moderate-to-high SNRs, where the structured prior enables more precise discrimination between active and inactive channel coefficients.

The superiority of STCS-DS over TCS-IID underscores the importance of modeling structural dependencies in the channel support. By capturing the clustered nature of non-zero coefficients in the angular domain and the sparsity pattern across delay taps, STCS-DS effectively leverages the additional information embedded in the channel's physical structure. The comparison with STCS-FS further demonstrates that the angle-delay domain representation provides a more natural and effective framework for exploiting channel sparsity than the angle-frequency domain alternative.

In summary, the EM-Turbo-CS framework, particularly when instantiated as the STCS-DS algorithm, delivers substantial improvements in channel estimation accuracy for FDD massive MIMO-OFDM systems. By combining the partially orthogonal sensing matrix design (partial DFT-RP) with structured sparsity modeling in the angle-delay domain, the approach achieves an attractive balance between estimation performance, computational efficiency, and fast convergence, making it well suited for practical implementation in next-generation wireless systems.

Table 14.5 System parameters for joint sensing and channel estimation Simulation

Parameter	Value
Number of OFDM subcarriers	$N = 1024$
Carrier frequency	3.5 GHz
Subcarrier spacing	$f_0 = 30\,\text{kHz}$
Total bandwidth	$B = N f_0$
Downlink pilot subcarriers	$\lvert\mathcal{N}_b\rvert = 32$
Uplink pilot subcarriers	$\lvert\mathcal{N}_u\rvert = 32$
BS antennas	$M = 64$
Time offset range	$\tau_o \in [-2/B, 2/B]$

14.6.2 Turbo-VBI for Joint Environment Sensing and Channel Estimation

14.6.2.1 Problem Setup and System Parameters

We consider a two-dimensional surveillance area of dimensions 100 m × 100 m, discretized with a grid resolution of 5 m. The BS is positioned at coordinates $[-50\text{ m}, 0\text{ m}]^T$, while the mobile user is located near $[50\text{ m}, 0\text{ m}]^T$ with a random positional offset. The prior distribution for the user location $\boldsymbol{p}_u = [p^x, p^y]^T$ is modeled as independent Gaussian

$$p^x \sim \mathcal{N}(50, \sigma_p^2/2), \qquad p^y \sim \mathcal{N}(0, \sigma_p^2/2), \tag{14.97}$$

with $\sigma_p^2 = 1$. This reflects realistic uncertainty in the initial user position estimate, as might be obtained from GPS or previous localization results.

The scattering environment comprises $K = 11$ radar targets and $L = 13$ communication scatterers, distributed in two clusters of different sizes to emulate the spatially clustered nature of real-world scattering environments. The key system parameters are summarized in Table 14.5.

14.6.2.2 Application of the Turbo-VBI Framework

To apply the Turbo-VBI algorithm to this joint sensing and channel estimation problem, we follow the modular architecture described in Sect. 14.4. The algorithm operates by iteratively exchanging extrinsic information between two specialized modules.

Factor Graph Partitioning. The factor graph of the joint distribution

$$p(\boldsymbol{x}^r, \boldsymbol{x}^c, \boldsymbol{\rho}^r, \boldsymbol{\rho}^c, \boldsymbol{s}^r, \boldsymbol{s}^c, \bar{\boldsymbol{s}})$$

is partitioned along the edges connecting the support variables $\boldsymbol{s}^r, \boldsymbol{s}^c$ with their corresponding precision factor nodes. Specifically, the cut is made between:

- The variable nodes $\boldsymbol{s}^r$ and the factor nodes $p(\boldsymbol{\rho}^r \mid \boldsymbol{s}^r)$,
- The variable nodes $\boldsymbol{s}^c$ and the factor nodes $p(\boldsymbol{\rho}^c \mid \boldsymbol{s}^c)$.

This partitioning creates two complementary subgraphs.

Subgraph $\mathcal{G}_{\text{A}}$: Contains the linear observation model together with the precision layers $p(\boldsymbol{\rho}^r \mid \boldsymbol{s}^r)$, $p(\boldsymbol{\rho}^c \mid \boldsymbol{s}^c)$ and the signal layers $p(\boldsymbol{x}^r \mid \boldsymbol{\rho}^r)$, $p(\boldsymbol{x}^c \mid \boldsymbol{\rho}^c)$, under an independent approximation for the support variables.

Subgraph $\mathcal{G}_{\text{B}}$: Contains the structured support prior $p(\boldsymbol{s}^r, \boldsymbol{s}^c, \bar{\boldsymbol{s}})$, which captures the 2D joint burst sparsity via the MRF model.

Module A: Sparse VBI Estimator. Module A operates on $\mathcal{G}_{\text{A}}$, performing sparse variational Bayesian inference under an effective independent prior. It takes as input the current observation $\boldsymbol{y}$ together with extrinsic messages received from Module B. These extrinsic messages, denoted as $\phi^r_{\text{A},q}(s^r_q)$ and $\phi^c_{\text{A},q}(s^c_q)$, are the messages from the factor nodes $p(\boldsymbol{s}^r \mid \bar{\boldsymbol{s}})$ and $p(\boldsymbol{s}^c \mid \bar{\boldsymbol{s}})$ to the variable nodes $\boldsymbol{s}^r$ and $\boldsymbol{s}^c$, respectively. They serve as independent Bernoulli priors for the support variables. Using these priors together with the linear observation model, Module A computes approximate Gaussian posteriors for the location-domain channel vectors $\boldsymbol{x}^r$ and $\boldsymbol{x}^c$, as well as Bernoulli posteriors for the individual support variables. The extrinsic messages passed back to Module B are obtained by removing the contribution of the incoming priors from these posteriors. These outgoing messages, denoted as $\phi^r_{\mathrm{B},q}(s^r_q)$ and $\phi^c_{\mathrm{B},q}(s^c_q)$, are the messages from the factor nodes $p(\boldsymbol{\rho}^r \mid \boldsymbol{s}^r)$ and $p(\boldsymbol{\rho}^c \mid \boldsymbol{s}^c)$ to the variable nodes $\boldsymbol{s}^r$ and $\boldsymbol{s}^c$, respectively.

Module B: Structured Support Inference via MRF. Module B operates on $\mathcal{G}_\mathrm{B}$, exploiting the two-dimensional clustered sparsity of the location-domain channels using the MRF prior introduced in Sect. 13.4.2. It takes as input the extrinsic messages $\phi^r_{\mathrm{B},q}(s^r_q)$ and $\phi^c_{\mathrm{B},q}(s^c_q)$ from Module A, which are interpreted as binary observations of the support variables s^r_q and s^c_q. Using these observations together with the joint support prior $p(\boldsymbol{s}^r, \boldsymbol{s}^c, \bar{\boldsymbol{s}})$, Module B performs sum-product message passing over the factor graph of the 4-connected MRF model to refine the estimates of the support variables. The detailed message-passing rules for this MRF, including the update equations for the beliefs and the computation of extrinsic messages, are provided in [12]. Interested readers are referred there for the complete derivation. The output of Module B consists of updated extrinsic messages $\phi^r_{\mathrm{A},q}(s^r_q)$ and $\phi^c_{\mathrm{A},q}(s^c_q)$, which are fed back to Module A as refined priors.

This iterative exchange between the two modules enables the algorithm to jointly exploit both the linear observation model (via Module A) and the 2D clustered sparsity of the location-domain channels (via Module B), leading to accurate joint sensing and channel estimation.

14.6.2.3 Simulation Results

The sensing matrix in this scenario is inherently ill-conditioned due to the structure of the location-domain dictionary and the undersampled nature of the measurements. Under such conditions, the Turbo-VBI algorithm is particularly well suited, as its variational Bayesian foundation provides robustness against matrix conditioning issues.

To validate the effectiveness of the proposed Turbo-VBI algorithm, we compare it against three baseline algorithms:

- **Orthogonal Matching Pursuit (OMP)**: A greedy algorithm that does not exploit any structural information and relies on a fixed grid.
- **Turbo-CS**: The Turbo-CS algorithm described in Sect. 14.2, which uses a support-based two-layer prior to incorporate the MRF structure.
- **Turbo-VBI with i.i.d. prior (Turbo-VBI IID)**: A variant of the proposed method that replaces the MRF prior with an independent Bernoulli prior on the support, isolating the contribution of spatial structure modeling.

We evaluate the proposed approach using two complementary metrics.

1. **Root Mean Square Error (RMSE)** for target and scatterer localization, quantifying the accuracy of position estimates.
2. **NMSE** for both radar channel and communication channel estimation, measuring the fidelity of channel reconstruction.

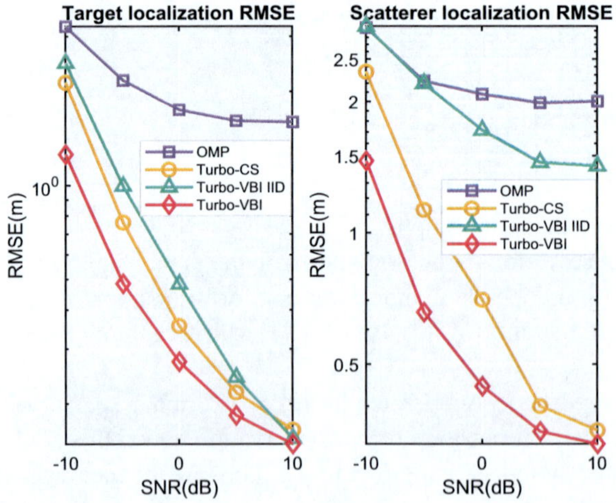

Fig. 14.7 RMSE of target and scatterer localization versus SNR

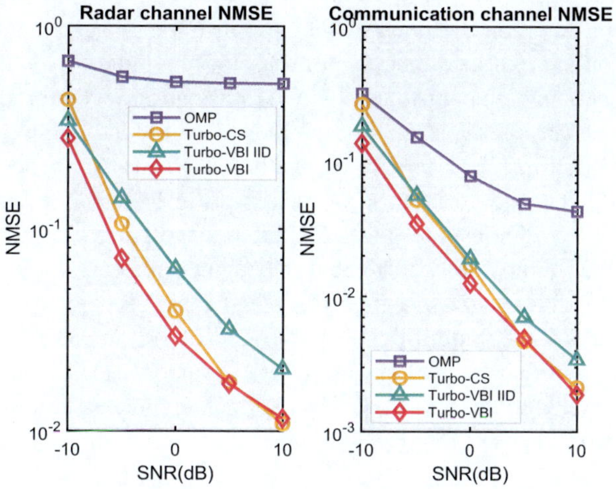

Fig. 14.8 NMSE of radar and communication channel estimation versus SNR

Figures 14.7 and 14.8 illustrate the impact of SNR on sensing and estimation performance across the evaluated methods.

Localization Accuracy: Figure 14.7 presents the RMSE of target and scatterer localization as a function of SNR. All algorithms except OMP exhibit the expected trend of improving performance with increasing SNR. The OMP algorithm, however, shows performance saturation in the high SNR regime. This limitation stems from its reliance on a fixed grid: even with perfect noise-free measurements, the estimation accuracy is fundamentally bounded by the grid resolution (5 m in this setup). In contrast, the Turbo-based methods, which incorporate dynamic grid refinement, can overcome this limitation and continue to improve as SNR increases.

Channel Estimation Accuracy: Figure 14.8 shows the NMSE performance for both radar and communication channel estimation. The proposed Turbo-VBI with MRF prior consistently outperforms all baseline methods across the entire SNR range, with the performance gains being particularly pronounced under low-SNR conditions. This advantage manifests in two key aspects: first, Turbo-VBI demonstrates superior robustness over Turbo-CS when handling ill-conditioned sensing matrices; second, the significant performance gain of Turbo-VBI with MRF over its IID counterpart underscores the critical importance of modeling the 2D joint burst sparsity through spatially non-stationary priors.

Collectively, these results demonstrate that the EM-Turbo-SBI framework, particularly when instantiated as Turbo-VBI with appropriate structured priors, effectively leverages domain knowledge to enhance both sensing and communication performance in challenging propagation environments. The modeling of structured sparsity emerges as a crucial factor in achieving these gains, enabling the algorithm to extract maximal information from limited measurements while maintaining robustness against system non-idealities such as ill-conditioned sensing matrices and timing offsets.

Summary

This chapter has introduced the EM-Turbo-SBI framework, an efficient and unified methodology for solving structured CS problems with a dynamic grid. The framework operates through two core iterative steps: the E-step performs Bayesian estimation of the sparse signal $\boldsymbol{x}$, while the M-step computes point estimates of the grid parameter $\boldsymbol{\theta}$ and other uncertain model parameters. To handle the M-step efficiently, an inexact block MM method is proposed, which alternates between updating each block variable using either closed-form solutions or simple gradient-based updates.

For the computationally challenging E-step, three variants of Turbo-SBI algorithms have been developed under a unified turbo inference architecture. This approach begins by partitioning the factor graph $\mathcal{G}$ into two complementary subgraphs: $\mathcal{G}_{\mathrm{A}}$, which encodes the linear observation model under an effectively independent sparse prior, and $\mathcal{G}_{\mathrm{B}}$, which captures the complex structural dependencies of the support vector $\boldsymbol{s}$ using decoupled binary/AWGN observations. This strategic decomposition renders Bayesian inference computationally tractable for each subgraph.

Two specialized modules are designed to exploit this graph decomposition. Module A performs sparse Bayesian inference over $\mathcal{G}_{\mathrm{A}}$ using an independent sparse prior and the linear observation model, while Module B executes sum-product message passing over $\mathcal{G}_{\mathrm{B}}$ under a structured sparse prior and decoupled binary/AWGN observations. These modules engage in an iterative information exchange, where extrinsic messages are computed by subtracting (in the log-probability domain) the incoming extrinsic message of the other module from the posterior, following the sum-product rule. This ensures that each module contributes unique information without redundant feedback.

Depending on the sparse Bayesian inference method employed in Module A, we have developed three distinct Turbo-SBI algorithms. Turbo-CS utilizes an LMMSE estimator in Module A, Turbo-AMP employs the AMP algorithm, and Turbo-VBI adopts the sparse VBI method. As analyzed in Sect. 14.5.2, each algorithm exhibits distinct operational characteristics: Turbo-CS demonstrates particular effectiveness with partially orthogonal sensing matrices; Turbo-AMP performs well under i.i.d. or weakly correlated sensing matrices, while Turbo-VBI offers superior robustness for general, irregular, or ill-conditioned sensing matrices, albeit at higher computational cost due to its requirement for high-dimensional matrix inversions and multiplications during posterior updates of $\boldsymbol{x}$.

Looking forward, Chap. 16 will introduce several low-complexity sparse VBI variants specifically designed to circumvent costly matrix operations, thereby addressing the computational challenges currently associated with Turbo-VBI while preserving its robustness advantages.

Homework Problems

14.1 Write down the detailed derivations for the LMMSE estimator in Eqs. (14.24) and (14.25), which correspond to Module A in the Turbo-CS algorithm. Your derivation should clearly show:

- The starting point from the linear observation model and prior assumptions.
- The application of the LMMSE estimation principle.
- All intermediate mathematical steps leading to the final closed-form expressions.

14.2 Prove that when the sensing matrix $\boldsymbol{A}$ is a partially DFT matrix, the posterior mean and variance in Module A of the Turbo-CS algorithm reduce to the simplified results in Eqs. (14.46) and (14.45), respectively.

- Clearly state the specific properties of the partial DFT matrix that enable this simplification.
- Show step by step how these properties lead to the diagonalized form of the covariance matrix.
- Discuss the computational advantages of these simplified expressions.

14.3 Consider a Turbo-AMP algorithm where the support vector $\boldsymbol{s}$ follows a Markov sparse prior:

$$p(\boldsymbol{s}) = p(s_1)\prod_{n=2}^{N} p(s_n|s_{n-1})$$

with transition probabilities:

$$p(s_n|s_{n-1}) = \begin{cases} (1-p_{01})^{1-s_n} p_{01}^{s_n} & s_{n-1}=0 \\ p_{10}^{1-s_n}(1-p_{10})^{s_n} & s_{n-1}=1 \end{cases}$$

where p_{10} and p_{01} determine the average cluster size and gap between clusters.

(a) **Factor Graph Analysis**:

- Draw the complete factor graph for the joint probability distribution.
- Partition the factor graph into two subgraphs according to the Turbo-AMP framework and draw both subgraphs.

(b) **Message-Passing Derivation**:

- For given transition parameters p_{10} and p_{01}, derive the complete message-passing algorithm for Module B of Turbo-AMP.
- Specify all message update rules and their computational forms.

(c) **EM Learning of Transition Parameters**:

- Derive an EM algorithm to estimate the unknown transition parameters.
- Specify how Turbo-AMP provides the required posteriors in the E-step.
- Derive the M-step update equations that maximize the surrogate function.

14.4 Using the provided MATLAB code for the Turbo-CS algorithm in the FDD massive MIMO-OFDM channel estimation application (Sect. 14.6.1):

- Modify the code to implement the Turbo-AMP algorithm.
- Compare the performance of Turbo-AMP and Turbo-CS algorithms in terms of:
  - Channel estimation accuracy (NMSE).
  - Computational complexity and running time.
  - Convergence behavior.
  - Robustness to different channel conditions.
- The MATLAB code for Turbo-CS is available at: https://github.com/biwcs-book/codebase.

14.5 For the joint environment sensing and channel estimation application in Sect. 14.6.2:

- Modify the provided code to handle unknown noise variance.
- Compare the performance of Turbo-CS and Turbo-VBI under imperfect noise variance knowledge.
- Analyze how the results differ from those in Figs. 14.7 and 14.8.
- Determine which algorithm demonstrates better robustness to unknown noise variance and explain why.
- The MATLAB code for this application is available at: https://github.com/biwcs-book/codebase.

References

1. A. Liu, G. Liu, L. Lian, V.K. Lau, M.-J. Zhao, Robust recovery of structured sparse signals with uncertain sensing matrix: a turbo-VBI approach. IEEE Trans. Wireless Commun. **19**(5), 3185–3198 (2020)
2. F.R. Kschischang, B.J. Frey, H.-A. Loeliger, Factor graphs and the sum-product algorithm. IEEE Trans. Inf. Theory **47**(2), 498–519 (2001)
3. J. Ma, X. Yuan, L. Ping, Turbo compressed sensing with partial DFT sensing matrix. IEEE Signal Process. Lett. **22**(2), 158–161 (2014)
4. L. Chen, A. Liu, X. Yuan, Structured turbo compressed sensing for massive MIMO channel estimation using a Markov prior. IEEE Trans. Veh. Technol. **67**(5), 4635–4639 (2017)
5. A. Liu, L. Lian, V.K. Lau, X. Yuan, Downlink channel estimation in multiuser massive MIMO with hidden Markovian sparsity. IEEE Trans. Signal Process. **66**(18), 4796–4810 (2018)
6. X. Kuai, L. Chen, X. Yuan, A. Liu, Structured turbo compressed sensing for downlink massive MIMO-OFDM channel estimation. IEEE Trans. Wireless Commun. **18**(8), 3813–3826 (2019)
7. Y. Teng, L. Jia, A. Liu, V.K. Lau, A hybrid pilot beamforming and channel tracking scheme for massive MIMO systems. IEEE Trans. Wireless Commun. **20**(9), 6078–6092 (2021)
8. P. Schniter, Turbo reconstruction of structured sparse signals, in *IEEE Conference on Information Sciences and Systems (CISS)* (IEEE, 2010), pp. 1–6

9. J.P. Vila, P. Schniter, An empirical-bayes approach to recovering linearly constrained non-negative sparse signals. IEEE Trans. Signal Process. **62**(18), 4689–4703 (2014)
10. J.P. Vila, P. Schniter, Expectation-maximization Gaussian-mixture approximate message passing. IEEE Trans. Signal Process. **61**(19), 4658–4672 (2013)
11. W. Xu, A. Liu, B. Zhou, M. Zhao, Successive Linear Approximation VBI for Joint Sparse Signal Recovery and Dynamic Grid Parameters Estimation," IEEE Trans.Wireless Commun. **24**(11), 9645–9659, (2025)
12. W. Xu, Y. Xiao, A. Liu, M. Lei, M.-J. Zhao, Joint scattering environment sensing and channel estimation based on non-stationary Markov random field. IEEE Trans. Wireless Commun. **23**(5), 3903–3917 (2024)
13. A. Liu, Y. Zhou, W. Xu, Subspace constrained variational bayesian inference for structured compressive sensing with a dynamic grid. IEEE Trans. Signal Process. (2025)
14. D.G. Tzikas, A.C. Likas, N.P. Galatsanos, The variational approximation for bayesian inference. IEEE Signal Process. Mag. **25**(6), 131–146 (2008)
15. L. Grippo, On the convergence of the block nonlinear Gauss–Seidel method under convex constraints. Oper. Res. Lett. **26**, 127–136 (2000). https://doi.org/10.1016/S0167-6377(99)00074-7
16. D.P. Bertsekas, Nonlinear programming. J. Oper. Res. Soc. **48**(3), 334 (1997)
17. A.P. Dempster, N.M. Laird, D.B. Rubin, Maximum likelihood from incomplete data via the EM algorithm. J. R. Stat. Soc. Ser. B (Methodol.) **39**(1), 1–22 (1977)
18. J. Vila, P. Schniter, Expectation-maximization Bernoulli-Gaussian approximate message passing, in *2011 Conference Record of the Forty Fifth Asilomar Conference on Signals, Systems and Computers (ASILOMAR)* (Pacific Grove, CA, USA, 2011), pp. 799–803
19. S. Rangan, P. Schniter, A.K. Fletcher, S. Sarkar, On the convergence of approximate message passing with arbitrary matrices. IEEE Trans. Inf. Theory **65**(9), 5339–5351 (2019)
20. Q. Guo, J. Xi, Approximate message passing with unitary transformation. arXiv preprint arXiv: 1504.04799,2015

15 Dynamic Sparse Bayesian Inference

This chapter extends the principles of sparse Bayesian inference introduced in previous chapters to dynamic environments where signals evolve over time. While earlier discussions focused on estimating sparse signals within isolated time frames, treating both signals and observations as random variables to compute posterior distributions, we now address scenarios where these quantities evolve as stochastic processes. The objective becomes sequential estimation: at each time step, we seek the posterior distribution of the current sparse signal conditioned on all observations accumulated up to that moment. To model the temporal evolution of sparsity patterns, we develop a time-varying Bernoulli–Gaussian Markov model that captures both instantaneous sparsity and temporal correlations. Within this probabilistic framework, we present a dynamic Turbo-CS algorithm designed for efficient sequential inference. The practical utility of this approach is demonstrated through a channel tracking application in 5G New Radio systems, where the algorithm must contend with real-world dynamic effects such as user mobility, frequency hopping, and various system imperfections. The chapter concludes by exploring extensions based on Turbo-AMP and Turbo-VBI methodologies, offering a comprehensive perspective on algorithmic options for handling time-varying sparsity.

15.1 Problem Formulation for Dynamic Sparse Bayesian Inference

15.1.1 System Model and Probabilistic Framework

Dynamic sparse Bayesian inference generalizes conventional sparse recovery to settings where both the signal of interest and the observations vary over time. Consider a discrete-time system with slots indexed by $t = 1, 2, \ldots$. At each time instant, we observe a measurement vector $\boldsymbol{y}^{(t)} \in \mathbb{C}^{M\times 1}$ that relates to an underlying sparse signal $\boldsymbol{x}^{(t)} \in \mathbb{C}^{N\times 1}$ through a linear observation model

$$\boldsymbol{y}^{(t)} = \boldsymbol{A}(\boldsymbol{\theta}^{(t)})\boldsymbol{x}^{(t)} + \boldsymbol{w}^{(t)}, \tag{15.1}$$

where $\boldsymbol{A}(\boldsymbol{\theta}^{(t)}) \in \mathbb{C}^{M\times N}$ is a sensing matrix parameterized by time-varying grid parameters $\boldsymbol{\theta}^{(t)}$, and $\boldsymbol{w}^{(t)}$ represents AWGN with variance $\sigma_w^{(t)}$. The collection of all uncertain parameters at time t, including grid parameters and other imperfection factors, is denoted collectively as $\boldsymbol{\xi}^{(t)}$.

To characterize the dynamic nature of the sparse signal, we adopt a time-evolving Bernoulli–Gaussian Markov model. This framework introduces two sets of latent variables at each time instant: a support vector $\boldsymbol{s}^{(t)} \in \{0, 1\}^N$ indicating which entries of $\boldsymbol{x}^{(t)}$ are non-zero, and a hidden amplitude vector $\boldsymbol{\beta}^{(t)}$ representing the magnitudes of those non-zero entries. The introduction of $\boldsymbol{\beta}^{(t)}$ decouples amplitude evolution from sparsity state transitions, allowing the model to maintain memory of amplitude values even during periods of temporary inactivity, a crucial feature for fully exploiting temporal correlations.

For notational convenience, we denote a sequence of quantities spanning times 1 through t as $\boldsymbol{x}^{(1:t)}$, with analogous notation for $\boldsymbol{s}^{(1:t)}$, $\boldsymbol{\beta}^{(1:t)}$, and $\boldsymbol{\xi}^{(1:t)}$. The complete probabilistic model factorizes as

$$\begin{aligned}
p(\boldsymbol{x}^{(1:t)}, \boldsymbol{\beta}^{(1:t)}, \boldsymbol{s}^{(1:t)}, \boldsymbol{\xi}^{(1:t)}) =& p(\boldsymbol{\xi}^{(1)})p(\boldsymbol{s}^{(1)})p(\boldsymbol{\beta}^{(1)})p(\boldsymbol{x}^{(1)}|\boldsymbol{\beta}^{(1)}, \boldsymbol{s}^{(1)}) \\
& \prod_{t'=2}^{t} p(\boldsymbol{\xi}^{(t')}|\boldsymbol{\xi}^{(t'-1)})p(\boldsymbol{s}^{(t')}|\boldsymbol{s}^{(t'-1)}) \\
& \times p(\boldsymbol{\beta}^{(t')}|\boldsymbol{\beta}^{(t'-1)})p(\boldsymbol{x}^{(t')}|\boldsymbol{\beta}^{(t')}, \boldsymbol{s}^{(t')}).
\end{aligned} \tag{15.2}$$

A. Liu and M.-M. Zhao, *Advanced Bayesian Signal Processing*, Textbooks in Telecommunication Engineering,
https://doi.org/10.1007/978-3-032-26656-9_15

This factorization reveals the temporal Markov structure: each variable evolves independently according to its own transition probability, with the sparse signal at each time depending only on the current support and amplitude.

15.1.1.1 Probability Model for the Sparse Signal

The relationship between the observed sparse signal and its latent components is modeled through a deterministic dependency: given the support and amplitude vectors, the signal is fully determined. This leads to the Bernoulli–Gaussian formulation:

$$p(\boldsymbol{x}^{(t)}|\boldsymbol{\beta}^{(t)}, \boldsymbol{s}^{(t)}) = \prod_{n=1}^{N} p(x_n^{(t)}|s_n^{(t)}, \beta_n^{(t)}) = \prod_{n=1}^{N} \delta(x_n^{(t)} - s_n^{(t)}\beta_n^{(t)}), \tag{15.3}$$

where $\delta(\cdot)$ denotes the Dirac delta function. Here, $\boldsymbol{s}^{(t)}$ governs the birth-death process of non-zero entries, while $\boldsymbol{\beta}^{(t)}$ captures the smooth evolution of their amplitudes. An important feature of this model is that the amplitude process evolves independently of the sparsity pattern; significant amplitudes may be associated with currently inactive entries, allowing these paths to become active in subsequent time slots with appropriate magnitude [2].

15.1.1.2 Probability for the Support Vector

In practical applications, the support configuration typically evolves slowly relative to the sampling rate. This temporal correlation makes previously estimated support information valuable as prior knowledge for current estimation. We model this using a Markov chain

$$p(\boldsymbol{s}^{(t)}|\boldsymbol{s}^{(t-1)}) = \prod_{n=1}^{N} p(s_n^{(t)}|s_n^{(t-1)}), \tag{15.4}$$

with transition probabilities defined as

$$\begin{aligned} p(s_n^{(t)} = 1|s_n^{(t-1)} = 0) &= \rho_{01}, \\ p(s_n^{(t)} = 0|s_n^{(t-1)} = 1) &= \rho_{10}. \end{aligned} \tag{15.5}$$

The parameters ρ_{01} and ρ_{10} characterize the degree of temporal correlation: smaller values indicate stronger correlation, meaning the support changes more slowly. These parameters can be learned automatically within an EM framework [2]. The initial distribution is set to the steady-state distribution of the Markov chain:

$$p(s_n^{(1)} = 1) = \lambda \triangleq \frac{\rho_{01}}{\rho_{01} + \rho_{10}}, \tag{15.6}$$

where λ represents the average fraction of non-zero elements in $\boldsymbol{s}^{(t)}$.

15.1.1.3 Probability Model for the Hidden Amplitude

The amplitudes of non-zero entries typically evolve smoothly over time, exhibiting a temporal structure that can be exploited for improved inference [5, 6]. We model this evolution using a Gauss–Markov process

$$p(\boldsymbol{\beta}^{(t)}|\boldsymbol{\beta}^{(t-1)}) = \prod_{n=1}^{N} \mathcal{CN}\left(\beta_n^{(t)}; (1-\alpha_\beta)\beta_n^{(t-1)} + \alpha_\beta\zeta_\beta, \alpha_\beta^2\gamma_\beta\right), \tag{15.7}$$

where $\alpha_\beta \in [0, 1]$ controls the degree of temporal correlation, ζ_β denotes the steady-state mean, and γ_β is the innovation variance. The steady-state variance of the process is $\frac{\alpha_\beta\gamma_\beta}{2-\alpha_\beta}$. The parameters α_β, ζ_β, and γ_β can be learned via the EM method. The initial distribution follows the steady-state Gaussian:

$$p(\beta_n^{(1)}) = \mathcal{CN}\left(\beta_n^{(1)}; \zeta_\beta, \frac{\alpha_\beta\gamma_\beta}{2-\alpha_\beta}\right). \tag{15.8}$$

15.1.1.4 Probability Model for Uncertain Parameters

The grid parameters $\boldsymbol{\theta}^{(t)}$ often exhibit slow temporal variation. We model this through a simple random walk

$$p(\boldsymbol{\theta}^{(t)}|\boldsymbol{\theta}^{(t-1)}) = \mathcal{N}\left(\boldsymbol{\theta}^{(t)}; \boldsymbol{\theta}^{(t-1)}, \gamma_\theta \boldsymbol{I}\right), \tag{15.9}$$

where γ_θ controls the rate of change. This formulation enables exploitation of temporal correlation in grid parameters, allowing high-resolution prior information from previous time slots to enhance estimation accuracy. Transition probabilities for other components of $\boldsymbol{\xi}^{(t)}$ depend on the specific application context, as illustrated in later examples.

15.1.2 Problem Formulation

Let $\boldsymbol{v}^{(t)} = \{\boldsymbol{x}^{(t)}, \boldsymbol{\beta}^{(t)}, \boldsymbol{s}^{(t)}\}$ denote the complete set of hidden variables at time t. Given the measurement sequence $\boldsymbol{y}^{(1:t)}$ and previous parameter estimates $\hat{\boldsymbol{\xi}}^{(1:t-1)}$, we pursue two fundamental objectives that together constitute a sequential estimation framework.

First, we seek the marginal posterior distribution of the current sparse signal:

$$p(\boldsymbol{x}^{(t)}|\boldsymbol{y}^{(1:t)}, \hat{\boldsymbol{\xi}}^{(1:t-1)}, \boldsymbol{\xi}^{(t)}) \propto \int_{-\boldsymbol{x}^{(t)}} p(\boldsymbol{y}^{(1:t)}, \boldsymbol{v}^{(t)}|\hat{\boldsymbol{\xi}}^{(1:t-1)}, \boldsymbol{\xi}^{(t)}), \tag{15.10}$$

where $\int_{-\boldsymbol{x}^{(t)}}$ denotes integration over all components of $\boldsymbol{v}^{(t)}$ except $\boldsymbol{x}^{(t)}$. This marginalization focuses inference on the current signal while properly accounting for uncertainty in other latent variables.

Second, we estimate uncertain parameters via MAP estimation:

$$\begin{aligned}
\hat{\boldsymbol{\xi}}^{(t)} &= \arg\max_{\boldsymbol{\xi}^{(t)}} \ln p(\boldsymbol{\xi}^{(t)}|\boldsymbol{y}^{(1:t)}, \hat{\boldsymbol{\xi}}^{(1:t-1)}) \\
&= \arg\max_{\boldsymbol{\xi}^{(t)}} \ln \int_{\boldsymbol{v}^{(t)}} p(\boldsymbol{y}^{(1:t)}, \boldsymbol{v}^{(t)}|\hat{\boldsymbol{\xi}}^{(1:t-1)}, \boldsymbol{\xi}^{(t)}) p(\boldsymbol{\xi}^{(t)}|\hat{\boldsymbol{\xi}}^{(t-1)}),
\end{aligned} \tag{15.11}$$

where $\hat{p}(\boldsymbol{\xi}^{(t)}) \triangleq p(\boldsymbol{\xi}^{(t)}|\hat{\boldsymbol{\xi}}^{(t-1)})$ serves as an effective prior propagating historical information forward.

A direct approach storing all observations for joint estimation quickly becomes computationally prohibitive as t grows. To address this challenge, we decompose the joint distribution as

$$\begin{aligned}
& p(\boldsymbol{y}^{(1:t)}, \boldsymbol{v}^{(t)}|\hat{\boldsymbol{\xi}}^{(1:t-1)}, \boldsymbol{\xi}^{(t)}) \\
& \propto \int_{\boldsymbol{\beta}^{(t-1)}} \sum_{\boldsymbol{s}^{(t-1)}} p(\boldsymbol{s}^{(t-1)}, \boldsymbol{\beta}^{(t-1)}|\boldsymbol{y}^{(1:t-1)}, \hat{\boldsymbol{\xi}}^{(1:t-1)}) \\
& \times p(\boldsymbol{s}^{(t)}|\boldsymbol{s}^{(t-1)}) p(\boldsymbol{\beta}^{(t)}|\boldsymbol{\beta}^{(t-1)}) p(\boldsymbol{x}^{(t)}|\boldsymbol{\beta}^{(t)}, \boldsymbol{s}^{(t)}) p(\boldsymbol{y}^{(t)}|\boldsymbol{x}^{(t)}, \boldsymbol{\xi}^{(t)}).
\end{aligned} \tag{15.12}$$

The exact posterior $p(\boldsymbol{s}^{(t-1)}, \boldsymbol{\beta}^{(t-1)}|\boldsymbol{y}^{(1:t-1)}, \hat{\boldsymbol{\xi}}^{(1:t-1)})$ is generally intractable due to cycles in the underlying factor graph. We therefore employ approximate posteriors $q(\boldsymbol{\beta}^{(t-1)})$ and $q(\boldsymbol{s}^{(t-1)})$ obtained via the Turbo-SBI algorithms developed in Chap. 14. This yields a computationally efficient factorization:

$$\begin{aligned}
& p(\boldsymbol{y}^{(1:t)}, \boldsymbol{v}^{(t)}|\hat{\boldsymbol{\xi}}^{(1:t-1)}, \boldsymbol{\xi}^{(t)}) \\
& \approx \sum_{\boldsymbol{s}^{(t-1)}} q(\boldsymbol{s}^{(t-1)}) p(\boldsymbol{s}^{(t)}|\boldsymbol{s}^{(t-1)}) \int_{\boldsymbol{\beta}^{(t-1)}} q(\boldsymbol{\beta}^{(t-1)}) \\
& \times p(\boldsymbol{\beta}^{(t)}|\boldsymbol{\beta}^{(t-1)}) p(\boldsymbol{x}^{(t)}|\boldsymbol{\beta}^{(t)}, \boldsymbol{s}^{(t)}) p(\boldsymbol{y}^{(t)}|\boldsymbol{x}^{(t)}, \boldsymbol{\xi}^{(t)}) \\
& = \hat{p}(\boldsymbol{s}^{(t)}) \hat{p}(\boldsymbol{\beta}^{(t)}) p(\boldsymbol{x}^{(t)}|\boldsymbol{\beta}^{(t)}, \boldsymbol{s}^{(t)}) p(\boldsymbol{y}^{(t)}|\boldsymbol{x}^{(t)}, \boldsymbol{\xi}^{(t)}),
\end{aligned} \tag{15.13}$$

where the predicted priors emerge naturally as

$$\hat{p}(\boldsymbol{s}^{(t)}) \triangleq \sum_{\boldsymbol{s}^{(t-1)}} q(\boldsymbol{s}^{(t-1)}) p(\boldsymbol{s}^{(t)}|\boldsymbol{s}^{(t-1)}), \tag{15.14}$$

$$\hat{p}(\boldsymbol{\beta}^{(t)}) \triangleq \int_{\boldsymbol{\beta}^{(t-1)}} q(\boldsymbol{\beta}^{(t-1)}) p(\boldsymbol{\beta}^{(t)}|\boldsymbol{\beta}^{(t-1)}). \tag{15.15}$$

This decomposition reduces the original multi-timeslot problem to a sequence of manageable single-timeslot problems. The MAP estimation simplifies to

$$\begin{aligned}\hat{\boldsymbol{\xi}}^{(t)} &= \arg\max_{\boldsymbol{\xi}^{(t)}} \ln \hat{p}(\boldsymbol{y}^{(t)}, \boldsymbol{\xi}^{(t)}) \\ &= \arg\max_{\boldsymbol{\xi}^{(t)}} \ln \int_{\boldsymbol{v}^{(t)}} \hat{p}(\boldsymbol{y}^{(t)}, \boldsymbol{v}^{(t)}|\boldsymbol{\xi}^{(t)}) \hat{p}(\boldsymbol{\xi}^{(t)}),\end{aligned} \tag{15.16}$$

with the effective joint distribution

$$\hat{p}(\boldsymbol{y}^{(t)}, \boldsymbol{v}^{(t)}|\boldsymbol{\xi}^{(t)}) \triangleq \hat{p}(\boldsymbol{s}^{(t)}) \hat{p}(\boldsymbol{\beta}^{(t)}) p(\boldsymbol{x}^{(t)}|\boldsymbol{\beta}^{(t)}, \boldsymbol{s}^{(t)}) p(\boldsymbol{y}^{(t)}|\boldsymbol{x}^{(t)}, \boldsymbol{\xi}^{(t)}). \tag{15.17}$$

This formulation enables efficient sequential estimation through three key mechanisms: temporal information propagation via predicted priors that encode historical evidence, computational tractability through approximate Bayesian inference using Turbo-SBI algorithms, and real-time implementation by focusing each estimation step on the current timeslot while properly accounting for temporal dependencies.

15.2 Dynamic Turbo-CS Algorithm

15.2.1 Algorithm Overview

Building upon the joint probability decomposition in (15.13) and employing the EM framework, the Dynamic Turbo-CS (DTurbo-CS) algorithm performs coordinated iterative updates between two core components at each time step until convergence, as elaborated below.

The **DTurbo-CS-E Step** uses predicted messages $\hat{p}(\boldsymbol{\beta}^{(t)})$ and $\hat{p}(\boldsymbol{s}^{(t)})$ propagated from time $t-1$, together with the current parameter estimate $\hat{\boldsymbol{\xi}}^{(t)}$ from the M-step, to compute an approximate posterior factorization $q(\boldsymbol{v}^{(t)}) \triangleq q(\boldsymbol{\beta}^{(t)})q(\boldsymbol{s}^{(t)})$. This inference leverages the effective joint distribution $\hat{p}(\boldsymbol{y}^{(t)}, \boldsymbol{v}^{(t)}|\boldsymbol{\xi}^{(t)} = \hat{\boldsymbol{\xi}}^{(t)})$ in (15.17) and the efficient message-passing architecture of the Turbo-CS algorithm for scalable approximate Bayesian computation. The variational factorization represents a practical compromise that maintains computational tractability while capturing essential statistical dependencies.

The **DTurbo-CS-M Step** constructs and optimizes a surrogate function for the MAP objective $\ln \int_{\boldsymbol{v}^{(t)}} \hat{p}(\boldsymbol{y}^{(t)}, \boldsymbol{v}^{(t)}|\boldsymbol{\xi}^{(t)}) \hat{p}(\boldsymbol{\xi}^{(t)})$ given the variational posterior $q(\boldsymbol{v}^{(t)})$. This maximization yields updated parameter estimates $\hat{\boldsymbol{\xi}}^{(t)}$ that reflect both current evidence and prior knowledge. The surrogate function approach transforms the potentially complex original objective into a more manageable form while preserving the essential optimization landscape.

Upon convergence at time t, the algorithm computes MMSE estimates of $x_n^{(t)}$ as the mean of the approximate marginal posterior $q(x_n^{(t)})$, providing both point estimates and associated uncertainty quantification. Crucially, the algorithm then propagates temporal information to the next time step through predicted messages:

$$\hat{p}(\boldsymbol{s}^{(t+1)}) \triangleq \sum_{\boldsymbol{s}^{(t)}} q(\boldsymbol{s}^{(t)}) p(\boldsymbol{s}^{(t+1)}|\boldsymbol{s}^{(t)}), \tag{15.18}$$

$$\hat{p}(\boldsymbol{\beta}^{(t+1)}) \triangleq \int_{\boldsymbol{\beta}^{(t)}} q(\boldsymbol{\beta}^{(t)}) p(\boldsymbol{\beta}^{(t+1)}|\boldsymbol{\beta}^{(t)}). \tag{15.19}$$

This iterative EM procedure, combined with systematic message passing across consecutive time steps, enables efficient sequential estimation while capturing the dynamic evolution of both sparse signals and system parameters. The predicted messages serve as informed priors for subsequent time steps, creating a continuous information flow that respects temporal dependencies while maintaining the computational advantages of the Turbo-CS approach.

15.2.2 DTurbo-CS-E Step

In the DTurbo-CS-E step, the Turbo-CS algorithm computes the approximate posterior distribution $q(\boldsymbol{v}^{(t)})$ by integrating multiple information sources: the current observation $\boldsymbol{y}^{(t)}$, propagated messages $\hat{p}(\boldsymbol{\beta}^{(t)})$ and $\hat{p}(\boldsymbol{s}^{(t)})$ from time $t-1$, and the given system parameters $\boldsymbol{\xi}^{(t)}$. This integration enables Bayesian inference that respects both current measurements and the temporal evolution of the sparse signal structure.

The computational efficiency stems from the Turbo-CS algorithm's modular architecture, which employs two specialized processing units with complementary roles. Module A functions as an LMMSE estimator incorporating the current observation along with extrinsic messages from Module B, effectively performing LMMSE estimation while accounting for structured prior information. Module B serves as a sparsity combiner integrating the extrinsic message from Module A with the prior messages $\hat{p}(\boldsymbol{\beta}^{(t)})$ and $\hat{p}(\boldsymbol{s}^{(t)})$ to exploit the dynamic sparsity structure encoded in the temporal model.

Through iterative information exchange between these modules, the algorithm fuses sparse prior information with observation data to produce refined Bayesian estimates of $\boldsymbol{x}^{(t)}$. This fusion process leverages both the statistical structure of the sensing process (handled by Module A) and the temporal dependencies in the sparsity pattern (handled by Module B), resulting in estimates more accurate and robust than what either module could achieve independently.

15.2.2.1 Module A: LMMSE Estimation

In Module A, the sparse signal $\boldsymbol{x}^{(t)}$ is estimated from the current observation using an LMMSE estimator incorporating a Gaussian prior $\mathcal{CN}(\boldsymbol{x}^{(t)}; \boldsymbol{x}_A^{pri(t)}, \text{diag}(\boldsymbol{v}_A^{pri(t)}))$, where $\boldsymbol{x}_A^{pri(t)}$ and $\text{diag}(\boldsymbol{v}^{pri(t)})$ represent the extrinsic mean and variance received from Module B. This Gaussian prior formulation encodes structured sparsity information in a computationally tractable form.

The LMMSE framework yields a closed-form solution with posterior distribution remaining complex Gaussian. The posterior covariance matrix and mean are given by

$$\boldsymbol{V}_A^{post(t)} = \left(\frac{\boldsymbol{A}^H \boldsymbol{A}}{\sigma_w^{(t)}} + \text{diag}\left(1/\boldsymbol{v}_A^{pri(t)}\right) \right)^{-1}, \tag{15.20}$$

$$\boldsymbol{x}_A^{post(t)} = \boldsymbol{V}_A^{post(t)} \left(\frac{\boldsymbol{x}_A^{pri(t)}}{\boldsymbol{v}_A^{pri(t)}} + \frac{\boldsymbol{A}^H \boldsymbol{y}^{(t)}}{\sigma_w^{(t)}} \right). \tag{15.21}$$

The extrinsic message passed to Module B is then computed as

$$\boldsymbol{x}_B^{pri(t)} = \boldsymbol{v}_B^{pri(t)} \left(\frac{\boldsymbol{x}_A^{post(t)}}{\boldsymbol{v}_A^{post(t)}} - \frac{\boldsymbol{x}_A^{pri(t)}}{\boldsymbol{v}_A^{pri(t)}} \right), \tag{15.22}$$

$$\boldsymbol{v}_B^{pri(t)} = \left(\frac{1}{\boldsymbol{v}_A^{post(t)}} - \frac{1}{\boldsymbol{v}_A^{pri(t)}} \right)^{-1}, \tag{15.23}$$

where $\boldsymbol{v}_A^{post(t)}$ is the vector of diagonal elements of $\boldsymbol{V}_A^{post(t)}$. This extrinsic extraction ensures Module B receives only new information derived from the linear observations, preventing redundant feedback of prior information.

15.2.2.2 Module B: Sparsity Combining

In Module B, the extrinsic mean $\boldsymbol{x}_B^{pri(t)}$ from Module A is interpreted within a virtual AWGN observation framework:

$$\boldsymbol{x}_B^{pri(t)} = \boldsymbol{x}^{(t)} + \boldsymbol{w}_B^{pri(t)}, \tag{15.24}$$

where $\boldsymbol{w}_B^{pri(t)} \sim \mathcal{CN}(\boldsymbol{w}_B^{pri(t)}; 0, \text{diag}(\boldsymbol{v}_B^{pri(t)}))$ represents virtual noise with variance given by the extrinsic variance from Module A. This interpretation transforms Module A's output into a form amenable to processing with structured sparsity models.

Under this observation model, the joint probability distribution associated with Module B becomes

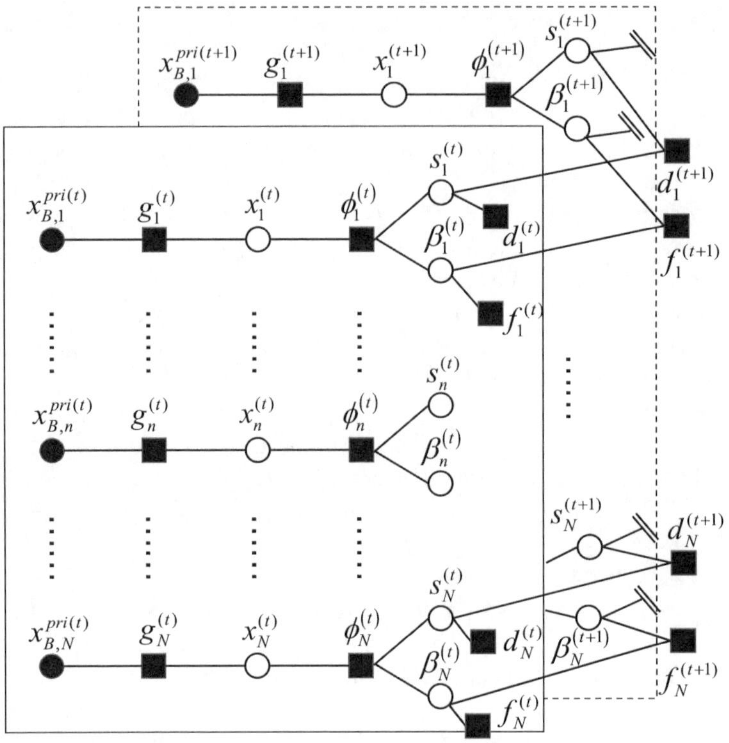

Fig. 15.1 An illustration of factor graph $\mathcal{G}_B$

Table 15.1 Factors, distributions, and functional forms in Fig. 15.1

Factor	Distribution
$d_n^{(t)}\left(s_n^{(t)}, s_n^{(t-1)}\right)$	$p\left(s_n^{(t)} \mid s_n^{(t-1)}\right) = \begin{cases} (\rho_{10})^{1-s_n^{(t)}} (1-\rho_{10})^{s_n^{(t)}} & s_n^{(t-1)} = 1 \\ (\rho_{01})^{s_n^{(t)}} (1-\rho_{01})^{1-s_n^{(t)}} & s_n^{(t-1)} = 0 \end{cases}$
$f_n^{(t)}\left(\beta_n^{(t)}, \beta_n^{(t-1)}\right)$	$p\left(\beta_n^{(t)} \mid \beta_n^{(t-1)}\right) = \mathcal{CN}\left(\beta_n^{(t)}; (1-\alpha_\beta)\beta_n^{(t-1)} + \alpha_\beta \zeta_\beta, \alpha_\beta^2 \gamma_\beta\right)$
$\phi_n^{(t)}\left(x_n^{(t)}, s_n^{(t)}, \beta_n^{(t)}\right)$	$p\left(x_n^{(t)} \mid s_n^{(t)}, \beta_n^{(t)}\right) = \delta\left(x_n^{(t)} - s_n^{(t)}\beta_n^{(t)}\right)$
$g_n^{(t)}\left(x_{B,n}^{pri(t)}, x_n^{(t)}\right)$	$p\left(x_{B,n}^{pri(t)} \mid x_n^{(t)}\right) = \mathcal{CN}\left(x_n^{(t)}; x_{B,n}^{pri(t)}, v_{B,n}^{pri(t)}\right)$

$$p(\boldsymbol{x}^{(t)}, \boldsymbol{s}^{(t)}, \boldsymbol{\beta}^{(t)}, \boldsymbol{x}_B^{pri(t)} \mid \boldsymbol{\xi}^{(t)})$$
$$= \hat{p}(\boldsymbol{s}^{(t)}) \hat{p}(\boldsymbol{\beta}^{(t)}) p(\boldsymbol{x}^{(t)} \mid \boldsymbol{\beta}^{(t)}, \boldsymbol{s}^{(t)}) p(\boldsymbol{x}_B^{pri(t)} \mid \boldsymbol{x}^{(t)}). \tag{15.25}$$

The corresponding factor graph $\mathcal{G}_B$, illustrated in Fig. 15.1 with detailed factor functions in Table 15.1, captures the complex statistical dependencies. Module B's primary objective is computing the posterior distributions $\hat{p}(x_n^{(t)} \mid \boldsymbol{x}_B^{pri(t)}, \boldsymbol{\xi}^{(t)})$ using sum-product message passing.

Message passing begins with input from Module A

$$\nu_{x_n^{(t)} \to \phi_n^{(t)}}(x_n^{(t)}) = \mathcal{CN}\left(x_n^{(t)}; x_{B,n}^{pri(t)}, v_{B,n}^{pri(t)}\right), \tag{15.26}$$

which carries observation information in computationally convenient Gaussian form. This is complemented by prior information propagated from time $t-1$, consisting of amplitude prior $\hat{p}(\boldsymbol{\beta}^{(t)}) = \prod_{n=1}^{N} \hat{p}(\beta_n^{(t)})$ approximated as Gaussian:

$$\hat{p}(\beta_n^{(t)}) \triangleq \mathcal{CN}\left(\beta_n^{(t)}; \mu_{out,n}^{(t)}, \Sigma_{out,n}^{(t)}\right), \tag{15.27}$$

and support prior $\hat{p}(\boldsymbol{s}^{(t)}) = \prod_{n=1}^{N} \hat{p}(s_n^{(t)})$ with

$$\hat{p}(s_n^{(t)} = 1) \triangleq \pi_{out,n}^{(t)}. \tag{15.28}$$

The marginal posterior distribution is computed through integration and summation over latent variables:

$$\begin{aligned}
&\hat{p}(x_n^{(t)}|\boldsymbol{x}_B^{pri(t)}, \boldsymbol{\xi}^{(t)}) \\
&\propto \int_{\beta_n^{(t)}} \sum_{s_n^{(t)}} \phi_n^{(t)}(s_n^{(t)}, \beta_n^{(t)}, x_n^{(t)}) \nu_{s_n^{(t)} \to \phi_n^{(t)}}(s_n^{(t)}) \\
&\times \nu_{\beta_n^{(t)} \to \phi_n^{(t)}}(\beta_n^{(t)}) \nu_{x_n^{(t)} \to \phi_n^{(t)}}(x_n^{(t)}),
\end{aligned} \tag{15.29}$$

where $\nu_{s_n^{(t)} \to \phi_n^{(t)}}(s_n^{(t)}) = \hat{p}(s_n^{(t)})$ and $\nu_{\beta_n^{(t)} \to \phi_n^{(t)}}(\beta_n^{(t)}) = \hat{p}(\beta_n^{(t)})$. From this posterior, key statistics are extracted as

$$x_{B,n}^{post(t)} = \mathbb{E}\left[x_n^{(t)}|\boldsymbol{x}_B^{pri(t)}, \boldsymbol{\xi}^{(t)}\right] = \int x_n^{(t)} \hat{p}(x_n^{(t)}|\boldsymbol{x}_B^{pri(t)}, \boldsymbol{\xi}^{(t)}), \tag{15.30}$$

$$v_{B,n}^{post(t)} = \mathbb{E}\left[\left|x_n^{(t)} - x_{B,n}^{post(t)}\right|^2 \middle| \boldsymbol{x}_B^{pri(t)}, \boldsymbol{\xi}^{(t)}\right]. \tag{15.31}$$

Finally, the extrinsic message back to Module A is

$$\boldsymbol{x}_A^{pri(t)} = \boldsymbol{v}_A^{pri(t)} \left(\frac{\boldsymbol{x}_B^{post(t)}}{\boldsymbol{v}_B^{post(t)}} - \frac{\boldsymbol{x}_B^{pri(t)}}{\boldsymbol{v}_B^{pri(t)}} \right), \tag{15.32}$$

$$\boldsymbol{v}_A^{pri(t)} = \left(\frac{1}{\boldsymbol{v}_B^{post(t)}} - \frac{1}{\boldsymbol{v}_B^{pri(t)}} \right)^{-1}, \tag{15.33}$$

where $\boldsymbol{x}_B^{post(t)} = [x_{B,1}^{post(t)}, \ldots, x_{B,N}^{post(t)}]^T$ and $\boldsymbol{v}_B^{post(t)} = [v_{B,1}^{post(t)}, \ldots, v_{B,N}^{post(t)}]$. This extrinsic extraction ensures Module A receives only incremental information about sparsity structure derived by Module B, excluding the observation information originally provided.

15.2.3 Temporal Message Propagation

Following convergence of the Turbo-CS algorithm at time t, we compute approximate posterior distributions for $\boldsymbol{s}^{(t)}$ and $\boldsymbol{\beta}^{(t)}$, then derive predictive messages for time $t+1$. This temporal propagation mechanism maintains continuity across time steps and enables tracking of evolving sparse signals.

15.2.3.1 Support Vector Message Propagation

Support vector propagation begins with computing the message from factor node $\phi_n^{(t)}$ to variable node $s_n^{(t)}$

$$\begin{aligned}
\nu_{\phi_n^{(t)} \to s_n^{(t)}}(s_n^{(t)}) &= \int_{x_n^{(t)}} \int_{\beta_n^{(t)}} \phi_n^{(t)}(s_n^{(t)}, \beta_n^{(t)}, x_n^{(t)}) \\
&\times \nu_{x_n^{(t)} \to \phi_n^{(t)}}(x_n^{(t)}) \nu_{\beta_n^{(t)} \to \phi_n^{(t)}}(\beta_n^{(t)}) \\
&= \pi_{in,n}^{(t)} \delta(1 - s_n^{(t)}) + (1 - \pi_{in,n}^{(t)}) \delta(s_n^{(t)}),
\end{aligned} \tag{15.34}$$

where the activation probability $\pi_{in,n}^{(t)}$ quantifies evidence from current observations:

$$\pi_{in,n}^{(t)} = \left(1 + \frac{\mathcal{CN}(0; x_{B,n}^{pri(t)}, v_{B,n}^{pri(t)})}{\mathcal{CN}(0; x_{B,n}^{pri(t)} - \mu_{out,n}^{(t)}, v_{B,n}^{pri(t)} + \Sigma_{out,n}^{(t)})} \right)^{-1}. \tag{15.35}$$

Here, $\mu_{out,n}^{(t)}$ and $\Sigma_{out,n}^{(t)}$ are the mean and variance from the effective prior $\hat{p}(\beta_n^{(t)})$, capturing historical amplitude information.

The approximate posterior for $s_n^{(t)}$, serving as the temporal message to the next time step, combines current evidence with prior knowledge:

$$\begin{aligned} q(s_n^{(t)}) &\triangleq \nu_{s_n^{(t)} \to d_n^{(t+1)}}(s_n^{(t)}) \\ &= \nu_{\phi_n^{(t)} \to s_n^{(t)}}(s_n^{(t)}) \nu_{d_n^{(t)} \to s_n^{(t)}}(s_n^{(t)}) \\ &= \hat{\pi}_n^{(t)} \delta(1 - s_n^{(t)}) + (1 - \hat{\pi}_n^{(t)}) \delta(s_n^{(t)}), \end{aligned} \tag{15.36}$$

with the combined probability given by

$$\hat{\pi}_n^{(t)} = \frac{\pi_{in,n}^{(t)} \pi_{out,n}^{(t)}}{\pi_{in,n}^{(t)} \pi_{out,n}^{(t)} + (1 - \pi_{in,n}^{(t)})(1 - \pi_{out,n}^{(t)})}, \tag{15.37}$$

where $\pi_{out,n}^{(t)} = \hat{p}(s_n^{(t)} = 1)$ is the effective prior. This Bayesian fusion optimally weights observational evidence against the temporal prior.

The predictive message for time $t + 1$ evolves this belief through the Markov transition model:

$$\begin{aligned} \hat{p}(s_n^{(t+1)}) &\triangleq \nu_{d_n^{(t+1)} \to s_n^{(t+1)}}(s_n^{(t+1)}) \\ &= \sum_{s_n^{(t)}} d_n^{(t+1)}(s_n^{(t)}, s_n^{(t+1)}) \nu_{s_n^{(t)} \to d_n^{(t+1)}}(s_n^{(t)}) \\ &= \pi_{out,n}^{(t+1)} \delta(1 - s_n^{(t+1)}) + (1 - \pi_{out,n}^{(t+1)}) \delta(s_n^{(t+1)}), \end{aligned} \tag{15.38}$$

where

$$\pi_{out,n}^{(t+1)} = \hat{\pi}_n^{(t)}(1 - \rho_{10}) + \rho_{01}(1 - \hat{\pi}_n^{(t)}). \tag{15.39}$$

For initialization at $t = 0$, we set $\pi_{out,n}^{(1)} = \lambda$ based on the expected sparsity rate.

15.2.3.2 Amplitude Vector Message Propagation

For amplitude propagation, the message from factor node $\phi_n^{(t)}$ to variable node $\beta_n^{(t)}$ integrates over signal value and support state:

$$\begin{aligned} \nu_{\phi_n^{(t)} \to \beta_n^{(t)}}(\beta_n^{(t)}) &= \int_{x_n^{(t)}} \sum_{s_n^{(t)}} \phi_n^{(t)}(s_n^{(t)}, \beta_n^{(t)}, x_n^{(t)}) \\ &\quad \times \nu_{x_n^{(t)} \to \phi_n^{(t)}}(x_n^{(t)}) \nu_{s_n^{(t)} \to \phi_n^{(t)}}(s_n^{(t)}) \\ &= \pi_{out,n}^{(t)} \mathcal{CN}(\beta_n^{(t)}; x_{B,n}^{pri(t)}, v_{B,n}^{pri(t)}) \\ &\quad + (1 - \pi_{out,n}^{(t)}) \mathcal{CN}(0; x_{B,n}^{pri(t)}, v_{B,n}^{pri(t)}). \end{aligned} \tag{15.40}$$

When $\pi_{out,n}^{(t)}$ is significantly less than 1, indicating uncertainty about path activity, the current observation may provide limited information about $\beta_n^{(t)}$. We employ a robust approximation to maintain stability:

$$\nu_{\phi_n^{(t)} \to \beta_n^{(t)}}(\beta_n^{(t)}) = \mathcal{CN}(\beta_n^{(t)}; \mu_{in,n}^{(t)}, \Sigma_{in,n}^{(t)}), \tag{15.41}$$

with parameters adapted based on activation confidence:

$$(\mu_{in,n}^{(t)}, \Sigma_{in,n}^{(t)}) = \begin{cases} (\epsilon, \frac{1}{\epsilon^2} v_{B,n}^{pri(t)}) & \pi_{out,n}^{(t)} \le T_{hr} \\ (x_{B,n}^{pri(t)}, v_{B,n}^{pri(t)}) & \pi_{out,n}^{(t)} > T_{hr} \end{cases}, \tag{15.42}$$

where T_{hr} is a threshold slightly less than 1 and ϵ a small positive constant. This approximation, validated through simulations and consistent with established practices [2], ensures robust performance under uncertainty.

The approximate posterior for $\beta_n^{(t)}$, propagating to the next time step, combines current and historical amplitude information:

$$\begin{aligned} q(\beta_n^{(t)}) &\triangleq \nu_{\beta_n^{(t)} \to f_n^{(t+1)}}(\beta_n^{(t)}) \\ &= \nu_{\phi_n^{(t)} \to \beta_n^{(t)}}(\beta_n^{(t)}) \nu_{f_n^{(t)} \to \beta_n^{(t)}}(\beta_n^{(t)}) \\ &= \mathcal{CN}(\beta_n^{(t)}; \hat{\mu}_n^{(t)}, \hat{\Sigma}_n^{(t)}), \end{aligned} \tag{15.43}$$

with parameters

$$\hat{\mu}_n^{(t)} = \frac{\mu_{in,n}^{(t)} \Sigma_{out,n}^{(t)} + \mu_{out,n}^{(t)} \Sigma_{in,n}^{(t)}}{\Sigma_{in,n}^{(t)} + \Sigma_{out,n}^{(t)}}, \tag{15.44}$$

$$\hat{\Sigma}_n^{(t)} = \frac{\Sigma_{in,n}^{(t)} \Sigma_{out,n}^{(t)}}{\Sigma_{in,n}^{(t)} + \Sigma_{out,n}^{(t)}}, \tag{15.45}$$

representing precision-weighted fusion of information.

The predictive message for time $t + 1$ incorporates temporal evolution through the state transition model:

$$\begin{aligned} \hat{p}(\beta_n^{(t+1)}) &\triangleq \nu_{f_n^{(t+1)} \to \beta_n^{(t+1)}}(\beta_n^{(t+1)}) \\ &= \int_{\beta_n^{(t)}} f_n^{(t+1)}(\beta_n^{(t)}, \beta_n^{(t+1)}) \nu_{\beta_n^{(t)} \to f_n^{(t+1)}}(\beta_n^{(t)}) \\ &= \mathcal{CN}(\beta_n^{(t+1)}; \mu_{out,n}^{(t+1)}, \Sigma_{out,n}^{(t+1)}), \end{aligned} \tag{15.46}$$

with predictive parameters

$$\mu_{out,n}^{(t+1)} = (1 - \alpha_\beta)\hat{\mu}_n^{(t)} + \alpha_\beta \zeta_\beta, \tag{15.47}$$

$$\Sigma_{out,n}^{(t+1)} = (1 - \alpha_\beta)^2 \hat{\Sigma}_n^{(t)} + \alpha_\beta^2 \gamma_\beta. \tag{15.48}$$

For initialization at $t = 0$, we set

$$(\mu_{out,n}^{(1)}, \Sigma_{out,n}^{(1)}) = \left(\zeta_\beta, \frac{\alpha_\beta \gamma_\beta}{2 - \alpha_\beta}\right), \tag{15.49}$$

based on the stationary distribution of the amplitude process.

15.2.4 DTurbo-CS-M Step

The MAP estimation problem for uncertain parameters $\boldsymbol{\xi}^{(t)}$ is addressed using the MM approach, extending the methodology of Chap. 14 to dynamic settings. We employ the EM framework to construct a tractable surrogate function approximating the challenging MAP objective $\ln \hat{p}(\boldsymbol{y}^{(t)}, \boldsymbol{\xi}^{(t)})$ around the current estimate $\dot{\boldsymbol{\xi}}^{(t)}$ as

$$u(\boldsymbol{\xi}^{(t)}; \dot{\boldsymbol{\xi}}^{(t)}) = \int q(\boldsymbol{v}^{(t)}; \dot{\boldsymbol{\xi}}^{(t)}) \ln \frac{\hat{p}(\boldsymbol{v}^{(t)}, \boldsymbol{y}^{(t)}, \boldsymbol{\xi}^{(t)})}{q(\boldsymbol{v}^{(t)}; \dot{\boldsymbol{\xi}}^{(t)})} d\boldsymbol{v}^{(t)}, \tag{15.50}$$

where $q(\boldsymbol{v}^{(t)}; \dot{\boldsymbol{\xi}}^{(t)})$ is the approximate posterior from the E-step conditioned on $\dot{\boldsymbol{\xi}}^{(t)}$, and

$$\hat{p}(\boldsymbol{v}^{(t)}, \boldsymbol{y}^{(t)}, \boldsymbol{\xi}^{(t)}) \triangleq \hat{p}(\boldsymbol{y}^{(t)}, \boldsymbol{v}^{(t)} | \boldsymbol{\xi}^{(t)}) \hat{p}(\boldsymbol{\xi}^{(t)}) \tag{15.51}$$

Algorithm 15.1 DTurbo-CS algorithm at time t

Input: $\boldsymbol{y}^{(t)}$, $\hat{\boldsymbol{\xi}}^{(t-1)}$, $\hat{p}(\boldsymbol{\beta}^{(t)})$, $\hat{p}(\boldsymbol{s}^{(t)})$, measurement matrix $\boldsymbol{A}$, noise variance $\sigma_w^{(t)}$.
Output: Approximate MMSE estimates: $\hat{\boldsymbol{x}}^{(t)}$.

Initialize $\boldsymbol{x}_A^{pri(t)}$, $\boldsymbol{v}_A^{pri(t)}$.
while not converge **do**
  E-step:
  %Module A:
  Update $\boldsymbol{x}_A^{post(t)}$, $\boldsymbol{V}_A^{post(t)}$,$\boldsymbol{x}_B^{pri(t)}$,$\boldsymbol{v}_B^{pri(t)}$ using (15.21), (15.20), and (15.22), (15.23).
  %Module B:
  Perform message passing over graph $\mathcal{G}_B$.
  Update $x_{B,n}^{post(t)}$, $v_{B,n}^{post(t)}$, $\boldsymbol{x}_A^{pri(t)}$, and $\boldsymbol{v}_A^{pri(t)}$using (15.30), (15.31), and (15.32), (15.33).
  Repeat Module A and Module B until convergence.
  M-step
  Construct surrogate function in (15.50).
  Update $\boldsymbol{\xi}^{(t)}$ using the inexact MM method.
end while
Pass the messages $\hat{p}(\boldsymbol{\beta}^{(t+1)})$ and $\hat{p}(\boldsymbol{s}^{(t+1)})$, as given in (15.38) and (15.46), to time $t+1$.

constitutes the effective joint probability density incorporating both likelihood and effective prior. The surrogate function lower bounds the original MAP objective while matching its value at the current estimate, ensuring that maximizing the surrogate guarantees improvement in the original objective. We then apply the inexact MM method detailed in Chap. 14, systematically updating $\boldsymbol{\xi}^{(t)}$ through block-wise optimization alternating between closed-form solutions and gradient-based updates tailored to each parameter block's structure.

Upon convergence of the iterative process between E-step and M-step, the algorithm obtains updated parameter estimates $\hat{\boldsymbol{\xi}}^{(t)}$ reflecting current understanding based on latest observations and historical information. Simultaneously, predictive messages $\hat{p}(s_n^{(t+1)})$ and $\hat{p}(\beta_n^{(t+1)})$ are computed following the derivations in Sect. 15.2.3, encapsulating evolved beliefs about support patterns and signal amplitudes. These predictive messages propagate to time $t+1$, serving as informed priors that initialize the next estimation cycle.

The complete DTurbo-CS algorithm at time t, summarized in Algorithm 15.1, provides a computationally efficient framework for dynamic sparse Bayesian inference that jointly estimates sparse signals and system parameters while maintaining temporal consistency through systematic message passing.

15.3 Application Example: Channel Tracking in 5G NR Systems

Massive MIMO-OFDM technology, a cornerstone of 5G NR systems, faces significant challenges in acquiring and maintaining channel state information under practical operating conditions. In TDD massive MIMO systems, the BS estimates uplink channels using sounding reference signals (SRS) and leverages channel reciprocity for downlink precoding. While TDD systems substantially reduce pilot overhead compared to FDD counterparts, several practical impairments complicate channel tracking in real deployment scenarios.

A primary challenge arises from channel aging effects caused by hopping SRS patterns, where SRS transmission occurs within bandwidth parts (BWP) that hop across different frequency ranges. Extended intervals between SRS transmissions in the same BWP exacerbate Doppler-induced phase rotation, rendering conventional approaches that assume quasi-static channels increasingly inadequate for high-mobility scenarios. Compounding these challenges are various system imperfection effects including residual time offset introducing cyclic shifts in delay-domain channel representations, and phase noise causing random phase variations that undermine temporal correlation. Additionally, cell-edge users experience low-SNR conditions due to limited mobile transmit power, creating a critical need for robust algorithms that maintain reliable coverage extension while operating near system sensitivity limits.

The dynamic sparse Bayesian inference methodology introduced in this chapter offers a unified framework simultaneously addressing these multifaceted challenges. This approach integrates robust temporal tracking through exploitation of structured time evolution of sparse massive MIMO channels, imperfection resilience via joint estimation of channel parameters and system impairments within an EM framework, and reliable low-SNR operation by leveraging the inherent noise immunity of

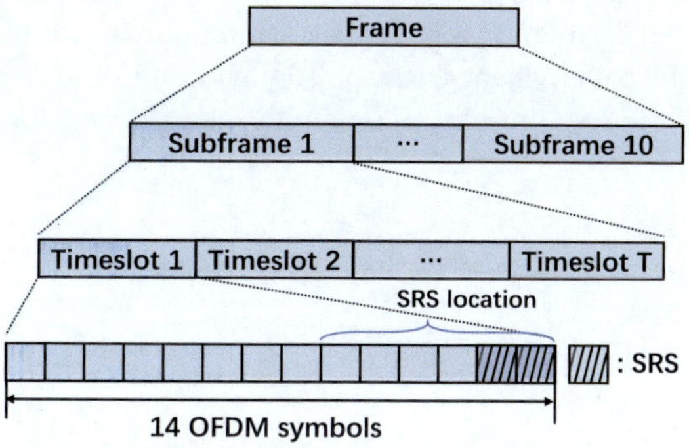

Fig. 15.2 An illustration of the frame structure

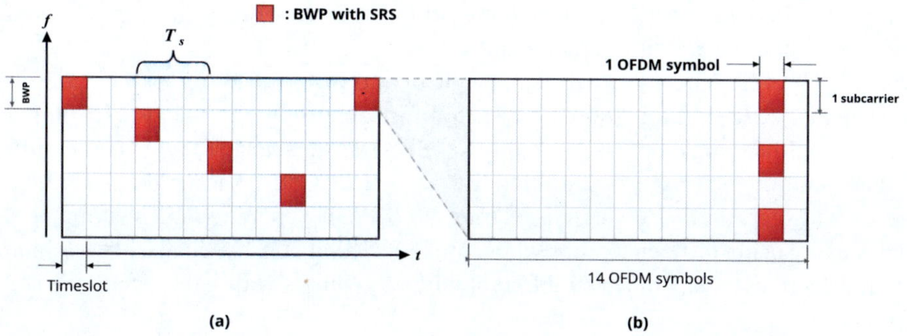

Fig. 15.3 An illustration of the hopping SRS pattern (adapted from Fig. 1 in [18])

Bayesian methods. Building upon our previous work in [18], the channel tracking scheme developed herein demonstrates how the dynamic Turbo-CS algorithm can be specifically adapted to handle the distinctive imperfections encountered in 5G NR systems. This practical implementation exemplifies the utility of our theoretical framework for solving real-world wireless communication problems, particularly under the challenging operational conditions of modern massive MIMO deployments where multiple impairment sources interact in ways that conventional methods cannot adequately address.

15.3.1 Frame Structure and SRS Pattern

Consider a TDD massive MIMO-OFDM system comprising one base station and multiple user equipments. The BS employs a uniform planar array with $N_r = N_x \times N_y$ antenna elements, where N_x and N_y denote antenna counts in horizontal and vertical dimensions, respectively. Each user equipment is equipped with N_t antennas configured as a uniform linear array.

The uplink frame structure, illustrated in Fig. 15.2, follows a systematic organization to support efficient communication. In the time domain, each frame contains 10 subframes, while in the frequency domain, the system bandwidth comprises N_{RB} resource blocks (RB) with each RB consisting of 12 subcarriers. Each subframe contains multiple timeslots, and each timeslot comprises 14 OFDM symbols carrying both data and reference signals.

To optimize the trade-off between pilot overhead and spectrum efficiency, SRS transmission follows a carefully designed pattern. SRS symbols are periodically transmitted in the last six OFDM symbols of selected timeslots with period T_S, ensuring regular channel sampling while minimizing interference with data transmission. The symbol configuration adapts based on active user load, with each SRS occasion utilizing 1, 2, or 4 consecutive OFDM symbols to balance estimation accuracy and resource utilization. SRS resources are efficiently shared among multiple users through appropriate frequency-domain and code-domain multiplexing schemes.

The SRS employs a frequency-hopping pattern to enhance power concentration and extend coverage, particularly beneficial for cell-edge users. As shown in Fig. 15.3a, the pattern completes h_p (e.g., $h_p = 4$) hops across the frequency band, allowing transmitted power concentration within narrower bandwidth parts compared to full-band transmission, significantly improving received signal strength at the BS. Within each BWP, SRS transmission follows a comb structure with symbols placed every N_c subcarriers, creating a sparse sampling pattern in the frequency domain. This discussion focuses on comb-2 mode ($N_c = 2$) with single OFDM symbol SRS transmission per timeslot, a commonly adopted configuration in practical 5G NR deployments.

15.3.2 Channel Sounding Reference Signal in 5G NR Systems

SRSs in 5G NR Release 16 are generated using cyclic-shifted Zadoff–Chu (ZC) sequences [10] as

$$r_{u,v}^{\alpha}(n) = e^{j\alpha n}\bar{r}_{u,v}(n), \quad n = 0, \ldots, P-1, \tag{15.52}$$

where $\bar{r}_{u,v}(n)$ is the base ZC sequence of length P, $u = 0, \ldots, 29$ denotes the base sequence group number, $v \in \{0, 1\}$ represents the sequence number within the group, and α is the cyclic shift factor creating orthogonal sequence variants. This structured design allows efficient generation of multiple orthogonal reference signals from a common base sequence, conserving system resources while maintaining signal quality.

The number of available base sequences per cell is strategically constrained by cell radius considerations to minimize inter-cell interference, making cyclic shift multiplexing essential for supporting multiple users within the same cell. 5G NR supports several orthogonal multiplexing methods for multi-user SRS transmission operating across different domains: code-domain multiplexing where multiple users employ different ZC sequences while sharing identical time-frequency resources; frequency-domain multiplexing across BWPs where users transmit SRSs in separate BWPs within the same OFDM symbol; frequency-domain multiplexing using different combs where users occupy distinct subcarrier sets within the same BWP; and time-domain multiplexing where users are allocated different OFDM symbols within the same time slot and BWP.

15.3.3 Physical Channel Model

In practical deployments, multiple user antennas are typically treated as virtual single-antenna users and assigned orthogonal SRS resources. Since channel tracking for users on orthogonal combs or bandwidth parts proceeds independently, we focus on the channel tracking problem for a single user antenna without loss of generality.

The channel frequency response at the n-th subcarrier during the t-th SRS symbol period is

$$\boldsymbol{h}^{(t)}[n] = e^{j\varepsilon^{(t)}} \sum_{l=1}^{L^{(t)}} x_l^{(t)} e^{j\varphi_l^{(t)}} e^{-j2\pi n f_s(\tau_l^{(t)}+\tau_0^{(t)})} \boldsymbol{a}_R(\theta_l^{(t)}, \phi_l^{(t)}), \tag{15.53}$$

where $L^{(t)}$ represents the number of propagation paths, f_s denotes subcarrier spacing, $x_l^{(t)}$ is complex path gain for the l-th path, $\theta_l^{(t)}$ and $\phi_l^{(t)}$ are azimuth and elevation angles of arrival, respectively, and $\tau_l^{(t)}$ is time delay of the l-th path. The model incorporates practical impairment factors: $\varphi_l^{(t)}$ characterizes Doppler phase rotation induced by user mobility, $\varepsilon^{(t)}$ represents random phase noise, and $\tau_0^{(t)}$ accounts for timing offset. For initialization, we set $\varepsilon^{(1)} = 0$, $\tau_0^{(1)} = 0$, and $\varphi_l^{(1)} = 0$ for all paths l.

Doppler phase evolution follows $\varphi_l^{(t)} = \varphi_l^{(t-1)} + 2\pi T_{SRS} f_{D,l}^{(t)}$, where $f_{D,l}^{(t)}$ denotes Doppler frequency shift associated with the l-th path at time t and T_{SRS} is the SRS symbol interval. For low-speed users with gradual acceleration, $f_{D,l}^{(t)}$ exhibits strong temporal correlation exploitable in tracking algorithms.

The BS's UPA response vector is

$$\boldsymbol{a}_R(\theta, \phi) = \boldsymbol{a}_x(\theta, \phi) \otimes \boldsymbol{a}_y(\phi) \in \mathbb{C}^{N_r \times 1}, \tag{15.54}$$

which, under half-wavelength spacing, takes the form:

$$[\boldsymbol{a}_x(\theta, \phi)]_{n_x} = \frac{1}{\sqrt{N_x}} e^{j\pi(n_x-1)\sin(\theta)\cos(\phi)}, \tag{15.55}$$

$$\left[\boldsymbol{a}_y(\phi)\right]_{n_y} = \frac{1}{\sqrt{N_y}} e^{j\pi(n_y-1)\sin(\phi)}, \tag{15.56}$$

for $n_x = 0, \ldots, N_x - 1$ and $n_y = 0, \ldots, N_y - 1$.

Within a bandwidth part containing P subcarriers, the received SRS signal model becomes

$$\boldsymbol{Y}^{(t)} = \mathrm{diag}(\boldsymbol{u})\boldsymbol{W}^{(t)}\boldsymbol{H}^{(t)} + \boldsymbol{N}^{(t)}, \tag{15.57}$$

where $\boldsymbol{u} \in \mathbb{C}^{P\times 1}$ is the transmitted frequency-domain SRS sequence, $\boldsymbol{W}^{(t)} \in \{0, 1\}^{P\times N}$ is a selection matrix determined by comb structure,

$$\boldsymbol{H}^{(t)} \triangleq [(\boldsymbol{h}^{(t)}[0])^T; (\boldsymbol{h}^{(t)}[1])^T; \ldots; (\boldsymbol{h}^{(t)}[N-1])^T] \in \mathbb{C}^{N\times N_r}$$

denotes the complete CFR matrix, and $\boldsymbol{N}^{(t)}$ is AWGN with element variance $\sigma_e^{(t)}$. For comb-2 mode in the first BWP, $\boldsymbol{W}^{(1)}(i, j) = 1$ for $j = 2i - 1, i = 1, \ldots, P$.

This formulation frames channel tracking as an extrapolation task, where the complete CFR across all bandwidth parts must be accurately estimated from limited observations within a single BWP.

15.3.4 Sparse Channel Representation

To develop a sparse representation in delay and angular domains, we define discrete grids over parameter spaces capturing essential dimensions of the wireless propagation environment. The delay grid $\mathcal{D} = \{\bar{\tau}_1, \ldots, \bar{\tau}_L\}$ contains $L \gg L^{(t)}$ points spanning $[-\frac{1}{4}T_d, T_d]$, where T_d represents maximum delay spread bound. Angular grids consist of $\mathcal{G}_\theta = \{\bar{\theta}_1, \ldots, \bar{\theta}_{N_x}\}$ with N_x azimuth points over $[-\frac{\pi}{2}, \frac{\pi}{2}]$ and $\mathcal{G}_\phi = \{\bar{\phi}_1, \ldots, \bar{\phi}_{N_y}\}$ with N_y elevation points similarly covering $[-\frac{\pi}{2}, \frac{\pi}{2}]$.

Under the assumption that true channel parameters align perfectly with these grids, the received signal model transforms into a structured sparse representation:

$$\boldsymbol{Y}^{(t)} = \mathrm{diag}(\boldsymbol{u})\boldsymbol{W}^{(t)}\boldsymbol{\Lambda}^{(t)}\boldsymbol{S}^{(t)}\boldsymbol{F}_d(\boldsymbol{H}_d^{(t)} \circ \boldsymbol{D}^{(t)})\boldsymbol{A}_R^T + \boldsymbol{N}^{(t)}, \tag{15.58}$$

where $\circ$ denotes Hadamard product. The constituent matrices play distinct roles in capturing different aspects of the channel behavior:

- **Phase noise matrix**: $\boldsymbol{\Lambda}^{(t)} = \mathrm{diag}(e^{j\varepsilon^{(t)}}, \ldots, e^{j\varepsilon^{(t)}})$ encapsulates the common phase noise affecting all subcarriers.
- **Timing offset matrix**: $\boldsymbol{S}^{(t)} = \mathrm{diag}(1, \ldots, e^{-j2\pi n f_s \tau_0^{(t)}}, \ldots, e^{-j2\pi(N-1)f_s\tau_0^{(t)}})$ models the linear phase progression caused by timing offset across the frequency domain.
- **Delay dictionary**: $\boldsymbol{F}_d = [\boldsymbol{a}(\bar{\tau}_1), \boldsymbol{a}(\bar{\tau}_2), \ldots, \boldsymbol{a}(\bar{\tau}_L)]$ contains steering vectors evaluated on the delay grid $\mathcal{D}$.
- **Angular dictionary**: $\boldsymbol{A}_R = [\boldsymbol{a}_R(\bar{\theta}_1, \bar{\phi}_1), \ldots, \boldsymbol{a}_R(\bar{\theta}_{N_x}, \bar{\phi}_{N_y})]$ contains steering vectors evaluated on the angle grids $\mathcal{G}_\theta$ and $\mathcal{G}_\phi$.

Together, $\boldsymbol{F}_d$ and $\boldsymbol{A}_R$ form the foundational basis for the sparse representation.

The sparse delay-angular domain channel matrix $\boldsymbol{H}_d^{(t)} \in \mathbb{C}^{L\times N_r}$ contains non-zero entries corresponding exclusively to active propagation paths, capturing inherent multipath sparsity. The Doppler phase rotation matrix $\boldsymbol{D}^{(t)} \in \mathbb{C}^{L\times N_r}$ with elements $e^{j\varphi_{l,n}^{(t)}}$ represents complex phase shifts from user mobility, where $\varphi_{l,n}^{(t)}$ denotes the Doppler rotation for the path corresponding to the (l, n)-th entry of $\boldsymbol{H}_d^{(t)}$.

To overcome resolution limitations of fixed grids, we introduce dynamic grid parameters adapting to true channel conditions: dynamic delay grid vector $\boldsymbol{\tau}^{(t)} = [\tau_1^{(t)}, \ldots, \tau_L^{(t)}]$, dynamic azimuth angle vector $\boldsymbol{\theta}^{(t)} = [\theta_1^{(t)}, \ldots, \theta_{N_x}^{(t)}]$, and dynamic elevation angle vector $\boldsymbol{\phi}^{(t)} = [\phi_1^{(t)}, \ldots, \phi_{N_y}^{(t)}]$. Collectively denoting $\boldsymbol{\Delta}^{(t)} \triangleq [\boldsymbol{\tau}^{(t)}, \boldsymbol{\theta}^{(t)}, \boldsymbol{\phi}^{(t)}]$ as the off-grid parameter vector, dictionary matrices become adaptive

$$\boldsymbol{F}_d(\boldsymbol{\tau}^{(t)}) = [\boldsymbol{a}(\tau_1^{(t)}), \ldots, \boldsymbol{a}(\tau_L^{(t)})], \tag{15.59}$$

$$\boldsymbol{A}_R(\boldsymbol{\theta}^{(t)}, \boldsymbol{\phi}^{(t)}) = [\boldsymbol{a}_R(\theta_1^{(t)}, \phi_1^{(t)}), \ldots, \boldsymbol{a}_R(\theta_{N_x}^{(t)}, \phi_{N_y}^{(t)})]. \tag{15.60}$$

This dynamic grid formulation enables super-resolution channel estimation by continuously adapting dictionaries to match true underlying channel parameters.

15.3.5 Markov Channel Model for Dynamic Sparsity

To capture temporal evolution of key channel parameters in dynamic environments, we propose a Markov channel model systematically representing time-varying propagation characteristics. The model centers on three fundamental parameter sets: the delay-angular domain channel vector $\boldsymbol{h}_d^{(t)} \triangleq \text{vec}(\boldsymbol{H}_d^{(t)})$ representing sparse DAD channel coefficients; the Doppler frequency offset vector $\boldsymbol{f}_D^{(t)} \in \mathbb{R}^{LN_r}$ with elements $f_{D,m}^{(t)}$ denoting Doppler shift for each DAD coefficient; and off-grid parameters $\boldsymbol{\Delta}^{(t)}$ encompassing dynamic delay and angle grid vectors.

This Markov channel model follows the time-evolving BG Markov structure introduced in Sect. 15.1.1, now adapted for channel tracking in massive MIMO-OFDM systems. The model comprehensively captures three critical aspects: evolution of dynamic sparsity patterns as active paths appear, disappear, or vary in strength; smooth temporal variation of off-grid parameters tracking continuous movement of scatterers and user equipment; and temporal correlation of Doppler frequency offsets reflecting physical constraints of user mobility and acceleration.

15.3.5.1 Probability Model for DAD Channel $h_d^{(1:T)}$

The BG probability model captures temporal correlation and sparsity:

$$\begin{aligned} p(\boldsymbol{h}_d^{(1:T)}|\boldsymbol{\beta}^{(1:T)}, \boldsymbol{s}^{(1:T)}) &= \prod_{t=1}^{T}\prod_{m=1}^{LN_r} p(h_{d,m}^{(t)}|s_m^{(t)}, \beta_m^{(t)}) \\ &= \prod_{t=1}^{T}\prod_{m=1}^{LN_r} \delta(h_{d,m}^{(t)} - s_m^{(t)}\beta_m^{(t)}), \end{aligned} \tag{15.61}$$

where support vector $\boldsymbol{s}^{(t)} = [s_1^{(t)}, \ldots, s_{LN_r}^{(t)}]^T$ describes multipath birth-death process, and $\boldsymbol{\beta}^{(t)} = [\beta_1^{(t)}, \ldots, \beta_{LN_r}^{(t)}]^T$ describes smooth amplitude evolution. The sparse Markov channel prior distribution is

$$\begin{aligned} &p(\boldsymbol{h}_d^{(1:T)}, \boldsymbol{\beta}^{(1:T)}, \boldsymbol{s}^{(1:T)}, \boldsymbol{f}_D^{(1:T)}, \boldsymbol{\Delta}^{(1:T)}) \\ &= p(\boldsymbol{f}_D^{(1:T)})p(\boldsymbol{\tau}^{(1:T)})p(\boldsymbol{\phi}^{(1:T)})p(\boldsymbol{\theta}^{(1:T)}) \\ &\quad \times p(\boldsymbol{s}^{(1:T)})p(\boldsymbol{\beta}^{(1:T)})p(\boldsymbol{h}_d^{(1:T)}|\boldsymbol{\beta}^{(1:T)}, \boldsymbol{s}^{(1:T)}). \end{aligned} \tag{15.62}$$

15.3.5.2 Probability Model for Channel Support $s^{(1:T)}$

Channel supports vary slowly over time. We use a Markov chain to model temporal correlation

$$p(\boldsymbol{s}^{(1:T)}) = \prod_{m=1}^{LN_r} p(s_m^{(1)}) \prod_{t=2}^{T} p(s_m^{(t)}|s_m^{(t-1)}), \tag{15.63}$$

with transition probabilities

$$\begin{aligned} p(s_m^{(t)} = 1|s_m^{(t-1)} = 0) = \rho_{01}, \\ p(s_m^{(t)} = 0|s_m^{(t-1)} = 1) = \rho_{10}. \end{aligned} \tag{15.64}$$

Smaller ρ_{10} or ρ_{01} indicates slower support variation.

15.3.5.3 Probability Model for Hidden Variable $\beta^{(1:T)}$

Path gain amplitudes evolve smoothly, modeled by Gauss–Markov processes [2, 3]

$$\beta_m^{(t)} = (1-\alpha_\beta)(\beta_m^{(t-1)} - \zeta_\beta) + \alpha_\beta \omega_{\beta,m}^{(t)} + \zeta_\beta, \quad m = 1, \ldots, LN_r, \tag{15.65}$$

where $\alpha_\beta \in [0, 1]$ controls temporal correlation, ζ_β is steady-state mean, and $\omega_{\beta,m}^{(t)} \sim \mathcal{CN}(0, \gamma_\beta)$ is i.i.d. circular white Gaussian perturbation. The joint distribution is

$$p(\boldsymbol{\beta}^{(1:T)}) = \prod_{m=1}^{LN_r} p(\beta_m^{(1)}) \prod_{t=2}^{T} p(\beta_m^{(t)}|\beta_m^{(t-1)}), \tag{15.66}$$

with $p(\beta_m^{(t)}|\beta_m^{(t-1)}) \sim \mathcal{CN}(\beta_m^{(t)}; (1-\alpha_\beta)\beta_m^{(t-1)} + \alpha_\beta\zeta_\beta, \alpha_\beta^2\gamma_\beta)$.

15.3.5.4 Probability Model for Doppler Frequency Offset $\boldsymbol{f}_D^{(1:T)}$

The probability model is

$$p(\boldsymbol{f}_D^{(1:T)}) = p(\boldsymbol{f}_D^{(1)}) \prod_{t=2}^{T} p(\boldsymbol{f}_D^{(t)}|\boldsymbol{f}_D^{(t-1)}), \tag{15.67}$$

with transition probability captured by Gauss–Markov processes

$$f_{D,m}^{(t)} = (1-\alpha_D)(f_{D,m}^{(t-1)} - \zeta_D) + \alpha_D \omega_{D,m}^{(t)} + \zeta_D, \quad m = 1, \ldots, LN_r, \tag{15.68}$$

$$p(\boldsymbol{f}_D^{(t)}|\boldsymbol{f}_D^{(t-1)}) = \prod_{m=1}^{LN_r} \mathcal{N}\left(f_{D,m}^{(t)}; (1-\alpha_D) f_{D,m}^{(t-1)} + \alpha_D\zeta_D, \alpha_D^2\gamma_D\right), \tag{15.69}$$

where α_D, ζ_D, $\omega_{D,m}^{(t)}$, and γ_D have similar definitions to the amplitude process parameters.

15.3.5.5 Probability Model for Off-Grid Vectors $\boldsymbol{\Delta}^{(T)}$

Delay and angle parameters vary smoothly over time, modeled as random walks [17]

$$\boldsymbol{\tau}^{(t)} = \boldsymbol{\tau}^{(t-1)} + \boldsymbol{u}_\tau^{(t)}, \tag{15.70}$$

$$\boldsymbol{\theta}^{(t)} = \boldsymbol{\theta}^{(t-1)} + \boldsymbol{u}_\theta^{(t)}, \tag{15.71}$$

$$\boldsymbol{\phi}^{(t)} = \boldsymbol{\phi}^{(t-1)} + \boldsymbol{u}_\phi^{(t)}, \tag{15.72}$$

where $\boldsymbol{u}_\tau^{(t)} \sim \mathcal{CN}(0, \gamma_\tau \boldsymbol{I})$, $\boldsymbol{u}_\theta^{(t)} \sim \mathcal{CN}(0, \gamma_\theta \boldsymbol{I})$, and $\boldsymbol{u}_\phi^{(t)} \sim \mathcal{CN}(0, \gamma_\phi \boldsymbol{I})$. The joint distributions factorize as

$$p(\boldsymbol{\tau}^{(T)}) = \prod_{m=1}^{L} p(\tau_m^{(1)}) \prod_{t=2}^{T} p(\tau_m^{(t)}|\tau_m^{(t-1)}), \tag{15.73}$$

$$p(\boldsymbol{\theta}^{(T)}) = \prod_{m=1}^{N_x} p(\theta_m^{(1)}) \prod_{t=2}^{T} p(\theta_m^{(t)}|\theta_m^{(t-1)}), \tag{15.74}$$

$$p(\boldsymbol{\phi}^{(T)}) = \prod_{m=1}^{N_y} p(\phi_m^{(1)}) \prod_{t=2}^{T} p(\phi_m^{(t)}|\phi_m^{(t-1)}), \tag{15.75}$$

with $p(\tau_m^{(t)}|\tau_m^{(t-1)}) \sim \mathcal{N}(\tau_m^{(t)}; \tau_m^{(t-1)}, \gamma_\tau)$, and similar distributions for angular parameters.

15.3.5.6 Adaptive Parameter Learning and Practical Advantages

The proposed Markov channel model features autonomous parameter adaptation for robust performance across diverse environments. It automatically adjusts transition probabilities ρ_{01} and ρ_{10} to match channel support variation rates and captures path amplitude changes through dynamic evolution of amplitude statistics governed by α_β, ζ_β, and γ_β. All parameters are learned directly from data via EM estimation, eliminating manual tuning.

The framework demonstrates intelligent speed adaptation: increasing ρ_{01} for high-mobility scenarios with rapid channel changes, while maintaining smaller transition probabilities for low-mobility environments with stronger temporal correlation. This combination of structured temporal modeling and data-driven learning creates an ideal framework for 5G NR systems in dynamic mobile environments, ensuring consistent performance while eliminating manual optimization burdens.

15.3.6 Outline of the Channel Tracking Algorithm

The channel tracking problem is formulated within a dynamic Bayesian inference framework by reformulating the received signal model as a standard CS problem:

$$\boldsymbol{y}^{(t)} = \boldsymbol{\Phi}^{(t)} \boldsymbol{h}_d^{(t)} + \boldsymbol{n}^{(t)}, \tag{15.76}$$

where the sensing matrix

$$\boldsymbol{\Phi}^{(t)} = [\boldsymbol{A}_R \otimes (\mathrm{diag}(\boldsymbol{x}) \boldsymbol{W}^{(t)} \boldsymbol{\Lambda}^{(t)} \boldsymbol{S}^{(t)} \boldsymbol{F}_d)] \boldsymbol{D}^{(t)} \in \mathbb{C}^{N_r P \times N_r L} \tag{15.77}$$

encapsulates combined effects of angular dictionary, comb structure, phase noise, timing offset, delay dictionary, and Doppler rotations. At each SRS symbol interval t, the algorithm simultaneously tracks both the time-varying DAD channel vector $\boldsymbol{h}_d^{(t)}$ and system parameters $\boldsymbol{\xi}^{(t)} \triangleq \{\boldsymbol{\Delta}^{(t)}, \boldsymbol{f}_D^{(t)}, \varepsilon^{(t)}, \tau_0^{(t)}\}$, representing a specific instance of the dynamic Bayesian inference problem that DTurbo-CS solves.

The inference process builds upon an effective joint probability model integrating temporal information from previous time steps:

$$\begin{aligned} \hat{p}(\boldsymbol{y}^{(t)}, \boldsymbol{v}^{(t)}, \boldsymbol{\xi}^{(t)}) \triangleq\ & \hat{p}(\boldsymbol{s}^{(t)}) \hat{p}(\boldsymbol{\beta}^{(t)}) p(\boldsymbol{h}_d^{(t)} | \boldsymbol{\beta}^{(t)}, \boldsymbol{s}^{(t)}) \\ & \times p(\boldsymbol{y}^{(t)} | \boldsymbol{h}_d^{(t)}, \boldsymbol{\xi}^{(t)}) \hat{p}(\boldsymbol{\xi}^{(t)}), \end{aligned} \tag{15.78}$$

where $\boldsymbol{v}^{(t)} \triangleq \{\boldsymbol{s}^{(t)}, \boldsymbol{\beta}^{(t)}, \boldsymbol{h}_d^{(t)}\}$. Temporal evolution of key parameters within $\hat{p}(\boldsymbol{\xi}^{(t)})$ follows physically motivated random walk models:

$$\hat{p}(\tau_m^{(t)}) \sim \mathcal{N}(\tau_m^{(t)}; \hat{\tau}_m^{(t-1)}, \gamma_\tau), \tag{15.79}$$

$$\hat{p}(f_{D,m}^{(t)}) \sim \mathcal{N}(f_{D,m}^{(t)}; (1-\alpha_D) \hat{f}_{D,m}^{(t-1)} + \alpha_D \zeta_D, \alpha_D^2 \gamma_D), \tag{15.80}$$

with similar Gaussian random walk models for angular parameters $\hat{p}(\theta_m^{(t)})$ and $\hat{p}(\phi_m^{(t)})$.

Based on this effective joint density, the channel tracking algorithm implements an iterative estimation framework alternating between expectation and maximization steps. The E-step, given messages $\hat{p}(\boldsymbol{s}^{(t)})$, $\hat{p}(\boldsymbol{\beta}^{(t)})$ from time $t-1$ and estimated parameter $\hat{\boldsymbol{\xi}}^{(t)}$ from the M-step, calculates approximate posterior $q(\boldsymbol{v}^{(t)}) \triangleq q(\boldsymbol{\beta}^{(t)}) q(\boldsymbol{s}^{(t)})$ using the effective joint PDF

$$\hat{p}(\boldsymbol{y}^{(t)}, \boldsymbol{v}^{(t)} | \boldsymbol{\xi}^{(t)} = \hat{\boldsymbol{\xi}}^{(t)}) = \hat{p}(\boldsymbol{s}^{(t)}) \hat{p}(\boldsymbol{\beta}^{(t)}) p(\boldsymbol{h}_d^{(t)} | \boldsymbol{\beta}^{(t)}, \boldsymbol{s}^{(t)}) p(\boldsymbol{y}^{(t)} | \boldsymbol{h}_d^{(t)}, \hat{\boldsymbol{\xi}}^{(t)}), \tag{15.81}$$

employing the Turbo-CS algorithm. The M-step, given $q(\boldsymbol{v}^{(t)})$, constructs a surrogate function for the MAP objective $\ln \hat{p}(\boldsymbol{y}^{(t)}, \boldsymbol{\xi}^{(t)})$, maximizes it, and obtains optimal solution $\hat{\boldsymbol{\xi}}^{(t)}$.

Upon convergence, the algorithm propagates essential information to time $t+1$ through predictive distributions:

$$\hat{p}(\boldsymbol{s}^{(t+1)}) \triangleq \sum_{\boldsymbol{s}^{(t)}} q(\boldsymbol{s}^{(t)}) p(\boldsymbol{s}^{(t+1)} | \boldsymbol{s}^{(t)}), \tag{15.82}$$

$$\hat{p}(\boldsymbol{\beta}^{(t+1)}) \triangleq \int_{\boldsymbol{\beta}^{(t)}} q(\boldsymbol{\beta}^{(t)}) p(\boldsymbol{\beta}^{(t+1)} | \boldsymbol{\beta}^{(t)}). \tag{15.83}$$

This comprehensive iterative EM framework enables robust channel tracking in 5G NR systems by jointly exploiting inherent spatial sparsity of massive MIMO channels and their temporal correlation across successive sounding intervals, while simultaneously adapting to practical system impairments. In addition, R-TST-MUSIC initialization can be employed

Table 15.2 System parameters

System parameter	Value
Carrier frequency	3.5 GHz
Bandwidth B	60 MHz
Subcarrier spacing f_s	60 kHz
BS antenna array $N_r = N_x \times N_y$	$64 = 8 \times 8$ (UPA)
User mobility	3 km/h (up to 20 km/h)
Number of delay grid points L	26
SNR	15 dB

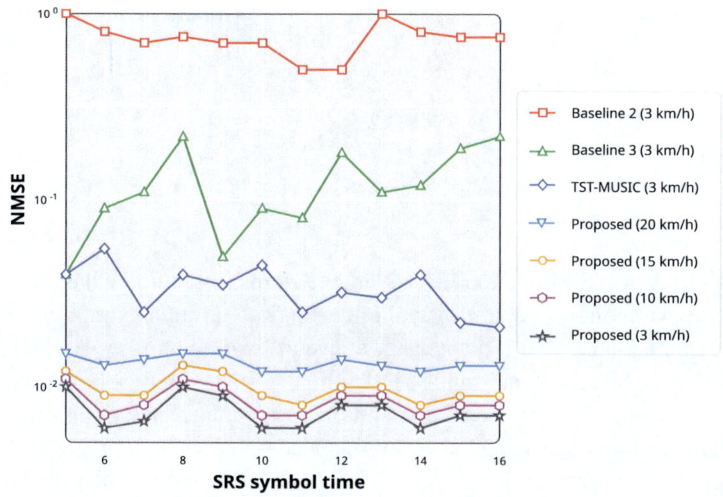

Fig. 15.4 NMSE of the full-band channel versus time for $h_p = 4$ (adapted from Fig. 11 in [18])

to accelerate the convergence speed. R-TST-MUSIC extends the TST-MUSIC algorithm from single-band/single-timeslot to multi-band and multi-timeslot scenarios, enabling coherent utilization of observations across different frequency bands and time slots. This provides a more accurate initial estimate, facilitating faster convergence of the DTurbo-CS algorithm. Details can be found in [18].

15.3.7 Simulation Results

We evaluate channel estimation performance through comprehensive numerical simulations, with key system configuration parameters summarized in Table 15.2. The experimental setup employs two SRS configurations: for a hopping pattern of $h_p = 2$, the bandwidth part spans 30 MHz with SRS sequence length $P = 250$; for $h_p = 4$, BWP bandwidth reduces to 15 MHz with $P = 125$. Channel realizations are generated using the QuaDRiGa toolbox [19] following the 3GPP R16 3D-UMa NLoS model, with performance averaged over 500 noise realizations and quantified using normalized mean square error.

To establish performance benchmarks, we compare against three baseline algorithms.

- **Baseline 1**: TST-MUSIC algorithm performing independent channel extrapolation at each timeslot [21].
- **Baseline 2**: OAMP-based channel tracking with time-domain extrapolation, estimating each BWP independently [22].
- **Baseline 3**: Proposed DTurbo-CS algorithm without imperfection compensation and off-grid updates (R-TST-MUSIC initialization is also employed to accelerate the convergence speed).

Figure 15.4 presents NMSE performance of full-band channel estimation over time across different algorithms. Results demonstrate that extrapolation-based schemes, specifically the proposed method and Baseline 3, achieve superior performance

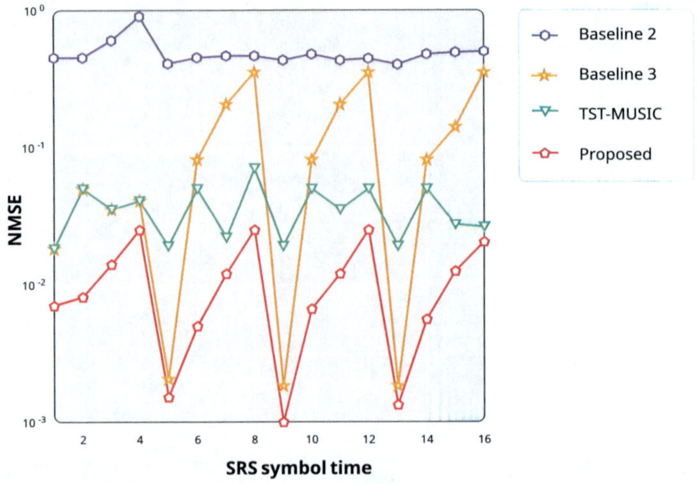

Fig. 15.5 NMSE of the first BWP channel versus time for $h_p = 4$ (adapted from Fig. 12 in [18])

compared to traditional channel tracking (Baseline 2). This advantage stems from two fundamental limitations of Baseline 2: It lacks genuine frequency-domain extrapolation capability and relies on time-domain extrapolation introducing approximation errors, while independent processing of each BWP prevents effective utilization of observational information across different bandwidth parts. Furthermore, the proposed scheme significantly outperforms both Baseline 3 and TST-MUSIC, underscoring the critical importance of imperfection factor compensation and off-grid channel modeling for accurate channel estimation.

We further investigate tracking performance under varying user speeds, also shown in Fig. 15.4. While TST-MUSIC exhibits minimal performance variation due to disregarding temporal channel correlation, other schemes experience performance degradation with increasing user velocity. Baseline performance is shown only at 3 km/h for clarity. Although NMSE of the proposed scheme increases with user speed, it maintains stable tracking behavior even at 20 km/h, delivering significant performance advantages even over baselines operating at the lower speed of 3 km/h.

Figure 15.5 illustrates NMSE performance for the first BWP channel across SRS symbol times, where the proposed scheme consistently surpasses all baselines. Both the proposed method and Baseline 3 exhibit periodic oscillations with period h_p, corresponding to frequency-hopping pattern of SRS across four BWPs. When SRS symbols are frequency-misaligned with the first BWP, channel extrapolation becomes necessary, leading to progressively degraded estimation accuracy with increasing extrapolation distance, evident between the fifth and eighth SRS symbols. Realignment at the ninth symbol abruptly reduces extrapolation distance, resulting in sharp NMSE improvement. Notably, Baseline 3 displays the most pronounced oscillations, revealing that uncompensated system impairments severely degrade extrapolation performance and validating the essential role of imperfection compensation in the proposed method.

Summary

This chapter addressed dynamic sparse Bayesian inference for recovering time-varying sparse signal sequences by proposing a time-evolving BG Markov model and developing a DTurbo-CS algorithm. The framework comprehensively exploits dynamic sparsity, encompassing both instantaneous sparsity at each time step and temporal correlation of support sets and non-zero coefficients across consecutive slots. Within this architecture, the cumulative effect of all past observations is encapsulated in messages propagated from preceding time slot, enabling decomposition of the dynamic inference problem into a series of independent sparse Bayesian inference subproblems. Each subproblem, characterized by a two-layer BG prior incorporating temporal messages, is efficiently solved using the Turbo-CS methodology. Practical applicability was demonstrated through a channel tracking application in 5G NR systems.

By employing different sparse Bayesian inference algorithms for per-slot sparse reconstruction, we can derive a family of dynamic methods including dynamic Turbo-AMP and dynamic Turbo-VBI. Dynamic Turbo-AMP uses the same time-evolving BG Markov model, with temporal messages serving as prior information conforming to a two-layer BG structure. For dynamic Turbo-VBI, we introduce a mean vector $\boldsymbol{\mu}^{(t)} = [\mu_1^{(t)}, \ldots, \mu_N^{(t)}]^T$ to capture temporal correlation of non-zero elements:

$$p(\boldsymbol{\mu}^{(t)}|\boldsymbol{s}^{(t)}, \tilde{\boldsymbol{\mu}}^{(t)}) = \prod_{n=1}^{N} \delta(\mu_n - s_n^{(t)}\tilde{\mu}_n^{(t)}), \tag{15.84}$$

where $\tilde{\boldsymbol{\mu}}^{(t)} = [\tilde{\mu}_1^{(t)}, \ldots, \tilde{\mu}_N^{(t)}]^T$ with $\tilde{\mu}_n^{(t)}$ denoting the mean of x_n conditioned on $s_n^{(t)} = 1$. Given mean vector $\boldsymbol{\mu}^{(t)}$ and precision vector $\boldsymbol{\rho}^{(t)}$, the conditional distribution of $\boldsymbol{x}^{(t)}$ follows:

$$p(\boldsymbol{x}^{(t)}|\boldsymbol{\rho}^{(t)}, \boldsymbol{\mu}^{(t)}) = \prod_{n=1}^{N} \mathcal{CN}(x_n^{(t)}; \mu_n^{(t)}, (\rho_n^{(t)})^{-1}). \tag{15.85}$$

To model temporal evolution of non-zero elements

$$p(\tilde{\boldsymbol{\mu}}^{(t)}|\tilde{\boldsymbol{\mu}}^{(t-1)}) = \prod_{n=1}^{N} \mathcal{CN}(\tilde{\mu}_n^{(t)}; (1-\alpha_\mu)\tilde{\mu}_n^{(t-1)} + \alpha_\mu\zeta_\mu, \alpha_\mu^2\gamma_\mu), \tag{15.86}$$

with parameters $\alpha_\mu, \zeta_\mu, \gamma_\mu$ defined analogously to previous models. Temporal messages from previous slots constitute a three-layer prior where the mean of $\boldsymbol{x}$ follows a Bernoulli–Gaussian distribution and the precision follows a Bernoulli–Gamma distribution, enabling derivation of a modified Turbo-VBI algorithm.

When temporal correlation of non-zero elements is corrupted by imperfection parameters or difficult to exploit, we may focus exclusively on support vector correlation by considering only the time-evolving Markov model $p(\boldsymbol{s}^{(t)}|\boldsymbol{s}^{(t-1)})$. In this simplified approach, temporal messages contain only binary prior information for the support, specifically $\hat{p}(\boldsymbol{s}^{(t)})$, providing a reduced-complexity alternative while maintaining essential temporal structure.

Homework Problems

15.1 Prove that the steady-state distribution of the Markov chain defined in Eq. (15.7) is given by

$$p\left(\beta_n^{(t)}\right) = \mathcal{CN}\left(\beta_n^{(t)}; \zeta_\beta, \frac{\alpha_\beta\gamma_\beta}{2-\alpha_\beta}\right).$$

In your proof, you should

- Start from the Markov chain dynamics specified in (15.7).
- Set up the steady-state condition for the distribution.
- Solve for the mean and variance parameters that satisfy the stationarity condition.
- Verify that the Gaussian distribution with the given parameters indeed remains invariant under the Markov transition.

15.2 (a) **Independent Hidden Values with Markov Support**: Consider the case where the support vector $\boldsymbol{s}$ follows the time-evolving Markov model in (15.4), but the hidden values $\boldsymbol{\beta}^{(t)}$ across different time slots are independent.

- Draw the factor graph for the joint probability distribution under these assumptions.
- Design a complete dynamic Turbo-CS algorithm for this scenario, specifying:
  - The message-passing rules between temporal frames.
  - The integration of support correlation information.
  - The estimation procedure for both support and signal values.

(b) **Bernoulli–Gamma–Gaussian Prior with independent Sparse Signal and Precision**: Now change the sparse prior to a three-layer Bernoulli–Gamma–Gaussian prior. Assume the support vector $\boldsymbol{s}$ still follows the time-evolving Markov model in (15.4), but both the sparse signal $\boldsymbol{x}^{(t)}$ and its precision $\boldsymbol{\rho}^{(t)}$ are independent across time slots. Design a dynamic Turbo-VBI algorithm for this case, including

- The variational distributions for all latent variables.
- The update rules that incorporate temporal support correlation.

15.3 Considering the application example in Sect. 15.3:

- Implement both the Turbo-CS and dynamic Turbo-CS algorithms using the provided MATLAB code.
- Compare their channel estimation performance in terms of:
  - Normalized Mean Square Error (NMSE) across different SNR conditions.
  - Convergence speed and computational complexity.
  - Robustness to time-varying channel conditions.
  - Ability to track sudden changes in channel sparsity patterns.
- Discuss the advantages and disadvantages of each algorithm, considering:
  - Modeling assumptions and their practical validity.
  - Computational requirements and implementation complexity.
  - Performance trade-offs in different operational scenarios.
  - Suitability for real-time applications.

The MATLAB code for the dynamic Turbo-CS algorithm is available at: https://github.com/biwcs-book/codebase.

References

1. J. Yu, X. Liu, Y. Gao, X. Shen, 3D channel tracking for UAV-satellite communications in space-air-ground integrated networks. IEEE J. Select. Areas Commun. **38**(12), 2810–2823 (2020)
2. J. Ziniel, P. Schniter, Dynamic compressive sensing of time-varying signals via approximate message passing. IEEE Trans. Signal Process. **61**(21), 5270–5284 (2013)
3. J. Yu, X. Liu, Y. Gao, X. Shen, 3D on and off-grid dynamic channel tracking for multiple UAVs and satellite communications. IEEE Trans. Wireless Commun. **21**(6), 3587–3604 (2022)
4. X. Liu, W. Wang, X. Song, X. Gao, G. Fettweis, Sparse channel estimation via hierarchical hybrid message passing for massive MIMO-OFDM systems. IEEE Trans. Wireless Commun. **20**(11), 7118–7134 (2021)
5. X. Zhu, L. Dai, W. Dai, Z. Wang, M. Moonen, Tracking a dynamic sparse channel via differential orthogonal matching pursuit, in *Proceedings - IEEE Military Communications Conference (MILCOM)* (2015), pp. 792–797
6. X. Zhu, L. Dai, G. Gui, W. Dai, Z. Wang, F. Adachi, Structured matching pursuit for reconstruction of dynamic sparse channels, in *Proceedings - IEEE Global Communications Conference (GLOBECOM)* (IEEE, 2015), pp. 1–5
7. F. Kschischang, B. Frey, H.-A. Loeliger, Factor graphs and the sum-product algorithm. IEEE Trans. Inform. Theory **47**(2), 498–519 (2001)
8. J. Ma, X. Yuan, L. Ping, On the performance of turbo signal recovery with partial DFT sensing matrices. IEEE Signal Process. Lett. **22**(10), 1580–1584 (2015)
9. L. Lu, G.Y. Li, A.L. Swindlehurst, A. Ashikhmin, R. Zhang, An overview of massive MIMO: benefits and challenges. IEEE J. Select. Topics Signal Process. **8**(5), 742–758 (2014)
10. *3GPP, TS 38.211 V16.7.0, Technical Specification Group Radio Access Network; NR; Physical channels and modulation* (2021)
11. T. Van Chien, E. Björnson, Massive MIMO communications, in *5G Mobile Communications* (Springer, Berlin 2017), pp. 77–116
12. X. Xia, H. Zhao, C. Zhang, Improved SRS design and channel estimation for LTE-advanced uplink, in *Proceedings IEEE 5th International Symposium MAPE EMC Technologies for Wireless Communications* (IEEE, 2013), pp. 84–90
13. O. Mehanna, N.D. Sidiropoulos, Channel tracking and transmit beamforming with frugal feedback. IEEE Trans. Signal Process. **62**(24), 6402–6413 (2014)
14. W. Guo, W. Zhang, P. Mu, F. Gao, H. Lin, High-mobility wideband massive MIMO communications: doppler compensation, analysis and scaling laws. IEEE Trans. Wireless Commun. **18**(6), 3177–3191 (2019)

15. H. Xu, L. Yang, Calibration of random phase rotation for multi-band OFDM UWB signals, in *Proceedings of the 44th Asilomar Conference Signals Systems and Computers* (2010), pp. 170–174
16. X. Wu, X. Yang, S. Ma, B. Zhou, G. Yang, Hybrid channel estimation for UPA-assisted millimeter-wave massive MIMO IoT systems. IEEE Internet Things J. **9**(4), 2829–2842 (2022)
17. C. Zhang, D. Guo, P. Fan, Tracking angles of departure and arrival in a mobile millimeter wave channel, in *Proceedings of the IEEE International Conference on Communications (ICC)* (2016), pp. 1–6
18. Y. Wan, A. Liu, A two-stage 2D channel extrapolation scheme for TDD 5G NR systems. IEEE Trans. Wireless Commun. **23**(8), 8497–8511 (2024)
19. S. Jaeckel, L. Raschkowski, K. Borner, L. Thiele, Quadriga: a 3-D multi-cell channel model with time evolution for enabling virtual field trials. IEEE Trans. Antennas Propag. **62**(6), 3242–3256 (2014)
20. 3GPP, Study on channel model for frequencies from 0.5 to 100 GHz (3 GPP TR 38.901 version 16.1.0 release 16) (2019)
21. Y.-Y. Wang, J.-T. Chen, W.-H. Fang, TST-MUSIC for joint DOA-delay estimation. IEEE Trans. Signal Process. **49**(4), 721–729 (2001)
22. L. Lian, A. Liu, V.K. Lau, Exploiting dynamic sparsity for downlink FDD-massive MIMO channel tracking. IEEE Trans. Signal Process. **67**(8), 2007–2021 (2019)

Advanced Variational Bayesian Inference Algorithms 16

In this chapter, we explore several advanced VBI algorithms designed to enhance and extend earlier VBI methods in multiple directions. First, to reduce the computational complexity of sparse VBI, we introduce an inverse-free VBI (IF-VBI) algorithm. Based on the MM principle, IF-VBI entirely avoids matrix inversion when updating the posterior mean of the sparse signal. Next, we present a subspace-constrained VBI (SC-VBI) algorithm that offers an alternative approach to managing complexity. Rather than eliminating matrix inversion entirely as in IF-VBI, SC-VBI replaces the high-dimensional matrix inverse with a low-dimensional subspace-constrained inverse, whose dimension equals the sparsity level of the signal. This strategy significantly accelerates convergence while minimizing computational overhead. Finally, we extend SC-VBI to the bilinear subspace variational Bayesian inference (BiSVBI) algorithm, tailored for bilinear sparse signal recovery problems. BiSVBI enables joint recovery of two sparse signals embedded within a bilinear observation model, further broadening the applicability of subspace-constrained variational inference. Building on these foundations, the chapter concludes with a discussion of advanced extensions, including successive linear approximation and super-resolution techniques, which push the boundaries of VBI toward next-generation wireless applications.

16.1 The Computational Challenge in Structured Sparse Recovery

16.1.1 Revisiting the VBI Update: The Matrix Inversion Bottleneck

We begin by revisiting the standard compressive sensing observation model that forms the foundation for much of our development:

$$\boldsymbol{y} = \boldsymbol{A}\boldsymbol{x} + \boldsymbol{w}, \tag{16.1}$$

where $\boldsymbol{x} \in \mathbb{C}^{N\times 1}$ represents a sparse signal to be recovered, $\boldsymbol{A} \in \mathbb{C}^{M\times N}$ is the sensing matrix, and $\boldsymbol{w} \in \mathbb{C}^{M\times 1}$ denotes additive white Gaussian noise. The noise is assumed to be independent and identically distributed, with each entry following a complex Gaussian distribution $w_m \sim \mathcal{CN}(0, \kappa^{-1})$. The noise precision κ is itself treated as a random variable with a Gamma prior $p(\kappa) = \Gamma(\kappa; c, d)$.

Under the three-layer independent BGG prior introduced in Chap. 13, the variational distribution for the sparse signal takes a Gaussian form:

$$q(\boldsymbol{x}) = \mathcal{CN}(\boldsymbol{x}; \boldsymbol{\mu}, \boldsymbol{\Sigma}), \tag{16.2}$$

with the posterior mean and covariance matrix given by the standard LMMSE expressions:

$$\boldsymbol{\mu} = \langle\kappa\rangle \boldsymbol{\Sigma} \boldsymbol{A}^H \boldsymbol{y}, \tag{16.3}$$

$$\boldsymbol{\Sigma} = \left(\langle\kappa\rangle \boldsymbol{A}^H \boldsymbol{A} + \mathrm{diag}(\langle\boldsymbol{\rho}\rangle)\right)^{-1}. \tag{16.4}$$

Here, $\langle\kappa\rangle$ denotes the posterior mean of the noise precision, and $\langle\boldsymbol{\rho}\rangle$ is the vector of posterior mean precisions for each component of $\boldsymbol{x}$. The diagonal structure of the prior precision term $\mathrm{diag}(\langle\boldsymbol{\rho}\rangle)$ reflects the assumption that the signal components are conditionally independent given their precisions.

The computation of $\boldsymbol{\Sigma}$ in (16.4) requires inverting an $N \times N$ matrix. For a dense matrix $\boldsymbol{A}$, this operation incurs $O(N^3)$ complexity, which becomes prohibitively expensive when N is large, a common scenario in modern applications such as massive MIMO channel estimation, where N can range from hundreds to thousands, or in high-resolution sensing problems where dense angular or delay grids are employed.

A. Liu and M.-M. Zhao, *Advanced Bayesian Signal Processing*, Textbooks in Telecommunication Engineering,
https://doi.org/10.1007/978-3-032-26656-9_16

Even when the matrix inversion lemma is applied to exploit the low-rank structure of $\boldsymbol{A}^H\boldsymbol{A}$ (with rank at most M), the complexity remains $O(N^2M)$, which is still burdensome for large N. Moreover, many practical sensing matrices do not possess the special structure (such as partial orthogonality) that would enable further simplification. This matrix inversion bottleneck motivates the search for more efficient alternatives.

16.1.2 Two Philosophical Approaches to Complexity Reduction

Two distinct philosophies have emerged in response to the computational challenge posed by the matrix inversion in VBI. While both aim to reduce complexity, they approach the problem from fundamentally different angles.

16.1.2.1 Relaxation-Based Methods: Modifying the Objective

The first approach modifies the objective function itself to enable simpler updates. Rather than working directly with the original evidence lower bound (ELBO), we construct a surrogate function that lower bounds it and admits closed-form updates without matrix inversion. This is the philosophy underlying the IF-VBI algorithm.

The key idea is to replace the quadratic form $\kappa\|\boldsymbol{y} - \boldsymbol{A}\boldsymbol{x}\|^2$ that appears in the likelihood with a majorizing function $g(\boldsymbol{x}, \boldsymbol{z})$ that has a diagonal Hessian. By introducing an auxiliary variable $\boldsymbol{z}$ and applying the MM principle, we obtain an iterative procedure where each E-step involves only diagonal matrix inversions and each M-step reduces to a simple assignment.

The primary advantage of relaxation-based methods is their exceptionally low per-iteration complexity. Once a majorizing matrix $\boldsymbol{T} = \lambda_{\max}(\boldsymbol{A}^H\boldsymbol{A})\boldsymbol{I}$ is constructed, where $\lambda_{\max}(\cdot)$ denotes the largest eigenvalue of a matrix, each iteration requires only $O(MN)$ operations, a dramatic reduction from the $O(N^3)$ or $O(N^2M)$ cost of standard VBI. However, this efficiency comes at a price. The construction of $\boldsymbol{T}$ requires computing the largest eigenvalue of $\boldsymbol{A}^H\boldsymbol{A}$, which for large N can be a costly one-time initialization. Moreover, the relaxation introduces a gap between the surrogate objective and the true ELBO, slowing down convergence. As such, IF-VBI may require many iterations to reach a stationary point, and the per-iteration progress is limited by the quality of the majorizing approximation.

16.1.2.2 Constraint-Based Methods: Imposing Structure on the Posterior

The second approach retains the original objective but imposes structural constraints on the variational posterior to reduce dimensionality. This is the philosophy underlying the SC-VBI algorithm.

The observation that motivates this approach is simple but powerful: the signal $\boldsymbol{x}$ is sparse, meaning that most of its components are zero. If we knew the true support $\mathcal{S}$ (the set of indices where $x_n \neq 0$), we could restrict our attention to the $|\mathcal{S}|$-dimensional subspace spanned by the corresponding columns of $\boldsymbol{A}$. Within this subspace, the matrix inversion would involve only a $|\mathcal{S}| \times |\mathcal{S}|$ matrix, reducing complexity from $O(N^3)$ to $O(|\mathcal{S}|^3)$, a dramatic saving when $|\mathcal{S}| \ll N$.

Since the true support is unknown, SC-VBI employs two levels of constraints to achieve efficient inference while preserving the original objective. First, an independence constraint forces $q(\boldsymbol{x})$ to have diagonal covariance, eliminating the need for full covariance matrix inversion in the variance update and reducing it to $O(N)$ operations. Second, a subspace constraint approximates the mean update within a low-dimensional subspace defined by the current estimate $\hat{\mathcal{S}}$ of the support, replacing the full $N \times N$ matrix inverse with a $|\hat{\mathcal{S}}| \times |\hat{\mathcal{S}}|$ inverse. A gradient-based refinement step follows the subspace-constrained update, correcting approximation errors and ensuring progress toward the true optimum of the original objective.

The key strength of constraint-based methods is that they work directly with the original likelihood, avoiding the costly eigenvalue decomposition to calculate $\lambda_{\max}(\boldsymbol{A}^H\boldsymbol{A})$ required by IF-VBI. By preserving the curvature information of the original objective, SC-VBI typically achieves faster convergence and better final accuracy. The per-iteration complexity is $O(MN + |\hat{\mathcal{S}}|^3)$, which remains manageable as long as the signal is highly sparse (i.e., $|\mathcal{S}| \ll N$). However, when sparsity is less pronounced and $|\hat{\mathcal{S}}|$ grows large, the cubic term eventually dominates, making constraint-based methods increasingly expensive.

16.1.2.3 Applicability and Algorithm Selection

The choice between relaxation-based and constraint-based methods hinges on problem-specific characteristics. Relaxation-based methods like IF-VBI are most attractive when $|\hat{\mathcal{S}}|$ is large and even a single $|\hat{\mathcal{S}}| \times |\hat{\mathcal{S}}|$ inversion is prohibitively expensive. In such regimes, the low per-iteration complexity of $O(MN)$ outweighs the cost of slower convergence, provided that the initial eigenvalue computation can be amortized over a sufficient number of iterations. These methods are particularly well suited for problems where the sensing matrix $\boldsymbol{A}$ is structured enough that $\lambda_{\max}(\boldsymbol{A}^H\boldsymbol{A})$ can be computed efficiently, or where the number of iterations required for convergence is not excessively large.

Constraint-based methods like SC-VBI excel in the opposite regime: when the signal is highly sparse ($|\mathcal{S}| \ll N$) and the cubic term $|\hat{\mathcal{S}}|^3$ remains manageable. In this setting, SC-VBI's faster convergence and avoidance of costly eigenvalue decomposition lead to significantly lower total computation time compared to relaxation-based alternatives. These methods are ideal for applications such as massive MIMO channel estimation, where the angular-domain channel is inherently sparse, and for high-resolution sensing problems where accurate parameter estimation requires preserving the original objective's curvature. As the sparsity level decreases and $|\mathcal{S}|$ grows, the advantage shifts back toward relaxation-based methods, highlighting the importance of understanding the underlying signal structure when selecting an appropriate algorithm.

The two approaches are not mutually exclusive. Hybrid strategies that combine elements of both are possible, such as using subspace constraints to accelerate the initial convergence of a relaxation-based method, or employing relaxation techniques to handle components outside the current support estimate. The algorithms developed in the following sections provide a foundation for exploring such combinations.

16.2 Relaxation-Based Complexity Reduction: Inverse-Free VBI

16.2.1 Overview of the IF-VBI Approach

The IF-VBI algorithm offers a fundamentally different strategy for addressing the matrix inversion bottleneck. Rather than attempting to compute the exact posterior updates more efficiently, IF-VBI modifies the objective function itself to create a surrogate problem that admits closed-form solutions without matrix inversion. This approach was originally developed in [1] for models with two-layer Gamma–Gaussian priors and later extended in [2] to accommodate the three-layer BGG prior that provides greater modeling flexibility.

The core idea is elegant in its simplicity: we construct a lower bound for the likelihood function that has a diagonal Hessian, then maximize this bound instead of the original ELBO. By introducing an auxiliary variable and applying the EM/MM principle, we obtain an iterative procedure where each E-step requires only the inversion of a diagonal matrix, an operation with $\mathcal{O}(N)$ complexity. The resulting algorithm achieves dramatic reductions in per-iteration computational cost.

Within the broader Turbo framework introduced in Chap. 14, IF-VBI serves as a drop-in replacement for Module A, the component responsible for compressive sensing under an independent prior. Module B, which handles structured sparsity through message passing over the support prior $p(\boldsymbol{s})$, remains unchanged. This modular design allows IF-VBI to leverage the same structured sparsity modeling capabilities as its predecessors (e.g., Turbo-VBI) while offering superior computational efficiency.

16.2.2 Compressive Observation and Sparse Prior Models

Consider the same compressive observation model $\boldsymbol{y} = \boldsymbol{A}\boldsymbol{x} + \boldsymbol{w}$ as in Sect. 16.1.1. To capture the sparsity structure of $\boldsymbol{x}$, we adopt the three-layer independent BGG prior introduced in Chap. 13. A binary support vector $\boldsymbol{s} = [s_1, \ldots, s_N]^T \in \{0, 1\}^N$ indicates which components of $\boldsymbol{x}$ are active ($s_n = 1$) and which are inactive ($s_n = 0$). Let $\boldsymbol{\rho} = [\rho_1, \ldots, \rho_N]^T$ denote the precision vector, where $1/\rho_n$ corresponds to the variance of x_n. The joint distribution factorizes as

$$p(\boldsymbol{x}, \boldsymbol{\rho}, \boldsymbol{s}) = p(\boldsymbol{s})\, p(\boldsymbol{\rho} \mid \boldsymbol{s})\, p(\boldsymbol{x} \mid \boldsymbol{\rho}), \tag{16.5}$$

with each component defined as follows.

Layer 1: Support Prior. The support vector follows an independent Bernoulli distribution:

$$p(\boldsymbol{s}) = \prod_{n=1}^{N} \lambda_n^{s_n} (1 - \lambda_n)^{1-s_n}, \tag{16.6}$$

where λ_n is the prior probability that component n is active.

Layer 2: Precision Prior. Conditioned on the support, the precision vector follows a Gamma distribution whose parameters depend on the activity state:

$$p(\boldsymbol{\rho} \mid \boldsymbol{s}) = \prod_{n=1}^{N} [\Gamma(\rho_n; a_n, b_n)]^{s_n} \left[\Gamma(\rho_n; \bar{a}_n, \bar{b}_n)\right]^{1-s_n}. \tag{16.7}$$

When $s_n = 1$, we choose a_n and b_n such that $\mathbb{E}[\rho_n] = a_n/b_n = \Theta(1)$, allowing x_n to take significant values. When $s_n = 0$, we enforce $\mathbb{E}[\rho_n] = \bar{a}_n/\bar{b}_n \gg 1$, driving the variance $1/\rho_n$ close to zero and effectively suppressing inactive coefficients.

Layer 3: Signal Prior. Given the precisions, the signal follows a complex Gaussian distribution:

$$p(\boldsymbol{x} \mid \boldsymbol{\rho}) = \prod_{n=1}^{N} \mathcal{CN}\left(x_n; 0, \rho_n^{-1}\right). \tag{16.8}$$

This hierarchical structure provides a flexible yet interpretable framework for sparse Bayesian inference, balancing modeling power with computational tractability.

16.2.3 Variational Problem Formulation

Let $\boldsymbol{v} \triangleq \{\boldsymbol{x}, \boldsymbol{\rho}, \boldsymbol{s}, \kappa\}$ denote the complete set of latent variables, and let $\boldsymbol{v}^k$ represent an individual variable in this set. The index set $\mathcal{H} \triangleq \{k \mid \forall \boldsymbol{v}^k \in \boldsymbol{v}\}$ enumerates these variables. In the variational Bayesian inference framework, we approximate the true posterior $p(\boldsymbol{v} \mid \boldsymbol{y})$ with a factorized distribution:

$$p(\boldsymbol{v} \mid \boldsymbol{y}) \approx q(\boldsymbol{v}) = q(\boldsymbol{x})\, q(\boldsymbol{\rho})\, q(\boldsymbol{s})\, q(\kappa). \tag{16.9}$$

Each variational distribution $q(\boldsymbol{v}^k)$ is obtained by maximizing the ELBO, which is equivalent to minimizing the KLD between $q(\boldsymbol{v})$ and the true posterior. The ELBO takes the form:

$$\begin{aligned} L(q) &= \int_{\boldsymbol{v}} q(\boldsymbol{v}) \ln \frac{p(\boldsymbol{y}, \boldsymbol{v})}{q(\boldsymbol{v})}\, d\boldsymbol{v} \\ &= \int_{\boldsymbol{v}} q(\boldsymbol{v}) \ln \frac{p(\boldsymbol{y} \mid \boldsymbol{x}, \kappa)\, p(\boldsymbol{x} \mid \boldsymbol{\rho})\, p(\boldsymbol{\rho} \mid \boldsymbol{s})\, p(\boldsymbol{s})\, p(\kappa)}{q(\boldsymbol{v})}\, d\boldsymbol{v}. \end{aligned} \tag{16.10}$$

As established in [13], a stationary point $q^*(\boldsymbol{v})$ can be reached by alternately updating each variational distribution while keeping the others fixed. For a given $q(\boldsymbol{v}^l)$ with $l \neq k$, the optimal $q(\boldsymbol{v}^k)$ satisfies

$$q(\boldsymbol{v}^k) = \frac{\exp\left(\langle \ln p(\boldsymbol{v}, \boldsymbol{y})\rangle_{\prod_{l\neq k} q(\boldsymbol{v}^l)}\right)}{\int_{\boldsymbol{v}^k} \exp\left(\langle \ln p(\boldsymbol{v}, \boldsymbol{y})\rangle_{\prod_{l\neq k} q(\boldsymbol{v}^l)}\right) d\boldsymbol{v}^k}, \tag{16.11}$$

where $\langle\cdot\rangle_{\prod_{l\neq k} q(\boldsymbol{v}^l)}$ denotes expectation with respect to all $q(\boldsymbol{v}^l)$ for $l \neq k$.

Applying this general update to $q(\boldsymbol{x})$ yields a Gaussian distribution with mean and covariance given by

$$\boldsymbol{\mu} = \langle\kappa\rangle \boldsymbol{\Sigma} \boldsymbol{A}^H \boldsymbol{y}, \tag{16.12}$$

$$\boldsymbol{\Sigma} = \left(\langle\kappa\rangle \boldsymbol{A}^H \boldsymbol{A} + \operatorname{diag}\left(\langle\boldsymbol{\rho}\rangle\right)\right)^{-1}. \tag{16.13}$$

The matrix inversion required in (16.13) incurs $O(N^3)$ complexity, which becomes prohibitive for large N. This bottleneck motivates the relaxation approach that follows.

16.2.4 Constructing a Relaxed Evidence Lower Bound

The key innovation in IF-VBI is the construction of a surrogate function that lower bounds the likelihood while admitting simpler updates. Specifically, we construct a lower bound for the likelihood $p(\boldsymbol{y} \mid \boldsymbol{x}, \kappa)$ using the following inequality:

$$\begin{aligned} p(\boldsymbol{y} \mid \boldsymbol{x}, \kappa) &= C \exp\left(-\kappa(\boldsymbol{y} - \boldsymbol{A}\boldsymbol{x})^H(\boldsymbol{y} - \boldsymbol{A}\boldsymbol{x})\right) \\ &\geq C \exp\left(-\kappa g(\boldsymbol{x}, \boldsymbol{z})\right) \triangleq F(\boldsymbol{y}, \boldsymbol{x}, \boldsymbol{z}, \kappa), \end{aligned} \tag{16.14}$$

where C is a normalization constant. The inequality follows from Lemma 16.1 in [1], and the function $g(\boldsymbol{x}, \boldsymbol{z})$ is defined as

$$\begin{aligned} g(\boldsymbol{x}, \boldsymbol{z}) \triangleq\, &(\boldsymbol{y} - \boldsymbol{A}\boldsymbol{z})^H (\boldsymbol{y} - \boldsymbol{A}\boldsymbol{z}) \\ &+ 2\mathrm{Re}\left\{(\boldsymbol{x} - \boldsymbol{z})^H \boldsymbol{A}^H (\boldsymbol{A}\boldsymbol{z} - \boldsymbol{y})\right\} \\ &+ (\boldsymbol{x} - \boldsymbol{z})^H \boldsymbol{T} (\boldsymbol{x} - \boldsymbol{z}), \end{aligned} \tag{16.15}$$

where $\boldsymbol{T}$ is a matrix satisfying $\boldsymbol{T} \succeq \boldsymbol{A}^H\boldsymbol{A}$. A computationally efficient choice is

$$\boldsymbol{T} = \lambda_{\max}\left(\boldsymbol{A}^H\boldsymbol{A}\right) \boldsymbol{I}_N, \tag{16.16}$$

which is a diagonal matrix.

Attention
Why $g(x, z)$ is a Majorizing Function?
The function $g(\boldsymbol{x}, \boldsymbol{z})$ majorizes the original quadratic form $f(\boldsymbol{x}) = (\boldsymbol{y} - \boldsymbol{A}\boldsymbol{x})^H(\boldsymbol{y} - \boldsymbol{A}\boldsymbol{x})$ because

1. **Upper Bound Property**: 2045For any fixed $\boldsymbol{z}$, $g(\boldsymbol{x}, \boldsymbol{z}) \geq f(\boldsymbol{x})$ for all $\boldsymbol{x}$, provided $\boldsymbol{T} \succeq \boldsymbol{A}^H\boldsymbol{A}$. This can be seen by examining the second-order Taylor expansion of $f(\boldsymbol{x})$ around $\boldsymbol{z}$

$$f(\boldsymbol{x}) = f(\boldsymbol{z}) + \nabla f(\boldsymbol{z})^H(\boldsymbol{x} - \boldsymbol{z}) + (\boldsymbol{x} - \boldsymbol{z})^H\boldsymbol{A}^H\boldsymbol{A}(\boldsymbol{x} - \boldsymbol{z}), \tag{16.17}$$

where $\nabla f(\boldsymbol{z}) = -2\kappa\boldsymbol{A}^H(\boldsymbol{y} - \boldsymbol{A}\boldsymbol{z})$. Comparing this with the expression for $g(\boldsymbol{x}, \boldsymbol{z})$

$$g(\boldsymbol{x}, \boldsymbol{z}) = f(\boldsymbol{z}) + \nabla f(\boldsymbol{z})^H(\boldsymbol{x} - \boldsymbol{z}) + (\boldsymbol{x} - \boldsymbol{z})^H\boldsymbol{T}(\boldsymbol{x} - \boldsymbol{z}), \tag{16.18}$$

we observe that replacing $\boldsymbol{A}^H\boldsymbol{A}$ with the larger matrix $\boldsymbol{T}$ in the quadratic term ensures $g(\boldsymbol{x}, \boldsymbol{z}) \geq f(\boldsymbol{x})$ for all $\boldsymbol{x}$.
2. **Touching Condition**: Equality holds when $\boldsymbol{x} = \boldsymbol{z}$, i.e., $g(\boldsymbol{z}, \boldsymbol{z}) = f(\boldsymbol{z})$.

The choice $\boldsymbol{T} = \lambda_{\max}(\boldsymbol{A}^H\boldsymbol{A})\boldsymbol{I}_N$ guarantees the majorization property while maintaining computational efficiency through diagonal structure. Since the exponential function $\exp(-t)$ is decreasing in t, $\exp(-\kappa g(\boldsymbol{x}, \boldsymbol{z}))$ consequently provides a lower bound for the original likelihood $\exp(-\kappa f(\boldsymbol{x}))$.

Substituting the likelihood lower bound into the ELBO expression yields a relaxed objective:

$$\widetilde{L}(q, \boldsymbol{z}) = \int_{\boldsymbol{v}} q(\boldsymbol{v}) \ln \frac{G(\boldsymbol{y}, \boldsymbol{v}, \boldsymbol{z})}{q(\boldsymbol{v})}\, d\boldsymbol{v}, \tag{16.19}$$

where

$$G(\boldsymbol{y}, \boldsymbol{v}, \boldsymbol{z}) = F(\boldsymbol{y}, \boldsymbol{x}, \boldsymbol{z}, \kappa)\, p(\boldsymbol{x} \mid \boldsymbol{\rho})\, p(\boldsymbol{\rho} \mid \boldsymbol{s})\, p(\boldsymbol{s})\, p(\kappa). \tag{16.20}$$

This relaxed ELBO retains the key property that $\widetilde{L}(q, \boldsymbol{z}) \leq L(q)$, with equality when $\boldsymbol{z} = \boldsymbol{x}$ and $q(\boldsymbol{v})$ is the true posterior. The introduction of the auxiliary variable $\boldsymbol{z}$ creates additional flexibility: by alternately optimizing over $q(\boldsymbol{v})$ and $\boldsymbol{z}$, we can monotonically increase the relaxed ELBO while never decreasing the true ELBO below its current value.

16.2.5 The IF-VBI Algorithm: An EM-Like Procedure

The IF-VBI algorithm maximizes the relaxed ELBO through an alternating optimization procedure that closely resembles the EM algorithm. Each iteration consists of two steps: an E-step that updates the variational distributions for fixed z, and an M-step that updates z for fixed variational distributions.

16.2.5.1 E-Step: Updating Variational Distributions

In the E-step, we hold the auxiliary variable fixed at its current value $z = z^{\text{old}}$ and update each variational distribution $q(\boldsymbol{v}^k)$ in turn. The update for $q(\boldsymbol{x})$ follows from retaining only the terms that depend on $\boldsymbol{x}$ in the relaxed ELBO, i.e.:

$$\begin{aligned}\ln q(\boldsymbol{x}) &\propto \ln F(\boldsymbol{y}, \boldsymbol{x}, z, \kappa) + \langle \ln p(\boldsymbol{x} \mid \boldsymbol{\rho})\rangle_{q(\boldsymbol{\rho})} \\ &\propto -\boldsymbol{x}^H \left(\langle\kappa\rangle \boldsymbol{T} + \text{diag}(\langle\boldsymbol{\rho}\rangle)\right) \boldsymbol{x} \\ &\quad + 2\Re\left\{\boldsymbol{x}^H \boldsymbol{A}^H \langle\kappa\rangle(\boldsymbol{y} - \boldsymbol{A}z)\right\} \\ &\quad + 2\Re\left\{\boldsymbol{x}^H \langle\kappa\rangle \boldsymbol{T} z\right\} \\ &\propto \ln \mathcal{CN}(\boldsymbol{x}; \boldsymbol{\mu}, \boldsymbol{\Sigma}),\end{aligned} \tag{16.21}$$

where $F(\boldsymbol{y}, \boldsymbol{x}, z, \kappa) = C \exp(-\kappa g(\boldsymbol{x}, z))$ is the relaxed likelihood factor. Completing the square reveals that $q(\boldsymbol{x})$ is Gaussian

$$q(\boldsymbol{x}) = \mathcal{CN}(\boldsymbol{x}; \boldsymbol{\mu}, \boldsymbol{\Sigma}), \tag{16.22}$$

with parameters

$$\boldsymbol{\mu} = \boldsymbol{\Sigma}\left(\boldsymbol{A}^H \langle\kappa\rangle(\boldsymbol{y} - \boldsymbol{A}z) + \langle\kappa\rangle \boldsymbol{T} z\right), \tag{16.23}$$

$$\boldsymbol{\Sigma} = \left(\langle\kappa\rangle \boldsymbol{T} + \text{diag}(\langle\boldsymbol{\rho}\rangle)\right)^{-1}. \tag{16.24}$$

The key computational advantage is now apparent: since $\boldsymbol{T}$ is diagonal, $\boldsymbol{\Sigma}$ is also diagonal, requiring only $O(N)$ operations to compute its inverse. The mean $\boldsymbol{\mu}$ involves only matrix-vector multiplications, avoiding the $O(N^3)$ cost of full matrix inversion.

The remaining variational distributions, $q(\boldsymbol{\rho})$, $q(\boldsymbol{s})$, and $q(\kappa)$, are updated using expressions similar to those derived in the Turbo-VBI algorithm of Chap. 14. For completeness, we summarize them here:

- **Precision update**: $q(\boldsymbol{\rho})$ follows a Gamma distribution with parameters updated using the current posterior means of $\boldsymbol{x}$ and $\boldsymbol{s}$.
- **Support update**: $q(\boldsymbol{s})$ follows a Bernoulli distribution with probabilities updated based on the current evidence and prior.
- **Noise precision update**: $q(\kappa)$ follows a Gamma distribution with shape $\tilde{c} = c + M$ and rate $\tilde{d} = d + \langle g(\boldsymbol{x}, z)\rangle$.

Let $q(\boldsymbol{v}; z^{\text{old}})$ denote the complete set of updated variational distributions obtained in this E-step.

16.2.5.2 M-Step: Updating the Auxiliary Variable

In the M-step, we hold the variational distributions fixed at $q(\boldsymbol{v}; z^{\text{old}})$ and optimize over the auxiliary variable z. The objective is the expected complete-data log-likelihood under the current variational distribution:

$$z^{\text{new}} = \arg\max_{z} \langle \ln G(\boldsymbol{y}, \boldsymbol{v}, z)\rangle_{q(\boldsymbol{v}; z^{\text{old}})}. \tag{16.25}$$

Computing the gradient with respect to z yields

$$\frac{\partial}{\partial z} \langle \ln G(\boldsymbol{y}, \boldsymbol{v}, z)\rangle = \langle\kappa\rangle(\boldsymbol{T} - \boldsymbol{A}^H \boldsymbol{A})(\boldsymbol{\mu} - z), \tag{16.26}$$

where $\boldsymbol{\mu}$ is the posterior mean of $\boldsymbol{x}$ from the E-step. Setting this gradient to zero leads to a remarkably simple update

$$z^{\text{new}} = \boldsymbol{\mu}. \tag{16.27}$$

The auxiliary variable is simply set to the current posterior mean estimate of the sparse signal. This update has an intuitive Bayesian interpretation: the expansion point for the majorizing function should be aligned with our current best estimate of $\boldsymbol{x}$, ensuring that the relaxation is tight where it matters most.

16.2.6 Convergence Behavior and Complexity Analysis

The IF-VBI algorithm, through its alternating E- and M-steps, generates a sequence of estimates that monotonically increases the relaxed ELBO $\widetilde{L}(q, z)$, ensuring convergence to a stationary point of $\widetilde{L}$. However, convergence to a stationary point of the original ELBO is not theoretically guaranteed.

The computational advantages of IF-VBI are substantial.

- **Per-iteration complexity**: Each E-step requires $O(MN)$ operations for the matrix-vector multiplications in $\boldsymbol{\mu}$ and $O(N)$ operations for the diagonal matrix inversion. Each M-step is essentially costless, involving only the assignment $z = \boldsymbol{\mu}$.
- **Memory requirements**: Only $O(MN)$ storage is needed for the sensing matrix $\boldsymbol{A}$, compared to $O(N^2)$ for methods that require storing the full covariance matrix.

However, these computational benefits come at a cost: the relaxation introduced by the majorizing function creates a gap between the surrogate objective and the true ELBO. This gap means that each iteration makes less progress toward the true optimum than a standard VBI update would, potentially requiring more iterations to achieve convergence. The number of iterations required depends on the conditioning of $\boldsymbol{A}^H\boldsymbol{A}$ and the quality of the majorizing approximation, poorly conditioned matrices may lead to looser bounds and slower convergence.

The overall efficiency of IF-VBI thus represents a trade-off: the dramatic reduction in per-iteration complexity must be weighed against the potential increase in the number of iterations required. For problems where N is very large and the matrix $\boldsymbol{A}^H\boldsymbol{A}$ is reasonably well conditioned, this trade-off is highly favorable, making IF-VBI an attractive alternative to standard VBI.

16.3 Constraint-Based Complexity Reduction: Subspace-Constrained VBI

16.3.1 Motivation: A Different Path to Efficiency

The IF-VBI algorithm presented in the previous section successfully eliminates large-scale matrix inversion through objective relaxation, but this computational economy comes at a price. The MM framework introduces a gap between the surrogate and true objectives, leading to slower convergence and potentially requiring many iterations to reach an accurate solution. For problems where fast convergence is critical, this trade-off may be unacceptable.

The SC-VBI algorithm, originally proposed in [3], offers an alternative strategy that strikes a different balance. Rather than modifying the objective, SC-VBI retains the exact variational updates but imposes structural constraints that reduce the dimensionality of the matrix inversion problem. The key insight is deceptively simple. Since the signal $\boldsymbol{x}$ is sparse, its energy is concentrated on a small set of active components. If we could identify this set, we could restrict all computations to the corresponding low-dimensional subspace, reducing an $N \times N$ matrix inversion to a $|\mathcal{S}| \times |\mathcal{S}|$ inversion, a dramatic saving when $|\mathcal{S}| \ll N$.

Of course, the true support $\mathcal{S}$ is unknown; it is precisely what we are trying to estimate. SC-VBI therefore employs an iterative approach that alternates between estimating the support and performing constrained inference within the current support estimate. The algorithm introduces two levels of constraints:

1. **An independence constraint** that forces $q(\boldsymbol{x})$ to have diagonal covariance, eliminating the need for full covariance matrix inversion in the covariance update.
2. **A subspace constraint** that approximates the mean update within the low-dimensional subspace defined by the current support estimate $\hat{\mathcal{S}}$, replacing the full matrix inverse with a much smaller one.

The approximations introduced by these constraints are not left uncorrected. A gradient-based refinement step follows each subspace-constrained update, ensuring that the algorithm progresses toward the true optimum of the original objective. As demonstrated in [3], this combination yields an algorithm that achieves an excellent balance among three key metrics: per-iteration complexity, convergence rate, and final estimation accuracy.

Like its predecessors, SC-VBI is designed for modular integration within the Turbo framework introduced in Chap. 14. The algorithm comprises two main components:

- **Module A**: The SC-VBI core module, which performs compressive sensing under an independent BGG prior while respecting the subspace and independence constraints.
- **Module B**: An unmodified structured sparsity inference module that handles complex prior distributions $p(s)$ through message passing, exactly as described in Chap. 14.

This modular design allows SC-VBI to leverage sophisticated statistical structures, such as the 2D-Markov priors used in channel estimation applications, while maintaining the computational efficiency that makes it suitable for large-scale problems.

16.3.2 Problem Formulation with Constraints

16.3.2.1 The Constrained Optimization Framework

We consider the same compressive observation model and three-layer independent BGG prior introduced in Sect. 16.2.2. Let $\boldsymbol{v} \triangleq \{\boldsymbol{x}, \boldsymbol{\rho}, \boldsymbol{s}, \boldsymbol{\kappa}\}$ denote the complete set of latent variables, with index set $\mathcal{H} = \{k \mid \forall \boldsymbol{v}^k \in \boldsymbol{v}\}$. The SC-VBI algorithm solves the following constrained optimization problem:

$$\mathcal{A}_{\text{SCVBI}}: \quad q^*(\boldsymbol{v}) = \arg\min_{q(\boldsymbol{v})} \int q(\boldsymbol{v}) \ln \frac{q(\boldsymbol{v})}{p(\boldsymbol{v} \mid \boldsymbol{y})} d\boldsymbol{v}, \tag{16.28}$$

$$\text{s.t.} \quad q(\boldsymbol{v}) = \prod_{k \in \mathcal{H}} q(\boldsymbol{v}^k), \quad \int q(\boldsymbol{v}^k)\, d\boldsymbol{v}^k = 1, \tag{16.29}$$

$$q(\boldsymbol{x}) = \mathcal{CN}\left(\boldsymbol{x}; \boldsymbol{\mu}, \text{diag}(\boldsymbol{\sigma}^2)\right). \tag{16.30}$$

Here, $\boldsymbol{\mu} = [\mu_1, \ldots, \mu_N]^T$ is the posterior mean vector and $\boldsymbol{\sigma}^2 = [\sigma_1^2, \ldots, \sigma_N^2]^T$ contains the posterior variances. The constraint (16.30) is the key innovation: by restricting $q(\boldsymbol{x})$ to have a diagonal covariance matrix, we eliminate the need for full covariance matrix inversion in the covariance update, reducing its complexity from $O(N^3)$ to $O(N)$.

This diagonal constraint is well justified in many practical applications. In massive MIMO channel estimation, for instance, $\boldsymbol{x}$ represents angular-domain channel gains, where different propagation paths correspond to physically independent scatterers. In image processing, wavelet coefficients often exhibit conditional independence given their support. Extensive numerical experiments in [3] confirm that the performance degradation from this constraint is negligible, while the computational benefits are substantial.

Since Problem $\mathcal{A}_{\text{SCVBI}}$ is non-convex, we seek a stationary solution according to the following definition.

Definition 16.1 (*Stationary Solution of SC-VBI Problem*) A factorized distribution $q^*(\boldsymbol{v}) = \prod_{k \in \mathcal{H}} q^*\left(\boldsymbol{v}^k\right)$ is a stationary solution of $\mathscr{A}_{\text{SCVBI}}$ if it satisfies all constraints and $\forall k \in \mathcal{H}$,

$$q^*\left(\boldsymbol{v}^k\right) = \arg\min_{q(\boldsymbol{v}^k)} \int \prod_{l \neq k} q^*\left(\boldsymbol{v}^l\right) q\left(\boldsymbol{v}^k\right) \ln \frac{\prod_{l \neq k} q^*\left(\boldsymbol{v}^l\right) q\left(\boldsymbol{v}^k\right)}{p\left(\boldsymbol{v} \mid \boldsymbol{y}\right)} d\boldsymbol{v}. \tag{16.31}$$

For $\boldsymbol{v}^k = \boldsymbol{x}$, the minimization is performed over the parameters $\boldsymbol{\mu}, \boldsymbol{\sigma}^2$ defining the Gaussian distribution $q(\boldsymbol{x}) = \mathcal{CN}\left(\boldsymbol{x}; \boldsymbol{\mu}, \text{diag}\left(\boldsymbol{\sigma}^2\right)\right)$.

16.3.2.2 Optimality Conditions for Constrained Distributions

For variables other than $\boldsymbol{x}$ (i.e., $\boldsymbol{v}^k \neq \boldsymbol{x}$), the optimal variational distribution follows the conventional VBI update rule:

$$q(\boldsymbol{v}^k) \propto \exp\left(\langle \ln p(\boldsymbol{v}, \boldsymbol{y})\rangle_{\prod_{l \neq k} q(\boldsymbol{v}^l)}\right), \tag{16.32}$$

where $\langle\cdot\rangle_{q(x)}$ denotes expectation with respect to distribution $q(x)$, and $p(\boldsymbol{v}, \boldsymbol{y})$ is the complete joint probability density function.

The update for $q(\boldsymbol{x})$ requires special treatment due to the diagonal covariance constraint. The following lemma characterizes the resulting optimization problem.

Lemma 16.1 (Stationary Solution of SC-VBI Problem) *For fixed distributions $q(\boldsymbol{\rho})$, $q(\boldsymbol{s})$, and $q(\kappa)$, solving Problem $\mathscr{A}_{\text{SCVBI}}$ reduces to the following parameter optimization:*

$$\min_{\boldsymbol{u},\boldsymbol{\sigma}^2} \boldsymbol{u}^H \boldsymbol{W}\boldsymbol{u} - 2Re\left\{\boldsymbol{u}^H \boldsymbol{b}\right\} + \sum_{n=1}^{N}\left(W_n\sigma_n^2 - \ln\sigma_n^2\right), \tag{16.33}$$

where we define

$$\boldsymbol{W} = diag\left(\langle\boldsymbol{\rho}\rangle\right) + \langle\kappa\rangle\, \boldsymbol{A}^H\boldsymbol{A}, \tag{16.34}$$

$$\boldsymbol{b} = \langle\kappa\rangle\, \boldsymbol{A}^H\boldsymbol{y}, \tag{16.35}$$

and W_n is the n-th diagonal element of $\boldsymbol{W}$. The optimal parameters are given in closed form as

$$\begin{aligned}\boldsymbol{\mu} &= \arg\min_{\boldsymbol{u}} \varphi(\boldsymbol{u}) \triangleq \boldsymbol{u}^H\boldsymbol{W}\boldsymbol{u} - 2Re\left\{\boldsymbol{u}^H\boldsymbol{b}\right\},\\ \boldsymbol{\sigma}^2 &= \left[W_1^{-1}, W_2^{-1}, \ldots, W_N^{-1}\right].\end{aligned} \tag{16.36}$$

Proof Given $q(\boldsymbol{\rho}), q(\boldsymbol{s}), q(\kappa)$, the KLD can be calculated as

$$\begin{aligned}\text{KL}(q \parallel p) &= \int q(\boldsymbol{v}) \ln\frac{q(\boldsymbol{v})}{p(\boldsymbol{v}\mid\boldsymbol{y})}\mathrm{d}\boldsymbol{v}\\ &= \langle\ln q(\boldsymbol{x})\rangle_{q(\boldsymbol{x})} - \langle\ln p(\boldsymbol{v},\boldsymbol{y})\rangle_{q(\boldsymbol{v})} + C\\ &= \langle\ln q(\boldsymbol{x})\rangle_{q(\boldsymbol{x})} - \langle\ln p(\boldsymbol{y}\mid\boldsymbol{x},\kappa)\rangle_{q(\boldsymbol{x})q(\kappa)} - \langle\ln p(\boldsymbol{x}\mid\boldsymbol{\rho})\rangle_{q(\boldsymbol{x})q(\boldsymbol{\rho})} + C,\\ &= \langle\ln q(\boldsymbol{x})\rangle_{q(\boldsymbol{x})} + \langle\kappa\rangle\left\langle\|\boldsymbol{y}-\boldsymbol{A}\boldsymbol{x}\|^2\right\rangle_{q(\boldsymbol{x})} + \left\langle\boldsymbol{x}^H\text{diag}\left(\langle\boldsymbol{\rho}\rangle\right)\boldsymbol{x}\right\rangle_{q(\boldsymbol{x})} + C,\end{aligned} \tag{16.37}$$

where C is a constant. Based on the constraint in (16.30), Problem $\mathscr{A}_{\text{SCVBI}}$ is equivalent to finding the optimal parameters of $q(\boldsymbol{x})$, denoted by $\left\{\ddot{\boldsymbol{\mu}}, \ddot{\boldsymbol{\sigma}}^2\right\}$, so that the KLD in (16.37) is minimized. The equivalent optimization problem is formulated as

$$(\ddot{\boldsymbol{\mu}}, \ddot{\boldsymbol{\sigma}}^2) = \min_{\boldsymbol{u},\boldsymbol{\sigma}^2} \mathcal{J}(\boldsymbol{u}, \boldsymbol{\sigma}^2), \tag{16.38}$$

where the objective function begins as

$$\begin{aligned}\mathcal{J}(\boldsymbol{u},\boldsymbol{\sigma}^2) = &-\sum_{n=1}^{N}\ln\sigma_n^2 - \left\langle(\boldsymbol{x}-\boldsymbol{u})^H\text{diag}(1/\boldsymbol{\sigma}^2)(\boldsymbol{x}-\boldsymbol{u})\right\rangle_{q(\boldsymbol{x})}\\ &+ \langle\kappa\rangle\left\langle\|\boldsymbol{y}-\boldsymbol{A}\boldsymbol{x}\|^2\right\rangle_{q(\boldsymbol{x})} + \left\langle\boldsymbol{x}^H\text{diag}(\langle\boldsymbol{\rho}\rangle)\boldsymbol{x}\right\rangle_{q(\boldsymbol{x})}.\end{aligned} \tag{16.39}$$

Substituting the definitions of $\boldsymbol{W} = \langle\kappa\rangle\boldsymbol{A}^H\boldsymbol{A} + \text{diag}(\langle\boldsymbol{\rho}\rangle)$ and $\boldsymbol{b} = \langle\kappa\rangle\boldsymbol{A}^H\boldsymbol{y}$, and expanding the expectations, the objective simplifies to

$$\begin{aligned}\mathcal{J}(\boldsymbol{u},\boldsymbol{\sigma}^2) = &-\sum_{n=1}^{N}\ln\sigma_n^2 - \boldsymbol{u}^H\text{diag}(1/\boldsymbol{\sigma}^2)\boldsymbol{u}\\ &+ \left\langle\boldsymbol{x}^H\left(\boldsymbol{W} - \text{diag}(1/\boldsymbol{\sigma}^2)\right)\boldsymbol{x}\right\rangle_{q(\boldsymbol{x})}\\ &- 2\text{Re}\left\{\left\langle\boldsymbol{x}^H\left(\boldsymbol{b} - \text{diag}(1/\boldsymbol{\sigma}^2)\boldsymbol{u}\right)\right\rangle_{q(\boldsymbol{x})}\right\}.\end{aligned} \tag{16.40}$$

We now evaluate the remaining expectations using $\langle\boldsymbol{x}\rangle = \boldsymbol{u}$ and $\langle\boldsymbol{x}\boldsymbol{x}^H\rangle = \boldsymbol{u}\boldsymbol{u}^H + \text{diag}(\boldsymbol{\sigma}^2)$, which leads to

$$\begin{aligned}\mathcal{J}(\boldsymbol{u}, \boldsymbol{\sigma}^2) = &-\sum_{n=1}^{N} \ln \sigma_n^2 - \boldsymbol{u}^H \operatorname{diag}(1/\boldsymbol{\sigma}^2)\boldsymbol{u} \\ &+ \boldsymbol{u}^H \left(\boldsymbol{W} - \operatorname{diag}(1/\boldsymbol{\sigma}^2)\right) \boldsymbol{u} \\ &+ \operatorname{Tr}\left(\left(\boldsymbol{W} - \operatorname{diag}(1/\boldsymbol{\sigma}^2)\right) \operatorname{diag}(\boldsymbol{\sigma}^2)\right) \\ &- 2\operatorname{Re}\left\{\boldsymbol{u}^H \left(\boldsymbol{b} - \operatorname{diag}(1/\boldsymbol{\sigma}^2)\boldsymbol{u}\right)\right\}.\end{aligned} \tag{16.41}$$

Finally, combining like terms leads to the following compact form:

$$\mathcal{J}(\boldsymbol{u}, \boldsymbol{\sigma}^2) = \boldsymbol{u}^H \boldsymbol{W} \boldsymbol{u} - 2\operatorname{Re}\left\{\boldsymbol{u}^H \boldsymbol{b}\right\} + \sum_{n=1}^{N} \left(W_n \sigma_n^2 - \ln \sigma_n^2\right),$$

which is clearly equal to the objective function in Problem (16.33). □

Three key insights emerge from Lemma 16.1.

1. **Diagonal Variance Structure**: The optimal posterior variances depend only on the diagonal elements of $\boldsymbol{W}$, yielding a computationally efficient update that requires no matrix inversion.
2. **Quadratic Mean Optimization**: The posterior mean $\boldsymbol{\mu}$ is given by the solution to a quadratic minimization problem. While this appears to require the matrix inversion $\boldsymbol{\mu} = \boldsymbol{W}^{-1}\boldsymbol{b}$, the subspace constraint method introduced next will avoid this computational bottleneck.
3. **Decoupled Updates**: The independence constraint decouples the variance updates across dimensions while maintaining coupling in the mean estimation through the off-diagonal terms of $\boldsymbol{W}$. This coupling is essential for accurate estimation, as it captures the correlations induced by the sensing matrix $\boldsymbol{A}$.

16.3.3 Subspace-Constrained Mean Update

16.3.3.1 The Core Idea

Lemma 16.1 establishes that the optimal posterior mean satisfies $\boldsymbol{\mu} = \boldsymbol{W}^{-1}\boldsymbol{b}$, which requires inverting an $N \times N$ matrix, the very operation we sought to avoid. However, the sparsity of $\boldsymbol{x}$ suggests a way forward: if we knew which components were active, we could restrict our attention to those dimensions. Let $\hat{\mathcal{S}}$ denote an estimate of the support of $\boldsymbol{x}$. We construct an approximation of the full posterior mean vector, $\boldsymbol{\mu}^s \in \mathbb{C}^N$, by setting its values on and off the estimated support $\hat{\mathcal{S}}$ separately as

$$\boldsymbol{\mu}^s_{\hat{\mathcal{S}}} = \langle \kappa \rangle \boldsymbol{W}_{\hat{\mathcal{S}}}^{-1} \boldsymbol{A}_{\hat{\mathcal{S}}}^H \boldsymbol{y}, \quad \boldsymbol{\mu}^s_{\hat{\mathcal{S}}^c} = \boldsymbol{0}, \tag{16.42}$$

where $\boldsymbol{\mu}^s_{\hat{\mathcal{S}}} \in \mathbb{C}^{|\hat{\mathcal{S}}| \times 1}$ is the LMMSE estimator for the components within the estimated support $\hat{\mathcal{S}}$, $\boldsymbol{W}_{\hat{\mathcal{S}}} \in \mathbb{C}^{|\hat{\mathcal{S}}| \times |\hat{\mathcal{S}}|}$ is the submatrix of $\boldsymbol{W}$ formed by selecting rows and columns indexed by $\hat{\mathcal{S}}$, and $\boldsymbol{A}_{\hat{\mathcal{S}}} \in \mathbb{C}^{M \times |\hat{\mathcal{S}}|}$ is the submatrix of $\boldsymbol{A}$ formed by selecting columns indexed by $\hat{\mathcal{S}}$. If $\hat{\mathcal{S}}$ accurately captures the true support, $\boldsymbol{\mu}^s$ provides an excellent approximation to the true posterior mean, and the complexity has been reduced from $O(N^3)$ to $O(|\hat{\mathcal{S}}|^3)$, a dramatic saving when $|\hat{\mathcal{S}}| \ll N$.

16.3.3.2 Support Estimation

The success of this approach depends critically on the quality of the support estimate $\hat{\mathcal{S}}$. We obtain this estimate through a combination of initialization and iterative refinement.

Initialization

For the first iteration, we can obtain an initial support estimate using prior knowledge or a simple algorithm such as orthogonal matching pursuit (OMP). In many applications, rough support information is available from physical considerations, for example, in channel estimation, the support corresponds to physically plausible angle-delay pairs.

Iterative Updates

In subsequent iterations, we update the support estimate based on the current posterior mean $\boldsymbol{\mu}$ using energy-based thresholding:

$$\hat{\mathcal{S}} = \{n : |\mu_n|^2 > \varepsilon\}, \tag{16.43}$$

where the threshold ε is chosen relative to the estimated noise power. A practical guideline, validated through extensive simulations, is

$$\varepsilon = (2 \text{ to } 3)/\langle\kappa\rangle. \tag{16.44}$$

This thresholding operation is computationally trivial, requiring only $O(N)$ operations.

16.3.3.3 Gradient-Based Refinement

The subspace approximation $\boldsymbol{\mu}^s$ may be inaccurate if the estimated support $\hat{\mathcal{S}}$ contains errors. To correct these approximations, we refine the estimate using gradient descent on the original objective $\varphi(\boldsymbol{u})$.

The gradient of φ is easily computed

$$\nabla\varphi(\boldsymbol{u}) = \boldsymbol{W}\boldsymbol{u} - \boldsymbol{b}. \tag{16.45}$$

Starting from an initial point $\boldsymbol{\mu}^{(0)}$, we perform B gradient steps

$$\boldsymbol{\mu}^{(i)} = \boldsymbol{\mu}^{(i-1)} - \epsilon^{(i)}\nabla\varphi(\boldsymbol{\mu}^{(i-1)}), \quad i = 1, \ldots, B, \tag{16.46}$$

where $\epsilon^{(i)}$ is a step size determined by the Armijo rule to ensure sufficient decrease of the objective function. The number of gradient steps B is a tunable parameter that balances per-iteration computational cost with convergence rate. In practice, small values of B (typically 1 to 5) suffice, as the subspace estimate already places us near the optimum.

16.3.3.4 Robust Initialization Against Support Errors

The subspace estimate $\boldsymbol{\mu}^s$ from (16.42) provides an excellent starting point when $\hat{\mathcal{S}}$ is accurate. However, when the support estimate contains errors, $\boldsymbol{\mu}^s$ may actually increase the objective function relative to the previous iterate. To guard against this, we implement an adaptive initialization procedure that selects the better of two candidates:

$$\boldsymbol{\mu}^{(0)} = \arg \min_{\boldsymbol{u}\in\{\boldsymbol{\mu}^s, \boldsymbol{\mu}^p\}} \varphi(\boldsymbol{u}), \tag{16.47}$$

where $\boldsymbol{\mu}^p$ denotes the posterior mean from the previous iteration. This adaptive selection offers several practical advantages.

- **Early iterations**: When $\boldsymbol{\mu}^p$ is still rough, $\boldsymbol{\mu}^s$ may provide a good initial point to accelerate convergence through effective subspace approximation.
- **Later iterations**: As the estimate converges, $\boldsymbol{\mu}^p$ becomes increasingly accurate and is often preferred over a new subspace estimate.
- **Automatic adaptation**: The algorithm automatically selects the better candidate without requiring manual intervention or parameter tuning.

Crucially, this selection guarantees $\varphi(\boldsymbol{\mu}^{(0)}) \leq \varphi(\boldsymbol{\mu}^p)$, ensuring that the objective function does not increase at the start of each iteration. Combined with the descent property of the gradient steps, this ensures monotonic decrease of the overall objective.

16.3.4 Updates for Remaining Variational Distributions

The updates for $q(\boldsymbol{\rho})$, $q(\boldsymbol{s})$, and $q(\kappa)$ are identical to those in standard VBI and the Turbo-VBI algorithm of Chap. 14. We summarize them here for completeness.

16.3.4.1 Update for $q(\rho)$

The posterior distribution of the precision vector factorizes as a product of Gamma distributions:

$$q(\boldsymbol{\rho}) = \prod_{n=1}^{N} \Gamma\left(\rho_n; \tilde{a}_n, \tilde{b}_n\right), \tag{16.48}$$

with parameters

$$\tilde{a}_n = \langle s_n \rangle a_n + \langle 1 - s_n \rangle \bar{a}_n + 1, \tag{16.49}$$

$$\tilde{b}_n = \langle s_n \rangle b_n + \langle 1 - s_n \rangle \bar{b}_n + \langle |x_n|^2 \rangle. \tag{16.50}$$

16.3.4.2 Update for $q(s)$

The posterior distribution of the support vector factorizes as a product of Bernoulli distributions:

$$q(\boldsymbol{s}) = \prod_{n=1}^{N} (\tilde{\lambda}_n)^{s_n} (1 - \tilde{\lambda}_n)^{1-s_n}, \tag{16.51}$$

where

$$\tilde{\lambda}_n = \frac{\lambda_n C_n}{\lambda_n C_n + (1 - \lambda_n)\bar{C}_n}, \tag{16.52}$$

with

$$C_n = \frac{b_n^{a_n}}{\Gamma(a_n)} \exp\left((a_n - 1)\langle \ln \rho_n \rangle - b_n \langle \rho_n \rangle\right), \tag{16.53}$$

$$\bar{C}_n = \frac{\bar{b}_n^{\bar{a}_n}}{\Gamma(\bar{a}_n)} \exp\left((\bar{a}_n - 1)\langle \ln \rho_n \rangle - \bar{b}_n \langle \rho_n \rangle\right). \tag{16.54}$$

16.3.4.3 Update for $q(\kappa)$

The posterior distribution of the noise precision is Gamma

$$q(\kappa) = \Gamma(\kappa; \tilde{c}, \tilde{d}), \tag{16.55}$$

with parameters

$$\tilde{c} = c + M, \tag{16.56}$$

$$\tilde{d} = d + \left\langle \|\boldsymbol{y} - \boldsymbol{A}\boldsymbol{x}\|^2 \right\rangle_{q(\boldsymbol{x})}. \tag{16.57}$$

16.3.4.4 Summary of Expectations

The expectations required in the above updates are

$$\langle \rho_n \rangle = \frac{\tilde{a}_n}{\tilde{b}_n}, \quad \langle \ln \rho_n \rangle = \psi(\tilde{a}_n) - \ln \tilde{b}_n,$$
$$\langle s_n \rangle = \tilde{\lambda}_n, \quad \langle \kappa \rangle = \frac{\tilde{c}}{\tilde{d}},$$
$$\langle |x_n|^2 \rangle = |\mu_n|^2 + \sigma_n^2, \tag{16.58}$$

where $\psi(\cdot)$ denotes the digamma function, the logarithmic derivative of the gamma function.

16.3.5 Convergence Analysis

The SC-VBI algorithm must overcome two fundamental challenges to ensure convergence. First, it relies on estimated support information, which may contain errors that could potentially cause instability. Second, the update for $q(\boldsymbol{x})$ employs an approximate optimization rather than an exact solution. Despite these practical compromises, the algorithm maintains strong theoretical guarantees, as established in the following theorem.

Theorem 16.1 (Convergence of SC-VBI Algorithm) *The sequence of variational distributions generated by the SC-VBI algorithm exhibits the following properties:*

1. *It monotonically decreases the KLD objective in Problem $\mathscr{A}_{\text{SCVBI}}$.*
2. *Every limiting point $q^*(\boldsymbol{v}) = \prod_{k\in\mathcal{H}} q^*(\boldsymbol{v}^k)$ constitutes a stationary solution of Problem $\mathscr{A}_{\text{SCVBI}}$.*

Proof The robust initialization mechanism (16.47) ensures that the initial point for the gradient refinement satisfies $\varphi(\boldsymbol{\mu}^{(0)}) \leq \varphi(\boldsymbol{\mu}^p)$, where $\boldsymbol{\mu}^p$ is the posterior mean from the previous iteration. The Armijo rule guarantees that each gradient step decreases φ, so $\varphi(\boldsymbol{\mu}^{(B)}) \leq \varphi(\boldsymbol{\mu}^{(0)})$. Thus, the update for $q(\boldsymbol{x})$ never increases the KLD objective.

For the other variables $(\boldsymbol{\rho}, \boldsymbol{s}, \kappa)$, the updates are exact minimizations of the KLD with respect to each factor, holding all others fixed. Each such update strictly decreases the KLD unless a stationary point has already been reached.

Since the KLD is bounded below by zero, the sequence of objective values converges. Standard results from block coordinate descent theory then imply that every limit point satisfies the stationary conditions in Sect. 16.3.2.2. For a complete rigorous proof, see Appendix B of [3]. □

The convergence analysis confirms that SC-VBI maintains theoretical soundness while achieving the computational efficiency necessary for practical large-scale applications. This combination of rigor and efficiency makes it particularly attractive for problems where both accuracy and speed are critical.

16.4 Extending to Bilinear Models: Bilinear Subspace-Constrained VBI

16.4.1 Bilinear Sparse Bayesian Inference

The algorithms developed in the preceding sections address problems where a single sparse signal is to be recovered from linear observations. However, many advanced signal processing applications involve a more complex structure: two sparse signals appear in a coupled form, with observations depending bilinearly on both. This section introduces the framework of bilinear sparse Bayesian inference, which addresses the challenge of recovering two sparse vectors, $\boldsymbol{\alpha} \in \mathbb{C}^{N_1\times 1}$ and $\boldsymbol{x} \in \mathbb{C}^{N_2\times 1}$, from such bilinear observations.

16.4.1.1 The Bilinear Observation Model

The general form of the bilinear observation model is

$$\boldsymbol{y} = \text{BL}(\boldsymbol{\alpha}, \boldsymbol{x}) + \boldsymbol{w}, \tag{16.59}$$

where $\text{BL}(\boldsymbol{\alpha}, \boldsymbol{x})$ denotes a bilinear function of the two sparse signals, and $\boldsymbol{w} \in \mathbb{C}^{M\times 1}$ represents additive white Gaussian noise with $w_m \sim \mathcal{CN}(0, \kappa^{-1})$. As before, the noise precision κ follows a Gamma prior

$$p(\kappa) = \Gamma(\kappa; c, d). \tag{16.60}$$

A fundamental characteristic of bilinear models is their conditional linearity, which is the key to developing tractable inference algorithms. When one of the signals is held fixed, the model becomes linear in the other.

Conditioned on $\boldsymbol{x}$: The model becomes linear in $\boldsymbol{\alpha}$:

$$\boldsymbol{y} = \boldsymbol{A}_\alpha(\boldsymbol{x})\boldsymbol{\alpha} + \boldsymbol{w}, \tag{16.61}$$

where $\boldsymbol{A}_\alpha(\boldsymbol{x}) \in \mathbb{C}^{M \times N_1}$ is an effective sensing matrix that depends linearly on $\boldsymbol{x}$.

Conditioned on $\boldsymbol{\alpha}$: The model becomes linear in $\boldsymbol{x}$:

$$\boldsymbol{y} = \boldsymbol{A}_x(\boldsymbol{\alpha})\boldsymbol{x} + \boldsymbol{w}, \tag{16.62}$$

where $\boldsymbol{A}_x(\boldsymbol{\alpha}) \in \mathbb{C}^{M \times N_2}$ depends linearly on $\boldsymbol{\alpha}$.

This conditional linearity property suggests an alternating estimation strategy: we can iteratively estimate one signal while treating the other as known, then switch roles. This is precisely the approach taken by the bilinear subspace-constrained VBI (BiSVBI) algorithm.

16.4.1.2 Hierarchical Priors for Coupled Sparse Signals

To model the sparsity of both signals, we employ the three-layer independent BGG prior for each. For the sparse vector $\boldsymbol{\alpha}$, we define

- A binary support vector $\boldsymbol{s}_\alpha = [s_{\alpha,1}, \ldots, s_{\alpha,N_1}]^T \in \{0, 1\}^{N_1}$, where $s_{\alpha,n} = 1$ indicates that α_n is active.
- A precision vector $\boldsymbol{\rho}_\alpha = [\rho_{\alpha,1}, \ldots, \rho_{\alpha,N_1}]^T$, where $1/\rho_{\alpha,n}$ represents the variance of α_n.

The joint distribution factorizes as

$$p(\boldsymbol{\alpha}, \boldsymbol{\rho}_\alpha, \boldsymbol{s}_\alpha) = p(\boldsymbol{s}_\alpha)\, p(\boldsymbol{\rho}_\alpha \mid \boldsymbol{s}_\alpha)\, p(\boldsymbol{\alpha} \mid \boldsymbol{\rho}_\alpha), \tag{16.63}$$

with the following components.

Layer 1: Support prior.

$$p(\boldsymbol{s}_\alpha) = \prod_{n=1}^{N_1} (\lambda_{\alpha,n})^{s_{\alpha,n}} (1 - \lambda_{\alpha,n})^{1-s_{\alpha,n}}, \tag{16.64}$$

where $\lambda_{\alpha,n}$ is the prior probability of activation.

Layer 2: Conditional precision prior.

$$p(\boldsymbol{\rho}_\alpha \mid \boldsymbol{s}_\alpha) = \prod_{n=1}^{N_1} \left[\Gamma(\rho_{\alpha,n}; a_{\alpha,n}, b_{\alpha,n})\right]^{s_{\alpha,n}} \left[\Gamma(\rho_{\alpha,n}; \bar{a}_{\alpha,n}, \bar{b}_{\alpha,n})\right]^{1-s_{\alpha,n}}. \tag{16.65}$$

Layer 3: Signal prior.

$$p(\boldsymbol{\alpha} \mid \boldsymbol{\rho}_\alpha) = \prod_{n=1}^{N_1} \mathcal{CN}\left(\alpha_n; 0, \rho_{\alpha,n}^{-1}\right). \tag{16.66}$$

The precision parameters are set following the same principles as in previous sections: when $s_{\alpha,n} = 1$ (active), $\mathbb{E}[\rho_{\alpha,n}] = a_{\alpha,n}/b_{\alpha,n} = \Theta(1)$, allowing significant values; when $s_{\alpha,n} = 0$ (inactive), $\mathbb{E}[\rho_{\alpha,n}] = \bar{a}_{\alpha,n}/\bar{b}_{\alpha,n} \gg 1$, forcing the variance toward zero.

Similarly, for $\boldsymbol{x} \in \mathbb{C}^{N_2 \times 1}$, we define a support vector $\boldsymbol{s}_x$ and a precision vector $\boldsymbol{\rho}_x$ with an identical three-layer BGG structure:

$$p(\boldsymbol{x}, \boldsymbol{\rho}_x, \boldsymbol{s}_x) = p(\boldsymbol{s}_x)\, p(\boldsymbol{\rho}_x \mid \boldsymbol{s}_x)\, p(\boldsymbol{x} \mid \boldsymbol{\rho}_x). \tag{16.67}$$

The explicit forms of these distributions mirror those for $\boldsymbol{\alpha}$ and are omitted for brevity.

16.4.1.3 Inference Goals

Given observations $\boldsymbol{y}$, our primary goals are to compute

1. **MMSE estimation of the sparse signals.**

$$\boldsymbol{\alpha}^* = \mathbb{E}[\boldsymbol{\alpha} \mid \boldsymbol{y}], \quad \boldsymbol{x}^* = \mathbb{E}[\boldsymbol{x} \mid \boldsymbol{y}]. \tag{16.68}$$

2. **MAP detection of the support vectors.**

$$s_{\alpha,i}^{*} = \arg\max_{s_{\alpha,i}} p(s_{\alpha,i} \mid \mathbf{y}), \quad s_{x,i}^{*} = \arg\max_{s_{x,i}} p(s_{x,i} \mid \mathbf{y}). \tag{16.69}$$

Exact computation of these marginal posteriors is intractable due to two factors: the coupled bilinear form creates dependencies between $\boldsymbol{\alpha}$ and $\boldsymbol{x}$, and the resulting factor graph contains loops that prevent exact message passing. The BiSVBI algorithm addresses these challenges through efficient approximate inference.

16.4.2 BiSVBI Algorithm

16.4.2.1 Problem Formulation with Constraints

The BiSVBI algorithm extends the SC-VBI framework from linear to bilinear observation models. Let $\boldsymbol{v} \triangleq \{\boldsymbol{\alpha}, \boldsymbol{x}, \boldsymbol{\rho}_{\alpha}, \boldsymbol{\rho}_{x}, \boldsymbol{s}_{\alpha}, \boldsymbol{s}_{x}, \kappa\}$ represent the complete set of unknown variables, indexed by $k \in \mathcal{H} \triangleq \{1, 2, \ldots, 7\}$. The BiSVBI algorithm solves the following constrained minimization problem:

$$\mathcal{A}_{\text{BiSVBI}}: \; q^{*}(\boldsymbol{v}) = \arg\min_{q(\boldsymbol{v})} \int q(\boldsymbol{v}) \ln \frac{q(\boldsymbol{v})}{p(\boldsymbol{v} \mid \mathbf{y})} d\boldsymbol{v}, \tag{16.70}$$

$$\text{s.t. } q(\boldsymbol{v}) = \prod_{k \in \mathcal{H}} q(\boldsymbol{v}^{k}), \quad \int q(\boldsymbol{v}^{k}) d\boldsymbol{v}^{k} = 1, \tag{16.71}$$

$$q(\boldsymbol{\alpha}) = \mathcal{CN}\left(\boldsymbol{\alpha}; \boldsymbol{\mu}_{\alpha}, \text{diag}(\boldsymbol{\sigma}_{\alpha}^{2})\right), \tag{16.72}$$

$$q(\boldsymbol{x}) = \mathcal{CN}\left(\boldsymbol{x}; \boldsymbol{\mu}_{x}, \text{diag}(\boldsymbol{\sigma}_{x}^{2})\right). \tag{16.73}$$

As in SC-VBI, the additional constraints on $q(\boldsymbol{\alpha})$ and $q(\boldsymbol{x})$ serve two crucial purposes:

1. **Computational Efficiency**: By restricting the covariance matrices to diagonal form, we avoid the complexity associated with full matrix inversion in posterior updates. This reduces the per-iteration cost from $O(N_1^3 + N_2^3)$ to $O(N_1 + N_2)$ for the covariance updates.

2. **Subspace Approximation**: The independence constraints enable the application of low-dimensional subspace methods to approximate the posterior mean updates for both signals, further reducing computational cost while maintaining accuracy.

Since Problem $\mathcal{A}_{\text{BiSVBI}}$ is non-convex, we target a stationary point defined as follows.

Definition 16.2 (*Stationary Solution of BiSVBI Problem*) A factorized distribution $q^{*}(\boldsymbol{v}) = \prod_{k \in \mathcal{H}} q^{*}\left(\boldsymbol{v}^{k}\right)$ is a stationary solution of $\mathcal{A}_{\text{BiSVBI}}$ if it satisfies all constraints and for every $k \in \mathcal{H}$:

$$q^{*}\left(\boldsymbol{v}^{k}\right) = \arg\min_{q\left(\boldsymbol{v}^{k}\right)} \int \prod_{l \neq k} q^{*}\left(\boldsymbol{v}^{l}\right) q\left(\boldsymbol{v}^{k}\right) \ln \frac{\prod_{l \neq k} q^{*}\left(\boldsymbol{v}^{l}\right) q\left(\boldsymbol{v}^{k}\right)}{p\left(\boldsymbol{v} \mid \mathbf{y}\right)}. \tag{16.74}$$

For $q(\boldsymbol{\alpha})$ and $q(\boldsymbol{x})$, the minimization is performed over the parameters $\left(\boldsymbol{\mu}_{\alpha}, \boldsymbol{\sigma}_{\alpha}^{2}\right)$ and $\left(\boldsymbol{\mu}_{x}, \boldsymbol{\sigma}_{x}^{2}\right)$, respectively.

16.4.2.2 Optimality Conditions

For variables other than $\boldsymbol{\alpha}$ and $\boldsymbol{x}$, the standard VBI update rule applies

$$q(\boldsymbol{v}^{k}) \propto \exp\left(\langle \ln p(\boldsymbol{v}, \mathbf{y}) \rangle_{\prod_{l \neq k} q(\boldsymbol{v}^{l})}\right), \tag{16.75}$$

where the expectation is taken with respect to all other variational distributions.

For the sparse signals $\boldsymbol{\alpha}$ and $\boldsymbol{x}$, the structural constraints modify the optimality conditions, as characterized by the following lemma.

Lemma 16.2 (Optimal Parameters under Bilinear Constraints) ***For*** $\boldsymbol{\alpha}$: *Given* $q\left(\boldsymbol{v}^{k}\right)$, $\boldsymbol{v}^{k} \neq \boldsymbol{\alpha}$, *the optimal parameters for* $q\left(\boldsymbol{\alpha}\right)$ *are*

$$\boldsymbol{\mu}_\alpha = \arg\min_{\boldsymbol{u}_\alpha} \varphi_\alpha(\boldsymbol{u}_\alpha) \triangleq \boldsymbol{u}_\alpha^H \boldsymbol{W}_\alpha \boldsymbol{u}_\alpha - 2Re\left\{\boldsymbol{u}_\alpha^H \boldsymbol{b}_\alpha\right\},$$

$$\boldsymbol{\sigma}_\alpha^2 = \left[\mathrm{W}_{\alpha,1}^{-1}, ..., \mathrm{W}_{\alpha,N_1}^{-1}\right]^T, \tag{16.76}$$

where $\boldsymbol{W}_\alpha = Diag\left(\langle\boldsymbol{\rho}_\alpha\rangle\right) + \langle\kappa\rangle\left\langle\boldsymbol{A}_\alpha^H\boldsymbol{A}_\alpha\right\rangle$, $\mathrm{W}_{\alpha,n}$ *is the n-th diagonal element of* $\boldsymbol{W}_\alpha$, *and* $\boldsymbol{b}_\alpha = \langle\kappa\rangle\left\langle\boldsymbol{A}_\alpha^H\right\rangle\boldsymbol{y}$ *with* $\boldsymbol{A}_\alpha$ *as a simplified notation for* $\boldsymbol{A}_\alpha(\boldsymbol{x})$.

For $\boldsymbol{x}$***:***

Similarly, given $q\left(\boldsymbol{v}^k\right)$, $\boldsymbol{v}^k \neq \boldsymbol{x}$, *the optimal parameters for* $q(\boldsymbol{x})$ *are*

$$\boldsymbol{\mu}_x = \arg\min_{\boldsymbol{u}_x} \varphi_x(\boldsymbol{u}_x) \triangleq \boldsymbol{u}_x^H \boldsymbol{W}_x \boldsymbol{u}_x - 2Re\left\{\boldsymbol{u}_x^H \boldsymbol{b}_x\right\}$$

$$\boldsymbol{\sigma}_x^2 = \left[\mathrm{W}_{x,1}^{-1}, ..., \mathrm{W}_{x,N_2}^{-1}\right]^T, \tag{16.77}$$

where $\boldsymbol{W}_x = diag\left(\langle\boldsymbol{\rho}_x\rangle\right) + \langle\kappa\rangle\left\langle\boldsymbol{A}_x^H\boldsymbol{A}_x\right\rangle$, $\mathrm{W}_{x,n}$ *is the n-th diagonal element of* $\boldsymbol{W}_x$, *and* $\boldsymbol{b}_x = \langle\kappa\rangle\left\langle\boldsymbol{A}_x^H\right\rangle\boldsymbol{y}$ *with* $\boldsymbol{A}_x$ *as a simplified notation for* $\boldsymbol{A}_x(\boldsymbol{\alpha})$.

Proof Sketch: The result follows from substituting the constrained Gaussian form into the KLD objective and solving the resulting optimization problem. The derivation closely parallels that of Lemma 16.1 for the linear case, with the key difference that the expectations involving $\boldsymbol{A}_\alpha$ and $\boldsymbol{A}_x$ now depend on the other signal's distribution.

16.4.2.3 Update for $q(\boldsymbol{\alpha})$

From Lemma 16.2, the variational distribution for $\boldsymbol{\alpha}$ is

$$q(\boldsymbol{\alpha}) = \mathcal{CN}\left(\boldsymbol{\alpha}; \boldsymbol{\mu}_\alpha, \mathrm{diag}(\boldsymbol{\sigma}_\alpha^2)\right), \tag{16.78}$$

with optimal parameters

$$\boldsymbol{\mu}_\alpha = \boldsymbol{W}_\alpha^{-1}\boldsymbol{b}_\alpha, \quad \boldsymbol{\sigma}_\alpha^2 = \left[W_{\alpha,1}^{-1}, \ldots, W_{\alpha,N_1}^{-1}\right]^T. \tag{16.79}$$

The computation of $\boldsymbol{\mu}_\alpha = \boldsymbol{W}_\alpha^{-1}\boldsymbol{b}_\alpha$ involves inverting the $N_1 \times N_1$ matrix $\boldsymbol{W}_\alpha$, which becomes computationally prohibitive for large N_1. Following the SC-VBI philosophy, we introduce a subspace-constrained approach that proceeds in four stages.

Stage 1: Subspace-Constrained Initial Estimate

Let $\hat{\mathcal{S}}_\alpha$ denote the estimated support of $\boldsymbol{\alpha}$ from the previous iteration. We define an approximate solution $\boldsymbol{\mu}_\alpha^s$ as

$$\boldsymbol{\mu}_{\alpha,\hat{\mathcal{S}}_\alpha}^s = \langle\kappa\rangle \boldsymbol{W}_{\alpha,\hat{\mathcal{S}}_\alpha}^{-1}\left\langle\boldsymbol{A}_{\alpha,\hat{\mathcal{S}}_\alpha}^H\right\rangle\boldsymbol{y}, \quad \boldsymbol{\mu}_{\alpha,\hat{\mathcal{S}}_\alpha^c}^s = \boldsymbol{0}, \tag{16.80}$$

where $\boldsymbol{\mu}_{\alpha,\hat{\mathcal{S}}_\alpha}^s \in \mathbb{C}^{|\hat{\mathcal{S}}_\alpha|\times 1}$ contains the elements corresponding to the estimated support, $\boldsymbol{W}_{\alpha,\hat{\mathcal{S}}_\alpha} \in \mathbb{C}^{|\hat{\mathcal{S}}_\alpha|\times|\hat{\mathcal{S}}_\alpha|}$ is the submatrix of $\boldsymbol{W}_\alpha$ indexed by $\hat{\mathcal{S}}_\alpha$, and $\boldsymbol{A}_{\alpha,\hat{\mathcal{S}}_\alpha} \in \mathbb{C}^{M\times|\hat{\mathcal{S}}_\alpha|}$ is the corresponding submatrix of $\boldsymbol{A}_\alpha$. If $\hat{\mathcal{S}}_\alpha$ matches the true support, $\boldsymbol{\mu}_\alpha^s$ provides an excellent initial point for minimizing $\varphi_\alpha(\boldsymbol{u}_\alpha)$.

Stage 2: Robust Initialization

To ensure robustness against support estimation errors, we select the initial point $\boldsymbol{\mu}_\alpha^{(0)}$ for gradient refinement from two candidates

$$\boldsymbol{\mu}_\alpha^{(0)} = \arg\min_{\boldsymbol{u}\in\{\boldsymbol{\mu}_\alpha^s, \boldsymbol{\mu}_\alpha^p\}} \varphi_\alpha(\boldsymbol{u}), \tag{16.81}$$

where $\boldsymbol{\mu}_\alpha^p$ is the posterior mean from the previous iteration. This adaptive selection provides important practical benefits: in early iterations, $\boldsymbol{\mu}_\alpha^s$ accelerates convergence through effective subspace approximation; in later iterations, $\boldsymbol{\mu}_\alpha^p$ becomes increasingly accurate and is often preferred.

Stage 3: Gradient Refinement

We then perform $B_\alpha \geq 1$ gradient steps to refine the estimate

$$\boldsymbol{\mu}_\alpha^{(i)} = \boldsymbol{\mu}_\alpha^{(i-1)} - \epsilon_\alpha^{(i)} \nabla_{\boldsymbol{u}} \varphi_\alpha(\boldsymbol{u})\big|_{\boldsymbol{u}=\boldsymbol{\mu}_\alpha^{(i-1)}}, \quad i = 1, \ldots, B_\alpha, \tag{16.82}$$

where $\epsilon_\alpha^{(i)}$ is determined by the Armijo rule to ensure sufficient decrease, and the gradient is

$$\nabla_{\boldsymbol{u}} \varphi_\alpha(\boldsymbol{u}) = \boldsymbol{W}_\alpha \boldsymbol{u} - \boldsymbol{b}_\alpha. \tag{16.83}$$

The parameter B_α balances per-iteration cost and convergence rate. This robust design ensures monotonic decrease of $\varphi_\alpha(\boldsymbol{u}_\alpha)$ and convergence to a stationary solution.

Stage 4: Support Update

The support estimate for the next iteration is obtained by energy-based thresholding:

$$\hat{\mathcal{S}}_\alpha = \left\{ n : |\mu_{\alpha,n}|^2 > \varepsilon \right\}, \tag{16.84}$$

where the threshold ε is chosen relative to the noise power as $\varepsilon = (2 \text{ to } 3)/\langle\kappa\rangle$. For the first iteration, $\hat{\mathcal{S}}_\alpha$ can be initialized using prior information or simple algorithms.

16.4.2.4 Update for $q(\rho_\alpha)$

Following the same derivation as in SC-VBI, $q(\boldsymbol{\rho}_\alpha)$ factorizes as a product of Gamma distributions

$$q(\boldsymbol{\rho}_\alpha) = \prod_{n=1}^{N_1} \Gamma\left(\rho_{\alpha,n}; \check{a}_{\alpha,n}, \check{b}_{\alpha,n}\right), \tag{16.85}$$

with parameters

$$\check{a}_{\alpha,n} = \langle s_{\alpha,n} \rangle a_{\alpha,n} + \langle 1 - s_{\alpha,n} \rangle \bar{a}_{\alpha,n} + 1, \tag{16.86}$$

$$\check{b}_{\alpha,n} = \langle s_{\alpha,n} \rangle b_{\alpha,n} + \langle 1 - s_{\alpha,n} \rangle \bar{b}_{\alpha,n} + \left\langle |\alpha_n|^2 \right\rangle. \tag{16.87}$$

16.4.2.5 Update for $q(s_\alpha)$

The posterior distribution of the support vector $\boldsymbol{s}_\alpha$ factorizes as a product of Bernoulli distributions:

$$q(\boldsymbol{s}_\alpha) = \prod_{n=1}^{N_1} (\check{\lambda}_{\alpha,n})^{s_{\alpha,n}} (1 - \check{\lambda}_{\alpha,n})^{1-s_{\alpha,n}}, \tag{16.88}$$

where

$$\check{\lambda}_{\alpha,n} = \frac{\lambda_{\alpha,n} C_{\alpha,n}}{\lambda_{\alpha,n} C_{\alpha,n} + (1 - \lambda_{\alpha,n}) \bar{C}_{\alpha,n}}, \tag{16.89}$$

with

$$C_{\alpha,n} = \frac{b_{\alpha,n}^{a_{\alpha,n}}}{\Gamma(a_{\alpha,n})} \exp\left((a_{\alpha,n} - 1)\langle \ln \rho_{\alpha,n} \rangle - b_{\alpha,n} \langle \rho_{\alpha,n} \rangle\right), \tag{16.90}$$

$$\bar{C}_{\alpha,n} = \frac{\bar{b}_{\alpha,n}^{\bar{a}_{\alpha,n}}}{\Gamma(\bar{a}_{\alpha,n})} \exp\left((\bar{a}_{\alpha,n} - 1)\langle \ln \rho_{\alpha,n} \rangle - \bar{b}_{\alpha,n} \langle \rho_{\alpha,n} \rangle\right). \tag{16.91}$$

16.4.2.6 Updates for $q(x)$, $q(\rho_x)$, $q(s_x)$, and $q(\kappa)$

The update equations for $q(\boldsymbol{x})$, $q(\boldsymbol{\rho}_x)$, and $q(\boldsymbol{s}_x)$ are entirely symmetric to those for $q(\boldsymbol{\alpha})$, $q(\boldsymbol{\rho}_\alpha)$, and $q(\boldsymbol{s}_\alpha)$. They follow the same subspace-constrained approach, gradient refinement, and parameter update formulas with the obvious substitutions. We omit the explicit expressions to avoid redundancy.

The posterior distribution of the noise precision follows the familiar Gamma form:

$$q(\kappa) = \Gamma(\kappa; \check{c}, \check{d}), \tag{16.92}$$

with parameters

$$\check{c} = c + M, \tag{16.93}$$

$$\check{d} = d + \left\langle \|\mathbf{y} - \mathbf{A}_\alpha \boldsymbol{\alpha}\|^2 \right\rangle_{q(\boldsymbol{\alpha})q(\boldsymbol{x})}. \tag{16.94}$$

16.4.2.7 Summary of Expectations

The expectations required in the updates are

$$\langle \rho_{\alpha,n} \rangle = \frac{\check{a}_{\alpha,n}}{\check{b}_{\alpha,n}}, \quad \langle s_{\alpha,n} \rangle = \check{\lambda}_{\alpha,n}, \quad \langle \kappa \rangle = \frac{\check{c}}{\check{d}},$$
$$\langle |\alpha_n|^2 \rangle = |\mu_{\alpha,n}|^2 + \sigma_{\alpha,n}^2, \quad \langle \ln \rho_{\alpha,n} \rangle = \psi(\check{a}_{\alpha,n}) - \ln \check{b}_{\alpha,n}, \tag{16.95}$$

where $\psi(\cdot)$ denotes the digamma function. The expectations $\langle \mathbf{A}_\alpha^H \mathbf{A}_\alpha \rangle$ and $\langle \mathbf{A}_\alpha^H \rangle$ are taken with respect to $q(\boldsymbol{x})$, and the expectations $\langle \mathbf{A}_x^H \mathbf{A}_x \rangle$ and $\langle \mathbf{A}_x^H \rangle$ are taken with respect to $q(\boldsymbol{\alpha})$. These can be computed efficiently given the diagonal covariance structure of the variational distributions.

16.4.2.8 Convergence Analysis

Despite the approximations introduced in the subspace-constrained updates for $\boldsymbol{\mu}_\alpha$ and $\boldsymbol{\mu}_x$, the BiSVBI algorithm retains strong convergence guarantees, as established in the following theorem.

Theorem 16.2 (Convergence of BiSVBI) *The BiSVBI algorithm generates a sequence of variational distributions that possesses the following properties:*

1. ***Monotonic Improvement****: The KLD objective in Problem $\mathscr{A}_{\text{BiSVBI}}$ decreases monotonically with each iteration.*
2. ***Stationary Point Convergence****: Every limiting point $q^*(\boldsymbol{v}) = \prod_{k \in \mathcal{H}} q^*\left(\boldsymbol{v}^k\right)$ of the sequence constitutes a stationary solution of Problem $\mathscr{A}_{\text{BiSVBI}}$.*

Proof sketch: The proof extends the convergence analysis of SC-VBI to the bilinear case. The key insight is that the updates for $\boldsymbol{\alpha}$ and $\boldsymbol{x}$ are each descent steps when the other signal is held fixed. The robust initialization mechanisms ensure that the starting point for each gradient refinement never increases the objective, and the Armijo rule guarantees sufficient decrease in each gradient step. The updates for the remaining variables are exact minimizations that further decrease the KLD. Since the KLD is bounded below, the sequence converges, and standard arguments from block coordinate descent theory imply that every limit point satisfies the stationary conditions. For a complete proof, see [19].

This theoretical foundation ensures that practitioners can employ BiSVBI with confidence in its reliability and mathematical soundness, even for challenging bilinear problems where exact inference is intractable.

16.5 Application I: Channel Estimation in Hybrid Beamforming Systems

16.5.1 Overview and Baseline Algorithms

This section demonstrates the practical application of the IF-VBI and SC-VBI algorithms to a concrete problem introduced in Chap. 13: channel estimation in TDD massive MIMO systems employing HBF. This application provides an ideal testbed for comparing the performance of different variational inference approaches, as it combines the challenges of high-dimensional sparse recovery with the need for computational efficiency in practical implementations.

We evaluate four algorithms, each representing a distinct approach to balancing accuracy and complexity.

Table 16.1 System configuration for channel estimation evaluation

Component	Specification
BS antenna array	UPA
Array dimensions	32 horizontal antennas × 72 vertical antennas
Total antennas	$N_r = 2304$
Compression ratio	$\frac{N_r}{N_{RF}}$, where N_{RF} is the number of RF chains
Elevation angle prior range	$-\frac{1}{6}\pi$ to 0 (practical deployment constraints)
Grid resolution	$N_1 = 32$ (azimuth AoA), $N_2 = 18$ (elevation AoA)
Pilot symbols	Random phase under unit power constraint

1. **EM-based Turbo-VBI (EM-Turbo-VBI)**: Originally proposed in [10, 13], this method combines the exact Turbo-VBI algorithm in its E-step with gradient-based updates for grid parameters in the M-step. It represents the gold standard in terms of accuracy, serving as a performance upper bound, but at the cost of high computational complexity due to exact matrix inversions.
2. **EM-based Turbo-CS (EM-Turbo-CS)**: Based on [11, 12], this approach employs the Turbo-CS algorithm in the E-step and gradient ascent for grid parameter estimation. It represents a state-of-the-art compressed sensing method, though its performance is known to degrade when the sensing matrix deviates from ideal properties.
3. **EM-based Turbo-IF-VBI (EM-Turbo-IF-VBI)**: This algorithm incorporates the IF-VBI method within the E-step, combined with grid parameter optimization in the M-step. It leverages the message-passing framework described in Chap. 13 to exploit the 2D clustered sparsity structure of angular-domain channels, offering reduced per-iteration complexity at the potential cost of slower convergence.
4. **Alternating estimation-based SC-VBI (AE-SC-VBI)**: Proposed in [3], this method departs from the traditional EM framework by updating all parameters simultaneously in a single timescale. It integrates three components:

   - **Module A**: SC-VBI for sparse channel estimation under the independence and subspace constraints.
   - **Module B**: Message-passing for structured support inference, exactly as in Chap. 14.
   - **Grid estimation**: Gradient ascent for refining the angular grid parameters.

   This unified update scheme eliminates the nested iteration structure of EM-based methods, potentially leading to faster convergence.

The comparative study that follows examines the trade-offs among estimation performance, computational complexity, and convergence behavior across these frameworks, providing practical insights for algorithm selection in real-world implementations.

16.5.2 Simulation Setup and Performance Metrics

16.5.2.1 System Configuration

We consider a realistic massive MIMO scenario with the system configuration summarized in Table 16.1. The wireless channel is generated using the 3GPP TR 38.901 standard channel model [9], which produces realistic multipath propagation characteristics essential for meaningful performance evaluation.

16.5.2.2 Performance Metric

The NMSE in decibels serves as the primary performance indicator

$$\text{NMSE (dB)} = 10 \log_{10} \frac{\|\hat{\boldsymbol{h}} - \boldsymbol{h}\|^2}{\|\boldsymbol{h}\|^2}, \tag{16.96}$$

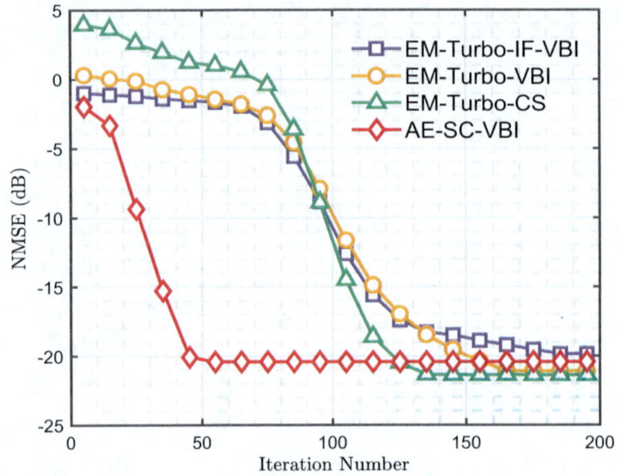

Fig. 16.1 NMSE of channel estimation versus the iteration number (adapted from Fig. 7 in [3])

where $\hat{\boldsymbol{h}}$ denotes the estimated channel and $\boldsymbol{h}$ represents the true channel. This metric inherently reflects the joint estimation accuracy of both elevation and azimuth angles; superior NMSE results indicate improved angle estimation across both dimensions.

All evaluated algorithms employ the 2D-Markov structured prior model introduced in Chap. 13, ensuring a fair comparison of their respective capabilities in exploiting channel sparsity structure.

16.5.3 Simulation Results

16.5.3.1 Convergence Behavior

Figure 16.1 illustrates the convergence behavior of the four algorithms at an SNR of 10 dB and a compression ratio of 4. The results reveal striking differences in convergence rates.

The AE-SC-VBI algorithm demonstrates substantially faster convergence than the three baseline methods, reaching steady-state accuracy in approximately half the iterations. This accelerated convergence can be attributed to its alternating estimation framework, which updates all variables simultaneously within a single timescale. In contrast, the EM-based approaches employ a nested optimization structure where the dynamic grid parameters are updated at a slower timescale than the other variables, creating a bottleneck that slows overall convergence.

Even when comparing algorithms with fixed grid parameters (i.e., disabling the M-step updates), AE-SC-VBI maintains a significant convergence advantage over EM-Turbo-IF-VBI. This performance gap highlights the effectiveness of the subspace-constrained matrix inversion technique in accelerating the optimization process without compromising estimation accuracy.

Importantly, all algorithms eventually achieve similar steady-state performance after convergence. This indicates that AE-SC-VBI provides substantial computational benefits during the convergence phase, exactly where it matters most in practical applications, while maintaining final estimation quality comparable to state-of-the-art methods. The combination of rapid convergence and competitive steady-state performance makes AE-SC-VBI particularly attractive for systems where computational efficiency and rapid channel acquisition are critical requirements.

16.5.3.2 Impact of Compression Ratio (Number of RF Chains)

Figure 16.2 compares the NMSE performance across different compression ratios at a fixed SNR of 10 dB. To ensure a fair comparison of computational efficiency, the total iteration time for EM-Turbo-IF-VBI was configured to slightly exceed that of AE-SC-VBI. Figure 16.3 complements this by quantifying the actual computational cost through CPU time measurements on a standard laptop with a 2.5 GHz processor. The results reveal several important trends.

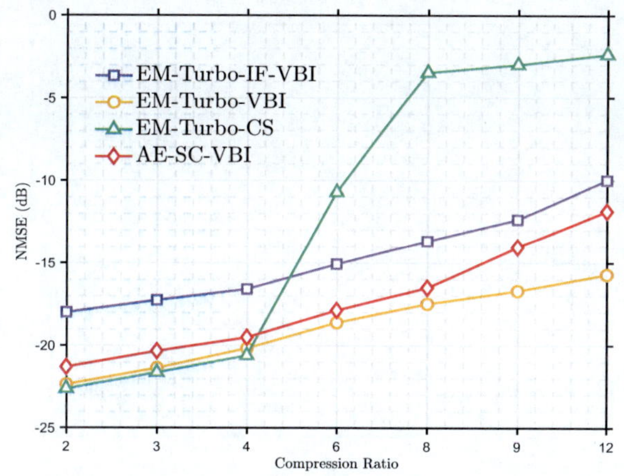

Fig. 16.2 NMSE performance of different channel estimation methods versus the compression ratio (Adapted from Fig. 8 in [3])

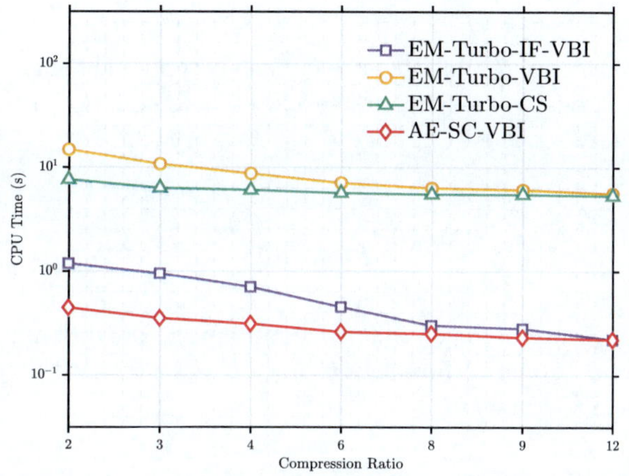

Fig. 16.3 CPU time of different channel estimation methods versus the compression ratio (adapted from Fig. 9 in [3])

Computational Efficiency:

- EM-Turbo-CS and EM-Turbo-VBI require significantly longer CPU time compared to both EM-Turbo-IF-VBI and AE-SC-VBI.
- AE-SC-VBI demonstrates substantially lower computational cost than EM-Turbo-IF-VBI, particularly at low compression ratios. This efficiency advantage stems from the reduced complexity of the subspace-constrained approach compared to the SVD operations required by EM-Turbo-IF-VBI at low compression ratios.

Performance Characteristics:

- EM-Turbo-CS achieves competitive performance at low compression ratios but experiences significant performance degradation when the compression ratio exceeds 4.
- All three VBI-based algorithms (EM-Turbo-VBI, EM-Turbo-IF-VBI, and AE-SC-VBI) maintain strong performance even at high compression ratios.
- AE-SC-VBI consistently outperforms EM-Turbo-IF-VBI across all compression ratios while consuming less computational resources.
- Remarkably, AE-SC-VBI achieves performance approaching that of the computationally intensive EM-Turbo-VBI algorithm, despite its significantly lower complexity.

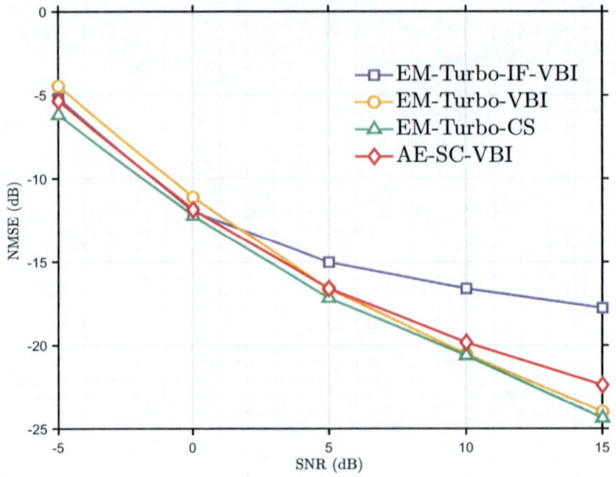

Fig. 16.4 NMSE performance of different channel estimation methods versus the SNR (adapted from Fig. 10 in [3])

These findings establish AE-SC-VBI as an algorithm that achieves an exceptional balance between estimation accuracy and computational efficiency, making it particularly suitable for systems where both performance and practical implementation constraints are important considerations.

16.5.4 Influence of SNR

Figure 16.4 presents the NMSE performance of the considered algorithms across different SNR levels, with the compression ratio fixed at 4. The results demonstrate distinct performance characteristics across the SNR spectrum. The proposed AE-SC-VBI algorithm exhibits several notable advantages:

- **High-SNR Performance**: AE-SC-VBI significantly outperforms EM-Turbo-IF-VBI in the high SNR region, demonstrating its superior capability to leverage improved signal quality for enhanced estimation accuracy.
- **Competitive Performance**: Across the entire SNR range, AE-SC-VBI achieves performance comparable to the high-complexity EM-Turbo-VBI and EM-Turbo-CS algorithms, despite its substantially lower computational requirements.
- **Robustness**: The AE-SC-VBI algorithm maintains consistent performance gains relative to EM-Turbo-IF-VBI while approaching the performance upper bound set by the more computationally intensive benchmarks.

These findings confirm that AE-SC-VBI effectively translates improved signal conditions into better estimation performance without incurring the computational overhead of traditional high-performance algorithms, making it particularly suitable for systems operating across diverse SNR conditions.

16.5.5 Influence of Number of Channel Paths

Figure 16.5 examines the influence of channel path number on estimation performance, with the compression ratio fixed at 8 and SNR at 5 dB. The number of channel paths, determined by the 3GPP TR 38.901 channel model, varies from 12 to 30 paths to represent realistic propagation environments.

The simulation results reveal two key observations:

- **Path Count Degradation**: All algorithms exhibit progressive performance degradation as the number of channel paths increases. This is expected since richer multipath environments present greater estimation challenges due to increased channel complexity.
- **Algorithm Comparison**: The EM-Turbo-CS algorithm demonstrates poor performance at this compression ratio (8), consistent with its previously observed limitations at higher compression ratios. In contrast, the AE-SC-VBI algorithm

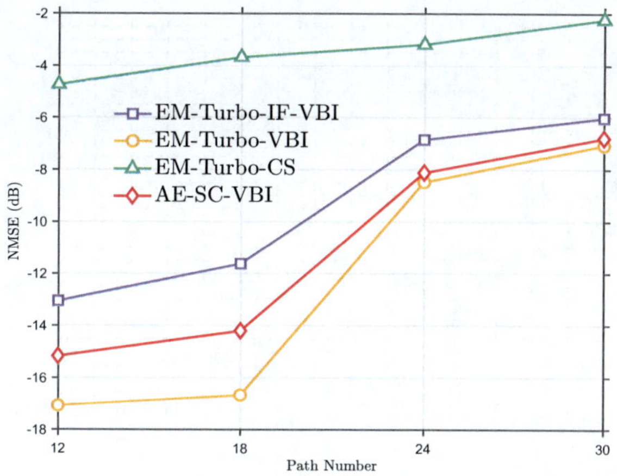

Fig. 16.5 NMSE performance of different channel estimation methods versus the number of paths (adapted from Fig. 11 in [3])

maintains a significant performance advantage over EM-Turbo-IF-VBI across all tested path counts, confirming its robustness to varying channel conditions.

These results underscore AE-SC-VBI's ability to handle diverse multipath scenarios while maintaining reliable estimation performance, further establishing its practical utility in realistic wireless environments.

16.6 Application II: Joint Sensing and Data Recovery in ISAC Systems

16.6.1 Background: ISAC and the Bi-Static Sensing Challenge

The evolution toward 6G wireless systems brings together two key technological trends: wideband OFDM waveforms and massive MIMO architectures. These advances enable communication signals with high resolution in both temporal and angular domains, creating new opportunities for ISAC systems. In ISAC, communication waveforms serve a dual purpose: they support high-rate data transmission while simultaneously enabling high-accuracy environmental sensing [15].

Within this framework, two fundamental ISAC scenarios have emerged [16], each presenting distinct technical characteristics and challenges. The first scenario involves a broadcast channel with mono-static sensing, where the BS functions as both radar transceiver and communication transmitter. Since the BS has complete knowledge of transmitted data symbols, sensing tasks become relatively straightforward due to full waveform knowledge.

The second scenario, a multiple access channel (MAC) with bi-static sensing, presents a far more complex situation. Here, the BS serves as both radar and communication receiver, while users act as both radar and communication transmitters. This configuration introduces significant challenges: the BS must perform joint environment sensing, channel estimation, and data recovery using only limited pilot signals and unknown data transmissions, all while contending with practical impairments such as timing synchronization errors and user position uncertainties.

This section focuses on this more challenging bi-static sensing scenario. We investigate a joint scattering environment sensing and data recovery problem in the uplink of a massive MIMO-OFDM ISAC system, adapted from [19]. The technical challenge involves developing algorithms that enable the BS to simultaneously:

- Detect and localize scatterers in the wireless propagation environment
- Estimate uplink communication channel for a reference user.
- Recover unknown data transmissions from the reference user.

The BS must accomplish these tasks by exploiting both limited uplink pilot signals and unknown data transmissions, making this an ideal application for the BiSVBI framework due to its ability to handle coupled estimation problems with structured sparsity.

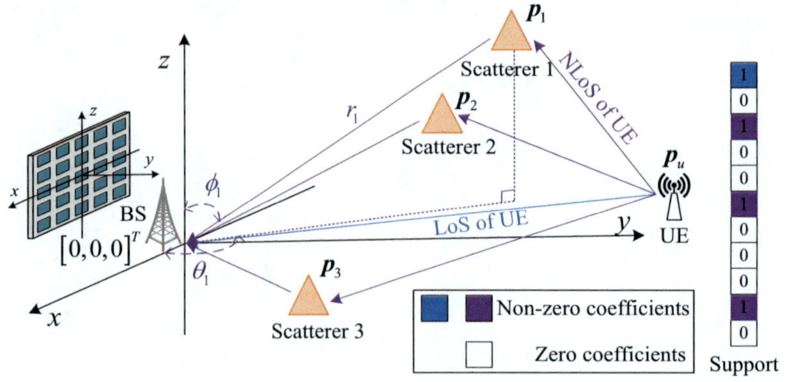

Fig. 16.6 Illustration of the system model, the location-domain channel, and its non-zero coefficients

16.6.2 System Model

16.6.2.1 Spatial Configuration and Geometry

We consider the uplink of a massive MIMO-OFDM ISAC system, where a BS equipped with a UPA of $M = M_x \times M_z$ antennas serves a single-antenna user while simultaneously sensing the scattering environment, as illustrated in Fig. 16.6.

In the spherical coordinate system, the BS array is positioned at $\boldsymbol{p}_b = [0, 0, 0]^T$. The user is located at $\boldsymbol{p}_u = [\theta_u, \phi_u, r_u]^T$, where θ_u, ϕ_u, and r_u denote the azimuth angle, elevation angle, and range, respectively. The environment contains L scatterers indexed by $l \in \mathcal{L} \triangleq \{1, \ldots, L\}$, with the l-th scatterer positioned at $\boldsymbol{p}_l = [\theta_l, \phi_l, r_l]^T$ within a 3D region $\mathcal{R}$. The BS possesses prior information about the user location through GPS or previous localization results.

16.6.2.2 Transmission Framework and Signal Model

The transmission framework comprises T_p pilot OFDM symbols and T_d data OFDM symbols, indexed by $t_p \in \mathcal{T}_p \triangleq \{1, \ldots, T_p\}$ and $t_d \in \mathcal{T}_d \triangleq \{1, \ldots, T_d\}$, respectively. Each OFDM symbol contains N subcarriers indexed by $n \in \mathcal{N} \triangleq \{1, \ldots, N\}$, with pilot subcarriers $\mathcal{N}_{t_p}^p \subseteq \mathcal{N}$ and data subcarriers $\mathcal{N}_{t_d}^d \subseteq \mathcal{N}$ allocated per symbol.

The signal model consists of two components: pilot signals and data signals. For pilot symbols, the received signal on subcarrier n of pilot symbol t_p is

$$\boldsymbol{y}_{t_p,n}^p = \boldsymbol{h}_n u_{t_p,n} + \boldsymbol{z}_{t_p,n}^p, \quad n \in \mathcal{N}_{t_p}^p, \tag{16.97}$$

where $\boldsymbol{h}_n \in \mathbb{C}^{M \times 1}$ is the channel vector, $u_{t_p,n}$ is the pilot symbol, and $\boldsymbol{z}_{t_p,n}^p \sim \mathcal{CN}(\boldsymbol{0}, \gamma^{-1}\boldsymbol{I})$ is AWGN.

For data symbols, the received signal on subcarrier n of data symbol t_d is

$$\boldsymbol{y}_{t_d,n}^d = \boldsymbol{h}_n x_{t_d,n} + \boldsymbol{z}_{t_d,n}^d, \quad n \in \mathcal{N}_{t_d}^d, \tag{16.98}$$

where $x_{t_d,n} \sim \mathcal{CN}(0, \sigma_n^2)$ is the data symbol with power σ_n^2, and $\boldsymbol{z}_{t_d,n}^d \sim \mathcal{CN}(\boldsymbol{0}, \gamma^{-1}\boldsymbol{I})$ is AWGN.

16.6.2.3 Channel Model

The channel model treats the LoS path as the 0-th path and the user as the 0-th scatterer ($\boldsymbol{p}_0 \triangleq \boldsymbol{p}_u$). The channel vector is expressed as

$$\boldsymbol{h}_n = \sum_{l=0}^{L} \alpha_l e^{-j2\pi n f_0 (\tau_o + \tau(\boldsymbol{p}_l, \boldsymbol{p}_u))} \boldsymbol{a}(\theta_l, \phi_l), \tag{16.99}$$

where α_l is the complex gain of the l-th path, f_0 is the subcarrier spacing, and τ_o is the timing synchronization error between user and BS.

The propagation delay $\tau(\boldsymbol{p}_l, \boldsymbol{p}_u)$ is given by

$$\tau(\boldsymbol{p}_l, \boldsymbol{p}_u) = \frac{\sqrt{r_u^2 + r_l^2 - 2r_u r_l g(\boldsymbol{p}_l, \boldsymbol{p}_u)}}{c} + \frac{r_l}{c}, \tag{16.100}$$

where c is the speed of light and

$$g(\boldsymbol{p}_l, \boldsymbol{p}_u) = \sin(\phi_u)\sin(\phi_l)\cos(\theta_u - \theta_l) + \cos(\phi_u)\cos(\phi_l). \tag{16.101}$$

The array response vector $\boldsymbol{a}(\theta, \phi)$ for a UPA with half-wavelength spacing follows the same structure as defined in previous chapters. Note that multiple-bounce NLoS paths can be equivalenced to single-bounce paths via virtual scatterers by solving the corresponding delay equation, a simplification that maintains model tractability without sacrificing accuracy.

16.6.2.4 The Joint Sensing and Communication Objective

The joint sensing and communication objective requires the BS to leverage both pilot and data signals to:

1. **Sense the scattering environment** by detecting scatterers and estimating their positions $\boldsymbol{p}_l$.
2. **Estimate and refine** the time offset τ_o and user location $\boldsymbol{p}_u$.
3. **Reconstruct** the channel vectors $\boldsymbol{h}_n$ for data detection.
4. **Recover** the transmitted data symbols $x_{t_d,n}$.

This creates a mutual enhancement loop: improved sensing yields better channel estimates, which enable more reliable data recovery, and recovered data provides additional information for refining sensing estimates. The BiSVBI framework is designed to exploit this synergy.

16.6.3 Bilinear Sparse Bayesian Inference Formulation

16.6.3.1 3D Location-Domain Sparse Representation

To enable efficient parameter estimation, we first obtain a sparse representation of the channel by discretizing the potential scattering region.

Angular Grid: We construct $Q_a = Q_1 \times Q_2$ angle grid points

$$(\bar{\boldsymbol{\theta}}, \bar{\boldsymbol{\phi}}) = \left\{(\bar{\theta}_{i,j}, \bar{\phi}_{i,j}) \mid i = 1, \ldots, Q_1, j = 1, \ldots, Q_2\right\}, \tag{16.102}$$

uniformly distributed in the transformed angular domain defined by

$$g_1(\theta, \phi) = \sin(\theta)\cos(\phi), \quad g_2(\phi) = \sin(\phi). \tag{16.103}$$

Specifically, the angle grid points are chosen to satisfy

$$g_1(\bar{\theta}_{i,j}, \bar{\phi}_{i,j}) = -\frac{Q_1 - 1}{Q_1} + \frac{2(i-1)}{Q_1}, \tag{16.104}$$

$$g_2(\bar{\phi}_{i,j}) = -\frac{Q_2 - 1}{Q_2} + \frac{2(j-1)}{Q_2}, \tag{16.105}$$

for $i = 1, \ldots, Q_1$ and $j = 1, \ldots, Q_2$.

Range Grid: We define Q_r uniformly spaced points $\{\bar{r}_1, \bar{r}_2, \ldots, \bar{r}_{Q_r}\}$ covering the operational range.

3D Position Grid: The complete 3D grid comprises $Q = Q_a \times Q_r$ total grid points $\{\bar{\boldsymbol{r}}_1, \ldots, \bar{\boldsymbol{r}}_Q\} \subset \mathcal{R}$, formed by combining angular and range coordinates.

To address the inevitable mismatch between discrete grid points and actual scatterer positions, we employ a dynamic grid $\boldsymbol{r} = [\boldsymbol{r}_1; \ldots; \boldsymbol{r}_Q]$ with $\boldsymbol{r}_q = [\theta_q, \phi_q, r_q]^T$, initialized using the uniform 3D position grid and then iteratively refined during estimation.

Based on the dynamic position grid $\boldsymbol{r}$, we define a dictionary matrix

$$\boldsymbol{A}(\boldsymbol{r}, \boldsymbol{p}_u) \triangleq \left[\boldsymbol{a}(\theta_u, \phi_u), \boldsymbol{A}^{\mathrm{NL}}(\boldsymbol{r})\right] \in \mathbb{C}^{M\times(Q+1)}, \tag{16.106}$$

where

$$\boldsymbol{A}^{\mathrm{NL}}(\boldsymbol{r}) \triangleq \left[\boldsymbol{a}(\theta_1, \phi_1), \ldots, \boldsymbol{a}(\theta_Q, \phi_Q)\right] \in \mathbb{C}^{M\times Q}. \tag{16.107}$$

The sparse representation of the channel vector is then given by

$$\boldsymbol{h}_n = \boldsymbol{A}(\boldsymbol{r}, \boldsymbol{p}_u)\boldsymbol{D}_n(\boldsymbol{r}, \boldsymbol{p}_u, \tau_o)\boldsymbol{\alpha}, \tag{16.108}$$

where $\boldsymbol{D}_n(\boldsymbol{r}, \boldsymbol{p}_u, \tau_o)$ is a diagonal matrix whose q-th diagonal element is $e^{-j2\pi n f_0(\tau_o+\tau(\boldsymbol{r}_q, \boldsymbol{p}_u))}$, and $\boldsymbol{\alpha} \in \mathbb{C}^{(Q+1)\times 1}$ is the location-domain sparse channel vector. Non-zero elements of $\boldsymbol{\alpha}$ indicate active scatterers at corresponding grid points.

16.6.3.2 Three-Layer Sparse Prior Model

To capture the sparsity pattern of the location-domain channel, we employ the familiar three-layer hierarchical Bayesian model:

$$p(\boldsymbol{\alpha}, \boldsymbol{\rho}, \boldsymbol{s}) = \underbrace{p(\boldsymbol{s})}_{\text{Support}} \underbrace{p(\boldsymbol{\rho} \mid \boldsymbol{s})}_{\text{Precision}} \underbrace{p(\boldsymbol{\alpha} \mid \boldsymbol{\rho})}_{\text{Sparse signal}}. \tag{16.109}$$

Layer 3: Signal Prior. Conditioned on $\boldsymbol{\rho}$, the sparse channel $\boldsymbol{\alpha}$ follows independent complex Gaussian distributions:

$$p(\boldsymbol{\alpha} \mid \boldsymbol{\rho}) = \prod_{q=0}^{Q} \mathcal{CN}\left(\alpha_q; 0, 1/\rho_q\right). \tag{16.110}$$

Layer 2: Precision Prior. The conditional distribution $p(\boldsymbol{\rho} \mid \boldsymbol{s})$ is

$$p(\boldsymbol{\rho} \mid \boldsymbol{s}) = \prod_{q=0}^{Q} \left(\delta(s_q - 1)\Gamma(\rho_q; a, b) + \delta(s_q)\Gamma(\rho_q; \bar{a}, \bar{b})\right), \tag{16.111}$$

where $\Gamma(\rho; a, b)$ is a Gamma hyperprior with shape a and rate b. When $s_q = 1$, α_q is active with variance $1/\rho_q = O(1)$, requiring $a/b = \mathbb{E}[\rho_q] = O(1)$. When $s_q = 0$, α_q is inactive with variance near zero, requiring $\bar{a}/\bar{b} = \mathbb{E}[\rho_q] \gg 1$.

Layer 1: Support Prior. The support vector follows an independent Bernoulli prior:

$$p(\boldsymbol{s}) = \prod_{q=0}^{Q} p(s_q), \quad p(s_q = 1) = \lambda_q. \tag{16.112}$$

Noise Precision Prior. The noise precision γ follows a Gamma distribution:

$$p(\gamma) = \Gamma(\gamma; c, d). \tag{16.113}$$

16.6.3.3 Bilinear Inference with Uncertain Parameters

For convenience, we define the pilot matrix and data matrix as

$$\boldsymbol{U}_{t_p} \triangleq \mathrm{BlockDiag}\left(u_{t_p,1}\boldsymbol{I}_M, \ldots, u_{t_p,N}\boldsymbol{I}_M\right), \tag{16.114}$$

$$\boldsymbol{X}_{t_d} \triangleq \mathrm{BlockDiag}\left(x_{t_d,1}\boldsymbol{I}_M, \ldots, x_{t_d,N}\boldsymbol{I}_M\right), \tag{16.115}$$

where $u_{t_p,n} = 0$ for $n \notin \mathcal{N}_{t_p}^{p}$, $x_{t_d,n} = 0$ for $n \notin \mathcal{N}_{t_d}^{d}$, and $\mathrm{BlockDiag}(\cdot)$ denotes block diagonalization.

Let $\boldsymbol{\xi} \triangleq \{\boldsymbol{r}, \boldsymbol{p}_u, \tau_o\}$ denote the collection of sensing parameters. Define the sensing matrix

$$\boldsymbol{\Phi}(\boldsymbol{\xi}) \triangleq \left[\boldsymbol{\Phi}_n(\boldsymbol{\xi})\right]_{n\in\mathcal{N}} \in \mathbb{C}^{MN\times(Q+1)}, \tag{16.116}$$

where $\boldsymbol{\Phi}_n(\boldsymbol{\xi}) \triangleq \boldsymbol{A}(\boldsymbol{r}, \boldsymbol{p}_u)\boldsymbol{D}_n(\boldsymbol{r}, \boldsymbol{p}_u, \tau_o)$.

Using the location-domain sparse representation, the received pilot and data signals on all available subcarriers become

$$\boldsymbol{y}_{t_p}^p = \boldsymbol{U}_{t_p}\boldsymbol{\Phi}(\boldsymbol{\xi})\boldsymbol{\alpha} + \boldsymbol{z}_{t_p}^p, \quad \forall t_p, \tag{16.117}$$

$$\boldsymbol{y}_{t_d}^d = \boldsymbol{X}_{t_d}\boldsymbol{\Phi}(\boldsymbol{\xi})\boldsymbol{\alpha} + \boldsymbol{z}_{t_d}^d, \quad \forall t_d, \tag{16.118}$$

where $\boldsymbol{y}_{t_p}^p \triangleq [\boldsymbol{y}_{t_p,n}^p]_{n\in\mathcal{N}} \in \mathbb{C}^{MN\times 1}$, $\boldsymbol{y}_{t_d}^d \triangleq [\boldsymbol{y}_{t_d,n}^d]_{n\in\mathcal{N}} \in \mathbb{C}^{MN\times 1}$, with analogous definitions for the noise vectors. For notational convenience, we include all subcarriers in the above expressions; however, in the actual algorithm, only the subcarriers with $n \in \mathcal{N}_{t_p}^p$ (for pilot signals) and $n \in \mathcal{N}_{t_d}^d$ (for data signals) are processed, as the remaining subcarriers contain no useful information for the sensing sensing the channel of the considered user, since they are allocated to other users.

For fixed sensing parameters $\boldsymbol{\xi}$ and data $\boldsymbol{x}_{t_d} = [x_{t_d,1}, \ldots, x_{t_d,N}]^T \in \mathbb{C}^{N\times 1}$, Eqs. (16.117) and (16.118) form a linear observation model for estimating $\boldsymbol{\alpha}$. For fixed channel $\boldsymbol{\alpha}$ and sensing parameters $\boldsymbol{\xi}$, Eq. (16.118) also forms a linear observation model for data recovery:

$$\boldsymbol{y}_{t_d,n}^d = \boldsymbol{\Psi}_n(\boldsymbol{\xi})x_{t_d,n} + \boldsymbol{z}_{t_d,n}^d, \quad \forall t_d, n, \tag{16.119}$$

where $\boldsymbol{\Psi}_n(\boldsymbol{\xi}) \triangleq \boldsymbol{\Phi}_n(\boldsymbol{\xi})\boldsymbol{\alpha} \in \mathbb{C}^{M\times 1}$.

Thus, the overall model is bilinear in $\boldsymbol{\alpha}$ and $\boldsymbol{x}_{t_d}$, with uncertain sensing parameters $\boldsymbol{\xi}$. Our primary goals are to estimate:

- The location-domain channel vector $\boldsymbol{\alpha}$.
- The data vector $\boldsymbol{x} \triangleq [\boldsymbol{x}_1^T, \ldots, \boldsymbol{x}_{T_d}^T]^T$.
- The support vector $\boldsymbol{s}$.
- The uncertain parameters $\boldsymbol{\xi}$.

From the complete observation $\boldsymbol{y} \triangleq [(\boldsymbol{y}_1^p)^T, \ldots, (\boldsymbol{y}_{T_p}^p)^T, (\boldsymbol{y}_1^d)^T, \ldots, (\boldsymbol{y}_{T_d}^d)^T]^T$.

For given $\boldsymbol{\xi}$, we seek the conditional marginal posteriors of $\boldsymbol{\alpha}$, $\boldsymbol{x}$, and $\boldsymbol{s}$. The uncertain parameters $\boldsymbol{\xi}$ are obtained via MAP estimation:

$$\begin{aligned}\boldsymbol{\xi}^* &= \arg\max_{\boldsymbol{\xi}} \ln p(\boldsymbol{y}, \boldsymbol{\xi}) \\ &= \arg\max_{\boldsymbol{\xi}} \ln \int_{\boldsymbol{v}} p(\boldsymbol{y}, \boldsymbol{v}; \boldsymbol{\xi})\, p(\boldsymbol{\xi})\, d\boldsymbol{v},\end{aligned} \tag{16.120}$$

where $\boldsymbol{v} \triangleq \{\boldsymbol{\alpha}, \boldsymbol{x}, \boldsymbol{\rho}, \boldsymbol{s}, \gamma\}$ collects all latent variables, and $p(\boldsymbol{\xi})$ is the known prior distribution of $\boldsymbol{\xi}$. The marginal posteriors $p(\boldsymbol{x} \mid \boldsymbol{y}; \boldsymbol{\xi})$ for the data can serve as soft information for digital demodulation and channel decoding.

16.6.4 The EM-BiSVBI Algorithm

The EM-BiSVBI algorithm solves the joint sensing and data recovery problem through an iterative procedure that combines BiSVBI with expectation–maximization.

16.6.4.1 Algorithm Structure

The algorithm alternates between two principal steps until convergence.

- **E-step (BiSVBI Estimator)**. Given the current estimate of sensing parameters $\boldsymbol{\xi}^{(t)}$ at iteration t, the algorithm computes approximate marginal posteriors for all latent variables using the BiSVBI framework:

$$q(\boldsymbol{v} \mid \boldsymbol{y}; \boldsymbol{\xi}^{(t)}) \approx p(\boldsymbol{v} \mid \boldsymbol{y}; \boldsymbol{\xi}^{(t)}), \tag{16.121}$$

where $\boldsymbol{v} \triangleq \{\boldsymbol{\alpha}, \boldsymbol{x}, \boldsymbol{\rho}, \boldsymbol{s}, \gamma\}$ encompasses all latent variables. This step follows the BiSVBI methodology detailed in Sect. 16.4, with the simplification that the data symbols $\boldsymbol{x}$ follow a Gaussian prior rather than a sparse prior, leading to streamlined update equations.

Table 16.2 Simulation configuration parameters

Parameter	Value
Base station position	$[0\text{ m}, 0\text{ m}, 0\text{ m}]^T$
BS antenna array	8×8 UPA ($M = 64$ antennas)
User equipment distribution	Random around $[100\text{ m}, 0\text{ m}, 0\text{ m}]^T$
User position prior	$p_u^x \sim \mathcal{N}(100, \sigma_p^2/2)$, $p_u^y \sim \mathcal{N}(0, \sigma_p^2/2)$, $\sigma_p^2 = 1$
Number of scatterers	$L = 6$ (randomly placed)
Subcarriers	$N = 1024$
Carrier frequency	3.5 GHz
Subcarrier spacing	$f_0 = 30$ kHz
Pilot subcarriers	$\lvert\mathcal{N}_{t_p}^p\rvert = 4$ (uniformly distributed)
Data subcarriers	$\lvert\mathcal{N}_{t_d}^d\rvert = 256$
Pilot symbols	$T_p = 1$
Data symbols	$T_d = 2$
Timing offset range	$\tau_o \in \left[-\frac{2}{B}, \frac{2}{B}\right]$, $B = Nf_0$

- **M-step (Parameter Update)**. The algorithm constructs a surrogate function for the MAP objective based on the current variational distribution and optimizes it with respect to the sensing parameters:

$$\boldsymbol{\xi}^{(t+1)} = \arg\max_{\boldsymbol{\xi}} Q(\boldsymbol{\xi}; \boldsymbol{\xi}^{(t)}), \tag{16.122}$$

where the surrogate function $Q(\boldsymbol{\xi}; \boldsymbol{\xi}^{(t)})$ is defined as

$$Q(\boldsymbol{\xi}; \boldsymbol{\xi}^{(t)}) \triangleq \int_{\boldsymbol{v}} q^{(t)}(\boldsymbol{v}) \ln \frac{p(\boldsymbol{y}, \boldsymbol{v}; \boldsymbol{\xi})}{q^{(t)}(\boldsymbol{v})} d\boldsymbol{v} + \ln p(\boldsymbol{\xi}), \tag{16.123}$$

where we use the shorthand notation $q^{(t)}(\boldsymbol{v}) \triangleq q(\boldsymbol{v} \mid \boldsymbol{y}; \boldsymbol{\xi}^{(t)})$ to denote the variational distribution at the t-th iteration. Since $Q(\boldsymbol{\xi}; \boldsymbol{\xi}^{(t)})$ is non-concave, we employ gradient ascent with adaptive step size selection:

$$\boldsymbol{\xi}^{(t+1)} = \boldsymbol{\xi}^{(t)} + \varepsilon^{(t)} \frac{\partial Q(\boldsymbol{\xi}; \boldsymbol{\xi}^{(t)})}{\partial \boldsymbol{\xi}}, \tag{16.124}$$

where the step size $\varepsilon^{(t)}$ is determined via the Armijo rule to ensure monotonic improvement of the objective function.

The alternating optimization between BiSVBI estimation (E-step) and parameter refinement (M-step) creates a virtuous cycle: improved channel estimates enable better data recovery, and recovered data provides additional information for refining sensing parameters. This mutual enhancement is particularly powerful in ISAC systems, where the coupling between sensing and communication is not a nuisance to be overcome but a feature to be exploited.

16.6.5 Simulation Results

16.6.5.1 Simulation Setup and Baseline Algorithms

We evaluate the proposed EM-BiSVBI algorithm through numerical simulations in a 3D coverage area of 100 m × 100 m × 10 m. The simulation configuration is detailed in Table 16.2.

We compare the proposed EM-BiSVBI algorithm against four baseline methods.

1. **LMMSE**: Linear minimum mean square error estimator, serving as a simple reference.
2. **Two-stage Scheme** [6]: Sequential channel estimation followed by data detection, representing conventional approaches that treat sensing and communication separately.

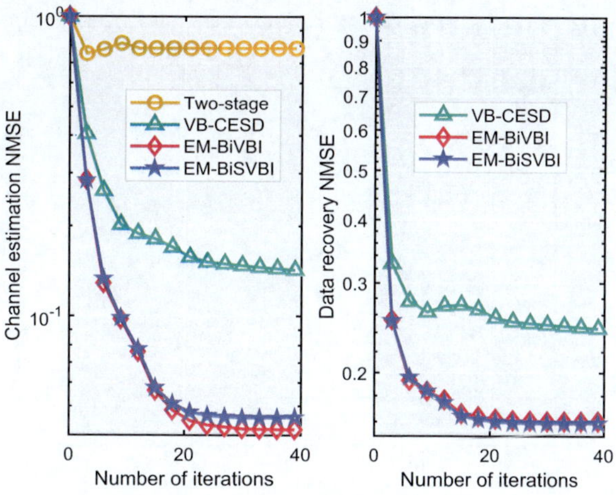

Fig. 16.7 Convergence behavior of different algorithms (adapted from Fig. 2 in [20])

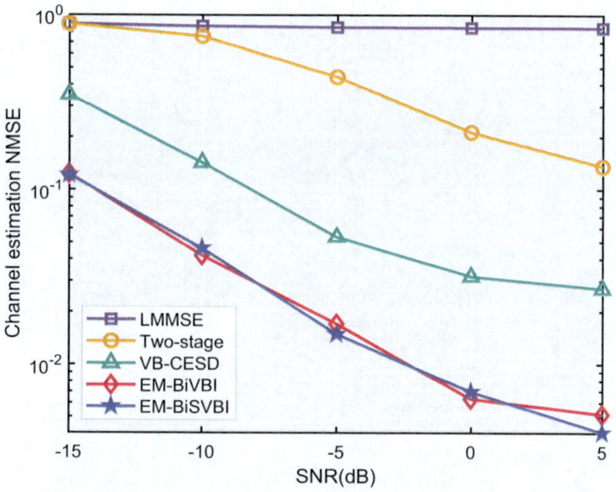

Fig. 16.8 Channel estimation NMSE versus SNR (adapted from Fig. 3 in [20])

3. **VB-CESD** [5]: Variational Bayes joint channel estimation and symbol decoding, a state-of-the-art method for joint problems.
4. **EM-BiVBI**: A variant of EM-BiSVBI that uses exact matrix inversions rather than subspace-constrained approximations, providing a performance upper bound for the subspace approach.

16.6.5.2 Convergence Behavior

Figure 16.7 shows the convergence behavior of different algorithms at SNR $= -10$ dB. The proposed EM-BiSVBI algorithm converges quickly, reaching a stationary solution within approximately 20 iterations. This convergence rate matches that of the more computationally expensive EM-BiVBI, demonstrating that the subspace-constrained approximations do not sacrifice convergence speed.

16.6.5.3 Channel Estimation Performance

Figure 16.8 evaluates channel estimation NMSE performance across different SNR levels. The joint estimation schemes (VB-CESD, EM-BiVBI, and EM-BiSVBI) achieve significant performance gains over the two-stage scheme and LMMSE. This improvement stems from the mutual enhancement between channel estimation and data recovery: better channel estimates improve data detection, and recovered data provides additional information for refining channel estimates.

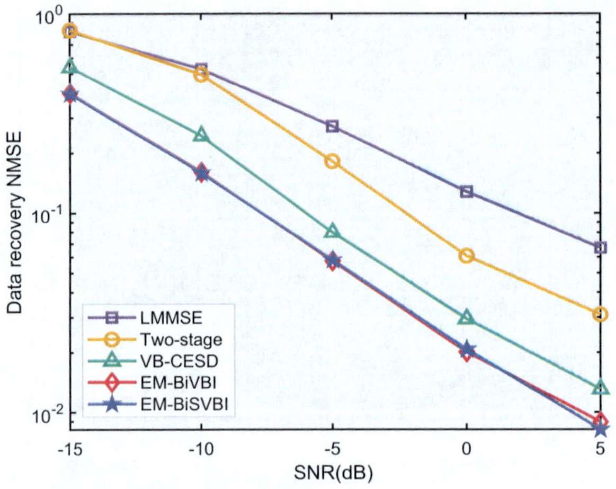

Fig. 16.9 Data recovery NMSE versus SNR (adapted from Fig. 4 in [20])

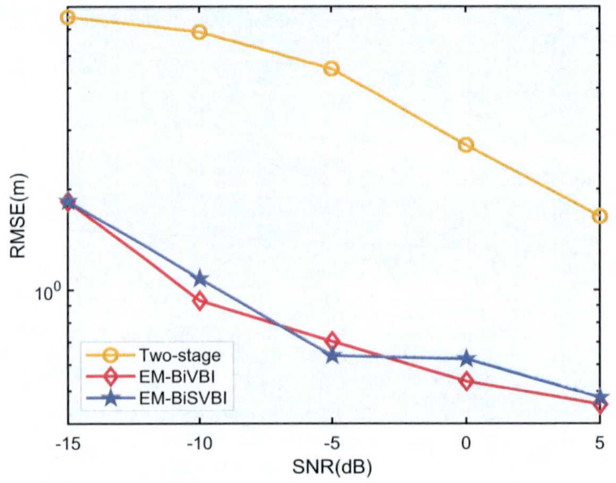

Fig. 16.10 Scatterer localization RMSE versus SNR (adapted from Fig. 5 in [20])

The proposed EM-BiSVBI performs better than VB-CESD, reflecting the advantage of 3D location-domain channel modeling over simpler angular-domain representations. Remarkably, EM-BiSVBI achieves performance very close to that of the more complicated EM-BiVBI, confirming that the subspace-constrained matrix inverse approach effectively reduces computational overhead without sacrificing accuracy.

16.6.5.4 Data Recovery Performance

Figure 16.9 presents data recovery NMSE performance. The same trends observed in channel estimation carry over to data recovery, with joint estimation schemes substantially outperforming sequential approaches. The gap between EM-BiSVBI and EM-BiVBI remains negligible across the entire SNR range, further validating the subspace-constrained approach.

16.6.5.5 Scatterer Localization Performance

Figure 16.10 shows scatterer localization RMSE performance. The ability to accurately localize scatterers is a key differentiator for ISAC systems. The joint estimation schemes achieve excellent localization accuracy, with EM-BiSVBI again matching the performance of EM-BiVBI. The VB-CESD and LMMSE, which do not explicitly model scatterer locations, are not evaluated for this metric.

16.6.5.6 Summary of Findings

The simulation results demonstrate that EM-BiSVBI effectively balances computational efficiency with estimation accuracy. Key findings include

- Joint estimation schemes significantly outperform sequential approaches, confirming the value of exploiting coupling between sensing and communication.
- 3D location-domain modeling provides tangible benefits over simpler angular-domain representations.
- The subspace-constrained matrix inverse approach reduces computational overhead without degrading performance, as EM-BiSVBI matches the accuracy of EM-BiVBI across all metrics.
- Fast convergence (within 20 iterations) makes EM-BiSVBI suitable for real-time applications.

These results position EM-BiSVBI as a promising solution for next-generation ISAC systems, where the tight integration of sensing and communication functionalities demands algorithms that are both accurate and computationally efficient.

Summary

This chapter has explored several advanced variational Bayesian inference algorithms that significantly extend and improve upon previous methods. Below, we summarize the key innovations and their connections to the broader landscape of Bayesian inference.

Complexity Reduction under General Sensing Matrices

Prior algorithms for structured sparse signal recovery often faced a fundamental trade-off between generality and computational efficiency. While Turbo-AMP and Turbo-CS achieved linear complexity, they were limited to specific sensing matrix structures. Turbo-VBI offered broader applicability but required computationally expensive matrix inversions with complexity $O(N^3)$ or $O(N^2M)$.

This chapter introduced two innovative approaches to address this challenge:

- **IF-VBI** [2]: Eliminates matrix inversion entirely by maximizing a relaxed evidence lower bound using the MM framework. While effective, this approach introduces additional MM iterations and requires an initial eigenvalue decomposition, potentially slowing down overall convergence.
- **SC-VBI** [3]: Replaces high-dimensional matrix inversion with low-dimensional subspace approximations, dramatically reducing computational overhead while maintaining convergence speed. By imposing independence and subspace constraints, SC-VBI preserves the original objective's curvature while achieving per-iteration complexity comparable to relaxation-based methods. The resulting Turbo-SC-VBI algorithm represents the first linear-complexity method for structured sparse recovery under general sensing matrices.

Bilinear Subspace Variational Bayesian Inference

Many practical applications, including ISAC, involve bilinear observation models where two sparse signals are coupled. The BiSVBI algorithm in [19] extends SC-VBI principles to this challenging domain by

- Formulating the bilinear sparse Bayesian inference problem as Kullback–Leibler divergence minimization.
- Imposing independence constraints on both sparse signals to avoid covariance matrix inversion.
- Employing subspace-constrained approximations for posterior mean calculation of the two coupled signals.
- Maintaining performance comparable to exact methods with substantially reduced complexity.

The algorithms introduced in this chapter lay a foundation for handling structured sparse signals. Building upon this foundation, several important extensions have emerged that push the boundaries of the VBI framework to address different application scenarios and challenges.

Successive Linear Approximation VBI (SLA-VBI)

Traditional EM-based compressed sensing methods studied in Chap. 14 suffered from slow convergence due to double-loop iterations and complex matrix operations. The SLA-VBI framework in [7] addresses these limitations by

- Providing Bayesian estimation of both sparse signals and dynamic grid parameters within a unified variational framework.
- Employing successive linear approximation to enable closed-form Bayesian inference.
- Incorporating MM principles to avoid matrix inversion, leading to an inverse-free variant referred to as IFSLA-VBI (Inverse-Free Successive Linear Approximation VBI).
- Offering flexibility to balance performance and complexity based on sensing matrix structure.

Notably, the subspace-constrained matrix inverse concept from SC-VBI can be further integrated into IFSLA-VBI to accelerate the MM iterations, demonstrating the complementary nature of these approaches.

Super-Resolution Variational Bayesian Inference

For applications requiring ultra-high resolution, such as channel extrapolation and precise wireless sensing, conventional CS/VBI methods face limitations in sparsity promotion and grid refinement. The emerging two-timescale tanh-VBI framework in [21] addresses these challenges through a novel alternating MAP estimation structure.

A key insight is that the traditional EM algorithm employed in earlier chapters follows a **multiple-E single-M (MESM)** iteration structure: multiple inner E-step iterations are performed between modules A and B before a single M-step update of the grid parameters. While this structure ensures stability, it also creates a bottleneck that slows convergence, particularly when high-resolution grid refinement is required.

The super-resolution VBI (SR-VBI) framework introduces a fundamentally different **single-E multiple-M (SEMM)** iteration structure. Here, **multiple-M** means that multiple grid update steps are performed within a single outer iteration, each update targeting only the currently active grid components identified by the sparse signal estimation module. This design is motivated by two key observations.

1. **Enhanced Resolution**: Frequent updates of the active grid components lead to finer granularity in parameter estimation, directly improving resolution capability. By updating the grid at a faster timescale than the sparse signal, the algorithm can progressively refine parameter estimates without waiting for complete convergence of the signal posterior.
2. **Computational Efficiency**: Since the number of active components in compressed sensing is typically small, updating only this subset incurs minimal computational overhead. The SEMM structure thus achieves faster grid convergence without the burden of processing the entire high-dimensional grid at each iteration.

The SEMM structure can be interpreted as a generalization of the EM framework where the timescale separation between E and M steps is reversed. While conventional EM prioritizes accurate posterior computation (multiple E-steps) before parameter updates (single M-step), SR-VBI prioritizes rapid grid refinement (multiple M-steps) based on current signal estimates, trusting that improved grid parameters will accelerate subsequent signal estimation. This reversal proves particularly effective in scenarios where the grid is dense and accurate parameter placement is critical for resolution.

Complementing this structure, SR-VBI employs two additional innovations:

- **Enhanced sparsity promotion using tanh distributions** that more accurately approximate ℓ_0-norm behavior than traditional Gaussian (ℓ_2-norm) or Laplace (ℓ_1-norm) priors. This stronger sparsity promotion is critical for achieving high resolution in dense grid scenarios where multiple components are closely spaced.
- **Fast-timescale grid refinement using BFGS optimization**, a quasi-Newton method that incorporates second-order curvature information to determine more effective search directions and step sizes. BFGS demonstrates enhanced robustness and reduced tendency to converge to suboptimal local minima, a particularly valuable property in the highly non-convex landscapes characteristic of high-resolution estimation problems.

The above advances collectively represent significant progress toward practical, efficient Bayesian inference for complex signal processing problems in next-generation wireless systems. By rethinking fundamental design choices, from constraint formulation to iteration structure, these algorithms expand the reach of variational methods to applications where accuracy, speed, and scalability are equally critical.

Homework Problems

16.1 Consider the application example in Sect. 16.5.
(a) **Effect of Gradient Update Steps (Parameter B)**:

- Implement the SC-VBI algorithm with different values of B (number of gradient updates in Eq. (16.46)), varying B from 0 to 10.
- Compare the convergence performance in terms of:
  - Convergence speed (number of iterations to reach a specified tolerance).
  - Final objective function value.
  - Computational time per iteration.
- Analyze how the hyperparameter B affects the trade-off between computational cost and convergence performance.
- Discuss the optimal range for B based on your experimental results.

(b) **Effect of Subspace-Constrained Matrix Inverse**:

- Modify the SC-VBI algorithm by removing the subspace-constrained matrix inverse step.
- Compare the convergence performance of the original SC-VBI and the modified version. Analyze the impact on:
  - Convergence stability.
  - Computational complexity per iteration.
  - Final solution quality.
- Discuss the importance of the subspace-constrained matrix inverse step and under what conditions it provides the most benefit.

The MATLAB code for this application example is available at: https://github.com/biwcs-book/codebase.

16.2 Prove Lemma 16.2. Your proof should include

- A clear statement of the lemma's assumptions and conclusions.
- A comprehensive derivation showing all mathematical steps.

16.3 Consider the application example in Sect. 16.6.

- Investigate how the number of pilot subcarriers $\left|\mathcal{N}_{t_p}^{p}\right|$ affects the system performance by
  - Varying the number of pilot subcarriers across a reasonable range.
  - Measuring the estimation accuracy for different pilot configurations.
- Analyzing the data detection performance under reduced pilot overhead. Specifically address the following questions:
  - What is the relationship between pilot subcarrier count and estimation performance?
  - Is it possible to correctly detect data when $\left|\mathcal{N}_{t_p}^{p}\right| = 0$ (no dedicated pilots)?
  - If data detection is possible without pilots, explain the underlying mechanism.
  - If not, explain why?

The MATLAB code for this application example is available at: https://github.com/biwcs-book/codebase.

References

1. H. Duan, L. Yang, J. Fang, H. Li, Fast inverse-free sparse Bayesian learning via relaxed evidence lower bound maximization. IEEE Signal Process. Lett. **24**(6), 774–778 (2017)
2. W. Xu, Y. Xiao, A. Liu, M. Lei, M.-J. Zhao, Joint scattering environment sensing and channel estimation based on non-stationary Markov random field. IEEE Trans. Wirel. Commun. **23**(5), 3903–3917 (2024)
3. A. Liu, Y. Zhou, W. Xu, Subspace constrained variational Bayesian inference for structured compressive sensing with a dynamic grid. IEEE Trans. Signal Process. 1–15 (2025)
4. J.A. Tropp, A.C. Gilbert, Signal recovery from random measurements via orthogonal matching pursuit. IEEE Trans. Inf. Theory **53**(12), 4655–4666 (2007)
5. S.S. Thoota, C.R. Murthy, Massive MIMO-OFDM systems with low resolution ADCs: Cramér-rao bound, sparse channel estimation, and soft symbol decoding. IEEE Trans. Signal Process. **70**, 4835–4850 (2022)
6. T. Cui, C. Tellambura, Semiblind channel estimation and data detection for OFDM systems with optimal pilot design. IEEE Trans. Commun. **55**(5), 1053–1062 (2007)
7. W. Xu, A. Liu, B. Zhou, M.-J. Zhao, Successive linear approximation VBI for joint sparse signal recovery and dynamic grid parameters estimation. IEEE Trans. Wirel. Commun. **24**(11), 9645–9659 (2025)
8. Y. Sun, P. Babu, D.P. Palomar, Majorization-minimization algorithms in signal processing, communications, and machine learning. IEEE Trans. Signal Process. **65**(3), 794–816 (2017)
9. E. Endovitskiy, A. Kureev, E. Khorov, Reducing computational complexity for the 3GPP TR 38.901 MIMO channel model. IEEE Wirel. Commun. Lett. **11**(6), 1133–1136 (2022)
10. A. Liu, L. Lian, V. Lau, G. Liu, M.-J. Zhao, Cloud-assisted cooperative localization for vehicle platoons: a turbo approach. IEEE Trans. Signal Process. **68**, 605–620 (2020)
11. J. Ma, X. Yuan, L. Ping, Turbo compressed sensing with partial DFT sensing matrix. IEEE Signal Process. Lett. **22**(2), 158–161 (2015)
12. A. Liu, L. Lian, V.K.N. Lau, X. Yuan, Downlink channel estimation in multiuser massive MIMO with hidden Markovian sparsity. IEEE Trans. Signal Process. **66**(18), 4796–4810 (2018)
13. A. Liu, G. Liu, L. Lian, V.K.N. Lau, M.-J. Zhao, Robust recovery of structured sparse signals with uncertain sensing matrix: a Turbo-VBI approach. IEEE Trans. Wirel. Commun. **19**(5), 3185–3198 (2020)
14. D.G. Tzikas, A.C. Likas, N.P. Galatsanos, The variational approximation for Bayesian inference. IEEE Signal Process. Mag. **25**(6), 131–146 (2008)
15. F. Liu, Y. Cui, C. Masouros, J. Xu, T.X. Han, Y.C. Eldar, S. Buzzi, Integrated sensing and communications: toward dual-functional wireless networks for 6G and beyond. IEEE J. Sel. Areas Commun. **40**(6), 1728–1767 (2022)
16. A. Liu, Z. Huang, M. Li, Y. Wan, W. Li, T.X. Han, C. Liu, R. Du, D.K.P. Tan, J. Lu, Y. Shen, F. Colone, K. Chetty, A survey on fundamental limits of integrated sensing and communication. IEEE Commun. Surv. & Tutor. **24**(2), 994–1034 (2022)
17. Y. Wan, A. Liu, A two-stage 2D channel extrapolation scheme for TDD 5G NR systems. IEEE Trans. Wirel. Commun. **23**(8), 8497–8511 (2024)
18. S. Ji, Y. Xue, L. Carin, Bayesian compressive sensing. IEEE Trans. Signal Process. **56**(6), 2346–2356 (2008)
19. A. Liu, W. Xu, W. Xu, G. Caire, Bilinear subspace variational Bayesian inference for joint scattering environment sensing and data recovery in ISAC systems. IEEE J. Sel. Areas Commun. **44**, 31–47 (2026)
20. W. Xu, A. Liu, W. Xu, M.-J. Zhao, G. Caire, Joint scattering environment sensing, channel estimation, and data recovery in ISAC systems. In: IEEE 36th International Symposium on Personal, Indoor and Mobile Radio Communications (PIMRC), Istanbul, Turkiye, pp. 1–6 (2025)
21. Y. Zhou, J. Li, W. Xu, A. Liu, Robust super-resolution compressive sensing: a two-timescale alternating MAP approach, pp. 1–10 (2025). arXiv:2508.07013

Bayesian Inference Based on Stochastic Optimization and Deep Learning 17

This chapter explores the integration of advanced stochastic optimization and deep learning frameworks to enhance Bayesian inference methodologies. As a representative framework, we examine particle-based variational distributions, which offer significant flexibility in approximating complex, multi-modal posterior distributions that arise in nonlinear estimation problems. Within this context, we introduce the stochastic particle variational Bayesian inference (SPVBI) algorithm, which addresses the exponential computational complexity of conventional particle-based methods by jointly optimizing both particle positions and weights. This joint optimization, enabled by stochastic successive convex approximation (SSCA), substantially reduces the number of particles required for accurate approximation while accelerating convergence. To further enhance efficiency, we apply deep unfolding to the SPVBI algorithm, transforming its iterative structure into a lightweight deep neural network with trainable parameters. The resulting learnable SPVBI (LSPVBI) network achieves end-to-end optimization through data-driven training, significantly reducing both iteration count and per-iteration complexity. Numerical simulations in a multiband WiFi sensing application confirm that this integrated approach yields substantial improvements in estimation accuracy while reducing computational cost.

17.1 Review of Particle VBI for Non-Convex Parameter Estimation

17.1.1 Problem Formulation Based on VBI Framework

We consider parameter estimation from nonlinear measurements corrupted by additive noise

$$y_m = f_m(\boldsymbol{\theta}) + w_m, \quad \forall m = 1, 2, \ldots M, \tag{17.1}$$

where $\boldsymbol{\theta} = [\theta_1, \ldots, \theta_N]^T$ represents the parameter vector, $f_m(\cdot)$ denotes a potentially nonlinear measurement function, y_m is the m-th measurement and $w_m \sim \mathcal{CN}(0, \sigma_w^2)$ models complex Gaussian noise. The joint likelihood function is given by

$$p(\mathbf{y}|\boldsymbol{\theta}) = \prod_{m=1}^{M} p(y_m|\boldsymbol{\theta}) = \prod_{m=1}^{M} \frac{1}{\pi\sigma_w^2} exp\left(\frac{-|y_m - f_m(\boldsymbol{\theta})|^2}{\sigma_w^2}\right). \tag{17.2}$$

When the functions $f_m(\boldsymbol{\theta})$ are nonlinear, the resulting likelihood landscape becomes non-convex, presenting significant challenges for conventional estimation methods.

Following Bayes' theorem, the posterior distribution is given by

$$p(\boldsymbol{\theta}|\mathbf{y}) \propto p(\mathbf{y}|\boldsymbol{\theta})\, p(\boldsymbol{\theta}), \tag{17.3}$$

where $p(\boldsymbol{\theta})$ represents the prior distribution. For computational tractability, the parameters are often partitioned into J blocks: $\boldsymbol{\theta} = \{\boldsymbol{\theta}_1, \boldsymbol{\theta}_2, \ldots, \boldsymbol{\theta}_J\}$. The marginal posterior for the j-th block requires integration over all other blocks, i.e.,

$$p(\boldsymbol{\theta}_j|\mathbf{y}) \propto \int p(\mathbf{y}|\boldsymbol{\theta})\, p(\boldsymbol{\theta})\, d\boldsymbol{\theta}_{\sim j}, \tag{17.4}$$

A. Liu and M.-M. Zhao, *Advanced Bayesian Signal Processing*, Textbooks in Telecommunication Engineering,
https://doi.org/10.1007/978-3-032-26656-9_17

where $\boldsymbol{\theta}_{\sim j}$ denotes all blocks except $\boldsymbol{\theta}_j$. This high-dimensional integration, combined with potential nonlinearities in $f_m(\boldsymbol{\theta})$, generally precludes closed-form solutions.

VBI addresses this intractability by approximating the true posterior with a simpler variational distribution $q(\boldsymbol{\theta})$. Under the mean-field assumption, this distribution factorizes across parameter blocks as

$$q(\boldsymbol{\theta}) = \prod_{j=1}^{J} q(\boldsymbol{\theta}_j). \tag{17.5}$$

The optimal variational approximation minimizes the KLD between $q(\boldsymbol{\theta})$ and the true posterior, leading to the following optimization problem:

$$\mathcal{P}_1 : \min_{q(\boldsymbol{\theta})} \int q(\boldsymbol{\theta}) \ln \frac{q(\boldsymbol{\theta})}{p(\boldsymbol{y}|\boldsymbol{\theta})\, p(\boldsymbol{\theta})} d\boldsymbol{\theta} \tag{17.6}$$

$$\text{s.t.} \quad q(\boldsymbol{\theta}) = \prod_{j=1}^{J} q(\boldsymbol{\theta}_j), \tag{17.7}$$

$$\int q(\boldsymbol{\theta}_j)\, d\boldsymbol{\theta}_j = 1, \forall j. \tag{17.8}$$

As Problem $\mathcal{P}_1$ is generally non-convex, we focus on designing efficient algorithms to find its stationary points. Two main challenges arise:

- The objective function often contains many suboptimal local minima due to the non-convexity and high dimensionality of the likelihood.
- Marginal posterior distributions are difficult to derive due to complex multi-dimensional integrals and unstructured distributional assumptions.

17.1.2 Particle-Based Variational Distributions

To capture multi-modal posterior distributions that arise in non-convex problems, particle-based VBI employs a discrete variational distribution with N_p particles

$$q(\boldsymbol{\theta}_j; \boldsymbol{w}_j) = \sum_{n=1}^{N_p} w_{j,n} \delta(\boldsymbol{\theta}_j - \boldsymbol{p}_{j,n}) \tag{17.9}$$

where $\{\boldsymbol{p}_{j,n}\}_{n=1}^{N_p}$ represent particle positions and $\{w_{j,n}\}_{n=1}^{N_p}$ are their corresponding weights. This approach offers several advantages.

- **Multi-Modal Representation**: As shown in Fig. 17.1, particles can capture multiple modes of complex posterior distributions.
- **Prior Information**: Initial positions can be sampled from the prior distribution $p(\boldsymbol{\theta}_j)$.
- **Adaptive Refinement**: Iterative weight updates enable progressive approximation improvement.

However, conventional particle-based VBI algorithm suffers from two primary limitations.

- **Static Particle Positions**: The algorithm updates only particle weights, not their positions, potentially requiring many particles for accurate approximation.
- **Computational Complexity**: Weight optimization involves summations with complexity growing exponentially with the number of particles and parameter dimensions.

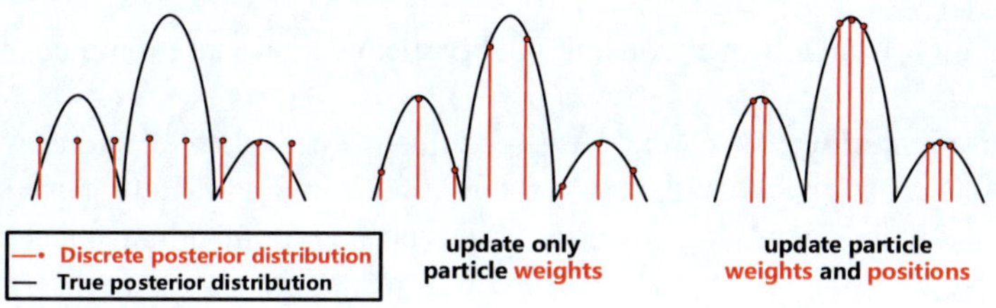

Fig. 17.1 Illustration of particle approximation

The following sections introduce stochastic optimization techniques that overcome these limitations by enabling simultaneous optimization of particle positions and weights, significantly improving computational efficiency and convergence properties.

17.2 Stochastic Particle Variational Bayesian Inference

17.2.1 Core Ideas and Problem Formulation

We now introduce the stochastic particle variational Bayesian inference (SPVBI) framework [5, 19], which enhances traditional particle-based variational inference by incorporating stochastic optimization techniques. Unlike conventional approaches that treat particles as static once initialized, SPVBI dynamically adjusts both the positions and weights of particles throughout the optimization process.

The method employs a discrete variational distribution constructed from a finite set of particles. For the j-th parameter block $\boldsymbol{\theta}_j$, this distribution takes the form

$$q(\boldsymbol{\theta}_j; \boldsymbol{p}_j, \boldsymbol{w}_j) = \sum_{n=1}^{N_p} w_{j,n}\, \delta(\boldsymbol{\theta}_j - \boldsymbol{p}_{j,n}), \tag{17.10}$$

where for each block j, $\boldsymbol{p}_{j,n} \in \mathbb{R}^{d_j}$ denotes the position of the n-th particle (with d_j being the dimension of $\boldsymbol{\theta}_j$), $w_{j,n} \in [0, 1]$ represents its corresponding probabilistic weight, N_p specifies the total number of particles per block, $\delta(\cdot)$ denotes the Dirac delta function.

The key advancement of SPVBI lies in its simultaneous optimization of both particle positions and weights $\{\boldsymbol{p}_{j,n}, w_{j,n} | \forall n = 1, \ldots, N_p\}$, in stark contrast to conventional particle-based VBI methods that only adjust weights while keeping positions fixed. This additional flexibility allows the variational distribution to adapt more effectively to the true posterior geometry.

Given this particle-based representation, the variational inference problem is reformulated as the following constrained optimization problem

$$\begin{aligned}
\mathcal{P}_2: \min_{\boldsymbol{p},\boldsymbol{w}} \quad & l(\boldsymbol{p}, \boldsymbol{w}) \triangleq \int q(\boldsymbol{\theta}; \boldsymbol{p}, \boldsymbol{w}) \ln \frac{q(\boldsymbol{\theta}; \boldsymbol{p}, \boldsymbol{w})}{p(\boldsymbol{y}|\boldsymbol{\theta})p(\boldsymbol{\theta})} d\boldsymbol{\theta} \\
\text{s.t.} \quad & \sum_{n=1}^{N_p} w_{j,n} = 1, \quad \epsilon \le w_{j,n} \le 1, \quad \forall j, n, \\
& \hat{\boldsymbol{\theta}}_j - \Delta\hat{\boldsymbol{\theta}}_j/2 \le \boldsymbol{p}_{j,n} \le \hat{\boldsymbol{\theta}}_j + \Delta\hat{\boldsymbol{\theta}}_j/2, \quad \forall j, n.
\end{aligned} \tag{17.11}$$

The constraints in this formulation serve three critical purposes:

- **Probability simplex constraint**: The conditions that weights sum to unity and are bounded above by one ensure that $\{w_{j,n}\}$ constitutes a valid probability distribution.
- **Numerical stability**: The strict lower bound $\epsilon > 0$ on each weight prevents numerical underflow in computations and maintains adequate particle diversity throughout optimization.
- **Prior-informed search**: The confinement of particle positions $\boldsymbol{p}_{j,n}$ to the hyper-rectangle defined by $\hat{\boldsymbol{\theta}}_j \pm \Delta\hat{\boldsymbol{\theta}}_j/2$ concentrates computational resources on regions of high prior probability, accelerating convergence.

Throughout this chapter, all vector inequalities are interpreted element-wise.

SPVBI provides significant improvements over conventional particle variational inference methods in three key aspects.

1. **Enhanced Approximation Capability**: Dynamic position adjustment enables accurate modeling of multi-modal and skewed distributions, a critical capability when the true posterior deviates significantly from simple parametric forms. The adaptive particles can effectively track evolving posterior modes during optimization, substantially reducing sensitivity to initial particle placement and preventing premature convergence to suboptimal solutions.
2. **Improved Computational Efficiency**: Coordinated position-weight updates enable faster convergence while reducing memory and computational requirements. By allowing particles to migrate toward regions of high posterior probability, the method requires fewer particles for equivalent approximation accuracy compared to approaches that only adjust weights. This particle efficiency makes SPVBI particularly suitable for high-dimensional problems where the cost of maintaining many particles would otherwise be prohibitive.
3. **Robust Optimization Properties**: The stochastic optimization framework inherent in SPVBI facilitates escape from local minima through its inherent randomness. The constrained search space, informed by prior knowledge, improves convergence stability by preventing particles from wandering into low-probability regions. The integration of prior information further enhances estimation reliability, particularly in scenarios with limited data.

The following subsections develop the parallel stochastic successive convex approximation (PSSCA) methodology that enables efficient solution of $\mathcal{P}_2$ and provide comprehensive analysis of the algorithm's theoretical properties and practical performance characteristics.

17.2.2 SPVBI Algorithm Design Based on Parallel SSCA

While updating particle positions accelerates convergence, the multiple summations in the objective function $l(\boldsymbol{p}, \boldsymbol{w})$ of Problem $\mathcal{P}_2$ remain computationally prohibitive. To address this fundamental challenge, the authors in [19] leverage the PSSCA framework [5, 6], a methodology specifically designed to handle the intractable integrals and summations that arise naturally in Bayesian inference problems.

17.2.2.1 Reformulation as a Stochastic Optimization Problem

The mean-field assumption, which postulates independence among parameter blocks, plays a crucial role in simplifying the variational posterior. Under this assumption, the joint distribution factorizes as

$$q(\boldsymbol{\theta}; \boldsymbol{p}, \boldsymbol{w}) = \prod_{j=1}^{J} q(\boldsymbol{\theta}_j; \boldsymbol{p}_j, \boldsymbol{w}_j). \tag{17.12}$$

This factorization yields a convenient decomposition of the entropy term. Because the logarithm of a product becomes a sum, the entropy integral separates across blocks

$$\begin{aligned} \int q(\boldsymbol{\theta}) \ln q(\boldsymbol{\theta})\, d\boldsymbol{\theta} &= \sum_{j=1}^{J} \int q(\boldsymbol{\theta}_j) \ln q(\boldsymbol{\theta}_j)\, d\boldsymbol{\theta}_j \\ &= \sum_{j=1}^{J} \sum_{n_j=1}^{N_p} w_{j,n_j} \ln w_{j,n_j} \triangleq C(\boldsymbol{w}). \end{aligned} \tag{17.13}$$

Here and in what follows, we use the shorthand $q(\boldsymbol{\theta})$ to denote $q(\boldsymbol{\theta}; \boldsymbol{p}, \boldsymbol{w})$. Expanding the full objective function $l(\boldsymbol{p}, \boldsymbol{w})$ using this decomposition gives

$$
\begin{aligned}
l(\boldsymbol{p}, \boldsymbol{w}) &= \sum_{j=1}^{J} \int q(\boldsymbol{\theta}_j) \ln q(\boldsymbol{\theta}_j)\, d\boldsymbol{\theta}_j \\
&\quad - \int q(\boldsymbol{\theta}) \left[\ln p(\boldsymbol{y}|\boldsymbol{\theta}) + \ln p(\boldsymbol{\theta})\right] d\boldsymbol{\theta} \\
&= C(\boldsymbol{w}) - \sum_{n_1=1}^{N_p} \cdots \sum_{n_J=1}^{N_p} \tilde{w} \left[\ln p(\boldsymbol{y}|\boldsymbol{\theta}) + \ln p(\boldsymbol{\theta})\right]\Big|_{\boldsymbol{\theta}=\tilde{\boldsymbol{p}}}, \qquad (17.14)
\end{aligned}
$$

where $\tilde{w} = \prod_{j=1}^{J} w_{j,n_j}$ and $\tilde{\boldsymbol{p}} = [\boldsymbol{p}_{1,n_1}^T, \ldots, \boldsymbol{p}_{j,n_j}^T, \ldots, \boldsymbol{p}_{J,n_J}^T]^T$.

Attention
Equation (17.14) reveals the core computational difficulty: the second term involves a summation over N_p^J terms. This exponential scaling with the number of variables J renders direct optimization of $\mathcal{P}_2$ intractable for any problem of practical interest. While the entropy term $C(\boldsymbol{w})$ decomposes nicely under mean-field, the expected log-likelihood and log-prior require summing over the entire Cartesian product of all particle sets, a combinatorial explosion that the PSSCA framework is specifically designed to circumvent.

To overcome this exponential complexity, we reformulate the objective as an expectation

$$
l(\boldsymbol{p}, \boldsymbol{w}) = \mathbb{E}_{q(\boldsymbol{\theta})}\left[g(\boldsymbol{p}, \boldsymbol{w}; \boldsymbol{\theta})\right], \qquad (17.15)
$$

where $\mathbb{E}_{q(\boldsymbol{\theta})}[\cdot]$ denotes expectation with respect to the variational distribution $q(\boldsymbol{\theta})$, and

$$
g(\boldsymbol{p}, \boldsymbol{w}; \boldsymbol{\theta}) = C(\boldsymbol{w}) - \left[\ln p(\boldsymbol{y}|\boldsymbol{\theta}) + \ln p(\boldsymbol{\theta})\right]. \qquad (17.16)
$$

This reformulation casts $\mathcal{P}_2$ as a stochastic optimization problem with a decision-dependent distribution: the random state $\boldsymbol{\theta} \sim q(\boldsymbol{\theta}; \boldsymbol{p}, \boldsymbol{w})$ is itself parameterized by the optimization variables $\boldsymbol{p}$ and $\boldsymbol{w}$. At iteration t, we denote $q^{(t)}(\boldsymbol{\theta}) \triangleq q(\boldsymbol{\theta}; \boldsymbol{p}^{(t)}, \boldsymbol{w}^{(t)})$ for brevity.

Attention
This reformulation is a pivotal conceptual step. It transforms an intractable deterministic problem into a stochastic one that can be approximated via Monte Carlo sampling. The decision-dependent nature, where the sampling distribution $q^{(t)}(\boldsymbol{\theta})$ evolves as we update $\boldsymbol{p}$ and $\boldsymbol{w}$, adds complexity but is precisely what enables the application of stochastic approximation methods. By sampling from the current distribution, we obtain noisy gradient estimates that guide the optimization, effectively replacing the exponential summation with tractable Monte Carlo averages.

17.2.2.2 Derivation of SPVBI Algorithm Based on the PSSCA Framework

The SSCA methodology [6] provides a systematic way to handle stochastic optimization problems. Rather than optimizing the original non-convex objective directly, SSCA iteratively constructs and optimizes a sequence of convex surrogate functions, as illustrated in Fig. 17.2.

For updating the particle positions $\boldsymbol{p}_j$ and weights $\boldsymbol{w}_j$ of the j-th variable $\boldsymbol{\theta}_j$, we adopt quadratic surrogate functions

$$
\bar{f}_{p_j}^{(t)}(\boldsymbol{p}_j) = \left(\boldsymbol{f}_{p_j}^{(t)}\right)^T \left(\boldsymbol{p}_j - \boldsymbol{p}_j^{(t)}\right) + \frac{1}{2\Gamma_{p_j}^{(t)}} \left\|\boldsymbol{p}_j - \boldsymbol{p}_j^{(t)}\right\|^2, \qquad (17.17)
$$

$$
\bar{f}_{w_j}^{(t)}(\boldsymbol{w}_j) = \left(\boldsymbol{f}_{w_j}^{(t)}\right)^T \left(\boldsymbol{w}_j - \boldsymbol{w}_j^{(t)}\right) + \frac{1}{2\Gamma_{w_j}^{(t)}} \left\|\boldsymbol{w}_j - \boldsymbol{w}_j^{(t)}\right\|^2. \qquad (17.18)
$$

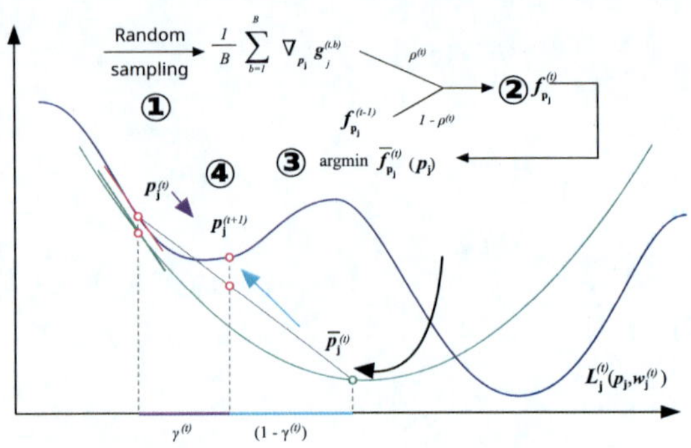

Fig. 17.2 Positional update in a single iteration of the stochastic successive convex approximation (Adapted from Fig. 3 in [19])

Here, $\boldsymbol{f}_{p_j}^{(t)}$ and $\boldsymbol{f}_{w_j}^{(t)}$ are unbiased estimators of the gradients $\nabla_{\boldsymbol{p}_j} l(\boldsymbol{p}^{(t)}, \boldsymbol{w}^{(t)})$ and $\nabla_{\boldsymbol{w}_j} l(\boldsymbol{p}^{(t)}, \boldsymbol{w}^{(t)})$, respectively. The parameters $\Gamma_{p_j}^{(t)}$ and $\Gamma_{w_j}^{(t)}$ are positive step sizes that control the curvature of the quadratic approximation.

These quadratic surrogates are carefully chosen for two reasons. First, they are strongly convex, ensuring that each subproblem has a unique global minimum. Second, they admit closed-form solutions that can be interpreted as gradient projection steps [5, 7], i.e.,

$$\bar{\boldsymbol{p}}_j^{(t)} = \text{Proj}_{\Delta p}\left(\boldsymbol{p}_j^{(t)} - \Gamma_{p_j}^{(t)} \boldsymbol{f}_{p_j}^{(t)}\right), \tag{17.19}$$

$$\bar{\boldsymbol{w}}_j^{(t)} = \text{Proj}_{\Delta w}\left(\boldsymbol{w}_j^{(t)} - \Gamma_{w_j}^{(t)} \boldsymbol{f}_{w_j}^{(t)}\right). \tag{17.20}$$

The projection operators $\text{Proj}_{\Delta p}$ and $\text{Proj}_{\Delta w}$ enforce the constraints from Problem $\mathcal{P}_2$: particle positions are confined to a prior-informed interval, and weights are projected onto the probability simplex.

To ensure stable convergence, we do not simply set the new iterates to these projected values. Instead, we employ a smoothing update that mixes the current iterate with the proposed update

$$\boldsymbol{p}_j^{(t+1)} = (1 - \gamma^{(t)})\boldsymbol{p}_j^{(t)} + \gamma^{(t)}\bar{\boldsymbol{p}}_j^{(t)}, \tag{17.21}$$

$$\boldsymbol{w}_j^{(t+1)} = (1 - \gamma^{(t)})\boldsymbol{w}_j^{(t)} + \gamma^{(t)}\bar{\boldsymbol{w}}_j^{(t)}. \tag{17.22}$$

Here, $\gamma^{(t)} \in (0, 1)$ is a decreasing step size that controls the update magnitude.

Attention

The PSSCA framework embodies a simple yet powerful idea: replace a difficult optimization problem with a sequence of easier ones. The quadratic surrogates are not merely convenient mathematical constructs. They have an intuitive geometric interpretation. Each surrogate is a local approximation of the original objective, constructed to be convex and easy to minimize. The gradient projection interpretation reveals that solving the surrogate is equivalent to taking a step in the direction of the negative gradient, followed by a projection back into the feasible set. The subsequent smoothing update with $\gamma^{(t)}$ acts as a memory mechanism, preventing oscillatory behavior and ensuring that the iterates evolve smoothly.

The heart of the algorithm lies in constructing unbiased gradient estimators $\boldsymbol{f}_{p_j}^{(t)}$ and $\boldsymbol{f}_{w_j}^{(t)}$. From the expectation form of the objective, the true gradients can be expressed as

$$\nabla_{\boldsymbol{p}_j} l(\boldsymbol{p}^{(t)}, \boldsymbol{w}^{(t)}) = \mathbb{E}_{q^{(t)}(\boldsymbol{\theta}_{\sim j})}\left[\nabla_{\boldsymbol{p}_j} g_j(\boldsymbol{p}_j^{(t)}, \boldsymbol{w}_j^{(t)}; \boldsymbol{\theta}_{\sim j})\right], \tag{17.23}$$

$$\nabla_{\boldsymbol{w}_j} l(\boldsymbol{p}^{(t)}, \boldsymbol{w}^{(t)}) = \mathbb{E}_{q^{(t)}(\boldsymbol{\theta}_{\sim j})}\left[\nabla_{\boldsymbol{w}_j} g_j(\boldsymbol{p}_j^{(t)}, \boldsymbol{w}_j^{(t)}; \boldsymbol{\theta}_{\sim j})\right], \tag{17.24}$$

where g_j isolates the terms that depend on the j-th block

$$g_j(\boldsymbol{p}_j, \boldsymbol{w}_j; \boldsymbol{\theta}_{\sim j}) = C(\boldsymbol{w}) - \sum_{n_j=1}^{N_p} w_{j,n_j} \left[\ln p(\boldsymbol{p}_{j,n_j}) + \ln p(\boldsymbol{y}|\boldsymbol{\theta}_{\sim j}, \boldsymbol{p}_{j,n_j})\right]. \tag{17.25}$$

To approximate these expectations, we generate a mini-batch of B samples

$$\{\boldsymbol{\theta}^{(b)} = [\boldsymbol{\theta}_1^{(b)}, \ldots, \boldsymbol{\theta}_J^{(b)}]^T, b = 1, \ldots, B\}, \tag{17.26}$$

drawn from the current product distribution $\prod_{j=1}^{J} q(\boldsymbol{\theta}_j; \boldsymbol{p}_j^{(t)}, \boldsymbol{w}_j^{(t)})$. Using these samples, we update the gradient estimators recursively

$$\boldsymbol{f}_{\boldsymbol{p}_j}^{(t)} = (1 - \rho^{(t)})\boldsymbol{f}_{\boldsymbol{p}_j}^{(t-1)} + \frac{\rho^{(t)}}{B} \sum_{b=1}^{B} \nabla_{\boldsymbol{p}_j} g_j^{(t,b)}, \tag{17.27}$$

$$\boldsymbol{f}_{w_j}^{(t)} = (1 - \rho^{(t)})\boldsymbol{f}_{w_j}^{(t-1)} + \frac{\rho^{(t)}}{B} \sum_{b=1}^{B} \nabla_{\boldsymbol{w}_j} g_j^{(t,b)}. \tag{17.28}$$

The per-sample gradients have explicit forms that reveal the underlying mechanics

$$\begin{aligned}\nabla_{\boldsymbol{p}_j} g_j^{(t,b)} &= \nabla_{\boldsymbol{p}_j} g_j(\boldsymbol{p}_j, \boldsymbol{w}_j^{(t)}; \boldsymbol{\theta}_{\sim j}^{(b)})\big|_{\boldsymbol{p}_j=\boldsymbol{p}_j^{(t)}} \\ &= \operatorname{vec}\left\{-w_{j,n}^{(t)} \cdot \nabla_{\boldsymbol{p}_{j,n}} \left[\ln p(\boldsymbol{p}_{j,n}^{(t)}) + \ln p(\boldsymbol{y}|\boldsymbol{p}_{j,n}^{(t)}, \boldsymbol{\theta}_{\sim j}^{(b)})\right]\right\}_{n=1}^{N_p},\end{aligned} \tag{17.29}$$

$$\begin{aligned}\nabla_{\boldsymbol{w}_j} g_j^{(t,b)} &= \nabla_{\boldsymbol{w}_j} g_j^{(t)}(\boldsymbol{p}_j^{(t)}, \boldsymbol{w}_j; \boldsymbol{\theta}_{\sim j}^{(b)})\big|_{\boldsymbol{w}_j=\boldsymbol{w}_j^{(t)}} \\ &= \operatorname{vec}\left[\ln w_{j,n}^{(t)} + 1 - \ln p(\boldsymbol{p}_{j,n}^{(t)}) - \ln p(\boldsymbol{y}|\boldsymbol{p}_{j,n}^{(t)}, \boldsymbol{\theta}_{\sim j}^{(b)})\right]_{n=1}^{N_p}.\end{aligned} \tag{17.30}$$

The step size $\rho^{(t)}$ is another decreasing sequence (discussed below), and we initialize $\boldsymbol{f}_{p_j}^{(-1)} = \boldsymbol{0}$, $\boldsymbol{f}_{w_j}^{(-1)} = \boldsymbol{0}$. By averaging both gradients and optimization variables over iterations, the algorithm ensures almost sure convergence to stationary points under standard regularity conditions.

Attention
The gradient estimators are the computational engine of the algorithm, and their structure reveals the underlying optimization dynamics. The position gradient $\nabla_{\boldsymbol{p}_j} g_j^{(t,b)}$ drives particles toward regions of high posterior probability: each particle's movement is weighted by its current probability $w_{j,n}^{(t)}$, with the direction determined by the gradient of the log-prior and log-likelihood. The weight gradient $\nabla_{\boldsymbol{w}_j} g_j^{(t,b)}$ balances two competing objectives: the $\ln w_{j,n}^{(t)} + 1$ term encourages entropy maximization (keeping weights diffuse), while the negative log-prior and log-likelihood terms favor concentrating probability on particles that explain the data well. The recursive update with step size $\rho^{(t)}$ is a common technique for variance reduction in stochastic approximation, smoothing out the noise from mini-batch sampling and promoting stable convergence.

The SPVBI algorithm updates all variables synchronously in parallel. An alternative non-parallel version updates variables sequentially: after updating one variable, new samples are generated from the updated distribution to estimate gradients for the next variable. Figure 17.3 compares these two schemes. The parallel structure, where all variables are updated simultaneously using samples from the same distribution, is particularly advantageous for deep unfolding as it creates a feed-forward computation graph amenable to automatic differentiation.

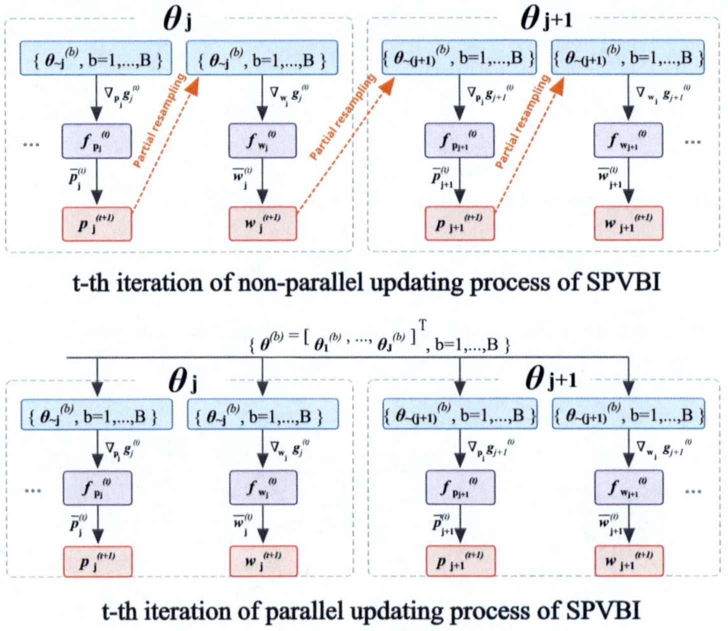

Fig. 17.3 Comparison of the parallel and non-parallel updating processes of SPVBI (Adapted from Fig. 4 in [19])

To ensure convergence, the step sizes $\rho^{(t)}$ and $\gamma^{(t)}$ must satisfy specific conditions drawn from stochastic approximation theory:

1. $\rho^{(t)} \to 0$, $\sum_t \rho^{(t)} = \infty$, $\sum_t (\rho^{(t)})^2 < \infty$;
2. $\lim_{t\to\infty} \gamma^{(t)}/\rho^{(t)} = 0$.

A typical choice satisfying these conditions is $\rho^{(t)} = O(t^{-\kappa_1})$ and $\gamma^{(t)} = O(t^{-\kappa_2})$, with $0.5 < \kappa_1 < \kappa_2 \le 1$, as commonly adopted in stochastic optimization [5]. Under these conditions, the SPVBI algorithm is guaranteed to converge to stationary points of Problem $\mathcal{P}_2$ [19].

Attention

These step-size conditions are not arbitrary mathematical technicalities; they embody fundamental principles of stochastic approximation. The first condition ensures that the gradient estimates are asymptotically unbiased (since $\rho^{(t)} \to 0$) while still allowing sufficient exploration in early iterations (since $\sum \rho^{(t)} = \infty$). The square-summability ensures that the cumulative noise from stochastic sampling remains bounded. The second condition, requiring $\gamma^{(t)}$ to decay faster than $\rho^{(t)}$, creates a two-timescale effect: the gradient estimators $\boldsymbol{f}^{(t)}$ evolve on a faster timescale than the optimization variables $(\boldsymbol{p}, \boldsymbol{w})$, leading to a quasi-static update that promotes convergence to a stationary point. The parallel update scheme, where all variables are updated simultaneously, aligns naturally with modern GPU architectures and is essential for efficient deep unfolding.

Upon convergence, the discrete distribution $q(\boldsymbol{\theta}_j)$ for each variable, represented by its weighted particles, provides an approximation to the true marginal posterior. From this particle representation, we can extract point estimates in two ways: the particle with the highest weight yields an approximate MAP estimate, while the weighted average of all particles gives an approximate MMSE estimate.

Algorithm 17.1 Parallel Stochastic Particle VBI Algorithm

Input: Measurement data, prior $p\left(\widehat{\boldsymbol{\theta}}\right)$, $\{\rho^{(t)}, \gamma^{(t)}\}$.
Output: $\boldsymbol{p}_j^*$, $j = 1, 2, \ldots J$.

Initialization: $\left\{\boldsymbol{p}_{j,n}^{(0)} \sim p\left(\widehat{\boldsymbol{\theta}}_j\right), w_{j,n}^{(0)} = \frac{1}{N_p}\right\}_{n=1}^{N_p}, \forall j$.
While not converged, do ($t \rightarrow \infty$)
  Parallel execution for all J variables:
    Generate a mini-batch of realization based on $q^{(t)}(\boldsymbol{\theta})$;
    Update $\mathbf{f}_{p_j}^{(t)}$ and $\mathbf{f}_{w_j}^{(t)}$, $\forall j$;
    Solve surrogate optimization problems;
    Update the particle position and weight of variable $\boldsymbol{\theta}_j$, $\forall j$;
end
Find the position $\boldsymbol{p}_j^*$, $j = 1, 2, \ldots J$ of the particle with the maximum weight $w_j^{(\max)}$ (i.e., the MAP estimation).

17.2.3 Algorithm Summary

Algorithm 17.1 summarizes the complete SPVBI framework, which introduces two major advancements over conventional variational inference methods.

- **Enhanced Model Flexibility**: SPVBI achieves enhanced modeling flexibility through its particle-based representation of the variational posterior. Unlike parametric approaches that impose rigid distributional forms (e.g., Gaussian or Gamma families), the particle representation can capture arbitrary posterior shapes, including multi-modal distributions that commonly arise in non-convex problems. This expressiveness mitigates the risk of converging to poor local optima and improves overall convergence behavior. The dynamic adjustment of both particle positions and weights introduces additional degrees of optimization freedom, leading to concurrent improvements in estimation accuracy, convergence speed, and particle efficiency.
- **Superior Computational Efficiency**: The framework achieves its computational efficiency by leveraging the PSSCA methodology [5, 6]. PSSCA directly addresses the intractability arising from high-dimensional integrals in the variational objective. By combining mini-batch stochastic sampling with iteration averaging, it enables efficient gradient updates without the need to exhaustively evaluate the entire probability space at each iteration. This approach reduces the per-iteration complexity from exponential to polynomial, making high-dimensional problems tractable.

The synergy of these two capabilities, flexible particle-based modeling and efficient stochastic optimization, renders SPVBI particularly effective for achieving accurate global posterior approximations in complex, high-dimensional, and non-convex estimation problems.

17.3 Deep-Unfolded SPVBI

17.3.1 Overview of the Deep Unfolding Technique

The SPVBI algorithm presented in the previous section represents a model-driven approach to Bayesian inference. By incorporating domain-specific knowledge, such as signal models and prior distributions, it offers theoretical performance guarantees and interpretable behavior. However, like many iterative algorithms, SPVBI faces practical challenges that limit its deployment in real-time applications. Chief among these are sensitivity to hyperparameter settings and relatively slow convergence rates, often requiring dozens of iterations to achieve satisfactory accuracy.

In contrast, conventional deep neural networks (DNNs) have demonstrated remarkable capabilities in automatic feature learning and function approximation [12]. Through end-to-end training on large datasets, they can learn complex mappings directly from data, often achieving fast inference through simple feed-forward computations. Yet this power comes with significant drawbacks: DNNs typically lack interpretability, offer no performance guarantees, and may exhibit poor generalization when deployed outside their training distribution.

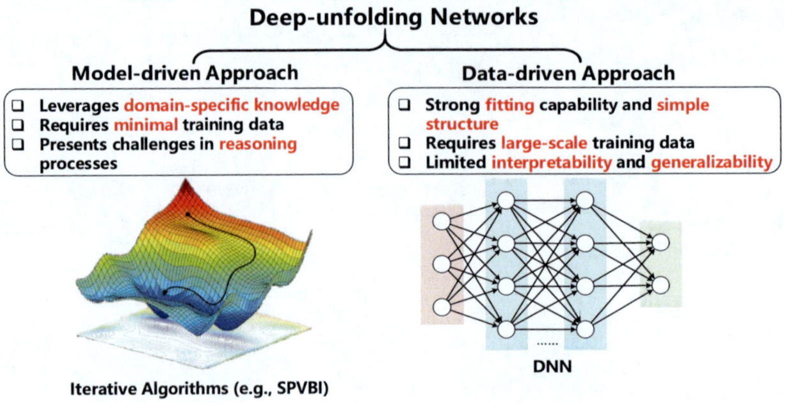

Fig. 17.4 Deep-unfolding networks: combining the strengths of iterative algorithms and DNNs

Deep unfolding [10, 11] emerges as a powerful synthesis of these two paradigms, aiming to combine the best of both worlds. As illustrated in Fig. 17.4, this methodology takes an iterative algorithm, in our case, SPVBI, and unfolds a fixed number of its iterations, denoted by T, into a layered neural network architecture. Each layer of this network corresponds to one iteration of the original algorithm, preserving its structural principles while introducing learnable parameters at key junctures.

Within this unfolded network, the core computations of the original algorithm are retained, but certain components, such as step sizes, projection operators, or even entire subroutines, are replaced or augmented with trainable parameters. These parameters are then optimized end-to-end using standard deep learning techniques (e.g., stochastic gradient descent) with respect to a task-specific loss function. The result is a hybrid architecture that inherits the algorithmic structure and interpretability of the model-based approach while leveraging data-driven optimization to achieve faster convergence and enhanced robustness.

The learnable SPVBI (LSPVBI) network, proposed in [19], embodies this principled integration. By applying deep unfolding to the SPVBI algorithm, LSPVBI aims to accelerate convergence while retaining the theoretical guarantees and interpretability of the original method. The transformation of an iterative procedure into a trainable network enables the algorithm to learn optimal internal parameters from data, achieving accurate inference in significantly fewer iterations than its non-learned counterpart. The following subsections detail the modular architecture of LSPVBI and the specific unfolding mechanisms that make this possible.

17.3.2 LSPVBI Network Architecture

The overall architecture of the LSPVBI network, depicted in Fig. 17.5, comprises T cascaded layers. Each layer corresponds to one complete iteration of the SPVBI algorithm, executing a full update of the variational distributions for all J variables. While every layer shares the same structural modules, they possess distinct trainable parameters, allowing the network to learn iteration-specific adjustments.

A key design decision in LSPVBI is the treatment of step sizes. Rather than treating all algorithm parameters as learnable, which would create an excessively large search space, we fix the non-critical step size sequences $\{\rho^{(t)}, \gamma^{(t)}\}$ to empirically validated values. Instead, we introduce a more targeted set of learnable parameters

$$\boldsymbol{\Lambda}_j^{(t)} = \left[\Gamma_{w_j}^{(t)}, \Gamma_{p_j}^{(t)}\right]^T, \quad \forall j = 1 : J, \ t = 1 : T. \tag{17.31}$$

This design choice is motivated by the dual role of $\Gamma_{p_j/w_j}^{(t)}$ in the algorithm:

- It acts as a step-size multiplier for the gradient estimates $\boldsymbol{f}_{p_j/w_j}^{(t)}$ in the projection steps (Eqs. (17.19) and (17.20)), thereby indirectly emulating the smoothing effect of $\rho^{(t)}$.

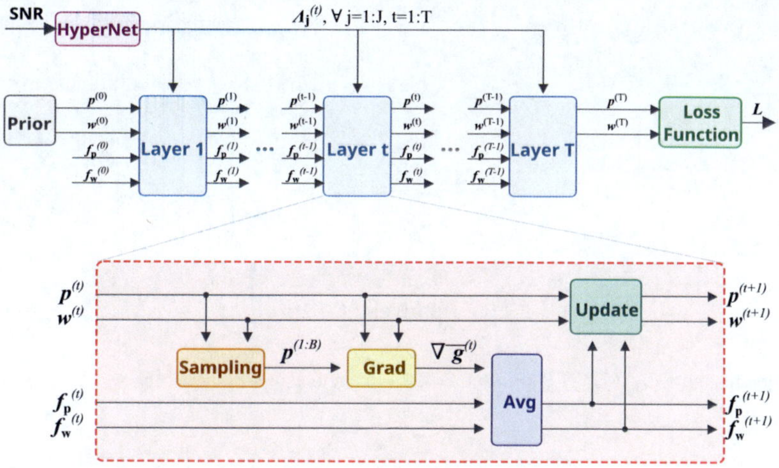

Fig. 17.5 Network architecture of the deep-unfolding SPVBI algorithm (Adapted from Fig. 5 in [19])

- It directly governs the magnitude of the candidate updates $\bar{\boldsymbol{p}}_j^{(t)}$ and $\bar{\boldsymbol{w}}_j^{(t)}$, contributing to a smoothing effect analogous to that of $\gamma^{(t)}$.

By focusing learning on these parameters, we strike a balance between expressiveness and trainability: the network gains the ability to adapt its behavior across iterations while maintaining a manageable number of trainable parameters. The particle positions $\boldsymbol{p}$, weights $\boldsymbol{w}$, and gradient estimates $\mathbf{f}_p, \mathbf{f}_w$, each of dimension JN_p, are propagated through the network layers. This propagation structure preserves the algorithmic state across layers, enabling the network to learn dynamics that unfold over multiple iterations.

17.3.3 Modular Architecture Design

Each layer in the LSPVBI network consists of four specialized modules, each responsible for a distinct computational task. This modular design facilitates both implementation and understanding, as each module's function aligns with a specific step in the original SPVBI algorithm.

17.3.3.1 Sampling Module

Accurate gradient estimation via Eqs. (17.27) and (17.28) requires sampling from the current discrete variational distribution $q^{(t)}(\boldsymbol{\theta})$. However, this sampling operation must be differentiable to support gradient backpropagation through the network, a requirement that rules out naive sampling approaches.

The Gumbel-Softmax technique [13] offers one solution, providing a differentiable approximation of discrete sampling by replacing random nodes with deterministic functions of Gumbel noise. Unfortunately, this approach suffers from high variance in both forward passes and backward gradients, particularly when combined with logarithmic transformations of probability values. As a result, accurately representing the target distribution typically requires a large number of samples, leading to substantial computational overhead.

To overcome these limitations, LSPVBI employs a deterministic sampling approximation using a lightweight, pre-trained neural network, as illustrated in Fig. 17.6. This sampling network takes as input the discrete distribution for the j-th variable, represented by its particle positions and weights $\{\boldsymbol{p}_{j,n}^{(t)}, w_{j,n}^{(t)}\}_{n=1}^{N_p}$, and deterministically outputs B sample values in proportion to the particle weights.

For example, consider an input distribution with weights [0.2, 0.3, 0.5] corresponding to particles at positions $\boldsymbol{p}_{j,1}^{(t)}$, $\boldsymbol{p}_{j,2}^{(t)}$, and $\boldsymbol{p}_{j,3}^{(t)}$. The DNN would output two samples of $\boldsymbol{p}_{j,1}^{(t)}$, three samples of $\boldsymbol{p}_{j,2}^{(t)}$, and five samples of $\boldsymbol{p}_{j,3}^{(t)}$, matching the underlying probability mass function. These samples are then shuffled across variables before gradient computation.

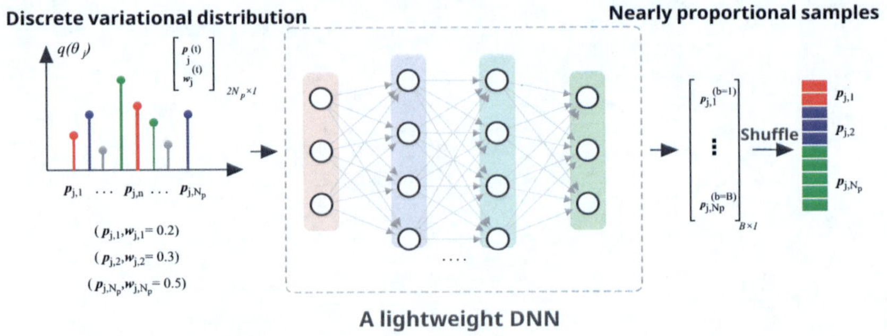

Fig. 17.6 An illustrative example of lightweight DNN approximation for discrete sampling (Adapted from Fig. 6 in [19])

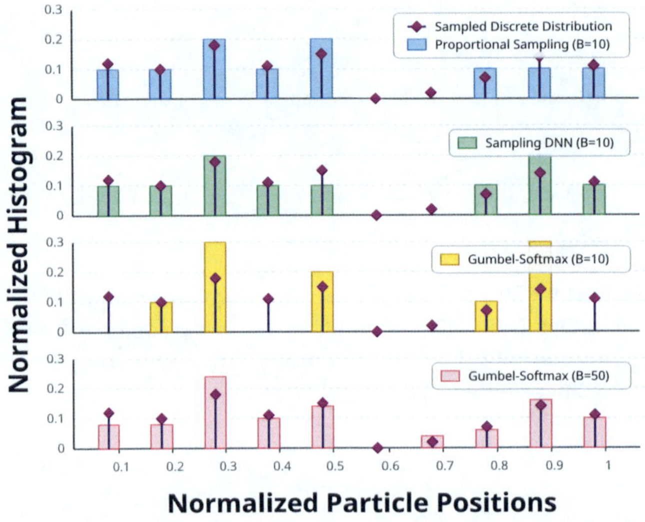

Fig. 17.7 An example demonstrating the sampling efficiency of different sampling methods (Adapted from Fig. 7 in [19])

Figure 17.7 demonstrates the sampling efficiency of this approach through a comparative example. Starting from a randomly generated discrete distribution with 10 particles (shown as stems), we generate samples using different methods and plot the resulting statistical histograms. The results reveal that the Gumbel-Softmax method requires approximately 50 samples to achieve a distributional approximation comparable to that obtained by our DNN-based sampler with only 10 samples. This fivefold reduction in sample count translates directly to computational savings, underscoring the efficiency of the proposed approach.

17.3.3.2 Gradient Calculation Module

With samples $\boldsymbol{\theta}_j^{(b)}$ generated for all j and b, the gradient calculation module computes average gradients with respect to particle positions and weights

$$\nabla_{\boldsymbol{p}_j/\boldsymbol{w}_j}\bar{g}^{(t)} = \frac{1}{B}\sum_{b=1}^{B}\nabla_{\boldsymbol{p}_j/\boldsymbol{w}_j} g^{(t,b)}. \tag{17.32}$$

The primary computational bottleneck in this step lies in evaluating the log-likelihood terms $\ln p(\boldsymbol{y}|\boldsymbol{\theta}_{\sim j}^{(b)}, \boldsymbol{p}_{j,n}^{(t)})$ appearing in Eq. (17.29) and their gradients in Eq. (17.30). These evaluations can be expensive, particularly for high-dimensional observations or complex likelihood functions.

To optimize this process, LSPVBI implements two key strategies.

- **Intermediate Result Reuse**: Across different particle evaluations, the variables $\boldsymbol{\theta}_{\sim j}^{(b)}$ remain fixed. Consequently, many intermediate results in the likelihood computation can be precomputed and reused, avoiding redundant calculations. This optimization is particularly effective when the number of particles N_p is large relative to the number of samples B.
- **Mini-batch Approximation for High-Dimensional Data**: For high-dimensional observations $\boldsymbol{y} = [y_1, y_2, \ldots, y_M]^T$, evaluating the full log-likelihood can be costly. Following [3], we employ a mini-batch approximation over a random subset of indices $\Omega \subset \{1, \ldots, M\}$ as

$$\ln p(\boldsymbol{y}|\boldsymbol{\theta}) \approx \frac{M}{|\Omega|} \sum_{m\in\Omega} \ln p(\boldsymbol{y}_m|\boldsymbol{\theta}). \tag{17.33}$$

This approximation reduces the per-sample gradient computation from $O(M)$ to $O(|\Omega|)$, with $|\Omega|$ typically chosen much smaller than M. The combination of these two strategies significantly reduces computational overhead per iteration, enhancing the scalability of LSPVBI to large-scale problems.

17.3.3.3 Gradient Smoothing Module

The gradient estimates obtained from the mini-batch samples exhibit inherent variance due to stochastic sampling. To mitigate this variability and promote stable training, the gradient smoothing module implements an exponential moving average on the gradient estimates, exactly as defined in Eqs. (17.27) and (17.28)

$$\boldsymbol{f}_{p_j}^{(t)} = (1-\rho^{(t)})\boldsymbol{f}_{p_j}^{(t-1)} + \frac{\rho^{(t)}}{B}\sum_{b=1}^{B} \nabla_{\boldsymbol{p}_j} g_j^{(t,b)}, \tag{17.34}$$

$$\boldsymbol{f}_{w_j}^{(t)} = (1-\rho^{(t)})\boldsymbol{f}_{w_j}^{(t-1)} + \frac{\rho^{(t)}}{B}\sum_{b=1}^{B} \nabla_{\boldsymbol{w}_j} g_j^{(t,b)}. \tag{17.35}$$

By incorporating historical gradient information through the recursive update, this smoothing operation effectively reduces estimation variance. The mechanism is particularly beneficial when a small sample size B is used for gradient estimation, as it tempers the high variability inherent in stochastic approximations without requiring additional samples.

17.3.3.4 Particle Update Module

The particle update module executes the final update of particle positions and weights, incorporating projection operations to enforce the constraints from Problem $\mathcal{P}_2$.

Position Update

Particle positions are updated according to

$$\boldsymbol{p}_j^{(t+1)} = (1-\gamma^{(t)})\boldsymbol{p}_j^{(t)} + \gamma^{(t)}\text{Proj}_{\Delta p}\left(\boldsymbol{p}_j^{(t)} - \Gamma_{p_j}^{(t)}\boldsymbol{f}_{p_j}^{(t)}\right). \tag{17.36}$$

Due to the box constraints in $\mathcal{P}_2$, the projection $\text{Proj}_{\Delta p}(\cdot)$ reduces to a simple clipping operation. For the i-th element $p_{j,i,n}$ of particle $\boldsymbol{p}_{j,n}$, it is defined as

$$\text{Proj}_{\Delta p}(p_{j,i,n}) = \begin{cases} \hat{\boldsymbol{\theta}}_{j,i} - \Delta\hat{\boldsymbol{\theta}}_{j,i}/2, & p_{j,i,n} < \hat{\boldsymbol{\theta}}_{j,i} - \Delta\hat{\boldsymbol{\theta}}_{j,i}/2, \\ p_{j,i,n}, & p_{j,i,n} \in [\hat{\boldsymbol{\theta}}_{j,i} - \Delta\hat{\boldsymbol{\theta}}_{j,i}/2, \hat{\boldsymbol{\theta}}_{j,i} + \Delta\hat{\boldsymbol{\theta}}_{j,i}/2], \\ \hat{\boldsymbol{\theta}}_{j,i} + \Delta\hat{\boldsymbol{\theta}}_{j,i}/2, & p_{j,i,n} > \hat{\boldsymbol{\theta}}_{j,i} + \Delta\hat{\boldsymbol{\theta}}_{j,i}/2, \end{cases} \tag{17.37}$$

where $\hat{\theta}_{j,i}$ and $\Delta\hat{\theta}_{j,i}$ define the center and width of the feasible interval for the i-th dimension. The derivative of this projection is straightforward

$$\nabla_{p_{j,i,n}}\text{Proj}_{\Delta p}\left(p_{j,i,n}\right) = \begin{cases} 1, & p_{j,i,n} \in [\hat{\boldsymbol{\theta}}_{j,i} - \Delta\hat{\boldsymbol{\theta}}_{j,i}/2, \hat{\boldsymbol{\theta}}_{j,i} + \Delta\hat{\boldsymbol{\theta}}_{j,i}/2], \\ 0, & \text{otherwise}. \end{cases} \tag{17.38}$$

This simple derivative facilitates gradient backpropagation through the projection operation.

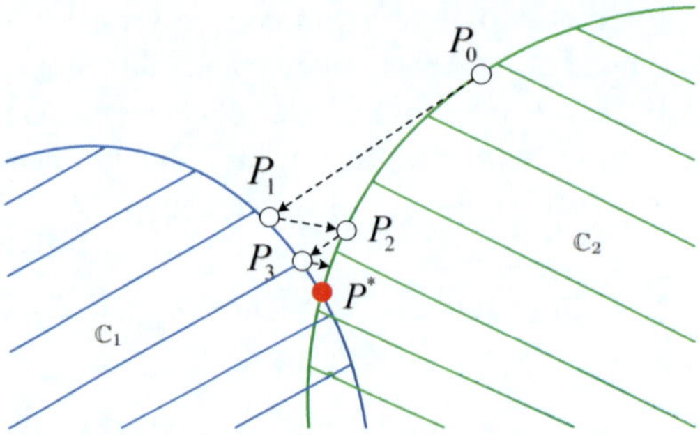

Fig. 17.8 Illustration of alternating projection

Algorithm 17.2 Alternating Projection Method.

Input: $\boldsymbol{y}_r = \boldsymbol{y}_{r+1} = \boldsymbol{w}_j^{(t)} - \Gamma_{w_j}^{(t)}\mathbf{f}_{w_j}^{(t)}$, $error = +\infty$, $\epsilon \rightarrow 0$.
Output: $\boldsymbol{y}_r$.

1: **While** $\left(error > 10^{-6}\right)$
2: $\quad \boldsymbol{y}_{r+1} = \max\left\{\left[\boldsymbol{y}_r - \frac{1}{N_p}\left(\mathbf{1}_{N_p}^T\boldsymbol{y}_r - 1\right)\mathbf{1}_{N_p}\right], \epsilon \cdot \mathbf{1}_{N_p}\right\}$
3: $\quad error = \left\|\boldsymbol{y}_{r+1} - \boldsymbol{y}_r\right\|_2$;
4: $\quad \boldsymbol{y}_r = \boldsymbol{y}_{r+1}$;
5: **end**
6: **Output** $\boldsymbol{y}_r$.

Weight Update

Particle weights are updated analogously

$$\boldsymbol{w}_j^{(t+1)} = (1 - \gamma^{(t)})\boldsymbol{w}_j^{(t)} + \gamma^{(t)}\mathrm{Proj}_{\Delta w}\left(\boldsymbol{w}_j^{(t)} - \Gamma_{w_j}^{(t)}\boldsymbol{f}_{w_j}^{(t)}\right). \tag{17.39}$$

The projection $\mathrm{Proj}_{\Delta w}$ enforces the simplex constraints (non-negativity and summation to one) and is implemented using an alternating projection method [14]. Specifically, we define two sets:

- $\mathbb{C}_1 = \{\boldsymbol{w}_j \in \mathbb{R}^{N_p} : \mathbf{1}_{N_p}^T\boldsymbol{w}_j = 1\}$ (sum-to-one), where $\mathbf{1}_{N_p}$ is an N_p-dimensional vector of ones.
- $\mathbb{C}_2 = \{\boldsymbol{w}_j \in \mathbb{R}^{N_p} : w_{j,n} \geq \epsilon, n = 1, \ldots, N_p\}$ (non-negativity).

As illustrated in Fig. 17.8, the algorithm alternately projects onto $\mathbb{C}_1$ and $\mathbb{C}_2$, typically converging to a valid simplex vector within 10 iterations. Algorithm 17.2 provides the complete alternating projection procedure.

The derivative of this projection can be approximated as

$$\nabla\mathrm{Proj}_{\Delta\boldsymbol{w}}(\boldsymbol{w}) = \left[\left(\boldsymbol{I}_{N_p\times N_p} - \frac{1}{N_p}\mathbf{1}\mathbf{1}^T\right)\mathrm{diag}\left(I\left(\boldsymbol{w}\right)\right)\right]^{\tau}, \tag{17.40}$$

where $\boldsymbol{I}_{N_p\times N_p}$ is an $N_p \times N_p$ identity matrix, $I(x)$ is an indicator function with threshold ϵ (taking value 1 when $x \geq \epsilon$ and 0 otherwise), and τ is the number of alternating steps. While not the most computationally efficient method, alternating projection provides a stable and differentiable procedure well-suited for gradient backpropagation in the deep-unfolding network.

17.3.3.5 Hyperparameter Network (HyperNet) Design

While LSPVBI demonstrates improved convergence and accuracy after training, its performance may degrade under test conditions that differ from the training environment. For instance, a network trained exclusively at SNR = 15 dB may perform poorly when deployed at SNR = 10 dB. This sensitivity to distribution shift is a common challenge in deep learning.

To enhance generalization across varying operating conditions, LSPVBI integrates a HyperNetwork (HyperNet) [16]. A HyperNet is a neural network architecture designed to generate or adapt the parameters of a target network based on input conditions. In our framework, we leverage the strong correlation between optimal LSPVBI parameters and environmental factors such as SNR.

Specifically, the HyperNet learns a mapping from the current SNR value to the full parameter vector $\boldsymbol{\Lambda} \in \mathbb{R}^{2JT\times 1}$ of the LSPVBI network. This mapping is implemented using a three-layer fully connected network

$$\boldsymbol{\Lambda} = \boldsymbol{W}_3 \cdot \text{ReLU}\left(\boldsymbol{W}_2 \cdot \text{ReLU}\left(\boldsymbol{W}_1 \cdot \text{SNR} + \boldsymbol{b}_1\right) + \boldsymbol{b}_2\right) + \boldsymbol{b}_3, \tag{17.41}$$

where $\boldsymbol{W}_i$ and $\boldsymbol{b}_i$, $i \in \{1, 2, 3\}$, are the trainable weight matrices and bias vectors of the HyperNet. During training, the HyperNet parameters are optimized jointly with the LSPVBI network parameters, learning to generate appropriate LSPVBI parameters for each SNR condition encountered.

In many parameter estimation scenarios, the number of variables J (e.g., number of propagation paths or targets) is not fixed a priori. To ensure that the unfolded network adapts to different values of J, we pre-train the model with a fixed, maximum expected number of variables J_0. During inference for a scenario with $J \leq J_0$, we simply set the initial states of the unused variables to zero

$$\left\{\boldsymbol{p}_j^{(0)}, \boldsymbol{w}_j^{(0)}, \boldsymbol{f}_{p,j}^{(0)}, \boldsymbol{f}_{w,j}^{(0)}\right\}_{j=J+1}^{J_0} = \mathbf{0}. \tag{17.42}$$

This simple strategy allows the trained network to be straightforwardly transferred to problems of smaller dimension without retraining, maintaining both flexibility and practical applicability.

17.3.4 Training Methodology

This subsection outlines the training procedure for the LSPVBI network. The training objective is to minimize the Kullback-Leibler divergence between the variational distribution after T iterations and the true posterior. This objective defines the loss function

$$Loss \triangleq \mathbb{E}_{q(\boldsymbol{\theta};\boldsymbol{p}^{(T)},\boldsymbol{w}^{(T)})}\left[g\left(\boldsymbol{p}^{(T)}, \boldsymbol{w}^{(T)}; \boldsymbol{\theta}\right)\right], \tag{17.43}$$

where $g(\cdot)$ is defined in Eq. (17.16).

The gradients of the loss with respect to the final layer's particle positions and weights are approximated via Monte Carlo sampling

$$\frac{\partial Loss}{\partial \boldsymbol{p}_j^{(T)}} \approx \frac{1}{B_{grad}} \sum_{b=1}^{B_{grad}} \nabla_{\boldsymbol{p}_j} g_j^{(T)}\left(\boldsymbol{p}_j^{(T)}; \boldsymbol{w}_j^{(T)}, \boldsymbol{\theta}_{\sim j}^{(b)}\right), \tag{17.44}$$

$$\frac{\partial Loss}{\partial \boldsymbol{w}_j^{(T)}} \approx \frac{1}{B_{grad}} \sum_{b=1}^{B_{grad}} \nabla_{\boldsymbol{w}_j} g_j^{(T)}\left(\boldsymbol{w}_j^{(T)}; \boldsymbol{p}_j^{(T)}, \boldsymbol{\theta}_{\sim j}^{(b)}\right), \tag{17.45}$$

where B_{grad} is the number of samples used for gradient estimation.

Using the chain rule, these gradients are backpropagated through the network to compute gradients for earlier layers $t = T - 1, \ldots, 1$ as

$$\frac{\partial Loss}{\partial \boldsymbol{p}_j^{(t)}}, \quad \frac{\partial Loss}{\partial \boldsymbol{w}_j^{(t)}}, \quad \frac{\partial Loss}{\partial \boldsymbol{f}_{p_j}^{(t)}}, \quad \frac{\partial Loss}{\partial \boldsymbol{f}_{w_j}^{(t)}}. \tag{17.46}$$

The gradients with respect to the trainable parameters $\boldsymbol{\Lambda}_j^{(t)} = [\Gamma_{w_j}^{(t)}, \Gamma_{p_j}^{(t)}]^T$ are then given by

$$\frac{\partial Loss}{\partial \boldsymbol{\Lambda}_j^{(t)}} = \frac{\partial Loss}{\partial \boldsymbol{p}_j^{(t+1)}} \frac{\partial \boldsymbol{p}_j^{(t+1)}}{\partial \boldsymbol{\Lambda}_j^{(t)}} + \frac{\partial Loss}{\partial \boldsymbol{w}_j^{(t+1)}} \frac{\partial \boldsymbol{w}_j^{(t+1)}}{\partial \boldsymbol{\Lambda}_j^{(t)}}, \tag{17.47}$$

where the component derivatives are

$$\frac{\partial \boldsymbol{p}_j^{(t+1)}}{\partial \Gamma_{p_j}^{(t)}} = \gamma^{(t)}(-\boldsymbol{f}_{p_j}^{(t)})\nabla \mathrm{Proj}_{\Delta p}, \quad \frac{\partial \boldsymbol{p}_j^{(t+1)}}{\partial \Gamma_{w_j}^{(t)}} = 0,$$
$$\frac{\partial \boldsymbol{w}_j^{(t+1)}}{\partial \Gamma_{p_j}^{(t)}} = 0, \quad \frac{\partial \boldsymbol{w}_j^{(t+1)}}{\partial \Gamma_{w_j}^{(t)}} = \gamma^{(t)}(-\boldsymbol{f}_{w_j}^{(t)})\nabla \mathrm{Proj}_{\Delta w}. \tag{17.48}$$

Finally, the gradients $\frac{\partial Loss}{\partial \boldsymbol{\Lambda}_j^{(t)}}$ are backpropagated through the HyperNet to update its parameters

$$\frac{\partial Loss}{\partial \boldsymbol{W}_3/\boldsymbol{b}_3} = \frac{\partial Loss}{\partial \boldsymbol{\Lambda}_j^{(t)}} \frac{\partial \boldsymbol{\Lambda}_j^{(t)}}{\partial \boldsymbol{W}_3/\boldsymbol{b}_3}, \tag{17.49}$$

$$\frac{\partial Loss}{\partial \boldsymbol{W}_i/\boldsymbol{b}_i} = \frac{\partial Loss}{\partial \boldsymbol{W}_{i+1}/\boldsymbol{b}_{i+1}} \frac{\partial \boldsymbol{W}_{i+1}/\boldsymbol{b}_{i+1}}{\partial \boldsymbol{W}_i/\boldsymbol{b}_i}, \quad \forall i = 2, 1. \tag{17.50}$$

For practical implementation, the network is trained using data generated from simulations or real-world measurements across diverse wireless environments, encompassing various SNR levels and delay path configurations. Standard deep learning techniques, including mini-batch averaging, stochastic gradient descent, and Adam optimization, are employed to optimize the parameters. Comprehensive implementation details, including specific network architectures, learning rates, and data augmentation strategies, can be found in [19].

17.4 Application Example: Multiband WiFi Sensing

17.4.1 System Model and Problem Formulation

To demonstrate the practical effectiveness of the SPVBI and LSPVBI algorithms developed in the preceding sections, we consider a multiband WiFi sensing scenario [8, 9]. This application is particularly well-suited to showcase the capabilities of particle-based variational inference, as it combines high-dimensional parameter spaces, nonlinear observation models, and practical hardware impairments, all characteristics that challenge conventional estimation methods.

As illustrated in Fig. 17.9, a single-input single-output (SISO) system employs known OFDM training signals across M non-contiguous frequency subbands to estimate the distance between a mobile node and a WiFi access point. The use of multiple bands provides enhanced resolution through both bandwidth aggregation and band-gap diversity, but also introduces additional complexity in signal processing.

The frequency-domain received signal for the m-th band, where $m = 1, 2, \ldots, M$ indexes the frequency bands, takes the form [9]

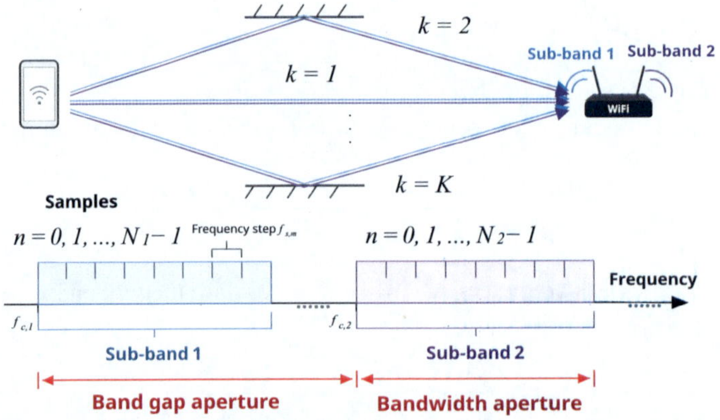

Fig. 17.9 Illustration of multiband WiFi sensing and spectrum distribution (Adapted from Fig. 2 in [19])

$$r_m^{(n)} = \sum_{k=1}^{K} \alpha_k e^{-j2\pi(f_{c,m}+nf_{s,m})(\tau_k+\delta_m)} e^{j\phi_m} + w_m^{(n)}, \tag{17.51}$$

with $n = 0, 1, \ldots, N_m - 1$ representing the subcarrier index within band m, where N_m denotes the number of subcarriers per band, and $k = 1, 2, \ldots, K$ indexing the individual scattering paths. The complex amplitude of the k-th path is denoted by α_k, while τ_k represents its time delay (with τ_1 corresponding to the line-of-sight path of primary interest for ranging applications). For each frequency band, $f_{c,m}$ and $f_{s,m}$ specify the initial frequency and subcarrier spacing, respectively. Hardware imperfections manifest through two additional parameters: ϕ_m captures the random initial phase that varies independently across bands, and δ_m accounts for the timing synchronization error arising from packet detection delay and sampling frequency offset. The term $w_m^{(n)} \sim \mathcal{CN}(0, \eta_w^2)$ represents additive white Gaussian noise.

Collecting all unknown parameters into a single vector yields

$$\boldsymbol{\theta}_{\text{ori}} = [\alpha_1, \ldots, \alpha_K, \tau_1, \ldots, \tau_K, \phi_1, \ldots, \phi_M, \delta_1, \ldots, \delta_M]^T. \tag{17.52}$$

The frequency-domain measurements across all bands and subcarriers are concatenated to form the observation vector

$$\boldsymbol{r} = \left[r_1^{(0)}, r_1^{(1)}, \ldots, r_1^{(N_1-1)}, \ldots, r_M^{(0)}, r_M^{(1)}, \ldots, r_M^{(N_M-1)}\right]^T. \tag{17.53}$$

Under the assumption of additive white Gaussian noise, the log-likelihood function follows from the product of individual likelihoods across all bands and subcarriers

$$\begin{aligned} \ln p\left(\boldsymbol{r}|\boldsymbol{\theta}_{ori}\right) &= \ln \prod_{m=1}^{M} \prod_{n=0}^{N_m-1} p(r_m^{(n)}|\boldsymbol{\theta}_{ori}) \\ &= -\sum_{m=1}^{M} N_m \ln\left(\pi\eta_w^2\right) - \sum_{m=1}^{M}\sum_{n=0}^{N_m-1} \frac{1}{\eta_w^2}\left|r_m^{(n)} - s_m^{(n)}\left(\boldsymbol{\theta}_{ori}\right)\right|^2, \end{aligned} \tag{17.54}$$

where $s_m^{(n)}(\boldsymbol{\theta}_{\text{ori}})$ denotes the reconstructed signal based on the current parameter estimates

$$s_m^{(n)}(\boldsymbol{\theta}_{\text{ori}}) = \sum_{k=1}^{K} \alpha_k e^{-j2\pi(f_{c,m}+nf_{s,m})(\tau_k+\delta_m)} e^{j\phi_m}. \tag{17.55}$$

This formulation presents a challenging high-dimensional, non-convex estimation problem. The likelihood function exhibits numerous local optima due to the high-frequency carrier terms, while the presence of unknown hardware impairments (i.e., the random initial phases ϕ_m and timing synchronization errors δ_m) further complicates the estimation landscape by introducing parameters that must be inferred alongside the physical quantities of interest. These characteristics make multiband WiFi sensing an ideal testbed for SPVBI and LSPVBI, as both algorithms are specifically designed to handle multi-modal posteriors and robustly estimate parameters under model uncertainty.

17.4.2 Performance Comparison

We now evaluate the performance of SPVBI and LSPVBI through comprehensive simulations. The experimental setup follows the parameters summarized below. Each variable is represented by $N_p = 10$ particles, with mini-batch size set to $B = 10$. The system operates over two non-adjacent frequency bands, each with 20 MHz bandwidth, centered at 2.401 GHz and 2.461 GHz, and a subcarrier spacing of 78.125 kHz. Channel impulse responses are generated using the QuaDRiGa platform in an indoor office scenario based on the 3GPP NR model [17]. The initial phase ϕ_m is uniformly distributed in $[0, 2\pi]$, while the timing synchronization error δ_m follows a Gaussian distribution $\mathcal{N}(0, 0.01\ \text{ns}^2)$. The step size sequences are configured as $\rho^{(t)} = 5/(5+t)^{0.9}$ and $\gamma^{(t)} = 5/(15+t)$, with initial values $\rho^{(0)} = \gamma^{(0)} = 1$. Unless otherwise specified, all experiments are repeated 400 times to ensure statistical reliability.

Figure 17.10 investigates the convergence behavior under different particle update strategies. When the particle count is reduced from 10 to 5, updating only the particle weights results in noticeably slower convergence. In contrast, jointly updating both particle positions and weights maintains nearly identical convergence speed and estimation performance. This observation confirms that dynamic position adjustment significantly enhances optimization flexibility, enabling the algorithm to preserve efficiency and accuracy even with a substantially reduced number of particles, a critical advantage for high-dimensional problems where particle count directly impacts computational cost.

Figure 17.11 compares the proposed LSPVBI against several baseline methods:

- **Weighted root-MUSIC (WR-MUSIC)** [18]: A high-resolution subspace-based method that provides coarse parameter estimates and serves as initialization for PSPVBI [7].
- **Black-box DNN**: A conventional data-driven deep neural network comprising fully connected layers with batch normalization and leaky ReLU activations, trained using the Adam optimizer. The network takes measurement vectors and prior node information as input and directly outputs delay estimates.
- **Ideal MAP**: A theoretical performance upper bound obtained by initializing gradient descent from the true parameter values, representing the best achievable performance under ideal conditions.

The results reveal that LSPVBI, enhanced by the HyperNet, adapts effectively to varying SNR conditions. Across the entire evaluated SNR range, LSPVBI achieves a RMSE closer to the ideal MAP bound than any of the competing methods. This superior performance confirms both the accuracy of the underlying probabilistic model and the effectiveness of the deep unfolding approach in learning optimized inference parameters from data.

17.4.3 Computational Complexity Analysis

Understanding the computational requirements of different algorithms is essential for practical deployment decisions. We analyze the complexity of each method considered in this study.

- **SPVBI Algorithm**: Through mini-batch sampling and minimization of quadratic surrogate objective functions, the per-iteration computational complexity of SPVBI is $O(2J(N_p B F_{\text{grad}} + N_p^3))$, where F_{grad} represents the cost of computing a single gradient, and the N_p^3 term accounts for the projection operations required to enforce constraints.
- **LSPVBI Algorithm**: The primary computational burden in LSPVBI lies in the gradient calculation module, with the complexity of other modules being negligible in comparison. The overall per-iteration complexity remains of the same order as SPVBI, but the network typically requires significantly fewer iterations ($T_2 < T_1$) due to its data-optimized parameters.
- **Black-box DNN**: The complexity of a fully connected network is determined by the sizes of its layers. Denoting the layer dimensions as L_0 through L_5, the forward pass complexity scales as $O(\sum_{i=1}^{5} L_{i-1} L_i)$.

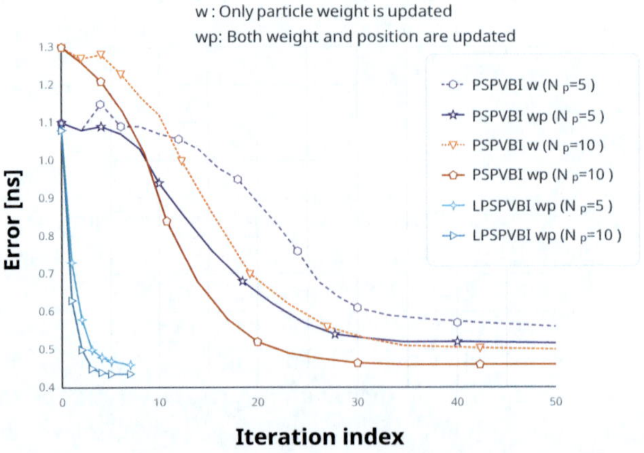

Fig. 17.10 Iteration error curves with different particle numbers (Adapted from Fig. 12 in [19])

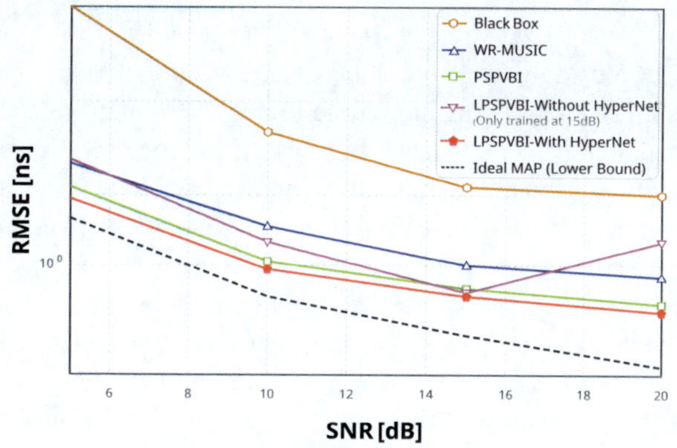

Fig. 17.11 RMSE of line-of-sight (LoS) delay estimation with respect to the SNR (Adapted from Fig. 13 in [19])

Table 17.1 Comparison of the complexity order for different algorithms

Algorithms	Complexity order	Typical values
WR-MUSIC	$O\left(MN_m^3 + M\left(2N_m - 1\right)^3\right)$	2.99×10^8
SPVBI	$O\left(T_1 2J\left(N_p B F_{grad} + N_p^3\right)\right)$	2.84×10^8
LSPVBI	$O\left(T_2 2J\left(N_p B F_{grad} + N_p^3\right)\right)$	5.68×10^7
Black Box	$O\left(\sum_{i=0}^{4} L_{i+1} L_i\right)$	2.47×10^7

- **WR-MUSIC Algorithm**: The WR-MUSIC algorithm's complexity arises primarily from subspace decomposition, which is $O(MN_m^3)$, and polynomial rooting, which scales as $O(M(2N_m - 1)^3)$.

Table 17.1 summarizes the computational complexity orders alongside representative numerical values computed for a typical parameter configuration: $M = 2$, $N_m = 256$, $J = 9$, $N_p = 10$, $|\Omega| = 64$, $B = 10$, $F_{\text{grad}} = 4500$, and layer dimensions $L_0 = 1028$, $L_1 = 3072$, $L_2 = 4096$, $L_3 = 2048$, $L_4 = 256$, $L_5 = 2$. The iteration counts are set to $T_1 = 35$ for SPVBI and $T_2 = 7$ for LSPVBI, reflecting the faster convergence of the learned approach.

The comparison of both performance and computational requirements demonstrates that the deep-unfolding approach not only achieves a substantial reduction in complexity, approximately a factor of five in typical configurations, but also realizes the full potential of the original model-based algorithm. This integration provides a more efficient inference framework without compromising estimation accuracy, effectively bridging the gap between iterative algorithms and deep learning paradigms for practical deployment in resource-constrained environments.

Summary

This chapter has systematically explored how advanced stochastic optimization and deep learning techniques can significantly enhance the design of variational Bayesian inference algorithms. The discussion began by addressing a fundamental challenge that arises in Bayesian inference: when the observation model is highly nonlinear, the resulting likelihood function, and consequently the posterior distribution, often exhibits a complex, multi-modal structure that cannot be adequately captured by simple parametric forms. In such scenarios, the standard Gaussian approximation, with its characteristic single mode and symmetric tails, becomes fundamentally inadequate for representing the true posterior's complexity. This limitation naturally motivates the exploration of more expressive variational families, among which particle-based representations stand out for their ability to model arbitrary distributions through collections of weighted samples.

However, conventional particle-based VBI methods, while offering the necessary flexibility, face several critical drawbacks that limit their practical utility. A primary issue is suboptimal particle placement: traditional approaches optimize only the particle weights while keeping their positions fixed at values determined by initial sampling. This static configuration often necessitates using a large number of particles to achieve accurate posterior approximations, which in turn slows convergence and increases memory requirements. Furthermore, these methods frequently suffer from computational intractability in high-dimensional settings, as the cost per iteration can grow exponentially with the number of variables due to the need to evaluate summations over all possible particle combinations.

To overcome these challenges, we introduced a progressive framework that systematically leverages modern stochastic optimization and deep learning techniques. The first component of this framework is the SPVBI algorithm, whose core innovation lies in jointly optimizing both particle weights and positions. This coordinated adjustment serves two purposes simultaneously: it enables a more efficient representation of the posterior by allowing particles to migrate toward regions of high probability, and it significantly reduces the number of particles required to achieve a given approximation accuracy. By integrating stochastic successive convex approximation with mini-batch sampling, SPVBI also achieves a substantial reduction in per-iteration computational complexity, transforming what would otherwise be an exponential-cost operation into a tractable stochastic optimization problem.

Despite these improvements, SPVBI may still require dozens of iterations to reach satisfactory accuracy in typical applications, a limitation that becomes particularly acute in real-time or resource-constrained scenarios. To further enhance efficiency, we employed deep unfolding, a methodology that reinterprets iterative algorithms as deep neural networks. In this framework, each iteration of SPVBI becomes a layer in a network, with learnable parameters introduced at key computational junctures. Complex, hand-designed operations are replaced with lightweight, trainable modules that can be optimized end-to-end using standard deep learning techniques. Through data-driven training, we obtained the LSPVBI algorithm, which substantially reduces both the number of required layers (iterations) and the computational complexity per layer compared to the original SPVBI, while often achieving superior estimation accuracy.

The deep unfolding methodology demonstrated in this chapter has broader applicability beyond SPVBI, offering a general pathway to enhance the efficiency and practical deployment of various advanced VBI algorithms. By bridging traditional Bayesian methods with modern deep learning, this approach preserves the interpretability and theoretical guarantees of model-based inference while leveraging the representational power and data-driven optimization of neural networks. The resulting framework opens new possibilities for scalable and adaptive probabilistic inference in complex, high-dimensional estimation problems where neither purely model-based nor purely data-driven approaches alone would suffice.

Future Research Directions: Bayesian Deep Learning for Wireless Frontiers

Throughout this book, we have explored a progression of model-driven Bayesian inference algorithms, from basic SBL to advanced Turbo and VBI frameworks, each designed to tackle the challenges of sparse signal recovery in structured environments. As we look toward the future, the convergence of these principled methods with data-driven deep learning, giving rise to the field of Bayesian Deep Learning (BDL), presents a paradigm shift. This fusion is not merely an incremental step but a strategic necessity for managing the escalating complexity and inherent uncertainties in next-generation wireless communication and sensing systems (e.g., 6G). This research trajectory is defined by a commitment to deep, principled algorithmic fusion, aiming to build trustworthy, interpretable, and intelligent systems with a clear pathway from theoretical advances to practical deployment.

1. Foundational Methodologies at the BDL Frontier

Current investigations focus on core algorithmic innovations that leverage the complementary strengths of both fields. Advanced hybrid frameworks are being developed to deeply integrate Bayesian principles with deep neural networks. Key thrusts include:

- **Deep-Unfolded Inference**: As exemplified in this chapter, deep unfolding transforms iterative algorithms like VBI into trainable deep networks. This approach retains the algorithmic structure and interpretability while enabling data-driven optimization of hyperparameters (e.g., step sizes) and even learned approximations of complex subroutines.

This principle can be extended to the entire Turbo-VBI family, creating a new class of ultra-efficient, learnable sensing and communication receivers.

- **Bayesian Neural Networks (BNNs) for Wireless**: Instead of using standard DNNs for black-box channel estimation or signal detection, BNNs can be employed to provide inherent uncertainty estimates. For instance, a BNN could predict a channel's impulse response along with a confidence interval, which is invaluable for robust link adaptation and autonomous decision-making in high-mobility scenarios.
- **Learning-Augmented Message Passing**: The message-passing algorithms central to this book (e.g., in EM-Turbo-SBI framework) rely on hand-crafted priors and statistical models. Future research can explore using deep learning to learn the optimal message update rules or to approximate intractable belief propagation in highly loopy graphs, combining the efficiency of message passing with the representational power of neural networks.

Building on these core methods, next-generation research is advancing scalable inference techniques for extremely high-dimensional models, a prerequisite for realizing digital twins and full-system level optimization. A prime example is the EM-TDAMP framework [42], which leverages expectation-maximization and turbo deep approximate message passing to simultaneously achieve efficient training and structured model compression of DNN. A parallel focus involves embedding robust uncertainty calibration directly into deep learning-based signal processing pipelines, ensuring that the confidence of a model aligns with its actual accuracy, a critical requirement for building trust in autonomous systems.

2. Enabling End-to-End Intelligent Integrated Sensing and Communications (ISAC) via BDL

The ultimate goal of ISAC is a single waveform that simultaneously and optimally performs both sensing and communication functions. This research direction applies BDL to create such cohesive and adaptive systems. Future work will focus on:

- **Unified Probabilistic Models**: Developing deep probabilistic models that can jointly infer communication channel states (e.g., for demodulation) and environmental sensing parameters (e.g., target location, velocity, and micro-Doppler signatures) directly from raw RF signals. This moves beyond separate estimation tasks to a holistic understanding of the radio environment.
- **Task-Oriented Semantic Communications**: Instead of focusing on bit-error-rate, future systems will optimize for the successful completion of a task (e.g., cooperative perception for autonomous driving). This can be achieved by designing communication systems that exchange probabilistic belief states (e.g., a distribution over object locations) rather than raw data. BDL provides the perfect toolkit for representing, compressing, and transmitting these rich belief states in a way that is inherently robust to channel impairments.
- **Bayesian Optimization for Resource Allocation**: The dynamic and uncertain nature of wireless environments makes resource allocation a challenging optimization problem. Bayesian optimization, a sample-efficient method for global optimization of black-box functions, can be leveraged to intelligently orchestrate resources like power, bandwidth, and beamforming vectors, directly maximizing network-level performance metrics under uncertainty.

3. Trustworthy and Distributed BDL for Network-Level AI

Scaling intelligent behavior from a single link to the entire network requires robust, distributed, and trustworthy AI. This research direction focuses on embedding BDL into the fabric of future wireless networks:

- **Federated Bayesian Learning**: In a multi-cell network, data is inherently distributed and privacy-sensitive. Federated learning enables collaborative model training without sharing raw data. By incorporating Bayesian principles into federated learning, we can create models that not only learn from distributed data but also provide well-calibrated uncertainty estimates, even with non-i.i.d data distributions across nodes. The federated EM-TDAMP framework [42] is an initial step toward reliable distributed inference over imperfect wireless channels.
- **Bayesian Digital Twins**: A digital twin is a virtual replica of a physical network. By constructing Bayesian digital twins, models that capture not just the expected behavior but also the uncertainty associated with their predictions, we can enable explainable analytics, proactive network management, and safe policy testing in a simulated environment before deployment.
- **Causal Bayesian Models for Network Analytics**: Moving beyond correlation, future network analytics will aim to understand causality. Causal Bayesian networks can model the intricate dependencies between network events, user

behavior, and configuration parameters, enabling root-cause analysis of performance degradation and ensuring that control actions lead to predictable and measurable improvements.

Through these interconnected research thrusts, BDL emerges not just as a tool, but as a foundational design philosophy for next-generation wireless systems. It provides the theoretical and practical framework for building networks that are not only efficient and high-performing but also trustworthy, adaptive, and truly intelligent—capable of navigating the complexities of the physical world and the demands of future applications.

Homework Problems

17.1 Compare the SPVBI algorithm with its learnable deep unfolding version (LSPVBI). In your analysis:

(a) Explain the fundamental mechanisms of both algorithms, highlighting how LSPVBI extends SPVBI through deep unfolding techniques.

(b) Discuss the advantages and disadvantages of each algorithm in terms of:

- Convergence speed and final estimation accuracy
- Computational complexity and memory requirements
- Robustness to hyperparameter settings
- Interpretability and theoretical guarantees
- Data requirements and implementation complexity

(c) Provide specific scenarios where each algorithm would be preferred and justify your recommendations.

17.2 Analyze how the mini-batch size B affects both SPVBI and LSPVBI algorithms:

(a) **Computational Complexity**:

- Derive the computational complexity of both algorithms as a function of B
- Explain how B influences per-iteration computation time and memory usage
- Compare the scaling behavior for both algorithms

(b) **Convergence Performance**:

- For SPVBI, analyze how B affects gradient estimate variance and convergence stability
- For LSPVBI, discuss how B influences training stability and final performance
- Identify the trade-offs between small and large batch sizes for each algorithm

(c) **Practical Guidelines**:

- Provide recommendations for selecting B in different practical scenarios
- Discuss any differences in optimal B selection between SPVBI and LSPVBI

17.3 Consider a simplified version of the application example in Sect. 17.4, with only a single scattering path and no non-ideal factors:

(a) **Algorithm Implementation**: Implement both PVBI and SPVBI algorithms for this simplified scenario, using the provided MATLAB codebase as a starting point: https://github.com/biwcs-book/codebase

(b) **Performance Comparison**: Compare the algorithms in terms of:

- Estimation accuracy and convergence behavior
- Computational efficiency and runtime

- Robustness to different SNRs
- Analyze the reasons for any observed performance differences

(c) **Theoretical Interpretation**:

- Explain how the simplified single-path scenario affects the relative performance of both algorithms
- Discuss whether your findings would generalize to more complex multi-path scenarios.

References

1. B. Zhou, Q. Chen, H. Wymeersch, P. Xiao, L. Zhao, Variational inference-based positioning with nondeterministic measurement accuracies and reference location errors. IEEE Trans. Mobile Comput. **16**(10), 2955–2969 (2017)
2. M.J. Beal, Variational algorithms for approximate Bayesian inference. Ph.D. thesis University of London, 2003
3. Q. Liu, D. Wang, Stein variational gradient descent: a general purpose Bayesian inference algorithm, in *Proceedings of Advances in Neural Information Processing Systems (NIPS)*, vol. 29 (2016)
4. B. Dai, N. He, H. Dai, L. Song, Provable Bayesian inference via particle mirror descent, in *Proceedings of the International Conference on Artificial Intelligence and Statistics (AISTATS)* (PMLR, 2016), pp. 985–994
5. Y. Yang, G. Scutari, D.P. Palomar, M. Pesavento, A parallel decomposition method for nonconvex stochastic multi-agent optimization problems. IEEE Trans. Signal Process. **64**(11), 2949–2964 (2016)
6. A. Liu, V.K.N. Lau, B. Kananian, Stochastic successive convex approximation for non-convex constrained stochastic optimization. IEEE Trans. Signal Process. **67**(16), 4189–4203 (2019)
7. Z. Hu, A. Liu, Y. Wan, T.X. Han, M. Zhao, A two-stage multiband delay estimation scheme via stochastic particle-based variational Bayesian inference. IEEE Internet Things J. **11**(11), 19632–19645 (2024)
8. Z. Hu, A. Liu, Y. Wan, T.Q.S. Quek, M. Zhao, Two-stage multiband Wi-Fi sensing for ISAC via stochastic particle-based variational Bayesian inference, in *Proceedings of the IEEE Global Communications Conference (GLOBECOM)*, pp. 5617–5622 (2023)
9. Y. Wan, A. Liu, Q. Hu, M. Zhang, Y. Cai, Multiband delay estimation for localization using a two-stage global estimation scheme. IEEE Trans. Wirel. Commun. 1–1 (2023)
10. A. Balatsoukas-Stimming, C. Studer, Deep unfolding for communications systems: a survey and some new directions, in *IEEE Workshop on Signal Processing Systems (SiPS)*, pp. 266–271 (2019)
11. J.R. Hershey, J.L. Roux, F. Weninger, Deep unfolding: Model-based inspiration of novel deep architectures. Comput. Sci. (2014)
12. P. Pan, Y. Zhang, Z. Deng, W. Qi, Deep learning-based 2-D frequency estimation of multiple sinusoidals. IEEE Trans. Neural Netw. Learn. Syst. **33**(10), 5429–5440 (2022)
13. E. Jang, S. Gu, B. Poole, Categorical reparameterization with gumbel-softmax (2016)
14. S. Boyd, J. Dattorro, Alternating projections. Lecture Notes of EeO (2003)
15. Y. Chen, X. Ye, Projection onto a simplex (2011). arXiv:1101.6081
16. W. Jin, H. He, C.-K. Wen, S. Jin, G.Y. Li, Adaptive channel estimation based on model-driven deep learning for wideband mmWave systems, in *Proceedings of the IEEE Global Communications Conference (GLOBECOM)*, pp. 1–6 (2021)
17. 3GPP, Study on channel model for frequencies from 0.5 to 100 GHz (3GPP TR 38.901 version 16.1.0 Release 16). Tech. Rep. (2020)
18. M. Pourkhaatoun, S.A. Zekavat, High-resolution low-complexity cognitive-radio-based multiband range estimation: concatenated spectrum vs. fusion-based. IEEE Syst. J. **8**(1), 83–92 (2014)
19. Z. Hu, A. Liu, W. Xu, T.Q.S. Quek, M. Zhao, A stochastic particle variational Bayesian inference inspired deep-unfolding network for sensing over wireless networks. IEEE J. Sel. Areas Commun. **42**(10), 2832–2846 (2024)
20. Y. LeCun, J. Denker, S. Solla, Optimal brain damage. Adv. Neural Inf. Process. Syst. **2**, (1989)
21. B. Hassibi, D. Stork, Second order derivatives for network pruning: optimal brain surgeon. Adv. Neural Inf. Process. Syst. **5**, (1992)
22. J. Frankle, M. Carbin, The lottery ticket hypothesis: finding sparse, trainable neural networks (2018). (arXiv:1803.03635)
23. H. Tanaka, D. Kunin, D.L. Yamins, S. Ganguli, Pruning neural networks without any data by iteratively conserving synaptic flow. Adv. Neural. Inf. Process. Syst. **33**, 6377–6389 (2020)
24. S. Scardapane, D. Comminiello, A. Hussain, A. Uncini, Group sparse regularization for deep neural networks. Neurocomputing **241**, 81–89 (2017)
25. S. Kim, E.P. Xing, Tree-guided group lasso for multi-response regression with structured sparsity, with an application to eQTL mapping (2012)
26. S. Kim, E.P. Xing, Tree-guided group lasso for multitask regression with structured sparsity. in *Proceedings of the 27th International Conference on Machine Learning* (PMLR, 2010), pp. 543–550
27. J. Li, Q. Qi, J. Wang, C. Ge, Y. Li, Z. Yue, H. Sun, OICSR: Out-in-channel sparsity regularization for compact deep neural networks, in *Proceedings of the IEEE/CVF Conference on Computer Vision and Pattern Recognition*, pp. 7046–7055 (2019)
28. K. Mitsuno, J. Miyao, T. Kurita, Hierarchical group sparse regularization for deep convolutional neural networks, in *2020 International Joint Conference on Neural Networks (IJCNN)* (IEEE, 2020), pp. 1–8
29. M. Barlaud, F. Guyard, Learning sparse deep neural networks using efficient structured projections on convex constraints for green AI, in *2020 25th International Conference on Pattern Recognition (ICPR)* (IEEE, 2021), pp. 1566–1573
30. X. Xiao, Z. Wang, S. Rajasekaran, Autoprune: automatic network pruning by regularizing auxiliary parameters. Adv. Neural Inf. Process. Syst. **32**, (2019)

31. S. Kuutti, R. Bowden, Y. Jin, P. Barber, S. Fallah, A survey of deep learning applications to autonomous vehicle control. IEEE Trans. Intell. Transp. Syst. **22**(2), 712–733 (2021)
32. G. Litjens, T. Kooi, B.E. Bejnordi, A.A.A. Setio, F. Ciompi, M. Ghafoorian, C.I. Sánchez, A survey on deep learning in medical image analysis. Med. Image Anal. **42**, 60–88 (2017)
33. D. Wang, W. Fu, Q. Song, J. Zhou, Potential risk assessment for safe driving of autonomous vehicles under occluded vision. Sci. Rep. **12**(1), 4981 (2022)
34. H. Wang, D.Y. Yeung, A survey on Bayesian deep learning. ACM Comput. Surv. (csur) **53**(5), 1–37 (2020)
35. L.V. Jospin, H. Laga, F. Boussaid, W. Buntine, M. Bennamoun, Hands-on Bayesian neural networks—a tutorial for deep learning users. IEEE Comput. Intell. Mag. **17**(2), 29–48 (2022)
36. M. Magris, A. Iosifidis, Bayesian learning for neural networks: an algorithmic survey. Artif. Intell. Rev. **56**(10), 11773–11823 (2023)
37. B. Hanin, Which neural net architectures give rise to exploding and vanishing gradients? Adv. Neural Inf. Process. Syst. **31**, (2018)
38. S. Hochreiter, The vanishing gradient problem during learning recurrent neural nets and problem solutions. Int. J. Uncertain. Fuzziness Knowl. Based Syst. **6**(02), 107–116 (1998)
39. A. Choromanska, M. Henaff, M. Mathieu, G.B. Arous, Y. LeCun, The loss surfaces of multilayer networks, in *Artificial Intelligence and Statistics* (PMLR, 2015), pp. 192–204
40. D.P. Kingma, J. Ba, Adam: a method for stochastic optimization, in *ICLR (Poster)*, ed. by Y. Bengio, Y. LeCun (2015)
41. C. Lucibello, F. Pittorino, G. Perugini, R. Zecchina, Deep learning via message passing algorithms based on belief propagation. Mach. Learn. Sci. Technol. **3**(3), 035005 (2022)
42. W. Xu, A. Liu, Y. Zhang, V. Lau, Bayesian deep learning via expectation maximization and turbo deep approximate message passing. IEEE Trans. Signal Process. (2024)
43. M. Rashid, M. Naraghi-Pour, Clustered sparse channel estimation for massive MIMO systems by expectation maximization-propagation (EM-EP). IEEE Trans. Veh. Technol. **72**(7), 9145–9159 (2023)
44. X. Bai, Q. Peng, A probabilistic model based-tracking method for mmWave massive MIMO channel estimation. IEEE Trans. Veh. Technol. **72**(12), 16777–16782 (2023)
45. T. Moon, The expectation-maximization algorithm. IEEE Signal Process. Mag. **13**(6), 47–60 (1996)
46. J. Ma, X. Yuan, L. Ping, Turbo compressed sensing with partial DFT sensing matrix. IEEE Signal Process. Lett. **22**(2), 158–161 (2014)
47. J.T. Parker, P. Schniter, V. Cevher, Bilinear generalized approximate message passing–Part I: derivation. IEEE Trans. Signal Process. **62**(22), 5839–5853 (2014)

Index

A

AMP, 149, 151, 153, 154, 156–159, 162, 164–168, 171, 207, 215, 216, 242, 246, 255–260, 270, 278
Angular reciprocity, 171, 179–182, 185, 186
Atom selection, 127, 128

B

Bayesian decision theory, 3, 9
Bayesian estimation, 11, 25–29, 35, 38, 260, 277, 334
Bayes' rule, 3, 9, 29
Bayes' theorem, 25, 54, 187, 337
Belief propagation, 41, 50, 52, 54, 167, 251, 357
Bernoulli–Gaussian Markov model, 281
Bias-variance tradeoff, 11, 23
Bilinear models, 167, 315
BiSVBI, 303, 316, 317, 320, 325, 327, 329–333
Block MM, 174, 175, 177, 242–244, 277
Burst LASSO, 101, 116–120, 124, 140

C

Channel estimation, 16, 18, 20, 60, 98, 101, 113, 116, 117, 119, 120, 124, 137, 149, 166, 171, 179, 180, 182, 185–187, 189, 213, 217, 219–223, 225–233, 238, 271–277, 279, 294, 297, 298, 300, 303, 305, 310, 312, 320–325, 330–332, 357
Channel tracking, 281, 290–294, 296–298
Classification, 3, 4, 10, 11, 22, 23, 57, 86
Compressive sensing, 22, 83, 85–87, 89–91, 93, 95, 98, 99, 101, 106, 113, 120, 127, 129, 130, 137, 138, 140, 146, 149, 166, 167, 171, 181, 185, 189, 190, 198, 201, 204, 205, 208, 213, 214, 217–220, 222, 223, 225–227, 233, 237–239, 241, 246–248, 253, 255, 256, 260–262, 269, 270, 272, 277, 296, 303, 305, 310
Computational complexity, 25, 33, 34, 57, 60, 64, 76, 127, 128, 149, 153, 184, 196, 197, 208, 209, 223, 238, 239, 250, 253–255, 260, 269–271, 279, 300, 303, 321, 335, 337, 338, 354–356, 358
Conjugate prior, 25, 28–30, 172, 208
Convergence analysis, 177, 267, 315, 320
Coordinate descent, 101, 107, 109, 110, 115, 116, 120, 124, 140, 189, 191–193, 195, 196, 267, 315, 320
CoSaMP, 127, 134–141, 149

D

Deep learning, 21, 189, 337, 346, 350, 352, 355–357
Deep unfolding, 337, 343–346, 354, 356, 358
Denoising function, 149, 156, 158–162, 164, 167
Direction-of-arrival, 57–60, 62–65, 68, 73, 76, 79, 198
Dynamic grid, 171, 172, 205, 206, 208, 213, 218–220, 222, 223, 225–228, 233, 237, 238, 241, 271, 272, 277, 293, 294, 322, 327, 334
Dynamic sparse Bayesian inference, 281, 290, 298

E

EM algorithm, 33–39, 104, 111, 259, 279, 308, 334
EM-Turbo-SBI, 241–243, 245, 247, 271, 272, 277, 357
E-step, 33–40, 111, 112, 199, 200, 205, 241, 242, 244, 245, 259, 264, 269, 277, 279, 289, 290, 304, 305, 308, 309, 321, 329, 330, 334
Estimation of Signal Parameters via Rotational Invariance Techniques (ESPRIT), 57, 64–70, 72–76, 78, 79
Evidence Lower Bound (ELBO), 189, 191–195, 304–309

F

Factor graph, 41–44, 46, 50–54, 150, 167, 218, 219, 225, 228, 236, 242, 245, 248, 250, 251, 254, 256, 257, 261, 262, 270, 273, 275–278, 283, 286, 299, 317
Factor nodes, 41, 43–45, 49, 50, 52, 150, 152, 153, 225, 248, 251, 252, 256, 262, 270, 271, 275, 276, 287, 288
5G NR, 60, 220, 290–292, 296, 298
Frequency Division Duplex (FDD), 98, 137, 171, 179, 180, 182, 185–187, 213, 219–221, 225, 271–274, 279, 290

G

Gaussian noise, 11, 16, 18, 22, 57, 124, 133, 137, 164, 180, 182, 196, 198, 203, 208, 248, 249, 262, 303, 315, 337, 353
Gaussian prior, 25, 27–30, 196, 209, 248, 249, 260, 270, 272, 273, 285, 300, 305, 329
Greedy algorithms, 127, 138, 139, 149, 276
Group LASSO, 101, 114, 116–120, 124, 140
Group sparsity, 177, 201, 204, 205, 215

H

Hybrid beamforming, 213, 220, 225, 226, 228, 320
Hyperparameters, 25, 111, 112, 140, 182, 187, 199, 200, 215–218, 235, 261, 269, 335, 345, 350, 356, 358

I

i.i.d. sensing matrix, 149, 260, 270, 271
Incoherence, 83, 85–87, 91, 93, 95, 98
Integrated Sensing And Communication (ISAC), 76, 189, 213, 229–232, 238, 272, 325, 326, 330, 332, 333, 357
Inverse-free VBI, 260, 271, 303–305, 307–309, 320–325, 333

J

Jacobian logarithm, 48, 52

K

KL divergence, 36, 39, 190

L

Latent variables, 33, 34, 39, 40, 111, 174, 200, 219, 224, 234, 245, 261, 264, 267, 281, 283, 287, 300, 306, 310, 329

A. Liu and M.-M. Zhao, *Advanced Bayesian Signal Processing*, Textbooks in Telecommunication Engineering,
https://doi.org/10.1007/978-3-032-26656-9

Least Absolute Shrinkage and Selection Operator (LASSO), 83, 91, 93, 101, 106–110, 112–120, 123, 124, 138, 140–143, 149, 150, 161, 162, 208, 213
Linear Minimum Mean Squared Error (LMMSE), 11, 20–22, 24, 99, 246, 247, 249, 250, 253, 260, 270, 272, 273, 278, 285, 303, 312, 330–332
ℓ_1-minimization, 16, 83, 88, 90, 93, 101, 113, 120, 123, 130
Loopy belief propagation, 41, 50, 52, 167, 251
Loss function, 3–5, 8, 9, 11, 12, 41, 42, 164, 346, 351
LS estimation, 16, 17, 22, 75, 127, 138
LSPVBI, 337, 346–356, 358

M
Majorization–minimization, 242–244, 254, 289, 290, 334
Majorization-minimization, 33, 37, 39, 40, 101, 104, 105, 120, 174, 175, 177, 181, 197, 290, 303–305, 309, 333, 334
MAP estimation, 11, 33, 38, 39, 41, 46–48, 52, 186, 189, 202, 241, 283, 284, 289, 329, 334, 345
Marginalization, 3, 5–10, 52, 283
Massive MIMO, 84, 86, 98, 101, 112, 113, 116, 119, 120, 124, 137, 138, 149, 171, 179–181, 185–187, 201, 203–205, 207–209, 213, 214, 217, 219–222, 225–230, 238, 271–274, 279, 290, 291, 294, 296, 303, 305, 310, 320, 321, 325, 326
Max-product algorithm, 46, 52
Mean squared error, 11, 15, 19–21, 23, 24, 42, 112, 124, 160–162, 164–166, 168, 176
Mean-field approximation, 190
Message passing, 41, 42, 44, 52, 149–153, 156, 157, 167, 216, 228, 241, 242, 245–254, 256, 257, 263, 269, 270, 273, 276, 277, 279, 284, 286, 290, 299, 305, 310, 317, 357
Minimum Description Length (MDL), 57–59, 61, 76
ML estimation, 12, 13, 15, 16, 25, 26, 28, 33–35, 38, 39
MMSE estimation, 11, 41, 151, 160, 272, 285, 316
M-step, 33–40, 111, 206, 208, 241–245, 254, 259, 264, 271, 277, 279, 284, 290, 296, 304, 308, 309, 321, 322, 330, 334
Multi-layer sparse prior, 213
MUltiple SIgnal Classification (MUSIC), 57–64, 66, 75, 76, 79, 296–298, 354, 355

N
Noise subspace, 57, 58, 63, 69, 76, 79
Nyquist rate, 83, 84, 93

O
Orthogonal Frequency Division Multiplexing (OFDM), 60, 62, 213, 219–223, 225, 229, 230, 238, 271–273, 275, 279, 290–292, 294, 325, 326, 352
Orthogonal Matching Pursuit (OMP), 127, 128, 130–135, 138, 140, 145, 146, 276, 277, 312

P
Parameter estimation, 3, 16, 41, 57, 73, 74, 76, 112, 171, 189, 190, 200, 208, 219, 222, 223, 227, 228, 233, 238, 239, 241, 243, 305, 321, 327, 334, 337, 351
Particle-based variational inference, 339, 352
Particle VBI, 189, 190, 194–198, 200–202, 208, 209, 224, 337, 358
Performance prediction, 149
Point estimation, 11, 22, 25, 29, 35, 264
Posterior distribution, 3, 11, 22, 25–30, 33, 35, 38, 39, 41, 42, 44, 46, 47, 50, 53, 110, 111, 150, 153, 157, 161, 174, 182, 186, 187, 189, 190, 196, 200–202, 208, 209, 217, 242, 246, 249, 252, 253, 257, 259, 264, 267, 281, 283, 285–287, 314, 319, 320, 337, 338, 355
Prior distribution, 3, 11, 12, 25–28, 38, 110, 111, 120, 152, 160, 161, 193, 199, 202, 215, 216, 218, 247, 268, 275, 294, 310, 329, 337, 338, 345

Q
Q-function, 33, 39, 111

R
Regularized Orthogonal Matching Pursuit (ROMP), 127, 130–132, 134, 135, 138, 140
Restricted Isometry Property (RIP), 83, 89–93, 96, 97, 127, 129–131, 134, 135, 138–140
Reweighted ℓ_1-minimization, 101, 120, 123
Risk minimization, 3, 8, 12
Root-MUSIC, 57, 60, 76, 79, 354
Rotational invariance, 57, 64, 66, 76, 166

S
Sequential estimation, 28, 30, 281, 283, 284
Signal processing, 3, 57, 64, 75, 76, 86, 112–114, 167, 198, 219, 231, 247, 315, 334, 352, 357
Signal subspace, 57, 58, 61, 64, 66, 67, 69, 70, 75, 78, 129
Soft-thresholding, 101, 108–110, 113, 115, 120, 121, 123, 143, 158, 161, 162, 164, 168
Sparse Bayesian learning, 171–175, 177–179, 181, 182, 185, 186, 189, 190, 198, 200, 208, 209, 215, 217, 356
Sparse prior, 22, 150, 153, 154, 156, 158, 160, 161, 165, 167, 168, 171, 173, 178, 186, 198, 213, 215–217, 224, 225, 227, 228, 234, 238, 239, 242, 245–248, 250, 251, 253, 255, 256, 261, 264, 270, 271, 273, 277, 278, 285, 300, 305, 328, 329
Sparse recovery, 90, 103, 113, 116, 128, 129, 131, 136, 156, 159, 172, 181, 205, 222, 224, 225, 227, 239, 257, 281, 303, 320, 333
Sparse VBI, 189, 190, 198–201, 203, 204, 207, 208, 246, 263, 264, 267, 268, 270, 271, 275, 278, 303
Sparsity, 83–86, 88, 90, 91, 93, 98, 101–106, 108–110, 112–114, 116, 119, 120, 124, 127, 128, 131, 134–136, 138–141, 145, 156–158, 160, 161, 166, 168, 171, 172, 176–178, 182, 183, 186, 187, 189, 198, 200, 201, 204, 205, 208, 209, 213–218, 222–225, 227–229, 233, 234, 236, 239, 241, 242, 247, 249, 251, 253, 256, 257, 259–261, 263, 267, 269–277, 281, 282, 285, 287, 288, 293, 294, 296, 298, 300, 303–305, 310, 312, 316, 321, 322, 325, 328, 334
Stagewise Orthogonal Matching Pursuit (StOMP), 127, 132–134, 139, 140
State evolution, 149, 159, 162–168, 227, 257, 259
Stochastic optimization, 337, 339–341, 344, 345, 355, 356
Stochastic particle VBI (SPVBI), 337, 339–341, 343–347, 352–356, 358
Stochastic Successive Convex Approximation (SSCA), 196, 337, 340, 341
Structured compressive sensing, 213
Structured sparsity, 101, 114, 119, 120, 124, 140, 166, 171, 186, 213–218, 223, 239, 241, 242, 247, 249, 256, 261, 263, 269, 272–274, 277, 285, 305, 310, 325
Sub-Nyquist sampling, 83
Subspace-constrained VBI, 260, 271, 303–305, 309–311, 315–321, 333–335
Subspace methods, 57, 76, 317
Successive Linear Approximation VBI (SLAVBI), 189, 190, 196–198, 208, 209
Sum-product algorithm, 41–43, 45, 46, 48, 50–52, 54, 252, 263, 270
Super-resolution, 213, 223, 227, 228, 233, 238, 294, 303, 334
Support estimation, 127, 139, 312, 318

T

Temporal correlation, 166, 167, 224, 281–283, 290, 292, 294–296, 298, 299

Time Division Duplex Hybrid Beamforming (TDD-HBF), 213, 220, 225

Time-varying sparsity, 281

Turbo Approximate Message Passing (Turbo-AMP), 241, 246, 255–262, 270–272, 278, 279, 281, 298, 333

Turbo Compressive Sensing (Turbo-CS), 241, 246–248, 250, 252–256, 260–262, 270–273, 276–279, 281, 284, 285, 287, 291, 296, 298–300, 321, 323, 324, 333

Turbo variational Bayesian inference (Turbo-VBI), 241, 243, 246, 260–262, 264, 267–272, 275–279, 281, 298–300, 305, 308, 313, 321, 323, 324, 333, 357

U

Uncertainty quantification, 25, 26, 29, 120, 140, 154, 171, 200, 257, 284

Uniform Linear Array (ULA), 60, 68–70, 72, 79, 86, 98, 112, 119, 137, 179, 185, 207, 214, 220–223, 238, 271–273

V

Variable nodes, 41, 43–45, 49, 50, 52–54, 150, 156, 256, 270, 275, 276, 287, 288

Variational Bayesian inference, 12, 21, 189, 194, 215, 217, 219, 241, 263, 275, 303, 306, 333, 334, 337, 339, 355

W

WiFi sensing, 337, 352, 353

Wireless localization, 189, 190, 200, 238

FSC
www.fsc.org
MIX
Papier | Fördert
gute Waldnutzung
FSC® C083411

Kolibri 360 GmbH
Ferdinand-Jühlke-Straße 7
99095 Erfurt, Deutschland
produktsicherheit@kolibri360.de